Systems Science: Theory, Analysis, Modeling, and Design

George E. Mobus

Systems Science: Theory, Analysis, Modeling, and Design

Springer

George E. Mobus
Faculty in Computer Science & Systems
Computer Engineering & Systems
University of Washington Tacoma
The School of Engineering and Technology (SET)
Tacoma, WA, USA

ISBN 978-3-030-93481-1 ISBN 978-3-030-93482-8 (eBook)
https://doi.org/10.1007/978-3-030-93482-8

This Springer imprint is published by the registered company Springer Nature Switzerland AG
The registered company address is: Gewerbestrasse 11, 6330 Cham, Switzerland

Preface

At the outset of writing this book, I had a reasonably clear vision of what I wanted to say for reasons I'll explain shortly. I should have easily wrapped up the manuscript in 2 years tops. The concepts had been accumulating in my head for a little more than five decades. And the overall scheme was really pretty straight forward, I thought. That scheme would be to draw from my first book with coauthor Michael Kalton, *Principles of Systems Science* (2015), for the "theoretical" basis of what I would be expanding in this book. Using the principles, I thought it would just be a matter of laying out the formula for *deep systems analysis* that I had developed in practice in the world of commerce; present my most recent research on an archetype model of a complex, adaptive, and evolvable system (CAES), which I had previously introduced at the 2015 Conference of the International Society for the Systems Sciences (ISSS); and wrap it all up with an example of how to analyze and design a more viable business system. But as the work proceeded, it became increasingly clear that this book would end up with much more than I had at first thought.

In a paper I presented to the 2016 Conference of the ISSS, coauthored with Kevin Anderson (2016), I introduced the notion of a universal lexicon and semantics for a language of systems that built upon systems dynamics (SD, Forrester, 1961) with its universal concepts of stocks, flows, and regulators, among others. This new language would significantly extend the lexicon to include things not generally found in SD, such as "process" as an abstraction for a subsystem. I also presented a set-theoretic formulation for the concept of *systemness* (see Chap. 4). During the question-and-answer portion of the talk, an attendee challenged me to produce a framework of justification for the terms and the mathematical handling. I made some clumsy attempt to do so off the top of my head, but not very successfully. Sometime after the conference, I started to realize the questioner had a very valid point. What were these "things" I claimed were part of systemness? How did I justify their inclusion in my language? What logic could I bring to bear to claim their reality and usefulness? Thus, started a dive into the rabbit hole of ontology.

That "quest" soon expanded my vision for what needed to be explained. I had a method for gaining deep understanding (an epistemology, explained in the Introduction). I had developed what seemed to be a pretty reasonable ontology starting from the basic substances of the Universe. And I had what looked like a means for translating the metaphysics into a solid, practical (or at least feasible) set of methodologies that would provide we humans with an approach to more deeply understand ourselves and our world, such that we might achieve true long-term sustainability. I realized that, with this revelation, showing how to design a single organizational structure like a business was, to be blunt, prosaic at best. Afterall, we humans, as a species and as a society, are currently in grave danger. Why not use all of the philosophy and methodology to tackle the real systemic problems we face? Why not consider the systems analysis of the human social system itself and use that analysis to design a sustainable human social system? That is, a system in which human beings can *be* human beings but the mechanisms within the system help guarantee the long-term viability of the whole of global society by placing reasonable constraints on the "component subsystems," us, to prevent self-destructive behaviors. This is clearly either a grandiose conception or, as I now think, a necessary endeavor. We have to figure this out. And I know of no other approach that could even begin to address the urgency with which we need to do so.

The present work has gone far beyond my initial vision. The Introduction explains the nature of deep understanding and the reasons why we need to strive for it in our knowledge of systems. Part I, then, covers the theoretical framework for obtaining that knowledge. It establishes the ontology of systems and provides the logical structure of systemness that permits us to examine any system and gain knowledge about it in a structured way (a system of system knowledge).

Part II, then, elaborates the methods of systems analysis that allow the discovery of the details of systems structures and functions while capturing that information into the system knowledgebase, a repository specifically structured to emulate the system of system knowledge.

Since our main concern is with extremely complex systems, like living beings, ecosystems, organizations, and societies, Part III examines the nature of CAESs and the archetypal sub-models that combine, recursively, to constitute these real systems.

Finally, Part IV brings the theory, methods, and archetype models together to explore how these very complex systems can (and should) be designed, with a special emphasis on the human social system that can exist in accord with the rest of the planet.

As I write this, the United States and parts of Europe are starting to recover from the COVID-19 pandemic. Unfortunately, many other parts of the world are not yet in that phase. For more than a year we have been self-isolating and wearing masks to prevent the spread of the novel coronavirus. We now have several vaccines which seem to be quite effective. The effort now should be to get more of these vaccines to places in the world where the virus is still running rampant. But those vaccines are the product of taking a systems approach to their development and distribution. Indeed, it can be argued that all of the disciplinary sciences are gradually coming around to grasping that they can no longer operate in silos of knowledge. They are seeing that a broader and transdisciplinary view of the world is required to tackle and manage the most complex problems that humanity and the world at large face.

There are many other challenges to consider. Global warming, soil degradation, potable water loss, social inequities, and so many more threats are systemic in nature, that is, have multiple complex and interacting causes with feedbacks. Humanity cannot afford to proceed with business as usual, that is, treating each problem as independent and linear.

Systems science, systems thinking, and using a systems approach to problem recognition and problem solution can provide deep understanding of the human (and world) condition and help us design a social organization in concert with the principles of systems. My hope is that this book will provide some insights into how this can be approached. Much more needs to be done, of course. What I offer here is only a glimpse of what is needed. But I sincerely hope it is a start.

Tacoma, WA, USA George E. Mobus
June 3, 2021

Acknowledgments

Systems science is only at best a partially recognized disciplinary subject in the modern world. Unlike the other sciences, such as biology or astrophysics, systems science functions as a kind of "meta" science. It is applicable to all of the sciences and yet has only a modicum of representation in academic departments. There are a few, very few, academic programs devoted to systems science, and even those that exist generally focus on only portions of the whole subject. They may center on systems thinking applied, for example, to management science. Or they may be organized as a "systems" component of an existing subject area. Systems engineering, for example, is often folded into industrial engineering or even mechanical engineering, or as an add-on to a classical science, for example, systems biology.

This is an unfortunate situation. Systems science should be a central topic in its own right with the other sciences and engineering programs building their disciplines on top of a systems perspective rather than adding the latter to the former.

I mention this by way of saying how incredibly fortunate I have been to meet and interact with souls who have had to adopt a traditional silo for their sustenance and yet persevered to maintain a systems perspective and developed a high degree of systems thinking in approaching their work. Hats off to so many colleagues who have, in a sense, kept the faith and have helped me come to understand the many varied, yet reasonable perspectives needed to start to grasp the full significance of understanding systems.

First, I would like to say thank you to the many members of the International Society for the Systems Sciences (ISSS)[1] for numerous wonderful conversations in the context of their annual conferences, but also in the Saturday morning seminars that we've been holding throughout the pandemic. These conversations have provided me with an extraordinary opportunity to present some of the more advanced ideas presented in this book, to get thoughtful feedback, and push me to hone my thinking on some extraordinarily complex topics. Special thanks to Peter Tuddenham, a past president of the ISSS who organized these meetings in an online format that

[1] See their website at: https://www.isss.org/home/. Accessed 6/3/2021.

kept interests in pursuing systems concepts alive and thriving, especially through the pandemic.

Next, I would like to thank the International Federation for Systems Research (IFSR)[2] for their sponsorship of a biannual program called "The Conversation" in which I participated with a fair number of people from the systems community. These conversations were open-ended and exploratory and served as incredible stimuli for me in developing many of the ideas in this book.

I could not begin to list all of the individuals who have affected my thinking as this book developed. My apologies to the many of you whose thoughts and insights I gained so much from. But I do need to acknowledge those individuals with whom I have had some of the most productive conversations, and some of whom provided many comments and critiques of parts of this book.

First, I want to acknowledge and thank my dear friend, coauthor, and colleague Mike Kalton. Mike has always given me a sense of direction when I've gone off on purely technical tangents. He helped craft our *Principles of Systems Science* book so that many more people, not necessarily steeped in quantitative details, could grasp the subject. I've had many people who have read the book comment on how readable it is. That is due to Mike's smoothening out any of my rough, technical prose. Both the systems community and I should always be grateful for his eloquence.

A fair number of individuals have had a direct influence on my thoughts. I'd like to thank Tyler Volk whose conversations on many of the topics, especially in Chaps. 3 and 4, have brought so much clarity to my understanding. For the last several years, we have had an ongoing conversation with Gary Smith, Hillary Sillitto, Tom Marzolf, Lynn Rasmussen, and Helene Finidori, which has been, often, exhilarating. I am especially thankful to have come to know Lenard Troncale, whose insights into many of the same areas that I consider in Chap. 3 have been both inspiring and innervating. We haven't always seen eye to eye on a variety of subjects, but our interactions have always been evocative and generative of better ideas.

There are many more wonderful thinkers in the systems community that I would acknowledge for their contributions to my grasp of the subject. I wish I had the space and time to list them all.

[2] See their website at: https://ifsr.org/. Accessed 6/3/2021.

Contents

Part I Foundations of Systems Understanding

About the Author

George E. Mobus is Associate Professor Emeritus of Computer Science & Systems and Computer Engineering & Systems in the School of Engineering and Technology at the University of Washington Tacoma. He received his PhD in computer science from the University of North Texas in 1994. His dissertation, and subsequent research program at Western Washington University (under National Science Foundation Grant No. IIS-9907102), involved developing autonomous robot agents by emulating natural intelligence as opposed to using some form of artificial intelligence. Mobus was awarded US Patent: #5,504,839, "Processor and Processing Element for Use in a Neural Network." He also received an MBA from San Diego State University in 1983, doing a thesis on the modeling of decision support systems based on the hierarchical cybernetic principles presented in this volume and in numerous papers. His baccalaureate degree was earned at the University of Washington (Seattle) in 1973, in zoology. He studied the energetics of living systems and the interplay between information, evolution, and complexity.Before completing his academic pursuits of a PhD, he had risen through the ranks of a small controls engineering company in Southern California, from software engineer to the top spot. The credit for success goes to the education he got in systems sciences during his MBA program and helping reshape many of the internals of the company that improved profitability and work conditions.

Chapter 1
Introduction

Abstract The problem of our age is how to go about understanding, deeply, the kinds of complex systems that we need to understand today. Our world is in trouble. Between suffering the unforeseen consequences of technologies that we thought were brilliant (when they were invented) and those of political decisions that seemed right (under the then circumstances), we seem to be witnessing complications in every aspect of societies, the environment, public health, and just about anything you can name.

The problem is complexity. Or, rather, the problem is one of managing complexity, so that we benefit and chaos does not overwhelm us. And in order to manage anything, you need to know it deeply. That is, you need to know how it works (and how it doesn't work at times) down to a level where interventions (if needed) can be used to keep the system stable.

We suggest that the pathway to deep understanding of any system, no matter how complex, is in front of us. Our guide to the pathway is to apply systems thinking and systems science to how we go about understanding systems. Many will argue that that is exactly what they do. And many people do so to one extent or another. Systems science has been developing for more than half a century and many thinkers have advanced concepts about systemness and many have used those concepts to pursue avenues in the sciences and engineering. There are a plethora of tools and methodologies developed around those concepts. But heretofore there has been no central focus on a set of comprehensive methodologies that could unify the various approaches to using systems science to gain deep understanding. There has not been a systemic systems approach. Many examples of disparate and sometimes contradictory views of systems science, or rather some part of systems science, will be provided throughout the book while still developing approaches to unification.

1.1 What This Book Is About

Fundamentally, this book addresses the problem of how to deeply understand the systems we encounter in the Universe or create from our own desires. There is a difference between deep understanding and more shallow understanding. This

G. E. Mobus, *Systems Science: Theory, Analysis, Modeling, and Design*,
https://doi.org/10.1007/978-3-030-93482-8_1

introduction will explore these differences and explain why deep understanding is preferable in every domain of knowledge. Briefly, though, superficial understanding is characterized by an ability to build an abstract model of a phenomenon and use that model to predict future behaviors under various conditions. Take a very simple system for example: a standard regression model can be constructed on the basis of correlations between two variables, one an independent variable like investment dollars and the other a dependent variable like returns. We can build this model by analyzing data from past experience. The regression curve can then be used to predict the value of the dependent variable given the value of the independent variable with some probability based on the correlation in the original samples. Correlation, however as the famous admonition goes, is not causation. Such models can only tell us what might happen (within certain error bounds) given the value of an independent variable. What is not known in this kind of model is what the causal relation is between the two variables. Shallow understanding is certainly better than no understanding (or guessing). But we know that shallow understanding can lead to surprises when systems do not behave as predicted. All too often the problem is that the original model failed to capture some internal causal mechanics that, perhaps only rarely, were operating to generate not understood aspects of the system.

At a slightly deeper level of understanding, we can build causal models based on the relations between macrostates of a dynamic system at a course-grained level of resolution. There needs to be an assumed mapping from microstates, which are not known in detail, to those macrostate variables of interest. Wolpert et al. (2017) describe this as a state space compression, that is, the macrostate variables are a compressed (in the sense of data compression) version of the aggregate of microstates. One example they give is of weather prediction based on measuring discrete points on the geography (weather stations) and using algorithms that map such measurements to larger phenomena such as pressure fronts or hurricanes. The latter are characterized as macrostates of the larger system whereas the former data points represent microstates. The point of having such mappings or models is to be able to predict future macrostates by also modeling the temporal evolution of the system at the macro level. The success of this procedure depends on how well the mapping from microstates to macrostates approximates the reality of the system.

Models based on observed relations between macrostate variables, based on the compression mapping from microstates to macrostates, are well developed, particularly with modern computational power. But they still are based on an inferred mapping process. The origins of these models can be traced back to work in the statistical mechanics of properties of, for example, gases where the macrostate variables of temperature and pressure could be related to the behavior of the gas particles in containment, possibly by knowing the average behavior of particles in collisions at average kinetic energies and in particular densities. Whenever a system, at the micro level, has a large number of degrees of freedom, it becomes impossible to work from the microstates to describe the behavior of the system. Even the notion of having knowledge of the initial conditions (at first observation) is fundamentally impossible. But because these, what we will call "simple," systems can be understood from their physics and because their overall behavior at the macro level is generally what interests us, this form of understanding and modeling is completely sufficient.[1]

The models we build are a direct reflection of the understanding we have about the system. They become an implementation of the knowledge we possess. Our concepts (in our brains) are models implemented in neural networks. When we build a model, we are making whatever knowledge we possess into a communication between the possessors of the knowledge and others. George Box (1976) used the aphorism: "All models are wrong, but some are useful," by which he meant that the models we build based on our best understanding of a system (or phenomenon) are not correct in every detail. However, models may be *sufficiently* correct to provide useful predictions. And by useful we mean actionable (e.g., weather predictions of rain give us cause to carry an umbrella).

This is all well and good as far as it goes. We do not disparage approximations that are useful in this sense. However, there is an argument that deeper understanding might be *more* useful. Science is, after all, an attempt to understand phenomena more deeply. Gregor Mendel's conceptualization of genetics was useful for a basic understanding of inheritance, but science dug deeper into the actual chemistry of genetic mechanisms, deoxyribonucleic acid (DNA), to better understand what was going on in the nature of genetic control over phenotypic characteristics and behaviors. We needed to have a deep understanding of genes in order to tie a number of phenomena together.

In this book we are interested in phenomena that transpire at higher levels of organization and complexity, all of which are composed from these simpler systems.[2] This is the domain of complex systems (CS), which includes living systems, human-derived artifacts, and social systems. The methods of modeling that have been so successful for simple systems (SS) start to break down in terms of providing veridical predictions (or anticipations) of future behaviors of such systems under varying environmental conditions. The mapping between microstates and macrostates needs to be based on a new way to obtain understanding. The accuracy and precision of computing the future behavior of extremely complex systems require a deeper understanding of the causal links between those microstates and the macrostates we observe.

Deep understanding involves obtaining knowledge of what goes on inside the phenomenon that produces the behaviors themselves. This is the meaning of systems. A system is composed of components that individually behave but in concert with the other components so as to produce the behavior of the whole. Knowing what those subsystems are and how they behave, essentially having models of them,

[1] This should not be interpreted to mean we have "perfect" knowledge/understanding of these systems. There are probably more details of such systems that physics may yet uncover as they continue to be studied. For example, we might do a better job of connecting the microstates with quantum-level laws.

[2] An excellent review of the nature of composition of more complex systems from simpler systems as the deep history of the Universe is by Tyler Volk (2017). His process of combogenesis is the story of how simple systems (like atoms) combine to form more complex composites.

leads to much better models of the system as a whole. Deep understanding provides a causal model of the system working. Henceforth, we explore the nature of obtaining deep understanding of complex systems and the construction of deep causal models as opposed to either statistical modeling (e.g., regression) or presumptive micro-to-macro mappings (e.g., as is often done in both dynamical systems and systems dynamics[3]).

Why this deep-understanding-based kind of model is superior may not be immediately obvious. The reason that deep understanding should be preferred over shallow understanding, even when the latter is adequate for many practical purposes, is that system behaviors can surprise us over longer time scales; the Universe is evolving. Things change and on all scales of space and time. This fact shows up most clearly in very complex dynamic systems such as living and supra-living systems (e.g., human societies or ecosystems). That means components of a system may change what they do by mechanisms that are just becoming clear to us. And if and when they do change it has an effect on the whole system that could not have been predicted by the shallow understanding models. A general rule is that the more complex a system is, the more potential there is for component subsystems to change.[4]

With deep understanding, coming from deeper analysis of components as subsystems, it is feasible in principle to construct models that can also demonstrate what we call *evolvability*. Such models might generate an ensemble of predictions or scenarios of anticipated behaviors of systems that would provide advanced warning about possible changes that we should know about. That is to say, deep understanding provides us with a path toward anticipation of different future states of affairs that might help us avoid threats or exploit opportunities. Armed with advanced knowledge of how systems might behave differently due to changes reduces the surprise factor.

Or deep understanding might simply provide us with the satisfaction of knowing better how the world works. For example, cosmologists seek deep understanding of the Universe not so much to exploit it, you can't very well control a super cluster of galaxies, but to better grasp the meaning of human existence within it.

Deep understanding in biology has led to the genetic revolution, the ability to decode genes and, now, even modify genes to achieve arguably practical purposes. All of the sciences achieve deep understanding through the approach known as reductionism, the process by which phenomena in one level of organization can be understood by knowing about phenomena at a lower level. For example, living systems (one level of organization) can be deeply understood if we take into account the chemistry that operates at the molecular level of organization. What happens in

---

[3] Dynamical systems are generally modeled as sets of differential equations integrated over time. Systems dynamics models are based on computer simulations of causal relations between macro-state variables.

[4] The capacity of complex systems, particularly ones with nonlinear internals, to undergo fundamental behavioral changes is called "evolvability." This topic will be taken up below in this Introduction and in several sections of the book.

the smaller scales of time and space (molecules) materially explains how living systems achieve their behavior at the larger scales of time and space (cells).

In this book, we will attempt to generalize the concept of reductionism from a systems science perspective to produce a methodology for gaining deep understanding of every kind of system. Our main concern will be directed toward what we call *concrete*, or real physical, systems. This is distinguished from *abstract* systems (see below). The reason for doing this is that the greatest needs for systems analysis (SA), modeling, and design are in relation to concrete systems, particularly human-designed or modified systems. We need to understand our world, our ecosystems, our organizations, etc. because we are constantly doing things that impact them and too often do not appreciate the consequences before doing so. Having a method to gain deep understanding before taking such actions might help reduce the numbers and severity of unintended consequences.

1.1.1 Talking About Systems

It will be important to understand a significant difference between systems science and the other sciences with respect to how the subjects are described. In the sciences, the subject of description is the specific substrate of the science. Biologists describe biological phenomena; chemists describe chemical phenomena. Their subjects are specific and their methods of description are particular to their domains. This is true even when the subject is a general pattern of phenomenon within that domain, such as when chemists derive laws of reaction rates or biologists expound on "laws"[5] of behavior in animals. When we read a paper reporting on a biological phenomenon, we expect to be told about the specific elements of the phenomenon, including names of the elements, the relations between them, and the dynamics of their interactions that collectively produce the phenomenon. We expect to be provided with specific measurements and analysis of the data relevant to the phenomenon itself. This is basically the same for all of the sciences and is just as much the case for various comparative studies between different representative elements, for example, between different species in biology, as for direct studies of single elements.

In systems science, the situation is quite different and we have a different way of talking about systems as phenomena.[6] The concepts of systems science apply to all

[5] The quotes here, and not on the chemists' laws, is a reminder that law-like phenomena seem to get rarer as we go from physics up the ladder of complexity to biological and psychological phenomena. There has been a long active debate within the sciences and the philosophy of science as to whether there really are laws in a strict sense, or emergent "rules" that govern the phenomena. See Unger and Smolin (2015), Unger's chapter 5 in particular.

[6] George Klir referred to Thinghood vs. Systemhood (2001). The former is the content of a science—what the thing is—while the latter is the set of properties that defines systemness and is applicable across all sciences.

domains of the sciences. When we describe a system phenomenon, we are not describing a specific phenomenon in a specific domain. Rather we are describing a general or meta-pattern of phenomena that is applicable across all domains. Going beyond mere comparison studies, or even analogous phenomena descriptions, we enter a domain of description that finds isomorphic relations that are true regardless of any specific subject domain. This general mode of description is different from how we approach descriptions in the sciences as a rule.[7] In Chap. 3, on the ontology of systems, we will be providing a guide to the generality of structural and functional patterns in systems science. In Chap. 4, we will develop these ontological elements into a language of systems that will provide us with a way to talk about systemness applied to specific phenomena in the sciences and engineering fields. We will assert (with evidence) that *systemese is a universal language that allows us to describe any concrete phenomenon regardless of the substrate's domain.*

Hence, this book will provide descriptions of general patterns, those that systems analysts will be able to use as guides to discovery of specifics within particular systems of interest. As much as possible we will also provide several examples of how the patterns pertain to several different phenomena in different knowledge domains. For example, we rely on examples from different areas of biology, physics, social sciences, and engineering to demonstrate the applicability of the patterns.

This way of talking about systems in general terms as the major mode of description may tend to put off domain scientists who are used to reading about the specifics of a phenomenon. To some it may even seem like a kind of "arm-waving" as a distraction from substance. But it is not. Systems science operates at a meta-science level, but we feel strongly a necessary level in order to better talk about concrete phenomena within any of the sciences. And as will become evident as we work through specific examples using the language of systems, how having a basic knowledge of systemness will facilitate transdisciplinary communications between the domain experts. Having systems language as a kind of Rosetta stone translator from domain to domain should significantly improve those communications, particularly with regard to the transfer of meaning (semantics).

1.1.2 The Nature of Systems Science

The study of systemness has a long history when viewed in retrospect.[8] Just as the way we talk descriptively about patterns that are isomorphic across different domains, the study of those patterns is the domain of systems science (c.f. Friendshuh

[7] Somewhat similar but still restricted descriptions occur in more general fields such as evolution theory, ecological theory, and reaction theory (to name a few), where the emphasis is on categories of phenomena. Ironically, however, these are the same areas that are currently morphing into systems sciences applied, for example, systems biology!

[8] See for example the excellent summary in Rousseau et al. (2018, Chap. 1).

and Troncale 2012). Consider a dominant pattern of organization that is found ubiquitously in nature and human-designed systems, a hierarchy of organization. Any system can be shown to be comprised of interconnected subsystems. And each of those will in turn be comprised of interconnected sub-subsystems. Each level in this hierarchy contains simpler elements or components (in Chap. 3, we will clarify the meaning of components). Eventually there is a level of the "simplest" components (also clarified in Chap. 3). So, the regress is not infinite.

Systems science is a transdisciplinary approach to discovery and characterization of the general nature and aspects of patterns, such as hierarchy of organization. It is transdisciplinary in that the systems scientist must examine a wide array of systems from many different subject disciplines in order to derive the isomorphic nature of the patterns of interest. Systems science seeks understanding of the nature of reality, which we believe is reflected in the fact that these patterns can be found so pervasive.

Systems science is concerned with the general theory of systemness.[9] It seeks to investigate the general nature of patterns that apply to all systems and derive a general theory that can, in turn, inform the disciplinary sciences as they explore their own subject domains. It should also be informative to engineers when designing complex systems. This is why we think of systems science as a meta-science.

That having been said, it is important to establish a basic understanding that what is going to be covered in this book is emphatically *not* presented as being a general systems theory (GST) as envisioned by von Bertalanffy (1968) or general systemology as described by Rousseau et al. (2018). Rather, our claim is much less grand in scope. In Chap. 3, we present *an* ontology that we believe covers the notion of systemness and the nature of an ontogenesis process that has produced the complex Universe we observe today. In Chap. 4, we present what we term a "*formal framework for understanding complex systems*" based on the ontological commitments made in Chap. 3. The aim of this framework is not to be taken as a general systems theory per se, but to guide the exploration of system properties in analysis and assist in designs of systems artifacts. It is possible that future research may arrive at a true General Systems Theory (labeled GST*, where the apostrophe, pronounced "star" signifies completeness) that will contain, at its core, some of what we are arguing in this book. Time will tell.

Chapter 2 will review a set of principles regarding what is known today about the patterns that recur in all areas of knowledge. This review comes from the author's previous book, with co-author Michael Kalton, *Principles of Systems Science*, published by Springer in 2015.

[9] Ludwig von Bertalanffy (1901–1972), a mathematical biologist, proposed the idea of a general systems theory (GST) and also the review presented in the reference in the previous footnote.

1.1.3 Concrete and Abstract Systems

It will be important to differentiate between concrete or real physical systems and abstract systems (Miller 1978; Ackoff 1971). Examples of concrete systems include animals, organizations, nations, and the whole Earth. Abstract systems are conceptual, that is, represented in a medium other than being embodied in a real physical, working structure and come in two flavors. Pure abstract systems are those, like the natural or real numbers, that exist as mental constructs that can be used generally for purposes such as measurement (in the other form of abstract system) or mathematics. Such abstract systems are applicable across a broad array of abstract systems of the second kind. The second form of abstract system is what we generally call a "model," of which there are several forms (see below). We will be concerned, in this book, with the relation between the concrete and second form of abstract systems, linked by the first form of abstract system. That is, models of concrete systems using abstract representations provided by the first form of abstract system. For example, an equation of motion that describes the progression of a pendulum's motion back and forth through an arc is an abstraction of the second form of a real pendulum. It is a model of a pendulum that captures its dynamics represented in the real number domain.

The distinction between abstract and concrete systems is a little less clear than suggested by this introduction. In Chap. 2, those subtleties will be explored more thoroughly. Anticipating that exploration, however, consider that even abstract systems take on a hint of concreteness when instantiated in the human mind. Indeed, all of our mental concepts could be considered abstract systems that, because they are based on real functioning neural firing patterns, are approachable as a form of concrete system!

As one example of an abstract system of the first kind, consider the theory of deterministic chaos. It can be invoked to explain a category of nondeterministic behavior in some physical systems such as the activities of the atmosphere we call weather. The models of chaos are important to our enterprise of understanding concrete systems, for certain. However, our focus will be on the internal aspects of concrete systems that give rise to chaotic behavior rather than the models that explore the concept of chaos (or any of the various forms of complexity generation). There are a large number of volumes that deal with such models (e.g., the Lorenz attractor), and we will be referencing some of them as might apply to understanding a specific example of concrete system.

The relation between concrete and abstract systems follows a straightforward pattern. Our objective is to show how to construct an abstract system, a model, from a deep understanding of a concrete system, using abstract systems of the first kind (mathematical constructs). Our approach is, however, not in line with the traditional use of modeling in the sciences and engineering. That is, the use of models has been seen as the way to come to understand systems outside of deep analysis. Based on some preliminary observations of a system's behavior (as a black box), a mathematical model is constructed and solved (run) to produce a result that is then tested against the behavior of the concrete system under experimentally controlled (or naturally occurring) conditions. The tradition is based on the notion that the model need only be constructed to answer questions of interest to the modeler and need not involve details considered to be outside the scope of those questions. As will be argued, this approach becomes increasingly problematic in the face of higher orders of complexity.

Complexity is the enemy of deep understanding in a timely fashion, unless the methods employed are systemic. We advocate that the program of science be turned around; that it is possible to use principled systems science to first obtain deep understanding and then construct models that will provide veridical predictions that can be used for design of complex artifactual systems and policies for governance of human affairs.

Our objective is to demonstrate how to come to deeply understand concrete complex systems and their environments first and then construct models from this understanding. What this book will show is a methodology based solidly on systems science theory that will lead to that kind of understanding. What will be demonstrated is that a deep understanding of a system leads to much better outcomes in terms of prediction or, in the case of engineered systems, better products.

Science, to date, has been the societal process of gaining deep understanding. But it has pursued a program of "discovery" that involves generating hypotheses regarding the behavior of interesting phenomena (in various domains such as chemistry or biology), constructing abstract models of these phenomena, for example, laws in physics and law-like relations in biology, and then making observations on the actual behavior of the phenomena in comparison to the mathematical solutions generated by the models. Upon discovering deviations of the models from the actual behavior (e.g., relativistic deviations from Newton's laws), science proceeds to dissect the phenomena to seek internal mechanisms, deconstruction for the purpose of discovering deeper understanding. In other words, science quite ordinarily starts with building models and then proceeds to open up the phenomenon to deeper inspection when the model fails to adequately predict the real system.

This formula—construct a model, test it, and dig deeper when the model fails—has been adequate in the history of the sciences. It works well enough when there is no hurry to gain deep understanding. However, there is reason to consider that we might not have the luxury of time anymore. In the second chapter, we will consider a new approach to gaining deep understanding that turns the program on its head. The main thrust of this book is to show how systems analysis, performed in advance of constructing models, can lead to deep understanding more rapidly. As will be argued, this approach is made possible by significant developments in our ability to decompose systems without destroying the inherent interconnections between component parts.

1.1.4 Systems and Complexity

For the last half of the last century and the first decades of this century people in many different disciplines, from the sciences and engineering to politics and governance to history studies, have been confronting increasingly complex issues.[10] The nature of the objects of their primary disciplines is now seen to be more complex than was historically true before World War II. The various disciplines, in fact, were established along the lines of silos such as they are organized in the universities because traditional reductionist methodologies operating on relatively simple objects of study worked amazingly well. Figure 1.1 shows a traditional hierarchy of major scientific disciplines and the root problem that runs through all of the sciences.

What the sciences are exploring and attempting to explain are phenomena that play out at each of these levels. Physics, for example, is concerned (among many other sub-subjects) with atoms and their structures and properties. Chemistry is concerned with their interactions. So chemical bonds (explained by physics) generate chemical reactions and molecular structures. Chemistry has a direct connection to physics in this regard but works with a new level of complexity that emerged from the atomic structures studied by physics.[11] The field of study called chemistry depends on the field of study called (atomic) physics just as the existence of molecules and chemical reactions depend on the properties and dynamics of atoms.

Biological phenomena depend on chemical phenomena. The study of biology now very much depends on the study of (organic and bio-) chemistry. And that same kind of relation proceeds at each level up the hierarchy. The gaps between the major subject boxes are filling in with multi-disciplinary sub-fields. As the sciences have matured, we find that knowledge from lower levels is reaching higher into the upper levels. For example, modern psychology now rests on the biology of the brain, but that in turn rests on specific chemical phenomena (e.g., neurotransmitter-receptor chemistry). The higher we go in these subjects the more we find we must understand the specific roles of phenomena that are studied at lower levels. This is a natural outcome of the fact of systemness; that systems are hierarchies of subsystems that, at some point, are based on the lower-level phenomena.[12] But it gets more

[10]Chapter 5, "Complexity," in Mobus and Kalton (2014) provides a general overview of various kinds of complexity but settles on Herbert Simon's hierarchy-based definition (see Sect. 5.2.2). A fuller development of the concept of "Simonian" complexity can be found in Sect. 4.3.3.7 in this volume.

[11]Physical chemistry is the interface between these disciplines, so is essentially transdisciplinary. Atoms emerged from the physical forces (strong, weak, and electromagnetic) during the emergence of the Universe (Big Bang).

[12]Throughout the book we use the term hierarchy as it is more commonly understood. However, the kinds of hierarchies we will be describing should be understood to actually be "holarchies," which are similarly structured hierarchies of "holons." These terms were proposed by Arthur Koestler in his book *The Ghost in the Machine* (1967). In the current book, the reader will recognize that what will be called "subsystems" are holons and the structural containment tree to be described in Chap. 4 is a holarchy.

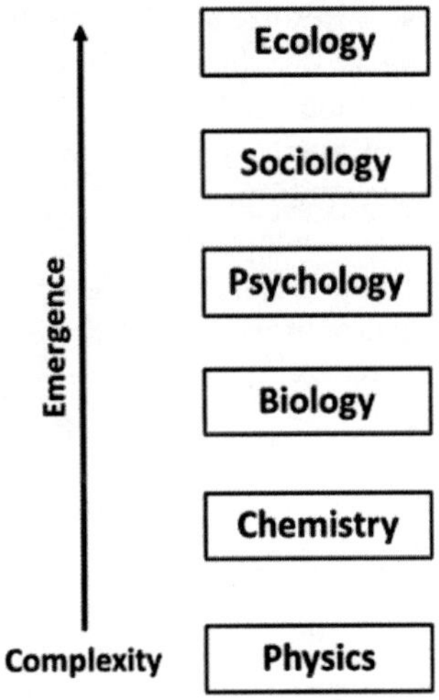

Fig. 1.1 There is a natural hierarchy of major science disciplines based on several aspects. The lower disciplines are fundamentally simpler to explore using the scientific process. Their objects of study are inherently simpler in terms of numbers of components. The emergence of greater complexity is explained in the text

complicated than that. As you go up the hierarchy of subjects (and phenomenal types), the levels below are themselves impacted by the kinds of actual structures and functions that do emerge and set patterns of phenomena at the higher level. One example to make this clear is the fact that all amino acids that form proteins in living systems have a left-handed chirality. Somewhere in the origin of life a presumed accident of nature broke the symmetry between left- and right-handed amino acid molecules being preferred (a form of symmetry breaking?) and it stuck in all subsequent evolution. Thus, causal arrows of influence point both upward and downward in the phenomenal processes contained in each of the boxes in Fig. 1.1.

At first, in the late nineteenth and early twentieth centuries, each of the sciences managed to stay within the bounds of its surface-level phenomena. But the methods of reduction were increasingly successful as instrumentation and models of phenomena improved.

Then, early in the twentieth century, people began to run into three aspects that started complicating the grasp of knowledge. One was the increase in structural and functional complexity of the objects under study as well as the complexities of how those objects operated in their environments (thanks to the aforementioned success of reductionist approaches but also to the fact that the objects of study were including more components at a given level). Another aspect was the realization that the behaviors of the objects being studied could not be described quite as simply as had been the case with simpler objects. Nonlinear dynamics and feedback structures started complicating the approaches to analyzing objects of study. The third aspect was the recognition that the objects of study often times spanned different fields or disciplines (those causal influence arrows going both up and down).

To really understand an object of study required more interaction between experts from different fields. But this is problematic because people in different disciplines often speak very different languages and have different approaches to study. Researchers in many fields began using what we call "systems thinking" to analyze the objects. The recognition that these more complex objects could not be understood in the standard disciplinary ways led to the explosion of new disciplines that arose to address these issues of complexity and interactions between disparate kinds of objects.

Fortunately for everyone interested in developing methods for dealing with complex objects and their environments, the advent of the digital computer coincided with the recognition of the problems associated with complexity and heterogeneity of component parts. Boosted with the development of the transistor and later the integrated circuit, computing became widely available and the capacity to build computational models of systems, as the complex objects of study came to be called, grew rapidly. Purely mathematical modeling, for example using systems of differential equations, was augmented rapidly by discrete recursive modeling, which allowed researchers to explore the dynamics of systems with feedback. The main difference between these approaches, however, was the requirement to know how a system worked from the inside in order to specify the internal behaviors and structures that gave rise to the whole system function. It was no longer possible to simply model a system based on data collected from observations of past behavior. That approach is still used, but still suffers the same problems described above. One needed a causal explanation of how the parts fit together and worked together.

There are three basic ways to approach modeling a system with complex structures and behaviors. The method applied to many kinds of systems, especially those where exploring the internals through dissection leads to loss of function, was to use a process called system identification, which looks strictly at the mathematical relation between inputs and outputs and builds a (usually) statistical function that can be used to "predict" behavior of the whole system under variations in the inputs. Such a model is often referred to as a "black box" since one cannot see inside the system to understand how it produces the results it does. As long as systems were seen to be relatively linear (or piece-wise linear) in their functions, this method would suffice to make predictions.

A second approach, used for example in biological work, is to take what one hopes is a representative sample of a population of similar systems, dissect them to find out something about what is going on inside, and use that knowledge to infer the input–output relations needed to, again, make predictions about behaviors under variations of the inputs. The latter can be determined by experiments on the still intact systems that can verify the understanding derived by the system identification approach. This kind of model is often referred to as a "gray box" since the inferences made between inner parts and outward behavior are still largely uncertain and require multiple observations and experiments to reduce that uncertainty.

A third approach that has been largely restricted to the engineering fields uses what we call "reverse engineering" or taking a complex machine apart to see what its parts are, and from a previous deep understanding of what the various kinds of parts do, work out the functional relations with fairly high certainty. The model is then called, as you have already guessed, a "white box.[13]"

This third approach leads to a system model with the highest level of capability to predict future behavior because it includes detailed understanding of what the components are, what relations they have with one another, and what independent behaviors can be expected from them (highly deterministic). Additionally, and perhaps most important, this kind of model allows the inclusion of nonlinearities resulting from internal amplifications and feedback phenomena. These are the aspects of systems that give rise to really interesting (and important) complexities that need to be understood for real anticipation (prediction).

Increasingly white-box-style analysis and modeling is being applied in the other sciences where the accumulation of lower levels of organization, thanks to reductionist methods, provides the same kind of knowledge of component behaviors as was the case for engineered systems. For example, the approach can be used in modeling and predicting the behavior of whole ecosystems since it is relatively easy to obtain information on the component plants, animals, fungi, bacteria, etc., and much is now known about species-determined behaviors as well as food webs and other interaction relations. It still takes a lot of work to obtain this information, but the approach of studying a system from the inside is feasible. Another example of this is when Information Technology (IT) specialists analyze the internal workings of an organization that is in need of information systems able to support management decisions.

However, the trend toward white box analysis of living systems in biology has made a tremendous amount of progress over the last several decades as well with the development of non-intrusive instrumentation such as functional magnetic resonance imaging (fMRI), ultrasound imaging, and now fairly spectacular abilities to see inside the brain and other tissues using "optogenetics."[14] These methods for examining the insides of opaque systems, while the latter are actually behaving, have given biologists dramatic abilities to pursue the same kinds of models that have been helpful in engineering and physics/chemistry since the last century.

Why is the ability to produce a white box model superior to gray box ones? It is the reduction in uncertainty that comes with what we have called "deep understanding" of the system. If we know more about how a system works on the inside, we can more reliably predict how it will behave given new environmental conditions.

---

[13] Alternatively, we call these "transparent boxes," and black boxes are referred to as "opaque boxes." The author prefers the latter terms since the former are too strongly associated with machine descriptions and the latter terms are more neutral with respect to substrate media (a living system, for example). Additionally, the latter terms are not associated with specific disciplines.

[14] A very good article in *Scientific American* reports on an important new imaging technology developed by one of the developers of optogenetics. This technology allows the imaging of the living connectome.

1.1.5 Deep Understanding

There are many different kinds of *knowledge*. We can know facts about things and relations between different things. But the epitome of knowledge is *understanding*.[15] This is the kind of knowledge that gives us the ability to anticipate future outcomes of processes and dynamic relations. Deep understanding comes from knowing not only what a thing is made of and how it is connected to other things in its environment, but how it works internally as well, which includes how its internal processes respond to changes in its environment. With understanding comes an ability to predict with some confidence that if the environment does something the system will respond in a particular way.

To be clear, all understanding comes from the knowledge we put into our models, be they purely mathematical, computer-based, or mental. The model is not the thing being modeled, but the intention is to obtain enough knowledge about how a system works to bring our models as close to representing the thing-in-the-world as realistically as possible. We will have much more to say about models, model construction, and the construction of a knowledgebase from which those models can be generated. But for now, recognize that we are pursuing the development of a method, a process, by which we can ultimately build more reliable models by gaining deep understanding of the internal workings of the system.

Deep understanding is more than just being able to anticipate behaviors. A predator has to anticipate the movements of its prey in order to be ready to pounce. This it accomplishes by having learned a set of causal relations between what is going on in the environment and what the prey normally does in response. If this is the dry season, the prey is likely to be gathered at a few remaining water holes so that is where the predator goes for dinner. The predator doesn't think about the prey being thirsty, its motivation for going to the water hole. It only has a model of what the prey does, its behavior, not what causes its behavior. This is effectively the same as the correlation-based regression model discussed above.

Rather, deep understanding goes beyond mere associations based on historical observations. It involves a deeper knowledge of causal mechanisms and motivations.

There are a considerable number of books covering knowledge discovery and construction methodologies. The traditional sciences, physics, chemistry, biology, psychology, etc., have long employed what is called the scientific method for this purpose. Other fields of inquiry have a variety of methods that work along the same lines as the scientific method but may not employ mathematics in the same way that the sciences do. History, for example, relies on the discovery of meanings in texts and historical artifacts. It is an interpretive method (i.e., involves some subjective thought) rather than a rigorous objective process. But it seeks the same sort of understanding about the human world that the sciences work toward.

---

[15] Several authors have considered higher levels of knowledge such as wisdom—the knowledge of how to use knowledge (Potter 1971)—and higher still is conscience or moral sentiment (Damasio 1994, p. 230). These higher forms of knowledge are relevant to high-order system patterns such as agency and governance, which will be discussed in Chap. 12.

The nature of understanding comes under the philosophical framework of epistemology (see that discussion in Chap. 3). Some of the major questions posed in this branch of philosophy are: What is knowledge? How do we humans gain knowledge? Why do we do so? There are many more, but this is the basic framework. This book will basically be about an approach to the gaining of deep understanding of complex systems.

1.1.6 Understanding Systems

How do we human beings succeed in living in the world? What does it mean to be successful? The evolutionary definition of success is "fitness," an organism's match of physical and behavioral characteristics meeting the demands of the environment in which it lives. A species is considered fit if its member individuals are fairly successful at staying alive and reproducing at least a replacement number of their kind. This species-level fitness is compared with any possible competing species that would usurp what is called the ecological niche. For example, if two different species of predators relied on the same prey for food, the one that was best able to catch the prey would tend to out-compete the other and, all other things being equal, eventually dominate the ecological niche.

Within the species individuals tend to have a range of variations regarding any traits or behaviors such that the population average defines the species fitness as above. Some individuals may have trait variations that reduce their average fitness relative to the species norm. Others might have variations that give them advantages above the population norm. In this case if there is a selection pressure, say from a competing species, and the individual is generally more fit, it will also tend to produce a larger proportion of offspring and help strengthen the species position in the niche.

Fitness, both species and individual, is a well-understood concept in biological evolution and is a core explanation, along with selection, for speciation or the long-term development of organisms. But, what about human beings? We are biological organisms. In what way are we fit?

There are a number of dimensions to this question. The dominant one is the fact that human beings have evolved to be able to adapt their environments to their biological needs or invent some technology that protects them from the vagaries of different environments. In other words, human beings have the ability to construct their own niches rather than simply fit into the existing conditions. We have succeeded in the evolutionary sense. And we have done so spectacularly. Almost too spectacularly, as it turns out.

Since there are myriad books on the subject of how clever humans are, how the various revolutions in technology, the Agricultural, the Metallurgical, the Industrial (I, II, and III), and the second agricultural (or "Green") revolutions have allowed us to literally cover the planet, and how doing so, along with the expansion of consumption, has turned around to put us in jeopardy (changed the fitness equation), that ground will not be covered here. The question we are really asking is: What characteristic of our thinking capacity has endowed us with the ability to do clever things? The answer to this, we think, is *our ability to understand how the world, or parts of it, works*. We have the basic capacity to understand the systems we encounter and the system in which we live. Our brains have the ability to observe phenomena carefully and postulate theories about causes and effects. We can build mental models of how things work and we have conscious access to those models to use them to anticipate the future (to some extent). We have, moreover, the capacity to test our models by carefully intervening in phenomena to see if our understanding is correct. We test our understanding through "experiments." The human brain is pre-equipped to do what amounts to science, at least an intuitive form of it.

This ability is what made it possible for humans to first exploit the effects of fire, to test the way in which stones could be chipped to produce a sharp edge, to discover the heat-preserving benefits of wrapping themselves in animal skins, and every cultural innovation that has emerged since.

It is the understanding of how things work, understanding the regularities in nature, which is at the root of human success as a species and as individuals. And it is the understanding of systemness, how things are interrelated and causally interact with consequences (e.g., feedback), that has made us so successful.

Unfortunately, we are not sufficiently equipped to understand broadly. We can, through our modern rigorous practice of science, understand very deeply but the vast majority of people are capable of specializing in only one or a few subjects. Polymaths are rare. Our sciences have operated for several centuries, quite successfully, by ignoring interactions between phenomena of interest with other phenomena that seem remote. We have been very successful at drilling down deeply in subjects but too often at the cost of understanding how the low-level phenomena relate to distant other kinds of phenomena. Sometimes we find out later that there actually are causal connections that were hidden from view, or at least not in our focus. Sometimes we find out because the interactions can have negative consequences. The connection between burning fossil fuels and climate change is a dramatic case in point.

We have also been extremely successful in exploiting what we have learned within various domains of knowledge. Engineering, in many different forms, involves taking the knowledge derived about an area and figuring out how to build technologies that serve our purposes. Historically, this has been realized within domains of phenomena such as mechanical or chemical or electrical. Engineers, trained in the formal methods (mathematics) of the domains have been able to invent and refine technologies that have improved the standards of living for large portions of the population.[16]

[16]Note that it is the political–social framework that prevents these advantages from being shared globally. There is no reason that people all over the world should not benefit from the standard of living provided by engineering, except for political machinations and the fact that the human population growth is out of the normal controls.

Once again there have been thousands of books (probably many more) on the subject of how engineering exploits understanding of phenomena, so we will not spend any time rehashing the obvious. But the point we wish to emphasize is, again, it is *understanding* that is at the root of this successful strategy of living.

1.1.7 Understanding Causal Relations

In all of this what is being understood is the causal relations that exist between events that occur and consequent events, or, equivalently, states that lead to other states. These relations are most often probabilistic, that is they are related statistically, not absolutely deterministically. The great successes of science and engineering have been in refining our ability to understand these stochastic relations and still do a very good job of predicting outcomes from varying initial conditions.

1.1.8 New Kinds of Causal Relations

For the entire history of science and engineering the kinds of causal relations that had been explored and understood have been what we call linear; that is, event *X* causes transition *Y*. And event *X* was caused by transition *W* that happened in the immediate past. We have learned how to trace linear causations backward (abduction) to discover what caused the observed event, and forward (deduction and induction), discovering what an event will cause in the future. We have learned how these causal relations are moderated by stochastic (probabilistic) factors like noise or nonlinearities that give rise to chaotic dynamics.

But now we are realizing this is not sufficient for complete understanding.[17] As we have gained greater understanding of how things work in our various disciplinary silos, we have also begun to realize how connected these things are across silos. Even within silos we have begun to realize that multiple phenomena at lower levels of organization contribute to the phenomena we observe at higher levels. For example, we have known for a long time that organic chemistry must be understood in order to grasp biochemistry, which, in turn, needs to be understood in order to grasp metabolism, which, in turn, needs to be understood in order to grasp cells… and so on. We've also known for a long while that organic chemistry is predicated on not just the peculiarities of carbon but the quantum properties of the electron shells of carbon atoms. Chemistry depends on physics.

[17] Complete understanding is a relative term! Complete suggests absolute, and science does not deal in absolutes (except possibly in temperature). The progression in science and engineering is based on "more complete" understanding, which characterizes a process of getting closer to the "truth" but not claiming to have arrived at the ultimate truth.

The sciences have been able to move ahead because it is possible to "abstract" the aggregate features of the lower levels of reality as we move up the hierarchy of physical organization. They have been able to act as if the lower levels are just "details" so long as those lower levels of phenomena acted in consistent ways. Fortunately, they mostly do.

What has started to become obvious is that as we move up the hierarchy of organization we also move up in terms of complexity. Phenomena that could not have been predicted strictly on the mechanics of the substrates start to emerge as those mechanics give rise to more complex structures. The origin of life problem is a case in point. In theory, we should be able to explain how metabolism in ancient cells originated and entered into biological evolution. In practice, we have not yet achieved that goal.

1.1.9 Complexity and Interactions

What we humans are coming to understand, though we don't understand it well enough yet, is that our whole world is a system, an extremely complex one. It is composed of myriad heterogeneous subsystems (see Principle 1 in Chap. 2). They are coupled through various cyclical flows (e.g., carbon cycle, hydro cycle) that are driven by massive energy flows through the planet. The subsystems are themselves composed of sub-subsystems that interact within their parent system with one another and some of them interact with other subsystems in the larger system. These interactions can be characterized also as flows of material, energy, or influence (information) that act to shape and regulate the structures and functions within.

In a very real sense, though difficult to perceive at times, everything is connected to everything else. Pull on this string, and it is likely that many things will result in different parts of the system changing in some way. Any change to one part of the system is likely to affect every other part of the system even if the effect is infinitesimal at the outer reaches. Some changes may be amplified by nonlinear processes and have significant impact even though rather minute in origin.

And this is where our problem arises. We humans are part of the larger Earth system. Whatever we do there will be some effect generated that affects the other parts of the system. We have grown in size and power as a collective force affecting a complex network of components. And, as we are now aware, it is too often in negative ways at the scale of our global civilization.

Even at the other end of the scale, at the local level, our technological wizardry is showing signs of fraying. Each new release of computing and communications technologies seems to have more bugs than the last release. Our attempts at producing artifacts like self-driving cars are beginning to demonstrate the limits of our ability to design super complex devices that are supposed to operate in a super complex environment. As this is being written there have been two deaths reported associated with such cars along with property damage. Those are unwanted interactions between one kind of complex system and others (human beings, the legal system, the economic system, etc.).

The simple fact is that we have reached a likely point of diminishing returns on the unununderstood increase in the complexity of our inventions. And the negative consequences are starting to show. That complexity arises from the desire to create new kinds of machines that combine diverse technologies in seeking additional functionality (the current buzz words are "cyber-physical systems" and "Internet of Things," IoT). For the most part, today, that means integrating computational and communications capabilities and the programs that exercise the mechanical parts of the machines. In turn, this means bringing together engineers from different backgrounds with different languages and different methodologies who must nevertheless work together to produce an advanced design. On top of that newer complexities have arisen. Environmental hazards, energy efficiency, material resource efficiency, recyclability, and many other factors have to be considered in designs.

Complexity is the bane of achieving understanding and designing new artificial systems. Systems science provides a direct way to grapple with complexity. The principles discussed in Chap. 2 provide a pathway to overcoming complexity in the pursuit of understanding. As discussed below, intuitions about systemness, what we call "systems thinking," is a necessary condition but insufficient when striving to overcome the vagaries of complexity. What is required is the discipline of systems science and the methods of systems analysis that are designed to work directly with complexity so as to manage it. The first step is to be clear on what we mean by a system and the principles of systemness. This will be the main task of the chapters in Part I.

1.1.10 Simple to Complex Adaptive and Evolvable Systems

Throughout the book we will be referring to four categories of systems based on degrees and forms of complexity. The degree of complexity is characterized by factors such as the number of elements and kinds of elements, and the number of levels in the structural/functional hierarchy as described in Mobus and Kalton (2014, Chap. 5). The forms of complexity are related but have to do with a system's capacity to interact with changes in its environment that deviate from operating norms. In general, the more complex systems have more capacity to cope with changing environments. Figure 1.2 shows these categories and provides some examples.

The four system categories are: simple systems (SS), complex systems (CS), complex adaptive systems (CAS), and complex adaptive and evolvable systems (CAES). We will have very little to say about simple systems (e.g., clocks or automobiles). We have some to say about complex systems (e.g., jet liners of space shuttles), which may be complicated but still not capable of the facility that seems to differentiate nonliving from living systems. That capability is an ability to adapt to changes in the environment that go outside of a nominal operating range. Adaptive change generally requires the system to alter its internal distributions of material and energy to compensate for changes. It is a key characteristic of living systems that can alter physiology or behavior to address external stresses brought on by

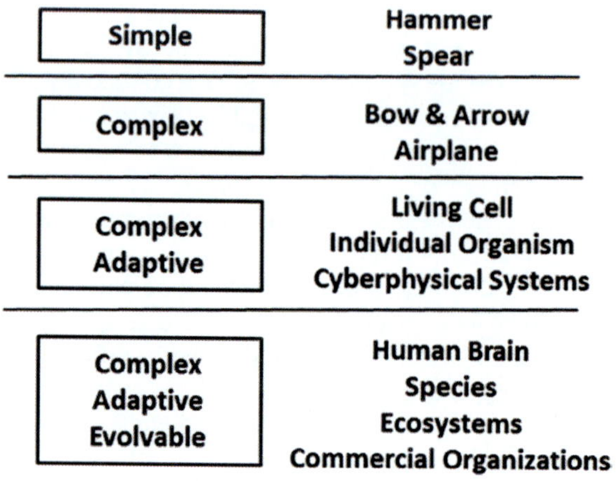

Fig. 1.2 Systems may be classified into these four categories according to measures of complexity based on the heterogeneity and number of parts as well as the numbers of levels of organization and by behavioral characteristics

environmental changes. Homeostasis is the paradigm example of an adaptive response.

Much has been written about CAS in a variety of arenas, such as complexity science. Most writers seemed not to distinguish between mere adaptivity in individual systems and evolutionary reconfiguration in more complex systems such as populations (species). In Mobus (2015), that distinction and a new class of complex systems were introduced and explained. Complex adaptive and *evolvable* systems (CAES) go beyond mere alterations in internal distributions and shifts in work processes for compensation of stresses. Evolvable systems are those that are able to adopt or eliminate functions (and the structures that perform them) de novo. That is, a system that has the capacity to incorporate a completely new mechanism for obtaining and processing a new material resource (e.g., a company starting a new product line), performing a new function, and perhaps finding a new purpose, has evolved. In the living world, pre-humans, individuals within a species were adaptive, but not evolvable. The species, or more generally the genus, as the system, was evolvable. Throughout the book we will refer to these categories of systems as they are pertinent to the theories or examples. It won't be until Chap. 10, however, that we will examine the full meaning of the categories. There we will elaborate on what we call model archetypes, starting with the CAS and CAES generic models. We will also explain the nature of several subsystem model archetypes, an agent, an economy, and a governance system that constitute the inner workings of the CAS/CAES archetypes.

1.2 A Systems Understanding Framework

An early approach to a knowledge framework can be found in Ackoff (1971). Ackoff calls for a "system of system concepts," noting that many terms and concepts used by systems thinkers were not sufficiently organized. The objective of this book is to develop a holistic system of systems understanding methodology based on the principles of systems science but for the purpose of doing system science and engineering. The same methodology, acting as a framework, applies to all of the sciences when they are directed at exploring the systemness within their individual domains.

1.2.1 Conventional Systems Understanding

Before proceeding to the outline of the framework proposed in this book, we should comment on the current widely practiced framework for understanding systems. The concept of modeling (especially mathematical and computer simulation) and systems understanding have gone hand in hand since the outset of the systems approach. In this framework the researcher starts with a hypothesis-like[18] conjecture about how a system works, what its relevant parts are, how they interact with one another, and what the relevant inputs and outputs are. They then construct a model using formal tools like a system of ordinary differential equations to describe the system as they conceive it. Using computing tools to "solve" the system output at some future time, given the inputs at time t_0,[19] they project the behavior of the system. Then it is necessary to observe the real concrete system in action to see if the predicted behavior is borne out in real life. Under the gold standard of scientific investigation, the researcher may be able to set up controlled experimental conditions to test the hypothesis. Otherwise, they need to hope for conditions that closely approximate the model inputs and record the behavior of the system as it reacts.

In any case the traditional approach is:

1. Make educated guesses about the nature of the concrete system and what are its important features.
2. Construct an abstract model of the system based on the suppositions made.
3. Run the model (do the computations) capturing the outputs as data.
4. Compare the model output with the concrete system's behavior to see if there is correspondence.
5. If yes, conclude that the hypothesized mechanisms are correct, otherwise tweak what you guess (educated, of course) is off in the model and run it again.
6. Repeat this process as often as necessary until the model behavior sufficiently replicates that of the concrete system.

This approach is really a reflection of the conventional scientific method and empiricism dating back to, for example, Galileo's attempts to grapple with how things fall to earth. Observe, hypothesize, test, and revise as needed. The history of science has been a gradual teasing out of details when the testing indicated more

---

[18] It is hypothesis-like, in that the conjecture says, in effect, the model I have created constitutes the important features of the real concrete system. If the two systems, the concrete and the abstract, both have the same behavior, within some arbitrary degree of accuracy and precision, then the hypothesis is *not* disproved. Otherwise, it is back to the drawing board.

[19] Most models require continuous input data as time passes.

knowledge was needed—further dissection was required to gain a better understanding of the phenomenon of interest.

The framework being proposed in this book turns this conventional approach on its head. Note that the decomposition of a phenomenon or system ends up taking place in any event, driven by the need to make the models better. The model, in this conventional framework, has been used as a spur to the action of deeper analysis more than a confirmation that our original guesswork was spot on. In what follows we suggest that the process of science, and systems science in particular, has reached a level of maturity in which we can effectively do the deep systems analysis (DSA) before necessarily producing models. In other words, the models we end up with are generated after we know what the system is and how it works at deeper levels. We are aided in this new approach by the fact that our knowledge of, for example, measurement theory, along with the state-of-art in advanced and inexpensive sensor technology, makes it possible to "instrument" systems in ways not achievable even a few decades ago. We have non- or minimally intrusive sensing methods (e.g., functional magnetic resonance imaging, fMRI) that allow deconstruction of concrete complex adaptive systems in ways never imagined previously. We've had advanced digital computing for many decades and so the ability to create models has been the main recourse for science and systems investigation. But now, for many systems of interest, we can analyze first and model second (and verify third is still a good idea).

One might be tempted to ask, then, what are models for if not to help us gain better knowledge? Well, they can still work in this way as needed. The process of functional/structural decomposition that will be described can still be spurred by models used as a tool when it is absolutely necessary to "guess" about mechanisms that we cannot easily directly analyze. However, much of what we really want from models is the ability to predict the future or, at least, project scenarios to aid us anticipate possible futures and make some decisions on actions that would be preemptive to prevent harm or take advantage of novel opportunities. With the proposed methodology, we will have models that are readily useful in this context with much less of the "development" phase shaping the model from the test-it-and-see approach.

A particularly important use of models that we will explore in later chapters is that they are used in the cybernetic problem of control, management, and governance (see Chap. 12). What we call "decision models" are used to process real-time data and make control decisions to correct errors. Such models cannot be based on guesses, no matter how educated they are. We must be able to generate veridical models early if they are to be used for this purpose.

1.2.2 Outline of the Framework

This framework is comprised of seven basic subsystems, four of them activities directly involved in obtaining system knowledge, one a governing process, one a monitoring process that provides information to the governing process, and one a core process, the knowledgebase, the structure of which links the activities together. Chapter 5 will provide a comprehensive introduction to these activities and structure. Here we will only introduce the ideas and a preliminary justification for why they are necessary for the entire enterprise of system understanding.

The first part is the activity called systems analysis (SA). This is the activity in which a system is first identified as an opaque-box[20] entity and then, using a structured analytic approach, converted into a transparent-box entity. Or rather a set of entities organized in a hierarchy of increasing details. The core structure (the second part) is a system knowledgebase, implemented in a hybrid of relational-object database technology along with Web technologies seen in examples like Google™ (for indexing and search) and Wikipedia (for editable content) to keep track of the system knowledge obtained through analysis. The third part involves the methods for generating various kinds of models from the knowledgebase for various uses. Most important among these are aids to mental modeling (i.e., diagrams) and building computer simulations for testing hypotheses. The fourth activity concerns the development of human-engineered systems, artifacts, procedures, and policies. These address the issues of systems engineering that apply to all kinds of complex designed systems from products to organizational systems to governance. Finally, all of these parts require that the concrete systems themselves, natural or engineered, be monitored on an ongoing basis with the predictions provided by the models (hypotheses) being verified by actual behaviors. These five will be expanded below.

However, the framework requires something more than a set of activities. Essential to the whole enterprise is the ability to use language to communicate the knowledge gained between multiple stakeholders or participants. In complex heterogeneous systems there will be many of these with many different (yet related) perspectives, and often from different disciplines using different languages. So, before any sort of mechanical framework can be developed and used, it is essential to consider what language will allow these multiple stakeholders to work together and gain the full value of understanding the system of interest (SOI). Chapters 3 and 4 will provide the explanation of the language that will make the understanding of systems feasible for all parties. Below is a brief expansion of the ideas of these five parts, but starting with the nature of a system language (SL).

1.3 Sharing Understanding

Understanding is a community process. No one human can possibly grasp every aspect of every phenomenon of interest. But, as a social enterprise, we should, at least in principle, be capable of deriving arbitrarily deep understanding of any phenomenon we encounter. Just as importantly we should be able to understand the

[20] We have adopted the term "opaque-box" to replace the conventional "black-box" terminology from reverse engineering. Opaque implies the possibility of transforming the opaque boundary to a transparent one.

connections between that phenomenon and all other phenomena as we expand our vision from a local to a universal scale.

Assuming a language that permits effective communications between members of the social system, the gradual buildup of multi-perspective understanding, along with methods for synthesizing seemingly different perspectives, through intersubjective integration provides a route to shared understanding.

1.3.1 Communications

Humans talk to each other. They use language to communicate ideas, facts, concepts, and meaning. Natural languages, which evolved as different cultural traditions spread out from Africa, resemble the speciation represented in phylogenetic trees in biology. And thus, people from different language groups have had difficulty communicating complex ideas, etc. Throughout history peoples from different backgrounds have had to work hard to learn translations of terms and interpretations of syntactical structures in order to share thoughts.

The same situation exists today for the various tribes of sciences. Each science has evolved different languages tuned to their individual disciplines. Even though English has become the dominant natural language of the sciences, the vast differences in terminology, meaning, and technical details of the objects of study in each discipline remain a drag on cross-disciplinary communication. Why this is problematic is that in today's world of complex objects of interest, multiple kinds of disciplines have to be called upon in order to study (analyze) them. The Human Genome Project is a case study in this problem. Calling upon scientists from computer science, genetics, chemistry, and several other support sciences, the project required that all of the participants basically learn a common language that borrowed terms and concepts from all of the disciplines. Geneticists had to become familiar with the ideas of computational complexity, while the computer scientists were called upon to learn the chemistry and structural aspects of DNA and genes. Initially this was challenging. Fortunately, DNA is organized as a string of "letters" that encode a "message." Computer scientists rapidly recognized the similarity between this and data structures with which they were intimately familiar and were able to develop efficient algorithms for deconstructing long strings of DNA to decipher genes. Thus, a sufficient commonality between genetics and computer science concepts allowed for the rapid deciphering of the genetic code.

It is possible that there is a similar Rosetta stone underlying the commonality among all science languages. In fact, this is the basis for why systems science can be considered as a meta-science. The claim, explained in Mobus and Kalton (2014) and in Chap. 2, is that systemness is universal, meaning that it is at the base of all objects and structures in the world. It follows that the language of systems should be a universal language that covers all phenomena (ideas and concepts) of interest. It could, therefore, provide a means for translating specific terms and meanings in one scientific language into any other language. An example will be provided below.

Chapters 3 and 4 are devoted to the development of a language of system. This is more than a language used to model systems. There are many of those that have been developed specifically to build abstract models of systems. Most have been designed to aid software and systems engineers capture their thoughts about what a to-be-designed system should be. A few, such as system dynamics (SD), were designed to capture abstract representations of existing systems (like the world—Meadows et al. 1972). All of these languages exist in order to produce abstractions of the real systems. Some of these abstractions can be turned into computer code and run in a simulation to produce an abstract representation of the behavior of the system model (which is presumed to represent the behavior of the real system, either an existing one or one yet-to-be built).

These languages serve an important purpose in allowing researchers or engineers to visualize the large-scale aspects of systems and their behaviors but do not produce the kind of deep understanding we described above. The models produced using these languages are simply great elaborations on the kinds of models discussed regarding shallow understanding. They are only as good at producing predictions or anticipations as the languages used permit and the modelers' skills allow. While this has proven good enough in many historical instances, it is not given that they will continue to be sufficient as we move inexorably into levels of complexity of systems that seem certain in the future. What is needed, in our opinion, is a language of system that is meant to actually describe in sufficient detail all aspects of a system of interest (and its environment). An abstract model might always be obtained from such a description if the language is sufficiently formal in its structure. At the same time, such a language permits many different disciplinarians to share their understanding as we try to do with natural language. Rather than reduce a system description to a set of state variables and functions alone, and those at a very abstract level for computational tractability, the language of system should be "speakable" as well as "viewable" as well as "computable." People should be able to talk to each other without over-abstraction causing the loss of details and meanings that support real understanding.

1.3.2 The Language of Systems

From another angle consider an interesting idea in linguistics. A number of linguists, psychologists, and philosophers of mind are proposing that the human brain has an internal, subconscious language that is essentially preprogrammed into the circuitry. They call it the Language of Thought (LoT) or "mentalese" (Fodor 1975; Pinker 2007b). The reasons for this proposal are too complex to go into here, but suffice it to say that there is a growing body of evidence that this is a valid notion. Let's go one step further to ask what are the symbols and syntax of this language? Assuming it is a universally possessed internal language—that is, all humans possess the same language—then that means there is already a basis for a common language. One more step. Suppose this mentalese evolved specifically to represent the world as humans found it.[21] What would be the most successful way to think about the world? If the world is a system of systems, then the language of thought should reflect this. It should be a language of systems, or *systemese*!

As it turns out modern functional neuroimaging is providing some preliminary support for this conjecture (see Mobus and Kalton 2014, Chap. 8, Sect. 8.2.5, and Chap. 9). All human beings recognize and talk about system concepts but using different terms and syntax and often having different surficial meanings. For example, the system concept of a "stock," a reservoir for a substance (matter or energy) or data, is universally used in many different contexts with seemingly different meanings. A retail store manager talks about her inventory of goods to be sold. A private citizen talks about the money he has in a bank account. An ecologist talks about the amount of water in a lake. An electronics engineer talks about the amount of charge contained by a capacitor in a circuit. All four use different words to describe the components of a system that temporarily "holds" a substance, along with associated notions of how the substance comes into the stock and how it goes out. All four have intuitions about the phenomenon of how much stuff is in the stock and that inflow and outflow rates work to determine this and it does not matter what spoken language they speak. A stock element of a system is a general and universal concept as is the dynamical behavior of its "level" due to difference in inflows and outflows.

The general claim is that the human brain of every individual on this planet is designed to grasp this general idea and all of the ideas of systemese. Chapters 3 and 4 will show how we derive a general systemese and give explicit (and English) names to them. Below we will introduce just a few systems concepts to make this notion more concrete.

The human brain does not just deal with names of things (nouns), relations (e.g., prepositions), and actions (verbs). It is also capable of quantification (and giving names to generalized forms) as well as constructing abstract relations between quantities (e.g., measurement, comparison, and calibration). That is, it attaches to concepts like inflow, stock, and outflow, magnitudes of those flows as per units of time. These may be qualitative, as in "that stream is bringing a lot of water into the lake." Or they can be more explicit, "the stream is deeper and wider after the rain so much more water is coming into the lake." Or it can be more explicit still, "the stream depth is twice as deep and twice as wide after the storm...." The brain has the ability to quantify measures, usually by comparison, and attach those to the phenomena of interest. Thus, communications consist of not only terms and syntax but also measures of magnitude (starting with sensory intensity measures) that add meaning to the terms. Handling quantification requires an addendum to ordinary language. We call it math.

[21] One implication of this notion is that some form of mentalese (systemese) is present in all brains throughout evolution. In other words, worms' very primitive brain-like structures process a very primitive version of systems, namely they are "aware" of limited environmental sources (e.g., food in the soil) and have sensory and processing ability to determine the internal states of their bodies and make response decisions that result in overall behavior.

What ties our quantification (concrete and abstract) to linguistic (concrete and abstract) representations is a kind of middle ground. As soon as humans evolved the language capability, they needed to also construct arguments to present to one another when making social decisions. Those arguments had to make sense in order for the species to achieve and maintain fitness in the late Pleistocene environment (relatively frequent climate shifts). Argumentation gives rise to logic that is valid reasoning forms such as deduction and useful heuristic forms such as induction, and abduction.

A full description of a system requires all three modes of thinking. Linguistic descriptions may be thought of as the skeleton onto which the flesh of mathematical and logical relations is attached for clarity and precision. Linguistic elements of language are most closely attached to our perceptual experiences. Thus, it forms the core of description. Mathematics and logic refine these descriptions, quantitatively and relationally.[22]

1.3.3 The Story of S: System Narratives

Communications is not a simple description of something for human beings. We do not share detailed descriptions. We share stories that tell of things that happened to the subject of the story, that tell of things the subject did, that tell of the subject's responses to situations, and so on. Human beings are storytellers. Indeed, every sentence that can be uttered is a micro-story. Every paragraph tells a scenario. A chapter tells a history. We don't just blandly describe the subject; we tell its story and that includes the story of what went on around the subject.

Systems thinking is effectively a form of third-person perspective in a narrative. The story is told from the vantage point of a quasi-omniscient narrator who can look down upon the subject and its environment and describe the unfolding story of its interactions with the other entities in its world.

One way to see this is to consider the story told by a system dynamics model using one of the popular SD languages, like STELLA. What the model simulation does is generate a story about what the values of key variables are over the course of the time of the simulation. The graphs that SD language produces are graphical representations of what happened as the story unfolded.

A really good language of systems must provide adequate descriptive power, but also adequate narrative power. Being able to tell us what the subject has done can tell us something about what we can expect the subject to do in the future.

---

[22] In addition, our vision processing module is interoperable with the linguistic module so that visual images play a role in describing models, e.g., flow diagrams, visual maps, etc. We will describe this more fully in Chap. 4.

1.3.4 The Mathematics of Systems

The formal language of science and engineering is often said to be mathematics. This has been demonstrated abundantly in all of the sciences and since what all sciences investigate is the systems in their domains of study, and all of the sciences use various levels of mathematics, it follows that all of mathematics is relevant to systems science. But the notion that mathematics is *the* language of science, engineering, and systems is too simplistic. For human understanding it requires linguistic, mathematical, and logical reasoning elements to describe things and situations. The linguistic components provide the semantic grounding and the logical and mathematical components provide two basic augmentations, the quantification of state variables (e.g., the level of substance in a stock of fixed volume) and a way to produce abstract models of the system. All three of these aspects are needed in order to talk about systems.

However, this book seeks to apply what is learned in systems science to a generalized methodology for gaining deep understanding of systems regardless of the domain of study. We employ the framework described above, the description of a universal structure that captures all of the aspects of systemness, to act as a form of skeleton upon which special forms of mathematics can be attached as needed for specific systems. For example, we will not be using differential equations to attempt to describe systems. However, specific elements of specific systems contained in our framework may have places for the use of differential equations during the building of specific models. Moreover, the logic of systemness is built into the framework. Where it might improve understanding of why a specific structure is the way it is, this logic will be explored more formally (e.g., temporal logic and its role in causal reasoning).

While a main objective of the author is to make systems science and these methodologies as widely accessible as possible, to help any educated person elevate their systems thinking and awareness, it will take more than just spoken/written language. We will be explicating all of the concepts and methods in this book in both language and mathematics. And we will see that the two go very much together toward gaining understanding. The level of math, however, is not so high as to exclude the average educated reader. Most high school graduates, these days, have had to take basic algebra and much of our mathematical formulations will be such. However, the mathematics that will be used to build our framework is called discrete math or sometimes discrete structures.

1.3.5 Discrete Mathematics

We will be using four interrelated topics within discrete math (there are many more topics typically covered in a course in discrete math). The term refers to the fact that the math itself is based primarily on the idea of working with integers as opposed to real numbers. We cannot avoid the real numbers entirely of course. The discrete math we will encounter is based on discrete "operations," which may or may not involve real numbers. But those operations are generally easier to grasp for most people, say as compared with the calculus!

We can get by with avoiding or even ignoring the continuum because of an interesting property of systems. Systems are composed of subsystems in a hierarchy of smaller scales of time and space. And what appears to be a continuous substance at one scale is actually found to be composed of discrete objects at a much smaller scale. For example, the water flowing in a stream appears to be a continuous substance at the scale of human perception. But if we could put a really powerful microscope on the water, we would discover that water is actually composed of independent molecules made of two atoms of hydrogen with one atom of oxygen. Water only appears to be a continuum from the perspective of ordinary human perception. The whole world that we humans deal with is composed of atoms and molecules, so, technically, if we choose our scale of measurement properly, we can work with purely integer values. Of course, this is totally impractical for many situations. Counting molecules of water or electrons passing a particular point during a particular time is really infeasible. So, we resort to mass averages (e.g., statistical mechanics) and real number representations (and calculus!). Most of the concrete systems we will be using as examples will require some real number representations but will not require calculus. For example, rates of flows will be couched in discrete time intervals (called a time series) even if the mass flowing during a discrete interval has to be represented by a real number with finite precision.

Readers unfamiliar with the subjects—set theory (including fuzzy set theory), graph theory (including flow networks), probability theory, and combinatorics—may want to spend some time studying the main results in these. We will not be *doing math*, in the sense of proving theorems or the like. We will, however, use these mathematical concepts in the construction of a basic and (we hope) universal definition of system that will provide a foundation for the language of system as well as the design of a knowledgebase in which to hold what we discover in a structured way, recoverable for a number of different purposes including building models.

Even if the reader is not certain of their level of understanding of the mathematical statements made, we are dedicated to providing enough explanatory verbiage that they will not feel they are missing anything essential to at least form a "better" understanding, if not a deep one.

1.4 Deep Systems Analysis (DSA)

Before anything about a system can be truly understood, it is necessary to do a thorough analysis. By thorough we mean deconstruction of the system and all of its subsystems down to the level where the components are so fundamental (and well understood) that no further deconstruction is needed. For very complex systems this can seem like a monumental, if not impossible, task. What this book is going to show is that, while the task is monumental in some cases, it is far from impossible. In fact, we will argue that it is absolutely necessary in order to assure success of the overall objective of deep understanding. Furthermore, in the case of human-engineered systems, it is absolutely mandatory for a successful deployment of the system upon implementation.

The term "systems analysis" has been used in a number of different contexts (Checkland 1999, p. 134). It conveys roughly the process of inquiry into what a system of interest is, or what a system of "desire" should be. The methods of inquiry have varied significantly from one domain (e.g., information systems) to another (e.g., policy design). Yet they all have two things in common. They are all based on a relatively loose kind of systems thinking, or, more appropriately, systems awareness. And they all attempt to go about the inquiry systematically, even if not guided by a principle-based approach. The systems analysis methodology introduced in this book attempts to base the process on the principles of systems science directly (Chap. 2). What we will consider is a procedure that works in two different contexts because it operates from the basis of a single set of principles. Namely, it provides a way of finding out *what exists* (science) and it provides an aid to thinking about *what needs to exist* (design). The latter context can be split into two major arenas: the engineering of physical systems and the design of human activity systems (Checkland 1999). Figure 1.3 shows the general relations between these contexts. Checkland makes a significant distinction between system types, in particular between engineered (hard) systems and organizational (soft) systems, and argued that the systems approach appropriate to each of those is quite different. The argument advanced in this book, however, is that even though there are differences in these systems, they are still systems in a general sense and so many of the methods given here, especially systems analysis, provide a unified approach—there need not

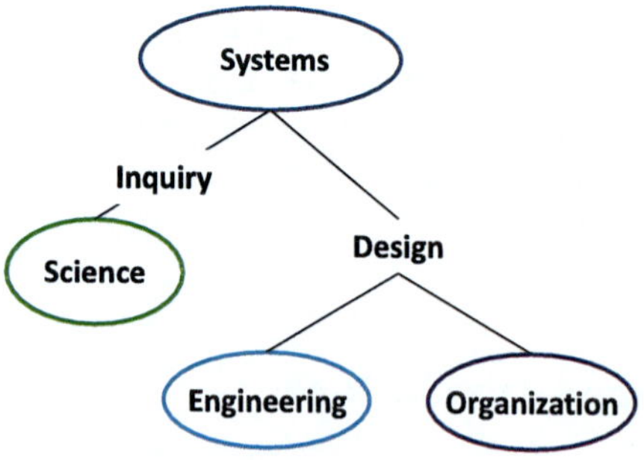

Fig. 1.3 Science (broadly defined) uses systems analysis to discover the nature of what already exists. Design seeks to produce something that needs to exist. Engineering is directed toward the design of machines, whereas organization is directed toward those systems that include human agents, such as the economy, businesses, schools. This classification roughly follows Checkland's (1999) classification of "natural," "hard," and "soft" systems

be different methods applied so long as the methods are based on system principles and not just ad hoc systems thinking.[23]

DSA is a top-down discovery process with bottom-up refinement that seeks deep understanding. More traditional engineering and organizational approaches have been based on a variety of methods, including a bottom-up, integration of modules, to a reliance on "requirements" of the users/customers that presume the latter know what they need. The typical engineering analysis (including software engineering) targets the technical details (and costs) of how to meet the requirements without ever necessarily asking whether those requirements are the right ones![24]

As will be explained in Chap. 5, and then thoroughly developed in Part II of the book, DSA involves several phases.

1.4.1 System Identification: Boundary Determination

The first phase involves actually identifying the system of interest (SOI) and establishing a boundary that demarks the system's insides from the rest of the environment. Sometimes this is as simple as identifying a living cell as the SOI and seeing the cell membrane as the boundary. More often this is a complicated task. As explained in Mobus and Kalton (2014) there are many kinds of boundaries, including conceptual boundaries that are often hard to recognize. In Chap. 6, we will cover some of the various techniques used to identify boundaries and "boundary conditions." Then, in Chap. 7, we will provide a few examples of kinds of boundaries in real systems and how to identify them.

Along with the determination of the boundary itself is the identification of all of the inputs and outputs that pass through the boundary. This means finding and quantifying (as best possible in this phase) the substances that enter and exit the system as well as the messages that it receives and sends, and the disturbances that, strictly speaking, are not inputs but may be found to affect the inner workings of the system when deconstruction is being done.

[23] Checkland's concern for differentiating was largely due to the attempts that were made in the 1960s to apply engineering approaches to the design of "soft" systems like companies or government agencies, where human factors could not be treated as parameters in an equation. At the time, systems engineering was still in a nascent state and a number of systems engineers and mathematicians were attempting to define systems from their perspective alone. This led to what this author considers some premature and limited notions of a definition of system (c.f. Wymore 1967, Chap. 2).

[24] As discussed in Chap. 6, this "starting point" of analysis, wherein some questionable user requirements are included in the initial specification of a system, has caused innumerable failures in projects in the past and has led modern systems and software development methodologies to be collectively known as "agile." These methods essentially abandon getting an upfront right specification of a system and instead adopt an iterative refinement method that includes having the user on the development team so that as the errors of design become apparent, the user member can change their specification on the fly.

Outputs include a system's overt behavior. Some systems move or generate forces all of which have to be accounted for in the analysis.

A major objective of this phase is to characterize the relation between inputs and outputs. To the degree possible given the complexity of the system and environment, the objective is to establish a quantitative function describing the output given the input. There are a number of techniques that will be discussed in Chap. 6 for accomplishing this. The importance of doing so is that this knowledge will provide guidance to the deconstruction phase.

1.4.2 Environment Analysis: Sources, Sinks, and Disturbances

Every real system, with the possible exception of the Universe as a whole, exists embedded within a larger supra-system. The Earth is embedded in the solar system and obtains its energy input from Sol. It receives a smattering of material inputs in the form of meteorites and comet dust (an occasional meteor of some mass). The biosphere is a subsystem within the whole Earth system (which we refer to as the "Ecos"—Greek for "home"). The human species and its cultures are a subsystem of the Ecos as well. A for-profit enterprise is embedded in a market/economic system. A child is embedded (we hope) in a family system. And thus, it goes.

No system exists as an isolated thing. If any such system did so, it would soon succumb to the second law of thermodynamics and cease to be a system at all—just a collection of bits and pieces in thermodynamic equilibrium. Thus, the understanding of any system depends on some understanding of its environment as well (what we will call the "embedding" environment).

The environment is comprised of other entities that provide resource and disturbance inputs (sources) and product/waste outputs (sinks) to/from the system of interest. Each and all of these need to be identified and their output/input flows quantified. Multiple sources could contribute the same kind of substance to an input to the SOI. Multiple sinks might absorb the outputs of a system. The relative contributions of all of them need to be identified and measured.

This is of fundamental importance. Every concrete, real, system is subject to the influences of inputs and outputs. Therefore, it is essential to identify all such source and sink entities and measure their rates of flow of material, energy, and messages in order to even begin understanding the nature of the SOI. Inputs other than disturbances (e.g., sudden changes in flows or thermal noise) are considered resources for the system, or otherwise they would not get the status of inputs. Similarly, outputs are either useful to other entities in the environment, or they are wastes that need to be absorbed and neutralized. For example, consider that every system involves a work process in which input energy at a high potential (and appropriate form—called exergy) is used to drive the work. The work might involve the transformation of matter, other forms of energy, or messages (e.g., computation) and will always result in the loss of some portion of that energy in the form of waste heat (the rest presumably used up in the work). Again, this is the second law of thermodynamics and absolutely must be taken into consideration in any truly deep understanding of systems. The heat must be transmitted to the environment and the total energy flow must account for it. Almost none of the existing modeling languages make provisions for this important aspect of concrete systems.

The first step in systems analysis is an analysis of the SOI's environment. We do not attempt to model the sources and sinks other than their outputs/inputs (usually via a flow function). As we will discuss in Chap. 6, there are times when accounting discrepancies are noted during subsequent systems analysis between the environmental flows and the internal system flows. At such times it becomes necessary to reverse the direction of SA and do what we will call a "supra-system analysis," that is including the source or sink in the boundary in order to resolve the discrepancy. An example of this can be found in the analysis of a supply chain for a manufacturing company in which the input of a specific part comes from a single source and thus the entity that supplies that part is more strongly coupled to the original SOI (the manufacturer) than initially understood. The single source supplier should probably be included within the boundary of the original SOI because it is acting as a resource obtainer for the manufacturer, that is, is actually a subsystem of the SOI.

1.4.3 Recursive System Decomposition

The general method of SA will be a top-down recursive search of the system-level tree. After the environmental analysis has established the main inputs and outputs of the SOI, the procedure opens up the opaque box to look at what is inside. It looks at the subsystems that comprise the whole system and determines the mapping of inputs to internal subsystems (those that work to obtain resources or deal with disturbances) and outputs from subsystems (those that export the products and wastes from within the SOI).

As will be shown in Chap. 3, subsystems of an original SOI may be treated (once discovered) as systems in their own right. That is, by focusing in on a subsystem as if it were an SOI, we can deconstruct it in the same way we do the original. The trick, here, is to treat the other subsystems as if they are the environment of the newly chosen SOI. We ignore their internal workings (for the time being) and simply concern ourselves with the mapping of inputs and outputs in exactly the same fashion as we did with the original SOI.

This constitutes a recursive method for analyzing the system from the top down to the level of components. The procedure used assures us that we account for all material/energy/message flows in accordance with the conservation laws and the second law of thermodynamics. By capturing these in the structured knowledgebase, we ensure that discrepancies will be flagged and provide a way to re-analyze the system to ensure we get it right.

Systems analysis as describe in this work is an expensive proposition. This will not sit well with managers or scientists concerned with short-term profit or budgets. Our justification for this intensive procedure, to be explained in Chap. 6, is summed up in an epithet often heard among working engineers, "Why is it there is not time to do it right the first time, but there is always time to fix it?" Profit-motive-driven (or cost-averse-driven) managers are often motivated to accelerate the designs of systems. Even budget-motivated scientists are driven to short-circuit analysis in order to produce a result within (or under) budget approved by a national funding agency, and on time. Yet with the methodologies in current practice, they invariably find themselves having to repeat or redo their work. We will argue that the problems being faced by large-scale and complex projects are the result of a failure to deeply understand a problem or requirement and thus lead to a failure of production of a final result.

1.5 The System Knowledgebase

With the definition of system to be worked out in Chap. 4, we will provide a structure of knowledge that emulates that of the human brain, and, so, is compatible with human communications of ideas. We will, in Chap. 8, explore the construction of a knowledgebase—a database with a specific structure for encoding knowledge.[25] This knowledgebase will be shown to emulate the knowledge of concepts held in the human brain and thus be able to provide a basis for communications between humans and between humans and machine (computational models). This is a major leap forward in the conceptualization of systems and models.

Fundamentally, the argument is that the human brain is designed to capture, and build, concepts that reflect the real world. We learn about the world from experience in it. The method by which this is accomplished is to encode percepts, concepts, and thoughts in neural networks. But those networks are direct representations of the real world. This will be explored in Chap. 4 in the context of the language of system, then further developed in Chap. 8 in describing the knowledgebase.

A major feature of the knowledgebase design is that the knowledge may be checked for internal completeness and consistency automatically. These checks will report any lack of adequate knowledge, thereby providing the systems analysts with guidance about what they need to go back and find. This is similar to the syntax checking that computer language compilers do to flag errors before trying to generate machine code.

1.6 Synopsis and Motivation

"Part I: Foundations of Systems Understanding" will provide, first, the theoretical underpinnings of systemness and why it can be understood. We start, in Chap. 2, with a review of the general principles or attributes of being a system that were

[25] The idea of a knowledgebase, as opposed to a mere database, comes from Principle 9, discussed in Chap. 2. Knowledge is organized and effective data—models, rather than just aggregations of data. Knowledgebases are accessed differently than databases—by association rather than by indexical search.

given in Mobus and Kalton (2014). Then, Chaps. 3 and 4 provide the developments of an ontology of systems, that is, the question of "what" exists in a systems framework, and a language of systems that will provide the basis for doing DSA and capturing the results in a knowledgebase. Chapter 5 then tries to situate what had been developed in the overall process of systems understanding. That is, we provide more details about how the different stages of the process, introduced above, take advantage of the theory of systems developed in Chaps. 3 and 4 (with the principles from Chap. 2 forming the backdrop).

In "Part II: Analysis for Understanding," we unpack the first two stages of the process, the analysis of a system and the knowledge that we acquire therefrom. Chapter 6 goes into the details of analysis, demonstrating the nature of systems analysis as generally a top-down recursive procedure using the system language (SL) that we developed in Chap. 4. We show how that language guides the process of analysis in that its syntax tells us what we should be looking for next as we construct an abstract model of the system. In Chap. 7, we will pause to let what we showed in Chap. 6 sink in. We provide several examples of the application of analysis to a wide variety of things we would call systems, each from different domains of knowledge, for example, biological or organizational. We will show how the process of analysis is carried out exactly the same in all cases, hopefully demonstrating its universality.

Chapter 8 then returns to the main thrust of the process. We take a much closer look at how the knowledgebase is designed in accordance with the mathematical definition given in Chap. 4 and is thus in conformance with the language structure developed in that chapter. It will be made clearer how the information captured in analysis, by being stored in a specific database architecture, becomes knowledge in the sense it will be defined in Chap. 3. In anticipation of that development, we will just say that knowledge and information are not the same thing even though they are related deeply from a cosmological ontology point of view.

The conclusion of Part II, Chap. 9, will be a single example of an incredibly complex system being analyzed. We will show how something as complex and "fuzzy" as the economy might be tackled using the methods thus far presented. This chapter holds some surprises for both seasoned systems scientist and traditional economists. We explore new ways to look at the economy as a system rather than as a free-form phenomenon.

In the next two chapters, "Part III: Archetype Models," we consider the nature of complex adaptive and evolvable systems (CAES), the most complex kinds we know of. Chapter 10 is devoted to an overview of the CAES concept. This archetype model is the integration and amalgamation of a number of historically important systems models that other researchers and thinkers have proposed in the past based on their observations of systems that were of interest to them, such as the Viable Systems Model (VSM) of Stafford Beer and Living Systems of James Grier Miller (and many other contributions) along with updated theories applicable to the sub-models of a CAES.

Chapters 11 through 13 present these sub-models. Chapter 11 is devoted to an updated understanding of Agents and Agency as the computational and decision processing elements in the next two sub-models. These are the Governance and Economy sub-models respectively. These are treated in more detail than is described in Chap. 10.

Finally, in "Part IV: Designing CAES Artifactual Systems," we address the practical uses of the process for deep systems understanding for producing complex systems artifacts. These include everything that humans invent and produce for use, including not just machines and cities, but procedures and policies; essentially anything that is the product of the human mind and did not arise from "natural" evolution. In Chap. 3, we describe this so-called natural evolution as the result of non-intentional auto-organization and selection by environmental factors. In Chap. 14, we introduce the notion that humans in their ability to imagine new organizations that are "intended" to solve problems or provide new capabilities transcend auto-organization and natural selection with intentional-organization and intentional-selection. What we typically call human culture, the aggregate of all artifacts along with beliefs, norms of behavior, and other mental properties of humans, has become a major part of the human environment, even more important, for most people today, than the so-called natural environment.

In Chap. 15, the process by which CAES can be brought into existence, through intentional-organization bolstered by systems engineering, resulting in artifacts that serve their intended purposes while minimizing unintended consequences will be spelled out. The latter aspect of human culture has been the source of many collapses of societies throughout history. Those societies were small and reasonably local as compared with the situation today in which cultures have global reach and impact.

There are growing signs that this cultural environment has developed in such a way that threatens human well-being. We will address this issue in the final chapter of the book in which we ask, can the use of deep understanding of the human social system (HSS) as an exercise in design and engineering produce a culture that is more conducive to both the human psyche and the rest of the natural world, and, if so, what would it look like in broad outline?

The last chapter, Chap. 16, is devoted to exploring the larger implications of designing systems that involve both human subsystems and the larger supra-system of the Ecos. Human beings and their social activities have already impacted the whole Ecos in ways we never would have imagined even a half century ago. The question now is how might systems science and the deep systems understanding process described in this book actually be put to the purposes of designing management subsystems to restore balance to the Ecos and make the human subsystem provide a purpose to that larger system.

Why should we consider the contents of this book? There are practical reasons in terms of doing better systems engineering. Some of this motivation will be covered in Chap. 2. There are purely intellectual reasons, in terms of the satisfaction of knowing how things work to a deep level. But as much as anything a more compelling motivation is that humanity is finally coming to realize that our world is in danger and the problems are systemic. We are just at the beginning of a process of understanding the systemness of the world. So far problems such as global warming and climate change and peak oil/energy and the role of neoliberal capitalism in driving these are just coming into focus in their own lights. But more and more, people are coming to realize that they are all interconnected. Thus, we have need of a scientific way to make those connections and grasp the various dynamics involved if we hope to begin solving these existential threats. Perhaps the process of gaining deep understanding described in these pages might help.

Part I
Foundations of Systems Understanding

The understanding of complex systems, how they work, what they do, how they interact with their environments, and much more, one must start with a general understanding of what a 'system' is! The first part of this book will seek to provide this understanding. There is a general understanding of what systems are; the term is used ubiquitously to indicate a 'something' that is a whole, but comprised of a multiplicity of components, often of heterogeneous composition and all parts interacting in patterned ways. This description applies to just about everything that exists. Everything is a system or a part of a larger system.

The nature of 'systemness' is the subject of a book by this author and coauthor Michael Kalton called *Principles of Systems Science* (Springer 2015), in which we provide a set of characterizing principles that are applicable to all complex systems and most of which apply to simple systems as well. The current book is a follow-on to that work in which we will expand on the concept of systemness. The background for this book will be found in the *Principles* book.

The first four chapters provide a summary of the main ideas as developed in the *Principles* book. They will expand those concepts presenting a system ontology and a language of systems that will be crucial to the process of understanding systems. These will provide foundational knowledge to be used throughout the rest of the book.

Chapter 2
Principles of Systems Science

Abstract Systems science, as a disciplinary basis for systems understanding and systems design, has had very mixed reviews with regard to its efficacy of methods and successes in application. A central reason for this is that there is not now a single discipline of systems science that is taught as a core topic and there is not a unified concept of systems science being practiced in the field. There is, however, a strong intuitive notion of systemness and many systems aspects and methods being applied to problem-solving. There is a strong notion that most of the major problems facing humanity are "systemic." *The Principles of Systems Science* (Mobus and Kalton 2014) was the first major attempt to reconcile and reunify the disparate aspects of systemness, showing how various aspects relate to one another. In this chapter we attempt to show how the principles outlined in that work may be applied to develop a single, consistent concept of a systems approach that is transdisciplinary and comprehensive. This leads to a set of methodologies (explicated in Parts II through IV) that will demonstrate how systems science can become a disciplined and principled way to tackle complex and important problems.

2.1 The Systems Approach

Within the sciences, engineering, management, and governance disciplines,[1] a relatively new approach to investigation and design has developed over the last several decades.[2] It is generally called the "the systems approach" to the discipline. What

[1] Through the rest of this book, we will use the term *scientist* in the very broad sense to mean someone who discovers new knowledge, be they facts, principles, or ways of doing things. We will use the term *engineer* to mean someone who designs new objects, processes, or social policies. Thus, someone who discovers a new market possibility in business would come under the term scientist, even though they are not necessarily using a strict scientific method. Someone who designs a new department to produce the new product is an engineer in this general sense.

[2] It is actually a bit difficult to put a definitive time frame on this development as it has entered the discourse of different sciences and engineering at different times. When the concept of a general systems theory was first emerging in the post-WWII period, several of the sciences, including

G. E. Mobus, *Systems Science: Theory, Analysis, Modeling, and Design*,
https://doi.org/10.1007/978-3-030-93482-8_2

exactly do we mean by the "systems approach" and how does it differ from what we ordinarily think of as science and engineering practices? This book will attempt to answer that question and, in addition, demonstrate how a strong systems approach can be actually implemented. We will explore what difference it can make to various disciplines, and why it is crucially important to pursue in order to gain deep understanding of our world, especially our social systems and cultures.

The concept of a system (Sect. 2.2 below) is a fairly easy one to grasp because our world—and the perceptions we have of that world—is organized into objects and interrelations between those objects, including ourselves as objects being in the world. Our brains evolved to grasp systemness because that is how the world is and grasping it makes us more fit evolutionarily. As a result, we intuitively understand the world (Chap. 4 will delve into this aspect). Systemness is a recognizable property of objects and their relations and it does not matter at what scale of space or time we are observing. It does not matter at what *level of organization* we look; we see systemness automatically. Elemental atoms are systems, as are living entities, as are societies. Thus, the application of systems science as an approach to practicing the other sciences promises to bring a new level of rigor to those studies beyond an intuitive recognition of systemness.

As Peter Checkland observed, "What distinguishes systems [as a subject of the study of systemness] is that it is a subject which can talk *about* the other subjects. It is not a discipline to be put in the same set as the others, it is a meta-discipline whose subject matter can be applied within virtually any other discipline" (Checkland 1999, p. 5, emphasis in the original). In other words, systems science is a meta-science, the findings of which may be applied readily in the other sciences. This is what we mean by a "strong" systems approach.

Currently, the need for a systems approach is intuitively understood by many scientists and engineers because they are undertaking work that involves high levels of complexity in terms of the numbers of interacting component parts and the heterogeneity of those parts. In the sciences, the subjects of study involve phenomena that are traditionally studied within a specific discipline like genetics or biochemistry but now combine in more complex levels where the knowledge and skills of several traditional disciplines are needed. They recognize that the traditional disciplinary phenomena cannot provide explanations for the complex ones that now take center stage. These cannot be understood through just one discipline alone; they require transdisciplinary approaches. Similarly, the kinds of engineering projects that are being tackled today cannot be handled by one engineering discipline alone. Nuclear power plants, space shuttles, and even commercial aircraft are not simple systems any longer. They are systems of systems (SoS)—complex objects that require the coordinated efforts of many engineering disciplines to design, produce, deliver, and monitor for performance and improvement. Similarly, large

management science, adopted aspects of systems science, as it was understood originally, quite readily, but possibly prematurely. Systems science has matured since then and the current discussions of the systems approach to understanding phenomena have re-emerged in multiple arenas in concert with the concepts of multidisciplinary research.

multinational corporations and non-governmental organizations (NGOs) have become extremely complex in their operations requiring well thought through organizational, policy, and procedural designs.

Unfortunately, intuitions about systemness are not nearly enough. While many scientists and engineers have been pioneering the use of systems thinking in pursuing their research and designs, the efforts have been largely ad hoc, not necessarily integrated, and not generally based on any central principles of systems science that could provide guidance to their efforts. Several examples from both the sciences and engineering provide a sense of this early, formative, but still immature notion of the systems approach.

To further clarify the issues involved, we will describe what is meant by system, system theory, system science, and other relevant concepts. Figure 2.1 shows these and their relations to one another.

We start with the basic concept of a system as a thing, a whole entity, and consider what that means. For this, we propose a theory of system as an object of study and develop a science of system to determine what attributes and characteristics systems have in general (i.e., make for a general system theory, von Bertalanffy 1968). System science leads us to discover these attributes, which, in turn, we posit must be found for anything we would call a system. The attributes, etc. taken together in concert we assert are principles of systemness, the properties of being a system (large gray box), elaborated in Sect. 2.3 below. Both system science and the principles contribute to our ability to design artifacts, from machines to organizational and social policies. They also contribute to providing deeper thinking about the various systems we find in the world naturally. System design and system thinking together guide actions in analysis and understanding of artifacts and natural

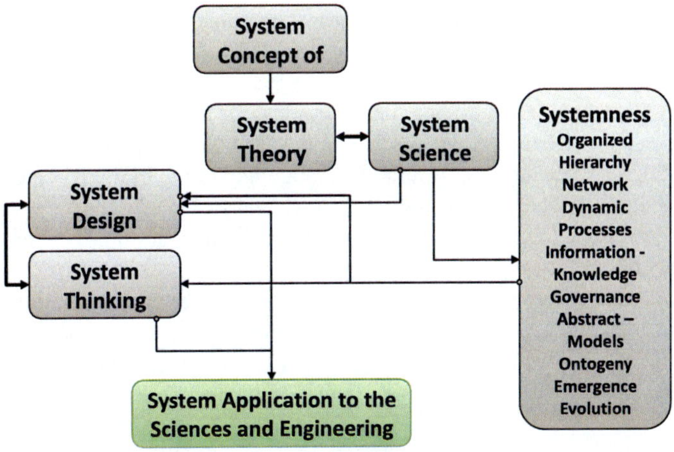

Fig. 2.1 The various areas of thought related to systems (gray) and how they influence the other sciences and engineering practice

systems in the world (concrete systems). And so, these contribute to the application of system concepts to the other sciences and engineering (green box).

This is our road map for applying system principles to understanding the world, including the things we invent.

2.1.1 Systems Intuitions, Thinking, and Sensibility

In the next two chapters we will be developing a "universal" language to be used in describing both qualitative and quantitative aspects of systems. The argument for the universality of this language is based on concepts developed in psychological linguistics having to do with how the human mind uses an internal language of thought (LoT)[3] to construct and think thoughts. Unlike public, natural languages that are acquired through learning, LoT is natural to the brain and is fundamentally the same in every human being. Like public languages, it has an internal syntax and symbolic representations of concepts. This subject will be further explained in Chap. 4. Public, natural languages are thought to translate into LoT representations and syntax for internal, subconscious processing. That is, what we say in our languages is the result of thinking in LoT translated to words and sentences in our learned public languages.

In this chapter, we introduce LoT to make an argument about the nature of systems thinking. That argument is that LoT, sometimes referred to as "mentalese," is actually the language of systems, or "systemese." We perceive systemness in the world because our brains are hardwired to do so. And our brains are so hardwired because the patterns of aspects of systemness are objectively in the world. Humans, indeed all living things, evolved to have this perceptual capacity at whatever level of organization (that is their econiche) they are embedded.[4] They succeed in interacting with the other relevant systems in their worlds by virtue of their own information processing capacity being based on systemness itself. Systemness can be thought of as the fundamental organizing framework for the Universe. Therefore, systemese is a basic template for describing the structures and functions of objects in the world. And because this is a built-in language, in the brain, its causal effectiveness is largely subconscious. Until, of course, the systemese sentences get translated into a public language—the one we hear inside our minds when we are consciously thinking.

When notions of systemness are brought into conscious awareness, that is when they are made explicit, we human beings are very quick to grasp how they apply. We can, without much effort, see that many kinds of objects in the world are connected with other objects, that they often have causal influences that follow time's arrow.

[3] A more complete discussion of LoT will be presented in Chap. 4.

[4] We ordinarily think of animals with brains as having perceptions of the world. But, plants, and even bacteria, receive messages from their environments and behave in response.

We can perceive readily that certain kinds of objects have internal processes and internal components that interact with one another to process inputs into outputs. This is how things work in the world so it is not at all surprising to find that our brains evolved to capture these aspects of the world. Our brains, in fact, build models of those things in the world that we observe—we call them concepts. And we use those models to make "predictions" about what will happen in the future.

All human beings, thus, have deep intuitions about systemness and how the world is composed of systems interacting with other systems, being part of larger systems. What we think happens in humans is that this internal LoT is linked with a specialist subsystem in the brain that encodes special "vocal" symbols and links these with the internal symbols of systemness. For example, every human being has an intuition about the nature of a container that receives input flows, stores whatever that flow consists of temporarily, and has output flows that lower the level of the stock. That basic concept can be translated into a gas tank in a car, or a bathtub, or a lake, or a parts inventory, or any other specific container that acts to temporarily store a substance. According to LoT theory the brain has a "primitive" representation of a element that can then be particularized for different real-world representatives. This mapping of physical instances in the world to a general concept in the mind is thought to be the basis for metaphor and analogy. More details on the mechanism and its workings will be found in Chap. 4.

2.1.2 *Examples of the Systems Approach Within Disciplines*

2.1.2.1 Systems Ecology

Pioneers in the science of ecology such as the Odum brothers, Howard and Eugene,[5] deeply understood the need to study ecological environments as systems. The term "ecosystem" denotes the idea that there are structures, part of the Earth's subsystems, which have identifiable and isolatable characteristics that determine or contribute to the biota that reside in those subsystems. Howard Odum developed a language of systems based on the flow of materials and energies through an ecosystem that could be mapped and modeled and that went far to explain how those systems behaved as a whole.

Energy flows have been formalized in the model of food webs but even more complex relations between species of plants, animals, fungi, and bacteria are now routinely studied in this field. The concepts of energy flow and network relations are keys to the study of systems ecology (Principles 4 and 3 below). However, so is the idea that ecosystems evolve over time. The concept of succession is also an important aspect of the systems nature of ecology (see Principle 6 below).

[5] Howard and Eugene Odum were brothers who pioneered the field of systems ecology (c.f. Odum 1983).

It is safe to say that systems ecology is one of the more advanced examples of the systems approach used in a science. The application of many other principles, as discussed below, to systems ecology is, however, somewhat spotty and not particularly integrated. For example, little work has been done on exploring the nature of information and knowledge encoding (Principle 7) as integrated with those described above. The field of ecological remediation touches on Principles 11 and 12 below, but only incidentally, intuitively.

Systems ecology is at the forefront of application of systems thinking to science. Other sciences are starting to follow suit. For example, psychology, sociology, and neurology are making forays into the systems approach as they expose the complexities inherent in their subjects. Even so, they are tackling these issues in a less than principled manner that will be suggested in this text.

2.1.2.2 System Dynamics

One of the most successful uses of the systems approach to understanding complex systems has been the development of the system dynamics (SD) modeling method. Developed by Jay Forrester at MIT, SD has provided a wonderful tool for analyzing and simulating complex systems (those having feedback loops internally). SD can be applied to an extremely wide variety of systems—ecological (Ford 2010), industrial (Forrester 1961), and global (Meadows et al. 1972, 2004). SD modeling allows a scientist or engineer to reduce a system of interest (SOI) to an abstract set of stocks, flows, and controls that represent the dynamical aspects of a system. Implicitly, these models also represent the networked nature of influences (flows of information) that lead to the overt behavior of the systems modeled.

2.1.2.3 Systems Management

How to organize and manage an enterprise is the subject of systems management. The enterprise is the system and the methods of analysis and design based on systems principles are applied in order to achieve an optimal design. Many ideas applied to management, especially operations and logistics management, derived from the field of operations research. Major early proponents of this field, and its implementation in management science, were Stafford Beer (1959, 1966, 1972), C. West Churchman (1960, 1968a, b), and Churchman along with Russel L. Ackoff and E.L. Arnoff (Churchman et al. 1957).

Another champion of systems thinking applied to human organizations was, as already mentioned, Peter Checkland (1999). An organization is an incredibly rich object for system study. It is, or can be, completely transparent to the analyst's probes. Subsystems and sub-subsystems are generally readily identifiable (e.g., departments). The behaviors of the various subsystems are mandated by their production functions, which are generally well known (e.g., accounting is standardized). And the main concern, from a systems perspective, is to carefully design and regulate the information systems that provide inter-departmental and division communications for the purpose of the cybernetic management functions (i.e., management decision-making). So important is this latter aspect that today the term "information system" is often taken as synonymous with the term "system" by itself. Whenever someone refers to "the system" they are likely referring to the information and knowledge infrastructure.

Recognizing that organizations are organic, evolving, and adaptive systems, Peter Senge (2006) developed the theory for how an organization's information and knowledge infrastructure is the basis for successful adaptation in changing environments. He embedded this concept within a systems thinking approach (the fifth discipline) to understanding organizations.

2.1.2.4 Systems Engineering

Any activity that involves designing a complex artifact is in the domain of an engineering practice. To a large extent the design of an enterprise, as described above, can be an engineering exercise. Chapter 14 will provide a general overview of the systems engineering process as it will incorporate the methodologies described in the following chapters.

Design is a creative process in which a system that has not previously existed, or existed in a desired form, is specified for construction. Engineering is the process by which design options are tested and improved based on principles (and laws of nature) so that the finally constructed system is capable of delivering its intended purpose, usually at a sustainable cost (in energy and material inputs) and over an extended time (its intended life cycle).

In today's world (e.g., the Internet of Things[6]) new class of artifacts that combine physical work with artificial intelligence (i.e., robots like self-driving vehicles) called cyber-physical systems can only be approached as complex systems requiring a systems approach to analysis and design.

It is no longer possible to just put a few mechanical or electrical engineers on the job to create these artifacts. In addition to the specialization of the traditional engineering approaches, a meta-engineering discipline is needed to coordinate and integrate the efforts of the disciplinary engineers. This is the role of systems engineering.

2.1.3 *Possible Issues with Current Practices*

As pointed out above using the systems approach, or systems thinking, in these and many other fields has been based on more intuitive and ad hoc methods. Systems intuitions turn out to be strong in the sense that the human brain is actually designed to perceive systemness (the next two chapters will develop this notion as the basis

[6]The Internet of Things (IoT) is a moniker given to the idea that many of our common appliances, washers and dryers, refrigerators, etc., will be given their own IP addresses and be accessible from central command centers in the home or from our smart phones.

for developing a "universal" language of systems that can be used by any human being to communicate ideas about systems). Hence, such intuitions are reasonably good, at least in the way they frame thinking about complex things. But this is only a start to using systems principles rigorously in order to obtain deep understanding of the systems of interest in any domain.

The biggest issue with current practices is that often time practitioners adopt a fairly limited framework for approaching their subjects. For example, it is often the case that researchers using systems thinking get stuck in one of the many facets of systems science and attempt to explore and explain phenomena from that facet alone. They may pursue modeling with system dynamics, or use network theory or information theory as their main focus of study. Of course, these facets need focal study at some point in developing a systems approach to the topics. However, they cannot be the sole set of conceptual and methodological tools used to gain deep understanding. *All of the principles of systems science have to be taken together in order to fully grasp the nature of phenomena.* Developing a theory based on only one facet, or principle, is inadequate for complete and deep understanding.

For example, dynamical behavior is, of course, an extremely important aspect of a system. And the ability to model the components that lead to the overt behavior is crucial to understanding the internal aspects of the system. But dynamics, although extremely important, is just one aspect of a system that needs to be taken into consideration when attempting to understand the whole system.

Granted that researchers and engineers tend to specialize in their preferred sets of concepts and methodologies to obtain a degree of expertise (and guarantee quality of their efforts) and that this is not necessarily a "bad" thing. Nevertheless, a complete systems approach has to link understanding of each of the principles to all of the others in order for scientists and engineers to claim holistic understanding. We will return to this issue in Part IV.

2.1.4 A Principled Systems Approach

The examples above demonstrate that scientists, managers, and engineers well understand the need to apply concepts of systemness to their subjects in order to grasp a more holistic understanding of the phenomena or what they are trying to achieve. Clearly the need for a systems theory that underlies and supports these approaches is demonstrated. What has been lacking is an actual set of principles that are integrated across the systems science spectrum of subjects that would provide a basis for pursuing a "true" systems approach. That is, an approach that examines every aspect of systemness in the system of interest, leaving no aspect uncovered so that a holistic understanding is the result.

In Sect. 2.5, we examine what can happen (and does happen all too often) when a "weak" systems approach is applied to the understanding and design of a system. Ironically, this system is in the category of systems from which the word itself became a household term, the world of information systems (computers and communications) deployed to improve the management of organizations. The field, today, is generally referred to as Information Technology (IT) since the use of computing and communications technologies covers much more than just management information systems. As we will demonstrate below, the analysis of IT systems is not as principled as practitioners have assumed it to be. It is, rather, a collection of believed best practices. In particular, the area of systems analysis is extremely weak, and as a result, too many designed and developed systems experience one or more project failures that materially affect the effectiveness and/or cost of the system. While many problems arise due to the nature of computer programming (and the inherency of errors in producing code), these are technically fixable. The other kinds of errors are found in the designs.

In this book, we will first review, briefly, the principles of systems science as described in Mobus and Kalton (2014), *Principles of Systems Science*, which is a comprehensive coverage of all currently understood aspects of system science. We will introduce all of those aspects and then, throughout this book, demonstrate how they are used to obtain a holistic understanding, a deep understanding, of the systems studied by the sciences and those to be designed by engineering. The rest of Part I will first establish the theoretical underpinnings, but the rest of the book is devoted to methodologies and examples of their application to real-world systems of great interest.

2.2 The Concept of a System

The concept of something being a system, or possessing the aspects of systemness, is, in one sense, simply a construct of the human mind, using systemese to describe the organized but complex phenomena it encounters. The brain treats things as systems because that is how it thinks. On the other hand, the model (concept) it is constructing is based on a reality in the physical world. We make a distinction between systems that are concrete, real things, and systems that seem ephemeral, conceptual, abstract, models that have their own kind of reality. As we discuss below, however, even the latter kind of system is a real thing. A concept in the mind, or a simulation model running on a computer, consists of real physical structures and processes that, through a process of abstract representation, are constructed so as to have isomorphic relations with the real things out in the world.[7]

[7] There is a long-standing debate among some systems theorists, those who hold that systems conceptualization is strictly a mental construct (constructivists) and those who hold that real things in the physical world are systems (realists). This debate, however, is very much like the old nature-vs-nurture debate that raged before the link between genetics and development (ontogeny) became clear. Both nature and nurture are at work in eliciting form and function in living things. Similarly, we hold that both constructivism and realism are describing one single reality but from two different perspectives.

2.2.1 Definition

In Chap. 4, we will provide a formal definition of system from which we derive a language of systems. This language is instrumental in pursuing understanding of systems through systems analysis and modeling. For now, let us look at an informal, qualitative, definition that will be sufficient to motivate the subjects covered till we get to Chap. 4. It is important to note that there are many "informal" definitions of systems that have been advanced over the years since the concept of things or processes as systems was formed in the early twentieth century. While some of these definitions sought parsimony, looking for the simplest definition, others sought completeness. Most, however, agreed on the main attributes of something being a system.

For our purposes, a concrete, "real," physical system is any identifiable object or entity that is composed of an internally organized set of heterogeneous components, each of which can be a subsystem, or, in other words, a system in its own right (recursive definition). Below we describe some other kinds of systems that on first pass don't seem to qualify as physical or concrete systems, for example, abstract or "conceptual" systems. However, we shall argue that on closer examination even these meet the definition as given here.

A system is bounded in some manner such as to maintain unity-hood and be distinguished from the rest of its environment.[8] Substances such as materials, energies, and messages can cross the boundary, either entering or leaving the system, and usually in a regulated fashion. Generally, not just anything can get inside or leave; only particular materials or energies are permitted. However, systems might be subject to unregulated disturbances of various sorts.

In general, systems process inputs to produce outputs (material, energy, messages, or forces). Inputs are received from sources in the environment and outputs go to sinks (both of which may be minimally modeled). Figure 2.2 shows a generic model of a "physical" system. Technically, Fig. 2.2 is itself an *abstract* system since it assigns names to the parts, for example, "boundary" is the name of the barrier that segregates the insides from the outsides of a real system. It stands for a concept, the act of "protecting" the internals of the system from external disruptions. Figure 2.2 depicts what we will call a "crisp" system. That is, it represents a system that has an identifiable boundary, environmental entities and which can be deconstructed in a straightforward way. Most systems in the world are not crisp. They are what we will call "fuzzy." That means it may be difficult (with whatever analytic tools are available) to define the boundary, the flows, or environmental entities, and the internal organization may be highly indefinite. Figure 2.3 provides an alternate

[8] A distinction between a real system and a model of a system needs to be made. Real, concrete systems have real, though sometimes fuzzy boundaries. Models of systems are mental constructs in which a considerable amount of abstraction has resulted in leaving some things out. The boundary of such a model system is often more a construction of the modeler's choices rather than a real boundary. This topic will be developed further in the chapters ahead.

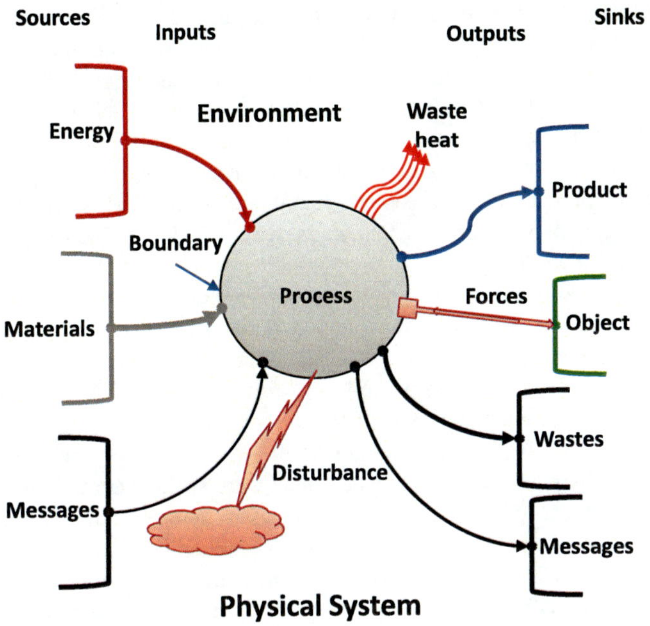

Fig. 2.2 This is an abstract representation of a generic "real" physical system. It is called a "process" because it receives inputs from its environment and does physical work on them to produce "product" outputs along with wastes (material and heat)

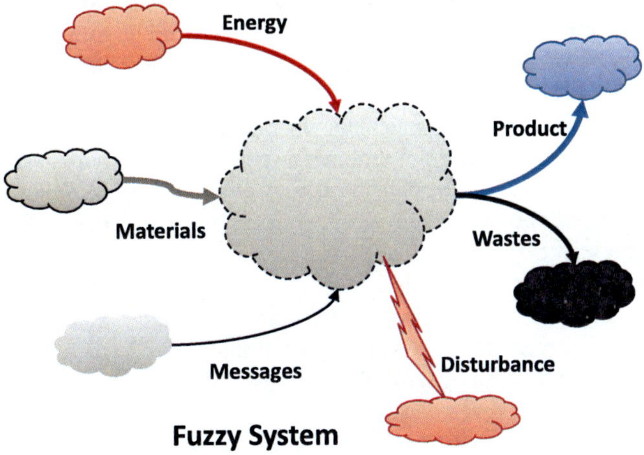

Fig. 2.3 When considerable uncertainty or ambiguity limits the capability for defining a system or its environment, then it needs to be considered as "fuzzy"

representation of fuzzy systems (and environments) that implies the difficulty of finding hard definitions. Fuzziness (which will include messy complexities, uncertainties, and ambiguities) is a major challenge in understanding systems, but, as will be argued in Chap. 6, this has always been the challenge in the sciences using reductionist methodologies to drill down to lower levels of organization (i.e., finding out how a system works and how it responds to its environment). In Chap. 6, we will show how we approach the analysis of a fuzzy system in terms of approaching the status of being a crisp system. In Chap. 4, we will provide the fuzzy set framework for making this possible.

Checkland (1999) has made a distinction between "hard" and "soft" systems, along these same lines. He notes that "human activity systems," that is, everything humans do in organizations and governance, are not definable as a concrete system (e.g., Fig. 2.2) due to the extraordinary number of additional factors such as value systems affecting decision processes. For the latter case he developed a "soft systems methodology" (SSM) that recognizes these complexities and eschews any attempt to treat human-based systems as if they were artifactual or simpler biological (natural) systems, yet still recognizing the systemness of such organized activities. It remains an open question, currently, as to whether the fuzzy system openness to be described in Chap. 4 will provide an effective middle ground between SSM and hard systems with respect to analysis and, for example, the design of public policy. In Chap. 7, and later in Part IV, we will show how the fuzzy aspects of the definition developed in Chap. 4 may be used, nevertheless, to analyze and understand social systems, or subsystems, thereof. Rather than treat systems as black and white (hard vs soft), our approach attempts to ameliorate the two extreme versions so that a consilience might provide greater insights at both ends of what we think is a spectrum. There is, we contend, no excluded middle here. Some systems will prove to have both hard and soft attributes, and in order to be truly integrative, we have to find a way to work with both simultaneously. After all, the human social system (HSS) is a subsystem of the natural Earth supra-system. Our whole existence depends upon our ability to understand the interactions between the hard systems of nature, the hard systems of technology, and the soft systems of human activities.

Real physical systems are the objects of study of the sciences and engineering, as defined broadly in footnote 1. They are real in the sense that they exist. They are physical because they are experienced via sensory perceptions or the senses via augmentation with instrumentation. The act of studying a physical system is accomplished with abstract systems such as the various symbols and names used in Figs. 2.2 and 2.3. Other abstract systems, such as mathematics, are used to describe the attributes, relations, and dynamical behaviors of those concrete systems.

2.2.2 *Systems That Seem Ephemeral*

In the Introduction, mention was made of abstract systems contrasted with concrete systems that are real, physical objects in the world. The term "ephemeral" is used here to suggest that abstract systems do not seem to have a physical embodiment.

Indeed, abstract systems are often treated as if they do not. But in reality, such systems do have a form of physical embodiment at times, for example when your brain is thinking about a concept or a simulation model is instantiated in a computer memory.

The term "abstract" does not mean that a system might be unembodied. Rather it means that the system has been representationally reduced to a set of parameters (e.g., variables) and mechanisms (e.g., functions) that can be computed either by brains or by computers producing results that resemble the real system. The ultimate abstraction is the naming of something (nouns), or some relation (preposition), or some action (verb) so that they are highly compressed for communications. Such abstractions have to have corresponding representations in the brain or computer memory that can generate an expansion to a less abstract version of the system.

2.2.2.1 Systems in the Mind

This specifically refers to mental models or conceptual models. That is, they are models of real or imagined systems that operate within the human brain.[9] Systems in the mind are instantiated in strongly interacting networks of neural clusters in the neocortex. When you see that house down by the lake with the red roof, your brain is activating circuits for a huge number of concepts—house-ness, redness, lakeness, nearness, and so on. Each of these contains multiple subsystem concepts. For example, the house concept includes walls, roof (which is connected to the redness concept), doors, windows. Each of these in turn is linked to concept clusters for more details, like texture of the siding or roof, shade of red, and so on. All of the features that constitute the scene are activated in multiple areas of the cortex, but are linked by mutually excitatory axons that reinforce all of the subsystems and subsubsystems to activating in what is now called "working memory," the aggregate of percepts and concepts that are currently active.

If this collection of percepts and concepts is consistently activated on multiple occasions they will be consolidated into another cluster of neurons (generally located further toward the front of the brain) that effectively code for the abstract concept; as discussed above, it is given a name. The latter is a long-term encoding of the house by the lake with the red roof instance. And when your mind recalls this cluster by activating it (for whatever reason), it, in turn, activates all of the perceptual and conceptual clusters that previously activated together when you were actually seeing the object and its environs.

This is a very brief description of an extremely complex process that goes on in the brain to represent systems in neural networks that can be formed "on-the-fly" and if properly reinforced can become permanent memories. The main intent here

[9] Would that there was space permitting a foray into the mind/brain arguments that dominate the psychology/neurology/philosophy worlds. We will not go that route. Here we simply claim that mental states and experiences (subjective) are the product of neurological processes, of clusters and networks of neurons working together to "represent" concepts in working memory that are, in turn, linked to abstraction concepts (words) that are processed sequentially to produce language.

is to show that the brain is very much able to record and use systemness in how it models the real world. It has the additional great ability to experimentally put multiple concepts together in ways not encountered in the real world naturally, like unicorns. This is the basis of our ability to invent tools and stories.

2.2.2.2 Systems in Symbolic Text

We invented the ability to represent mental models or images in symbols and icons. We can write down a sequence of symbols that represent, even if arbitrarily mapped, our spoken languages. Once the conventions of the mapping are agreed upon among participants, we can start recording and recalling systems that are etched into an appropriate medium. We can write and read sentences so that the systems in our minds can be conveyed not only by spoken words but also by the conveyance of the media itself. A message written on a piece of papyrus that designates farmer Ahab as owning 13 jars of wheat in the granary records a system relation between the farmer Ahab and the jars of wheat and the granary. That piece of papyrus could just as easily be given to another person in order to transfer the claim for the 13 jars under Ahab's mark to another person.

Language captures descriptions of systems in the mind, which in turn capture systems in the world. But language is a very fuzzy system. Words and sentences can be ambiguous. They can mean slightly different things to different participants in a conversation.

In order to circumvent the problems with simple language descriptions, scientists and engineers have turned to structured texts with well-defined terms. For example, engineers employ standardized forms called "specifications" along with diagrams to convey descriptions that are readily understood by anyone who is knowledgeable in the standard. The collection of texts and diagrams is a system for expressing a concrete system.

Mathematics is a step further in abstraction but also in disambiguation. For example, measurements of system attributes, by reliable measuring devices, can be represented much less ambiguously in numbers. Moreover, relations between attributes (and their numeric representations) can be highly unambiguous. Mathematics (and its sister, logic) still depend on symbolic representation, but the symbols chosen are generally quite crisp. The relations can be represented in operation symbols that have unambiguous meaning. Through operations of deduction one can prove a sequence of applied operations always produces a particular and specific outcome.

When we are trying very hard to be careful and precise in our meanings we turn to mathematics and logic.

It is important to recognize that categories of mathematical symbols belong to an abstract system. It is also important to realize that this is a static system as long as the symbols live only on paper. In order to be dynamic (to prove new theorems, for example), it is necessary for the systems to be encoded again into the minds of people who can then manipulate those symbols according to the rules of behavior adopted for the particular system.

2.2.2.3 Systems in Computers

Combining text, graphics, and mathematics, it is now possible to construct system models in computer memories and to simulate the dynamics of the system. Computer simulations of many different systems are now quite common and in fact indispensable in modern technology, governance, management, and sciences in general. We will have a considerable amount to say about this approach to abstract systems in Part III.

2.2.3 Systems Equivalencies

Each type of system is characterized by its location in the sense that each type exists in a different medium. Even so there is a form of equivalency between them. That is, a system in the world is represented as a system in the mind and, in turn, is represented as a system in text, or a system in a mathematical object with its rules of transformation. That mathematical system can be represented as (or converted to) computer codes. The system in the world is ultimately reduced to a system in a computer memory.

Figure 2.4 shows a diagram of this equivalency between systems as represented in different media. It is an extended version of the treatment in Mobus and Kalton (2014, Chap. 1).

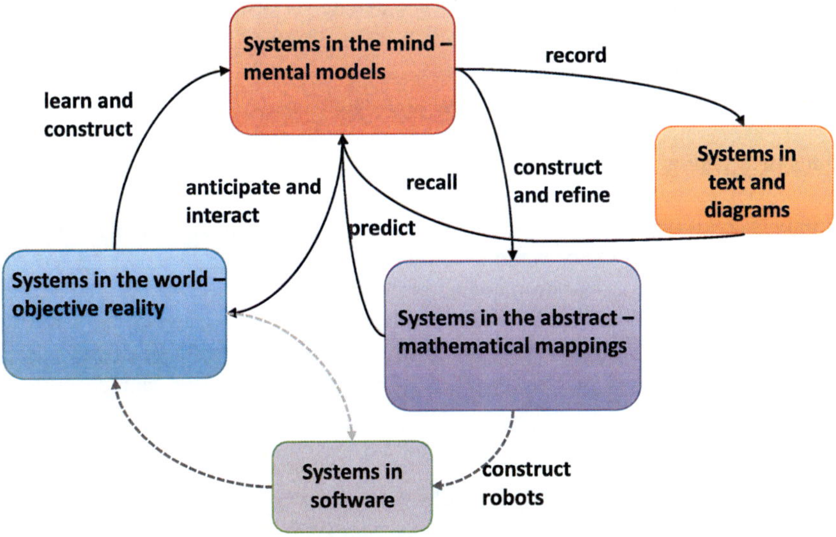

Fig. 2.4 Systems can be found represented in different media. (After Mobus and Kalton (2014), Fig. 1.1 with addition of systems in text)

Systems located in other media, for example, in mental models, are just as real as what we ordinarily think of as systems in the world (out there). As explained in Mobus and Kalton (2014), mental models are actually physical embodiments of abstract systems in neural tissues! Concepts are encoded in brain engrams, patterns of activation in neural clusters across the neocortex. These encodings, when activated in working memory, have all of the physical characteristics of physical systems as described above. Moreover, as we will explain in the next chapter, these physical embodiments of concepts constitute a symbol system. The brain's design imposes a syntax on how such symbols are processed as thoughts and to expression in language.

In a similar way, we can argue that systems encoded in mathematical formalisms only take on a life (have dynamics) when being active within the brain of a mathematically astute person. The mathematical formulas are, after all, concepts within the brain encoding mechanisms and thus, too, become real physical systems. The same argument applies to systems in text. A mind constructs the language equivalent of a system being described (see Chap. 4 for a full explanation) and records it in text. Another mind can read and interpret the text constructing a mental model of the system that then, once again, becomes a real physical system in neural tissues.

If a system can be translated to a mathematical representation, it can be translated into computer codes and run as a simulation in a computer memory. As with representations operating in a brain, being thus real physical systems, a system being simulated in a computer has all of the same attributes of a real physical system too.

There are, of course, significant differences in terms of systems themselves, as they are in the real physical world, and systems represented in the other media. Systems in other media are only models or abstractions of the real systems. They suffer from "loss of information" in being reduced to representations. Mental models are probably the least "reduced" because of the way the human brain can encode the representations—being able to represent levels of detail as needed to make judgments and predictions about the real system's behaviors.

What we want to emphasize here is that all of the various supposed ephemeral kinds of "systems" are, in reality, real physical or concrete systems when one considers that their only real life is either in a brain or in a computer memory and simulation. They all require organization, boundaries, inputs of matter and energy, and output behaviors and wastes when in brains or computers (e.g., waste heat). When systems are recorded as text or mathematical formulas, those recordings are passive representations. When concepts in the brain are not brought into working memory, they too are just recordings waiting to be brought to life. But while alive, they are as subject to the principles of systems as any so-called real physical system.

2.3 Principles of Systems Science Reviewed

The term "principle" is used here in its sense of a regular consequence of nature operating according to laws. For example, in nature, at all levels of organization from fundamental particles up through super clusters of galaxies the bits (like atoms) have tendencies to come together (be it the strong force or gravity) or be forced apart. They have "personalities," like valence shells in atoms that may either be attractive or repulsive. It doesn't matter if we are talking about real atoms (elements) or people, or societies. Each of these has characteristics that either facilitate interactions or cause repulsion. In the former case, some new, higher-order structure with emergent properties and/or behaviors comes into being. The principles invoked here are: (1) formation of networks of components, (2) hierarchical structure, (3) complexity increase, and (4) dynamics resulting from new configurations (constraints and degrees of freedom). But these are just the obvious ones. In systems science, we learn that all of the principles may be at work at once.

In Mobus and Kalton (2014), we identified 12 basic principles that apply variously to all systems.[10] Some apply more to very complex systems, but all 12 apply to the most complex ones. Here is an abbreviated list of the 12 principles covered in Mobus and Kalton (2014).

1. Systemness: Bounded networks of relations among parts constitute a holistic unit. Systems interact with other systems, forming yet larger systems. The Universe is composed of systems of systems.
2. Systems are processes organized in structural and functional hierarchies.
3. Systems are themselves and can be represented abstractly as, networks of relations between components.
4. Systems are dynamic on multiple time scales.
5. Systems exhibit various kinds and levels of complexity.
6. Systems evolve to accommodate long-term changes in their environments.
7. Systems encode knowledge and receive and send information.
8. Systems have governance subsystems to achieve stability.
9. Systems contain models of other systems (e.g., simple built-in protocols for interaction with other systems and up to complex anticipatory models).
10. Sufficiently complex, adaptive systems can contain self-models.
11. Systems can be understood (a corollary of #9)—Science.
12. Systems can be improved (a corollary of #6)—Engineering.

Below we provide brief descriptions of these as they appeared in that prior book as a review or to provide some background explanations. These principles will be used throughout this book as the basis for various methods, but particularly for the whole enterprise of understanding real systems.

Figure 2.5 organizes these principles according to relations between them. Evolution is the overarching principle that determines the long-term unfoldment of the other principles. The core principles apply to even simple systems. The operational principles apply to all systems but in simple systems are not as readily

[10] We neither claimed that these were all of the major principles nor even that they were *the* major principles, only that they seemed to capture what we felt was the spectrum of systems science. In addition to these 12 "major" principles there are numerous sub-principles, some of which we explicitly identify in various chapters in the book.

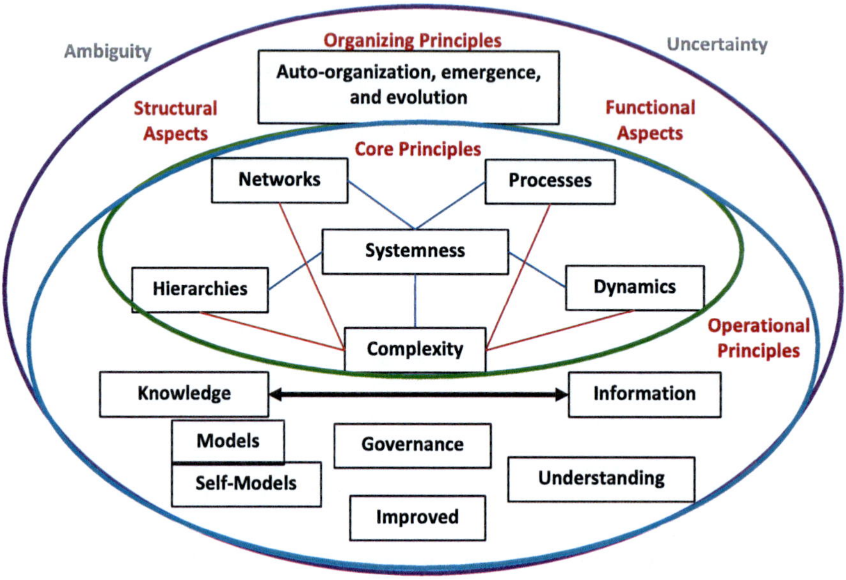

Fig. 2.5 A system of system science principles

identified. Since this book relates to understanding very complex systems, all of these 12 will apply.

The purpose of having a set of principles that apply to systems in general is so that we know what we are looking for when we analyze a system of interest (SOI).

A note on Principles 11 and 12: These are couched as corollaries to Principles 9 (models within systems) and 6 (evolution). They are special cases of the lower-numbered principles. In humans, mental models of the things in the world are what we mean by understanding (in principle). For intentional agents, like birds that build nests or humans that design spacecraft, the system in which they operate can be evolved (improved) for a purpose. This latter case will occupy a fair portion of this book.

The rest of this section is borrowed from Mobus and Kalton (2014, Chap. 1)[11] to provide readers with an introduction to the main principles we introduced there. In each subsection title we include the chapter reference in Mobus and Kalton (2014) as "M&K Chapter(s) *x*," where *x* is the chapter number.

---

[11] These paragraphs are quoted from Mobus and Kalton (2014) but have some additional information not in the original. No attempt to identify such has been made as it seems immaterial to the intentions of this book.

2.3.1 Principle 1: Systemness (M&K Chapter 1)

As shown in figure 1.3, this is the *core of the core principles*, meaning that it is connected to all of the other principles. This principle asserts that everything observable in the Universe is a system and establishes the idea that the core principles are to be found at every scale of space and time in one form or another.

The observable Universe is a system containing systems.[12] Every system contains subsystems, which are, in turn, systems. The Universe may be the only instance of a closed system, which establishes the upper bound on systemness. If Smolin (1997) and others postulating "Loop Quantum Gravity" (LQG) are correct, then "monads" existing at Planck scales of time and space may be the smallest objects that meet the qualifications of systemness.[13] It is possible to imagine monads as the smallest entities that qualify as systems (i.e., have inputs and outputs and interrelations with other monads), and, hence, provide the lower bound on systemness. All things that humans observe between these two extremes, from Planck scale to the "size" of the whole Universe, are systems (see Chap. 3, Sect. 3.4).

Systems of all kinds are found to be composed of components, which may themselves be subsystems (i.e., systems in their own rights). Systemness is a recursive property in which, starting at any arbitrary mid-level scale, one can go upward (seeing the initial system of interest as a subsystem of a larger system) or downward (analyzing the components and their interrelations[14]). This principle guides both analysis, in which systems are understood by breaking them down into components and the functions of components, and synthesis, in which the functioning of systems is understood in terms of their place in a larger relational web.

For our purposes, we will not be much concerned with the very largest scale (cosmology) of the Universe, nor the very smallest scale of the presumptive monads of LQG. But every scale in-between that we can observe will be fair game for understanding systemness.

---

[12] Smolin (1997) and Primack and Abrams (2006).

[13] Monads are "entities" invented by the late seventeenth-, early eighteenth-century German philosopher, mathematician, and physicist Gottfried Leibniz who posited them as the simplest objects of substance. Smolin (1997) relates monads to strings in string theory.

[14] One might think there is a potential problem with infinite regress in this definition. We address this in Principle 2 and later in the book. There are reasonable stopping conditions in both directions of analysis. Practically speaking, however, most systems of interest will not require approaching those conditions.

2.3.2 *Principle 2: Systems Are Processes Organized in Structural and Functional Hierarchies (M&K Chapter 3)*

Since all components and their interactions exist only as processes unfolding in time, the word "system" and the word "process" are essentially synonymous (see Chap. 3). We often use the word when wishing to denote a holistic reference to an object considered as an organized relational structure.[15] When we use the term, we are usually denoting the internal workings of an object that take inputs and produce outputs. Even systems that seem inert on the time scales of human perception, for example, a rock, are still processes. It is a somewhat different way to look at things to think of rocks as processes, but at the atomic/molecular scale, inputs like water seepage and thermal variations. Cause the component molecules and crystals to change. The rock's output, while it still exists, is the shedding of flakes (e.g., of silica) that end up as sands and clays in other parts of the environment. So, in order to understand the organized structure of the Earth, the geologist must study it as a process, not just structure!

A hierarchy is a layered structure in which the lowest layer is constituted of the simplest components in the structure, and usually the numbers of components are large compared with other layers. In general, also, the time constants for dynamics of layers lower in the structure are much smaller, i.e., things happen faster. The next higher layer consists of subsystems composed of components from the lower layer in which component interactions within the subsystem are stronger than interactions between components in other subsystems. The subsystems have effective boundaries or concrete boundaries. This layering and the composition of subsystems taken from the next lower level are usually represented by a tree structure (in the graph theoretic sense, see Chap. 4, Fig. 4.6 for an example).

The hierarchical nature of system structures has long been recognized (Simon 1998; Koestler 1967). As process, functional hierarchies correspond with the structural hierarchical architecture of systems. Hierarchies are recognized as the means by which systems naturally organize the work that they do as a response to increasing complexity (the increase in the number and kind of components at any one level in the hierarchy). Analytical tools that decompose systems based on these hierarchies are well known, especially in reductionist science. But also, when we attempt to construct a system that will perform some overall function for us, we find it is best to design it as a hierarchy of components integrated into working modules, which, in turn, are integrated into meta-modules. The notion of hierarchy will become especially important when we take up the question of coordination and control in our discussion of cybernetics.

[15] Arthur Koestler (1905–1983) used the term "Holon" to describe this fundamental structuring of a whole composed of parts that are themselves wholes. He used the term "Holarchy" to refer to the hierarchical relation between the system as a whole and its subsystems. See: Koestler (1967).

2.3.3 Principle 3: Systems Are Networks of Relations Among Components and Can Be Represented Abstractly as Such Networks of Relations (M&K Chapter 4)

Systems are networks of components tied together via links representing different kinds of relations and flows. This principle ties several other principles together. Namely, Principles 9 and 11 have to do with how we can create models of systems in the world with systems in the mind, or systems in the abstract. The emerging network science (Barabási 2003) provides us with a range of formal tools for understanding systems. For example, graph theory, in discrete mathematics, provides some powerful tools for examining the properties of networks that might otherwise be hidden from casual observations.

For example, Fig. 2.6 shows the existence of a node type, the hub, that was not understood until the application of network theory to several example networks. A "hub" is a node that is strongly connected to many other nodes in such a way that it provides a kind of bridge to many other nodes (depending on the direction of connectivity—the two-way connections in this figure represent the more general case).

Another powerful way to use graph and network theories, very closely related to one another, is the "flow graph," also called a "flow network." In a standard graph, the links represent relations and directions of influence. In a flow graph, the links show a single direction of influence but the influence is carried by a flow of a real substance, i.e., matter, energy, or informational messages. In these cases, the rate and magnitude of the flow are considerations and need to be represented in some fashion. Typically, we use numeric and textual labels to identify those flows. More abstractly, as in Fig. 2.7, they can be represented by the thickness of the arrows showing direction.

These kinds of graphs and the networks represented have been used to analyze so many kinds of systems to date that they have become an essential tool for the pursuit of systems science. Grounding in network and graph theoretical methods is thus very helpful. Even if the quantitative methods of graph theory are not fully made explicit, it is still an invaluable conceptual tool to know how to qualitatively characterize systems as networks of interacting components and to provide detailed descriptions of the nature of the links involved in order to provide a "map" of the inner workings of a system.[16]

[16]The relationship between a network and a map should be really clear. The word "map" is used generically to refer to any graphic representation of relations between identified components. A map of a state or country is just one example of such a network representation, as the network of roads that connect cities, etc.

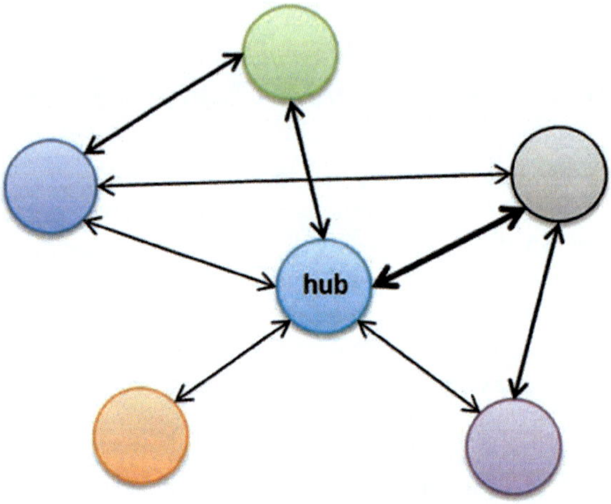

Fig. 2.6 A network of components (nodes) can abstractly represent interrelations by links (edges) in a graph structure. Interaction strengths are represented by arrow thickness, but could be represented by numerical labels. This is a bi-directional graph meaning that the relation goes both ways, for example, like electromagnetic force. In this graph the node labeled "hub" is connected to all other nodes, so it would be expected to play a key role in the functional dynamics of this network

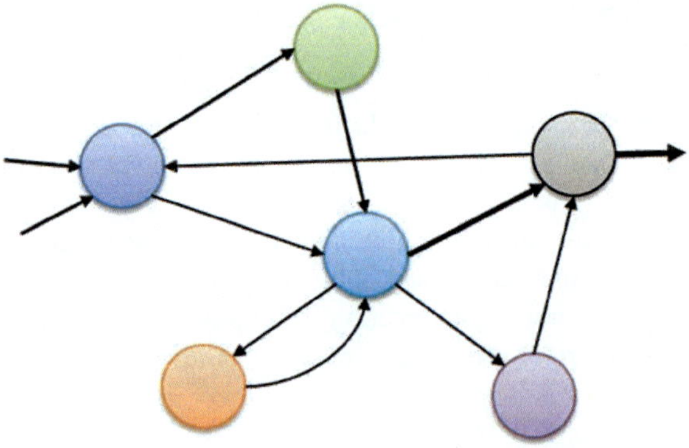

Fig. 2.7 The same network as represented in Fig. 2.6 is here represented as a "flow network." The arrows are unidirectional indicating that the net flow is toward the node with the arrow point. Flow networks provide additional mathematical tools for analyzing the dynamics of a system. Here we have added inputs and outputs in conformance with the idea that such a network is a process with an overall function (outputs given inputs). Again the "volume" of a flow is indicated by the thickness of the arrow for simplicity

2.3.4 *Principle 4: Systems Are Dynamic over Multiple Spatial and Time Scales (M&K Chapter 6)*

Dynamics (or overt behavior) refers to how the processes operate or change inputs into outputs over time. In the most general sense, the lower the level of resolution in space dimensions, the smaller the resolution in time scales relevant to dynamics. At very small spatial scales (e.g., molecular) such processing proceeds in the micro- and millisecond time scales. At somewhat larger spatial scales, say at the level of whole cells, the time constants might be given in deciseconds (one-tenth of a second). On still larger spatial scales processes might be measured in seconds and minutes. On geological spatial scales geophysical processes might be measured in centuries or even millennia. What about the Universe as a whole?

We sometimes find that critical processes that operate over sufficiently different time scales can have hidden negative consequences for the system as a whole. Systems constantly adjust themselves by feedback loops, but when interdependent components operate with feedback loops of different temporal scales the system may become unstable.[17] In understanding what goes wrong and leads to disruption and collapse of function in natural and human-built systems, we generally find dynamical mismatches at the root. For example, the fast economic payoff for clear-cutting forests or harvesting fish by factory trawlers is not in itself scaled to match the reproductive cycles of trees or fish. Systems science explicitly calls for attention to dynamics at all time scales in which conflicts could threaten the sustainability of the system. In those cases where sustainability is desirable, we look for ways to find "harmony" among the different levels of system composition (the hierarchy).

2.3.5 *Principle 5: Systems Exhibit Various Kinds and Levels of* Complexity *(M&K Chapter 5)*

Complexity, like network science, is really one characteristic of *systemness*. But since the complexity of systems is a critical attribute in understanding why a system might behave as it does or fail to behave as might be expected, complexity science has emerged as a subject standing on its own (See Mitchell 2009). Chapter 5 in Mobus and Kalton (2014) discusses the nature of complexity in more detail. Specifically, we have adopted the view expressed by Herbert Simon (1998) that complexity is a measure derived from several attributes of system structure. Systems are described as hierarchically organized networks of strongly interacting modules at any level in the hierarchy. Each module is, itself, also a subsystem and thus a hierarchically organized network of strongly interacting modules. That is, systems are nested structures. This structure is repeated down to a level where we recognize

[17] For an extensive analysis of the dynamics of development and collapse in human and natural systems, see Gunderson and Holling (2002).

what we will call atomic components (see Chaps. 3 and 4). Ultimately the complexity of any identified system is based on the total number of components, number of kinds of components, and the attributes of networks within a level and between levels (i.e., the degree of modularization).

For our purposes we adopt this intuitive definition of complexity. A full explication of complexity would require a full volume on its own. Numerous other authors have tackled the problem with varying degrees of success. Most of modern complexity science has focused on several interesting phenomena in which various systems demonstrate complex behavior in spite of their particular dynamics being based on relatively simple rules. These views of complexity are important when considering the generation of complexity in real systems, but we maintain that the final manifestation of complexity is embodied in the structures and functions of the systems themselves. Understanding the generative functions that give rise to complex structures is clearly needed (see Chap. 3) but not the same as an index measure of the complexity of a system.

Later, when we consider the transformation and evolution of systems (Chap. 3), we will see that systems become more complex by virtue of broadening and deepening the hierarchy and new functionality and unexpected potentials may emerge. But complexity also carries a price: some of the more important findings in complexity science, such as deterministic chaos, self-organized criticality, and catastrophe theory, have shown us that complexity and non-linearity can, themselves, be sources of disruption or failure.

Human societies and institutions present one of the most trenchant examples of the trade-offs of complexity. Joseph Tainter (1988) has put forth a very credible argument that as societies become increasingly complex as a result of trying to solve local problems, only to create bigger problems, the marginal return (e.g., in stability) decreases and even goes negative. This phenomenon is linked with the collapse of many historical civilizations (Scott 2017), such as the Roman Empire, and causes some social scientists today to voice concerns regarding the trajectory of our modern technological civilization. This aspect will be important to keep in mind when we discuss the issues of governance of complex systems.

Note for Principles 6–12 The following principles, those outside of the core principles in Fig. 2.5, apply mainly to more complex systems, especially those described as "complex adaptive systems" (CAS). The meaning of this term will be made clear later.

2.3.6 Principle 6: Systems Evolve (M&K Chapters 10 and 11)

In many ways, this principle, shown in a box under "Organizing Principles" in Fig. 2.3, is itself composed of several sub-principles and is the most overarching of them all. Indeed, it can be reasonably argued that the complexity of the systemness we find in the Universe is an outcome of evolution. All systems can be in one of three situations. They can be evolving toward higher organization, maintaining a steady-state dynamic,[18] or decaying. The principle that systems evolve is based on the systemic effects of energy flows. If there is an abundance of inflowing *free* energy, that which is available to do useful work, then systems (as a general rule) will tend toward higher levels of organization and complexity (see Principle 8 below). Real work is needed to maintain structures and to create new, compound structures. When the energy flow is diminished, the Second Law of Thermodynamics (entropy) rules, and instead of the uphill climb to higher order and complexity or the energy-demanding maintenance of complex order, a process of decay sets in and systemic order deteriorates toward random disorder.

2.3.7 Principle 7: Systems Encode Knowledge and Receive and Send Information (M&K Chapters 7 and 8)

Information and knowledge are most often thought of as pertaining to systems in the mind, a subset of systems. Another way of looking at it, however, finds them in the operation of all systems as they move into a future with possibilities already shaped by the present state of the system. This approach usefully grounds knowledge and information in systemic structure, which not only *is* as it is but *means something* for any possible reaction to events as they unfold. That is, the system by its very structure "knows" how to react. From this foundation in physics, we will be able to more clearly trace the emergent systemic differences in modalities of knowledge and information as biological and psychological life evolves from the original matrix of physical and chemical systems. This will allow a more careful differentiation of the mental way of possessing knowledge and processing information from the physical way, making it clear that the way living organisms hold knowledge and process information is a more complex, evolved form of doing something every system does. The subjects of information and knowledge and their duality will be discussed in Chap. 3.

2.3.8 Principle 8: Systems Have Regulatory Subsystems to Achieve Stability (M&K Chapter 9)

As systems evolve toward greater complexity the interactions between different levels of subsystems require coordination. At a low level of complexity, cooperation between subsystems may emerge as a matter of chance synergies, but more complex systems need more reliable mechanisms of control to coordinate the activities of

[18] Steady-state does not imply a constant level of some measure. Steady-state systems may as readily be characterized by statistical properties that are stationary but not static.

multiple components. Thus control, typically exercised through feedback processes linked with specialized subsystems, becomes an important issue in any discussion of the function of both fabricated and evolved systems. When we take up cybernetics in Chap. 12, we will see how complex logistical coordination is achieved through the development of control hierarchies (multiple controllers require another layer of coordination among themselves!). And then the question will reemerge in an even more challenging form when we discuss the reproductive ability that marks the emergence of life, where not just coordination but accurate copying of the entire system pushes the control question to new levels.

2.3.9 Principle 9: Systems Can Contain Models of Other Systems (M&K Chapters 8 and 13)

We are all aware of the function of mental models, how the image of how someone looks aids in meeting with them, how the map modeling the street layout enables us to navigate the city, or how the blueprint guides the construction of the building. However, modeling occurs not just with minds but in all sorts of systemic relations where one system or subsystem somehow expects another. Thus a piece of a puzzle models inversely the shape of the piece that will fit with it, and, in a similar way, molecules, by their shape and distribution of charges, model the molecules with which they might interact. In general, systems encode in some form models of the environment or aspects of the environment with which they interact, though this modeling element of functional relationships is realized in many different ways and levels in different sorts of systems.

2.3.10 Principle 10: Sufficiently Complex, Adaptive Systems Can Contain Models of Themselves (M&K Chapters 7, 8, 9 and 13)

Adaptive systems such as living organisms can modify their models of an environment to adapt to changes, or simply for greater accuracy (i.e., learning). Creatures capable of having mentally mediated roles and identities include models of themselves, and these likewise may involve greater or lesser accuracy. For humans, as we shall see, the intertwined models of the world and of themselves become structured into their societies and inform the way societies interact with the environment. Systems science reveals the dynamics by which such models are shaped and supplies a framework within which to critique their validity. Insofar as inaccurate models contribute to dysfunctional interaction between society and the environment, systems science thus offers an especially valuable window on the question of sustainability.

2.3.11 *Principle 11: Systems Can Be Understood (a Corollary of #9, M&K Chapters 12 and 13)*

As discussed above, science is a process for explicating the workings of systems in the world, and it has very recently been turned to a better understanding of systems in the mind as well. It has moved our understanding of these systems to new levels by employing formal systems in the abstract. As these formal systems mature and are fed back into mental models arising from experience, we humans can develop better understanding of how things work both in the world and in our own minds. We will never reach an end to this process of understanding systems and some levels of systems may continue to elude us, but in principle systems function in terms of relational dynamics, and this is an appropriate object for human understanding.

The reason we call this principle a corollary of Principle 9 is that the understanding comes from the efficacy of the models we hold of the systems we study. Science is the paradigmatic example. As a social process, science seeks to characterize and model natural phenomena by a piecewise approximation process. The models are improved in terms of accuracy and precision as well as predictive capacity over time. Models are sometimes found to be incorrect and so are abandoned in pursuit of better models. Alchemy evaporated as chemistry arose. In the end, efficacy is the test of the explanatory power of the models. This is what is meant by "understanding" something. When you understand you can make predictions, or at least project scenarios that can then be tested. Then, the accuracy, precision, and explanatory power of the models can all be assessed and, according to Principle 8, using information feedback for self-regulation, the models can be further improved (or found wanting and abandoned).

The human capacity to learn, especially in abstract conceptual models, is an individualistic form of Principle 11. Whereas science builds formal models in the abstract (and increasingly in software), we individuals construct knowledge in our neural networks. Knowledge is just another word for model in this sense. Our brains are capable of building dynamic models of systems in the world and using those models to predict or spin scenarios of the future state of the world given current and possible conditions. We have a strong tendency to anticipate the future, and in so doing we are relying on subconscious models of how things work in order to generate plausible explanations, both of what has happened and what might happen in the future. When we say we learn from our mistakes we are, just as in the case of science, correcting our models based on errors fed back to our subconscious minds where the construction takes place.

When we say that systems can be understood, then, we are referring to our ability to function successfully through the guidance of models in the mind that correlate with relevant features of systems in the world. A model is not identical with the object it models, so our understanding of a system is not identical with the system itself, and therefore is never final. A given model can always be carried further, and another perspective yielding an alternative model is always possible. And this takes us to our twelfth and final principle.

2.3.12 Principle 12: Systems Can Be Improved (a Corollary of #6, Chapter 12)

If one has the boldness to assert that something is an improvement, they are likely to meet the familiar counter, "Who's to say?" If I say it's great to have a new highway, someone else can always bring up sacrifice of land, air quality, noise, or others of a virtually unlimited (and equally systemic) number of ways in which the improvement might also be considered a degradation. Systems science will furnish a framework for thinking through these issues. Principle 6 notes that with available free energy, systems can evolve to higher complexity with emergent new properties. But this is not to say the dynamics that ratchet up complexity automatically lead to improvement. Quite the contrary, increased complexity can also lead to instability and collapse. And then again, who's to say that in the big picture stability is better than collapse?!

We will frame the systemic question of improvement in terms of function. Dynamic systems in their operation necessarily produce consequences. This is their functioning. And in the study of auto-organization and evolution we find that functioning changes as complexity increases. But unless this causal functioning somehow *aims* at some result, all results are equal and the notion of improvement has no basis, no metric. So a systems account of evolution will have to take up the question of when and how causal functioning gets to the condition where the operation of the system can be observed *to aim* selectively at some kind of result. Although the aim is hard to identify in the pre-life universe, the world of life is full of such processes. How then does evolution ramp up to start working in terms of *improved* function, the selection of the fittest?

We, our metabolisms, and the entire world of life operate and organize in an ongoing process of looking for what fits and better fits our varied aims and purposes. And, of course, all those aims and purposes are not perfectly harmonized, as both individual lives and whole systemic sectors intersect with vectors that include tensions, competitions, and sometimes outright contradiction. The success of the predator is the failure of the prey. So, with a narrow focus one can improve either the hunting of the predator or the elusiveness of the prey. At a more inclusive systemic level, improvement would have to look for the best dynamic balance, since too much success on either side, while good for the individuals involved, would have negative consequences at the level of the species' well-being. And improving an integral ecosystem in which myriad intersecting competitions are woven into a dynamically shifting mutual fit would be a yet more daunting challenge. And the same goes for social systems, where clarity regarding individual lives or narrowly focused issues is much easier than the endlessly contested visions of what constitutes an improved society.

2.4 Systems Science

Systems science is a meta-science. This is implied in Sect. 2.1 and Fig. 2.1. That is, it crosses all of the traditional science domains to address patterns of physical reality that are found in all of those domains. The examples mentioned above about attraction and the formation of networks are as true for quarks as for atoms, as for molecules… as for galaxies and everything in between in terms of scale of space and time. Such patterns are found in physical systems, chemical systems, biological systems, social systems, and large-scale planetary systems (i.e., biospheres, atmospheres). The objective of systems science is to explicate these patterns in such a way that the knowledge of them can be applied in any domain to help the domain specialist discover the patterns in their particular subject of interest. The principles of systems science listed above are the starting points for aiding domain specialists do their particular brand of science.

Systems science attempts to organize system knowledge in a rigorous, structured manner, just as all sciences do for their domain knowledge. The principles given above are a first (possibly clumsy) attempt to provide a starting framework for this effort. We will be refining the use of systems science to understand the world in this book.

Systems *thinking* is a more informal way of thinking about the world systemically. Everyone, to some degree or another, has intuitions about the workings of the world they live in and observe. For some people this naturally includes understanding connections and interactions—connecting the dots, as we say. It includes grasping how causal relations can be circular, how feedback can drive the dynamic behavior of some systems. A few people have strong intuitions that allow them to make good decisions about how to work with the rest of the world for everyone's benefit. Unfortunately, not everyone does. Systems thinking comes in shades of gray. People range from systems-oblivious to strong awareness that they are observing systems and are part of a larger system. The objective of systems science is to make explicit what systems thinkers know implicitly with the hope that doing so will improve the systems thinking of everyone through education.

In the next chapter on Systems Ontology we will provide a small sidetrack on how the human brain evolved to perceive systemness in the world. It is actually an expansion from Principles 9 and 10 regarding how a system (the brain) can contain models of other systems (everything out there) including a model of the self (some portion of what's in here that gives rise to the feeling of "I").[19] In turn, this is related to Principle 7 regarding information (communications) and knowledge (models).

---

[19] Damasio (2000).

2.4.1 "Problems" to Be Solved

Ultimately, we need strong systems thinking and the tenants of systems science in order to understand the problems with which we contend. That understanding is mandatory because sometimes what we think is a problem to be solved is not really a problem at all, not when understood as part of a larger system in which it is embedded. For example, human societies often find themselves faced with conflicts with one another. The problem to be solved is how to destroy the enemy before they destroy us. In the context of the immediate cause of conflict, this might seem like a reasonable approach. We have enlisted the aid of scientists to invent more destructive weapons. But to what end? When viewed in the larger context of history and the Earth's natural systems, our little wars are immensely destructive not just of the other guys but of a lot of our environment and our own psyches. The real problem to be solved is the question of why conflicts arise and what steps might be taken to avoid those situations. Historians and sociologists, of course, grapple with such questions. But so far as I know, they have not done so using the kind of systems analysis being described in this book, and usually not even strong systems thinking.

It is ironic that many of the tenants of systems science emerged from the efforts of the Allies and Germans alike in World War II. Concepts from cybernetics (control), information theory, and computing were germinated during the war. But so was the atom bomb. The global politicians were caught in the worst form of non-systems thinking—not understanding how the world as a whole is a system. The scientists employed were involved in domain-specific development. Yet out of that, a number of scientists, already sensitive to the ideas of systems, were able to bring many of those ideas together in a unified concept of systemness.[20] Led by biologists like Ludwig von Bertalanffy, and philosophers like Alfred North Whitehead, who grasped a vision of how these newer concepts fit into general patterns of organization and dynamics, many of the scientists and thinkers came out of the war effort with a realization that humanity had to literally see the bigger picture if it was to survive the twentieth century and beyond. Systems science emerged from a wonderful collective effort among many great minds in the late 1940s and early 1950s.[21]

---

[20] I will be using this term frequently even though you will not find it in any standard dictionary. Systemness is the collective properties of a "thing" that make it a system. Specifically, things that have the properties covered as the "core" principles above, as a minimum, display systemness. The term "systemic" refers to a property or situation that is part of a whole system, so it does not quite work to encapsulate the constellation of properties of a system. Another term that is sometimes used in the same way as systemness is "systemicity." It too is hard to find (perhaps in some British-based dictionary).

[21] It is quite typical in books on systems science to venerate as many of the great minds who founded the discipline in this time period. In Mobus and Kalton (2014), we acknowledged many of these people as they figured specifically into one or another sub-discipline. However, in this book we will not be spending much time recounting where the various ideas came from and lauding the founders. It is not for lack of respect and gratitude; it is for space efficiency! Histories of systems science and the scientists who helped found the area can be found in Warfield (2006) among others.

And then something typical happened. After the war, it was recognized that scientific research had played a significant role in helping the Allies win. Vannevar Bush, a prominent engineer who had been head of the US Office of Scientific Research and Development (OSRD) during the war, wrote an influential memo that recommended forming research funding agencies that would direct money to research universities to support innovative research to keep the United States dominant in the world of science and technology. The National Science Foundation came out of this idea. And academia was changed forever. When you dangle money in front of administrators, they come up with incentives to get the academics to play ball. That incentive was tenure and promotion. In order to win tenure, young academics were encouraged to obtain grants to do very specific research in their domains. Part of the problem was that this included getting published in recognized journals within the specific domains. As a result of this emphasis on domain-specific research (to keep your job and get promoted), systems science became disintegrated into specialized subfields like complexity science, cybernetics, information theory, and so on. In spite of the fact that each of these used the word "system" extensively (because their subject matter was obviously embedded in systems!), the various sub-disciplines became specializations just like all of the other sciences.

This situation persisted throughout the remainder of the twentieth century and into the start of the twenty-first century. Those who continued on in the tradition of an integrated whole science of systems found themselves on the periphery of academia. There were concerted efforts to keep systems science alive—interestingly one of the most successful areas for systems science was in business schools where management scientists recognized the benefits of systems thinking. But on the whole, it remained fractionalized and the parts only loosely coupled.

That situation started to change in the latter part of the twentieth century when technologies like the Internet began to play a significant role in economic life. The impetus was complexity. Not only were technologies getting extremely complex but the whole fabric of our social systems was too. Globalization, enabled by computing and communications networks along with the availability of cheap fuel for shipping goods, created an additional layer of complexity that made it clear that the world really was a system. With the realization of how humanity was affecting the ecosystems of the world and, in fact, the whole world with climate change,[22] the notion that disciplinary sciences could address our issues began to crumble.

The watchword of today is "sustainability." People everywhere are asking what do we need to do to solve our energy problems, our food security problems, our ecosystems and mass extinction problems, and the list goes on. They are implicitly recognizing that all of these global and massive problems are interrelated. Their

---

[22] Many scientists are calling for a formal renaming of the current era (at least since the advent of the Industrial Revolution) from the Holocene to the Anthropocene due to the global geologically embedded effects of human activity.

implicit systems thinking is kicking in and they are beginning to realize they need systems solutions.[23]

Thus, it seems it is time to re-integrate the various sub-disciplines that make up systems science and use this science to tackle problem-solving in a systemic way. Systems engineering will be the application of a holistic systems science to design. This book will walk the reader through this process. We will apply holistic systems science to the understanding of complex systems and the problems that they suffer so as to find holistic solutions—the kind that minimize unintended consequences.

Note that while these complex global problems are an immediate and compelling reason to employ formal systems science, it isn't just problem-solving in the defensive sense that compels us to turn to systems science. Technological progress today comes from the integration of systems that are comprised of subsystems, or what is now called "systems of systems" (SoS). The "smart power grid," the electric power distribution systems that bring electricity from power plants to our homes and businesses, is being investigated as a necessary solution to upgrading the current (quite antiquated) grid, to accommodate the intermittency problem with alternative energy sources such as solar and wind power, and to improve the efficiency of electricity consumption at the end user. This grid will involve so much more than relays and wires. It will involve sophisticated computing and communications subsystems for coordination of distribution. The same is true for a large variety of progressive developments in man-made artifacts such as transportation, communications, and food production.[24] All of the most interesting developments, for example, self-driving cars and trucks, involve significantly greater complexities as compared with late-twentieth-century technologies. As such, and noting Principles 5 and 6 in which we will find that unanticipated behaviors often emerge from very complex combinations of heterogeneous subsystems, using rigorous systems science to analyze these "solutions" and model their behavior so as to catch these behaviors in advance of unleashing them on the world is becoming an essential part of engineering.

2.4.1.1 Complexity

Beyond any doubt the most obvious cause of social and ecological problems is the sheer complexity of our globalized civilization.

For example, according to Worldwatch Institute the average American plate of food has traveled between 1500 and 2500 miles from farm to consumer.[25] The number of steps in growing, transporting, processing, retailing, etc. are unbelievable. Or, how many people today can repair, or even tune up, their own automobile. Look

[23] *Science*, 356:6335, April 21, 2017, is a special issue featuring articles on "Ecosystem Earth" in which many of these exact questions are prominently featured (Vignieri and Fahrenkamp-Uppenbrink 2017, and related articles).

[24] It is true that many of these developments are in part driven by the need to address carbon pollution, but they are also desirable developments even if we did not have this climate change sword of Damocles hanging over our collective necks.

[25] See: http://www.worldwatch.org/globetrotting-food-will-travel-farther-ever-thanksgiving

under the hood of a modern car! Thirty years ago, it was still possible to do a certain amount of maintenance on one's car. Fifty years ago, it was ordinary for people to service their automobiles or even repair much of the engine, transmissions, or bodies. Cars were simpler. There were no computer chips to deal with.

Today it is almost impossible to do any kind of project like Mt. Rushmore's sculptures of four presidents. Why? Because the number of environmental impact studies that would need to be done make such an undertaking a non-starter (in recognition that we humans tended to destroy natural habitat in such projects)! In a sense this is actually progress. We humans will not be directly responsible for the demise of a species of fish or beetle just to satisfy some hubristic urge. But the complexity of regulations also inhibits projects that might not result in any degradation just because of the added expenses of doing the studies.

Then there are the stories of complex artifacts such as baggage handling systems in major airports that were catastrophic failures due to poorly designed and implemented software.

Software systems are among the most prominent forms of complexity in today's societies. And the failures of software development projects are among the costliest results. However, the same root of the problem and the same sorts of project failures are found in many different works of human design. Below, in Sect. 2.5, we provide a few more examples along with explanations of how failure to employ rigorous systems methodologies (based on rigorous systems theory) leads to these problems.

2.4.1.2 Material and Energy Consumption

Every system receives material and energy inputs for processing to produce products and waste outputs. If the system is in balance with its environment it means that the resources of material and energy are available in sufficient quantities to meet the system's needs. It also means the products and wastes will be absorbed by the environment (e.g., there are customers that use the products, and sinks that can recycle the wastes) at the rates of production. Systems that grow larger than their balance point (or what is called the carrying capacity of the environment) will extract more resources than can be replaced (either recycled or renewed). They will dump more products or wastes into the environment than can be readily absorbed, thus poisoning the environment for themselves and other systems. We know this to be the case from empirical studies of both natural and human systems.

Systems that continue growing beyond their natural limits also tend to waste or inefficiently use the resources they extract so they attempt to increase their extraction rate in order to compensate. We humans are guilty on all counts with the consequences that we have severely polluted our world and are depleting our resources more rapidly than they can be replaced. Resource extraction and output disposal are systemic problems that involve not only our desires for wealth and clean air but a deep understanding of just what the carrying capacity of the whole planet is relative to human consumption. We are just beginning to get a handle on these issues and we hope that with the aid of systems science we will have a better set of tools with which to do so.

2.4.1.3 Governance for Sustainable Existence

It is tempting to pronounce a newly understood principle to add to the list of 12 above. However, it is possible to consider this as a sub-principle within Principle 8—systems have subsystems for regulating their behavior. The sub-principle is that systems serve a purpose within their embedding supra-system. This means that the system, as a subsystem of the supra-system (Principle 1 regarding hierarchies of systems within systems), is strongly coupled with (generally) multiple other subsystems from which they get resources and deliver products of value to the others (who are "customers"). The whole supra-system persists over time as long as all of the interacting subsystems are doing their "jobs." And so long as that is the case the whole supra-system and its subsystems are sustainable.

The notion of serving a purpose (i.e., producing products of use in the system) is behind the evolution of internal regulation of sub-subsystems that keep the subsystem doing the "right" things. Such systems that succeed in serving their purpose are "fit" in their environments (the other subsystems).[26] Another term for systems that have such internal regulatory or governance subsystems is "purposive." That is, they are motivated to behave in what appears to be a purposeful manner—that is, they appear to have their own purposes. In fact, their purposes are just the dual of the purpose they serve. A good example of this is the role of a top carnivore in a food web. Their apparent purpose is to find and eat prey. In fact, they are serving the purpose of culling (returning a certain amount of prey biomass to the environment in the form of excrement that will be recycled by bacteria) and population control in the ecosystem. The governance system involved is a combination of internal drives within each individual carnivore coupled with feedback loops through the food web itself. If the carnivores cull too many prey, the latter numbers decline, leading to less food for the carnivores and their numbers eventually also decline. The nutrients returned to the soil may boost primary production (plant growth) providing sustenance to a then growing population of prey animals. Finally, this increase in population means the predator population can start to recover.

Governance mechanisms involving a hierarchy of cybernetic sub-subsystems are what make possible subsystems like carnivores or businesses fulfilling their purposes, both their internal goal-directed purposes and their larger purpose within the whole supra-system. In Chap. 12, we will see how this works in more detail. We will also show how these mechanisms constitute the ideal of a distributed decision system. Even though we talk about a hierarchy of structure and function, this is not the same as a command-and-control version of governance.

---

[26] In Mobus and Kalton (2014), Chap. 10, "Emergence," Sect. 10.3.2.4, we describe the process by which simpler components interact with one another to form composites that must then survive a selective environment. Those configurations and any behaviors that result from them that do survive are said to have emerged. The fact of their survival shows that they are fit in that particular environment. In Sect. 10.2.1.4, "Fit and Fitness," we had given the criteria for being fit in an environment.

Chapter 10 will go deeply into the nature of governance and management subsystems that are required for all complex adaptive (CAS) and complex adaptive and evolvable systems (CAES), which will be discussed throughout this book. Every such system is exposed to challenges from environmental changes to internal wear and tear or entropic decay processes that can produce substantially suboptimal behavior. Any such system must remain fit by compensating or adjusting for these challenges so that a central problem for them is the constant monitoring of the environment and self along with the decision processes and activation capabilities in place to maintain themselves in the face of these challenges.

This requires an internal governance subsystem composed of decision agents that collect data from their specific domains of control, process the data to determine if any actions are required, and if so, what, and then issue command signals to actuators to carry out those actions.

No system is sustainable without a well-functioning governance subsystem. At various places in the book we will be referring to, what for us is the most important system on the planet, the human social system (HSS), which at present has global dynamics but no real global governance subsystem in place to regulate the internal actions. We do not really control our population growth or resource depletion behaviors in order to be in balance with our environment—the rest of the planet. Moreover, most of our national governments are actually not really very functional from the systems perspective on governance. One of our greatest challenges in systems science will be to bring to the fore where our more common ideas of governance (and political process) deviate from systemic ideals and suggest ways in which these can be corrected for the sake of the HSS sustainability.

2.4.2 *Systems Science as a Meta-Science*

Though the application of systems science to engineering and development of human-built artifacts, including our own systems of governance, will have a huge benefit for humanity and the rest of the world, the sciences can benefit from its application as well.

Already most of the sciences have branches, some mainstream, that are called "systems <subject>," where <subject> is a placeholder for name of the science. Systems biology is, perhaps, the most prominent. Many of the original founders of systems science were primarily biologists (Miller 1978; von Bertalanffy 1968; and for a general overview, Capra and Luisi 2014). However, people from diverse areas, such as management theory, were very fast to take up the systems perspective in understanding management science and especially understanding management information systems in terms of information theory and cybernetics (Beer 1966).

Most of the sciences that have adopted some aspects of systems science to their methodological approaches (i.e., systems modeling) have benefited from the insights gleaned from doing so. In systems ecology, for example, H.T. Odum (1994) used a systems modeling methodology to gain a deeper understanding of how energy flows through an ecosystem worked to shape and sustain that system.

However, there is much more than methods for modeling that could be gotten from applying rigorous systems science to any science. With the principles described above, along with others that derive from them as described in Mobus and Kalton (2014), scientists would know a priori that their objects of inquiry would be subject to considerations of network and hierarchical organization, internal and external communications, regulated energy and material flows, and so on. Basically, they can address their subjects more holistically than has been the general practice up until the present century. System theory should aid in better understanding theories in the sciences. For example, social scientists are already making great strides in understanding social dynamics from a systems perspective, using network theory, auto-organization theory, game theory, and many other aspects from systems science. Presently, they have developed their own methodological approaches (similar to the biologists), and have, in fact, contributed to a better understanding of some systems theory issues applied to human beings and societies. It stands to reason that if they work at connecting all of the principles of systems to their subjects they will gain new, more general insights into human social systems.

2.4.2.1 Understanding the Universe

The goal of the sciences (both natural and social) is to come to understand the way the world works. The "world," as used here, means the Universe and us as well. This statement is not an argument in favor of scientism, the view that science is the "only" way to gain knowledge. It is simply an observation of how the work of the sciences has proceeded over the last several centuries. One cannot argue against the successes thus far achieved (without denying some kind of objective reality). Our grasp of physical reality today is a simple fact, validated by our ability to produce artifacts, machines, medicines, media, and all aspects of culture.

So, without denying that there may be other forms of knowledge about realms of some reality that is not, strictly speaking, physical, we start with the presumption that insofar as physical reality is concerned, so far, the sciences have provided humanity with a working understanding of how things work.

Up until quite recently the work of science has proceeded along a path based on a mostly analytical methodology. The version of the scientific method taught in schools reflects the philosophical approach to gaining understanding called "methodological reductionism."[27] This is, essentially, the idea of finding explanations for

[27] Not to be confused with a philosophical posture, also called reductionism, in which it is asserted that all phenomena can ultimately be "explained" by lower-level phenomena. What systems science has lent to this notion is the concept of emergence, in which higher-order phenomena are found that could not have been predicted just from the properties or behaviors of the constituents a priori. Methodological reductionism recognizes the need to describe the higher-order phenomena first and then look for underlying phenomena that a posteriori can be seen to explain the higher-order phenomena. There is an extensive literature on emergence but, of course, we recommend Mobus and Kalton (2014) for a concise overview.

phenomena (or objects) in terms of component phenomena (or objects). Analysis is the process of decomposing (or deconstructing) something to find out what it is made of and trying to understand its workings based on the composition found. For example, living cells are composed of biochemical molecules, which are, in turn, composed of atoms (the mnemonic CHNOPS—carbon, hydrogen, nitrogen, oxygen, phosphorous, and sulfur—captures the main elements), which, in turn, are composed of subatomic particles, etc. The behaviors of atoms are based on the behaviors of the subatomic particles (e.g., the formation of various kinds of bonds based on the properties of electrons in their orbitals around the nuclei).

But as science has explicated the way the world works it has run into some interesting problems with its methods. It isn't enough to explain how life is based on chemistry. Life has properties and behaviors that knowledge of chemistry alone cannot explain from below. For example, where does the motivation of living systems to survive and replicate come from? Biologists observe this living mandate but there is nothing in chemistry, even organic chemistry, which would necessarily predict that cells would strive to survive and reproduce.

The properties and behaviors of living systems are emergent from interactions between extremely complex organic chemistry subsystems. They have to be observed in action (in situ) to be appreciated. After that, deconstruction of the metabolic and reproductive processes can show how the biochemical processes produce these phenomena. There is no known way to go the other way, from biochemical to phenomena.

The biological grasp of what biochemistry does (and means) is a clear example of a systems approach to understanding.[28] Methodological reductionism is a necessary step toward understanding but only when predicated on a holistic model of the phenomena of interest. That is, the system must be appreciated in its whole behavior before trying to grasp how the system works from the inside. We shall be attempting to apply this lesson from biology to the full range of sciences. Systems science offers an approach to understanding how low-level (organization) phenomena translate, through emergent properties and behaviors, to more complex yet understandable higher-order phenomena.

Understanding, however, is not as straightforward as might be supposed. There are actually two levels of what we call understanding. There is the kind in which we can make simple predictions of what might happen in terms of a system's future behavior. And then there is the kind where we can explain *why* a system might behave the way it does.

[28]Which may help explain why the greatest and most comprehensive advances in understanding systems came from biologists!

2.4.2.2 Shallow Understanding

What do we mean when we say we understand something? There are at least two senses implied by the word "understanding." Both carry the idea that we can predict the future behavior of a system or predict that instances of an observation of a system's state will be within particular bounds. But such prediction can be based on a "shallow" understanding of the system. It derives from many prior observations of the system's behavior under varying conditions (environmental) along with a recording of the history that allows us to make predictions about behavior under conditions not yet observed. We treat the system as a "black box"[29] wherein we know nothing about how the system transforms its inputs into output behaviors. We just know that it does. Observational sciences, those that rely on observation of natural phenomena and use statistical methods like linear regression analysis to capture some behavioral tendency, are often stuck with this sort of understanding.

Often a shallow understanding of a system is all we can expect to get. For example, when trying to understand the behavior of a particular animal species we can only watch and record the conditions of the environment and the animals' responses to it. We have to collect large amounts of data and analyze it using these statistical tools to see if a pattern emerges that can lead to prediction. But being able to successfully predict future behavior based on past observed behavior in observed environments does not guarantee that those predictions will hold when a completely new kind of environment presents itself. The temporal window of observation may have been too short to accommodate all of the possible scenarios and hence fail to produce a correct prediction under what might be rare conditions.

Experimental sciences are not limited to merely observing. They can conduct designed condition experiments that extend the notion of a temporal window and rare conditions. For example, a behavioral biologist might be interested in how a particular species of ant reacts to a lower temperature than has commonly been observed in the wild. Setting up that experiment would be relatively simple and yield data that would extend the inferences about behaviors' correlations with conditions.

Even so, the observations of behavior remain in the realm of shallow understanding. What would be nice to know is how that ant species adapts internally to low temperatures (if it can). And to do that you need to do some dissection.

2.4.2.3 Deep Understanding

If you really want to understand something, you have to find out *how* it works, how its behavior under different conditions derives from those inner workings, and be able to predict its reactions to various kinds of inputs on "first" principles, i.e., based

[29] The term comes from the engineering world and is used to describe a machine, the internal mechanisms of which are unknown. All that can be discerned is its behavior.

on kinetics or internal dynamics. When you can do that last part, you can claim to have a basic understanding of the thing. Doing a good job on the first two, finding how it works and how it behaves, brings you a long way to the last—prediction. Being able to say how something works generally means knowing something about its component parts and how they interact with one another, especially the roles they play in transforming inputs to behavior and outputs. Observing and collecting data regarding behavior over time and over a wide distribution of inputs is still important so as to confirm one's predictions based on this deeper understanding. The more thorough the job one does the deeper one's understanding goes.

But really deep understanding comes from continuing to "deconstruct" the internal components themselves. This is the framework of systems analysis that we will be exploring (forgive the pun) in depth in Chap. 6.

We must take a moment and explain the use of the word "deconstruct" in the previous paragraph. To deconstruct something means to take it apart to evaluate its internal components. It does not necessarily entail *separating components* from one another (as in dissection) and, in fact, can include the notion of keeping track of the connections between components so as not to disrupt the organization. Consider an analogy that might be useful. Suppose a biologist wants to examine a single-celled organism. She might first put the organism under a microscope low-power lens to observe it from the outside. She might take pictures of the organism as it swims about in its medium. She might observe its behavior as she introduces various compounds into the medium to see how the organism reacts to them. She might take chemical samples of the medium to see what sorts of molecules the organism is exuding. But at some point, when she has a good idea of what the organism does, she decides to put it under a higher-power lens that allows her to see inside the organism without disrupting it. She sees a multitude of organelles, tiny subsystems that carry on the internal work in the organism. She can catalog all of the various kinds, get some accounting of their numbers, and perhaps see some connectivity between them. Then she shifts to an even higher-power lens to look at more details; perhaps now an electron microscope has to be brought in. One by one she examines the structure and functions of the organelles. She might note how mitochondria take in low-weight carbohydrates and pump out adinosine tri-phosphate (ATP) molecules. Still not disrupting the cell, she can see how outputs from one kind of organelle are inputs to other kinds.

This is the sense in which we use the word "deconstruct" and the process of "deconstruction." The objective is to not disrupt the workings at finer levels of resolution, but to preserve the whole system's integrity as much as possible in order to understand how the parts' activities contribute to the whole. Later we will be elucidating the process of structural/functional deconstruction in systems analysis and show how its aim is explicitly to maintain functional organization of the system being analyzed. This is different from many traditional reductionist approaches (mentioned above) where the methodologies may leave a system disrupted because of a belief that knowing what the parts are is all you need to know to explain the whole. We have learned that this belief is in error most of the time.

Understanding the world and the things in it, like ourselves, our society, our complex technologies, and the environment, is the principal endeavor of humanity as it seeks to shape its world in wise ways. Not knowing how things work is dangerous. If we act on ignorance instead of understanding, we risk unintended consequences. Humanity is facing that situation right now in the form of global warming and climate change that will dramatically alter our environment in ways that may involve existential threats. Acting with shallow understanding gets us into trouble. It takes deep understanding to advance our situation keeping within the limitations imposed by the natural world—the laws of physics and chemistry, for example.

Deep understanding has been the purpose of science. Science is a process, a self-correcting process that works to increase our understanding of the world and everything in it. It operates primarily by observing and measuring, by finding mathematical or formal descriptions of phenomena, by dissecting phenomena into components, and by modeling phenomena so as to make predictions.

However, science has been operating in a fractured, almost haphazard way, though it has been evolving toward greater organization due to the way the results of science are accumulated and organized. Individual scientists observe various phenomena and investigate their often small piece of the puzzle. Science has focused on a reductionist approach to gaining understanding. This is a historical phenomenon that resulted from starting from a base of ignorance of things, and a limited ability for individual scientists to envision a larger or more holistic picture of how their phenomenon of interest fit into that larger picture.

As the sciences have evolved (or matured), from ignorance to greater understanding, an interesting new phenomenon has emerged. More and more, the sciences are turning to system *thinking*, a system *perspective*, and a concern for interactions between numerous phenomena and subsystems. They are turning more and more to a systemic approach to gaining understanding and are recognizing that the real basis for deep understanding is not mere reductionist exposure of the components but a concern for how those components interact with one another to produce the larger behavior of whole systems.

Systems science is about understanding the systemness of things in the world (as depicted in Fig. 2.1). It is a science that explores the nature of being a system, of classifying kinds of systems, and developing the methodologies for applying principles of systems science to the other sciences and engineering as they become more systemic. This book is about the latter aspect.

2.5 What Goes Wrong

In this final section, we examine some of the issues that arise from not following a principled approach to system understanding. That is, what kinds of problems do we experience when we attempt to design and build an artifact, or generate a policy, or understand a complex scientific problem when we are not using a whole systems approach to the work? Here we will briefly take a look at a few examples of areas where the work of systems analysis and design are not based on the principles of systems science but on systems intuitions and more casual systems thinking. Along the way, we will point out the issues that arise and discuss how a failure to take a particular principle, or set of principles, into account can lead to less than desired results.

2.5.1 The Software Development Process

2.5.1.1 A Bit of History

Our first example comes from the Information Technology (IT) arena. Information systems, as used in organizations primarily for the assistance they provide in management decision-making, were understood to be complex very early on. During the early 1970s large corporations, particularly banks, were beginning to adopt computers to automate routine account management functions. The use of computers, what were called "mainframe" machines, spread rapidly as the tools for building large software systems were developed.

For example, the COmmon Business Oriented Language (COBOL) had been designed initially in the late 1950s and standardized in 1968.[30] With that standardization and the construction of compilers by mainframe manufacturers, a cadre of specialists called "systems analysts" arose to help translate the management needs of business workflows into computer programs that would automate those workflows. Early popular programs were developed for the organizations' accounting needs, such as payroll, accounts receivable, and accounts payable. Over a relatively short period of time, other business needs were automated, especially inventory management. The success of these early information systems can be attributed largely to the fact that the underlying workflows that needed managing had been in practice, literally, for centuries. The accounting practices in standard business were so well understood that the translation task was generally straightforward. Paper-based processes were replaced with computer-based ones more or less directly. Analysts did not have to actually deeply analyze the accounting procedures as if they were "new" applications. The designs were simply assumed from the existing paper-based procedures. In that environment, the only problems that generally arose were the result of programmers making mistakes while coding, leading to logic errors (bugs) in the programs. The most insidious bugs were those that only showed up on rare occasions due to particular but uncommon inputs. They were often not detected in pre-deployment testing and only showed up during field operations. The software industry (both third-party and in-house developers) got used to maintaining their code, which actually meant fixing bugs when they surfaced. On the whole,

[30] COBOL was an early English-like programming language, often attributed to the work of Grace Hopper. It was designed specifically to support the development of business information management computing support.

this was manageable. The costs were believed to be offset by the increased productivity gained using automation.

Once traditional accounting processes were automated the role of systems analysts came to be associated with helping accounting managers decide what reports (data summaries) they wanted to get from the now-computerized data files. The analysts were needed because they understood what could be done through the report generation languages and how to access the various files needed to process the data. Systems analysis became a process of local problem-solving—one report at a time. The generation of a report only depended on the location (file) and disposition (age) of the data and not on the whole system. This practice may have been the origin of practices that persist to this day—to do analysis by asking users/stakeholders what they want from some particular "new" system, rather than do a true systems analysis of the whole system as context for the development and design of processes to achieve the goals.

Problems started to emerge as the technology became more complex. Complexity also affected the nature of the business processes being automated. Distributed mini-computers were employed to manage different aspects of business workflows. For example, manufacturing companies sought means of monitoring and controlling the flow of materials from inventory through manufacturing to finished goods. Marketing and sales wanted to track customers' orders and deliveries. These kinds of operations are non-standard across industries, and even within industry types there are many variations and specializations of needs. Hence the nature of IT systems began to include heterogeneity of sub-processes. In addition to structural complexity increasing, the distribution of computing nodes while still needing cross-organizational cooperation led to the need to solve problems in concurrency—the coordination, synchronization, and communication of state information between processes.

2.5.1.2 IT Development Failures

Information Technology projects conducted by businesses and government agencies tend to be extremely complex, involving hundreds to thousands of "users," managers (decision makers), and affecting numerous stakeholders such as customers or clients, and suppliers. They involve computing and communications equipment and the software to make them work. The main source of trouble seems to be in software development. Software has unique system component properties that make its development subject to many sources of problems.

For the most part software development is conducted in planned projects by project managers whose job is to integrate all of the components into a working subsystem. Those projects are managed with the intent of producing and delivering the software modules for the economic benefit of the organization and stakeholders. For now, we report a sobering fact that has been plaguing the software development industry ever since the level of complexity (above) rose beyond a straightforward payroll system. Most projects suffer some kind of failure that costs in wasted effort, lost revenues, and other economic consequences.

Projects don't generally have a binary result—success or failure. However, we can categorize project results into those categories with shades of gray. Success or failure potentials can be measured along several dimensions. Reasonable metrics to use include project costs, delivery schedule, and final performance (compared with requirements). By these criteria, most software development projects experience some degree of failure. According to a report published by the Standish Group from a 1995 survey of IT managers an average success rate (on-time, on-budget) was a paltry 16.2% with less than 50% of originally specified functionality delivered.[31] According to this report, in 1995, approximately $81 billion dollars were to have been spent by American companies and government agencies for projects that were likely to have been canceled before completion. This amount is compared to approximately $250 billion spent each year on software development. And this was 1995!

Have things improved since the end of the last century? Ten years later a report was posted on the Institute of Electrical and Electronic Engineers (IEEE) Spectrum site, "Why Software Fails: We waste billions of dollars each year on entirely preventable mistakes," by Robert N. Charette (2005):

> This year (2005), organizations and governments will spend an estimated $1 trillion on IT hardware, software, and services worldwide. Of the IT projects that are initiated, *from 5 to 15 percent will be abandoned before or shortly after delivery as hopelessly inadequate.* Many others will arrive late and over budget or require massive reworking. Few IT projects, in other words, truly succeed.[32] (emphasis added)

In the same article Charette gives the following list of "reasons" that projects fail:

- Unrealistic or unarticulated project goals.
- Inaccurate estimates of needed resources.
- Badly defined system requirements.
- Poor reporting of the project's status.
- Unmanaged risks.
- Poor communication among customers, developers, and users.
- Use of immature technology.
- Inability to handle the project's complexity.
- Sloppy development practices.
- Poor project management.
- Stakeholder politics.
- Commercial pressures.

[31] The Standish Group web site: http://www.standishgroup.com/ (accessed October 26, 2016). The group report: Chaos at https://www.projectsmart.co.uk/white-papers/chaos-report.pdf (accessed October 26, 2016). There have been dissenting voices to the results reported then (see this Dr. Dobbs report by Scott W. Ambler, The Non-Existent Software Crisis: Debunking the Chaos Report, February 04, 2014 http://www.drdobbs.com/architecture-and-design/the-non-existent-software-crisis-debunki/240165910 [accessed October 26, 2016]).

[32] See: http://spectrum.ieee.org/computing/software/why-software-fails (accessed October 26, 2016).

This list actually applies to more than just software development projects. In Chap. 5, we will provide a sketch of the systems analysis process and use items from this list to demonstrate how a full analysis based on systems principles would greatly reduce the negative impacts of failures. For now, note that the first three items above all point to a simple fact—ignorance of what the system would or should be, and not bothering to gain that understanding before commencing the design. But it goes deeper than not understanding the IT system requirements. Most information systems that are delivered but fail to produce adequate results do so because the real system isn't just the IT subsystem. It is the underlying work process that is being served by the IT system. It is impossible to understand the requirements of an IT system without understanding the system being served and all of its subsystems. This will be thoroughly explored in Chap. 6.

And what about today (as of 2015)? The IEEE Spectrum has posted an interactive website that allows users to explore the costs of failures globally.[33] Basically, it appears that the industry has not learned anything from these failures, perhaps a result of the multifactor aspects of the complexity of projects and their environments. But upon post-crash analysis (Charette 2005), most or all of the problems that caused the failure were in some sense preventable.

It is the contention of this book that following a more rigorous systems approach than is typically done would uncover most of the problems before a significant amount of money is thrown away.

In truth, the main form of systems analysis that is practiced in all of the various forms of project management relies on a weak questioning of the users, stakeholders, managers, etc. This is called requirements gathering, and even when it is backed by supposed quantitative measures it suffers one major flaw. Usually and commonly the users, et al., *do not actually understand their own requirements* nor do they have a language with which to articulate them if they did. In Chap. 5, we provide a few examples of situations where users' claims for what was required were later shown to be incorrect after the development work had already gotten underway. It is interesting to note that the software development project management community has reacted to these kinds of examples by designing methodologies that attempt to work around this problem. So-called agile and iterative methods have evolved in which users are actually considered part of the development team and the systems are developed in modules quickly allowing users to actually see the results and, most importantly, change their minds at an earlier stage in development. This approach has helped to some degree but in terms of dollars lost, these methodologies have only shaved about ten percentage points off of the experiences of the "heavy planning" approaches of the 1970s and 80s.

[33] See: http://spectrum.ieee.org/static/the-staggering-impact-of-it-systems-gone-wrong (accessed October 26, 2016).

2.5.2 Systems Engineering Process

Systems engineering involves quite a lot more than just designing an information system alone, however many of the approaches used in systems engineering are similar to software engineering.[34] Today we design whole physical operations, machinery, labor, and management subsystems together. Large-scale products like supercolliders and jumbo jets involve every conceivable engineering input to bring everything together in an integrated and fully functioning whole. This means that not only do the engineers have to concern themselves with developing software, they also have to develop every physical and "mental" aspect of the system. They have to coordinate the efforts of mechanical, electrical, materials, chemical, and sometimes many more engineering experts. The scale of complexity in these efforts is many times greater than an IT project alone.

One of the more famous examples of a massive system design failure was the baggage handling system at the Denver International Airport (DIA). This subsystem of a megaproject was supposed to speed up baggage handling and save labor costs. The system was supposed to serve all of the airlines using DIA, which was itself a huge modern airport, designed for expansion. It involved many mechanical sub-subsystems (e.g., the conveyor belts and their drive motors, switching equipment) as well as the computer controls that were supposed to route the movements of baggage from airline to customers or from service desks to the correct flights quickly and efficiently.

Unfortunately, the system never worked. In several post-mortem analyses many of the problems given in the Charette list above were found to be operative. Megaprojects are large, complex systems that require multidisciplinary-based engineering. Their purposes are to solve really large problems such as integrated transportation, energy production and distribution, and healthcare. They have become increasingly popular globally and they have proven to be extremely costly in terms of cost overruns and late completions (Flyvbjerg et al. 2003). They reflect the same sets of problems encountered in software projects but with massively greater risks and costs. As the global problems facing humanity grow and require mega-megaprojects to tackle solving them, the need to learn how to approach them is extreme. This is especially true given the decreasing resources that are available to the world (Meadows et al. 2004).

[34] For the rest of this book, we will consider all such engineering processes as systems engineering, i.e., software engineering is a subset of systems engineering.

2.5.3 The Sciences

It would not be entirely correct to claim that science "goes wrong." But it is correct to claim that science goes slow. The process of science has been guided by principles evolved from the earliest thoughts of empiricists such as Roger Bacon (circ. 1219/20–circ. 1292). The discipline of the scientific method and the process of doing repeated experiments (or making repeated observations), even when occasionally faltering, tend to cause the convergence of understanding over time. There is nothing particularly wrong with this process. Except that we might not have the luxury of time to let things work themselves out. Moreover, we've already argued that the complexities associated with phenomena of modern interest have reached a point where simple single disciplinary approaches cannot always make progress.

We assert that the methods described in this book are as appropriate for the scientific process as for the design process. Indeed, the systems analytic procedures, which preserve the holistic relations needed to reconstruct the system of interest in model form, are just starting from a higher-level view of the whole process. The sciences are already moving toward systems approaches and this is particularly the case for interdisciplinary investigations.

The systems approach as developed in this book adds a higher-level principled integrating framework for all of the sciences, natural as well as social. In Chap. 9, we will take an in-depth examination of a scientific approach to economics, typically thought of as a social science and therefore not subject to the same kinds of methodological disciplines as the physical sciences. We will show one approach that is capable of replicating more quantitative methods for economic questions because it follows the systems approach to the subject. We would like to put to rest the myth that social sciences are "soft" because their subject matter involves human beings. The principles of systems science are used to demonstrate that as we go up the hierarchy of complexity and levels of organization, the same rules apply even though they manifest in far more complicated ways.

We believe and will argue that using the systems approach as an umbrella framework for all of the sciences will make scientific investigations more efficient.

2.6 What Could Go Right

The main contention of this book is that all human activities aimed at gaining knowledge, and particularly understanding, will benefit from applying the principles of systems science and the methods described in this book. At the very least, the knowledge gained will be "pre-integrated" into the larger body of knowledge of the world by virtue of tying the system to the sources and sinks of its inputs and outputs, its environment. For example, we will demonstrate in Chap. 9 how analyzing economic phenomena as a system can provide a stronger (and more scientific) linkage with other subsystems of the human social system such as the education subsystem, which is supposed to be a beneficiary of the former.

Similarly, we will have examples of how systems engineering, if pursued under the framework of the principles of systems, will result in much higher-quality products, services, and policies for designed systems. This comes in the same form as for the sciences. The system of interest is automatically and firmly tied to its environment. The principles of systems science will ensure that the designed system will fulfill its purpose—it will be *fit* in its environment.

In the background of all of this is the idea that doing things right the first time will reduce overall costs in the long run. Doing so will help make sure resources will be conserved and available to other efforts in the future. It will reduce operating costs, in the case of designed systems, and curating costs in the case of natural systems. Structuring knowledge for ease of access and making sure that knowledge is complete will reduce search time (and costs) in the future. Finally, being able to, at the push of a button, generate various models for visualization and simulation provides ready access to many views of a system of interest and provides ways to test hypotheses about the behavior of the system under conditions not necessarily observed in nature or designed originally.

2.7 Proceeding with Principled Methods

Throughout history, humans have demonstrated time and again that once basic principles of nature were learned and applied, the quality of knowledge gained subsequently improved substantially. This book will attempt to apply the principles discussed above, and the general theory of systemness, to a methodological approach to furthering the knowledge we have of how the world works. Such knowledge is becoming vital to our very existence as entities on this planet. Our hope is that what is offered here will be seen as beneficial to the process of gaining understanding and will be adopted, in some form, in time to make a difference.

Chapter 3
System Ontology

Abstract Given the physical history of the Universe as we understand it now, there are a number of questions we need to address having to do with how "things" come into existence and how is it that things seem to be getting more complex over time, going from the fundamental particles and energies that seem to have been created in the Big Bang to systems like the Earth and its complex "spheres." What exists is a matter of record. How the Universe proceeded to evolve toward greater complexity and organization is another matter. The things are organized into categories based on similarities and differences. That organization points to a hierarchy of complexity. In this chapter, we will propose an ontology of systemness that accounts for the organization. We say "an" ontology because there have been many attempts throughout human history to establish such an organization. This chapter focuses on an ontology that is guided in formation by the principles of systems science covered in the last chapter. Again, this is not a first such attempt. We will acknowledge a few more recent attempts in this chapter but will not present a history of the developments in ontology as this can be found in many sources.

In all such ontologies the starting point is a metaphysical perspective, a point of view or world view. As metaphysics, one has to assert some basic "substances" that are universal in nature and make commitments to derive all that exists from those. We start the chapter making such commitments at the most fundamental level of existence and then derive all else from there. The process that leads to increasing organization and complexity is essentially a universal evolution, which is an evolution theory that is not just about living systems. We call it "Ontogenesis" and describe the mechanisms (like variation, combination, emergence, and selection) that generate new things and new levels of organization.

We will also dip into one of the other main branches of metaphysics, epistemology of systems—the knowledge basis of systems understanding and a brief look at how we can use systems knowledge to obtain understanding of all systems. This is done in preparation for elaborating a systems language (names of the things and their dynamic relations) to be pursued in the next chapter.

G. E. Mobus, *Systems Science: Theory, Analysis, Modeling, and Design*,
https://doi.org/10.1007/978-3-030-93482-8_3

3.1 What Exists

Ontology is the study of what exists. Its objective is to identify and name those categories of "things" that are found in reality.[1] It is not concerned with specific instances of things themselves, e.g., your pet dog. Rather it is the "about-ness" of things, e.g., your dog shares many characteristics with every other dog, a group of characteristics we might call "dog-ness." It involves things that are general in nature, not specific instances of those things. For example, a biological entity that has some level of volition, it moves in order to obtain food, is an animal. Animals, defined by a set of properties (such as volition), exist. There are numerous instances of animals, from simple flat worms to whales. The concept of an animal is a category[2]; ontology deals with categories' given names for identification purposes. More general categories sit atop a tree graph, subcategories are linked below. The implication of the linkage is that the lower category is a type (or kind) of the higher category. A good example of an ontological category tree is the one implied by the animal example—a phylogenetic tree.[3] The root of the tree is the most general category, e.g., "Animal" (kingdom) with all of the kinds of animals (phyla, e.g., vertebrates) forming nodes linked to it. Each of these then has subsequent sub-subcategory nodes (classes, e.g., mammals), and branching further down (see footnote 2) to yet lower levels (less generality). The lowest category in this tree is the species.[4] So, we have within the tree the logical relation, "is-a-kind-of," from the bottom up—a pet dog is a kind of canine, is a kind of mammal, is a kind of vertebrate, is a kind of animal.

The philosophical study of ontology delves into questions such as: "How do we know these characteristics define a category?" Or: "Is this characteristic particular

[1] We need to clarify that ontology is not just a list of things that exist, it is a guide to fundamental existence (the most basic things that exist) and to how other more complex things composed of these can come into existence. We are concerned with a claim of not only what basically exists (e.g., quarks) but also what can be generated into existence from those few basics.

[2] We will be using concepts from a field of mathematics called category theory. However, we will not delve into the more arcane aspects of the field. In Chap. 4, we will be developing a mathematical description of systems using set and graph theories. Those familiar with category theory will grasp the derivations. Our position, for the present, is that category theory is a little "too" abstract for the development of a theory that can be grasped by the average reader for whom this book is intended—for readers interested in knowing how to apply category theory to the structures we will be exploring.

[3] Note that phylogenetic trees generally are represented as branching upward. At the risk of confusing the reader, this differs from a conceptual tree, which inverts the relations in having the root above and branching downward. There is no real difference in these representations. They come from different disciplinary origins (i.e., biology vs. logic)—yet another example of how systems theory is meta-theory for science!

[4] The term species is not quite as cut and dried with respect to categories. There are subspecies, and hybrids that can cause a little confusion. In the next chapter, we deal with the notion of such fuzzy situations using fuzzy set theory.

to this category?" Or better still: "Is this a valid characteristic, and if so, why?" For the most part we will leave these kinds of questions to the philosophers.[5]

The concept of "an" ontology is adopted in the field of information science primarily for the purpose of categorization and logical linking as in the above example of "is-a-kind-of." In the computational world of information science, the purpose of what is called "domain ontology" is to represent knowledge in a *computable form*. That is, the representation should be able to be manipulated with time-efficient algorithms to derive new knowledge. The approach taken in this book is something between the philosophical and the computational ones. The reason will become much clearer toward the end of Chap. 4 and as we use the language of system developed there to work with concrete systems. Basically, we assert that the language/ontological representations need to be a hybrid between algorithmic computation and natural language so that the ideas of systemness can move freely between the world of computer simulations and human understanding.

One thing that needs to be stated clearly, at this point, is that "an" ontology cannot make a claim of truth—that the things and relations declared to be existing truly are the things and relations that truly do exist. Or more to the point, these are the only things that exist and matter. Any ontology is just a "best guess" based on hopefully unbiased observations. What is being claimed is not ultimate truth but a *commitment* to a set of categories and relations that are treated as complete and internally consistent. Any metaphysicist is free to make other commitments and then show how those lead to our phenomenal experience of reality. For once commitments have been made, it is what we do with the sets subsequently to build descriptions of the world. Once the commitments are made, we can then turn attention to the nature of knowledge and how it is acquired, to epistemological pursuits. We begin this chapter with a few commitments to what we believe is the fundamental underlying nature of reality and how that nature gives rise to the things and relations we actually find in the Universe (ontogenesis). We then use the logic of ontogenesis (Sect. 3.3) to derive the systems that have come into existence in the history of the Universe, how those systems give rise to new, more complex systems, and that produces all of the things that exist today (that we know of).

The purpose of this chapter is to explore and develop a system ontology in the service of developing a language of systems in the next chapter. The terminology developed in this chapter will form the basis of a lexicon and syntax of this language (Chap. 4) and establish the semantics of that language, its meaning with respect to

[5] Consider, for example, the concepts put forth by philosopher Roberto Unger (Unger and Smolin 2015) who asserts that ontology in philosophy is ultimately an empty concept. He holds that everything in the Universe is subject to change, including the "laws" of nature, time being the only inclusive aspect of the Universe as a whole. What he proposes instead is a "proto-ontology," with three basic "attributes" (section starting on page 239). The first is an existence postulate—something has to exist. The second is a plurality postulate—there has to be more than one kind of anything. The third postulate is that everything that exists must be connected causally. These three postulates are fundamental to systemness.

descriptions of systems in the real world. The language will be used to build our knowledge of the world and how it works.

Before developing a means of saying what exists we need to consider several higher-order concepts that shape our thinking about systems. These concepts are at once phenomenological, that is, we can perceive their reality, and metaphysical, that is, they exist outside, above, and beyond our merely phenomenological treatments of things that exist.

The first is what we will call the "category" of things. Categories are abstract types of things. A common example of categories is the phylogenetic tree of life. The binomial nomenclature method of naming species is predicated on a hierarchy of types. We humans are a type of primate. Primates are a type of mammal, and so on.

The second concerns the reappearance of certain patterns of forms and relations throughout the hierarchy of organization—the levels of organization[6] from subatomic particles and forces to atoms, to molecules, to geophysical and biological systems, and so on. There are a set of archetype patterns that recur at all of these levels. These have been described as "isomorphic"[7] in that they seem to possess the same formal structures and functions whenever they occur. For example, all systems can be described as networks or sets of objects/components that have strong interactions with one another and can be represented in graph theoretical terms. This form can be found at all levels of organization in nature and human system designs (c.f. Mobus and Kalton 2015, Chap. 4).

3.1.1 Categories and Type Hierarchies

Below we will argue that the nature of a process we call *cosmological ontogenesis*[8] naturally produces a hierarchy of elaboration of patterns we recognize as categories (see Sect. 3.2.1.3 for an introduction). As mentioned above, humans are a kind of primate. There are, extant, many concurrent versions of primates that are the leaf nodes in an evolutionary tree, having arisen from a common primordial primate archetype. Humans are contained within the category of primate (actually in the category of hominin, which includes our cousin humans such as *Homo neanderthalensis*). Primates are a *type* of mammal. The terms category and type are somewhat interchangeable with caution.[9] A category is a generalized grouping of systems that

[6] For a graphical view of levels of organization, along with an explanation of their dynamics, see Volk (2017).

[7] One pattern is isomorphic with another when there is a one-to-one mapping of features or functions between them.

[8] The term ontogenesis comes from biology where it means the emergence of form. We are adopting the phrase "cosmological ontogenesis" to name the process whereby forms of new systems are produced and develop.

[9] In the dictionary they are considered symmetrical synonyms. The explanation given by WikiDiff does little to clear up the semantics and usage.

have a subset of similar characteristics. All mammals, for example, have hair and the females possess mammary glands for lactation and feeding the young. If you are asked to "categorize" an animal that has hair and mammary glands, you would put it in with mammals. At the same time, mammals are a type of animal (and animal is a category). So, the distinction between type and category can be difficult. Our position is that we will not try to make a distinction explicitly. What is important is to recognize that types represent evolved individuations from categories. The latter are recognized by us humans from a process of inductive learning whereby we find representatives possessing common traits and lump them together into said categories.

3.1.2 Recurring Patterns at All Scales

We find, in the elaboration of what exists in the Universe, repeating patterns of process and form at each level of organization. For example, we will find that certain work processes such as "combining inputs" or "temporary storage of substance" are repeated at molecular or societal levels of organization. As mentioned above, the use of networks is replicated at all levels of organization. We will argue below, in the section on ontogenesis, that this is a natural consequence of the way systems are composed at the different levels of organization.

There are a set of universal patterns of structure and function that will be found at whatever level of organization we explore. In this book we will use the concept of an archetype model or a generalized (and abstract) model of process that can be shown to be applicable to any level of organization with appropriate elaborations. As an example, consider the notion of an *economy*. In later chapters we will demonstrate that the concept of an economy is a universal pattern of how energies and materials interact to produce stable, sustainable complex adaptive systems (CAS). Those systems emerge from the same process of ontogenesis we have been alluding to. They seem to be natural consequences of the process. In anticipation of our explication, we assert that metabolism is a fundamental form of an economy and that what we call "the" economy of human society is nothing more than an expanded version of metabolism at an obviously larger scale of organization.

3.1.3 Origins of Things: A Universal Process

Where do hierarchies of organization come from? Why are cells complex organizations of biochemical processes (e.g., metabolism)? Why are communities complex organizations of humans, who are, in turn, complex organizations of cells? Indeed, how do "things" come into being and, over time, the complexity of things increases? Our answer is that a recursive process of auto-organization, emergence and selection, and evolution (variation and selection) operates from the simplest level of organization (quantum fields) through successive levels of organization (typified by the interests of the sciences, i.e., subatomic, atomic, molecular) to produce the Universe we see today (Mobus and Kalton 2015, Chaps. 10 and 11; Volk 2017; Morowitz 2002; Smith and Morowitz 2016; Bourke 2011).

When the Universe was "born" it was a super-hot mixture of fundamental particles and energies (we do not really know what these were: strings, monads?).[10] Somehow, as the Universe expanded (through "inflation," then "ordinary" expansion) and cooled, matter condensed. Matter interacted with energies (photons) and having inherent interaction potentials formed structures under the influences of gravity, strong and weak forces, electromagnetic attraction/repulsion, and radiation. From the very beginning the Universe engaged in a process of *ontogenesis*—the process of combining lower-level entities to form complexes that could then interact at a higher scale—that led to the origin of living forms (at least on one planet we know of for certain; see Sect. 3.3). Living forms continued the process of ontogenesis to form multicellular forms, and then societies. Human societies created cultures, a combination of biological and technological complex systems.[11]

Universal evolution appears to be driving matter and energy toward greater levels of complexity—this in seeming opposition to the nominal account of the increase in entropy. In fact, so long as stars produce energy flows that affect disequilibria (i.e., energy gradients) in planetary systems, there will be a drive toward greater organization. Locally, entropy decreases as life evolves elaborately. Globally, i.e., the whole Universe, entropy increases as it should. There is no contradiction. High potential energy does work to increase organization at the microscopic level of life on a planetary surface even while the Universe tends toward equilibrium (Morowitz 1968).

3.2 What Is "a" System Ontology, and Why Does It Matter?

We have claimed that systemness, the "property" of being a system, as described by the 12 principles in Chap. 2 and to be formally defined in the next chapter, is the fundamental organizing principle of the Universe. This is a very audacious claim on

[10] This version of the cosmic origin story, currently favored by scientific cosmologists, is the current best guess about how the Universe we see today originated approximately 13.5 billion years ago, and developed over time to the present. As has been the case throughout the history of the sciences it is the current paradigm, but questions have been raised that raise the possibility that it is not quite the right story.

[11] Tyler Volk (2017) has described a clear progression of increasingly complex stages on what he calls the "Grand Sequence" from free quarks to our current states of civilizations. He calls the process "Combogenesis" wherein ontological elements within any stage (e.g., quarks, nucleons, atoms, molecules, prokaryotic cells, etc.) interact with one another forming combinations that effectively become the new entities at the next stage. So quarks combine to form nucleons and the latter combine with electrons to become atoms, and so on up the sequence. This is the same process as described in Mobus and Kalton (2015), and again later in this chapter. See Sect. 3.3.2.2 below.

its face. So, it will be prudent to examine the basis of this claim because it is fundamental to everything that follows.

As one way to see how this claim comes about, we will describe the process of auto-organization of smaller, simpler system objects into larger, more complex systems with emergent properties and behaviors. Once stable properties and behaviors have come into being, the cycle of auto-organization and emergence at the next higher level of organization proceeds. This is the grand cycle of universal evolution that occurs on all levels of size and time. This is what we call ontogenesis, the evolution of forms of organization that become progressively more complex over time. The Universe that we observe today, especially our own world, came into being by this process.

If the concepts being explored by cosmology and quantum physics are reasonably close to reality, then the cycle would seem to "start" at some fundamental scale (the Planck scale?) of space and time with pre-particle entities (Unger and Smolin 2015). These have attractive and repulsive interactions that lead to primitive structures (perhaps quarks and leptons) with emergent properties and behaviors. Those, in turn, led to interactions resulting in the particles of the Standard Model (or something reasonably close to it). And then everything from that point on is, as we say, history. This description may recount the bootstrap of physical reality. However, much research is needed to be done to reveal it. For our purposes we will need to assume that something like this had to have happened to get the whole process started.[12]

Regardless of the details at the quantum level, the fact is that the Universe has evolved into a very complex and complicated place. Here on our planet Earth that evolution has included the emergence of life, complex multicellular plants and animals, social structures, human beings, and civilizations. The latter includes a remarkable non-biological aspect—complex artifacts, or technology and cultures. Our starting point for understanding the Universe is already problematic. We exist in extreme complexity (as compared with atoms or subatomic particles). We have a major task ahead of us to grapple with that complexity, to understand how the world works.

The main topic of this book is the description and explanation of a *methodology* (set of methods) that can be used to acquire knowledge of how things in the Universe work. Further claimed is that this methodology rests on understanding "things" as *systems*, that systems interact with other systems and in doing so form a larger, encompassing *supra-system*, that systems are composed of subsystems (down to an atomistic level), and that this hierarchical structure encompasses everything from the Universe as a whole (closed) system down to the uncountable number of fundamental particles and energy packets that have auto-organized to form the

[12] Parenthetically, we harbor a suspicion that a key to understanding subatomic phenomena (quantum world) might very well end up being helped by systems science! But also, large-scale phenomena like gravity. It is amusing to think that a resolution to the ultimate amalgamation of quantum and gravitational theories might have a root in systems thinking!

complexity of the Universe that we (as systems that can contain models of other systems) observe in the present.

The claim(s) thus made is (are) unprovable and, more importantly, un-falsifiable as it (they) stand. This puts the whole issue of systemness into the realm of philosophy, metaphysics, as a starting point. Systems science begins at the point where falsifiability of hypotheses becomes possible. That is after we have established a set of "axioms"[13] generated from the claim of systemness and then set out to test whether this or that aspect of a specific system of interest (SOI) fails, upon proper testing, to be falsified. The program for generating axioms from the principles of systemness is beyond the scope of this work. However, we do need to ground the rest of the book in a philosophical treatment to show the justifications for what follows.

3.2.1 System Philosophy

Following the tradition of philosophies, particularly those of the Western World (e.g., ancient Greeks),[14] we will discuss the general ideas of systemness as a metaphysical domain, followed by briefs on what counts as a philosophical ontology and epistemology. This chapter will use this philosophical starting point to transition to what might be called a more practical version of ontology (and epistemology) as it relates to doing systems science. The term "ontology" has been appropriated by the computer science/information sciences communities to mean a set of terms that name what exists in a domain of interest (e.g., diseases in the domain of medical science). The main use of domain ontologies is, for example, providing keyword tagging for documents, especially graphics that contain objects for which the name applies (e.g., X-rays of lungs with cancerous tissues). World Wide Web developers have proposed a standard for use of such ontologies in what is called the "Semantic Web."

In this chapter, we will develop a set of terms that cover *objects*, *relations*, *actions*, *transformations*, *quantifications*, and *categorizations* that constitute at least a beginning of a system ontology.[15] In the next chapter, we use that ontology, along

[13] The term is in quotes to note that these are more like assumptions than logical or mathematical axioms in the rigorous sense. However, their treatment as true axioms is not precluded (indeed it is expected).

[14] Michael Kalton, coauthor of *Principles of Systems Science*, is a philosopher who has studied Eastern philosophies/religions. He has demonstrated that many aspects of these philosophical traditions are very much systems concepts.

[15] One of the reviewers of an earlier draft of this chapter introduced the author to a more universal approach to an upper ontology, the General Formal Ontology (GFO). After a review of its approach it appears that there are many aspects of GFO that map to the system ontology developed here. It is certainly conceivable that both approaches have many aspects in common because both are attempts to find the ultimate nature of what exists. One substantive difference, however, is that the GFO started out with an assumption of formality, using first-order logic. The system ontology

with a formal definition derived from the principles and reflecting the ontological development presented below to produce a language of systems. This language will then be the basis for everything else that follows, namely the extraction of knowledge about systems in the world through systems analysis, the modeling of systems for various purposes, and the generation of system designs.

But first we start with the metaphysical aspects of systemness as a basis for what we do.

3.2.1.1 Metaphysics

Systems as we will be describing are real things in the real world of material and energy. By definition they are physical. This even applies to "thoughts" in the mind, which we now have solid evidence are patterns (spatial and temporal) of synchronized neural excitations of particular (though fuzzy[16]) clusters that encode representations of sub-patterns across the neocortex of the brain.[17]

The position being taken here might be termed "practical physicalism," or a philosophical stance that only matter and energy exist and all phenomena can be known in terms of these substances and their *interactions* (however, see below for an extension of this position with respect to information and knowledge[18] as "nonphysical" substances having causal powers). It is "practical" in the sense that the stance is taken because all we seem to be able to actually say we know objectively is that which we can measure by physical means. It is thus somewhat meaningless in developing a philosophy of physical systems to admit to other substances at the base of our phenomenal Universe. We do not deny other possibilities but take this stance for the simple reason that we have nothing we can say, systemically or otherwise, about some other existence. On the other hand, we will soon introduce other aspects of reality that we will call "ethereal substances." Like matter and energy these are

---

presented in this chapter started with an assumption that imposing such a formality too soon might foreclose the exploration of interesting ideas that cannot be immediately captured in a formal language of that sort. The use of formalism (set and graph theories) in the next chapter is used primarily to capture descriptions of systemness without the restrictions imposed by an axiomatic system. We believe that at some point such a system will be developed but hope it will be after there is a broad understanding of the general terrain.

[16]The patterns of clusters contributing to any concept are under constant revision due to learning and forgetting attributes at the periphery of a concept. For example, the concept of a "house" may need considerable modification after one sees a geodesic dome-shaped dwelling.

[17]This view rejects the Cartesian dualism, matter versus mind. What happens in the mind can be explained by the physical activities in a person's brain. Using functional magnetic resonance imaging (fMRI) methods, neuroscientists are able now to view the brain in the act of thinking specific thoughts. Some kinds of thoughts, or activations of concepts, appear to be found in similar locations in the brains of different people.

[18]Throughout this book and in Mobus and Kalton (2015), we differentiate between information and knowledge. The former is seen as the content of a message whereas the latter is seen as the content of a structure, See below, main text, for explanations.

real in the sense that they participate in causal relations with the former two, i.e., constitute their interactions. These are "information" and "knowledge" (see Fig. 3.1).

Once we find ourselves in the realm of physical phenomena, we can pose scientific questions and hope to answer them via the process of science. But questions such as "What is matter?" or "What is energy?" or "What makes them interact as they do?" are on the edge of metaphysics (meta: beyond or above). An even more difficult-to-contemplate question is: "Why do they exist?" These kinds of questions are beyond the capabilities of science to answer. They are currently solidly in the realm of metaphysics.

Systems metaphysics is based on a worldview as previously described. We simply assume that the Universe is a closed system.[19] That is, the Universe has all of the properties of a system except the input and output of anything. Further we assume that the Universe came into being in a state of ultimate simplicity (in the Big Bang) and has been evolving toward higher levels of internal organization (at least in quasi-isolated pockets of space-time) and complexity through the interchanges of substances between emerging subsystems. Our main piece of evidence for this proposition is the existence and state of affairs on the Earth. We have at least one example of an evolved complex system (c.f. Morowitz 2002).

The metaphysics of systems can be summarized by looking at the philosophical ontology and epistemology of a systems-based worldview.

The starting point is that the Universe is composed of matter and energy, that both come in discrete packets. Matter can be thought of as compressed energy. We

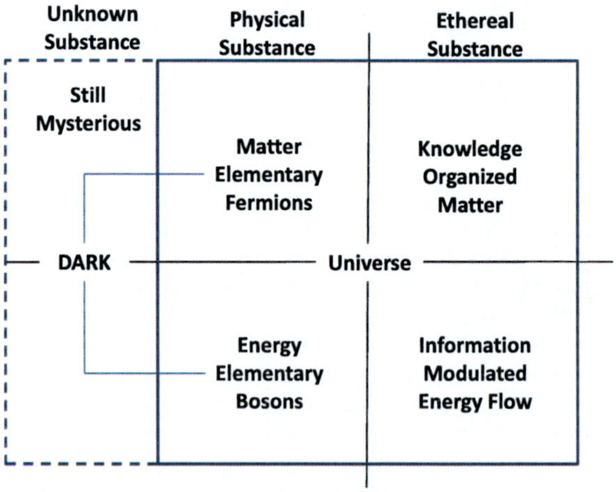

Fig. 3.1 These are the claimed substances that exit (have ontological status). The substances in the solid boxes have understood relations and interactions. The dotted-lined box indicates that our knowledge of these substances is still developing

[19]We have to assume since as things stand, we have no access to the boundary of the Universe due to the finite speed of light and the distance to the presumed "edge" of the Universe. The assumption is considered reasonable even under so-called "multi-verse" interpretations of cosmology (Unger and Smolin 2015).

know that the two substances are interchangeable.[20] Let us assume (as one of our axioms) that at the Big Bang, matter was created out of some of the energy, itself contained in the singularity described by the standard model in cosmology.[21] There are many additional aspects of this "creation" such as time, space, the four fundamental forces, and randomicity that are important in a full description but go far beyond the scope of this book, so will be left to another time. We will simply assume them.

3.2.1.2 Cosmological Ontology

As a first step in delineating what exists from the standpoint of systems, we need to establish the overall context. Our planet is a system in the Cosmos. It is a subsystem of the Solar System and that is a subsystem of the Milky Way Galaxy. We find systems at increasing scales of size and time (c.f. Smolin 1997). The Earth is our local "global" system, mostly closed (at present) to material flows but luckily very open to energy flow, in particular solar radiation coming in and waste heat radiated out to space. In order to begin the construction of a system ontology, we need to at least examine aspects of a cosmological ontology as it supports and constrains what we can say about systems.

Figure 3.1 shows a conceptual framework for the basic constituents of reality as we currently understand it from physics. In addition to matter and energy the figure shows "substances" discovered and characterized in the twentieth century, information and knowledge. It also shows the somewhat speculative situation with regard to what are called dark energy and matter. These substances are real enough by virtue of their observed causal impact on the long-term dynamics of the Universe (Clegg 2019; Primack and Abrams 2006; Smolin 1997). At this time all we can say is that dark matter is thought to be responsible for keeping galaxies from flying apart and dark energy is thought to be responsible for accelerating the expansion of the Universe as a whole. They are shown next to regular matter and energy with dotted-lined boxes to indicate their still uncertain status.[22] Because of this status we will proceed with what we do know about regular matter and energy as they participate in the emergence of systems.

Our metaphysical claim is that the entire Universe is comprised of these substances interacting with one another such that structures composed of matter and energy evolve toward greater organization over time through a process of

[20] Of course, thanks to Albert Einstein's famous equation: $E = mc^2$.

[21] Not all cosmologists are in agreement about the "pre-history" of the Universe. See Unger and Smolin (2015).

[22] At this writing there is scant evidence for dark matter being some exotic kind of particle or dark energy being some form of propelling energy working against gravity. In fact, there are alternative theories that might turn out to be more correct, *modified gravity* rather than matter, and *quintessence* rather than energy.

auto-organization, emergence, and subsequent selection[23]—a process we have dubbed *ontogenesis* (see below).

This despite a seemingly countervailing process of randomness that appears universally. In reality this randomness is a result of the interactions of particles and lower potential energies that result in unpredictable changes in behaviors of the particles observed from a more macroscopic perspective. Thermal noise and various other sources of uncertainty, such as nonlinear interactions driven by energy flows, result in a stochastic process of combination as well as a stochastic environment. Without this stochasticity, combinations of particles (the term being used generically to represent an aggregate of independent objects at any level of organization) would not explore the space of possible combinations. Also, the stochasticity of the surrounding environment of the combinations can act to break up such combinations that are "weakly bound" so that the particles can be recycled. Only the most stable combinations in any particular environment will persist—their internal bonds are stronger than the forces that bombard them. This can be described as a search through the space of stable configurations,

A second and ultimately more important process is at work as well. That is, the propensity for energies to equilibrate throughout the available space; that is, the tendency toward energy equilibrium over time. This tendency, known as the Second Law of Thermodynamics, means that systems, even if they are stable against the vagaries of their environments, can, nevertheless, lose organization—become more disordered—and their components dissociate by losing energies that formed the bonds that held the components together.

This provides a bridge to physics and a cosmological process of ontogeny (development). The origin and evolution of systems in the Cosmos is cosmogony (or cosmogeny[24]). Once the existence of fundamental particles and energies was established, presumably in the Big Bang, the development of increasing levels of organization (i.e., particles to assemblies, their interactions, to supra-assemblies) took over. This describes an upward spiraling process of auto-organization wherein components at a given level of organization, say atomic elements (nuclei of hydrogen and helium), would combine in nucleosynthesis, to produce more complex systems of protons and neutrons—the heavier elements. After the Cosmos cooled sufficiently, nuclei could attract electrons to form atoms with the emergent properties that allow yet further combinations in the form of molecules. We now enter firmly the realm of increasing complexity and the emergence of new properties and behaviors.

[23] Mobus and Kalton (2015), Chaps. 10 and 11) explicate these processes as they pertain to how matter and energy interact, mediated by information to produce more complex dissipative structures that, in turn, interact at a new level of organization. The degree to which such structures are dissipative is a measure of the entity's "knowledge" of the interaction. See also the sections below regarding ontological information and knowledge.

[24] The suffix, "-geny" derives from the same root as "generate" and "gene," referring to birth and development. The term "cosmogony" appears to be a derivative therefrom. C.f. the online etymology site: https://www.etymonline.com/word/-geny for reference (accessed January 8, 2018).

Auto-organization is a result of the capacity of some fundamental units of matter to possess interaction potentials of varying kinds. Namely, these units may interact by binding (attraction) or mutually repelling each other, using or giving up energy (e.g., photons) as appropriate. Moreover, we posit that these interactions are determined by geometry; the potentials exist only at certain points on the "shell" of the matter. An atom is a model of this matter–matter interaction relation. Atoms (elements) are already systems, meaning they are composed of simpler subsystems (e.g., quarks, organized as subsystem components of protons and neutrons, and electrons). Our assumption is based on the idea that we will see this pattern of matter–energy–matter interactions (with the structures of matter exhibiting "personalities" that condition the kinds and dynamics of interactions) as we go "up" the organizational hierarchy.[25]

But remember this is metaphysics at this stage. We are not compelled to "prove" this conjecture, only to make clear what we are conjecturing because of our position that systemness obtains therefrom. The happy fact is that from atoms, and chemical interactions, up the hierarchy we are clearly dealing with the interactions of matter and energy and the "ordinary" sciences. Below we will review the roles of information and knowledge in this interaction and as they pertain to the on-going process of evolution. For completeness, Fig. 3.2 shows some of the relations that exist between matter, energy, knowledge, and information.

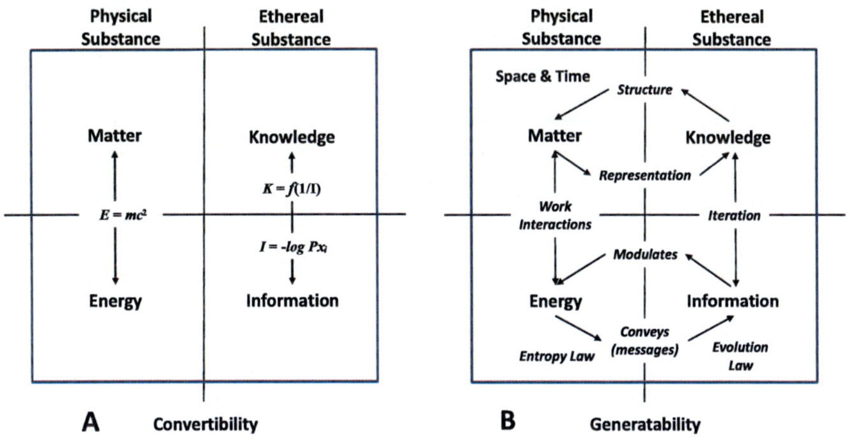

Fig. 3.2 (**a**) Matter and energy are inter-convertible according to Einstein's formula. Information and knowledge are similarly inter-convertible (see text). (**b**) The relations between information/knowledge and energy/matter constitute the cause of dynamics and the evolution of all structure

[25] What we are describing is reminiscent of the Marx–Engels theory of dialectical materialism. See: Spirkin, Alexander (1983) by Progress Publishers. https://www.marxists.org/reference/archive/spirkin/works/dialectical-materialism/index.html (accessed October 28, 2016).

In Mobus and Kalton (2015, Chap. 7), the subjects of information and knowledge are explicated. Here we summarize the main points covered in that chapter.

3.2.1.2.1 Ontological Information

The term "information" is used so broadly in everyday language that it is often misused, even by those who work in a field related to information theory (e.g., computer science).[26] Partly this is because those working in a field have developed a culture of mutual understanding of their subject in which they all know what meaning should be attached to any particular instance of the use of the term by virtue of the context in which it is used. Outside of that culture, other cultures or laypersons do not share this contextual distinction so the real meaning can get lost.

A particularly difficult case of confusion in the definition of information comes from the field in which the definition was born, communications. In communications theory we deal with a sender, a channel, a receiver, and an agreed upon coding that allows the sender to represent into the channel a message that conveys the current state of the sender, or if we are normally talking about human communications something the human intends to send. Communications engineers talk about encoding the information into a signal/message but what they really mean is encoding into the message what they *think* will be informational to the receiver. The sender cannot know a priori what state the receiver is in and therefore whether the message will, in fact, inform the receiver, meaning conveying news that the receiver did not previously have. Information, as we will see, is not defined by the sender or its actions in coding a message, but by the state of the receiver when the message arrives.

For another example, information is often used when referring to a subject that is actually about "knowledge" (explained in the next section below). Information and knowledge are *not* one and the same in a technical sense, but are often interchanged in casual use. Take the case of genetic coding in deoxyribonucleic acid (DNA); we often say "the information contained in our genome…" for example. The actual codes (nucleotide triplets, called codons, corresponding to specific amino acids) are embodied in the structure of the gene and are, technically, *knowledge*, in the sense of knowledge of how to construct living systems. The sequences are the result of evolutionary learning[27]—variability tested by natural selection. Many authors indiscriminately refer to the genes as "information" because they know that the gene is

[26] For a thorough and extensive review of the complications involved in the usages of the term, see the *Stanford Encyclopedia of Philosophy* review at: https://plato.stanford.edu/entries/information/ (accessed January 10, 2020).

[27] The use of the word "learning" is a bit risky in that it is usually associated with what animals do mentally. However, it refers to any change in a structure resulting from the information exchanges going on at an appropriate time scale. In the case of evolution, a mutation can cause a change in the phenotype that can result in different form or behavior, essentially an informational message to the rest of the environment. The latter, then, has an opportunity to act upon that new structure, positively or negatively. If positive then the species has "learned" something that will be retained.

the source of a message (messenger ribonucleic acid [mRNA]) that will be sent out into the extra-nuclear cytoplasm where it will be decoded by a ribosome to construct a polypeptide (or protein). The message is informational to the ribosome as will be explained shortly.

In this work we follow the basic insight of Shannon (Shannon and Weaver 1949) that information is the *measure* of uncertainty regarding the state of a message, or the next "symbol" to be received in a message stream, which is a property of the receiver (e.g., the ribosome) and not of the sender (the DNA). The DNA codes, read and interpreted by DNA sequencing, as in the Human Genome Project, are informational to the gene researchers in the same way the message carried by mRNA is informational to the ribosome. When the genome is first sequenced, the researchers cannot know with better than a 1 in 20 chance what the next codon will be (unless they already knew what the amino acid sequence of the protein was so that they could make a prediction—but it was not until after the genetic sequences were discovered that researchers could match up genes and proteins more rigorously). Hence the sloppy use of the word information when describing the actual structure of the gene.

The measure of information in a message is strictly a function of the properties of the receiver—the entity that uses the message—to instigate a change in its own structure based on that measure. This is the essence of learning in its most fundamental form. A message is received that reduces uncertainty in the receiver, which, in turn, results in a modification of its own structure, thus producing a new level of knowledge in it.[28] The inter-conversion relation between information and knowledge is shown in Fig. 3.2a. The formula for information, I,[29] comes from Claude Shannon's (Shannon and Weaver 1949) formulation of the amount of information conveyed by the receipt of the next symbol (from an ensemble of symbols possible) in the message (Mobus and Kalton 2015, Chap. 7, Quant Box 7.2).[30]

Another mistake made in the use of the word information in common usage is to intermingle the notion of "meaning" with information. Meaning is actually quite separate from the measure of information. Meaning comes from the a priori

[28] In Mobus and Kalton (2015), it is explained that messages are carried in energy streams that are modulated in one form or another to encode the content. The energy variations in the message flow (the code) then may be amplified by the receiver so that effective energy flow can be directed to the site of work on the receiving structure so that it is modified based on the informational content of the message received. This is the resolution to the difference between information and meaning that Shannon and Weaver (1949) pointed out. Meaning derives from knowledge, information derives from improbability.

[29] Typically, information theory authors use the symbol H rather than I as we do. The reason has to do with the resemblance of the formula for information that resembles that of thermodynamic entropy as noted by Ralph Hartley in Hartley 1928.

[30] Shannon got his inspiration from the work of Ralph Hartley (1928) who developed the idea that the members of a set (of symbols, say) represented an amount of uncertainty proportional to the size of the set. This notion closely aligns with that of physical entropy and so in subsequent developments of formal information theory, the terms message entropy and information have become confusingly intertwined!

arrangement of senders and receivers that involves the interpretation of messages. This in turn depends on the prior agreement on the symbols, rate of message flow, and other factors. The receiver is predisposed to act (make internal changes) on the message. However, the information measure of the message determines how much action ensues at the receiver. Meaning tells the receiver *what to do*. Information tells the receiver *how much to do*. In Chap. 12, we introduce a macro-model of governance, the Hierarchical Cybernetic Governance System (HCGS), in which we will further demonstrate the nature and role of information in forming and maintaining structures and functions in systems. Chap. 10 in Mobus and Kalton (2015) covers this material as well.

The ontological status of information derives from the fact that messages have causal power to change a physical structure. This is, however, a strange kind of causal power. It is not a result of the intentions of a transmitter/sender, but rather it is based on the ignorance of the receiver. This ignorance is a function of the structure of the receiver relative to other possible structures that could obtain with the accomplishment of work. That is, the receiving structure already has the capacity to alter itself so as to better facilitate the dissipation of energies, to become more thermodynamically stable. It will only do so, however, as a result of receiving messages that trigger such a reaction. Real work is accomplished in making these alterations, so energy is consumed and heat dissipated. Afterward, the structure is a priori prepared to dissipate future flows of energies without having to do additional work. Information is the capacity to cause changes in receiving systems that better prepare that system for future messages (minimize surprise). The change in the structure of the receiver that reflects this preparation is what we call knowledge.

3.2.1.2.2 Ontological Knowledge

As noted above many authors conflate the notion of information with what we are calling here *knowledge*. Whereas information is the measure of uncertainty reduced when a message is received by a competent recipient (i.e., able to decode the modulations appropriately), knowledge is the a priori structure (which channels energy flows) of the recipient. The latter will undergo changes in structure in proportion to the "amount" of information in the message. Figure 3.3 provides a diagrammatic depiction of the interrelation between information in a message and knowledge gained.

According to ancient Vedic literature,[31] "Richo aksare parame vyoman," "Knowledge is *structured* in consciousness (Rig Veda I.164.39 as translated by Maharishi Mahesh Yogi, 1974)." The key word here is "structured." The theory of knowledge, ordinarily the subject of epistemology (discussed below), is rising to the level of science with the recognition that knowledge is encoded in the actual physical

[31] The Vedic texts include four volumes of hymns said to embody the Veda, knowledge, orally transmitted from the 2nd millennium BCE. The Rig Veda is thought to be the oldest.

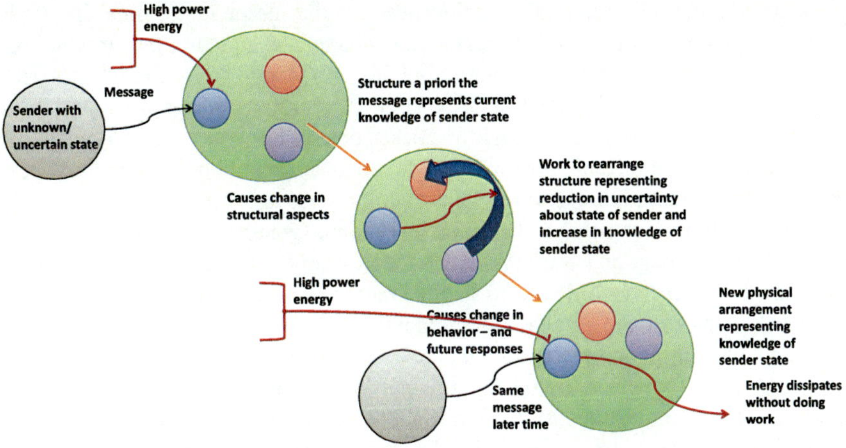

Fig. 3.3 This figure shows the time course events when a recipient receives a message from a sender. A sender encodes a message based on its current state, which may be different from the recipient's knowledge of the sender's state as coded in the recipient system's internal structure. The message would therefore contain an amount of information representing the degree and direction of difference. The interpreter (blue small circle) uses that information to direct a work flow modifying the internal structure. If at a later time the very same encoded message is received again it does not contain information because there is no difference and the energy that would have gone to do additional work is simply dissipated, or conserved for future use

structures of systems,[32] and thus acquires ontological status (Principle 7 in Chap. 2). Knowledge is acquired through experiences that a system has. Internal structures change in organized work processes in response to the informational qualities of those experiences. Once changes in the structure have been made the result is that the system has internalized the experiences and if exposed to similar conditions in the future will receive less information from the same situation in those subsequent experiences. This description applies to all systems starting with atoms and up through the hierarchy of organization and complexity. According to the Vedic tradition, this is the essence of consciousness and so everything in the Universe is conscious (see the next section on panpsychism). What we call human consciousness is just a complex version of universal consciousness. What we claim here is much less expansive. We do claim that information and knowledge, and their interactions, do have ontological status.

[32]There is a somewhat widespread misapprehension about the nature of knowledge, especially among biologists who assert that information is an emergent phenomenon in living (and possibly prebiotic) systems as opposed to a fundamental substance. Many of these authors also prefer to use the term "information" more broadly than seems appropriate. For example, some use the adjectives "latent" and "semantic" to name what we are calling here, "knowledge." For example, see Grisogono (2017) for arguments for the emergence (vs. primacy) of information with the advent of life and the conflation of "semantic" information with what we call here, ontological knowledge.

Leaving aside the nature of consciousness for the moment, we note that it is nevertheless a phenomenon of the brain. Knowledge is encoded in the structure of the brain (as one example). The brain contains knowledge from three basic processes—evolution, development, and experiential learning—along with a facility for persistence: memory (Coen 2012).[33] These correspond with the time domain of information generation and knowledge production. They could be renamed as "species learning," "maturation learning," and "autonomous agent learning."

In Fig. 3.2b we see some of the transformational (generation of) relations. The right side of the diagram shows that information receipt generates knowledge construction in the receiver. This means that the receiving system possessed an a priori low expectation of the messages communicated from sender to receiver. Since the "amount" of information in a message is inversely proportional to that expectation, the receiver has to respond by reconfiguring its own structure to become more dissipative (Mobus and Kalton 2015, Chap. 7). That is, it is better able to pass through the energy flows that are the result of information (perhaps through amplification) so that next time the same message is received it is less likely to cause a similar amount of change in the structure.

What Fig. 3.2b shows is the iterative cycle of information–knowledge construction mediated by the roles of matter and energy. In essence the configuration of matter (structure) holds the knowledge of the system. Energy conveys the message from the sender to the receiver. Usually a very small amount of energy (and possibly matter) is involved in the message. What makes it a message is the modulation imposed on the flow by the sender. That modulation encodes the symbols (a set of patterns) into the message flow.

In the figure, the term "Entropy Law" refers to the fact that work needs to be accomplished in order to change the structure and to maintain it if it is to be retained. Forgetting is a loss of structure resulting from non-maintenance. This phenomenon can be found in living and supra-organic systems quite readily. The term "Evolution Law" is the reciprocal phenomenon. Where work is done to maintain newly acquired structure (or where the structure is highly stable under the extant environmental conditions), the change in structure has a ratcheting effect. In essence, the gaining of knowledge is progressive so long as suitable energy flows through the system (c.f. Morowitz 1968).

The concept of learning and knowledge encoding is not limited to living beings and certainly not limited to human beings. Any time one system sends a message that is received by a second system, and that second system is altered by its receipt, learning has occurred. Knowledge has been increased.

This is true across all scales of space and time. It begins when the Universe cooled sufficiently to allow, for example, quarks to begin signaling each other through the mediation of gluons (the strong attractive force). Quarks that end up

[33] In the above example of a ribosome receiving information from the DNA, it is using that information to build a polypeptide structure. However, it, as a system, does not really learn anything since once the polypeptide transcription is complete the polypeptide is released to wander off into the cytoplasm. The ribosome has no memory of what it did!

moving toward one another, reacting to the information value (say based on separation distance and relative direction), undergo a structural combination (three quarks to a proton, for example) that constitutes the knowledge of the subatomic particle. As will be seen later, this is the beginning of ontogenesis, but it also illustrates that as soon as the constituents of the early Universe could form structures and react to signals, information and knowledge attained their ontological positions.

At the scale of atoms, we see photons being the mediators of information. The outer electrons are perfectly "happy" buzzing around in their orbitals, the structural aspect of atoms. But the receipt of a photon of the right energy can boost an electron to a higher-energy state causing it to step up an energy level (or even completely away from its atom). The atom as a whole is now in a new configuration and will react, say by exposing a net positive charge to its environment. Eventually the electron will fall back to its normal place emitting another photon as a message to the Cosmos. Of course, when atoms are aggregated closely together, they have multiple different ways to communicate with one another, like the sharing of electrons to form covalent bonds in molecules (another instance of ontogenesis!). Each atom achieves a new configuration (knowledge) and behaves differently as a result.

The pattern of message receipt and reconfiguration (even if marginal or miniscule) embodies the gaining of knowledge (even if fleeting) and repeats at every level of organization in the Universe.

One further note, on the nature of knowledge as structural configuration; it does not generally persist. Entropy prevails in the end. Except for the configuration of quarks in a proton, which appears to be eternal, all systems at higher levels are subject to decay or destruction. Atoms are relatively stable, except if radioactive, but molecules are subject to many chemical and physical conditions that can change them. Protein molecules degrade with time at physiological temperatures and need replacing in living systems. DNA molecules are among the most long-lived biomolecules, which is why we can recover genetic knowledge from Neanderthal bones. But even they have a tendency to occasional degrading (this is especially the case when DNA is being copied). In human-scale terms, our memories, which are configurations in neural networks, the strength of connections between neurons participating in an engram, do tend to fade with time unless reinforced through recall from time to time. In terms of the situation shown in Fig. 3.3, it would amount to a slow decay back toward the initial state with the emission of waste heat. Mobus (1994) provides a mathematical model of "forgetting" or memory trace fading in neural networks as a perquisite for learning new associations in a nonstationary world. Paradoxically, the entropy law is what makes it possible to learn new things and build new knowledge.

3.2.1.2.3 Panpsychism

Recall the last section opened with a quotation from the ancient Vedic texts, the Upanishads, regarding knowledge, structure, and consciousness? There has been an on-going debate, mostly among philosophers of science but also some physicists, that consciousness is a property of the Universe, indeed probably the key property. That is, every physical thing in the Universe possesses some level of consciousness.[34] The interplay between matter, energy, information and knowledge suggested in Fig. 3.2 hints at what might be called a micro-panpsychism or a concept of how the most fundamental physical substances can be considered conscious. If we define consciousness broadly as an ability to receive messages and change our internal configuration in accordance with the amount of information contained in the message and especially if that results in a change in behavior of the whole system, then it would be true that consciousness is possessed by the most fundamental units of reality. And, if the most fundamental units, say Leibnitz's or Smolin's (1997) monads, were able to receive information and change their behavior then it seems reasonable that the composition of more complex entities (quarks, nucleons, atoms, molecules, etc.) would possess more elaborate forms of that property. It would then be reasonable to argue that the consciousness that is so evident in living systems is actually rooted in a basic capacity of elements in the Universe to "sense" and change in accordance with what they sense.

We might not be able yet to say anything about strings or monads, but we do know that nucleons are able to change in response to weak interactions (W and Z bosons conveying messages) that can result in nuclear transmutations. Atoms can be ionized by the receipt of photons knocking out outer electrons (and potentially leading to chemical bonding). Change in structure due to "unexpected" energetic flows, messages, starts at the lowest known levels of material organization. So, it is not a stretch to claim that the given definition of consciousness is operative at every level in the complexity structure of the Universe. It becomes really interesting in the realm of living things, and even more so in the realm of sentient beings such as ourselves.

How the fundamental units of reality combine and construct higher levels of organization, complexity, and capacity to construct knowledge is a central concern of ontology or ontological development, here termed *ontogeny*.

3.2.1.3 Ontogeny: A First Pass

Ontogeny (also ontogenetic process) is the process of something becoming or coming into existence[35] and then furthering development. The central question here is: How do systems of increasing complexity arise out of the four (or six) fundamental

[34] Levels of consciousness vary among researchers. A simple model goes like this: an entity is conscious (that is aware) of its environment; an entity is conscious of itself and its environment; an entity is conscious of its being conscious of both self and environment; an entity is conscious of being conscious in the last sense; an entity is conscious of a larger environment and its place within it—transcendent consciousness.

[35] The term ontogeny is generally associated with biological development, i.e., embryonic or fetal development. However, the term can also refer to any development process in which a new, more complex form (and function) develops from more primitive or simpler forms (and functions).

substances in the first place? In other words, where and how do systems originate and develop into their stable forms and functions? How does complexity increase over the history of the Universe?

This question had little meaning in the time before Charles Darwin put forth an evolutionary development process. For most, existence was simply a given; God created heaven and Earth and all therein in exactly the forms witnessed. Scientist had begun to question the firmness of this stance before Darwin but it was Darwin who provided a mechanism for biological evolution in which new characteristics, in the form of new species, could emerge from existing forms.[36] Biological evolution, especially with the addition of Mendelian genetics and the theory of genetic mutation, goes a long way in explaining how biological systems at the level of species and genera change over time, but it does not directly address the problem of origins, or how things got started, nor does it address the "apparent" trajectory of evolution, that is, producing more complex biological systems over the course of time. In the realm of biology, we need to examine the origin of life problem, the emergence of living cells from nonliving chemical reactions, in order to consider how life got started prior to neo-Darwinian evolution. Then we have to address the issue of why biological evolution, over the course of time, has led to increasingly complex life forms, and ultimately to human intelligence and consciousness.[37]

Today we now also understand the process of embryonic/fetal development as an unfolding of potential, coded in the genes, but influenced by environmental factors that have the effect of turning on or off genetic expressions at key points. There is a new amalgam of neo-Darwinian evolution with embryonic development (known as Evo Devo: Carroll 2006a, b) that has gone a long way to explain biological ontogeny much better than Ernst Haeckel's recapitulation theory.

From the systems standpoint, we need to understand how do systems originate (biological or not) and how do they evolve to achieve a stable existence in which they can be observed to behave in particular ways. The solution to this question is found in the nature of auto-organization, emergence, and contextualized selection on what emerges. We assert that there is a universal form of evolution, of which biological evolution is just one example (Mobus & Kalton, Chaps. 10 and 11). This is the process, composed of those three sub-processes working in concert, that, starting at the most fundamental level of matter and energy, creates new objects or entities at a new level of organization. We call this process Ontogenesis. Once the new entities are formed, they are subject to ambient conditions that select for or

[36] Alfred Russel Wallace developed a very similar thesis at the same time as Darwin, but the priority of publication goes to the latter.

[37] Of course, not all species became "more complex" over time. The Earth is still full of less complex forms such as prokaryotes, single-celled critters, worms, etc. Only a few evolutionary lines lead to greater complexity in terms of body forms, organs, and behaviors. However, the increase in complexity in those few forms also contribute to the overall complexity of the whole planet. For example, the human invention of antibiotics has worked back on prokaryotes to force the selection of antibiotic resistant strains. The Ecos is thus more complex overall.

against various configurations. Those that remain stable can then interact with each other in new ways. And the process starts over again at this new level.

Briefly (from Mobus & Kalton, previous reference), before systems as such exist there are components—entities or particles with various "personalities"[38]—contained within a volume of space that, under the influence of energy flows, mix and interact (Morowitz 1968, 2002) to form more complex structures.[39] Some of these structures will prove to be stable under the ambient conditions (a kind of selection process) and demonstrate new, emergent, behaviors. They become, in effect, new components capable of new interactions at a new, higher level of organization.[40] This is how new things come into existence. And once a new level of organization obtains, the *ontogenic* cycle starts again to produce the next level of organization. There are components at each level that have potentials for interactions (e.g., atoms, molecules, cells) and energy flows that supply the capacity to do the work of combination (as well as dissociation), and when combinations are made, they are selected for their stability in the ambient conditions. The history of the origin of life will be written in this scenario. So, too, will the origins of eukaryotic cells, multicellular organisms, and societies (Bourke 2011; Volk 2017). Biological evolution accounts for the micro-steps of biological development. But the emergence of species, genera, and all of the subsequent divisions of life as well as the nature of social systems comes from a universal process of auto-organized subsystems, the emergence of higher-order structures and functions, and the ultimate selection of that which works under the prevailing conditions. This is universal ontogeny.

The process of ontogeny is a natural consequence of the interplay between matter, energy, information, and knowledge, which is the final product of the process. Knowledge is encoded in the structures of matter, bound by energy, and stable in the face of testing by on-going energy flux. Systemness is the natural consequence of ontogenesis. The construction of higher-order systems from lower order, but tested ones, gives rise to the principle of systemness (Mobus and Kalton 2015, Chap. 1). It is what produces the hierarchy of physical forms and interactions that constitute the world. It gives us a Universe of interactions, a network of entities that behave. In Sect. 3.3, we will expand the explanation of the ontogenic cycle.

[38] As explained in Mobus and Kalton (2015), the term "personality" refers to interaction potentials expressed outwardly from the boundary of a component. These potentials provide the points on the boundary where components may connect (see "Interfaces" below). Components may have multiple potentials and thus combine with multiple other diverse components to form structural complexes. These personalities are also referred to as affordances elsewhere.

[39] Morowitz also emphasized the geometric aspects of energy flow in which energy enters the system at one point and exits, as heat, at another point, leading to convective cycles that further act to organize the structure, providing another dimension to the selective environment.

[40] Note, however, that not all new entities are going to be stable in light of the ambient environmental conditions. For example, just after the Big Bang there were a plethora of particles created (from monads?) that could exist in the tremendous energy flux that was ambient. As the Universe expanded, however, it cooled and many of those particles were not stable configurations. Today we can recreate some of them in high-energy particle accelerator experiments, but they rapidly decay into more stable constituents.

3.2.1.4 System Epistemology

The bulk of this book is actually about system epistemology or more specifically how we humans gain knowledge of systems. Using the to-be-developed ontology (this chapter) and a formal definition of system (developed in Chap. 4), we will generate an organized structure that is modifiable in the sense of learning described above. This structure, which we will implement in a database, the structure of which is similar to our brains, will be predisposed to receive information about a specific system, the system of interest, generated from the act of analysis (in the broad sense). We call this structure a knowledgebase. In Chap. 5, we introduce the mechanics of how system information will be generated, how it results in the encoding in the knowledgebase structure, and how that knowledge may be "recalled" and used for multiple purposes, such as refining its own structure, testing hypotheses (through models), and constructing/generating designs for physical implementations. The latter will include not just engineered products, normally associated with systems engineering, but organizations and organizational processes, and policy recommendations for public action (e.g., water distribution policies for regions experiencing severe drought).

The key to system epistemology is recognizing that the gaining and storing of knowledge in the knowledgebase is accomplished using a language of system that is completely comprehensible to human beings. In the next chapter where we develop this language, we show how the language of system is already grasped subconsciously by humans and thus becomes accessible consciously by making the nature of the language explicit through systems science and the ontology (terms) translated into a spoken language (in this case English).

Thus, the knowledgebase, an artifactual system, can converse with a human mind (actually many minds) through a language facility to share knowledge about other systems. We use systems science to build a system of communication between machines and people that will, hopefully, enrich our totality of knowledge.

3.2.1.4.1 The Role of Modeling

All systems are initially observed from the "outside." That is, the scientist has access to monitoring the inputs and outputs (behavior) of the system but does not immediately have access to internals of the system that generate that behavior.[41] The first task of gaining knowledge is to produce a formal description of the behavior in the form of a model of the system that can be used to generate the same behavior as witnessed in the system itself. Very complex and/or fuzzy bounded systems may require considerable observation before the researcher arrives at the set of variables

[41] This is even partially true of systems in which humans are components, such as an organization or the economy (see Chap. 9). The scientific approach requires objectivity, which can be gotten, in theory, by the scientists removing themselves from the system and pretending to be an outside observer.

that adequately describe the system's behavior. Once a model's results conform to the behavior of the system of interest, empirically determined, the scientist may claim to have a certain degree of understanding of the system.

More often than not, however, the first several attempts at models will not produce conforming results and the model builder will seek to resolve the discrepancies by looking for unrecognized variables and/or presumed internal relations between variables or other factors (e.g., rate constants) that might cause the model to perform better. The traditional approach to scientific epistemology has been along these lines. Observe the phenomenon, develop a hypothesis regarding how the system works, build a model with the "right" variables and relations, and then run the model to see if its behavior conforms to the real phenomenon. If it does, write a paper and submit it to a journal!

In the history of the sciences, however, this has not been a completely satisfying process, as mentioned in the Introduction. Ultimately models fail to provide the kind of knowledge that we really seek, that is causal knowledge—what makes the system behave as it does. Whether due to changes in the environments of the systems, insufficient precision or accuracy in the solutions, or, in the case of evolvable systems, to some alterations inside the system boundaries that change the transformation functions that are hidden from view, models only based on input–output behaviors can become obsolete. Or, the scientists are just curious about what is really going on inside the system (see below)! Unless we know what the causal mechanisms within the system are that transform the inputs to outputs, and know how those mechanisms can react to changes in the inputs, say, we are always in jeopardy of not being able to equate the models we build with real veridical knowledge of the system.

Models, in combination with modelers, are thus, themselves, evolvable systems. They tend to get more complex as scientists seek to explain more about complex phenomena than the original model could provide. More than that, the discrepancies between models and the real systems provide the motivation for opening up the black box and attempting to deconstruct the internals. In other words, science has proceeded to explore phenomena more deeply as a result of not getting the models "right" in the first place.

We can understand this tradition in terms of the tools that the sciences have had to work with. This includes the kinds of sensors that were available for monitoring systems, mostly from the outside. They had to "infer" the internals based solely on the observed behaviors (or they may have settled on simple correlation when no hypothesis of causality was available). And then they had to take great care to make sure their observations were complete and accurate. The use of models, whether statistical correlations between variables, or sets of ordinary differential equations, or, more recently, system dynamics (SD) and agent-based methods, has been considered a "strong inference" method for saying we understand a system. When the models fail to predict new or different phenomena they need to be changed, or guide the eventual deconstruction of the system.

This has been the typical pattern in the sciences: develop a hypothesis about how something works based on its behavior, generate a model to test the hypothesis (especially if normal experimental approaches cannot be used), collect additional observations for analysis, and see if the model helped predict the system’s future behavior. Today this methodology is pervasive thanks to the ease of building models for computer-based simulation. Many scientists race to construct a model as early as possible and then spend the next several years tweaking the models until they seem to produce the appropriate outputs.

The problem with this approach is that a model is not necessarily the same thing as understanding the system itself. It is reasonable to say it is understanding *something* about the system, but that is all one can say with any kind of certainty. Yet, for many sciences, this has been the best that could be done until someone proceeded with deconstructions that clarified the internal mechanisms and produced better causal models.[42]

Let us be very clear. A mental concept is a model implemented in dynamical neural structures. Any person’s conceptualization of the world and the things in it is, in fact, a mental model of that world. Computational models are, effectively, extensions of mental model, formalized and encoded in a modeling language (mathematics or something like systems dynamics) but, nevertheless, are just models of the systems of interest in the world. Models built on deeper knowledge of causal processes within the systems are generally better for capturing the predictive or anticipated behavior of the systems and this is what the sciences strive for.

3.2.1.4.2 Inquiry

In the description of information and knowledge from the prior section one might conclude that the communication act is all “set up” prior to any messages being sent and results in knowledge encoding. However, this is not always the case. Just above we described knowledge acquired by the method of analysis. In a sense there is no “sender” of messages in the case of inquiry into a subject. The act of gaining knowledge is initiated by the actions of the one seeking to gain knowledge. A scientist proposes a hypothesis and then does experiments to (ideally) refute it if it does not correspond to the truth of the matter. The messages being received originate in a phenomenon that the scientist is actively observing. There is no sender encoding the message as described above in a strict sense. The scientist, in setting up the experiment and then observing the results, has constructed the sender (the experimental medium) and designed the code (measurements).

Experimental science is not the only possibility. Many sciences have to rely on observations of naturally occurring phenomena and employing various methods, mostly statistical inference and modeling, in order to obtain their observations. Again, there is not an a priori communication connection between a sender (in this case the phenomenon) and the receiver (the observer). However, in both of these

[42] In Chap. 9, we will see a specific example of a science that has relied on modeling heavily as it attempted to explain very complex phenomena. We will also show how the approach can now migrate toward that being put forward in this book.

cases there is a flow of information from a source (experiment or phenomenon) to a receiver (scientist), usually via instrumentation. The scientist, as the receiver in this case, has an a priori expectation associated with the results of the observation and whatever reduction in uncertainty of the outcome constitutes information. The scientist gains knowledge and conveys messages (via journal articles) that are informational to a broader audience. The knowledge acquired is encoded into structured literature where it is accessible by future receivers.

What is interesting about this situation is that the motivation for the messages derived from experiments or just observations is the curiosity of the scientist. The scientist wanted to understand something and set up conditions whereby they could get the information necessary to construct new knowledge. Where does this curiosity come from? Humans are by no means the only animal to demonstrate curiosity-driven exploration. Usually, such explorations can be related to searches for food or other rewards but they are undirected (foraging for example). But in humans, exploration and inquiry seem more opportunistic; that is, there is no direct benefit sought, but just some information that might come in handy some day! Humans have been labeled "informavores!"

3.2.1.4.3 Knowledge Gets Refined and Expanded

Principle 12 in Chap. 2 says that systems can be improved. This has many versions, but in the present case we are referring to the improvement of knowledge, which is, in fact, a system of encoded forms. In fact, we are referring to Principle 9, which relates to systems containing within their structures *models* of other systems. In the human brain these mental models are what we mean by concepts. And concepts can not only be learned, but also they can be improved with additional experience (c.f. Carey 2009, for an explication of the nature of concepts). They can be refined (finer resolution, correction, etc.) and they can be expanded (incorporate more perspectives).

Learning in this sense is motivated by discrepancies between what a currently held concept (model) predicts and what actually happens. Predictive ability is the key to fitness in a changing or nonstationary and uncertain environment. Ergo, if a concept currently held is failing in some way to make sufficiently accurate predictions, the mind is motivated to find out why.

This is what is behind the scientific quest for more and better knowledge and it is what is behind individuals being curious. There is reason to suspect that many mammals and birds have this same facility (as in the cat's curiosity being a source of loss of one of nine lives).

In the next chapter, we will examine the idea that the brain contains a natural template of system structure and function that it applies to observations of the real world and uses to make primitive predictions, as when a newborn is starting life. The real-world systems that are observed deviate in multiple but normal ways leading to the triggering of learning. The template model is used to construct a specific instance model, which then will be refined and expanded as the young person gains experience with that kind of system. When real systems are encountered that deviate more significantly from already encoded models, this is a trigger to cause the brain to create a new category of system and start building from the prototype template anew.

3.2.1.4.4 Learning Systems’ Learning Systems

No this is not a typo. Here we refer to the idea of systems that can learn actually going through the process of learning other systems (improve their models of other systems; Principle 9 in Chap. 2). This is the basic model of systems epistemology. Systems of sufficient complexity (like our brains) are capable of constructing models of all other systems, including supra-systems, by improving and expanding on a fundamental “system” template. Our effort will be to show how this basic model can be replicated in an artifactual system (the knowledgebase fed by systems analysis).

One remaining question is: “Where did the template system model come from in the first place?” The answer, one that would require several volumes to explain in detail, is the evolution of brains. Animals have to behave, move, react to, and otherwise negotiate a dynamic world of systems. Brains evolved from simple reactive sensor–decider–actor subsystems to the complexities of the human brain by virtue of selection favoring those brain structures that were better able to learn and work with systems in the world. Even the simplest sensor–decider–actor brains (in flat worms for example) are themselves input–process–output devices and that is the fundamental template definition of a system! By virtue of having a primitive brain, a flat worm possesses a model of self and a primitive model of its world (e.g., mud as a medium/food).

This being so, we are now in a position to tackle the problem of developing a universal ontology of system. That is, we will need to identify all of the existing elements that make a system what it is. The simple input–process–output model is a good starting place for our understanding, just as it was a good starting place for the evolution of brains and for the development of knowledge in an individual human being.

3.2.1.4.5 Leveraging Isomorphic Structures and Functions

Systems science studies the way in which systemness applies to all kinds of systems at all levels of organization and complexity (e.g., see Sect. 3.3.1.2). We can recognize objects or entities, explained later, as systems whether they are microscopic (atoms, molecules, single cells), mesoscopic (multicellular organisms, human beings, societies), or macroscopic (planets, solar systems, galaxies), because all of these share common properties or aspects that all systems have (what makes for systemness). Such properties/aspects are embodied in the structures and functions of system parts and the fundamental nature of these are *isomorphic* across scales and complexities (iso—same; morphic—of shape). In this chapter, we will see the basis for this in real system components. As a brief example, consider the concept of a container or a discernable boundary that holds some "substance." We will return to this example again. Atoms have boundaries, even if fuzzy.[43] Living cells have boundaries (cell membranes), physical asset inventories have boundaries. And solar systems, in similar fashion to atoms, have fuzzy boundaries. They are all physically different in terms of the substance of makeup. A cell membrane is made from lipoproteins whereas the walls of a house can be made of bricks and mortar. Even so, both delineate a containment—something is inside and everything else is outside.

Many different elements of systems at all scales have easily identifiable "isomorphisms" in terms of the forms they take and the functions they perform. The collection of these elements is the set of isomorphic structures and functions that constitute all systems. Another example is the network structure of all systems. All systems are composed of a set of components and linkages between those components. Thus, all systems may be described as a network structure (Principle 3 in Chap. 2). Network structures are ubiquitous throughout nature and as we will see in the next chapter constitute a main element of systemness.

The process of coming to understand specific systems, no matter how complex they might be, involves using this set of isomorphic elements, along with a "grammar" for composing them into meaningful structures, to discover the deep nature of the real system of interest. In the next chapter, we will introduce these elements in more detail along with the language of system composition and the grammar of composition. It is precisely because these elements of systemness are a constant feature of all levels of organization and complexity, because they are isomorphic across and within these levels that we can hope to realize a deep understanding of complex systems.

3.2.2 Examples of Developing a System Ontology

The desire to construct an ontology that captures the essence of human thinking is not at all new. For example, Gottfried Leibniz (1646–1716), who we will refer to below, pursued a "*Characteristica universalis*," or a universal language of science, that would unify all of the pursuits of the sciences.

In this section we will briefly look at several representative but more modern, methodological approaches used to develop a system ontology. These have been and are being employed by several researchers in systems science, linguistics, and systemic applications in the sciences. We should hasten to note that not all of the work that has been done is couched in terms of "ontology" per se. Most of the work done to date has been, rather, the exploration of frameworks, lexicons, and semantics aimed at grasping an organization of systems knowledge that can be used to

[43] We will be returning to the notion of fuzziness in Chap. 4. For the moment we can use the intuitive notion of fuzziness.

further the cause of systems inquiry. We offer this sample of work as, nevertheless, examples of attempts to establish a common language of what exists based on a conceptualization of systemness. We would assert that these examples are oriented toward the same goal even if perspectives are somewhat dissimilar. Therefore, with deepest respect for the efforts examined here we hope the authors (those still alive) will permit us to aggregate their insights into this concept of ontological development. This is only a small survey of approaches to developing the ontology of systems. We believe that this is a fruitful area of open research. What we present here is just the beginning (but hopefully an extremely useful one). The objective of the development methods is to define a clear and concise *framework* (next section) for then examining system-related phenomena to assess their ontological status.

The key questions asked in developing such a framework revolve around discovering common features and patterns of features that are found to be true in all systems, or at least those that we have a deep interest in. The principles given in Mobus and Kalton (2015) and replicated in Chap. 2 are examples of this kind of study. The listed principles in Chap. 2 are, however, only the most general features and patterns. Many more are believed to exist and will be the subject of much research in the coming decades (c.f. Rousseau et al. 2016). As presented below, the ontology framework is based largely on the principles, but there have been several additional important influences.

3.2.2.1 Miller's Living Systems Theory

James Grier Miller (1978), in his monumental tome, *Living Systems*, did not address the idea of an ontology as such; however, he outlines 12 critical system concepts relevant for life and a number of terms referring to commonalities across kinds of living systems (e.g., from cells to societies). Table 3.1 provides the list of top-level concepts from his Chap. 3. Each one of these has a number of sub-concepts (details and sub-subconcepts!). This list is similar to the Ontology framework in Sect. 3.4.

Missing from Miller's list is the concept of *knowledge* as separate from but related to information as shown in Figs. 3.1 and 3.2, and explained above in the section on Cosmological Ontology. Otherwise, the reader will already recognize some of these items from our prior discussion and will encounter more as we proceed. Another difference between Miller and Mobus and Kalton (2015) is that the former includes the term "steady state" as having principle ontological status whereas the latter establish a general concept of "dynamics" of which steady state dynamics are only one kind. The reason is likely to be that Miller was only concerned with *living* systems that continued living over time, which would require a steady state dynamic condition, at least so far as the then state of knowledge of what being alive meant. Other biologists tackling the distilling of principle terms, such as Len Troncale (see below), have included more general forms of dynamical systems such as chaotic processes.

Along with these basic concepts Miller listed 20 critical subsystems of a living system (page xix of the preface, Table P-1), and a select number of components

Table 3.1 Miller's basic concepts include what we are calling a system top-level ontology for living systems (complex adaptive and evolvable systems)

Miller's basic concepts (Chap. 3)
Space and time
Matter and energy
Information
System
Structure
Process
Types
Level
Echelon
Supra-system
Subsystem and component
Transmission in concrete systems
Steady state

(Table P-2, pp. xx–xxiii). His approach was to examine all of these across multiple levels of living system organization, e.g., cells to supranational systems. Even though Miller was concerned with living systems as a biologist and medical scientist, he recognized the life-like characteristics of human-based organizations.[44] So, most of the subsystems/components are found in all levels.

Miller grouped his critical subsystems into three categories: subsystems that process matter, energy, and information; subsystems that process only matter and energy; and subsystems that process only information. A fourth quasi-category holds a single critical subsystem—the boundary. A few representatives in categories 2 and 3 are listed here (with notes on relations to what will be covered later in this book in square brackets). Examples of category 2 include: "Ingestors" [later in this book to be seen as acquisition processors], "Distributors" [as the name implies], and "Producers" [later to be seen as sub-subsystems]. Category 3 examples include: "Input transducers" [later to be seen as message transducers, e.g., amplifying receivers], "Channel and nets" [later to be seen as flow networks], "Memory" [as the name implies], and "Decider" [later to be seen as an agent subsystem (Chap. 12)]. His category 1 also contains a single element that combines categories 2 and 3 along with the boundary element—that is, a "Reproducer" [later to be seen as a specialized work processor for repair and propagation]. All of these critical subsystems are to be found in all levels of Miller's hierarchy of systems [later to be seen as levels

[44] In Chap. 9 and again in Chap. 12, we will look at the economy and its governance as a form of exosomatic metabolism, thus reflecting Miller's notion of a supra-level of organization that is still reflective of a living system.

of organization] from a single cell (lowest level of organization in his vision) through organisms to societies and supranational systems.[45]

3.2.2.2 Troncale's System Process Theory

Lenard Troncale (Troncale 1978a, b; Friendshuh and Troncale 2012) and co-workers have been working on the concept of systems as processes (systems process theory [SPT] and systems of systems process theory [SoSPT], see below). He and his co-workers have been leading an effort to build a true science of systems through the rigorous examination of a large number of processes or transformations and relations between elements that give rise, for example, to stability, structures, or products.[46]

System-level processes (originally called "systems field axioms"), and how they interact, called "linkage propositions," act to produce higher-order processes. The latter leads to what he calls "Systems of Systems Process Theory" (SoSPT) reflecting the hierarchy of organization from simple relations (at the bottom of the hierarchy) to much more complex multiples of processes and realized linkage propositions at the higher levels.

These workers emphasize the importance of the concept of "process," which they define as, "...a series of steps of change through which a set of objects proceeds" (Friendshuh and Troncale 2012, p5). They argue that the concept of process is so important because change (transformation of objects) is what the world is fundamentally about. But not just any change; they emphasize that changes that lead to organization or structures are the basis of systems and should, thus, be the main subject of systems science. Hence, they look for fundamental process archetypes that are "isomorphic" across many if not all systems that come under study. In other words, they are looking for processes that underlay all of nature. These would be the ontological elements of systems. Friendshuh and Troncale (2012, p6) explain isomorphisms:

> A process is isomorphic if its abstracted, generalized "form" or "pattern" can be found at many scales in many phenomena (Gr. iso- = same as, equal; morpho- = form).

---

[45] At the time of Miller's writing and up to the present, there really are no examples of supranational systems that would meet his criteria. Neither the United Nations, nor the North Atlantic Treaty Organization (NATO) or any other such organization is vested with decision-making authority. State powers that approximate a supranational, such as the former Soviet Union and the current European Union, were and likely are unstable arrangements. The former already has disintegrated, and the latter seems on the brink as of this writing.

[46] Troncale has noted many similarities between his own work and that of Miller mentioned above. Both researchers had very similar goals and methodologies by which to arrive at general theory. See Troncale (2006).

Some examples of processes that are isomorphic are[47] (from Friendshuh and Troncale 2012):

- Boundaries
- Competition vs. Cooperation
- Cycles and cycling
- Emergence/Origin
- Feedback
- Hierarchy
- Networks

The ontology that emerges from the Troncale et al., work consists of major system states and progressions, such as system "form" and system "linkage," each with its own set of "isomorphic processes" and cross-connected through the linkage propositions that are paths of influence between the isomorphies.[48] For example, Troncale defines a system state as "origin." In this state he defines the isomorphy of "boundary conditions". In other words, the system comes into existence as a system when a boundary appears that distinguishes it from the background environment in which it is embedded. The boundary condition isomorphy is linked with the "input/output" isomorphy that is part of "system linkage." In other words, there is a direct relation between inputs and outputs with boundary conditions established earlier in the sequence of system ontogeny (to be discussed below). This very important relation will be developed more explicitly in what follows.

Troncale also points out a basic problem with ontologies. Ontologies are built with words, and words are notoriously changeable in meaning. (Troncale 1993). The selection of a lexicon that is truly representative of an ontology is not a trivial problem.

3.2.2.3 INCOSE: Systems Science Working Group Effort

The International Council for Systems Engineering (INCOSE) has had difficulty describing a discipline of systems engineering owing, in part, to a poor consensus on what systems science actually is and what parts of systems science should inform the engineering of the products systems engineers produce. And this comes from a poor consensus on what a system is! Within the ranks of INCOSE a number of systems engineers recognized this failing and set out to correct it.[49] The INCOSE

[47] Many of Trocale's processes correspond with what Mobus and Kalton (2015, p8) call principles of systemness.

[48] The term "isomorphy" is used mostly in biology where it means, loosely, similarity of features between species. The term is not generally found in other fields. Technically it takes the form of a noun as opposed to the more general term, isomorphic, an adjective.

[49] It was also motivated by a growing number of systems engineering projects that had "failed" in one sense or another and the growing awareness that systems engineering, unlike other engineering professions that were solidly based on physics or chemistry, had no scientific base to build upon.

Systems Science Working Group (SSWG)[50] was formed to address this problem. They sought alliances with existing systems science organizations such as the International Federation for Systems Research (IFSR)[51] and the International Society for Systems Science (ISSS)[52] to begin the process of constructing definitions for systems, systems science, and, eventually, a basis for systems engineering. As part of that effort, members of the INCOSE SSWG have been working on defining an ontology of systems that could provide a basis for inter-group communications about systemness (though they do not use that term explicitly).

It is a little cart-before-the-horse to have engineers define a scientific subject area in this way. However, given that the need is extremely great (systems engineering is going to happen regardless of the state of a system ontology!) it is not entirely unprecedented. And, as it happens, many high-level systems thinkers have been involved in the effort so there is a high degree of expectation that the work will produce useful insights. As of this writing, the authors' own efforts are being invited to be incorporated in the effort. So, this chapter and the next will become part of the process within INCOSE to build a universal ontology (and language) of systems.

3.2.2.4 Natural Semantic Meta-Language

The idea that there might exist a common lexicon (of concepts) pointing to a fundamental ontology has been explored by Anna Wierzbicka and Cliff Goddard (Wierzbicka and Goddard 2014) in terms of a minimal set of semantics that are common across "all" natural languages and are used as a base set of terms to define other terms in a language. They have identified a set of terms, called "semantic primes," that they claim are common to all investigated languages (meaning that cross-language interpretation is straightforward). There are words in every language that are used to express these prime concepts, e.g., "mother" (in English) is found in all studied languages.

The idea that there might be a meta-language containing universal concepts lends some weight to the idea that mentalese exists and is subconsciously being used in thought. Further, that there are universal concepts suggests that the proposition that the human brain comes hardwired with a set of templates for those concepts—that they are not a product of cultural indoctrination—provides support to the more refined proposition that the human brain is wired to perceive systemness.

[50] See the INCOSE SSWG web site: https://www.incose.org/ChaptersGroups/WorkingGroups/transformational/systems-science for more information (accessed October 26, 2018).

[51] See the IFSR web site: http://www.ifsr.org/ for more information (accessed January 26, 2018).

[52] See the ISSS web site: http://www.isss.org/home/ for more information (accessed January 26, 2018).

3.2.2.5 System Dynamics (SD)

One of the most well-known systems modeling methods was developed originally by Jay Forrester at the Massachusetts Institute of Technology (MIT). Forrester sought to construct a computer modeling language that would be general enough to encompass any arbitrary dynamic system process (Radzicki and Taylor 2008). His original targets were related to management of organizations and economics, but the language he developed could be applied to a wide variety of systems, such as eco-systems, and engineering projects. DYNAMO, his early language, used a lexicon based on the idea that all systems involve flows of matter, energy, and messages, or, more generically, just flows of stuff and influence. It also involves the states of reservoirs or "stocks" of stuff that rise and fall relative to the flow rates of inputs versus outputs. He defined regulators or "valves" that controlled the rate of flows as well as sensors and control decisions that would change the settings on the valves.

Subsequent versions of SD have continued to view the Universe as being comprised of stocks, flows, regulators, and influences determined by decision functions. Newer languages with graphic front-ends have been developed to help researchers build complex models of systems in these terms. Donella Meadows in *Thinking in Systems* (2008) provided a summary of SD lexicon, syntax, and semantics for building models out of the simple ideas of stocks and flows.

Indeed, a significant number of issues involving system dynamics can be handled using this ontology. Just about everything you see is composed of stocks of something and flows into and out of those stocks. The generality of the language (and its basic ontology) allows it to be useful in any number of systems models. A great advantage of SD is that it is immediately translatable into computer code for simulation.

On the other hand, there are still many aspects of systemness that are beyond mere dynamics. As important as dynamics are, they are not the end all of system modeling or understanding. In the last decade, work has been done on combining agent-based modeling languages with SD in order to produce more realistic models of systems involving cognitive processes (instead of mere decision functions) and social interactions. Thus, the ontology of SD is being combined with the ontology of cognitive science and sociology to arrive at a much richer definition of reality.

3.2.2.6 H.T. Odum's Energese

Howard Odum is recognized as one of the fathers of systems ecology. He made an incredibly significant contribution to that field but also to our general understanding of how the whole world works with his introduction of the study of energy flows in ecological systems.

His focus was on how energy enters the systems (through primary producers capturing solar energy and producing biomass) to how that energy is then distributed among the participant species in the ecosystem, and finally dissipates to the embedding environment (heat transfer to exiting water and radiation to the atmosphere).

Odum recognized that these principles of energy flow applied equally to the human economy (Odum 2007). He developed an implicit ontology and lexicon called "energese." What he recognized and used to great effect is, in Len Troncale's terms, the isomorphic relations between energy flow-through in ecological systems and that in human economic systems. An economic system is, after all, a process for doing useful work to produce wealth for humans to exploit (see Chaps. 9 and 13). This is no different than primary producers making biomass for exploitation by the consumers in an ecosystem (Fig. 3.4).

3.2.2.7 WordNet

George A. Miller (Miller 1995), at Princeton University Department of Psychology, started a lexical database of English words that are linked semantically. The resulting network can be searched for meaningful relations (they call "synsets") that have deep conceptual structure. From the WordNet website:

> WordNet® is a large lexical database of English. Nouns, verbs, adjectives and adverbs are grouped into sets of cognitive synonyms (synsets), each expressing a distinct concept. Synsets are interlinked by means of conceptual-semantic and lexical relations. The resulting network of meaningfully related words and concepts can be navigated with the browser. WordNet is also freely and publicly available for download. WordNet's structure makes it a useful tool for computational linguistics and natural language processing.[53]

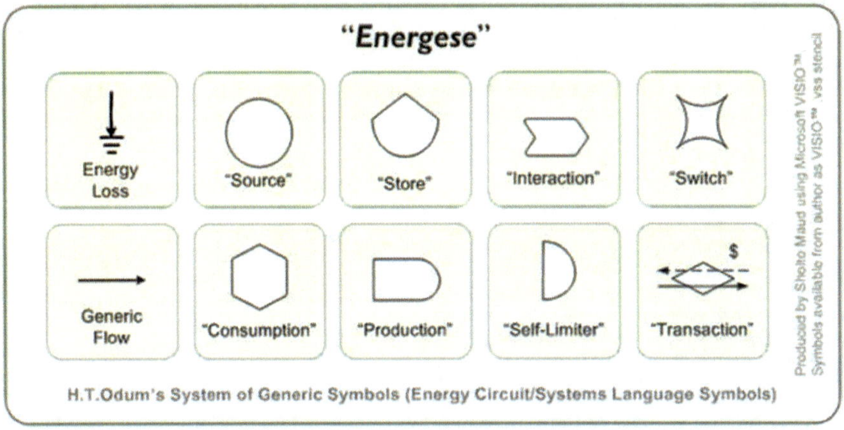

Fig. 3.4 H.T. Odum's energese lexicon and iconography used to model both ecological and human economic systems (1984)

[53] See the WordNet website at: http://wordnet.princeton.edu/wordnet/ (accessed November 3, 2017).

Using WordNet's online search tool for the word "container" links to full hyponym words like bag, basket, bowl, and bin, and many others, including vessel, are produced. A subsequent search led to "reservoir" and a search for the inherited hypernym (superordinate inheritance) produced the display below (Fig. 3.5).

Hyponyms are subcategories of a more general class (according to Dictionary.com).

What the synsets in WordNet demonstrate is how terms used in English are semantically related and how they are organized in a category hierarchy from most general sense to specific meanings. A reservoir holds something as ethereal as talent or as physical as drinking water. It has some sense of an associated quantity of that something (e.g., a talent agency has 13 actors). It is an entity that performs a particular function (holding, at least temporarily). Thus, a reservoir is essentially what we called a stock in Chaps. 3 and 4. The term stock did not, surprisingly, relate to reservoir in this sense in WordNet. One might question then if it is advisable to continue to use the word stock as our semantic primitive. This is an open question that will deserve more investigation as the practical tools deriving from the use of systemese are developed.

3.2.2.8 Strong Movement Toward a General (Universal) Ontology of System

In all the above cases we see the researchers making concerted attempts to accomplish two things. The first is the establishment of a framework for understanding systemness in terms of what sorts of "things" exist and participate in the patterns of objects being systems. The second is to use that framework to generate terms that can be interpreted universally as having the semantics we attach to the things, performing their functions, within the patterns of systemness.

There is clearly a desire to derive such a set of terms. The underlying motive would seem to be the desire to have a common language of systems that would describe the systemness ("domain of system knowledge" according to Klir 2001) into which the particular terms of disciplinary terms (domain of thing knowledge) could be mapped. With good reason.

reservoir inherited hypernym
- S: (n) reservoir (a large or extra supply of something) "a reservoir of talent"
  - direct hypernym / inherited hypernym / sister term
  - S: (n) supply (an amount of something available for use)
- S: (n) indefinite quantity (an estimated quantity)
  - S: (n) measure, quantity, amount (how much there is or how many there are of something that you can quantify)
    - S: (n) abstraction, abstract entity (a general concept formed by extracting common features from specific examples)
      - S: (n) entity (that which is perceived or known or inferred to have its own distinct existence (living or nonliving))

Fig. 3.5 The inherited hypernym display from WordNet shows the inheritance of the concept of a reservoir from the concept of an entity. It acquires traits of "measure" and "supply" along the way. Source: http://wordnetweb.princeton.edu/perl/webwn?o2=&o0=1&o8=1&o1=1&o7=&o5=&o9=&o6=&o3=&o4=&r=1&s=reservoir&i=1&h=10000#c (accessed November 3, 2017)

At the base of this motivation is the need, becoming more perceived as time goes on, to have a true basis for transdisciplinary communications. Efforts such as presented in this chapter and by the workers described above will continue and likely strengthen as this need deepens.

3.3 Ontogenesis: The Origin and Development of Systems

In a previous Sect. (3.2.1.3), we provided an introduction to the notion of ontogeny of things that exist. We asserted that auto-organization, emergence, and selection—the ontogenic cycle—accounted for the existence of things in the Universe. The mechanisms by which this is accomplished are the subject of Mobus and Kalton (2015), Chaps. 10 and 11). The argument we now advance is that this is the basis for systemness evolving in the Universe: the organization of matter, the levels of organization, and the increases in complexity of systems as new levels emerge. That is, the process of ontogeny in the Universe is responsible for the fact that all systems are constructed of subsystems, which, in turn, are themselves systems amenable to further decomposition until we reach a most fundamental "particle" level.[54] What we find in the Universe today is a very complex structure of systems and their subsystems. The Earth is a system comprised of numerous subsystems (see Chaps. 9 and 12 for example). This complexity evolved through the ontogenic cycle driven by the availability of high-potential energy flows (e.g., from the Sun) through the Earth structure, doing work to complexify[55] the system, and radiated as low-potential waste heat to outer space (Coen 2012; Morowitz 1968, 2002; Schneider and Sagan 2005).

The Earth, indeed, the whole Universe, is what it is today because electromagnetic energy has been flowing from stars through intermediate systems, planets like the Earth for example, to deep space since the gravitational formation of bodies undergoing nuclear fusion reactions[56] capable of radiating these energies (stars). On this planet we observe systems comprised of subsystems that are, themselves, comprised of subsystems (i.e., sub-subsystems), down to the level of subatomic particles precisely because the pattern of systemness is the natural outcome of ontogeny.

[54] A most fundamental particle or most "atomic" particle is posited at the quantum level. Loop quantum gravity theory (see Smolin 1997; Unger and Smolin 2015, for background) and string theory both seek to explain how things come into being based on such fundamental particles or strings. These subjects are far beyond the scope of this book. Our conception of such particles and their interactions with quanta of energy is based on inference and the need for an ultimate stopping condition for downward recursion of the system–subsystem relation more fully explained in the next chapter.

[55] This term implies that the work of combination of subsystems and components through auto-organization increases the Simonian complexity of the whole supra-system.

[56] Nucleosynthesis is, itself, a version of ontogenesis in which lighter atomic nuclei are brought together to form heavier nuclei.

3.3.1 *The Ontogenic Cycle*

The ontogenic cycle is the process whereby "things" that exist at a level of organization interact with one another. Some things interact strongly, for example forming strong bonds or mutual dependencies that will hold them together in space over time.

3.3.1.1 The Basic Process

We recapitulate the descriptions of auto-organization, emergence, and selection that underlay universal evolution from Mobus and Kalton (2015). Figure 3.6 is a simplified view of the process as presented in that source by adding the explicit flow of energies that drive the cycle and adding a third loop representing a different and significant mode of organizing, we call "intentional-organization," which we will encounter in Part 4 of the book in chapters on artifactual systems and their design by human intention.

The figure depicts three loops, from left to right, that represent ontogenesis in three different regimes (or dynamic realms, as Volk 2017, terms them). The innermost loop is labeled pre-biological and entails ontogenesis of systems from the subatomic level of organization up to and including the pre-biological chemical level. Upon the emergence of life (described by Smith and Morowitz 2016, as a major phase transition of matter) we enter the realm of biological evolution that

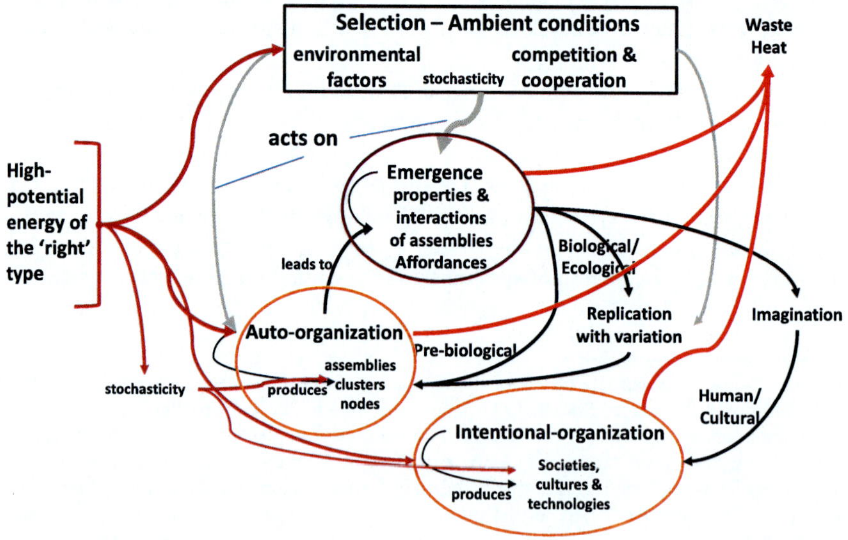

Fig. 3.6 The ontogenic cycle is depicted. This cycle produces ever more complex systems from simpler systems as long as there is a flow of high-potential energy from a localized source through the system and out to a general sink, as is the case for the Earth system. (Figure modified from Mobus and Kalton (2015), Fig. 10.3, p. 471))

inserts a new sub-process, the methods of replication with variation (genetic mutation, for example) to the overall process of emergence followed by a new round of auto-organization. This significantly expands the range of possible variant entities (e.g., species) and thus the opportunities for auto-organization to generate many new kinds of entity relations, including symbiotic and ecological (e.g., food webs). This new level of organization offers entities with vastly increased numbers of affordances, and, hence, greater- or longer-term stability for the new systems emerging. Finally, once hominids evolved through the operation of this biological loop, and transcended ordinary consciousness—that of the state of the body, of the state of the immediate environment, and the state of the coupling of these two (Mobus 2019) and mere signaling communication—a new kind of organizing process comes into play. Humans invent things that they can imagine and intend to construct and use; they are conscious of possible future worlds (Mobus 2019). This latter process, here identified as "intentional-organization," will be explored further in Part 4, as mentioned. We introduce it here in preparation for that discussion.

The flow of high-potential energy[57] through a potential system, that is one that is not particularly organized but contains "particles" or individuated entities with complimentary affordances that have the potential to form interactions requiring the doing of work, drives the auto-organization, or realization of those potentials, in the form of new assemblies of particles, i.e., particles form selective bonds, or in the case of the biological loop, species relations. In the case of intentional-organization, this is realized in the actual work of humans inventing and constructing new cultural artifacts (see Chap. 14). Since work is being done, the energy eventually flows out of the system as waste heat when a steady-state condition is achieved. The figure also depicts a smaller flow of energy that is not necessarily driving work directly but contributes instead to stochastic disturbances or the "shuffling" of the particles to increase the opportunities for disparate kinds to interact. Note that in the auto-organization process this stochasticity is a relatively important component of the process (e.g., in the chemical system, this is in the form of thermal vibrations and variations in momenta of the atoms, such as Brownian motion). In the intentional-organization loop, stochasticity plays a role, to be described further in Part 4 as the novelty element in imagining designs, but much less so as in the pre-biological and biological loops.

The most notable version of this process, to provide an explicit example, might be the chemical interactions between molecules that were thought to be involved in the origin of life on Earth (Morowitz 1968, 1992, 2002; Schneider and Sagan 2005). During this phase ambient conditions (such as temperature, salinity, pH, and other physical forces in the case of pre-life chemistry) tend to select for more stable

[57] High potential energy refers to a specific form of energy that is capable of coupling with specific forms of work within the potential system. Different forms of energy will couple with bonding activities in the different levels of organization. At subatomic levels, the energies are inherent in the fundamental nuclear and electromagnetic forces. At the molecular level, energies may involve thermal modes or electromagnetic quanta (i.e., photons). At the social level, these same forms of energy are now operating in the form of behaviors such as eating and replicating.

assemblies and disrupt lesser stable ones. If the geometry of the boundary conditions is favorable, the flow of energy through the system will also drive convective, mixing, and sorting cycles that might favor the interactions between these new assemblies. The assemblies have different interaction potentials than their components had possessed before forming bonds. For example, larger organic molecules may have very different bonding potentials for exposed valence shells when some of the valence opportunities have already been subsumed under the formed bonds. Some of these new potentials are "stronger" than would have been the case for a "naked" atom. They also can exhibit very different behaviors. More complex molecules may show affinities for other molecules that would not have been present in the pre-molecular state. This is the phase of emergence—new structures and new behaviors emerge from these new potentials.

Lest the reader think this general pattern of the ontogenic cycle only applies to atoms and molecules, think about social situations. There are many instances where strangers are gathered into a new potential social situation, like a new organization (the containment boundary), where they enter a phase of "getting to know one another." People have personalities and have affinities for certain other personalities in forming relations of trust, friendship, or being repelled by some. The organization may have an "official" organization network, but here we can see the "unofficial" social network auto-organize. The resulting relations, in particular the bonds that are forged, provide the bound participants with much more social power than any of them might have had as individuals.

Consider the rings of Saturn, or, for that matter, the orbits of planets around the Sun. On the scale of the solar system, we see interactions mediated by gravitational coupling that led to new structures. Planets like Earth are condensed and structurally organized aggregations of dust, water, and gases that form new organizations during planetary evolution. The final (or semi-final) states of the solar system and planets demonstrate considerable organization through the selection force of gravity (e.g., the various geo-layers from the Earth's core to its outer atmosphere based on densities of the materials) pulling mass toward the planetary center versus the flow of high-temperature heat outward toward the surface and into deep space.

Just as important as the emergence of new structures and behaviors, the stable assemblies represent a new level of complexity, i.e., complexity of the whole system has increased. The prebiotic chemical "soup" underwent a phase change when self-replicating cells emerged and has increased in complexity ever since with the rise of eukaryotic cells, like paramecia, then multicellular organisms, and eventually up to human societies. The Earth, as a planet, is far more complex as a whole, than any one society. Figure 3.7 depicts the spiraling upward over time as the result of the continuous application of the ontogenic cycle. The width of the spirals represents increasing complexity. The spiral shape represents the auto-organization and emergence of new more complex assemblies through time. As long as there is a flow of free energy, i.e., energy in a form that can drive the useful work of forming new combinations of assemblies at any given level, the spiral will continue to be upward and wider. This increasing complexity is corresponded with the increasing levels of

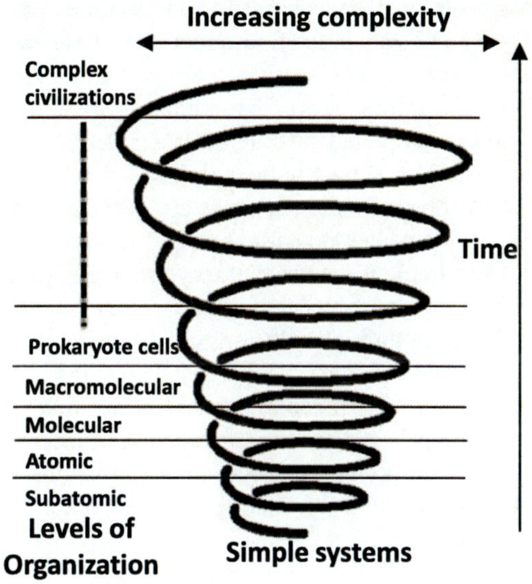

Fig. 3.7 The continuous pass through the ontogenic cycle produces a spiral of ever-increasing complexity as the general ontology of the Universe

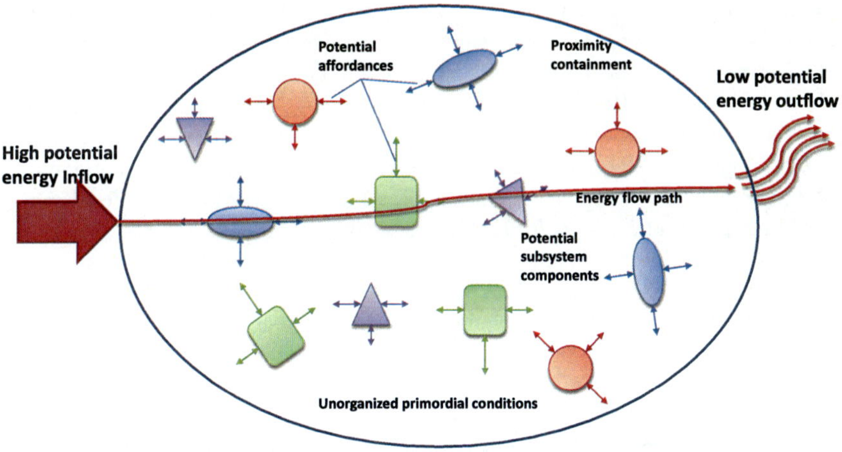

Fig. 3.8 The primordial conditions before the onset of the ontogenic cycle just at the onset of auto-organization

organization that we have witnessed on our planet (after Volk 2017; c.f. Morowitz 2002; Troncale 1978a).

Here we present a summary of the mechanisms involved in the ontogenic cycle as far as auto-organization is concerned (as mentioned above we will consider intentional-organization later in the book). Figure 3.8 shows the situation prior to the onset of auto-organization. This shows schematically a multiset of components (possibly themselves subsystems as described in the next chapter). Each component possesses what we have called "personalities" that make them unique kinds of entities. The objects in the figure represent these entities and each has a set of bidirectional arrows representing their potentials to interact with other components (potential affordances). These entities are shown contained in some fashion that ensures their potential for interactions (later we will discuss boundaries and boundedness). Also shown is the existence of a high-potential energy source providing energy input into the container at a particular position. The figure also shows the dissipation of low-potential energy (not shown are the actual source and sink entities for simplicity) and a pathway for energy to flow through the system.

The state of the system at this primordial situation is that the components may interact, say through collisions or bumping; we assume motion degrees of freedom. The situation shown can represent a wide range of entities in the levels of organization in Fig. 3.10. It can represent, for example, the simple molecules available for interactions leading to the construction of macromolecules, the precursors of living systems. Or it can represent individual human being brought together in a social situation in which personalities (as the word is normally used) have the potential to interact.

This model of auto-organization follows closely and was inspired by the work of Harold Morowitz (1968) and amplified by Smith and Morowitz (2016). Morowitz developed the concept of how energy flow, under particular conditions of containment and configuration, could act to organize the system in a molecular environment. He was one of the premier workers in the field of biogenesis (origin of life).

We use the term "affordance" in a very general way. More specifically, mutual affordance means that two entities are able to couple unidirectional flows (of material, or energy, or messages), which will be explained later in this chapter and more thoroughly in the next chapter; for convenience these two-way flows are here represented as bidirectional arrows for brevity. An example is the coupling of a positive charge from a proton with the negative charge of an electron. Another is a dialog between two people.

What we see in Fig. 3.8 is a "potential" or latent system. Each entity's potential affordances are matched to one extent or another with affordances on one or more of the other entities, by type. That is, each entity has a potential to bond[58] with some other entities under the right circumstances. Principally that means an availability of the "right" kind of energy (also known as exergy) in the energy flow to do the work of forming the bond. Another "right" kind of energy might involve that needed to move the entities about within their containment. In the work Morowitz did, he noted how aqueous molecular systems (under conditions of temperature and pressure) respond to heat flow by forming convective cycles, assuming the waste heat can escape the system as shown in the figure. This keeps the potential system at relatively constant temperature and pressure. Moreover, the convective cycles carry the

[58] For completeness we should also understand that some affordances may involve repulsion and not just attraction. To keep the explanation simple, we are only considering attractive forces here.

entities about in a stochastic manner making the probability of finding matches higher than equilibrium chance alone.

Though the process of auto-organization can most easily be pictured for molecular formation, as the entities in Fig. 3.8 represent atoms of different kinds with different configurations for valence electrons, it is important to note that this picture represents the general dynamics of auto-organization and is valid for all levels of organization. For potential systems higher in the levels of organization, the same phenomenon can be seen but with differences in the details of containment, configuration, and what the specific form of energy flow is involved. For example, in the case of people brought together at a social meeting, the containment might involve the event (like a wedding) and a room. The energy flow is clearly much more complex involving both the energy each person uses to move about, which they brought into the situation from the food they consumed at home, and things like the temperature controls for the room.

From the initial primordial state, we now consider the actual process of ontogenesis, the combining of these entities as a process and how it leads to more complexity and increasing levels of organization (see Sect. 3.3.1.2).

Figure 3.6 shows three pathways from emergence back toward the next phase of auto- or intentional-organization. One pathway, the purely physical, involves auto-organizations of things such as nucleons, atoms, and molecules, but also covers auto-organization (via the gravitational force) of stars, planets, and galaxies, in other words, pre-life systems. The second, middle pathway routes through the special kind of evolution that applies to living systems. Mechanisms that can be viewed as generators of unbounded diversification, such as mutation and duplication, create opportunities for new living systems to emerge. The third, right-most pathway will be covered later. Figure 3.9 provides a pictorial representation of the auto-organization pathways. Here we see the entities from Fig. 3.8 beginning to interact and follow a time course of such interactions, thus becoming components of a system. These are shown as bonding at the mutually compatible affordances (Fig. 3.9).

The figures on the left side (labeled A) represent a time sequence (blue arrow labeled Time) of pre-life auto-organization. In this figure we represent a shortened sequence leading to biological evolution, but, independent of scale, the various objects could represent quarks, or nucleons, or atoms, or molecules (or planets, or stars, etc.). In that sequence higher complexity is obtained when various entities with different personalities and having different, but compatible, potential affordances (thin two-headed arrows) combine owing to mutual attraction to produce realized affordances (thick black two-headed arrows). In this sequence we see the emergence of increasing complexity by "accretion" at the same level of organization. The combinations may lead to the emergence of new possible affordances or, as shown, residual affordances from the original components and newer combinations (middle object in A). The success, of course, depends on the ambient conditions that add or subtract energy to or from the process, as shown above and detailed in Mobus and Kalton (2015), Chap. 10). Those conditions select for stable configurations and select against weak bonds. The third object shows another round of auto-organization that leads to a chemical system sufficiently able to exploit the

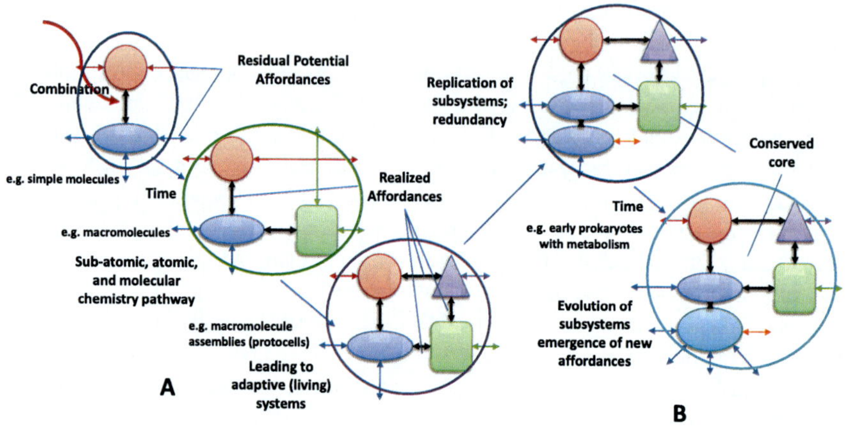

Fig. 3.9 Ontogenesis: (**a**) the pre-life pathway to greater complexity; (**b**) the evolutionary pathway to greater complexity of living systems through biological evolution. The surrounding ovals in the A sequence represent effective boundaries where the connections between entities as components create a unity. The boundaries indicated in the B sequence may represent actual boundaries like cell membranes. This figure depicts just the formation of stable structures, those having significant duration given the milieu in which they formed. A more complete picture would show many unstable formations (at a specific time) hovering around the stable ones that will later disaggregate and many components that have not been incorporated into more complex structures owing to the specific conditions of the milieu

energy disequilibria and to adapt to minor shifts and retain stability. This is the beginning of ancient metabolism (Smith and Morowitz 2016) and probably coincident with the building of the first genetic code. At some point subsequent these systems acquired an ability to replicate themselves and internal parts of themselves (the two similar blue ovals in the upper object in B). Replicated internal components may have the ability to evolve differently from the original component.[59] We have now entered the pathway of biological evolution where when such events occur the redundant part is free to evolve into something new, sometimes also evolving new affordances (bottom object in B).

Atomic auto-organization proceeded along a fixed pathway by merely combining nuclei via nucleosynthesis. Residual affordances took the form of increasing positive electric charge, as protons were fused to form higher atomic number. New affordances took the form of creating additional (potential) electron shells with increasing "slots" for electrons to occupy when nuclei captured electrons (in cooler conditions). These new affordances of atoms, valence shells, not just nuclei, created the potential for chemical interactions between atoms, e.g., covalent or ionic bonds. Of course, when a valence shell has all of its electrons and the atom is electrically

[59] In Chap. 10, we will address the issues of what is required within a complex, adaptive, and evolvable system to implement these mechanisms.

neutral, the atom system is essentially closed to further interactions; no residual affordances are available at the boundary, except under very extreme conditions.

Chemical evolution (and here we can include planetary construction) begins to introduce more elaborate combinations of atoms and under the right circumstances leads to molecular mixes of sufficient complexity and organization that they live. We suppose that the process of ontogenesis continues apace. Biological evolution eventually produced our clever species that has proceeded to change the dynamic of ontogenesis by introducing intentions into the combination business (and into the business of reshaping material resources for advantage). They create systems for their use and combine with a spectacular array of those artifactual systems (see Chap. 14) to form a multiplicity of cultures, organizations of societies with seemingly unlimited affordances. At this writing there has been some hint of these cultures further combining by a process of blending. You can find a McDonalds™ almost anywhere in the world.

In the following section, we sketch the elaboration of structures, forms, and functions in a hierarchy of complexity as ontogenesis produced the Universe we witness today.

3.3.1.2 Levels of Organization

The Universe, as we find it contemporarily, is rather complex in terms of heavenly bodies (stars, planets, galaxies, and even more exotic bodies). But in addition, there are islands of exceeding complexity (at least one that we are sure of) on planets with living systems.[60] This complexity pertains to the fact that there exist nested levels of organization. Atoms might seem relatively "simple" compared with molecules, especially organic molecules. In turn molecules are relatively "simple" compared with living cells, prokaryotes like bacteria. And bacteria (or cells in a multicellular organism) are much "simpler" than a tree or a goat. When we recognize that all organisms are participants in larger-scale ecosystems, including geophysical aspects as well as a large-scale stable set of other living species, we can see that the whole Earth is exceedingly complex.

Figures 3.10, 3.11 and 3.12 show some relations between "things" that show progressively greater complexity as new levels of organization (to which they belong) emerge. We will discuss the progressive origins of these things below as well as the levels of organization and emergences. Figure 3.10 presents the purely physical (i.e., subatomic and atomic) entities that end up producing both large-scale entities like galaxies and small-scale entities like gases and rocky bodies through chemical interactions.

The figure shows the transition from nonliving to living systems through stages that involve the co-evolution of primitive ecosystems (believed to be primitive

---

[60] Recall that the kind of complexity we are talking about is that suggested by Herbert Simon, a hierarchically organized object composed of subsystems and those composed of smaller subsystems, what he called a "partially decomposable" system.

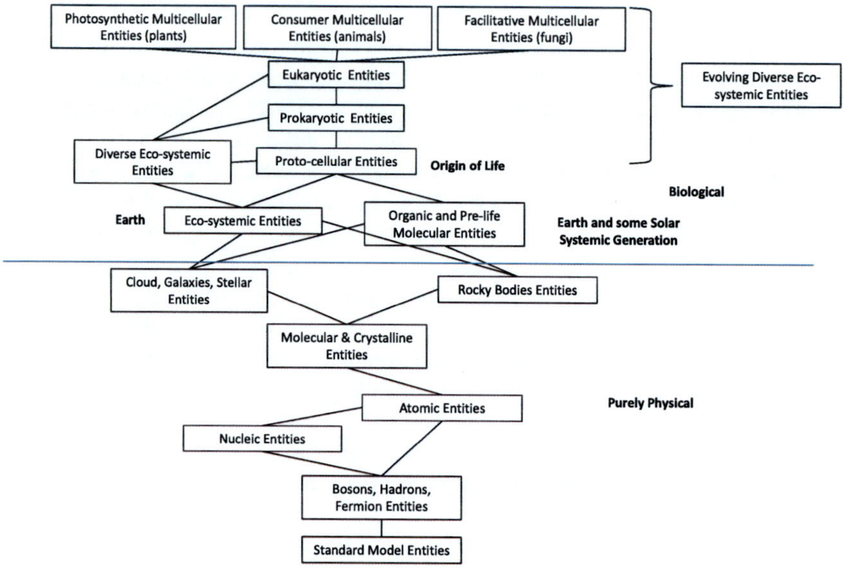

Fig. 3.10 Levels of Organization starting with the lowest levels—the "Purely Physical"—and showing the generation of the biological levels. Note that this does not reflect a simple linear evolution, but that multiple kinds of subsystems have emerged and co-evolved over time. The boxes are labeled with the kinds of "things" that exist at any given level. Those things tend to continue to exist at higher levels although things like "Proto-cellular Entities" at the origin of life stage may have been consumed by living organisms after the "Prokaryotic Entities" stage emerged. Also see below sections under 3.3.2 "Other Takes on Ontogenesis"

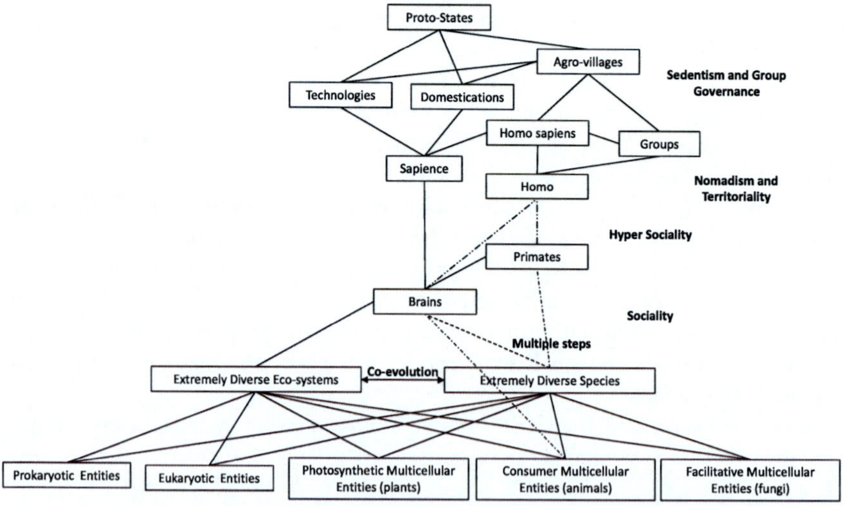

Fig. 3.11 The continuation of emergence of levels of organization after the emergence of life. Living cells rapidly diversified through the Darwinian mechanism of descent with modification (i.e., gene mutations). The great Linnaean Kingdoms (along the bottom) originated during the earliest era of life as a result of modification and selection (discussed below). The various dashed lines represent progressions that include many more intermediate steps that have been left out. Also, the diagram only follows the Animalia line of evolution, being very anthropocentric

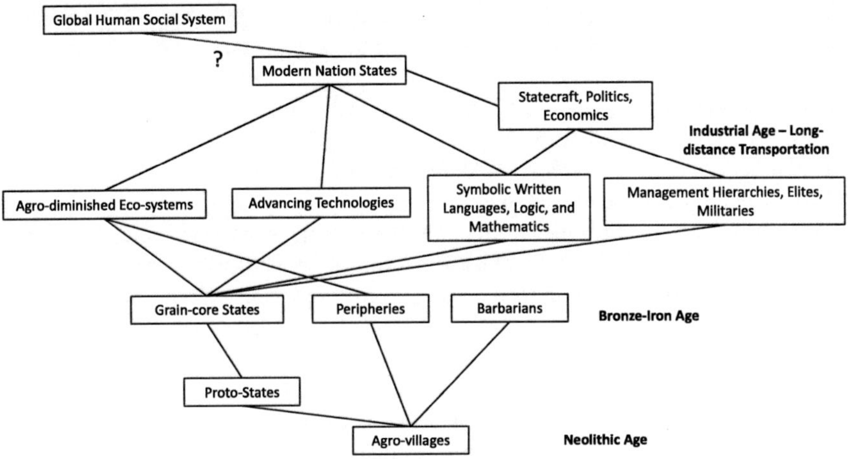

Fig. 3.12 The levels of organization for humans grow quite complex but can be summarized by the historical "ages" that paleontologists and historians consider. This characterization of the Neolithic and Bronze–Iron ages is derived from Scott (2017). The highest level of organization is represented by a world in which the whole of humanity is integrated into a single "Global Human Social System," which will be addressed in Chaps. 7 and 9, and the Postscript

thermal vents in the new oceans: Smith and Morowitz 2016). The events that had to occur for the origin of living matter are only now beginning to be understood. Morowitz (1992) considers the origin of metabolism and some primitive form of it, providing clues as to what sorts of chemical interactions must have been transpiring that would produce a substrate for full-on living metabolism in true cellular structures.

The exact mechanisms involved in the origin of life, the processes of auto-organization and autocatalysis that led to the first biological entities (prokaryotic cells), are still unknown but many clues have been discovered (Smith and Morowitz 2016). But once living systems emerged in the ancient Earth, a new capacity for self-maintenance emerged as well, the capacity for individual cells to adapt within boundaries to changes in their extant environment. We now enter the realm of the complex adaptive system (CAS) insofar as individual organisms are concerned. But something even more interesting co-emerges with living cells. They reproduce and form populations of cells. They transmit their genetic materials via copies to descendants, but sometimes with minor errors in the code. Most of the time the errors are either harmful and kill the individual descendant, or they may be neutral, having no substantive effect on the phenotype of the offspring and so are invisible to selective forces. What this does do, however, is introduce a completely new kind of system in the form of a population of individuals that share the same basic genetic code but for a few errors in copying here and there—a species. And the population constitutes a new kind of system. It, taken as a whole, is a complex, adaptive, and *evolvable* system. Not only are individuals capable of adapting within limits to short-term changes in environments, but also, due to some individuals possessing mutations that, by chance, give them the capacity to become adaptive to larger changes, the species gains the ability to survive on the whole when things really do change substantially. The capacity to evolve modified and more fit phenotypes, when things change, gives the species resilience in the face of change and makes it long-term sustainable.

So, with the origin and development of life, we go into the realm of the complex adaptive and evolvable systems (CAES) mentioned in the Introduction, again in Chap. 2, and to be refined in Chap. 6.

It is the CAES that intrigues us the most. Each of us, even as individual biological entities, turn out to be CAES, at least in terms of our ability to "learn" new ideas and behaviors. But so are our social systems, our organizations, and, increasingly, our artifactual systems (Chap. 14).

The horizontal blue line in Fig. 3.10 marks the point at which systems types transitioned from simple and merely complex to complex adaptive and then complex adaptive and evolvable systems. In this book we will be mostly concerned with CAS and CAES instantiations, and, in particular, those at the high end of the chart in Fig. 3.12.

Ontogenesis has continued unabated from the production of simple systems like atoms, complex systems like planets, complex adaptive systems like biological individuals, and complex adaptive and evolvable systems like species, human individuals (the capabilities of their neocortex), ecosystems, and human social systems like tribes and agro-villages (Fig. 3.11).

Each turn of the cycle traces out the spiral increase in complexity pictured in Fig. 3.7, focused on the Earth as a whole (and presumably many other planets in the Universe able to support living systems). We will refer to the whole Earth system as the Ecos[61]—the home. The human species (as with all species of plants, animals, fungi, bacteria, etc.) are subsystems of the Ecos. The human social system is a particularly interesting subsystem of the Ecos. We shall have very much more to say about it in the following chapters.

Figure 3.11 continues the evolution of higher levels of organization within the biological domain, but focuses mostly on the line of animal evolution, particularly the development of complex nervous systems leading to complex behaviors. Of course, of significance in this evolution is the origin and evolution of human beings. The figure represents a number of unnamed stages with dashed lines. We jump up to *Homo sapiens* from Primates (in general) through a number of species of pre-*Homo* and early *Homo* (e.g., *erectus*).

Sapience is defined as the cognitive ability of humans to form communications networks using complex symbolic representations (words and sentences—language). But it also involves the hyper-social capacity to cooperate (with more than just altruistic actions), to think both systemically (see Chap. 4) and strategically (into the future). And it involves moral reasoning. So far as we know, these

[61] The Ecos is an alternative and broader name for the whole Earth system (which includes Luna and Sol having major input influences on what happens on the surface of the Earth) not in opposition to the popular term, Gaia (Lovelock and Margulis 1974), but as a way to situate Gaia as the physiology of a larger phenomenon (c.f. Volk 1998).

capacities evolved in the human line of great apes, though we cannot be certain which in the several predecessor species they first became operative.

One of the earliest indications of rising sapience in the Homo genera was the reliance on small group interactions, extended families, and fission–fusion dynamics (Sapolsky 2017, pp. 429–430) leading to "bands" or extended groups that obtained stability, leadership hierarchies, and a cohesion that supported what is called group selection (Sober and Wilson 1998; Wilson and Sober 1994; Wilson and Wilson 2008; Wilson 2013) in which competition between groups but cooperation between members of groups led to increased fitness. That is, in groups that excelled in cooperation internally, their ability to outcompete other groups led to a higher rate of success in exploiting resources and reproduction.

Homo sapiens proceeded to invent new technologies in stone, wood, and bone tools, along with improved clothing and shelter construction. The process of domestication of plants and animals (e.g., a wolf-like ancestor to a dog) also led to more reliable food sources. In particular the domestication of various grains and living in uniquely abundant ecosystems that promoted sedentism (areas like the Fertile Crescent in the Middle East and the Nile River Valley) gave rise to increasing group sizes and new kinds of social organization that ultimately put emphasis on the logistical management of planting fields and irrigation. A new level of organization emerged involving humans, tools, other living organisms, and a new relation to territory—the agro-village (Scott 2017, c.f. Volk 2017, Chap. 13).

Early agro-villages entered into a number of kinds of relations with each other and with the remaining hunter-gatherer-herding cultures that still lived around them (Scott 2017). Some of these relations involved trade and mutually beneficial interactions (e.g., a source of "fresh" genetic material). Some were more aggressive and destructive. It seems a lot depended on the weather! Climate variations over the centuries determined crop successes, and needs to raid other villages' grain stores. Basically, the Mesolithic and Neolithic ages, right into the Bronze–Iron age saw a fluctuation of proto-state formations followed by dissolutions (due likely to plagues and wars, ibid).

In Fig. 3.12, we pick up the continuing story of the ontogenesis of complex human social systems from Agro-villages and Proto-States to what Scott (2017) calls "Grain-core States" such as Mesopotamian Ur and the united kingdom of Egypt to the rapidly developing cultures of humans.

A lot goes on in terms of the auto-organization and emergence of new forms once early states come into existence and show some relative stability. Writing, logic, geometry, and mathematical reasoning come into existence in the context of managing the grain-core. Management hierarchies (see Chap. 12) emerge. So does religion and its coupling to those management hierarchies—various forms of what is called "statecraft".

Money is invented as a form of "promise to pay" backed by real commodities. Those promises are increasingly taken over by the evolving nation states. Coinage is invented as a form of messaging—the state promises that you can use this coin to buy commodities that you need; trust us!

The modern nation state, indeed the whole global system of nation states interacting with one another, has been the result of emerging entities engaging with each other and forming coupling interactions that are both super complex and yet wholly understandable in light of the ontogenic cycle. Below, we review the major components of that cycle.

3.3.1.3 Energy Flow: A Prerequisite to the Generation of Structures and Levels of Organization

The key proviso of the ontogenic cycle is the fact that energy is needed in any physical, chemical, biological, and social/ecological system in order to do work. That is, to bring material components together, to force or induce them to interact (e.g., bind or communicate), and to provide an on-going source of binding energy, there needs to be an input of the right kind of energy from an external source as well as an external sink to which "spent" energy can dissipate.

Harold Morowitz (1968) provided an exquisite model of how energy flow through an unorganized chemical system provided the motive force to facilitate the auto-organization of that system, to produce combinations of elementary things that would then give rise to a new level of organization, prebiotic molecules. His approach was to focus on the processes that gave rise to the origin of living systems (see also, Smith and Morowitz 2016), the prebiotic regime of complex molecular formation that occurs as a result of the "right" kind of energy (i.e., light photons or heat differentials) flowing into a highly constrained geometric "container" at a particular point, and then out at a diametrically opposite point as waste heat. What Morowitz showed is that given the right energy flows and geometry, complex molecular forms would emerge that could then interact further.

Morowitz was (1927–2016) an eminent scientist dedicated to strict rules of empiricism. In a personal meeting with this author, he was unwilling to leap to non-empirical implications of his energy flow principle beyond its role in pre-biological chemistry (he later would actually venture out into other domains, c.f. Morowitz 2002). Yet, what systems science deals with is patterns and after many years of examination it is clear that the notion of energy flow is at the heart of driving increasing complexity. We propose to call this aspect of ontogenesis the Morowitz principle.

3.3.1.4 Auto-Organization: Creating Structure

At any level of organization, one finds elemental entities. Whether these be atoms or individuals in a society, the fact is that the world contains individualized entities or objects that have the capacity to interact with other (perhaps diverse) entities in either attractive or repulsive interactions. Potential systems, that is, systems that are composed of multiple independent individuals of various types prior to the flow of energy, generally will have some form of random or semi-random distributions, subject only to the chance interactions (think of a container of a mixture of gases at thermal equilibrium). Every type of entity will possess what we called a "personality" (Mobus and Kalton 2015), or the range of interaction potentials or affordances exposed to the other entities. These may be outward exposed potentials such as valence shells on atoms that have already formed bonds with other atoms in molecules and then proceed to form more complex molecules (e.g., polymerization of repeating chains of smaller molecules). However other kinds of potentials may develop as a result of increased complexity and geometry. Following the molecular story, once proteins (a heteromer of amino acids) evolved they took on folding into new shapes that gave them abilities to manipulate other kinds of molecules, such as carbohydrates.

The personalities of human individuals are extremely more complex and diverse, of course. By personalities, here, we do not just mean the standard psychological personality profile, e.g., extroverted versus introverted. We mean the whole range of behaviors and appearances that are perceived by other humans. Some are going to be perceived as attractive and followed by potential formation of bonds (friendships, romance, trading relations, etc.). Others may put off any such relations, be essentially repulsive. These are clearly not forces like nuclear or electromagnetic/electrostatic, but they are real and it can be argued are derivatives of those simpler forces. Furthermore, any bonds forged between individuals (including those between parents and children) require the direction of biophysical energy toward maintenance.

Attractions lead to aggregations, groupings. Repulsions lead to increasing distances between entities and separation of these entities.[62] Combining through attraction creates new entities at what we have called higher levels of organization. Repulsions between these new entities lead to competitions and sorting. That is, the new entities are engaged in pulling together and pushing apart dynamics that tend to group attractive entities into interactions.

These patterns of interactions mediated by forces and energy flows should not be thought of as merely analogous. The energy flows in biological groups are exactly the same as in, for example, cellular metabolism, but now channeled through more complex mechanism (psychology and brain physiology). To insist that they are different is a grave category error that only perpetuates the belief that somehow humans are different from the rest of nature and beyond the laws of physics and chemistry (not to mention the laws of biological systems).

Auto-organization operates at every level of organization bringing diverse personalities (that is, objects with various affordances) into proximity and interactions that, if stable under ambient conditions, are the basis for the next higher level of organization as depicted in Figs. 3.8, 3.9 and 3.10. Attraction (e.g., exothermic and

[62] Unless the energetic environment includes forces that can overcome the repulsive force; for example as occurs in nucleosynthesis inside a massive star where the effects of gravity and pressure overcome the electrostatic repulsion between two positively charged nuclei, bringing them close enough together such that the residual strong nuclear force causes them to fuse. On the scale of human interactions, the requirement for economic viability frequently forces parties to cooperate to achieve a common goal.

exergonic reactions) or forced interactions (e.g., endothermic or endergonic reactions) along with the sorting and organizing influences of repulsion under the excitatory influences of an energy flow (and that energy's interactions with the material forms of the entities in the system) produce a stochastic mixture of higher-order structures that can then interact and form a new level of organization. This is an ongoing and recursively generative process that, as we track it from the simplest subatomic entities (i.e., quarks, electrons, and photons), meaning considering the history of the Universe, has led to the kinds of complex structures (living systems and societies) that we observe today.

3.3.1.5 Emergence: New Structures and New Interactions

Auto-organization is fundamentally a randomized process of combination and sorting (or segregating). It is *chance*[63] because there is no a priori organization in the distribution of entities throughout the volume of a potential system. The Big Bang saw to that. Entities at a level of organization tend to be distributed in a random, mixed manner. Energy flow and auto-organization then tend to generate randomized combinations of more complex entities. Energy flow drives segregating movements and material flows, like convective cycles. But in the end, new structures are formed and these have new potentials for interactions with each other. The example of the emergence of enzymes (or ribozymes in RNA world) that have shapes that facilitate catalytic reactions in other molecules in metabolism that enabled the emergence of life processes is a case in point (Smith and Morowitz 2016).

But auto-organization often gives rise to many more emergent entities, with new structures and new interaction potentials, than can achieve stable relations at a given level of organization. Nature is continually blindly experimenting with new forms to see what they can achieve as new systems. Because the emergent new forms are the result of chance encounters at the lower level of organization, not all of them are necessarily fit to contribute to the new level. In fact, most new experiments are doomed to failure. Only a few stable configurations of structure and their potential interactions are destined to survive a fundamental testing and, thus, provide the entity components that can enter the next cycle of ontogenesis. The environment that is extant for any given level of organization is a harsh taskmaster.

3.3.1.6 Selection

Every new structure, with its new interaction capacity, is a blind experiment (remember the "chance" part of emergence?). Structural configurations have to be subjected to tests like thermal stability (at the ambient temperature, can the entity exist?).

[63] As in *Chance and Necessity* (Monod 1971); the role of chance or stochastic process in evolution.

They have to be stable against other environmental conditions (such as pH, pressure, social norms).

Consider the situation for the most elemental particles, the quarks. In the earliest phase of the Universe, shortly after the Big Bang, in an extreme temperature, all of the various generations of quarks (I up/down; II charm/strange; III top/bottom) could exist and potentially interact to form exotic kinds of hadrons. But as the Universe cooled radically (as it expanded radically) only the up/down quarks proved stable enough to form realized hadrons (protons and neutrons).

Or consider the conditions of the early Earth and primitive oceanic thermal vents where conditions may have acted to select for the stability of co-factor molecules that would eventually contribute to primitive metabolism (Smith and Morowitz 2016).

Selection is, of course, a major aspect of Darwinian evolution. There are several candidate selection processes that are reasonably well understood: natural selection, or the selection of the phenotypes most capable of surviving and reproducing; sexual selection, or the capacity of certain body forms to be more attractive to potential mates; and multilevel selection, in which capabilities at the genetic, phenotypic, and group level of form and behavior act to provide advantage or otherwise to systemic entities (genes, individuals, and groups, respectively).

We view selection as an all-encompassing process in which the context or environmental conditions operate on emergent forms to destroy those that are susceptible (i.e., denature proteins that are not stable) and leave in place those that are thermodynamically or physiologically, or sociologically, stable. The survivors are free to interact and start the next phase of ontogenesis; to produce the next level of organization.

The result of selection is the basis for the next round of ontogenesis. That which proves effective, and survives within the context of the environment created by the level of organization, participates in the next round of auto-organization. The ontogenesis of new levels of organization thus emerges and the Universe experiences new levels of complexity. Hence, Figs. 3.6 and 3.7 and the framework they invoke.

3.3.2 Other Takes on Ontogenesis

This view of increasing complexity is neither new nor unique to our conceptualization. In fact, many authors have examined this apparent trajectory of the evolving Universe. In this section we examine a number of previous or current concepts of how the Universe is evolving toward increasing complexity and, especially, as represented by the evolution of life and humanity on the planet Earth.

3.3.2.1 The Noosphere: Teilhard de Chardin

Pere Teilhard was a Jesuit priest living in the early twentieth century who completely bought into the theory of Darwinian evolution applied to humanity. For de Chardin, the mind of humanity represented a new level of organization above the mere animalistic capacity for cognition.

The mind of humans was a result of a continuing process of evolution (set in motion by the Christian god). But it was fundamentally different from the sentience of prior creatures (particularly the great apes). Teilhard's objective was to reconcile mental evolution with the Christian theology. He recognized that human mentation was apart from our ancestral apes but did not develop a firm empirical theory for how this was so.

Even so, Pere Teilhard provided some instructive insights with respect to how humans represented to new level of organization that involved mentation beyond basic ape-biology. Leaving his metaphysics and theology aside, he did give us a sense of humans as a new level of organization—that our species represented an emergence from the mere biological.

3.3.2.2 Combogenesis

Tyler Volk has coined a term to describe the process of forming combinations of simpler things to produce more complex things—Combogenesis (Volk 2017).[64] He describes how the process first kicks off in the early Universe (right after the Big Bang) with the combining of subatomic particles (quarks) to form nucleons (protons and neutrons) as the ultra-hot Universe began to cool down upon expansion.[65] Nucleons and electrons could then combine to form the early light atoms (e.g., hydrogen and helium). As gravity condensed gaseous clouds to form the earliest stars, combogenesis went to work in their interiors through nuclear fusion to produce heavier nuclei. Supernovae then produced even heavier nuclei that, when ejected into space, formed the atoms, which we know from the Periodic Table.

Volk then describes a "Grand Sequence" of transitions in which combinations at the molecular level led to the formation of planets and the origin of living matter (see also Smith and Morowitz 2016). The stages in this sequence represent major transitions in a hierarchy of complexity (i.e., from prokaryotic to eukaryotic cells, to multicellular organisms, to societies of multicellular organisms, and particularly human societies).

[64] Volk's work is perhaps the most similar to the results reported above. The author and Volk developed very similar conceptions of how the Universe got more complex over time. We have been in close communications to see if an integration of the two sets of ideas is possible.

[65] The picture is similar to the cooling of a hot gas in a volume that can expand. The expansion gives rise to fewer particle collisions per unit time, which is what we recognize as a lowering of the temperature of the gas.

As with the ontogenesis cycle described here, combogenesis is acted upon by evolutionary processes. He does not include physical process selection, but once into the realm of living things—first the process of biological evolution and then, when humans enter the picture, cultural evolution—the role of selection becomes obvious.

In our reading of Volk's work, it seems that we and he have independently arrived at a concept of universal generativity for complexity and levels of organization. This is not really remarkable in that the seeds of thinking this way have been long germinating (Bourke 2011; Coen 2012; Miller 1978; Maynard Smith and Szathmáry 1998; Morowitz 1968, 1992, 2004; Prigogine and Stengers 1984; Troncale 1978a).

3.3.2.3 Emergence by Continuous Integration and Diversification

Another perspective on how hierarchies of increasing complexity have emerged from a bottom-up process of things interacting at one level and then producing objects at a new higher level is provided by Len Troncale (1978a). The integration aspect seems to be fundamentally the same idea of combogenesis (above) and auto-organization of ontogenesis. Different entities with different interaction potentials actualize those potentials to form more complex entities at a new level of organization.

Troncale's vision includes a phase of diversification or variation generation that seems influenced by the process in Darwinian evolution in which species diverge from a common ancestor in incrementally subtle ways. The mechanisms for generating diversity depend on the level of organization. In biology this is Darwin's famous decent with variation, now known to be due to mutations and chromosomal cross-over. It depends on making many copies of a genome wherein some configurations may be different (mutations) and lead to phenotypic variations that can then be tested in the crucible of selection.

What seems different between Volk's vision of a grand sequence of discrete stages and Troncale's model is that the latter sees the process as continuous and on-going, whereas the former sees what look like discrete jumps. Volk does offer that there might be a place in the model for micro-combogenesis events that, at a larger scale of resolution, might reasonably appear to look like discrete steps.

What seems important here is not whether emergent new things and behaviors are due to a continuous or discrete stage process, but rather that both these views along with the others mentioned all point to an obvious (but often contested) trajectory of universal evolution going from the simple (as in the Introduction) to the complex, to the complex adaptive (early life, cells, and individuals), to the complex adaptive and evolvable (life, genera, and societies). The details of how this trajectory is supported by ontogenesis will be the subject of intense research in the coming years.

3.4 An Ontological Framework

In Sect. 3.2.1.2, "Cosmological Ontology," we established a basic background ontology, claiming that all that exist in the Universe, or at least that part of it accessible informationally by we humans, ultimately are formed from matter and energy, organized by the influences of information and resulting in representation of knowledge. Under the influence of the ontogenic cycle above, the Universe has evolved from a primordial, simple state to what we observe today, a high level of organization and complexity that includes living beings such as ourselves capable of understanding that organization. We will take this as a starting point for "that which evolves," that is, all the "things" we find in the Universe.

In this section, we will provide an ontological framework that asserts that what can exist in this evolving Universe, made of matter and energy, organized by knowledge and information, is systems. Put differently, matter and energy interact according to the influences of knowledge and information to produce systems at all scales of space and time, at all levels of emergent organization, and complexity.

The claim, made in Chap. 2, Principle 1 (Systemness, and supported by the ontogeny arguments above), is that everything is a system, meaning that all things in existence are organized with system attributes and are, themselves, subsystems of larger supra-systems, up to the Universe as a whole (which we take to be the only true closed system). This claim must be counted as a conjecture as much as a premise at present. However, it is arrived at by induction over the examination of many actual concrete systems in nature, as indicated in the above Sect. 3.2.2, on "Examples of Developing a System Ontology." The framework offered here is as much a guide to future research in systems science as an outline of ways to view systems ontology.

3.4.1 The Framework, Its Purpose, and Its Justifications

A framework is a set of guides and organizing principles used to elaborate the details of a complex subject. This book is about a framework for understanding systems through the application of the principles in Chap. 2—its main structure being the method of analysis and knowledge capture. The framework we offer here is an approach to finding and defining a set of unique objects, relations, and actions that are found in all systems.[66] These objects, etc., are at once both concrete and abstract. They are concrete in that each is found to be a real thing. They are abstract in the sense that we are applying symbols, names, and icons, for example, to denote the real object, etc.

[66] Troncale (2006) proposes a similar notion of a framework for his SSP—system of system's processes. Troncale does not call his approach an "ontological" framework, but one can find significant relations between it and that given here.

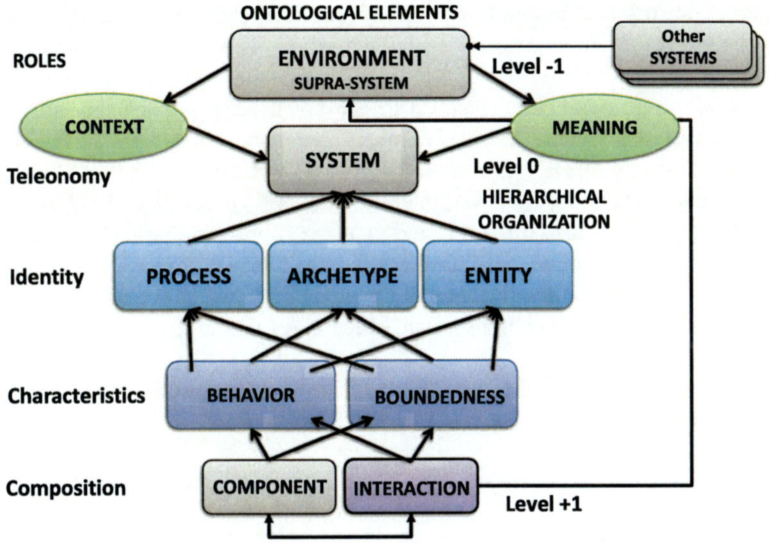

Fig. 3.13 A general framework for the system ontology identifies the relations between upper ontological elements. Gray elements are the main elements, i.e., the ENVIRONMENT/SUPRA-SYSTEM in which the SYSTEM is embedded and the COMPONENT(s) that comprise the system are the main structural elements. Ovals, CONTEXT and MEANING, are logically imposed on the relation between the environment and its system component. Light blue rounded corner rectangles (PROCESS, ARCHETYPE, and ENTITY) are Identity attributes of any system. Dark blue rectangles (BEHAVIOR and BOUNDEDNESS) derive from the way a system's components interact with one another to form its internal structure/function. INTERACTIONS among internal components and between entities (Other SYSTEMS) in the environment involve exchanges or flows of matter, energy, and messages (or influences)

The framework for the system ontology is shown in Fig. 3.13. This may be called a conceptual model in that it demonstrates not only the ontological commitments but also how they are organized. The figure is a graph representation of elements (nodes) and their relations (edges) along with category names and organizing labels (levels).

Elements (nodes in the directed graph) in all capital letters denote terms that will be used in the top-level ontology of systems. These are the things that exist by virtue of the Universe organizing as it does, through auto-organization, emergence, and evolution (Mobus and Kalton 2015, Chaps. 10 and 11), which we are now calling the ontogenic cycle in Sect. 3.3.1. These are the core things that we will look for in our analysis of systems regardless of the level of organization or complexity. As we will show in the next chapter, for example, the element labeled COMPONENT at Level + 1 may be readily seen as a subsystem of the system at Level 0. That component/subsystem would then be subject to the same analysis imposed by the framework in the figure. It will become the system of interest and be its own Level 0. The analysis simply reapplies the conceptual framework with the previous SYSTEM becoming the new ENVIRONMENT.

The framework is composed of three *aspects*. The first aspect, the nodes in the graph, is the set of ontological elements themselves. The second, signified by the words on the left side of the graph, is the ROLES that the items in that region (horizontal level) of the graph play. The third aspect shows the higher–lower relation that frames the system hierarchy; the level numbers are relative positions in the hierarchy.

Starting at the top, the ENVIRONMENT encases and defines both the CONTEXT and the MEANING of the system of interest (SOI). This is designated as Level − 1, or one level up.[67] It is also often referred to as the supra-system. Level 0 is that of the SOI, the system that we seek to understand and model. Level + 1 is one level down from the SOI meaning the level of organization in which we find the internal components and their interactions—that which gives rise to the SOI behavior and defines its boundary.

In the next section we will elaborate on all of the elements, relations, and categorizations illustrated in Fig. 3.13.

3.4.2 *The Aspects*

3.4.2.1 The Ontological Elements of the Framework

In the section below, "The System Ontology", we will define and describe these elements and their expansion into sub-elements that exist. Here we provide a brief explanation of the major elements in the framework in preparation for that expansion.

3.4.2.1.1 ENVIRONMENT

In this structure the top or highest element is the environment. This is the supra-system that encloses the system of interest (see Fig. 3.4). Environments are, by definition, more complex than the SOI. This is because that larger system contains the SOI, which is then a subsystem, as well as other systems interacting with the original SOI. Hence, the complexity of the environment includes the complexity of the various subsystems.

The environment is what gives context and meaning to the SOI. Both have to be considered in order to understand the roles, particularly that of *purpose* (see below).

[67] The author recognizes the irony of calling a Level − 1 as *up*. This somewhat unfortunate condition is a result of adopting the computer science way of looking at "tree" structures, which are inverted from our ordinary understanding of trees. The root is at the top and the level number is zero. To go down the tree level is to increase the level number. Thus, to go back to something before the presumptive root means to go in a negative direction. What can one say? Conventions are conventions after all.

3.4.2.1.1.1 CONTEXT

This refers to the set of conditions that obtain at a time instant relevant to the dynamics of the SOI. The environment in an evolving Universe is continually changing in so many various complex ways, so contexts also change.[68] The fundamental context is that of the various other systems with which the SOI interacts in an on-going way. These are the sources of inputs and the sinks for outputs across the system boundary.[69] Other sources, not necessarily recognized as an identifiable nature, may affect a system through what we will call a "disturbance."

Changes in context will occur in several ways. The first way is when a source or sink entity changes its behavior (e.g., the rate of supply or receiving) from a long-term average (if such ever existed). This is one form of what is technically called non-stationarity. We will be examining this phenomenon extensively. A second kind of change involves the appearance or disappearance of sources or sinks. A loss of a resource source or a shut off of a sink changes the context in which the SOI operates. This too is a kind of non-stationarity that can change slowly or rapidly. Both of these changes are part of the uncertainty factors that affect all concrete systems.

One source of uncertainty we will not be concerned with is that of purely random action of chaos in the vernacular sense. True randomness could only be achieved in closed system where there is absolutely no kind of gradient operating, e.g., an adiabatically isolated container holding simple gas near equilibrium and far from any gravitational influences. While such systems might exist in an idealized abstraction, they are technically impossible. What has come to be called deterministic chaos, however, is not of this same sort of randomness. Supra-systems may have chaotic dynamics, meaning that one or more of their internal causal relations between subsystems (including the SOI) may operate on a chaotic attractor; this still counts as a form of organization. And that organization has implications for the organization and sustainability of the SOI.

---

[68] This pertains primarily to the Universe as it appears to exist today, 13.5 billion years after the Big Bang. The environment of the first particles (e.g., quarks) and photons, indeed all of the elements of the Standard Model, started out in extreme temperature in which these elements likely existed as independent non-binding or interacting entities. But the Universe rapidly cooled and as the temperature came down elements began to interact and form the first systems—protons, neutrons, electrons, and neutrinos, with their various interactions mediated by the bosons. Yet later, cooling led to the formation of atoms that could then interact through chemistry. Thus, one could say that the environment for these primary systems is no longer changing; thus they are not compelled to make further adjustments.

[69] We have used the term boundary in a rather loose way so far. As will be explained in Sect. 3.4.2.1.2.5, below, a boundary is an effective demarcation between what is in the system and what is not. It is realized in many ways and in many forms described there and in the next chapter.

3.4.2.1.1.2 MEANING

The concept of "meaning" has always been problematic in philosophical terms. The ideas are tied up with things like human values, which are extremely variable and very often inconsistent even within a single individual.

In our system ontology, we consider MEANING to describe a set of conditions that have a trinary affective influence on a system. We use the term "valence" to designate one of three influences: positive, negative, or neutral.[70] The changes described in the context could have neutral effect, in which case the system does not need to do anything differently; indeed it is free to choose a next state randomly. But some context states can be negative with respect to how the conditions will affect the system (e.g., part of a negative feedback loop). For example, a radical change in average temperature in a climate zone will require some response from the species affected or they will go extinct. Other contexts might be viewed as positive in that they support the system, as when the average temperature returns to that which was the norm for the species. An important aspect of this interpretation of meaning is that it is the "handshaking" between an environment and the system in terms of interpreting messages (information) and computing decisions for action (if such is possible).

We assert that virtually all meaning ultimately resolves down to one of the three valences and the magnitude of those changes that give urgency or motivation for action by the system. This assertion is based on work done by Antonio Damasio (1994) in the realm of human brain functions and decision-making and by observations widely made in animal behavior. As well, the author has demonstrated this working in a robot's brain (Mobus 2000).

Throughout the book, we will provide examples of the context and meaning of an environment's effect upon a system.

Note in Fig. 3.7 that we show an arrow from the INTERACTION element (at the bottom) going all the way back to the ENVIRONMENT element. This represents closing the loop between the SOI and the ENVIRONMENT via its interactions with the latter. If those interactions are inappropriate given the CONTEXT situation, then the SOI will incur a negative valence from the ENVIRONMENT. Conversely, if the interaction is appropriate, then the valence will be positive. If the interaction is neutral, then the valence will also be neutral.

3.4.2.1.2 SYSTEM

This ontological element is quite obviously the core of the whole enterprise. There are three ways to view a system of interest: as a process, as an object, and as an entity.

[70] In psychology, the term valence refers to the degrees of badness (and repulsion) or goodness (and attraction) with no reference to neutral. We are using the term in a broader, though inclusive, sense.

3.4.2.1.2.1 PROCESS

All systems within the Universe are open to input and output flows of at least one of: material, energy, or messages (which are special forms of material/energy flows). Systems process or transform the inputs into outputs. The collective effect of the process is also called the "function" of the system. The processes inherent in the bits and pieces of a rock, strong chemical bonding, produce its qualities of hardness and stability over time (soft sandstone notwithstanding). The processes of digestion in your alimentary track function to acquire nutrients for your body's maintenance.

3.4.2.1.2.2 TYPE

The ontogenic cycle (Sect. 3.3.1) as it has proceeded over the history of the Universe, from simple, to complex, to complex adaptive, and to complex adaptive and evolvable systems, has generated a plethora of diverse real systems as variations on common themes. These systems share a basic set of attributes that puts them in what we call a category. For example, the evolution of the geosphere has produced an incredible variety of lakes (a category of entities) on almost all continents. These lakes vary in volume, depth, surface area, and other important features. But they are all recognizable as lakes. They all serve the same function of acting as reservoirs for water. Similarly, all mountains are recognizable as mountains, being in the category of mountain. Within the tree of life, the relations and derivations are even more telling. Life has evolved into many different genera from some common ancestor, forming a complex tree-like structure. There is a continuity of genotypic information such that we can trace current living things back to their predecessors (in most cases). But the elaboration of the tree of life also created a set of categories, captured in the nomenclature system that designates, for example, phyla, classes, families, etc. down to the species level.

We are using the term "type" here in a very general way. It can refer to a category, which is usually found in a hierarchical structure, or it can refer to what we are calling an "archetype" (see Chap. 10 for an elaboration on the concept of archetypes). Generally, the word is used to designate a derivation from abstract classification hierarchies. As such it places the system of interest within a larger evolutionary framework that is within a diversity-generating ontogenic process. Its use here, in the framework, will become clearer in the next chapter when we introduce the mathematical concept of a system's history.

Type corresponds with the way mental models are formed in the human brain.

3.4.2.1.2.3 ENTITY

An entity is a concrete system that does something actual, having causal effect on other entities or objects in its environment. Concrete systems are real and identifiable as unique, at least within a minimal category. That is, a system as an entity is an actual thing in itself and not something like a category (though it belongs to a type). The idea of a "house" is a concept and a category of objects that exist in the world. "My house" is an actual member of this category and particular in existence. Models of systems are abstract versions of real systems that are more like categories in action. On the other hand, a specific model instantiation, say in a particular computer memory, is an entity for the duration of its run.

3.4.2.1.2.4 BEHAVIOR

All systems display activity on some time scale. They interact with other systems by virtue of properties of their boundaries and their components (below).

3.4.2.1.2.5 BOUNDEDNESS

All systems are delineated from their environments by virtue of *coherence* of the components. We can talk about this as providing an "effective" boundary. In some systems this will manifest as a physical barrier demarking the insides of the system from its environment, such as the case of cell membrane. In other cases, there may not appear to be any such structure. The atmosphere is "bound" to the planet by gravity but we do not conceive of a physical boundary keeping the gases enveloping the planet. Indeed, some of the lighter gases such as hydrogen may be light enough to escape the atmosphere if struck by an energetic photon. The effective boundary of the atmosphere is the gravitation field gradient where the pull of gravity is no longer able to keep the lightest gases from escaping.

This type of boundary belongs to a class of fuzzy boundaries, to be explained more fully below in Sect. 3.4.2.2.3, "Attributes," and in Chap. 4. Other characteristics of boundary conditions, such as porosity, will also be explained.

Boundedness is a property that, in the Universe of today, generally arises as a result of electromagnetic interactions between molecules or gravitational force for massive objects. But we can also count the bonds of love between people as a form of binding force that keeps, for example, families together. In this case the electromagnetic force is operating on several levels. It is behind the electrochemical operations in the brains and it mediates message flows between people. They do not feel a physical force per se, but they do feel the effects of physical forces at play in their mental states and communications. Boundaries in social systems are extremely fuzzy and generally porous. People come and go and are members of a given system based on, for example, time of day or season of the year. Other membership relations are also possible, e.g., members of a religious persuasion based on commonly held beliefs.

Boundaries need not be crisp or clearly physical entities in themselves. For example, some boundaries are implications of the strength of interactions between components being such that the entity has a degree of "apartness" from other entities in the environment (see Fig. 3.8). Take a molecule as an instance. The boundary is provided by the chemical bond strengths between atoms and the lack of available bonding sites on the molecule. CO_2 is a good example. The four bonding sites available on the carbon atom are occupied by the two bonding sites on each of the oxygen atoms, thus leaving no additional bonding sites. The bonds established are quite stable so the molecule is established as an entity and will only interact with other molecules through physical collisions (i.e., temperature related). That particular molecule is especially good at absorbing photons at characteristic wavelengths and becoming more agitated as a result.

Boundaries, when not completely explicit, can be determined by the relation between internal interaction strengths and densities in the network of subsystems compared to interaction strengths and number with external entities. Figure 3.14 shows a depiction of a system where the boundary is "effective" as a result of the bindings between internal subsystems that are strong and dense as compared with interactions between a few components and some external entities. A graph-theory-based analysis of the network depicted in the figure would reveal the "clique" of gray nodes, which, in turn, provides a strong argument that the components inside the blue oval constitute a delineated system even though no explicit physical boundary can be seen.

The notion of a boundary as a natural element is still contentious in schools of systems thinkers. We define a boundary to exist when the sum of binding forces from within a given system act to establish long-term cohesion among the component parts of the system. Colloquially, "the family that plays together stays together."

The Boundary "Problem"

Over the last few years members of the systems science and system engineering communities have been holding regular discussions for the purpose of coming up with an agreed definition of systemness. One of the difficulties that have emerged from these discussions has been the recognition of little agreement on the nature of boundaries. Engineers, coming from a tradition of being the ones who define what a system is (i.e., they have to design a system to perform a given function), tend to prefer the notion that boundaries are more-or-less arbitrary, or, rather, that we

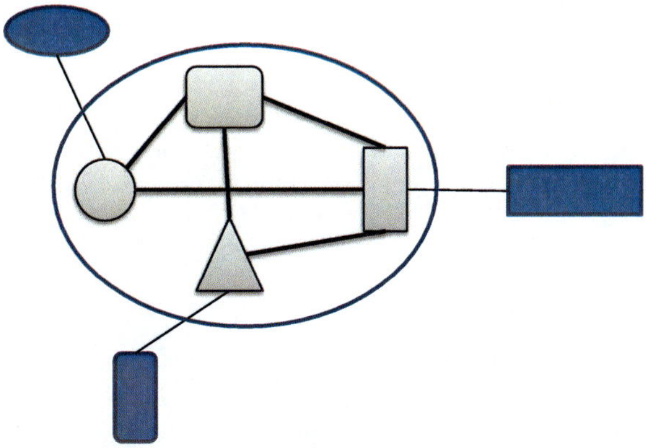

Fig. 3.14 The density and strengths of interactions (links) between components that are "inside" the system are greater than those between external entities and a few components of the system. The effective boundary (blue oval) distinguishes between the system and its environment. Thicker undirected lines represent stronger links

humans determine what the boundary of a system is. They then transfer this sentiment to the existence of boundaries in natural systems as well, some claiming that boundaries (indeed systemness itself) are a construction of the human mind. Some systems scientists, especially the ones more inclined to pure formal definitions of systems, agree, thinking that the whole notion of a system is a mental construct (and, of course, the mathematical representation of a system *is* a mental construct—see Chap. 4). But the majority of natural scientists and especially those who do practice systemic thinking about their subjects, tend to hold the opposite view; those systems are clearly bounded even when the boundary is based on internal forces and interactions as opposed to a constraining physical container.

The boundary "problem" is one of world views rather than an actual problem in finding the so-called boundary conditions, the conditions that pertain to demarking a system from its environment. Where it becomes problematic for systems scientists who are trying to do analysis of real systems is when they are faced with finding those conditions and clearly differentiating the insides from the outsides of a system. A formal approach to the problematique is to use the concept of a fuzzy boundary, borrowed from fuzzy set theory. As will be shown in Chap. 4, a fuzzy boundary can be defined in which each component member of a system has its own membership function that can be based on location in space and/or time. This means a member component may be inside the system boundary to some degree at certain spatial coordinates and at certain times. At all other locations and times, the component is outside the system.

While there remain significant philosophical debates on the nature of boundaries and systems, in reality these debates have not seriously hindered progress toward understanding systemness itself. The main problem tends to be with the methods and tools used to resolve the question of what is a system, natural or engineered. For the most part, this means how one goes about constructing a model of a system for purposes to be discussed in Chap. 11. One of the things this book seeks to show is how this difficulty can be mitigated.

3.4.2.1.2.6 COMPONENT

All systems contain components that internally operate (behave) to produce the process that the system entails. Components may themselves be systems, or atomic, i.e., irreducible.

3.4.2.1.2.7 INTERACTION

All systems interact with other systems to some extent or another. There are no isolated systems. Interactions result from properties of their boundaries and processes. Fundamentally, all physical components/subsystems possess interaction potentials mediated by the electromagnetic and/or gravitational forces. Chemical reactions (exchanges of electron), radiative interactions, mechanical interactions, and convective cycles mediate the interactions between systems at all levels of organization above the atomic. Even love is mediated by chemical interactions (pheromones and brain receptors!).

At all levels of organization, the entities that exist possess interaction potentials mediated by physical forces but possibly transmitted over long distances. Electromagnetic phenomena, especially photon transmitted, can be used to transmit messages at low energy costs over tremendous distances. And in fluidic media such as air and water, mechanical waves, when modulated effectively, can also transmit information that causes changes in the receiving system. Pheromones along with words of endearment, transmitted either by electromagnetic (writing) or mechanical (sound waves), can seal a bonding quite effectively.

Another term used to describe interactions is "relations." An interaction is a relation, but the latter term captures the static situation more than the dynamic one. There are a number of kinds of relations that are temporary or transitory. They can be logical or situational, such as positional in space and time ("in front of," "before"), or social ("dominant," "submissive"). An interaction may incorporate a relation, but should do so as part of a functional description; that is, the relation should be time-dependent at least but likely to depend on other factors. For example: a train engine can be "in front of" the trail of cars it is pulling as long as it is going from point A to point B, after which it is decoupled. The "in front of" relation only exists under certain functional conditions.

3.4.2.2 Roles

All of these elements play various roles within the framework.

3.4.2.2.1 Teleonomy

Referring to the interactions between the environment and the SOI (Level − 1 to Level 0 in Fig. 3.3): The term "teleonomy" is used to designate something like a "purpose." Purpose is a highly problematic concept, both philosophically and practically. Do systems have a purpose? Does the environment have a purpose? Does either *serve* a purpose? Purpose implies intention. Pre-Darwin, the organization of the world was easily explained by the intentions of God. Post-Darwin the situation has gotten more complicated.

Systems exist and persist in their environments as a result of performing some kind of *function* that is beneficial, ultimately to the supra-system. This is the meaning of "fitness." As long as a system fulfills its function and that function is of net benefit to the embedding supra-system, the system gets to enjoy the receipt of beneficial resource inputs. In other words, there is a net benefit (profits minus costs) to both the supra-system and the system and thus, the system sustains.

A system performs its function in the sense that it produces outputs, some of which are products in that they are inputs to other entities in the environment that need those inputs as resources, while others might be actions, behaviors, that contribute to the success of other entities in the environment. The output of a system might be a physical substance or it might be a behavior. In either case that output is of value to some other entity or entities in the environment and, thus, a benefit to the supra-system as a whole.

Consider an ecosystem. This is a semi-closed system composed of multiple species of plants, animals, bacteria, fungi, and the geophysical substrates on which it is built. It is also dependent on a pattern of climatological regularity. Ecosystems achieve a more-or-less (quasi-) stable dynamic, called a climax state, in which the interactions between all of the various species act to mutually constrain one another. In this system, the purpose of a carnivore is to constrain the population of a prey species; the purpose of the latter is to convert photosynthetic (primary) production to animal biomass for the predator. The plants that feed the primary consumer have the purpose of converting solar energy into preliminary biomass. The predator's purpose is to eventually return the biomass back to the bacterial colonies that return the nutrients to the soils so that the next generation of plants can fulfill their purpose. Every biological system depends on every other biological system for its existence and the steady-state flux of energy and material exchanges through the whole ecosystem maintain the organization of that system.

The purpose of a system is to "fit" into its supra-system in the sense that it produces actions or products that keep the whole supra-system in dynamic stability. The supra-system's purpose is to channel flows of resources to those subsystems that fit this objective. Subsystems that contribute to the long-term stability of the supra-system are favored by virtue of the fact that other subsystems will produce resources that they need as long as they produce resources that other subsystems need.

Thus, purpose is a mutual support phenomenon. You scratch my back and I will scratch yours. Or, I will scratch species A's back, they will scratch species B's back, etc. until species Z scratches my back. Subsystems that do not do so are ultimately not fit in the evolutionary sense and will be selected against in the long run. The environment of an SOI is a system seeking dynamic equilibrium so long as energy flows through it. An SOI that does not meet the purposes of a supra-system (environment) will be selected against in the long run.

A system can be said to have a purpose in terms of the environment. This does not necessarily mean that the system is purposeful (has its own purpose for existing) but if it does its purpose should be aligned with the "needs" of the environment in order for there to be any kind of long-term sustainability for the SOI. Context is the current state of the environment with respect to those interactions it has with the system. It affects the system from moment to moment and the system must behave in such a way that it accommodates that context. In the totality of interactions between an SOI and its environment, misalignments will invariably harm the SOI. Nature bats last, and always wins. Even if the environment is stressed by the actions of the SOI, the former will counter and select against the SOI in the long run. When the variables of the environment reach certain critical values, they create a stressful situation for the SOI.

Meaning is based on the imposition by the environment of its requirements for the SOI, i.e., gives meaning to the existence of the SOI. The SOI should be viewed as a subsystem of the larger environment taken as a supra-system. Thus, the SOI at Level 0 must behave in a way that fulfills the requirements imposed by the environment. This is the meaning of natural (and other forms of) selection in biological evolution, for example. This is the meaning of market factors in the economy as another example. Biological systems like species that fail to provide a service to the larger ecosystem niche, will be selected against in the long run. Those that provide a useful output (e.g., meat for carnivores or food for herbivores, or are keystone species[71] in specific ecologies) will generally benefit from the environment providing useful feedback. Carnivores work to keep populations of herbivores in check (a regulatory function) and receive food for doing so. Herbivores often spread seeds or manures and, similarly, receive food in the process. Companies that produce a product or provide a service that consumers desire are rewarded with sales and profits. Purpose comes from this matching of subsystem output to other subsystems' inputs in a way that is mutually beneficial through feedback loops (both positive and negative). Overall balance and stability of the whole supra-system depends on these kinds of matched interactions.

Environments have a tendency to change on longer time scales than that of the overt behavior of the SOI. This is the basis for the SOI to change (its composition and behavior) to remain in compliance with the new(er) requirements. This is what adaptation and evolution are about.

Living and supra-living systems (e.g., organizations) are certainly purposeful in doing what they have to in order to stay alive (operate) and procreate (profit/grow). They are adaptive and evolvable systems that can reconfigure their internals to meet new challenges.

3.4.2.2.2 Qualities

This refers to the set of aspects that confer systemness on a "thing." Every concrete system is, at once, a process, an object, and an entity. The process is a set of active internal transformation activities that are required for the system to obtain and manipulate resource substances and produce output substances (and forces). The process gives rise to the identity of the kind of entity an observer perceives. Objecthood entails the system being a unified, whole, identifiable body. All of its parts move in an orchestrated way.

[71] A keystone species is one that plays a critical role in the sustainability of an ecosystem. Their influence on the structure and function of the system is such that removal of representatives of the species from an ecosystem has been shown to cause a major change in the abundance of various other species.

3.4.2.2.3 Attributes

We attribute systemness to an entity (or process) based on the fact that what is perceived by another system (we humans for example) is the behavior of a bounded object. When we cannot know directly what is going on inside the object, we can still infer something is happening inside by virtue of sensing what is going in from the outside and what is coming out from within, or, in other words, crossing the boundary, and this includes the responses to forces as well as the production of forces applied to entities in the environment.

The behavior of the whole system is a function of the combined behaviors of its internal components (and cannot be predicted on the basis of individual component behaviors). It is recognized that the behavior "belongs" to the SOI by virtue of an effective boundary that demarks the system from the rest of its environment.

3.4.2.2.4 Composition

The boundary of the system keeps the contents in and the rest of the environment not needed as resources out. A system is composed of components that have interactions with one another. They are linked by various ways, forces, and flows of substances. As we will see in the next chapter, but start to get a hint of here, is that a component of a Level 0 SOI can be, itself, a system. That is, it has the ontological status of a system, making it a subsystem of the original SOI. By a recursive rule of hierarchical organization (see Eq. 3/4 in the next chapter), such a subsystem may be remapped into this same framework and treated as a system with an environment (the original SOI as supra-system) and composed of internal components and interactions giving rise to its behavior. This remapping is the basis for the process we will introduce in Chap. 5 as functional/structural deconstruction.[72]

3.4.2.3 Levels

The concept of levels in a hierarchy is quite natural. In this framework, we define a three-level structure depicted in Fig. 3.15. We label the SOI as Level 0 (L 0) as it is the focus of our investigations. The environment of the SOI is labeled Level − 1, as in it is one level out from the SOI. The explanation for this comes shortly. Similarly, the components within an SOI are assigned Level + 1.

---

[72]The term "decomposition" is often used to mean the same operation. In mathematics this is well understood. Unfortunately, I have found that many naïve students or laypersons think of corpses rotting when the word is used, or at least that thought seems to lurk subconsciously, perhaps biasing their understanding of what we are doing, or creating a barrier of mild disgust that blocks understanding. We prefer the term "deconstruction" since it evokes a mechanical operation of taking something apart and it sounds better than "dissection."

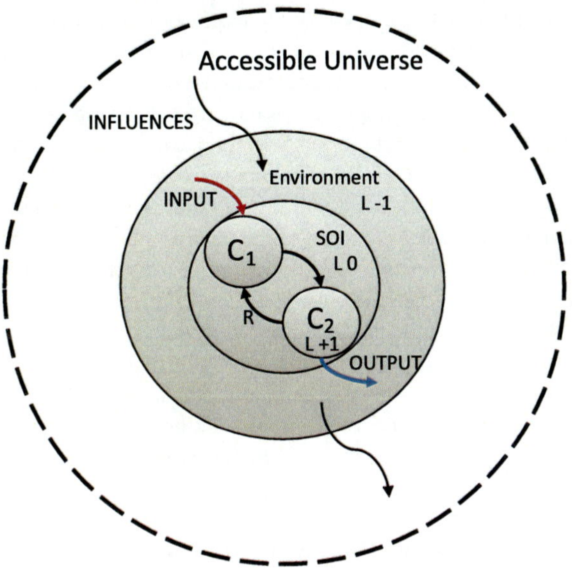

Fig. 3.15 This depicts a "map" of system hierarchy (see Fig. 3.14). Level 0 is the SOI. Components (C1 and C2) along with relations (arrows labeled R) constitute the next level down, L + 1. The environment is the next level up, L − 1

Some possible confusion arises from this use of plus and minus symbols. It is logical, however, from the standpoint of a graph theoretical tree, which has its root at the top and branches downward in levels. The root is generally said to be at Level 0 and the branches to lower nodes are at Level 1; those at the next level down (!) are at Level 2 and so on. We will be using tree structures and graph theoretical math in formulating the structure of system representation in the next chapter, so it will become clear how this scheme works in practice. The assignment of −1 to the level of the environment will come into play when we are doing functional and structural deconstruction on levels further down in the hierarchy of composition.

3.4.3 Using the Framework

3.4.3.1 Applying the Principles at Level − 1 in the Framework

We are now in a position to attempt an organization of ontological elements using the framework and examining various proposed elements from the numerous methods discussed above and examined elsewhere. Our starting point is to use Level − 1, the ENVIRONMENT, and Level 0, the existence of a SYSTEM, with the role of CONTEXT and MEANING, focusing our attention on the Purpose of the system. Taken along with the principles of organization, hierarchy (especially that the environment is a supra-system), network, and behavior, we establish the existence of other systems as entities in the environment that interact directly with the system of interest through the flows of matter, energy, and messages. The entity-hood of the system is established by the existence of a boundary that acts to regulate the inflows and outflows and these, measured at points in time, establish the CONTEXT of the environment relative to the SYSTEM. The inputs to the system must necessarily be construed as resources (except for disturbances) and the system must necessarily have at least one output that is of benefit to the environment. Benefit to the environment has the same characteristics as benefits to the system from the context. The tri-valent values discussed earlier apply mutually to the environment and the system. We conclude that the value produced by the system must benefit some entity in the environment that through some feedback loop through the environment comes to support the continued availability of critical inputs to the system.

These are some of the objects, relations, and actions that have ontological status as a result of applying the principles to candidates within the framework.

3.4.3.2 Applying the Principles at Level 0 in the Framework

We next address the qualities of Level 0, the SOI, by examining the attributes that make an SOI what it is. An SOI is, by definition, a system. Our examination of Level − 1 established the input/output relations with environmental entities, so the SYSTEM status of the SOI is given. Also established, if the output(s) of the SOI are transformed aspects of the inputs (e.g., some of the high-potential input energy appears at the output as waste heat), is that the system contains a PROCESS. Work must be accomplished within the SOI so that transformations are established. Finally, ENTITY-hood is given by locating the system in time and space relative to the other entities in the environment. In some environments, as supra-systems, a multiplicity of similar entities may constitute subsystems. That is, an SOI may be replicated within the environment multiple times. However, thanks to the fact that no two physical objects can occupy the same space at the same time, it is possible to establish the unique characteristics of a specific SOI by affixing the time and place of its existence. Note that the first element, time, is considered always sequential. However, space is established by fixing a coordinate system on the whole environment and specifying the coordinates of a given SOI. This operation is only sometimes necessary when doing a micro-scaled analysis, as will be demonstrated in Chap. 6.

The transition to the application of Level 1 begins with establishing the existence of the BOUNDARY and BEHAVIOR of the SOI. Boundaries establish the points of contact/interaction between the SOI and the other entities in the environment. The boundary also provides the frame of observation for the behavior of the SOI with respect to its position in the environment and the interactions it has with the other entities. The external or public behavior of an SOI is observed at the boundary first.

3.4.3.3 Applying the Principles at Level 1 in the Framework

The full establishment of the boundary and behavior of an SOI precedes and leads to the exposure of the internals of the SOI. The latter now becomes a new ENVIRONMENT in that the existence of COMPONENTs and their INTERACTIONS constitute the establishment of the internals of the original SOI. We now move the index up so that the new Level − 1 is the entire contents of the original SOI and the new Level 0 applies to all of the subsystem components to be found there. And the new Level 1 will be reapplied to each of those that are not atomic.

The differences between Tables 3.2 and 3.3 indicated the key to understanding systems through analysis by deconstruction. Whereas in the original Level − 1 we started with unmodeled sources and sinks, at Level 1 we designate the components of the original Level 0 as subsystems and (possibly) atomic processes (those that need no further deconstruction). Systems are composed of subsystems, some of which are already known (e.g., atoms are already well understood as systems). This will be of help in deciding how to proceed in analysis in Chaps. 5 and 6.

3.4.4 Upper Ontology

What we have accomplished at this point is the construction of an "upper" ontology or the base concepts that are used to derive all other concepts of that which exists in the real world. The framework we employed has provided us with a basic vocabulary of highest-order categories of those things that exist. The work ahead, then, is to construct the "lower" ontology for systemness, e.g., identifying specific categories of things that have specific kinds of causal effect in the world.

Table 3.2 Examples of that which has existence (ontological status) at Level − 1 in the framework

Existence of	
Objects (Substance)	System of Interest, Sources, Sinks, Channels, Material, Energy, Messages
Relations	Source-Of, Sink-For, Relative Position in Space, Relative Activity in Time
Actions (Work)	Applies Force, Moves, Accepts-From, Exports-To, Reacts-To

Table 3.3 Examples of that which has existence (ontological status) at Level 1 in the framework

Existence of	
Objects	Subsystems, Components, Atomic Processes, Channels, Material, Energy, Messages, Stocks/Reservoirs
Relations	Source-Of, Sink-For, Relative Position in Space, Relative Activity in Time
Actions (work)	Applies Force, Moves, Accepts-From, Exports-To, Reacts-To

Note that basically the same ontological items are repeated at this level. Sources and Sinks are replaced by subsystems and components

3.4.5 *System Domain Ontology*

By the contention that all things are systems we can begin to recognize specific things that perform different, yet still general, kinds of functions in complex systems. For example, we can differentiate INTERACTIONS based on the idea of a flow of influence and further differentiate those flows based on the real-world attributes (BEHAVIOR) of energy, matter, and messages. These latter are attributes of interactions at every level of organization in the Universe, so are not just attributes of disciplinary domains, but of all systems. In the former arenas we will make further distinctions, for example of different kinds of energies, e.g., electricity versus gravitational potential. In the next section we develop the system ontology.

3.4.6 *Discipline Domain Ontologies*

Every domain of knowledge has its own peculiar identifications for things that are systemic. This is the key to understanding how to apply system science and, specifically, systems analysis to the understanding of domain-specific systems. Biology has its own terminology, but those terms can ultimately be derived from systems terms (the systems ontology). The same is the case for all the other natural and social sciences. We contend it is also the case for the humanities, though this is a bit more difficult to show.[73]

3.5 The System Ontology

Thus, we will now develop the ontology of systems and derive a terminology that is applicable to any kind of system at any level of complexity. This ontology will be the basis for the development of the lexicon of a system language (SL), to be developed in the next chapter.

3.5.1 *Background Categories*

These categories are not substances as such, but rather used to establish relative locations and relative times.

[73] Consider, for example, a painting as a model of perceptions had by the artist. Even abstract art is representational. Moreover, those representations "speak" to the observer of the work. If this were not the case then there would be little or no value in art as a means for human minds to connect.

3.5.1.1 Space

The existence of SPACE is problematic. In Newtonian physics, space is an absolute, a kind of lattice of locations. In Einsteinian, (General Relativity) space is defined by the distance relations between objects. In this book we will use the notion of LOCATION-RELATIVE-TO, having identified a frame of reference. In most cases this will be a Cartesian frame with the center chosen as the center of mass of a system. This is strictly for measuring relative distances in some appropriate metric/scale, which will be identified on a case-by-case basis.

Systems as real things in the Universe have extent, that is, a volume occupied in space. In the next chapter, when we turn to the discussion of real systems and their boundaries, we will make more explicit what this means. In addition to the extent of a system, there is the issue of space between systems, especially with respect to the "distance" that substances must travel between systems. As we will see, when multiple systems communicate and transfer substances between one another over an extended time we are really talking about a supra-system, a system of systems that constitute the supra-system being, itself, a system and thus, the spatial extent of all of these subsystems plus the distances between them form the extent of the supra-system.

3.5.1.2 Time

Time could also present special problems. Time scales vary with the level of organization. Things go more quickly in smaller-scale subsystems; time constants are smaller relative to the larger supra-system. The shortest time interval for the lowest subsystem level is taken as the basic clock tick for the system as a whole. Ideally, time constants that are integer multiples of this smallest time constant (Δt in discrete time) can be applied to subsystems higher in the hierarchy. This is advisable when one desires to run simulations of a higher level of organization as an abstraction of the lower levels. This will be discussed in Chap. 10.

Systems exist over some time scale. As with the extent of a system in space, they have extent in time as well. Indeed, the very notion of a system having duration means that during its lifetime it is recognized as an organized whole thing. But systems do not last forever, with the possible exception of protons and sub-nucleon particles.[74] Indeed, the Universe as a whole may not be eternal. The evidence is strong that it will either evaporate or ultimately contract back into a singularity (perhaps exploding in another Big Bang).

All systems above protons have a lifespan. Though it is difficult to make any causal claims at this time, there is reason to believe that lifespans, when lived in full

[74] Protons may or may not be stable indefinitely. Some indications are that protons do decay with a half-life of about 10^{32} years. As of this writing no one has detected a verifiable proton decay event. See the HyperPhysics article: http://hyperphysics.phy-astr.gsu.edu/hbase/Particles/proton.html for background (accessed April 5, 2020).

measure, are related to the workings of entropy. Systems eventually simply fall apart. On the other hand, more likely scenarios involve systems being disrupted by other systems as when galaxies collide or a prey animal is eaten by a predator.

3.5.1.3 Space-Time

As is now understood, space, with its three dimensions, and time, as a fourth dimension, constitute a singular framework for placing entities in relation to one another. In this book we will not dwell on the implications of a system's position in space-time per se since that will always be relativistic with respect to other systems and will be particular to the kinds of systems with which we will be concerned. So, for our treatment, we will assume that systems have space-time extent, a location and time during which the system may be observed and recorded (as covered in Chap. 6). It is simply noted that this aspect of systemness, that it does occupy space for an amount of time, is a given.

3.5.2 Framework Elements

3.5.2.1 ENVIRONMENT

Every SOI is embedded in a larger system, the supra-system. Generally, the SOI will interact directly with only a few other subsystems of the same level of organization status and/or atomic components (see below). The environment contains all of the sources of inputs (matter, energy, and messages) and the sinks for outputs (same substances). It also contains all of the channels (or fields) through which flows occur. The entities of the environment connect directly with the boundary of the SOI.

3.5.2.2 SYSTEM: The Root Category

The SOI is the entity that we are most interested in understanding. It can be viewed as the root of a tree (inverted) with branches to nodes representing the subsystems at Level 1. Figure 3.16 shows the basic tree structure of a deconstructed system.

The environment is shown as another upward directed tree with the sources and sinks as the nodes (Level − 1). The complexity of the system is indexed by the number of levels of subtrees. Each subsystem of the SOI, at Level 1, is, itself, the root of a tree. Only the tree rooted at S3 is shown in the figure. At Level 2, the sub-subsystems of S3 are shown. Similarly, S3.4 constitutes a sub-subtree root for

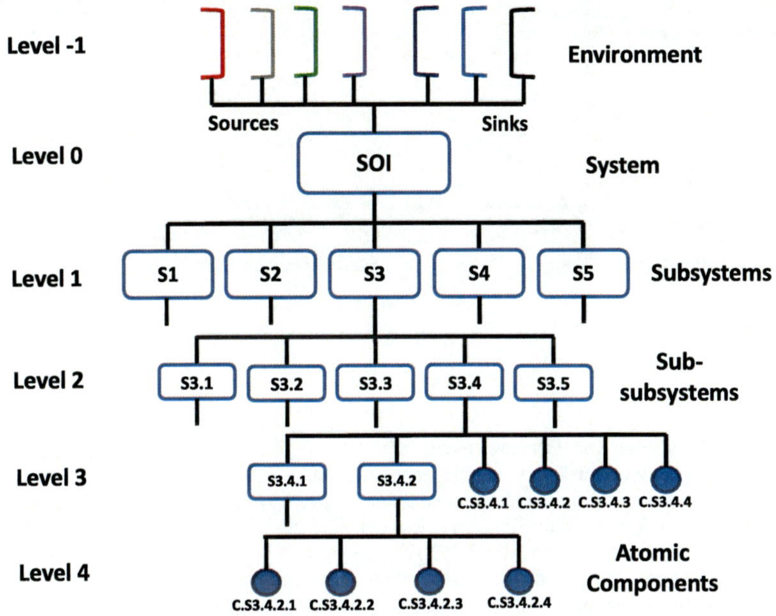

Fig. 3.16 The Environment-SOI-Subsystems hierarchy represented as a tree structure shows the levels of organization in a complex system. The labels in the various nodes below the root node (SOI) are used to distinguish each subsystem or atomic component and preserve the ordering of the tree. Branch lines that terminate are presumed to have children but are not shown. Note that Level 3 shows a mix of both subsystems and atomic components. This tree is unbalanced in that node S3.4.2 is not a leaf node and has additional children atomic components

children at Level 3. Level 3 in this figure contains some atomic components (see explanation below) mixed with additional children sub-sub-subsystems.[75]

The relation between the depth of a hierarchical tree and the concept of system complexity is explained more completely in Mobus and Kalton (2015), Chap. 5).

3.5.2.2.1 PROCESS

A process is a system that performs a transformation on its inputs to produce outputs that are different in form, quantity, or organization. The atomic work processes of Figs. 3.17 and 3.18 are archetypes. Systems are generally speaking compositions of atomic work processes and so accomplish more complex processes.

[75] The extending of prefix "sub" with hyphens for each level down in a tree is used here only for example purposes. In the future we will drop this kind of notation with the understanding that "subsystem" means a child node of some sub-root in a tree like this. Where necessary, the level of the subsystem will be given using the kind of dotted integer notation shown in the figure to make it explicit.

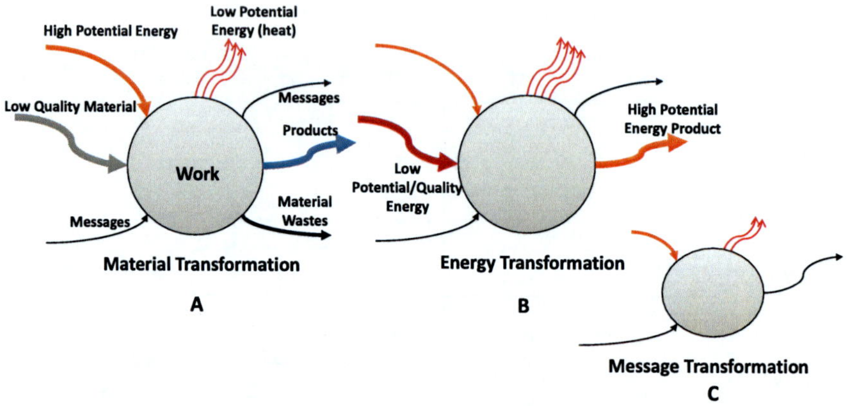

Fig. 3.17 All processes transform low-quality material, energy, or messages into high-quality versions of the same. Work processes require the input of high-potential energy to drive the work itself. In doing work, according to the Second Law of Thermodynamics, some of the energy does not accomplish work, but is transformed to a low-potential form—waste heat. The work produces a "product" or high-quality material, or high-potential energy, or messages of greater use (higher information content) downstream. Material transformations involve some material waste products while energy transformations produce a greater proportion of waste heat

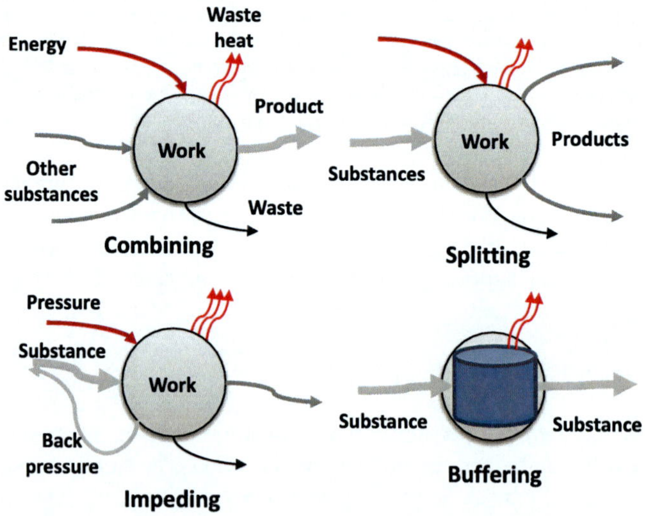

Fig. 3.18 Subsystems can be treated as atomic components when their work processes are simple and already understood

Every real system imports material and/or messages and energy. It does work on the imported matter, energy, or messages, converting some portion of the energy into a lower form—waste heat. The work performed will depend on the details of the system (or subsystem) but will involve some form of transformation or translation (moving). In general, the imported material, energy, or messages will be transformed from a high-entropy form into a lower-entropy form, having balanced the entropy equation by virtue of the lost heat. Messages are processed for their information content. This may involve work done in receiving, transducing, and interpreting the message for its information content. Further work done by an "information processor" would include conversion of the information to knowledge for storage in the structure of the system. It may also result in encoding the information for retransmission.

Material transformations involve processes that reduce the entropy of the material, such as refining or shaping matter for a new use. The work could be mechanical, electrical, thermal (high temperature), or chemical. In all cases energy is consumed to change the material from its "raw" form into some form that is going to be more useful to clients or to the system itself (e.g., repairing structures internally). In most cases there will be some form of transport of the material from one location to another. One major form of transport is exporting products or wastes through the boundary.

Energy transformations involve work done on a low-grade (low potential or quality) energy form to boost its potential or quality (with some loss as heat) so that the energy flow becomes, product-like, available to other work processes. Examples include refining oil to produce gasoline, or transforming water flow into electricity.

In all three cases (Fig. 3.17), there are two types of high-quality energy inputs. One type is the operations energy (as shown in the figure), which must flow into the process in real time. The second type is the energy that was consumed constructing the energy work process (equipment) originally, plus periodic energy consumption for maintenance of the equipment. The former is, essentially, the operating costs, in high-potential energy. The second is, essentially, the investment of high-potential energy to make it possible to do the on-going work.

As per Fig. 3.16, showing the subsystem hierarchy, each one is also a sub-process that is responsible for a smaller portion of the whole system process. Atomic components, at levels of organization above atomic nuclei, are also processes but perform the smallest unit of work at that level. Components, including atomic components, are discussed below.

Non-atomic processes are usually mixtures of various work processes, having multiple material, or energy, or message inputs and multiple outputs. Atomic processes are defined below.

3.5.2.2.2 OBJECT

Every real system has physical extent in an appropriate frame of reference. It takes up space and has duration. No two objects may occupy the same space at the same time.

There is, however, a situation in which some systems can "seem" to occupy the same space. This occurs when the boundary of an object is fuzzy, that is, the boundary may soften or morph for some period of time. One of the best, and most dramatic, examples of this is when a single person (or a group of individuals) moves from one system in which they participate to another. For example, an individual goes to work in an office in the morning—they become a component in the business system. All of the individuals who arrive around the same time, in essence, expand the boundary of the business system. At the same time that they become part of the business system, they have ceased to be directly participating in their own family systems (at least in body if not mind). The same argument applies to a person's involvement in various out-of-the-home activities. When someone is shopping, they are participating in the "consumer" part of the economic system. When they are working (at the office) they are participating in the labor part of the economic system. An individual is a multi-capacity component of many different systems. What makes it even more complicated is that a single individual's mind can be involved in many different sub-processes, effectively, through subconscious mental activities. A father at work can be worried about a sick child at home while working at the office. However, no human can effectively actually be engaged in multiple activities at the same time (no matter what teenagers may think about their abilities to "multitask").

The fact that boundaries may morph, or be fuzzy, in this sense they are no less real. But the issue of time needs to be accounted for. In the example of people going to work, each individual can be thought of as part of a flow (of an agent component—see below) between delineated subsystems. That is, for example, people going to work are outputs from the "human" subsystem flowing to input to the economic subsystem.

3.5.2.2.3 ENTITY

A system, through its behaviors, will affect other entities in its environment (level of organization in the hierarchy). This is what we call "entity-hood." In this perspective, entities affect each other, which is the same as saying subsystems interact with one another and, by virtue of their output/input flows, affect each other's behaviors. All entities are actors regardless of how simple they may be. More complex entities have more latitude in making decisions and more impact on other entities by virtue of their degree of agency.

3.5.2.2.3.1 Actor

In a very real sense even a dead body is an actor. The very act of decay sends various chemical compounds into the environment and affect other entities. Every atom is an actor. Every system, by virtue of being a process with inputs and outputs, is also an actor.

3.5.2.2.3.2 Agent

A special pattern of an information processing subsystem is a decision maker. The decision maker takes in information from the environment, processes this information in the context of a decision model, and generates a "decision." The latter is output that is coupled with an effector subsystem to cause behavior to occur. The nature of agents and agency will be covered in Chap. 12.

BEHAVIOR

Every system interacts with its environment in some fashion. What it does in reaction to forces or chemical interactions constitutes its behavior. This is the changes in space or composition that a system undergoes over time in response to those interactions. Even a pebble being buffeted in a mountain stream can be said to have some kind of behavior. The degree to which its internal constituent minerals retain their bonds and the overall shape of the pebble even in the face of the buffeting can be counted as "doing something" even though we prefer to think of it as being inert. Behavior in this context may be just the dynamics of a solid object reacting to physical forces.

Things get more interesting for systems such as the Earth's mantle, which continuously undergoes convective streaming and pushes continental plates around the crust. This too is behavior, but motivated by energy flows internal to the system.

Living systems are similarly motivated from internal energy flows that are even more highly constrained resulting in complex behaviors that are goal directed.

BOUNDARY

A boundary demarks the difference between the inside of a system and the outside, or environment. A boundary exists whenever components of a system are restrained in time and space to a particular location. Materials, energies, and messages may penetrate the boundary, but the components that interact with one another to make the system what it is, are kept together so that they will consistently interact.

Boundaries need not be separate objects or components though they certainly can be. For example, a cell membrane is a distinct and special substance that surrounds the cell interior allowing the influx of molecules that are (generally) good for the cell's metabolism, and keeping out foreign molecules that would interfere with normal operations. And the membrane facilitates the expulsion of by-products that could be harmful to the normal operations of the insides.

Other kinds of boundaries may simply be implied by operational forces and interrelations between components that are much stronger than those between internal components and others in the environment. For example, the congregation of a church does not need a specific building (walls and roof) to maintain it as a system. The fellowship and bonds of shared beliefs are psychological forces that keep the congregation structurally and functionally unified. Of course, in this last example, the boundary is porous in that members may come and go over time. Moreover, it is fuzzy in that different members may have slightly different degrees of membership, say, based on their particular understanding of a bit of dogma.

Boundaries may or may not be solid, easily identifiable enclosures. However, everything that we could call a system has a boundary that makes it capable of maintaining system-hood over time.

3.5.2.2.4 COMPONENT

Every object elements within the system boundary, including the interfaces, channels, stocks, sensors, and regulators, are components of the system. At higher levels of organization in the SOI, components will be found to be complex objects themselves, or, in other words, subsystems that will require further white box analysis (see Eq. 4.1 in the next chapter).

Atomic components are those subsystems that need no further deconstruction in order to understand the SOI internals. What makes them atomic is that they involve a minimum of inputs and outputs and their transformative work is easily handled as an opaque box. Four simple work processes are shown in Fig. 3.18. The word substance, as used in these figures, can mean any of material, energy, or messages. However, the forms operating on material inputs are shown. These all conform to the principle that all systems are processes that involve work being done, requiring energy flow, and transform the shape, purity, rate of flow, or dynamics of materials. Simple work includes: combining two substances into a more complex substance (a product) with the loss of some heat and waste substance, splitting a single substance flow into two (or more) products with some loss of energy and substance,[76] impeding a flow or slowing the rate of flow with a consequent back-pressure, and buffering a flow, used to smooth out the flow volumes over time.

Atomic components are most generally found at the lowest level of organization in the hierarchy. However, at lower levels it is possible to find mixes of atomic and complex components, the latter being subsystems that will still need to be deconstructed. The hierarchy tree, in such a case, is unbalanced.

Figure 3.19 shows some additional work processes that are atomic. Copying takes a patterned input substance and an un-patterned one, outputting the original input (think of it as a template) and a copy of the pattern in the second output (plus some waste from imprinting the pattern). Propelling is work done to push a substance against a gradient, like a pump pushing water through a pipe. Sensing is somewhat similar in that it responds to a modulated force or energy flow, where the modulation is a kind of pattern. It outputs a modulated energy flow, usually very low power, which encodes the modulation (or variation over time) in the applied force. Amplification is related. It adds energy (power) to a weak but modulated energy input to produce a "copied" modulated high-powered energy flow output. A

[76] Note that in this scheme a product is any configuration that is more usable by downstream processes than the input substance would have been.

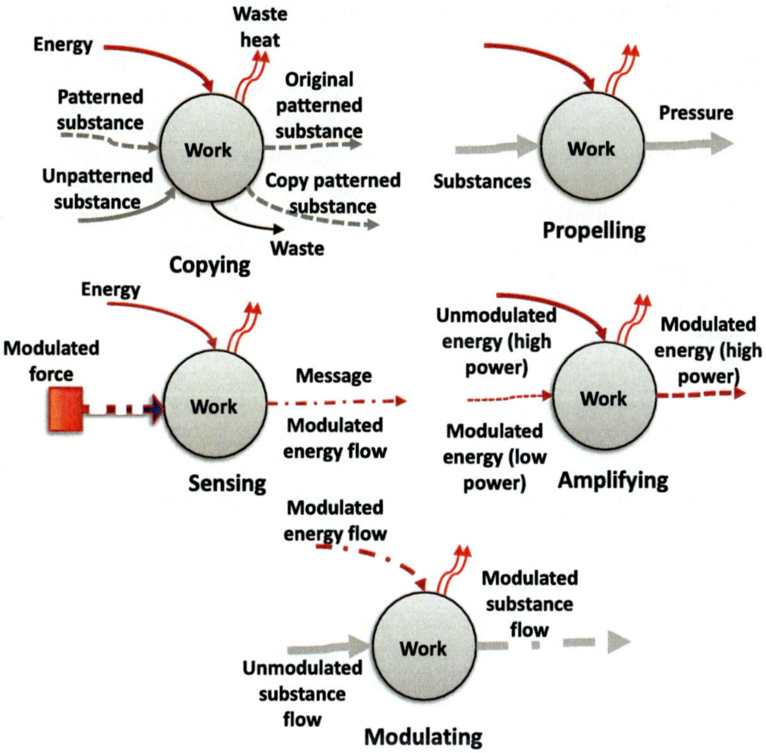

Fig. 3.19 Additional atomic processes are shown

modulator is in this same category in applying a modulated energy flow (signal) to a substance flow to produce a modulated output substance flow. Note that when the input substance is a force or energy flow, we have the sensor or amplifier effect.

The term "component", however, can apply to a subsystem within a system. As above we reserve the term "atomic component" to specify a leaf node in the deconstruction tree. Otherwise a component should be viewed as having internal structure that needs further deconstruction to elucidate its structures and behaviors. This is the basis, to be developed in the next chapter, for the recursive structure of systems. Component definitions apply at Level + 1 in our framework. But once all components in a given system are known, that system becomes the new Level – 1 insofar as further analysis of the components is concerned. Each component, in turn, becomes the system of interest where its environment includes all of the other former components in its supra-system (the old Level 0). The net result of this is the realization of a tree structure as shown in Fig. 3.8.

3.5.2.2.4.1 Personalities

Components of a system express different potentials for interactions with other components, what we have also called potential affordances. In other words, components have "kinds." They are differentiated by the fact that all components of systems are subsystems having internal structures that produce unique (to the kind) behaviors and interaction potentials. For example, elemental atoms have differing kinds of bonding potentials due to the effects of the Pauli Exclusion Principle and the atomic weight of the nucleus (leading to different valence shell levels). Amazingly, different people have different personalities that have varying amounts of attractiveness or repulsion to other personalities. Both kinds of personalities lead to interactions between entities!

3.5.2.2.5 INTERACTION

There are various kinds of interactions between components and between components and entities outside the original Level 0 (i.e., in the original Level − 1). The interactions between internal components are generally characterized as flows of substance, either channelized (through channels) or broadcast as fields. Material flows may be through constrained channels, like pipes, or through circulations such as convective cycles. Energy, similarly, may be delivered through channels or through general fields. Either way the laws of conservation and degradation (or diffusion) have to be observed. Interactions occur because the component receiving the flow is affected by it. The existence of complex systems depends on the fact that one component's output is a resource input to another component allowing the latter to fulfill its purpose, i.e., perform its intended work.

There are several aspects associated with interactions (see the higher-order aspects given above). These are related to several principles listed in Chap. 2. They have ontological status from the standpoint of being found in every system.

3.5.2.2.5.1 Networks (Principle 3)

The interactions between components in a system at a particular level of organization form a network, generally a flow network for different substances and influences. Relational networks are also part of the interactions, i.e., some components (nodes) in the system may have key roles in the system relative to other components. For example, upstream components in a supply chain can influence the behavior of downstream components by virtue of the quality of their products. Every such network can be modeled (Principle 9) as an abstract graph where the components are node (edges) and the links (edges) are the relational connections (Mobus and Kalton 2015, Chap. 4).

The components of every system (every kind of system) form networks of relations.

3.5.2.2.5.2 Complexity (Principle 5)

As described in Mobus and Kalton (2015), Chap. 5), complexity is a measure related to the degree to which a system has both a multiplicity of components (subsystems) at any given level of organization (below) and the number of such levels forming a hierarchy. Every system may be characterized by this measure. Simple systems (e.g., elemental atoms) have very low measures of complexity relative to a living cell, which contains all of the complexity of elemental atoms, their various molecular compositions, and their interactions in complex molecular interactions, e.g., enzymatic reactions.

Complexity might be better characterized as a property rather than a "thing." However, as we will see in coming chapters, this property is fundamental to systemness and thus we claim has ontological status as something that exists.

3.5.2.2.5.3 Hierarchies (Principle 2)

As with the related aspect, *Complexity*, the hierarchical structure of systems is a reality owing to the nature of the levels of organization that emerges from auto-organization (Principle 6). Hierarchies of systems and subsystems, etc. are also called "holarchies," a term suggested by Arthur Koestler to describe how a whole system is comprised of subsystems that are themselves essentially each one wholes in their own rights. No nonhierarchical system is known. Even so-called "flat" organizations turn out to have implicit hierarchical (power) structures if not hierarchical work processes. No mailroom clerk gets to decide if and when they will distribute the mail!

3.5.2.2.5.4 Messages (Principle 7)

Messages are specialized versions of energy (and often time material) flows. They are characterized by using very little energy in their transmission from a source to a sink. They are generally pushed out (actively transmitted) and received passively. Messages are flows that are modulated in a way that encodes symbols (e.g., frequency, amplitude, digital) that are a priori recognized by the receiver. They take little energy to propagate signals (inject into a channel or broadcast) and are most often amplified at the receiving end so as to have effect on the receiving system (i.e., result in work being done to modify the structure of the receiver).

Messages are used by active transmitter systems to influence receivers, to inform them and generally to control future behavior by receivers. However, receivers might also be receptive to naturally occurring signals, such as reflected light or naturally generated sounds or emanated molecules (taste and smell). Living systems have evolved to use ambient messages to observe and perceive their environments. They can observe entities that are not themselves actively sending messages but whose behaviors nevertheless transmit information.

3.6 Putting It All Together

3.6.1 Terminology

The point of an ontological study is to determine the existence of objects, their relations with one another, their effects on one another, and their overall structure in time and space. The second point is to assign terms (symbolic names) to these objects, relations, and actions, and to determine legitimate connections between the terms, in other words, the syntax of a language. That will be the task tackled in the next chapter where we elaborate an initial lexicon and syntax of the system language.

This work requires establishing a hierarchy of abstractions, such as the idea that there is a universal concept of a movement or "flow" of something (material, energy, or message) that sits atop a Space-Time coordinate structure (a Newtonian version) or a set of distinctive kinds of channels (a relativistic version). A flow of material is distinguished from a flow of energy.

3.6.1.1 The Natural Language Version of Terms

We will be expressing our ontological terms in English. The choice should be completely arbitrary but was made for practical reasons. Most scientists today, regardless of their native language, speak English in conferences. Most journals in the Western world are in English. We will leave it to others proficient in other native languages to translate these terms to their language as needed. In the next chapter we introduce the idea that every human on this planet has an internal inherent mental language that is the same for everyone regardless of their public (spoken) language. The terms in this mental language (lovingly called "mentalese") are at the core of natural public language, i.e., every word in every language has its root meaning buried in the subconscious use of mentalese. Thus, if this conjecture is correct (see the ideas of Natural Semantic Meta-language above and to be further explored in the next chapter), then the selection of a specific public language should be immaterial since every term in any one language is directly translatable into any other public language.[77]

Thus, the terms in our ontology of system will attempt to be English words that represent the most abstract conceptualization of the system term relating to the systemese concept.

The lexicon of a language of system to be developed in Chap. 4 will be given in English. In chapters following Chap. 6, we will see several example systems using the derived English terms specific to the domain of the systems of interest. The

[77] Not to imply that word-for-word translation between any two languages is trivial. Sometimes it takes a phrase in one language to convey the semantics of a single term in another language. However, here we are talking about systemness concepts, which are innate in all human minds, so the translation of those concepts between languages should be fairly straightforward.

language of systems may be viewed as a kind of Rosetta stone, providing a means of translating terms in one domain to terms in another domain; this may help provide a basis for transdisciplinarity.

Terms used in quantification and mathematical descriptions of functions, for example, are, at this point, relatively universal in the realm of discourse. Mathematics is its own language, as is logic. Visual images, e.g., icons representing the various terms, may or may not be culturally neutral. This is an area that requires more research. For example, the use of a barrel icon, for the term "stock" may not convey the meaning in all cultures. Another term, "store," is used by Odum (2007) to refer to the same concept. In some natural languages the translation of store might be more appropriate than stock. We will leave it to linguistic experts to make that determination. We do not think anything vital is going to be lost in this English text by using stock.[78]

3.6.1.2 A Basic Example of the English Ontology of Systems

Finally, we list here a set of terms along with their hierarchical relations derived from using the framework described in Sect. 3.4. This list is not meant to be exhaustive, but it includes some of what we believe to be the most useful concepts in understanding concrete complex systems.

- Environment
  - Entities

    Source
    Sink
    Disturbance
  - Flows [MOVES-FROM-TO]

    Input

    - Material
    - Energy

[78] There is some difficulty in talking about this particular term since there needs to be a distinction made between the substance of the stock and the container in which it is held. The stock value is a measure of the amount of the substance, e.g., volume or weight for material, or charge in a capacitor for electrons. The confusion comes from the fact that a holding condition (e.g., a boundary) involves some kind of container having a capacity measure. Further complicating the distinction is the fact that some such containers are fuzzy. For example, a herd of cattle while feeding in a free range is not physically contained in the conventional sense. When they are herded into a corral they are in a container. In both cases we can talk about the stock of cattle in terms of the number of heads. In the free-range condition, the cattle are still "contained" but now by the geographical distribution of edible grasses. And too, many ranchers do erect fences to keep their cattle from getting too far away. The author is deeply indebted to Hillary Sillito for pointing out this important distinction.

  - Message

  Output

  - Material
  - Energy
  - Message

- System
  - Boundary

    Type (e.g., porosity, fuzziness)
    Interfaces

    - Receiver (of inputs)
    - Exporter (of outputs)
  - Subsystem

    Component

    - Sub-subsystem

      Object
      Entity
      Agent

    Interactions
    Relations
  - Behavior
  - Hierarchy
  - Complexity

These basic concepts of what exists will be expanded throughout the book, but will be given more definitive status in the next chapter.

3.6.2 Accounting for Everything

This chapter has presented a pass at defining what exists insofar as systemness is concerned. The use of a framework in Sect. 3.4 sought to provide a means for producing a full accounting of the attributes of systemness but as with all human endeavors, something is bound to have slipped through unnoticed. There are some "things" in our Universe that have been missed. Our framework approach, based as it is on the principles identified in Chap. 2, will surely be found wanting at some time in the future. It may need revision or even a complete overhaul. But something like it will be needed to guide the identification of all of the unique things, interrelations, and actions that do exist.

What we hope will happen is that this ontology will be used to guide the process of understanding complex systems and when it fails or finds inconsistency those who understood the use of the framework will go back to that point and rethink as needed how the framework might have failed to determine whatever that something that is missing. The danger is that researchers may be tempted to simply add onto the ontology in an ad hoc manner without really working through the framework and finding out how to fix it. This would be unfortunate because just like in the sciences, guided by the scientific method, for example, improvements in the process of discovery of new domain knowledge are made to the scientific method itself, e.g., adding double-blind experiments helps diminish biases. The framework presented in Sect. 3.4 is to system ontology what the scientific method is to normal science. If it lacks something as shown by the discovery of an ontological object that did not come out of the framework, then the framework itself needs updating and improving.

3.7 Summary of Ontology and Ontogeny

This chapter has been an extensive examination of the concept of an ontogeny of systems starting from a cosmological level and developing the systemness properties and attributes that apply universally to all "things" as we find them. All systems, we have asserted, are based on a few fundamental patterns of organization and behavior that work together to make a system a system. We have also examined the nature of the ontogenic cycle, which accounts for how systems emerge from lower levels of organization and complexity. We claim that all system components, at higher levels of organization, recapitulate the fundamental work processes depicted in Figs. 3.18 and 3.19. These represent atomic processes that are found throughout the natural and artificial world and their patterns are simply elaborations of higher-level assemblies of lower-level components.

It remains to now use the discovered concepts of things and their terms to develop a structured description of all systems at any level of organization and complexity. Then using our terminology and such a structured description, we intend to define a system language (SL) that can be used in analysis to obtain a deep understanding of any system of interest (and its environment) as well as be a basis for generating simulation and static models of those systems. This language will provide a single common way to describe all systems regardless of their specific disciplinary content.

Chapter 4
A Model of System and a Language to Represent It

Abstract Human beings have an innate capacity to recognize systems and reason about their dispositions and interactions. This capacity is not explicitly experienced in conscious awareness but is embodied in a subconscious language of thought, which we assert is "systemese." That is the brain is already aware of systemness and how to think subconsciously about systems. Systems science attempts to make this implicit language explicit, a language of systems that can be used in the process of understanding complex systems. To accomplish this goal, we need to establish a formal framework and definition of system and from this, elaborate the lexicon, syntax, and semantics for expressing systemness in the context of real-world systems. We propose a mathematically based definition of system based on the ontological framework developed in the last chapter. Finally, we use the semantics, syntax, and lexical elements from the ontology and the model of the human systems to propose a formal (but extensible) language of system.

4.1 In This Chapter

In the last chapter, we developed a set of terms and relations representing the elements of systemness. It was shown how systems come into existence in the real universe by a process of ontogenesis that started with the Big Bang and fundamental particles (matter-like entities) and forces (energy-like) mediated by information and constructing structures that encode knowledge about the history of things. That process did not stop with, for example, the creation of atoms. Indeed, we suspect it continues today so long as free energy is available to do the work needed for construction.

What we saw is that systems at any "intermediate" scale (excluding the lowest quantum level and the Universe as a whole at the extremes of scale) are comprised of subsystems precisely because the process of ontogenesis is one of the compositions. The subsystems are "components" in the construction of the more complex super-system representing a next higher level of organization.

In this chapter, we turn to the use of ontology in devising the one key tool we need in order to express all of the elements of systemness, a *language* of systems.

G. E. Mobus, *Systems Science: Theory, Analysis, Modeling, and Design*,
https://doi.org/10.1007/978-3-030-93482-8_4

This language is very special as a tool for communications between humans, and between humans and machines. That is, the language of systems possesses both the characteristics of a natural language for expressing systems and the characteristics of a formal language (both mathematics and computation) making it useful in constructing computer models for simulation of the systems we are gaining deep understanding about.

What we will consider, then, are aspects of the human linguistic capabilities, including cognition, that make it possible for people to communicate ideas about the systems of interest and the terms examined in the last chapter can be deployed for this purpose. Our interest in this aspect is more than just the fact that we can string some words together in a correct syntax and construct a thought having semantic content and pragmatic context. We will examine a radical notion regarding the source of human public language (that which we speak and hear) and a more primitive language that is subconsciously used. Linguists and philosophers of language call this the "Language of Thought" (LOT) or "mentalese." We have proposed, and will develop the concept further in this chapter, that mentalese is, in reality, *systemese* (Mobus and Anderson 2016). That is, the language we use for thought is based on systemness. This includes native or innate concepts and built-in mental computations enacting the constructs of systems and their behaviors. There is sufficient evidence coming out of cognitive science that humans are born with these innate concepts and computations and that some of them are evolutionarily quite old (Carey 2009). This makes sense. Animals with brains, including humans have evolved in a world that is full of systems that all share the elements of systemness. Those animals that evolved mechanisms for perceiving and conceiving of those systems as systems (i.e., objects, flows, and agency) would fare better in the fitness game. Systemese includes archetypal concepts of system elements and a set of rules for composition/decomposition. We will outline some of the simpler elements in this chapter. In Part III, we cover more complex archetype models used in construction of complex adaptive and evolvable systems such as *agent* and *exchange*.

The systemese hypothesis leads us to the idea that our natural languages are extensions of the fundamental elements. We have ways to apply different names to the elements based on particulars of how the system is constructed. The names we give things are grounded in a deep semantics based on their roles in the system model framework of Chap. 3. For example, a container (of a stock) can be a woven basket, a clay jar, or a skin pouch used to carry necessities. It can, of course, have many different forms and names. But the purpose is always the same—it temporarily holds a stock of something.

Once we have a clear idea of how language works to communicate ideas about systemness we turn to a more direct way to use the ontology. We will provide a formal definition of system that incorporates the elements developed in the previous chapter. This definition, framed in set and graph theoretic terms, provides a structure that applies to all systems, though some elements may be missing or "null" in simple systems. The structure then acts as a guide to conducting analysis (Chap. 6) and capturing relevant information about the system of interest in a knowledgebase

(Chap. 8). That knowledgebase becomes the source of constructing system simulation models and artifact specifications (Part IV).

The chapter then provides an example of what a formal language of system would look like. The language would use all of the human linguistic modules along with several other processing modules to describe the concept of the system, its model, in sufficient detail that ambiguity and uncertainty are minimized. The language can also be used to construct simulations of systems operating in variable environments—telling the system's story—letting the modelers test hypotheses about the system they might not otherwise be able to examine empirically.

4.2 The Communication of Ideas

Any attempt to produce a language of systems needs to be based on our understanding of human cognition and the nature of public language. Our goal is to link how people communicate ideas about the world to one another and the ability to describe the structures and functions of systems. Additionally, our goal is to formalize the language in such a way that humans can communicate with computers (and vice versa) efficiently and effectively. So, our first task is to consider how ideas are communicated.

Ideas are mental states that involve the composition of several concepts into single modules of thought. Sentences spoken in public languages express ideas. It is possible for ideas to have some non-trivial importance in the context of other thoughts (beliefs) and the environmental situation (perceived). Such ideas may become permanent or semi-permanent memories accessible in the future. They are, in effect, systems of concepts that may participate in further compositions. As such, these seemingly non-material systems are the result of the very same ontogenesis as described in the last chapter. We can say this because the reality of mentation is that it is very much involving material subsystems—neurons and networks of neurons. Concepts, beliefs, and ideas are all represented in neural networks in the brain. The details of this very physical process are still a bit fuzzy but are beginning to be elucidated. For our purposes for the present, it is enough to point out that the formation of concepts and so forth is the very same ontogenic process but operating in a cryptic medium, the brain.

Time and space do not permit a full accounting of mentation here. What follows is a brief outline of the major aspects as related to conscious thought and language. This will be followed by a deeper mental capacity for LOT which underpins the conscious thoughts and communications of them with other humans and with machines.

4.2.1 *Representations of Concepts, Beliefs, and Ideas*

The various contents of thought, concepts, beliefs, and ideas are constructed in neural representations in the brain, primarily in the neocortex, the outer rind of the cerebral cortex where it is thought that learned representations reside in complex hierarchical networks (Abdou et al. 2018; Alkon 1987; Arsenault and Buchsbaum 2015; Brodt et al. 2018; Carey 2009; Deacon 1997; Huth et al. 2012; Kraus and Cheour 2000; Mobus 1994; Scalaidhe et al. 1999; Seung 2013; Sporns 2016; Squire and Kandel 2009).[1] Representations in the neocortex are supported in learning (including modifications with ongoing experience), in maintenance, and in recall (bringing to working memory) by deeper, older brain systems.

In addition to the learned representations that are generally available to conscious thinking, there are an array of inherent, native and tacit representations that are either operative in infants or develop under genetic control during maturation (Carey 2009). These are available to a core cognition "engine" that underlies the ability of an infant to begin constructing the learned representations. An example is the simple concept of something being an "object" (as discussed in the ontological framework in the prior chapter). An object has features such as occupying a particular space at a particular time and some degree of solidity and coherence. All of these features are automatically perceived and integrated such that a singular material entity is perceived. This capacity is in-born and is evolutionarily old.

4.2.1.1 Concepts

The mind holds units of thought that correspond with things in the world and we call them concepts. Cognitive scientists and linguists have studied how human beings possess, construct, or form concepts through learning processes, and how they then use concepts to compose ideas that can be shared through language with other humans.

4.2.1.1.1 Conceptual Hierarchies

There are actually several ways in which there are hierarchies of concepts. One way is in terms of degree or amount of content. Low-level concepts are like thought primitives, small and pertaining to some detail. Take the concept of "hair," related to mammals. One can perceive or think of a single hair or of an aggregate of hairs

---

[1] The architecture of biological neural network representations is not the same as so-called "connectionist" models, artificial neural networks that rely on distributed representations, all 'synaptic' connections in a multi-layered network contribute to representation (cf., Rumelhart and McClelland 1987). This is in direct contradiction to the nature of biological neural representations, which are accomplished in hierarchical feature-percept-concept subnetworks. See, especially, Mobus and Kalton 2015, Chap. 8, Sect. 8.2.5, Biological Brain Computation.

(fur). Fur is another concept based on or using the concept of multiple single hairs. The fur of, say a dog, is one "feature" of dogness. Other features include body form, shape of head, and teeth. The concept of a dog, then, is composed of these various featural sub-concepts properly arrayed in a space-time context. The hierarchy of concepts runs from small detail features to the whole concept of, in this case, a dog.

The dog concept participates in another kind of hierarchy, one of phylogenetic ancestry of natural kinds. A dog is a canine; a canine is a mammal; a mammal is an animal. This hierarchy is a more sophisticated concept and it is not entirely clear that the human mind forms it. An alternative theory of kind relatedness is that of prototype learning, an inductive process whereby, for example, all exemplars of dogs encountered, and identified by an "authority" helps a child form a prototypical set of features and forms that constitute the "eigen-dog"[2] that is used then as a basis for pattern matching against new encounters of non-typical examples. Supposedly, in a similar vein, encounters with other types of mammals, like cats, accompanied by authoritarian exposure of the way the other forms are related to the dog form creates a higher-level "eigen-mammal," thus a hierarchy of "eigen-forms" of animate objects.

4.2.1.1.2 Innate vs. Learned Concepts

As mentioned above, human beings, and presumably many other kinds of animals, come with built-in, or innate concepts of the most important things in the world that they will need to interact with from an early point in their development. Some of these, such as the "object" concept mentioned above, are already at work in very young infants. There is evidence that the concepts of "causality," "agent," "agency," and "intentionality" are at work at an early age as well (cf., Carey 2009). We come with an ability to differentiate between inanimate and animate objects. These innate concepts are what we will call "archetypes." They are not inductively constructed in the same way eigen-patterns are. They represent basic models of important things in the world upon which the developing child can begin the process of learning particulars and sorting into kinds.

What we don't come with is specific concepts of particular objects and agents. These have to be constructed through experience with the world. It seems newborn babies, for example, have the ability to locate two oval shapes in their near visual field upon which they immediately fixate. That is, they attend to their mother's face. Other features of faces are probably also innately represented but in the form of "sLOTs to be filled in." A nose and mouth have a pre-ordained relation with the eyes and the infant rapidly captures their mother's facial features filling in those sLOTs. All of which is highly motivated by the fact that the mother is the source of sustenance and comfort.

---

[2] An eigen-image is an image that, in essence, averages the most relevant features needed to identify an example image as being of that type. It draws on large sets of the images selecting the relevant features through a somewhat complicated mathematical procedure.

Starting with archetype concepts, which we also call "mental models," every human being must construct increasingly complex concepts of the things they encounter in the world. Then sometime after their first year of encounters with objects and agents, they begin to form another kind of concept, an abstract representation of the "name of things and their interaction relations." They begin the process of acquiring language. They learn words as placeholders for the more elaborate image concepts representing the actual features of the actual things, or at least those features they had, at that point, been able to capture. They learn words attached to actions that objects and actors undergo, the verbs.

And they learn the rules for composing these concepts into ideas, the syntax of their native language.

4.2.1.2 Ideas

Sentences that we create in the syntax of our native language convey ideas or concepts that relate things and actions in the world. These are micro-narratives of what happens, or in the case of human imagination, what could happen. There are libraries full of works on linguistics so we will not dwell on the details of communication of ideas here, other than to recount the fact that human ideation, as reflected in spoken/heard language is capable of nested recurrent structures. Sentences such as "Bob believes Jane believes he likes her," express how one person can possess a model/concept of another person's mind, called "theory of mind," or "folk psychology" in the cognitive science literature (recall Principle #9 in Chap. 2). That sentence also reveals another important aspect of human cognition. The use of the word "believes" reflects on the fact that people are not just conscious (of things in the world) but of their own thoughts (recall Principle #10). They are "self-conscious."

4.2.1.3 Beliefs

We will not have a great deal to say about beliefs (as opposed to concepts) other than to recognize that what one believes about the nature of a conceptual object or agent is, or can be, quite idiosyncratic, path dependent on the sequences of experiences one has with the subjects of concepts. For example, if one had a bad experience with a vicious dog once in childhood, they may harbor a suspicion for all dogs (and a fear of dogs in general simmering in the background). They would "believe" that dogs are potentially dangerous. One would need to have, nevertheless, a concept of vicious (threatening) behavior, which may also be innate. So, beliefs are one way to connect concepts not necessarily for the purpose of communicating an idea but as a way to form context around concepts and ideas. This, along with perceived current conditions surrounding a communication, forms what we call the pragmatics of language and thought.

4.2.2 *Names of Things—Abstraction and Language*

Names of things, relations, and actions are special abstract concepts. They are formed in auditory processing space and are composed of phonemes produced by auditory feature detectors. Words, in general, are heard and learned long before they become mapped to the premotor cortex for speech production. The neural subnetworks encoding the words must also be associated with the neural symbols actually representing the concepts. The latter are considered iconic, or having "shapes" that are directly related to features of the thing in the world, or action of things in the world. Thus, there are two representations related to the things in the world, an iconic image of the thing and an abstract symbol that is encoded in auditory/aural space. The latter has no necessary iconicity with respect to the thing in the world. It is based on the particular combinations of phonemes used to name the thing in whatever the native language happens to be. This is the reason that different spoken/heard (and signed) languages can exist in the first place. They can all have different names (different words) for the same object, say, but the semantics are determined by the iconic image representation which is invariant among languages. And this is why it is possible to learn to speak and hear foreign languages because the semantics of different sounding words is fundamentally the same. This goes for written and read language as well.

Words are abstract and mostly arbitrary representations in the brain. But they are associated, through learning, with the actual iconic image of the concept representation in memory. Humans can create other abstract symbols, letters to represent phonemes (or combinations thereof) and written sequences of letters to represent the words. So, visual processing is linked with auditory/aural processing. Numerical processing (which does not appear to be innate, but has antecedent processes that address concepts like "more" or "several") is similar to linguistic processing in using external symbols related to words, for example, the number word sequence, and abstract written symbols, the numbers, 1, 2, 3, etc. Figure 4.1 shows the various modules involved in communications of concepts and ideas. These modules intercommunicate as described above through a low-level language of icon and abstraction representations and syntax. Low level means subconscious and "primitive." Each module is responsible for converting the semantics of this low-level LOT into high-level public language representations—words and visual symbols.

4.2.3 *Language of Thought and Innate Concepts*

Philosopher of mind, Jerry Fodor (1975) advanced the notion that ideas and complex concepts are mediated by a lower-level set of more primitive concepts and a specific syntax that when operating creates those higher-level thoughts, the ones in conscious thinking. Other linguists and philosophers have endorsed that notion or something similar (Pinker 2007a, b). Given that representations in neural tissue

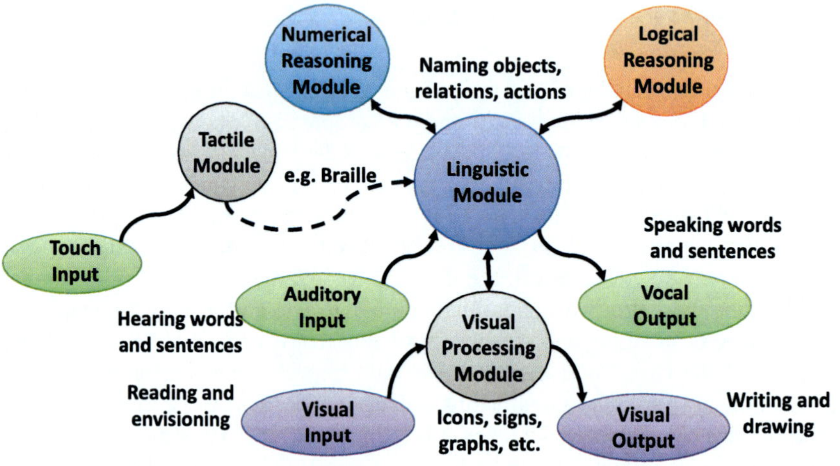

Fig. 4.1 Various processing modules communicate between each other coordinating external communications

have symbol-like qualities, are organized synchronous firing of specific subnetworks of neurons, then a low-level compositional computation that would organize the low-level concepts into higher-level ones according to an innate syntax makes sense.

The use of a LOT engine for production of complex concepts starts with the innate archetype concepts discussed above, even before it is used to connect those concepts as ideas (and possibly beliefs). That is, many developmental concepts, for example, *mother*, are formed before the infant even begins to grapple with spoken word understanding let alone production. The innate archetype concepts may act as scaffolding or as templates. When a particular concept of object or agent is being learned (a particular toy, person, or word) the particular features that particularize that entity are appended or combined (possibly with a "copy" of the archetype) in memory. This, too, is accomplished by the LOT as this process occurs in prelinguistic toddlers.

As originally conceived, LOT is a computational theory of mind (CMT) where computation is taken to be an algorithmic manipulation of strings of "symbols" (something Fodor seemed to have both asserted and denied!) That neural representations of concepts (things in the world) are very clearly consistent subnetworks of firing patterns constructed either from innate concepts or from learned ones, they can be construed as being symbol-like (cf., Abdou et al. 2018; Brodt et al. 2018). That there is some internal syntax for combining these symbol-like firing patterns into composite patterns in a kind of sequential stimulation, one can say that the brain does indeed compute an internal language to produce constructs that make sense, that is, have semantic content and fulfill pragmatics.

What the LOT thesis has not yet accomplished is to say exactly how a human mind becomes full of meaningful concepts and then results in public language constructions—thoughts. Neither the LOT nor the "Origin of Concepts" thesis has satisfactorily explained what Stephen Harnad (1990) has called the "Symbol Grounding Problem" or how concept representations, symbols, come to have the semantic contents connected with the things they represent in the real world. The "Systemese Hypothesis" seeks to provide such an explanation.

4.2.4 The Systemese Hypothesis

How many and what kinds of archetype models are innate for humans? There are a few that have been well studied along with their role in helping launch the learning of more elaborate and complex concepts as children mature, what Susan Carey (2009) refers to as "Quinean bootstrapping," which refers to the process Willard Van Orman Quine suggested for the learning of concepts. It is persuasively argued by Carey and others that these archetypes came into existence as "hardcoded" networks in the brain through evolution. As species evolved to deal with more complex environments and life histories, and with more complex other entities in their social and extant environments they would need more kinds of archetypes that could be composed, in the LOT process, into more complex meta-archetypes. The latter might be thought of as complex primitives. For example, above we used the newborn's propensity to learn the face of its mother starting with a very limited archetype of two ovals separated at just the right distance apart and a "placeholder" slot for something like a nose and one for a mouth. But what the infant is also doing is learning the concept of a face. It is exposed to other adults and older siblings, for example, and begins to construct an eigenface that allows it to recognize the face concept. It is also learning a more general concept of faces of other kinds of agents with eyes, noses, and mouths, but with very different configurations. Neuroscientists have now found a hierarchy of neural circuits, and even specific neurons (Scalaidhe et al. 1999) that play a very important role in a child's overall concept of an agent, that of associating a face-like structure with an intentional agent. Humans and many non-human animals have been shown to have special neural circuits for recognizing a "generalized" face regardless of species, the specific facial structure of specific species, and conspecific individuals within one's own species.[3]

Facial recognition circuitry, archetype face models, and the construction of eigenfaces as prototypes for faces provided evidence that there is thinking going on at a pre-conscious level. This is taken as evidence for confirming some version of the LOT hypothesis. But the question remains as to how many more archetypes might be needed in order that the human mind can develop to hold the numbers and kinds of concepts it typically does. And what is the nature of the syntax of this LOT?

[3] It is also the case that the great apes are good at recognizing the faces of members of other great ape species. It takes a long and intimate exposure to one another, say in the case of a primatology field station, for learning the kinds of important features in the other species' faces that differentiates individuals. But apes recognize humans and vice versa reasonably easily.

The *systemese hypothesis* is advanced to suggest answers. This hypothesis, put simply, is that the lexical elements of the language, the basic concepts, are precisely those elements of systems devised in the ontology of systems. We shall claim that, at least for human beings, the semantic elements we enumerated in Chap. 3, such as sources, sinks, and flows are represented in some primitive form in innate neural representations.

An example may help. Consider the sentence describing an action:

"John caught the ball thrown by Jim," translates into the agent system, the abstract name of which is "John," receives the ballistic[4] flow of an object, the name of which is "ball" from a source, another agent, the name of which is "Jim." The receiving part of the "John" system is an extended interface called a "hand" using a protocol called a "glove." A toddler observing this action does not necessarily know the names of the things involved, but already understands the roles of sources, systems-as-sinks, flows of substance, and even interfaces (receivers) and specific characteristics of the receiving act (protocol) because recognizers for these things are already available in the brain.

This hypothesis may be considered speculative, of course. But the existence of some kind of LOT along with an examination of a few of the known innate concepts lends strong support for the idea that LOT is system syntax and those core concepts are system archetypes.

The argument for why this hypothesis deserves more examination by cognitive scientists (and neuroscientists) is the evolutionary argument; having innate recognizers of systems elements would increase the fitness of the possessor. It would provide efficient pattern recognition machinery and efficient/effective pattern learning of newly encountered instances of unfamiliar systems that, nevertheless, resemble a given archetype. This is because, as argued in the last chapter, the Universe is a system of systems and subsystems recursively structured down to the most fundamental substance entities. It would be useful to come equipped with mechanisms for rapidly recognizing the systemness of things in the world, at least on the mid-scales of time and space that our sensory apparatus permit sensing.[5]

Our immediate interest in the systemese hypothesis is in devising a public language (using English names of things, though this is not required) of systems that is based on the ontology framework and embodies the systems structure/function syntax so that human beings can describe systems and their behaviors to other humans and to machines. Both humans and machines (computers) can form representations of the described systems and achieve a more consistent communication between systems in the world, systems in the mind, systems in the abstract, and systems in computer code.

---

[4]There are reasons to believe that ballistic-form paths need to be learned from experience but the launching event (from the source) and receiving event (at the system acting as a sink) are innately recognized (Carey 2009).

[5]And with instrumentation such as telescopes and microscopes, etc., extending the range of sensing to larger and smaller scales, respectively.

The objective of systemese is to build a model of a system that corresponds with systems in the world and to translate the internal language of thought into the public language of systems. Let's consider one example of this process. We have already considered the system role of a "stock" of substance (or data) as being a fundamental concept. A stock of something grows larger or smaller depending on the rate of inflow versus outflow of the substance. And the dynamics of this phenomenon seem completely intuitive. But we have glossed over a critical consideration. Stocks are invariably bounded; they are contained and containers have capacity attributes determined by physical dimensions.[6] You cannot force more of a non-compressible fluid, like water, into a closed container, like a boiler, than it can hold while the outlet is closed, so the inflow is stopped when it reaches capacity. So, stocks and containers have a tight coupling (with flows) that must be represented in the language of thought.

That language must include an innate representation of containment and a model of a generic or archetypical container (as mentioned above). It includes an attribute of capacity, which is a variable. Now when a child encounters a real container in the world, they do not have to wonder what on earth it is, their archetype model provides the answer—the syntax and the semantics. The brain builds a higher-level archetype using the sensory attributes of the particular container. Say a child encounters a woven basket that has nothing in it. The sensory inputs and perceptual models handle the nature of the weaving, the shape, and size. Without having to be told, the child will instinctively understand that something goes inside and can be "contained" there temporarily because the container archetype model module contains that concept innately. What the child needs to be told is what the name of this thing is. And this can be any publicly spoken word in whatever language is native to the child. What it is called in public language doesn't matter so long as everyone uses the word consistently in reference to this container.

Thus, the child learns a specific instance of an archetype object, attaches the needed perceptual attributes, and links that to a name of a type of thing—a basket in this example. What is likely going on in the brain is that a new concept cluster in the prefrontal cortex is linked to the container archetype module and the sensory and perceptual modules participating in the perception of the basket are then associatively linked to the same cluster. Multiple encounters with such a container are probably required to strengthen the links. This new cluster comes to represent the concept of a basket-container, with the archetype providing the needed bootstrapping basis for constructing the specific type archetype model.

At the same time, an auditory cortex cluster is formed to encode the heard word and that cluster is linked, by association, to the basket-container image. The child learns the word "basket" and knows what it is (a container). Moreover, the container

[6]At first glance the stock of air in the atmosphere might not seem to fit this description. But the key notion here is containment, not the physical container as a closure. Air is contained on the Earth's surface by the gravitational pull from the planet. The same is basically true for the global ocean. This kind of containment, while valid, is not perceived by ordinary human senses in the same way containment of, say, some water is contained in a cup.

archetype includes the concept of putting something in, containing for a time, and then taking something out (inflows, stocks, and outflows!).

In sum, then, systemese consists of a set of innate archetype concepts (models) that evolved to reflect the systems (and components of systems) that exist in the environments of animals with which they have to interact. These concepts come with built-in syntactical rules for combination that preserve the semantics and pragmatics of the situations in the world. All animals with brains of any complexity "think" in their version of systemese, and this is not a conscious process. Humans, on the other hand, have an extraordinarily complex environment—complex systems with which to interact—and accordingly have a much more sophisticated set of innate concepts with which to construct more complex thoughts. Moreover, humans possess a supervening concept encoding mechanism that associates specific auditory patterns (speech) with complex constructions, and these become the names given to things, relations, and actions—language.

This is how the human brain/mind builds models and constructs ideas for communications. Our objective now is to replicate this notion of a systemese, archetype concepts and innate syntax, in a formal way so that we can build a communicative language of systems. Chapter 15, which describes a new approach to systems engineering, will revisit the notion of using systemese to construct models, not in the mind but in a machine computable form, which is very much the same process going on in the mind. When a human constructs a spoken sentence in the manner described above, they are, in essence, engineering a design for a thought. We will propose, there, that systems engineering is exactly the same process—or should be. The concepts of language and modeling developed in this chapter will come full circle to provide a method for engineering complex systems. Just as the language of thought gives rise to the public language of communication in order to install a model/thought/idea in another person's head, so too, the same, but explicit, systemese will be shown to give rise to complex system designs.

This program starts with providing a formal definition of system.

4.3 A Formal Definition of System

We are now ready to develop a formal definition of a system that, along with the ontological commitments of the last chapter, will be the basis for producing a language for systems. This language will be used to guide analysis of systems, since the definition tells us what we should be looking for, and to build models of systems at various levels of abstraction.[7]

[7] There is a point we need to be clear about. What is being presented here is, itself, a concept about what a system consists of, how it is composed. It is not being represented as *the* "general theory of systems," though it might be a candidate for that title. It is based on having made the ontological commitments from Chap. 3 and following them to their "logical" conclusions. That being said, we are fairly certain that if there is a general systems theory, as posited by von Bertalanffy, for exam-

The definition is given in three complimentary forms: verbal, graphical, and mathematical. All three forms provide views of the system definition that provide access to stakeholders from different backgrounds. The mathematical definition is needed in order to create an abstract representation of the system definition that can be directly applied to creating a language of system (hereafter called SL).

4.3.1 Verbal

The lexicon of SL is taken from the primitive and derived elements in Chap. 3. All verbal descriptions use those lexical elements as the skeleton of meaningful statements. For example, in describing a subsystem that processes material inputs (with energy) to produce a product output we would simply say something like: "Process M takes in materials A and B from sources 1 and 2 along with energy E from source 3 to make product Z with waste product X going to sinks 5 and 6 respectively, at an efficiency of 68%." Additional verbal descriptions would include the rates of flow of materials A and B, energy E as well as those of product Z and waste X. The heat output can be automatically determined given these flows and the fact that the work efficiency of the process is 68%. Another statement that particularizes these would be to identify (name) the materials and products, such as: "Material A is iron." At a still deeper level, we could describe the variations in rates of all inputs and outputs due to various disruptions or things like diurnal cycles: "Material A comes in discrete pulses, one mass of delivery each 24 hours during an interval between 2:00 and 4:00 pm." A complete system description can be given in a paragraph of such statements. Upon deconstruction of the system to find its internal workings, each subsystem and internal flow would generate its own sub-paragraph.

The verbal descriptions start with the lexical elements acting as placeholders for specific items, such as material A is a placeholder for "water" in the above example. With each additional statement, the system description gets refined. As the deconstruction of the system proceeds, sub-paragraphs are added, each describing the subsystem at a lower level of detail.

Figure 4.2 shows a conceptual model of the verbal description above. We show the relations between the main conceptual elements. The sentence demonstrates

ple, then it is likely to look something like this. Many theorists have attempted to define systemness in a formal way and suggested that their definition constituted the general theory. Some of them have shared common approaches (e.g., using set theoretical language) yet after nearly six decades, no universal agreement over what exactly "system" means has emerged in the scientific literature. The reader may recall from the discussion in the Preface regarding the way we "talk" about systems science that we regard the latter as more of a meta-science than just another kind of science. Which means that systems science theories are not ordinary theories at all, but metaphysical theories. Ergo, any definition of system must simply accept some ontological (and epistemological) commitments and then get on with it. Only in successes with usage over an extended time will the veracity (or acceptance) of a definition be turned into a meta-theory, i.e., a general theory of systems.

how we can verbally describe a system owing to the semantic and syntactical relations drawn from the underlying systemese.

4.3.2 Graphical

Following the age-old dictum that "A picture is worth a thousand words," SL provides a graphical way to express systems. This has actually been done in most modeling languages such as Stella (for system dynamics), and SysML. Various kinds of diagrams can be produced to capture the system as a model. Figure 4.3 provides a graphical version of the system paragraph from above.

A graphical representation such as in Fig. 4.3 can be adorned with the names of the elements and quantitative aspects. It can be used in an animation of the simulation of the model generated from the knowledgebase.

Graphical models of systems help people grasp the essential relations in the system very quickly. In fact, the analysis support tool for using SL in analysis would include a drag-and-drop graphic user interface that would allow them to do the analysis and capture the data in a user-friendly manner.

4.3.3 A Mathematical Structure Defining a System

A formal language is based on the existence of a formal structure into which elements of the language fit (see Pragmatics section below). For example, a computer programming language is based on a formal structure involving arithmetic, logic, and conditional flow control (e.g., IF-THEN) used to describe data processing algorithms. These are realized in an actual computer architecture based on the theory of

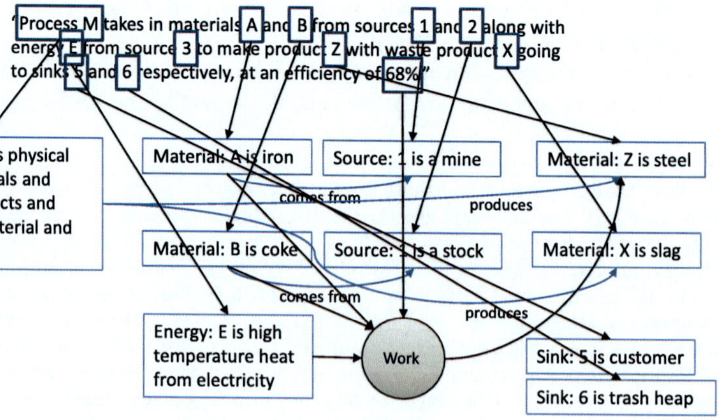

Fig. 4.2 Verbal descriptions have relations that map to the system language

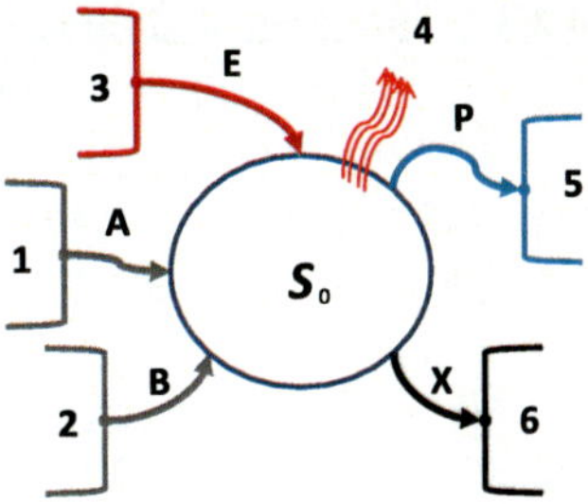

Fig. 4.3 A graphical representation can be built from the verbal description given above. S_0 is the system of interest, lettered arrows represent flows. Numbered entities represent sources or sinks. All of these will be explained presently

computation (e.g., the Turing Machine formalism and the von Neumann architecture).

The following definition is proposed as a starting point for developing a formal definition of system. The development of this approach was inspired originally by Klir (2001) although he, by his own claim, was a radical constructivist, whereas this work is inclined toward a realist interpretation. Another similar approach was that of Wymore (1967). Both of their works were, however, devoted to a purely mathematical approach to defining system and then using the definition and the math to explore the mathematical implications. However, the approach taken in this book is quite different in purpose. We start with a principle-based definition and then apply mathematics as a way to provide a *structure* for "holding" the details of a system description. We will not be *doing* math as much as *using* math.

Deriving a definition of system from the principles and the ontology, a system $\boldsymbol{S}$ is a 7-tuple:

$$S_{i,l} = C, N, G, B, T, H, \Delta t_{i,l} \tag{4.1}$$

where i and l are indexes. The index i is a subsystem index and the index l is the level of organization in the system-subsystem hierarchy.[8] Both are 0 for the initial system of interest, $\boldsymbol{S}_{0,0}$ is the designated SOI. Recall from the previous chapter that the ontological framework specified that the SOI be designated as level 0. This will make sense in this definition as we show how the system definition leads to a natural way to deconstruct the component/interaction level (+1) in the framework. Several later chapters in the book will describe the application of these equations (and the knowledgebase described in Chap. 8) to even more complex adaptive and evolvable systems (CAESs) described in Part III.

[8] As we will see later, we will actually incorporate both indexes into a single coding scheme that will give both the level in the hierarchy and the component index using a 'dotted' numbering scheme. This will be conducive to providing key values for the knowledgebase schema to be covered in Chap. 8.

4.3.3.1 Structural Skeleton

C is a set of components and their "type" along with membership functions in the event the set is fuzzy, that is, the components may have partial inclusion.

$$C_{i,l} = \left\{ \left(c_{i.1.l}, e_{i,1,l}, m_{i.1.l}\right), \left(c_{i.2.l}, e_{i,2,l}, m_{i.2.l}\right) \cdots \left(c_{i.k,l}, e_{i,k,l}, m_{i.k.l}\right), \cdots \left(c_{i.n,l}, e_{i,n,l}, m_{i.n.l}\right) \right\}_l \quad (4.2)$$

is the set of components at level l and i is the component index from the level above (if any). The components of $C_{i,l}$, for example, $(c_{i.k,l}, e_{i,k,l}, m_{i.k,l})$ use the dotted integer index that keeps track of the lineage of a component. That is, $i.k$ is the kth component belonging to the ith component in the level above (i.e., $l-1$). The $e_{i,k,l}$ are to indicate equivalence classes where applicable. An equivalence class is determined by some set of criteria which are found common among multiple components. For example, in a living cell there are numerous organelles divided into specific types such as ribosomes and mitochondria. All ribosomes share common features and are thus equivalent in those features even if not in terms of locations within the cytoplasm. An alternative approach to Eq. 4.2 would be to use integer numbers to represent, say, the average count of a particular class. The $m_{i.k,l}$ are membership functions for fuzzy sets. A component might be a member of a given system only partially or only part of the time. If the set is crisp, then all $m_{i.k,l}$ are equal to 1. Figure 4.4 shows the construction of the set $C_{0,0}$ with three components, one of which is a multiset.

Note that this is not the standard formulation for a fuzzy set. Traditional fuzzy sets are defined as a set and a membership function that applies to all possible members. That is a fuzzy set is defined as a pair, (C, m) where C is the set of components and $m{:}C \rightarrow [0,1]$ is a function mapping a member of C to the interval [0, 1] representing the degree of membership. In the above formulation, each member component has its own membership function.[9] This is called "member autonomy." It allows for members to be individually evaluated for membership based on their particular characteristics. For example, various high-weight proteins and organelles in living cells are prevented from leaving the interior of the cell by the properties of the membrane; they are always members of the cell system. On the other hand, water and various low-weight molecules and ions can transport across the membrane depending on their individual characteristics and the membrane transport mechanisms. Therefore, the high-weight molecules, etc. have a membership function that always returns 1 whereas the membership functions of other molecules depend on their particular properties and those of the membrane.[10]

---

[9] An alternative approach, one more conducive to some kinds of mathematical treatment, would be to set the membership functions up as a function space or a set of functions that map components with the same index into the system $S_{i,l}$.

[10] Under most circumstances these kinds of components, that is, water molecules, are lumped in various ways, for example, osmotic pressure or concentration for non-water molecules.

Standard fuzzy set theory, along with fuzzy logic, addresses a form of non-certitude regarding the status of objects in a set that appears at odds with the usual notion of probability. Both fuzzy set theory and probability theory map this non-certitude onto the real number-based range. There are long-standing arguments in the mathematical arena on whether fuzziness or probability is the better representation of a lack of certainty about the status of things in real systems. Kosko (1990) puts it this way:

> Fuzziness describes event *ambiguity*. It measures the degree to which an event occurs. Randomness describes the uncertainty of *event occurrence*. An event occurs or not, and you can bet on it. At issue is the nature of the occurring event: whether it itself is uncertain in any way, in particular whether it can be unambiguously distinguished from its opposite. [italics in the original]

Further he distinguishes: "Whether an event occurs is 'random.' To what degree it occurs is fuzzy." The distinction is important. Whether a component is currently in a particular system or not is subject to ambiguity of some kind. In some circumstances it might be best characterized by a probability distribution, for example, $Prob\{c \in S\}$ is given by a classical frequency-based mapping onto [0,1]. But in other cases, whether c is a member of S is not either-or but simply ambiguous. The resolution of this conundrum is going to be dependent on the system and the components. There are two complicating factors that will need to be dealt with. The first is the fact that there are situations in which the component is a member of multiple systems, seemingly simultaneously. This is clearly an ambiguity of the fuzzy kind. We will examine several cases of this condition. For example, a human being seems to be simultaneously a member of a family and a member of an organization (see Chap. 9). This is resolved in recognizing that the component can only do one task at a time and must therefore time multiplex between processes (systems). The second complication comes from trying to describe the collective behaviors of components that have this property. As with the concept of temperature or pressure in gasses, this is resolved by resorting to probabilities. We look at the aggregate likelihood of components being members of one system or another.

Figure 4.4 is a graphical representation of a system with three components in its $\boldsymbol{C}$ set. The two views shown provide different ways to think about systems and their subsystem components. The "tree view" (A) shows the structural aspects of hierarchical organizations. The "map view" emphasizes the nestedness and the topological arrangement of components. It can also show some aspects of heterogeneity among components (e.g., different colors and sizes).

Multisets are allowed. In the figure component $c_{0.2,1}$, with the tilde, represents a set of components with many instances of that type of component. For example, it could represent the water molecules in a cell. Later examples of how this is used will make it clear.

Components of a system that are not multisets or atomic components (as discussed below) may themselves be subsystems, that is, having sufficient complexity to warrant further deconstruction. That is:

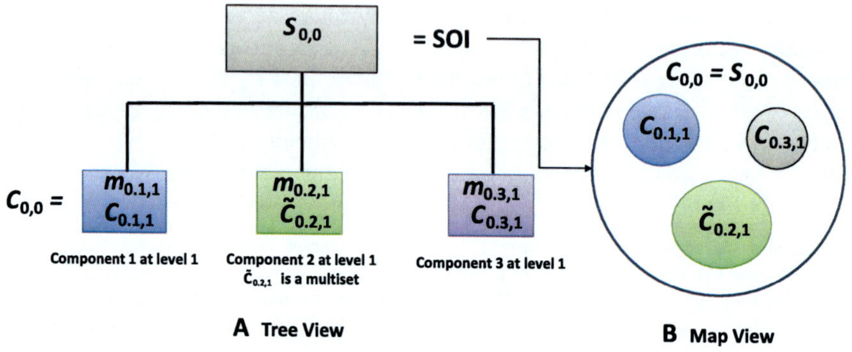

Fig. 4.4 The structure of a system that is composed of three component subsystems. $c_{0.2,1}$ is a multiset. (**a**) The tree view of the system of interest shows the compositional hierarchy down one level. (**b**) The map view shows the components in $C_{0,0}$ (which is the same as $S_{0,0}$) organized within the boundary of the SOI

$$c_{i.j,l} = \begin{cases} S_{i.j,l+1} & \text{if component is complex} \\ c_a & \text{if component is atomic} \end{cases} \tag{4.3}$$

is the *i*th component treated as a new system of interest at the $l + 1$ level in Eq. 4.3 that describes the recursive structure of system structural hierarchies. The dotted index, *i.j*, is used to maintain the global position of the subsystem component in the original SOI. Later we will drop the leading "0." index from the level 0 SOI since it is the same for all subsystems. Subsequently, the *i* index will designate the level *1* component number. As the number of levels increases (downward) the dotted number index will extend accordingly. For example, the 4th component of the 2nd component at level 2, itself being the 3rd component at level 1 would be designated 3.2.4 (0.3.2.4 truncated).

Equation 4.3 defines a tree structure rooted in the original SOI at level 0. The index *i* designates the branch of the tree, $0 \leq i \leq n$, where *n* is the branching factor. This same scheme continues down the tree where the dotted notation extends the *i* index. Of course, the *l* index is redundant in that the number of dots in the dotted index actually encodes the level in the tree. We include it for the sake of explicitness (to keep the reader from having to count dots!)

The recursion cannot go on forever, obviously. Eventually the tree must have leaf nodes. What stops it? We have identified several stopping conditions, some semi-formal, others a matter of choice by the analysts. As an example of a semi-formal stopping rule, we use the "simplest process rule." This means that a component, $c_{i.1,l}$, $l \gg 1$, needs no further deconstruction because it is doing work by either merely combining two inputs to produce a single output (has a simple transformation function), or it is splitting one input into two outputs. This applies to material, energy, and messages alike. It also requires that there are no internal decision rules beyond the transformation function. Other atomic-level work includes impeding a flow or propelling a flow. All four of these simple work processes involve the consumption of energy and the loss of energy as waste heat. A fifth simple component is a "raw" stock being used simply as a buffer and without regulating controls. These atomic processes are shown in Fig. 4.5.

Informal stopping conditions include a judgment that the component's inner workings are already well known and specified outside of the system deconstruction. For example, transistors or ATP molecules do not need further deconstruction as components since their internal structures and specifications for their behaviors are given. Similarly, an organic molecule in a biophysical system need not be further deconstructed. Figure 4.6 depicts a system deconstruction tree resulting from the recursion. The tree view is easier to see the hierarchical relations of system-subsystems-sub-subsystems. A map view becomes unwieldy but is possible to use as a representation. One would simply see ovals inside of ovals. Later, as we present more examples of actual systems, we will see ways to use both representation views as needed. The idea that we can treat any complex component at any level in the hierarchy as a system in its own right, according to Eq. 4.3, supports a way to semi-isolate any component and develop representations for it.

The tree shown in Fig. 4.6 is just the skeleton of the system hierarchy of systems and subsystems. Figure 4.8, will show a more complete tree including subsystems and flows, to be covered next.

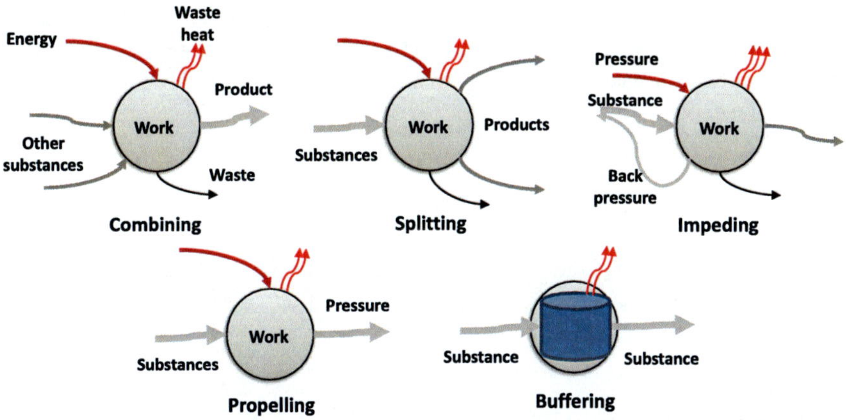

Fig. 4.5 The formal atomic processes used in the "simplest process rule" for stopping the recursive deconstruction procedure involve either work on the inputs (e.g., combining, splitting, impeding, or propelling), producing output changes (kind or rates), or a passive buffering (stock). The various arrows represent the flows that will be explained below

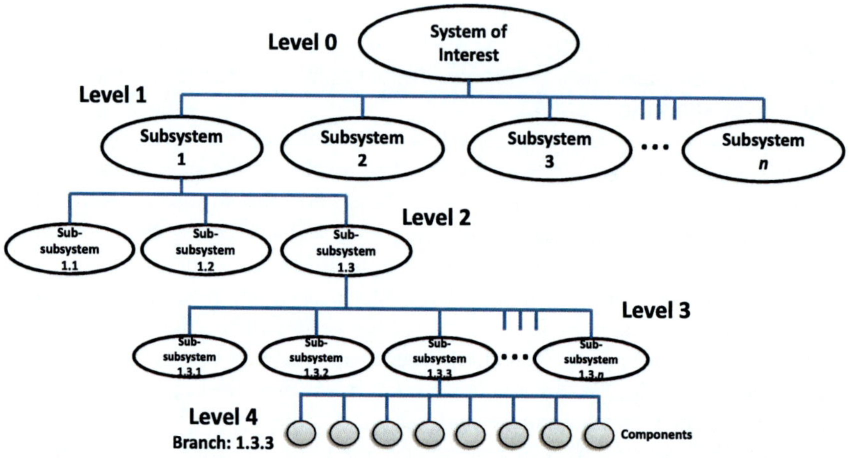

Fig. 4.6 Equation 4.3 expanded through system deconstruction procedures gives rise to a tree structure. Branch 1.3.3 (where the pre-appended "0." has been removed) ends the recursion with a set of atomic components. Note that the various subsystems are represented in this tree view by ovals rather than boxes as in Fig. 4.4a. Either kind of closed shape can be used

4.3.3.2 Interactions

The structural skeleton established thus far has only explicated the organization of subsystems (complex components) and sub-subsystems (complex or atomic components) in a hierarchy of scale. What needs to be established now is the relations between components within the system and between some components within the system and entities outside the system that constitute the system's environment of sources and sinks.

4.3.3.2.1 Between Components Internally

Equation 4.4 is a graph (from graph theory) that defines the interactions between all of the components in *C*.

$$N_{i,l} = C_{i,l}, L_{i,l} \tag{4.4}$$

N is a graph with vertices, $(c_{i.k,l}, m_{i.k.l}) \in C_{i,l}$, and directed edges, $(e_{i.k,l}, cap_{i.k,l}) \in L_{i,l}$. Edge $\boldsymbol{e_{i.k,l}}$ is the vertex pairs, $(\boldsymbol{c_{i.k,l}}, \boldsymbol{c_{i.o,l}})$, where $\boldsymbol{k \neq o}$ and the direction is assumed from $\boldsymbol{k}$ to $\boldsymbol{o}$.

N is generally a flow network through which real substances are moving from one node (component) to the next with causal influence. The term, *cap*, is a function, $cap_{i.k,l}$: $C_{i,l} \times C_{i,l} \rightarrow \mathbf{R}_{\infty}$, giving a capacity describing the flow rates. Rate functions are determined by one or more of the atomic processes from above (as in Fig. 4.3). The actual function will be generally complex in that flows are usually

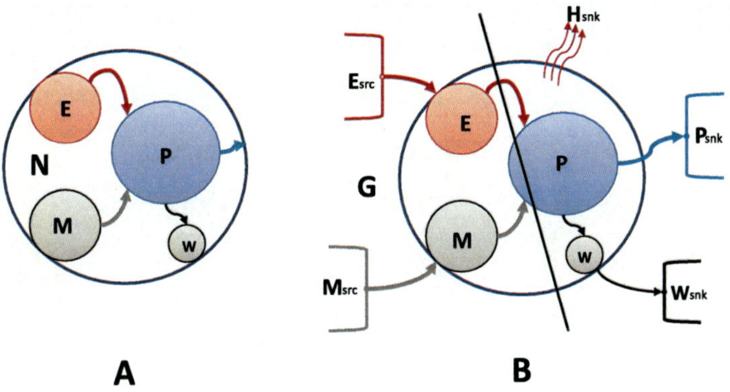

Fig. 4.7 (**a**) A set of nodes (components within the system) and edges or flows between them. This is the *N* graph described in the text. (**b**) The flows from environmental sources, the open rectangles, and to environmental sinks (right side of the figure). This is graph ***G***. The line in (**b**) is the graph partition or "cut." Nodes: *E* energy obtaining process, *M* material obtaining process, *P* production and exporting process, *W* waste removal process. In (**b**): E_{src} energy source, M_{src} material source, H_{snk} heat sink, P_{snk} product sink, W_{snk} waste sink

fluctuating as a function of several different factors. Alternatively $cap_{i.k,l}$ may simply specify the max flow rate possible. The difference will be determined in context.

N captures the internal flows within the system, that is, between subsystem components. Figure 4.4a shows a graphic representation of a four-component system with flows of matter and energy. Note that we treat both flows of actual materials/energy/and messages through channels (e.g., pipes) and phenomena such as application of forces or diffusion (i.e., fields) as generalized flows, where the appropriate equations are used to differentiate in the models.

4.3.3.2.2 Between Environment and Components of S

G is a bipartite flow graph defined as:

$$G_{i,l} = \left(C'_{i,l}, Src_{i,l}\right), \left(C''_{i,l}, Snk_{i,l}\right), F_{i,l} \tag{4.5}$$

where:

$C'_{i,l}$, $C''_{i,l} \subset C_{i,l}$ are the subsets of components within $\boldsymbol{C_{i,l}}$ that receive inputs from the source elements $\boldsymbol{e_{i.k,l}} \in \boldsymbol{Src_{i,l}}$ and send outputs to the sink elements $\boldsymbol{e_{i.j,l}} \in \boldsymbol{Snk_{i,l}}$ respectively (see Fig. 4.4b and Eq. 4.2). $\boldsymbol{Src_{i,l}}$ and $\boldsymbol{Snk_{i,l}}$ are sets of the nodes situated in the environment. In the figure they are shown as open rectangles (the colors are meant to suggest different categories of sources and sinks). This is because they are unmodeled in terms of their internal workings as elements in the original SOI environment; they are only encountered as members of the

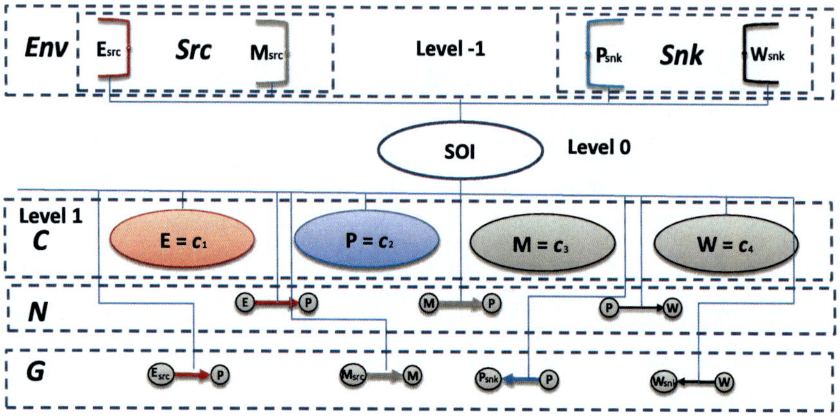

Fig. 4.8 A more complete tree view of the system depicted in Fig. 4.7 shows the subsystems and both internal and external flows. We also show the system of interest environment and the sources and sinks for the external (***G*** network) flows. Note that flows are edges in a graph and are labeled with the source node and the sink node with the direction of the arrow indicated. Radiated waste heat and the environmental sink are not included in this figure

environment and their internal details cannot be known.[11] Together the sets $\boldsymbol{Src}_{i,l}$ and $\boldsymbol{Snk}_{i,l}$ are a superset, the environment. In certain contexts we will talk about the tuple $E_{i,\,l} = \langle Src_{i,\,l}, Snk_{i,\,l}\rangle$ as being the environment of component i at level l.

$\boldsymbol{F}_{i,l}$ is the set of directed flow edges as was the case for $\boldsymbol{N}$ above. Edges are of the form: $(f_{i.\,k,\,l}, cap_{i.\,k,\,l}) \in F_{i,\,l}$ ($\boldsymbol{cap}_{i.k,l}$ is the capacity function from above). Edge $\boldsymbol{f}_{i.k,l}$ is the vertex pair, $(\boldsymbol{e}_{i.k,l}, \boldsymbol{c}_{i.o,l})$, $\boldsymbol{e}_{i.k,l} \in \boldsymbol{Src}_{i,l}$ and $\boldsymbol{c}_{i.o,l} \in \boldsymbol{C}'_{i,l}$, the subset of subsystems in $\boldsymbol{C}$ that are responsible for obtaining inputs from the environmental sources, or $(\boldsymbol{c}_{i.o,l}, \boldsymbol{e}_{i.k,l})$, $\boldsymbol{e}_{i.j,l} \in \boldsymbol{Snk}_{i,l}$ and $\boldsymbol{c}_{i.o,l} \in \boldsymbol{C}''_{i,l}$, the subset of subsystems in $\boldsymbol{C}$ that are responsible for expressing outputs to the environmental sinks. As above, $\boldsymbol{k} \neq \boldsymbol{o}$ and the direction is assumed from $\boldsymbol{k}$ to $\boldsymbol{o}$. Nodes $\boldsymbol{e}_{i.k,l}$ specify those in the **e**nvironment (sources and sinks) relative to the SOI.

Figure 4.8 provides a tree view of the entire system, thus far defined. It shows the environment entities (Fig. 4.7b), the internal subsystems (components), and the internal and external flows.

4.3.3.3 Boundary

Up to this point the definition of system should resemble those given in most accounts whether in mathematical form or not. Languages for modeling systems, such as system dynamics (SD) utilize the same sets of elements as described so far but not in this mathematical form. At this point we start to depart from historical

[11] Later we will see this "rule" is modified as we go deeper into the original SOI and deconstruct to lower components.

accounts of system definitions as will be explained along the way. The first departure involves the concept of a boundary. If you recall from the last chapter, we claimed that boundedness is a real element (has first-class ontological status). Boundedness constructs effective boundaries to systems.

This situation is not without controversy. Workers who are primarily interested in modeling systems have maintained that there are no real boundaries but what the modeler chooses (Meadows 2008, pages 95 and 97). This is a reasonable claim when the focus of attention is on the replication of certain dynamics of "parts" of a whole system. In such cases, the boundary of the simulation model, not the system, is being chosen to keep the problem tractable. On the other hand, some systems clearly have identifiable boundaries such as cell membranes, skin, or walls.

Our position is that all systems are kept intact by virtue of some kind of internal binding that produces an "effective boundary." The distinction between what is "inside" a system versus what is outside depends on this effective boundary. Systems that do not have physical boundaries nevertheless have distinct subsystems that interface with the entities in the environment that supply resources or act as sinks for the wastes produced by the system. Another way to think of these kinds of boundaries is that they are a result of the interactions between internally bound special component subsystems, interfaces, to be described shortly, and those external entities. Recall Fig. 3.8 in the last chapter.

For purposes of analysis and design, we will make boundaries explicit elements of a system definition.

The boundary, $\boldsymbol{B}$ in Eq. 4.1, at level l, then is a tuple. That is:

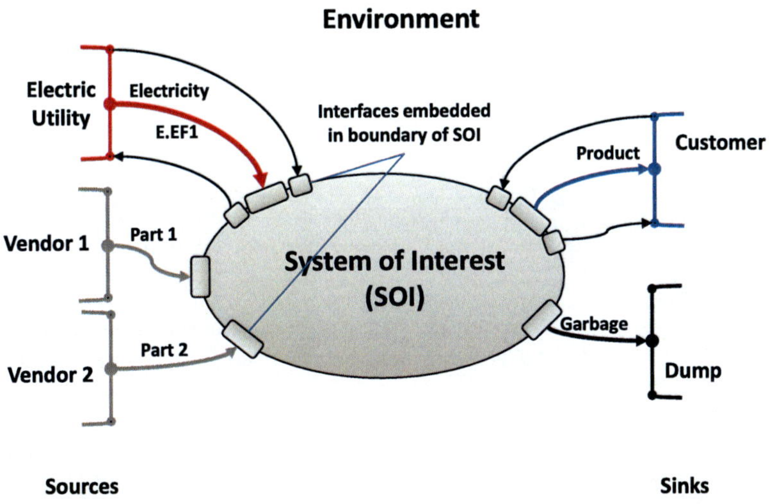

Fig. 4.9 Interfaces are associated with the boundary of a system. Here they are shown as round-edged rectangles that penetrate the boundary and act as pass-ways for inputs and outputs

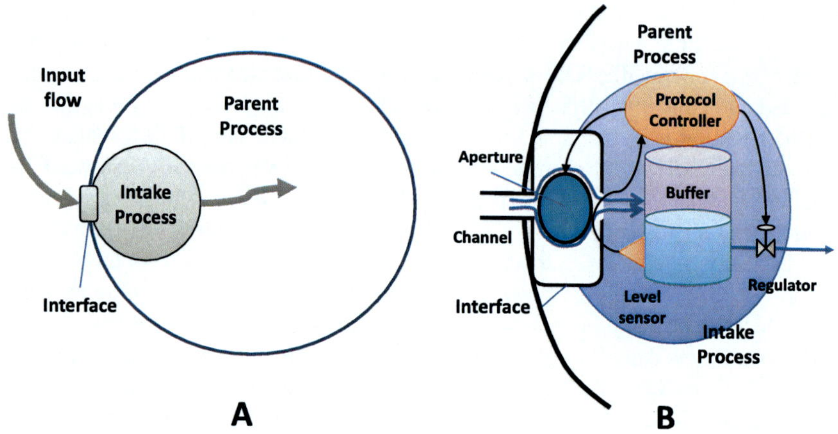

Fig. 4.10 Interfaces are represented generically (**a**) showing the general form of the interface between the outside world and, in this case, an intake process that then supplies the substance flow to the parent process (or SOI). (**b**) Provides some details of a more complex interface, for example, an active membrane pore on the surface of a cell. The aperture is the main channel control device (here shown as a "ball valve"). Many interfaces are subsystems (processes) in their own rights and so will have additional "equipment" devoted to regulating the flow of the substance. Other elements in the figure (e.g., sensor, buffer, and regulator) will be explained below

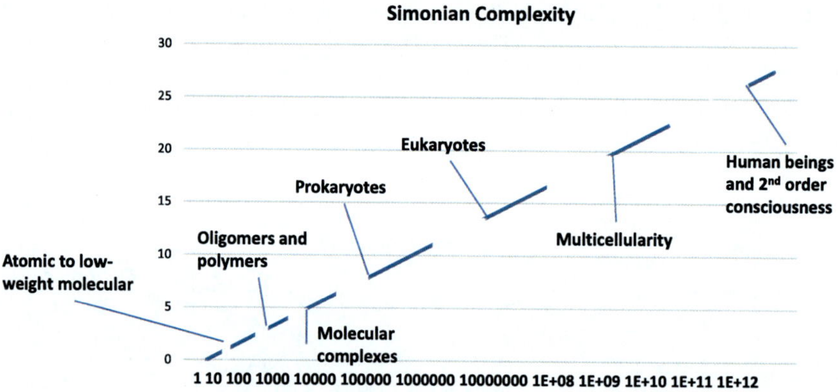

Fig. 4.11 As Simonian complexity explodes up the levels of organization (and complexity), the logarithm of that value rises linearly and gaps (probably exaggerated) in the measure are seen at the major transitions, where the graph line (blue) appears indicates a range of complexities within the various domains

$$B_{i,l} = P_{i,l}, I_{i,l} \tag{4.6}$$

where $\boldsymbol{P}$ is the set of properties and the second set, $I_{i,\,l}$, is the set of *interfaces*. The exact form of $\boldsymbol{P}$ is still an object of research. At present it includes such properties

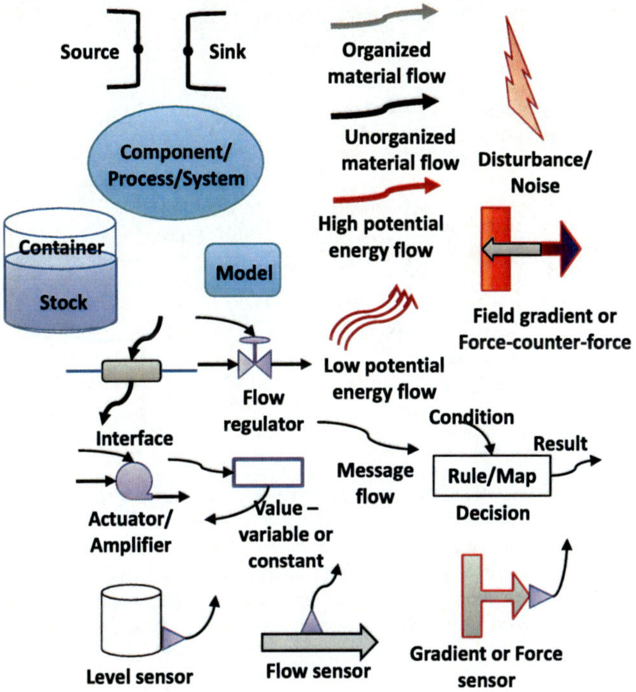

Fig. 4.12 An error sensing mechanism resulting from the composition of four atomic work processes. The circuit can be used by, for example, a homeostat, a cybernetic feedback control

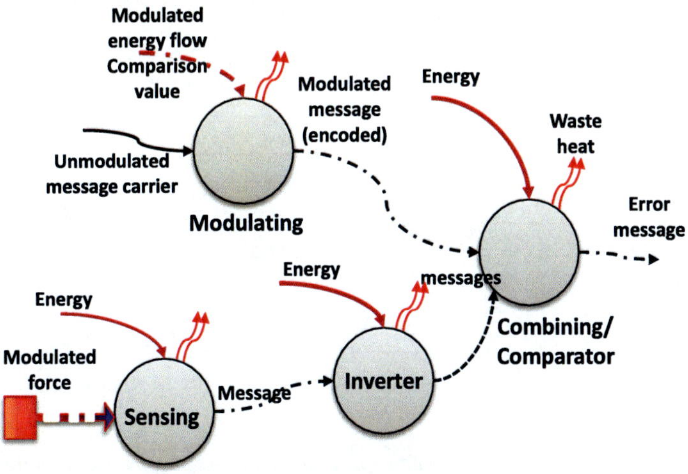

Fig. 4.13 These are some of the icons used to represent elements of the system language

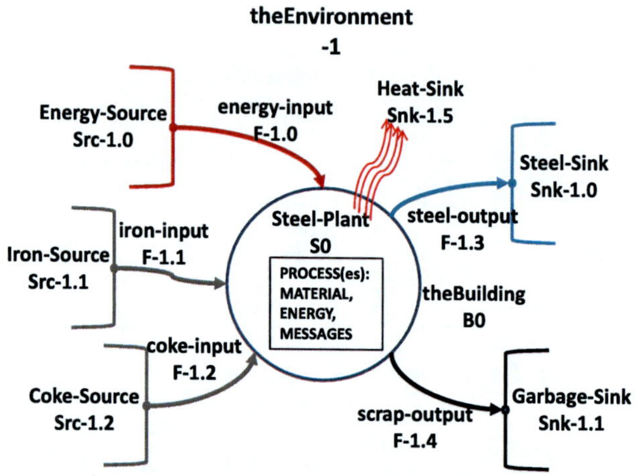

Fig. 4.14 A simplified example of the use of SL to describe the SOI, a steel manufacturing plant. This figure shows the SOI and its environment. Note that labels (names of objects) are user-defined (using prescribed characters as in most programming and markup languages), but may follow some determined style requirements. As in most programming languages, the strings are case-sensitive. The identification codes, for example, Src-1.0, on the other hand, follow strict conventions that will be explained briefly in the text and discussed more thoroughly in Chap. 6

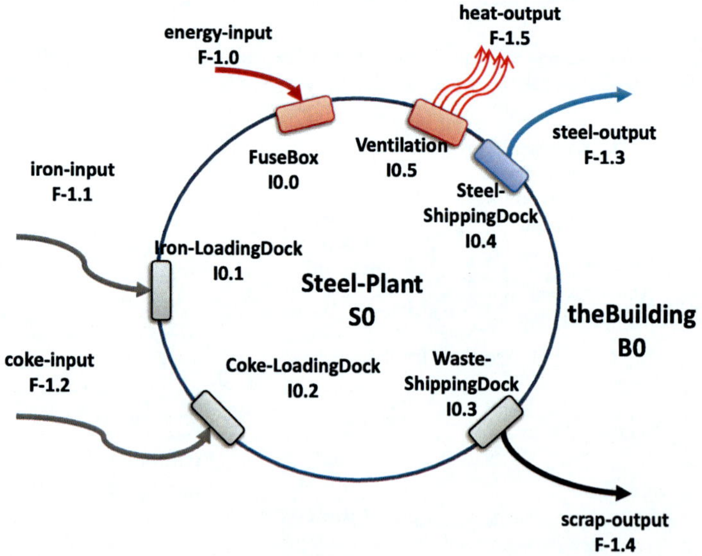

Fig. 4.15 Analysis of the boundary identifies the main material and energy interfaces for importing resources and exporting products and wastes

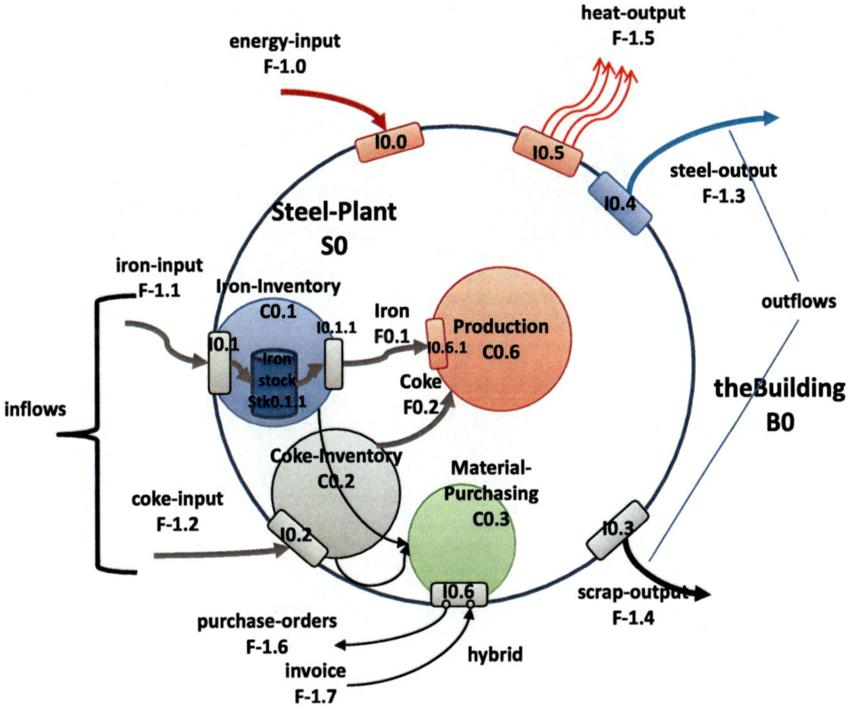

Fig. 4.16 Further analysis of the SOI as a transparent-box reveals internal components and flows. A new component has been added, the Materials-Purchasing (C0.3) subsystem. Its interface with the supplier demonstrates the situation when a message interface communicates bidirectionally, both receiving and sending. It also indicates that a single environmental entity (the supplier) also sends and receives. Moreover, the supplier is a source and a sink simultaneously

as porosity (0 being completely non-porous) and "perceptive fuzziness," meaning the degree to which it is easily perceived; 0 being able to identify and locate in space the separator between inside and outside any number greater than 0 but less than 1 being the degree to which a physical phenomenon corresponding to "keeping the insides in and the outsides out."

Boundaries can be hard physical structures such as a cell wall (in plants and bacteria) or a structure resulting from competitive forces such as the phospholipid bilayer membrane of living cells that results from its own unique internal chemistry. The boundary of an ecosystem is very fuzzy and very porous, both perceptually and in the technical sense. Some members of an ecosystem may transit in and out at various times but usually through particular portals (e.g., game trails). Nevertheless, ecologists agree that there are boundary conditions that provide an internal milieu suitable for the species that live there. The climate conditions, mostly regulated by geographical features surrounding that of the ecosystem, provide a supportive home for those species adapted particularly for them.

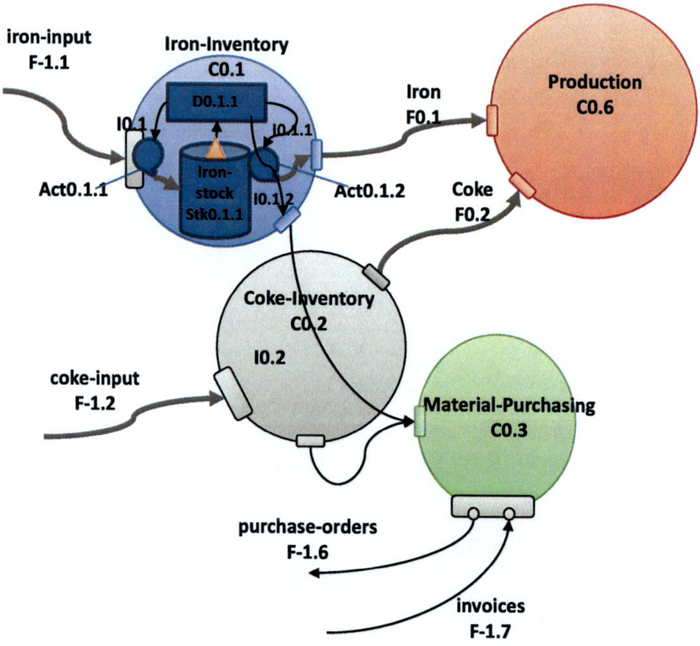

Fig. 4.17 Further expansion and decomposition applied to the Iron-Inventory shows it to be a more complex process including actuators receiving iron-input and putting it into the Iron-stock (Stk0.1.1). A decision process monitors the level of inventory (orange triangle) and decides when to order more stock (black arrow coming out of I0.1.2 going to Materials-Purchasing)

By analogy with the set $\boldsymbol{C}$ above, the set of interfaces embedded in the boundary are components in the boundary subsystem and are themselves subsystems. That is, every $\boldsymbol{r}_{i,l} \in \boldsymbol{I}_{i,l}$ is an $\boldsymbol{S}$ itself but including what we call a protocol object, φ. That is:

$$r_{i,l} = \left(S_{i,l+1}, \varphi\right) \tag{4.7}$$

There is a reason for treating interfaces as different from other system components due to their special role in crossing boundaries. Interfaces do not typically alter the substance of the flow, that is, do not transform the substances as a process does to create products.[12] In the case of interfaces as subsystems, the φ object is called a protocol which is an algorithm for letting the flow across the boundary in an ordered fashion. The receiving dock of a manufacturing company has a *procedure* that receives packages, checks the manifest, checks the material being received,

[12]With the exception of an interface containing a "filter" protocol. Filters are designed to let the desirable substance through but impeding whatever might be undesirable. Some filters may require periodic replacement if they get clogged up with material substance, or they might be flushed periodically to some waste flow. Other filters such as those used to filter electric flow frequencies dissipate the "waste" continuously.

and sends it on to the inventory subsystem. A neuron's postsynaptic membrane allows ions to pass into or out of the cytosolic compartment through special pores that are activated by the transmission of neurotransmitter. Most interfaces will be found to involve both a mechanical (or electrical) pass-through control and a message processing component that controls the pass-through control.

A generalized model of interfaces on the boundary of an SOI is shown in Fig. 4.9. The flow arrows from sources are shown going into an interface component; arrows coming out of output interfaces go toward the sinks. Interfaces are special kinds of regulators, usually allowing passage in one direction only, and only under the right conditions of the protocol.

Figure 4.10 expands on the nature of interfaces through an example. Each interface provides a connection to the channel of the flow in the environment on its external side. Internally it is interfacing with a subsystem of the SOI that is either a receiver or an exporter process (as in Fig. 4.10b).

The biggest single mistake that system scientists (or scientists exercising systems thinking) make is to ignore or trivialize the concept of an interface (and its protocol). Examples of how this can lead to trouble in adequately accounting for system behavior will be provided in the chapters ahead. Suffice it to say that many mistakes in system analysis and design result from the inattention to interfaces and their protocols.

As a rule of thumb, the interfaces for more complex systems tend to be more complex themselves. A simple opening in the wall of a hut functions as a door, but it allows anyone (or anything) to walk through.[13] An actual hinged door, with door knobs inside and out, and an internal lock mechanism (keyed from the outside) in a modern (and more complex) structure is a bit more complex and has a controlled entry/exit protocol. Entries and exits from airport departure/arrival gates have become exceedingly complex with separate entry and exit paths required by security needs.

4.3.3.4 Transformations

$\boldsymbol{T}$ is the set of transformation rules for the subsystems in $\boldsymbol{S}$. That is, for each $\boldsymbol{c}_{i,l} \in \boldsymbol{C}_{i,l}$ there is a formula, equation, or algorithm, $\boldsymbol{t}_{i,l}$, that describes the transfer function of that component for transforming inputs to outputs. These may be expressed in any suitable form, such as ODEs or computer codes.

The inputs and outputs are the same flows as represented in the $\boldsymbol{G}$ graph (above). Specifying $\boldsymbol{T}$ for the level 0 SOI will be exceedingly difficult initially since it involves many different inputs and outputs (for complex systems). As we will show in Chap. 6, however, it isn't necessary to make a full specification of the transformation at the start. For purposes of analyzing natural and designed systems both, all

[13] This is an example of a bidirectional interface with a nearly non-existent protocol, unless you consider that if the hole is not very large you might need to duck your head each time you pass through.

that is needed is a rough approximation of the transformation, used as a placeholder that suggests its nature. This is, effectively, an abstract model of the transformation that will be refined as the deconstruction process continues. One of the advantages of following this framework is that the higher-level transformations get refined by discovery of the lower-level ones. Once more is known about the internal transformations of subsystems and their combined effects, the higher-level model can be made more rigorous (Principle 12 in Chap. 2). This recursive improvement tracks the deconstruction all the way down to the leaf nodes in the system tree.

Note that this approach varies from many systems' engineering methods in which an attempt is made to specify the system output (with great specificity) as the first step. This inclination derives from device engineering practices where the function of a device can be a priori specified and the device engineered to produce that result. In systems engineering the final output(s) of the system are generally far too complex to have a perfect specification upfront. When system engineers believe they are obtaining such a specification (e.g., in requirements gathering), they are often faced later with discontinuities between expectations and actualities that are highly disruptive to design projects. With the approach of deep analysis promoted in this book (top-down deconstruction of transformations and bottom-up refinement) the engineering process will be following a formal procedure that they often end up following informally anyway!

4.3.3.5 Memory

H in very complex systems could be a super complex object that records the history of the system, or its record of state transitions, especially as it develops or evolves. For example, brains learn from experience, and as such their internal microstructures change over time. This is called memory and the current state of ***T*** can be based on all previous states. Some simple systems, like atoms for example, may have a NULL ***H***; that is there is no memory of past states and future states depend only on the current state and current inputs. As just mentioned, on the other hand, brains (and indeed all biological systems) have very rich memories. ***H*** augments ***T*** and all variables associated with elements in ***N*** and ***G*** in that it records traces of the changes in these variables over time. A simple version of ***H*** would be the time series data of all state variables of the system averaged over some appropriate time window. This too is an area of research to pursue. The best model for ***H*** would be the human brain, particularly the neocortex, where memories are encoded, stored, and retrieved for use (Mobus 1999).

An example of emulating the way brains develop memories and its possible relation to the system model as captured in the global knowledgebase must wait until we get to Part II. The human brain is the quintessence of a way to build models of systems (as discussed above). There will be numerous comparisons between the use of this mathematical framework and the workings of the brain as ways that understanding is captured.

For our purposes here, however, we will consider more formal methods for capturing history in a usable way. Consider, for example, the use of a recorded time series of state measures. Let H at time t be defined as a set of measures (a list of variables in the system),

$$H_t = [v_1, v_2, v_3, \ldots, v_i, \ldots, v_n]_t \tag{4.8}$$

At each time instance the variables of the system are measured with an appropriate instrument and recorded. The time series of H_t sets provide a set of snapshots of the state of S at each time increment. For example, the profit and loss statement of a corporation is an annual snapshot of the corporation's most essential state variables. In the simplest case H_t is record made every Δt unit. It is what we call a data stream. However, just having a record of the data is not very useful. Ordinarily, just as with the profit and loss statement, we look for patterns in the data after processing it in some fashion.

4.3.3.6 Time

Finally, the last element in $\boldsymbol{S}$ is $\boldsymbol{\Delta t}$, a time interval relevant to the level of the system of interest. The time interval is familiar to those who work with discrete-time simulations. In general, higher levels in the hierarchy of organization have larger Δts; the activities take longer than those at lower levels. Δt is generally an integer multiple of the lowest level time constant that is deemed relevant for a particular system. In discrete-time simulation, it is the time step over which the model of that level is computed.

4.3.3.6.1 Time Indexing

For all systems at any level in the structural hierarchy an index of time step, t, is used to count the amount of time in Δt units for that level between events or state changes. Under ordinary circumstances (e.g., when not trying to model infinite time) there will be an additional parameter used to designate time units in a larger period. Thus Δt might be replaced by a tuple, $\langle \Delta t, x \rangle$, where x is the integer count of a single cycle.

4.3.3.6.2 Cyclic Intervals

Real systems are embedded in supra-systems that undergo cyclical behavior. The Earth rotates on a diurnal cycle, tides rise and fall, seasons come and go and come again. Some cycles are regular intervals, the general meaning of the term "periodic."

Others, such as the tides, are "quasiperiodic" meaning that they are irregular but nevertheless repeating, just over varying intervals.

In systems that undergo periodic or quasiperiodic behavior, the element Δt can be replaced by a "clock" or "quasi-clock" function that counts Δt units until that count reaches a limit and the counter is reset to 0. A quasi-clock has a secondary function that generates the limit number (which is not a constant) according to the phenomenon leading to the quasiperiodicity. Admittedly, this is an unsettled area requiring more research. But models of tides based on the major parameters such as position of the sun and moon have been developed. There are instances when the periodicity of a cycle may be varied in a constrained-random fashion, using, for example, the bounded Monte Carlo method.

4.3.3.7 Considering Very Complex Systems

Complexity is a very difficult concept and the word itself has many different senses. Intuitively most people have the notion that if something has many parts (i.e., subsystems) and many interactions (i.e., flows) between the parts and many interactions with its environment, then it is complex.

There are many different ways to characterize the complexity of a system. Here we investigate two complementary approaches. The first looks at what we can consider "structural" complexity using the above-mentioned intuition. The second approach looks at dynamic or "behavioral" complexity, that is, how are the behaviors of the complex system in relation to their environmental interactions. Both include consideration for the number of states that a system can be in, but behavioral complexity looks only at the states of interactions with the environment.

4.3.3.7.1 Simonian Complexity

We have adopted a definition of structural complexity derived from Herbert Simon's description of a near decomposability of a system (Simon 1998, 209).

Simon's description of systems as hierarchies of modular units (what we called components or subsystems above) gives rise to a metric that can be used to characterize the complexity of real systems. We develop that metric below.

There are two fundamental aspects that go into defining complexity: structural complexity and functional complexity. One deals with the hierarchy of modules and submodules and the other deals with what each module does. The former is exposed by the structural decomposition process outlined in Chap. 6. The latter is much more difficult to estimate. For our purposes here, we will use the concept of a state space to approximate the metric of functional complexity.

Imagine taking a reading on every flow (connection) and every reservoir in a system and all of its subsystems every Δt instance. The state, σ_i, of the system where, i, is the index of the set of possible states, $\boldsymbol{S}$, and σ_i is an element of that set. The instantaneous measure of all of these dynamical elements at time t defines the system as being in state i at time t, or $i = \sigma_t$. Since by definition a system is an organized set of parts (components and subcomponents) the number of possible states is constrained and so can be represented mathematically, at least in principle, by a Mealy finite-state machine (or automaton). For systems where multiple state transitions may occur from any one state to a successor state and where these are stochastically chosen, the representation is that of a non-deterministic finite-state machine with statistically determined transition probabilities on each node out link. The number of transitions possible in the entire state space is $\boldsymbol{T}$, a list of the pairs of from and to states. The Mealy FSM takes into account the system's interactions with the environment or context. The state transitions are determined by a combination of inputs to the system and the system's current state at time t.

As a reasonable approximation of the *size* and complexity of the state space, we can use the number of states, $\mathbf{S}$, in the space (number of nodes in the finite automaton) and the number of transitions, $\boldsymbol{T}$.

$$|\mathcal{S}| = f\left(|S| + |T|\right) \tag{4.9}$$

The function, f, is as yet unspecified but assumes some kind of scaling factor.

We can then define Simonian complexity measure as a sum of the number of components and subcomponents down to and including leaf nodes (atomic components) along with the sum of interactions in the $\boldsymbol{N}$ networks within each component module, the size of the boundary as a list of interfaces, approximating the number of inputs and outputs, and size of the state machine.

$$\mathbb{C} = \ln\left(\left(\sum_{L}^{l=0}\sum_{I}^{i=1}\left[\left[|C_{i,l}| + |N_{i,l}|\right] + |B_0|\right]\right) + |\mathcal{S}|\right) \tag{4.10}$$

This function sums the number of all components and relations at each level in the decomposition hierarchy and the number of affordances with external entities or number of interfaces on the boundary and takes into account the state space size. The log function compresses the numerical value of the complexity measure.

The working hypothesis is that while **C** is a continuous variable, we should see gaps between values for lower complexity systems and higher complexity ones that mark phase transitions. For example, we should see a, perhaps small, gap between elements and molecules and a larger gap between macromolecular assemblies and whole cells. In fact, the hypothesis is that as we ascend the hierarchy of organization, we will see larger and larger gaps between the complexity measure of lower domain systems and higher domain ones as the complexity of higher-domain systems explodes exponentially. Figure 4.11 shows this basic idea.

It is a research challenge to compute the Simonian complexity of various systems as shown in Fig. 4.11. To the degree it might be done, we expect to observe some kind of discontinuity (e.g., the gaps shown) that demarks the transition from a lower complexity domain to a higher one.

4.3.3.7.2 Behavioral Complexity

Behavioral complexity can be far more difficult to characterize compared with structural complexity. The behaviors of a system are related, clearly, to the number of states the system might be in as described above. However, there are possibilities of transitions from internal states that are not easily modeled by finite state machines, even probabilistic machines. Among these possibilities is behavior resulting from internal non-linearities, resulting, for example, in chaotic behaviors. That is, if one captures time series data on the externally displayed states of a system, they describe a chaotic attractor basin in phase space.

4.3.3.8 Ontogenesis of Greater Complexity

Ontogenesis, recall from Chap. 3, is the general process that brings components together to form more complex entities. Here we provide a somewhat specific example of how components as "atomic" work processes (from Figs. 3.18 and 3.19) are composed through linkages of outputs to inputs to produce a functional and useful object. Figure 4.12 shows the end result of an ontogenic linkage that results in a device capable of computing an error message, the net value of a difference between a reference signal (e.g., setpoint) and a measured value.

In this example we take four different atomic work processes, sensing, an inverter, a modulator, and a combiner, here being used as a comparator. Refer to Figs. 3.17, 3.18, and 3.19 to the atomic processes. The sensor takes in energy to do the work and transduces a modulated force, such as pressure or temperature producing a modulated message encoding the force parameter. The inverter takes as input a message and does what its name implies; it inverts the sense of the message code (for numerical values it effectively changes the sign of the value). It too uses energy to do this work and output a message. The combining process receives two inputs. In this case, the combining process is for messages, in other words, a computational process. It receives the output message from the inverter and a message from the modulating process which has transduced a modulated energy input onto a message carrier which is output to the other input to the combiner. The latter does the combining, which in the case of messages is a superposition of the two messages resulting in either a null message output or a signed output corresponding to the amount of error.

This whole assembly can now function as an error detector as long as the major inputs, the modulated force and the modulated energy flow, are commensurate. That would mean that the final output of an error message had some veracity in the context of a controller agent that could take a decision and activate whatever power it had to counter the error.

This model is extremely general. It could model the creation of a feedback loop at the molecular level, say in the lead up to the origin of life. Or it could represent the engineering design of an error detection circuit in a nuclear power plant control. In any such case, if the inputs match the outputs and the work processes perform their functions appropriately, the circuit's use leads to stability and continuation of the larger system in which it is embedded. If not, better luck next time!

In ontogenesis, as described in the previous chapter, luck was a factor in bringing components into coupling proximity and in the right ordering in auto-organization. We saw that a mixing and sorting regimen such as occurs in a convective cycle in a gravitational or electric field increases the chances for, say the sensor and the inverter to couple since their input/output interfaces match and once coupled they are reasonably stable awaiting the proximity of a comparator to the inverter's output interface. There is a kind of "rachet effect" that keeps promising couplings stable for at least a while until the next component joins the group. One explanation for this is that each of these work processes gets an input of high-potential energy that has to be dissipated properly in order for the system to not overheat, as it were. Once two processes are suitably coupled, they create a better pathway for this energy to do so, thus lowering the total energy (specifically the thermal modes) and raising the likelihood that they will be more stable. Add another, or several more processes and this stability ratchet increases.

Of course, the process is purely stochastic for auto-organization, with only slight biases provided by the mixing-sorting regimen. But even for intentional-organization (design) as we will see in Chap. 15 chance still plays some role in getting the right components together in the right way at the right time.

We now turn attention to the problem of how to use our ontological framework, understanding of ontogenesis, and our formal framework for describing systemness to create a way to describe all systems, but particularly able to describe even the most complex systems we encounter. We need a language.

4.4 Toward a Language of Systems

Armed with the mathematical structure described in Eq. 4.1 and subsequent equations, along with the ontological commitments in Chap. 3, we are in a position to design a formal language of systems (SL). With such a language we will be able to describe, in principle, any system in any domain, for example, biological or sociological. The language uses generic terms to represent various elements of systemness, for example, nouns like "process" to embody a system that does work and "flow" to embody the movement of materials, energies, and messages. Its syntax is based on allowable patterns or relations, for example, a work process that varies a flow (e.g., a valve or resistor) cannot be put inside a stock; it has to be inserted into the flow.

Ultimately SL's semantics describe the systemness of a specific kind of system, for example, a biological cell. The generic lexicon is translated into domain-specific terms. For example, a vacuole (an organelle that holds stuff) is a kind of reservoir. The semantics of SL (of systemness) provides the fundamental organizing aspects of all systems in any domain.

4.4.1 Semantics

SL is a "description" language. It is used to describe structures, relations, functions, and the other items in the ontology of Chap. 3. This is different from a typical programming language, which is designed to "describe" algorithms, or sequences of steps in a data manipulation process. An algorithmic description language (i.e., programming language like JavaScript) is incorporated into SL to specify the operations of behaviors of the various elements, such as interface protocols. Relations between elements are built into the syntax (see below).

4.4.1.1 Lexical Elements

We translate elements from the ontology (Chap. 3) into operative terms to be used in constructing descriptions. Each lexical element has a corresponding graphical representation. Figure 4.13 provides a few examples of graphical icons used to represent the lexical elements for a visual programming interface or display.

We envision a graphical user interface tool that allows a user to drag one of these from a palette of icons to a workspace to assemble a system construct. The tool would enforce the syntax as well as capture associated relevant data, such as the domain-specific names and types associated with the generic names and types. For example, a process/system object would be given a domain-relevant name, for example, "theCompany." The tool would prompt the user-analyst/designer to fill in a table of associated data (see example below).

4.4.1.2 Descriptions

Descriptions are structured statements that provide domain-specific knowledge about the elements and their relations. Descriptions include table-organized descriptors, which can be general for the lexical type, for example, a material flow includes descriptors for what kind of material, how much, receipt rates and variances, timing, etc. Descriptors may be specialized as needed through "user-defined" extensions.

A description includes the element type identifier, a name identifier, and the descriptors. Some elements include lists of sub-elements in their descriptors. For example, a description of a boundary includes descriptors for boundary properties (e.g., fuzzy, porous) and a list of interfaces that provide receptors or sources for flows from/to the environment of that system.

4.4.1.2.1 Components

Table 4.1 shows an example of a description for a component. Component types and sub-types are predefined (by the system language and by user domain-specific extensions—example below). The pull-down menu (blue arrows) can be used to select one or more of these options.

This is only an example, but it shows the mechanism for capturing domain-specific information about an element. Similar popups would be used for every element type in the lexicon. Thus is established that the translation of systemness attributes to the domain subject.

4.4.1.2.2 Flows

Network relations are implicit in several ways. The flow paths are implied in the sender-receiver links, for example, a source's output links to an interface's receiver. The structural tree is implied in the identification numbering scheme using the dotted integer described above.

4.4.1.2.3 Behaviors

Behaviors or dynamics are implemented using embedded scripts. For example, the above flows are associated with scripts that run algorithms for simulation of the flow rates specific to the kind of flow. The transformations (***T*** in Eq. 4.1) are expressed in transfer functions or simulation programs for the relevant component.

Table 4.1 An example of a component description that pops up when a user drags a component (or subsystem) into an existing SOI image

Component	
Type:	
Sub-type:	
ID code:	
Name:	
Description:	
Structural specification reference:	
Functional specification reference:	
Reference:	

4.4.1.3 Syntax

The syntax of SL is two-dimensional. The structure of a system being described is governed by the mathematical definition given above. The syntax determines that, for example, a flow must come from a source and go to a sink, whether from environmental sources and sinks to boundary interfaces, or between components and stocks internal to the SOI. There are no other options. Similarly, flows and stocks can be sensed (rate and level), but processes cannot be sensed in the same way. Everything is determined a priori by the ontological commitments.

4.5 Example

As mentioned above, SL is a description language; it has more in common with document description languages like hypertext markup language (HTML) used to describe web pages for display in browsers than with conventional programming languages. Combined with a scripting language like JavaScript, which is used to specify behaviors within the web page environment, a markup-like language is quite powerful in providing form and function in models. SL has much in common with HTML, or more correctly with the Extended Markup Language (XML), which is a superset of HTML. Below we provide a very simplified example of a system described in a markup language we'll refer to as sysXML to give some idea of how this computational platform can be used to do so. We should emphasize that this exercise is just a preliminary concept of how SL might be implemented and is as much a playful exploration as a serious example. Research into how SL might be realized continues. Should the reader not care to get bogged down in technical details of a formal language, this section may be skipped without serious loss of grasp of the concept.

Figures 4.11, 4.12, 4.13 and 4.14 provide examples of a system of interest in SL graphics along with some of the captured data, that is, labels. Listings 4.1, 4.2, 4.3 and 4.4 show the sysXML output from the captured data as organized in the knowledgebase after analysis. The system is analyzed following the procedures given in Chap. 6, in a top-down decomposition with the analytical engine capturing the results into the knowledgebase format that will be described in more detail in Chap. 8. From there, it is possible to generate a systems dynamics-like simulation model and produce a human- (and machine-) readable sysXML specification as shown below.[14]

[14] We should point out that there are a number of ways to represent a model in a computer readable form. For example, the knowledge might be used to generate a straightforward program in a language such as C++. However, we feel that the representation of a system should be human readable and understandable. Extensions of XML are rapidly becoming the language of choice for many fields where knowledge sharing among humans and machines is done. XML started out to be used

The example is of a very simplified steel manufacturing plant that takes in electric energy, iron, and coke as its main resources and produces steel for sale to customers as well as some scrap (waste materials) for disposal and LOTs of waste heat. In Fig. 4.14 we start with situating the SOI, Steel-Plant, in its environment as per the process detailed in Chap. 6.

Our model SOI is overly simple but, hopefully, instructive of both the analysis process and the use of the language to decompose opaque-boxes in a principled manner. We start with the identification of the SOI (recognizing its boundary) and the analysis of its environment.

We are keeping this example as simple as possible so that the reader may keep track of the process and what data is collected during it to produce the results shown in the sysXML listings below. In Chap. 7, we will look at how the language assists in the analysis of several example CASs and CAESs, and then in Chap. 9, we will provide a much more realistic example with more details after consideration of the knowledgebase in Chap. 8.

All of the major elements of the system and environment are designated with prefix codes, for example, "F" for flow or "Src" for source. These elements are labeled with names that will be registered in the knowledgebase. For example, the environmental entity "Coke-Source" is identified as "Src-1.2," meaning it is a source object in the environment (recall that the environment is at level "−1." That specific entity is number 2, hence it has an "id" of "Src-1.2," using the dotted notation for ids. The flow from this entity is "coke-input" with an identification of "F-1.2." Note that the text strings chosen for identification (names) are somewhat arbitrary and up to the users. They are case-sensitive, but beyond some basic rules for formation, are totally dependent on the user (who is the analyst).

Listing 4.1 shows the sysXML version of the description of the SOI and its environment. Note that all of the listings given are meant only to convey the general approach and not meant to be taken too seriously, especially the examples of constants (all caps).

Listing 4.1 The XML output from the knowledgebase of the description of the Steel-Plant system. This shows the opening of the description (an SOI with id of S0) and its main environment (level-1) sources, sinks, and flows.

---

to represent documents (like static web pages) but soon evolved to be able to represent any kind of system model one might like. Thus, we adopted this approach for our development.

```
<SOI   name="Steel-Plant"   id="S0"   type=PROCESS   type=CAES
subtype=MATTER subtype=ENERGY subtype=MESSAGE>
   <environment name="theEnvironment" id="-1" evolvable=TRUE>
        <sources>
                    <source  name="Energy-Source"  id="Src-1.0"
type=ENERGY subtype=ELECTRICITY inflow_id="F-1.0"
                evolvable=TRUE>
                <description>"This is the vendor who supplies
us with electricity."</description>
              <source_model name="electricity_delivery_sched-
ule" type="JavaScript"></source_model>
            </source>
        <source name="Iron-Source" id="Src-1.1" type=MATERIAL
subtype=IRON inflow_id="F-1.1"
                evolvable=TRUE>
                <description>"This is the vendor who supplies
us with iron."</description>
                   <source_model name="iron_delivery_schedule"
type="JavaScript"></source_model>
            </source>
         <source name="Coke-Source" id="Src-1.2 type=MATERIAL
subtype=COKE inflow_id="F-1.2"
                evolvable=TRUE>
                <description>"This is the vendor who supplies
us with coke."</description>
                   <source_model name="coke_delivery_schedule"
type="JavaScript"></source_model>
            </source>
        </sources>
        <sinks>>
           <sink name="Steel-Sink" id="Snk-1.0" type=MATERIAL
subtype=STEEL outflow_id="F-1.3"
                evolvable=TRUE>
                  <description>"This is the customer who buys
our steel product."</description>
                    <sink_model name="steel_delivery_schedule"
type="JavaScript"></sink_model>
            <sink>
         <sink name="Garbage-Sink" id="Snk-1.1" type=MATERIAL
subtype=GARBAGE outflow_id="F-1.4"
                evolvable=TRUE>
               <description>"We export trash and wastage from
the steel production."</description>
```

```
                    <sink_model name="waste_delivery_schedule"
type="JavaScript"></sink_model>
              </sink>
              <sink name=ATMOSPHERE id="Snk-1.2" type=HEAT out-
flow_id="F-1.4" evolvable=FALSE>
                  <description>"Radiate waste heat to the atmo-
sphere."</description>
                    <sink_model name="heat_production_schedule"
type="JavaScript"></sink_model>
              </sink>
          </sinks>
      </environment>
```

XML code uses *tag* elements to denote the structural element being described. We have defined the tag "SOI" to represent the start of a system description document. Tags are demarcated by the "<" and ">" characters and the end of a description is given with the tags "</", ">." The other entities included within the brackets are called attributes. Here the majority of attributes are "name," "id" (meaning the identification number in prefix dotted number format), "type" and "subtype." The latter two attributes may be present multiple times since elements might have more principle and subordinate characteristics that need to be identified. The words in all capital letters are predefined attribute values that largely follow the ontology of the last chapter. We suspect that all readers will have no trouble interpreting the tags and their contents (which are for the most part simple strings, but also additional tags that are part of the content of superior tags—which are indented to show their subordination).

In our approach the first major element to be described is the environment, between the <environment> and </environment> tags.[15] The description follows directly from Eqs. 4.1 and 4.5, describing the external sources, sinks, and disturbances (the latter have been left out in this simple example). Note the name is pretty obvious (theEnvironment) and would be more nuanced in a regular analysis of a real system.

Figure 4.14 represents what a user/analyst would see on a screen using the analytical engine we will describe in Chap. 6. They would drag elements to their positions and be prompted for the names/identifications of the elements along with other attributes as shown in Table 4.1. Later, the system would produce the code in Listing 4.1.[16]

[15] Tag elements like environment are not case sensitive. So, the programmer could have as easily spelled it with an upper-case E to show its importance.

[16] As will be covered in Chap. 6, the items used in the construction of a model come from a palette of items and are automatically connected according to the rule of the syntax, i.e., a flow entity dragged to source-interface would automatically be connected to both.

Note the repetition of source and sink elements as well as the flows from/to them. We included flows declared within the source or sink entities, named inflows and outflows consistent with whether something is a source or a sink—some such entities may actually be both sources and sinks, in which case they are designated as a "hybrid" object,[17] not shown in this figure.

Figure 4.15 depicts the boundary of the SOI with its various interface entities. We have eliminated the sources and sinks from the figure because they were captured in the previous figure (analysis), but we have included the flows that connect with the interfaces. Note that at this point the analysis is still a kind of opaque-box in that the internals of the SOI are completely unknown (for the present). All that can be said at this point is that the flows of materials, etc. from the environmental sources and to the environmental sinks are captured.

Listing 4.2 provides the sysXML listing of the boundary analysis. The process, as described in Chap. 6, selects a point on the boundary where a flow enters or exits. At this point, the internal details of the interface may not be understood, but the fact that an interface is essential to control the flows into or out of the SOI is enough to locate it on the boundary. In this example, the inflow of energy, flow F-1.0, is found to enter the building at a major power junction box (the "FuseBox"), designated interface I0.0. The assignment of .0 is because it is the first element found in the boundary. Elements may be either numbered from .0 on or .1 on depending on which conventions are declared by the organization doing the analysis.

The analysis proceeds around the building finding where materials, energies, and messages enter or leave. For example, it is determined that there is a special loading dock designed to accept iron shipments, I0.1. In Fig. 4.15, we have purposely left out an important, but not immediately obvious interface that will be discovered later. It is a messages-only interface, which is an often overlooked one in real life. More will be said about this in Chap. 6.

The type of the boundary is designated as CONCRETE, meaning it is a real physical structure (a building with walls, doors, etc.). The porosity attribute is a fuzzy linguistic variable, VERY_LOW, meaning that materials and energies cannot easily get in or out except through the interfaces. Messages, however, might easily get in or out through non-specified interfaces—a very usual problem in human organizations! Porosity of boundaries is still a rich area of research, as are some other attributes not discussed here. Non-concrete boundaries have different attributes. For example, a boundary may exist only because the internal component interactions are stronger than those between internals and elements in the environment. Thus, the boundary is real but not thought of as concrete.[18]

Listing 4.2 shows interfaces with the <protocol></protocol> tag elements already designated, for example, "@receive_electricity." At this stage of analysis, these are

---

[17] Some interfaces are also designated as hybrids when the protocol involves the exchange of messages between the source/sink and the interface. See Fig. 4.17 for an example.

[18] As an example, from biology, in plants the cells have exterior "walls" composed of cellulose, and considered concrete in this sense. In animals and plants, the actual cell is contained within a cell membrane, which is created by dynamic chemical interactions between molecules available in the cytoplasm. These molecules interact to form membrane structures quite naturally so the cell membrane (and those of organelle internally) are self-organizing, self-producing structures. Thus, they would better be of a type we might call DYNAMIC.

generally just stub names, the functions have not yet been determined, necessarily. The <recievesFrom> and <sendsTo> tags (not case sensitive, just using capital letters for readability) have to correspond with the type of interface. In the example below, we are dealing with strictly input and output interfaces (not hybrids). In a hybrid type both tags may be included in the content of the <interface></interface> tags.

Listing 4.2 Boundary description (note the indentation level is the same as the <environment></environment>) tags.

```
<boundary     name="theBuilding     id="B0"     type=CONCRETE
porosity=VERY_LOW adaptive= TRUE
    evolvable=FALSE>
    <interfaces>
          <interface name="FuseBox" id="I0.0" type=RECEIVES
adaptive= TRUE>
            <recievesFrom>Src-1.0</recievesFrom>
            <protocol>@receive_electricity</protocol>
             <description>"Receives the electricity from the
grid and connects it to the power distribution
               system."</description>
    </interface>
   <interface name="Iron-LoadingDock" id="I0.1" type= RECEIVES
adaptive= TRUE>
        <recievesFrom>Src-1.1</recievesFrom>
        <protocol>@receive_iron</protocol>
         <description>"Receives the iron shipments from ven-
dors. Protocol includes moving iron
               supplies into inventory."</description>
    </interface>
   <interface name="Coke-LoadingDock" id="I0.2" type= RECEIVES
adaptive= TRUE>
        <recievesFrom>Src-1.1</recievesFrom>
        <protocol>@receive_iron</protocol>
         <description>"Receives the coke shipments from ven-
dors. Protocol includes moving coke
               supplies to inventory."</description>
        </interface>
             <interface  name="Steel-ShippingDock"  id="I0.4"
type=EXPORTS adaptive= TRUE>
            <exportsTo>Snk-1.0</exportsTo>
```

```
            <protocol>@ships_steel</protocol>
            <description>"Prepares steel shipments according
orders and places them ready for
                    pickup."</description>
        </interface>
        <interface name="Waste-ShippingDock" id="I0.3" type=
EXPORTS adaptive= TRUE>
            <exportsTo>Snk-1.1</exportsTo>
            <protocol>@waste_disposal</protocol>
             <description>"Processes wastes into garbage and
recyclables and puts them in appropriate
                    containers for pickup."</description>
        </interface>
        <interface name="Ventilation" id="I0.5" type= EXPORTS
adaptive= TRUE>
            <exportsTo>Snk-1.5</exportsTo>
            <description>"Removes excess heat from the build-
ing and pushes it to the
                    exterior."</description>
            <protocol>@cooling</protocol>
        </interface>
    </interfaces>
</boundary>
```

Note that in the attributes we list the following: porosity=VERY_LOW adaptive=TRUE evolvable=FALSE. Porosity or the ability for foreign things to get into the system or things that are part of the system to get out in an uncontrolled way (i.e., no interface) needs to be specified. We're using the English phrase, "very low" as what is known as a linguistic value. The attribute, porosity, (a variable) can take on any number of linguistic values, such as "very high" (LOTs of holes), or "medium." The number of possible values depends on the exact nature of the boundary. The use of linguistic values in this context is another indication of treating boundaries as fuzzy in the technical sense.

The "adaptive" variable here is indicated with a true or false value. The significance of this is that the boundary has some internal dynamics that allows it to modify its operations within the adaptive limits. For example, one or more of the loading docks may be expanded by borrowing some inside floor space temporarily to accommodate a larger than ordinary shipment. Research on the implementation of adaptivity is still developing, but indications are that more than just a binary value may be needed (similar to porosity). Adaptability has to be built into the programming code, for example, the strings not in double quotes but preceded by an @ symbol. The protocol tags in each of the interfaces are functions in a suitable language, here we anticipate something like JavaScript.

A system element that is capable of evolvability means that element might be modified during simulation. Again, evolvability, like adaptivity, needs to be implemented in the computer code associated with the dynamic elements. For example, if the boundary had been designated as evolvable (TRUE) then the function(s) needed to implement this would appear in another tag, <evolution>@evolution_program</evolution>, coming just before the <sources> tag. The boundary defined in Listing 4.2 could have been evolvable if, for example, a mutation occurs in the porosity attribute, or in one of the interfaces' protocols, meaning that the function code itself has been mutated (a somewhat similar idea as used in genetic algorithms).

In Fig. 4.16 we begin the process of exposing the internal subsystems of the Steel-Plant SOI. In this depiction, we have identified four internal subsystems of interest, three associated with the environment-boundary interfaces and one, the production shop that interacts with the three.

The third internal process, the Material-Purchasing process was discovered during analysis of one or the other of the loading dock interfaces. There are internal message flows such as Invoices or shipping documents that are interchanged between the purchasing function and the receiving functions. Thus, the newly discovered subsystem needs to be captured. But the purchasing function also interacts with vendors through a variety of messages that are both incoming and outgoing. Its interface is thus a hybrid and so are the sources (i.e., they receive messages from the purchasing function that regulates the shipping of material).

In addition to showing these four subsystems, one, the Iron-Inventory has also been decomposed to expose level 2 sub-subsystems. The scenario for the iron flows is: Receipt of shipments of iron from the vendor (F-1.1) through interface (I0.1), the Iron-LoadingDock in Fig. 4.15, from there into the iron-stock (Stk0.1.1). When it is time to make steel, iron is withdrawn from the stock in batches (F0.1) and moved to Production (C0.6) through the interface (I0.1.1) and interface (I0.6.1). See Fig. 4.16 for more details.

The discovery of the Material-Purchasing subsystem happened when analyzing the Iron-Inventory management sub-subsystem (D0.1.1). Figure 4.17 provides more details about what goes on inside the inventory room. We show active (pump shapes) work being done to move the iron into and out of the stock. These work processes are managed by an agent (decider) who also monitors the inventory levels (orange triangle sensor) and sends a purchase request to the Material-Purchasing office (C0.3). These messages transit interfaces (I0.1.2 and a receiver in C0.3, as yet unnumbered).

A complete transparent-box analysis would identify the other subsystems and internal flows at level 1 and give rise to Listing 4.3.

Listing 4.3 This listing does not show the whole system but does show how the components, flows, and other objects in Eq. 4.1 are represented.

```
<components>
<!-- component elements in this section are actually refer-
ences to systems elements later in the file -->
    <component>"C0.0"</component>
    <component>"C0.1"</component>
    <component>"C0.2"</component>
    <component>"C0.3"</component>
    <component>"C0.4"</component>
    <component>"C0.5"</component>
    <component<"C0.6"></component>
    <component<"C0.7"></component>
</components>
<flows>
      <flow name="Iron" id="F0.1" type=MATERIAL subtype=IRON
units=TONS>
        <source>"C0.1"</source>
        <sink>"C0.6"</sink>
        <capacity>"10/hr"</capacity>
    </flow>
</flows>
<transformation>@monthly_steel_production</transformation>
<!-- The history of the firm is reflected in the financial and
cost accounting records -->
<history>@accounting_system</history>
<delta_t>MONTHLY</delta_t>
</SOI>
```

Only one flow is shown in the listing to save space. All of the internal flows would have similar structure. The tag "<!--" and closed with "-->" denote a comment in XML.

The last several elements of the model, "transformation," "history," and "delta_t" complete the system model description of Eq. 4.1. The transformation description is a computer code called @monthly_steel_production which simulates the average monthly production schedule of the plant. The history is captured in the firm's accounting records. Those records would need to be replicated within the namespace of this model! Finally, delta_t is the time step over which a discrete simulation would cover.

Note that the components list only contains the identification references of the subsystems exposed by the first level decomposition. This is because of the recursion of Eq. 4.3. Listings 4.1, 4.2 and 4.3 cover just the structures and functions of the SOI and its environment. In order to describe these subsystems, we need to provide listings for each one. We do this in Listing 4.4. Each discovered subsystem has its own <system></system> tag at the same indentation as the SOI listing. We use these tags instead of the SOI tag, all of the components now being treated as systems in their own right. When we reach the level of non-decomposable elements (the atomic elements) we use an attribute, ATOMIC to signal the interpreter that there are no more elements other than transformation and delta_t.

Listing 4.4 This listing shows how the subsystems are treated. In this listing we decompose the Iron-Inventory, identified as C0.1 in Fig. 4.17.

```
<system    name="Iron-Inventory"    id="C0.1"    alt_id="S0.1"
type=PROCESS type=COMPLEX
    subtype=MATERIAL subtype=IRON adaptive= TRUE>
    <parent>"S0"</parent>
    <environment>
        <sources>
            <source id="Src-1.1" inflow_id="F-1.0">
            <description>"This is the vendor who supplies us
with iron."</description>
                <source_model name="@iron_delivery_schedule"
type="JavaScript">
            </source>
        </sources>
        <sinks>
            <sink id="C0.6" outflow_id="F0.1">
           <description>"C0.6 is the production process where
the iron is made into steel."</description>
             <sink_model name="@steel_requirements_schedule"
type="JavaScript"></sink_model>
            </sink>
        </sinks>
    </environment>
          <boundary  name="theIronInventoryRoom"  id="B0.1"
type=CONCRETE porosity=EXTREMELY_LOW
        adaptive= TRUE>
        <interfaces>
            <interface id="I0.1.1" type=RECEIVES>
                  <interface  name="Iron-Batching"  id="I0.1"
type=EXPORTS adaptive= TRUE>
                <exportsTo>C0.6</exportsTo>
                <protocol>@batch_iron</protocol>
                <outflow>"F0.1"</outflow>
              <description>"Creates batches of iron for pro-
duction and puts them in a holding area for
                  moving into the production process C0.6."</
description>
            </interface>
```

```
        </interfaces>
    </boundary>
    <components>
        <component>"D0.1.1"</component>
        <component>"Stk0.1.1"</component>
    </components>
      <transformation>@stocking_and_distributing_iron_inven-
tory</transformation>
    <history>@recieving_and_kitting_records</history>
    <delta_t>WEEKLY</delta_t>
</system>

<system>      name="IronInventoryManagement"      id="D0.1.1"
type=AGENT subtype=COORDINATION
    adaptive= TRUE>
    <parent>"C0.1"</parent>
<!-- This body contains the managerial functions for the Iron
Inventory subsystem -->
</system>

<system name="IronInventoryStorage" id="Stk0.1.1"  type=STOCK
type=MATERIAL subtype=IRON
    adaptive= TRUE>
    <parent>"C0.1"</parent>
<!—This body would contain any details that the stock might
have. -->
</system>
```

We will wait until Chap. 6 to explain the other details of these listings (which comprise a single sysXML document) and several others that arise during analysis and information capture.

4.6 Conclusion

In this chapter, we have covered a number of closely related topics surrounding the concept of a language of systems. We started by describing systems thinking. First, we considered the way in which some people can think explicitly about systemness. This has been the standard way of looking at it. But we then claimed that there is another way to look at thinking itself as implicit systems thinking. That is, our brains already communicate among various representational modules based on the kinds of systems constructs explored in the last chapter. Many linguistics researchers and philosophers of mind believe that every human brain has a subconscious language of thought that supports the symbolic module interchanges/integrations. We suggested that this is actually "systemese," the mental language.

At that point, we focused on the communication of ideas with so-called public languages. We asserted that public—natural—languages are auditory translations of ideas captured in mental models using systemese.

The progression from thinking to language led further to consideration for how we could develop an explicit system language to take systemese public, as it were. We expressed the desire to have a language which was both machine and human readable so that people could communicate explicit systems descriptions (both structure and function) with one another and with computers (which could then run simulations from those descriptions). Such a language needed to take on characteristics of both natural and formal languages. Natural languages are extensible (adaptive) and evolvable. Formal languages have a strict syntax and semantics. This led to the notion of a formal definition of system that followed from the ontology. In Sect. 4.3, we provided the equations that defined the elements of a system and how they relate to one another.

Finally, we previewed our candidate system language (SL). We previewed the process of systems analysis with an example of a relatively simplified physical plant that makes steel using the graphical lexicon and syntax of systemness. And we showed the text of a system model that could be generated from the visual model. The text version (using sysXML) can be read by a display and simulation interpreter (similar to an active web page with embedded scripting).

Now that we have a language, we turn to a somewhat more detailed examination of the entire understanding process showing how the various pieces operate and fit together. That will be the subject of Chap. 5.

Chapter 5
An Introduction to the Process of Understanding Systems

Abstract Recalling that all systems are processes at some scale, the inverse is true, that all processes are systems. Understanding complex systems will require a system of inquiry, a systematic process. In this chapter, we provide an overview of the whole system understanding process, *presented as a system*. As outlined in the Introduction, this process is comprised of seven major sub-processes, each of which can be further deconstructed into sub-sub-processes (an example is given in Fig. 5.1 below). After introducing the whole system and developing the initial deconstruction the chapter will provide a general description and brief explanation of each of these major sub-processes. Each will be more fully explained, with examples of how the processes work, in Parts II, III and IV.

5.1 From General Systems Principles and Theories to Actual Systems Knowledge

Let us utilize the principles of systems science discussed in Chap. 2 to design a process of analysis and synthesis (maintaining a holistic knowledge) that will allow us to understand complex systems as described in the Preface and Introduction. This system of analysis and synthesis is the product of using the process that is to be summarized in this chapter and elaborated in the balance of the book. That is, reflexively, this process is a product of itself. To explain: The author has developed this process over years of studying real-world systems using elements from this process without having formalized their relations in the way developed in this book. Only after completion of the *Principles of Systems Science* book (Mobus and Kalton 2014) did the author realize that organized appropriately, these various methods, that is, deep analysis, knowledgebase construction, and generating/testing models from the knowledgebase, constituted a whole process or a system for understanding systems in its own.

This experience is a microcosmic version of the story of humanity evolving the formal systems we have for gaining increasingly veridical knowledge about parts of the world—the sciences and math. We begin with that story because it is instructive

G. E. Mobus, *Systems Science: Theory, Analysis, Modeling, and Design*,
https://doi.org/10.1007/978-3-030-93482-8_5

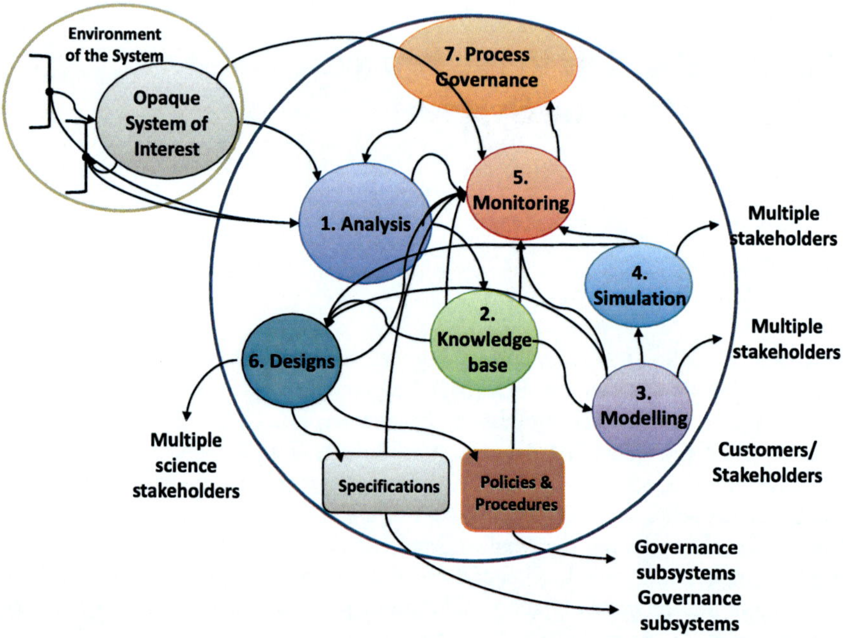

Fig. 5.1 The system for coming to deep understanding of other systems is comprised of the seven sub-processes (described in the text), including a governing process. The latter will be developed further in Chap. 12. The opaque concrete system, whether existent or to be designed, along with elements of its environment, is the input to the process. Various products are produced and distributed to customers and stakeholders. The arrows showing flows are of messages that convey information so long as there is a disequilibrium between the knowledgebase (and its outputs) and the analyzed system. Note that the "specifications" and "policies & procedures" are product outputs from Design. Specifications pertain to engineered systems, while policies/procedures pertain to other governance subsystems

with respect to how we can gain systems knowledge by the process described here. But unlike knowledge gained within a single disciplinary silo, systems knowledge is transdisciplinary allowing us to tackle the most complex kinds of systems we want and need to understand.

5.1.1 *From Observing to Characterizing to Modeling*

As described in the Introduction, we human beings have been most successful in our ability to interact with all of the various environments on the planet by virtue of our ability to more deeply understand the systems with which we interact in those environments. We can anticipate the near future given our knowledge of how things work and prediction of how they will behave. For most of human prehistory and even into the Bronze Age, humans could use their natural, intuitive, capabilities to understand their world based on the language of thought, which, as asserted in Chaps. 2 and 4, is actually systemese. As long as humans could observe the systemic aspects of the entities and processes in their environment, they could infer regularity of behaviors that served for anticipating the future. It also accounts for early humans who left Africa to adapt fairly quickly to very different environments. Systemness is the same everywhere even though the systems may seem very different superficially.

But the very capabilities that served so well as intuitive thinking produced a world of increasing complexity in which mere intuitive thinking would not lead to deep (enough) understanding, at least for individuals. Specialization, already incipient in tribal life (see discussion of the origins of economic culture in Chap. 9), became increasingly necessary because the breadth and depth capacity of the individual brain is limited (as discussed in the Introduction, polymaths are quite rare).

Science and math/logic were invented (or discovered through trial and error at first) to begin formalizing the process of gaining much deeper understanding of the world. Formalization consists primarily of developing the general pattern recognition of mathematics and logic so that a set of patterns could be applied to multiple different specific domains. Math and logic were elevated to methods of finding and exploiting patterns according to generalized sets of rules that did not depend on any particular substrate system. The scientific method (and the science process) is formalizations for acquiring direct knowledge within substrate domains through the application of mathematics and logic to generate theories and hypotheses, and hypothesis verification or falsification through empirical testing. These formalized processes have worked amazingly well despite the fact that many of the domain-specific methodologies were developed in a somewhat ad hoc or independent (from one another) approach. It wasn't until philosophers of science, like Karl Popper, explored the epistemological aspects of modern science in general that an overview of the common approaches to all sciences began to emerge.

The sciences and engineering processes have advanced significantly with the maturing of formal methods. One consequence has been a significant jump in the complexity of the subject domains. Biology has gone from a descriptive discipline (naturalist studies) to one dependent on mathematical modeling (population studies, protein folding, and many others). The term "systems biology" is used to describe the new mathematically oriented approaches to getting deep understanding in the domain. The same phenomenon is taking place in the so-called hard sciences as well as the social sciences. Models are being developed to gain deep understanding of many phenomena. But there is more to systems <subject> than just mathematical modeling.

There is a growing realization that there are many patterns of organization and behaviors that transcend the boundaries of disciplines. For example, the infamous logistic (S-shaped) curve associated with population growth is found to be useful in many different domains where some value is increasing, at first exponentially positive, and then after some time, turns exponentially negative.[1] Many "growth" processes can be modeled using this formulation with appropriate choices for the parameters.

Early pioneers of general systems theory recognized that there were general principles about nature that appeared to operate in all explored domains, which they termed "isomorphies" (Troncale 1978a, b) and realized that all of these phenomena display the properties of systemness, hence, their start at codifying aspects of systemness under the rubric of General Systems Theory (GST, von Bertalanffy 1968). At present, there are still many "versions" of what is purported to be a GST; many researchers have rekindled the vision of the pioneers and seek some consensus on a general set of principles that would constitute a science of systems (systemology, cf., Rousseau et al. 2016). The author along with co-author Michael Kalton attempted to collect the range of identified concepts of systemness and organize them into a principled framework in Mobus and Kalton (2014).

Though many would argue that we are still in the early stages of developing a consensus of what the principles are and how they should be organized, there has been significant progress made in that direction. In this author's view, we have enough background now that allows us to begin mapping the understood principles of a GST onto real systems (Checkland 1999, p. 9). We present here a methodological framework for accomplishing this task.

5.1.2 *All Sustainable Processes Are Systems*

The process of understanding a system deeply (from principle #11) is shown to be a system! This follows from principle #2 summarized in Chap. 2. Processes are defined as sustained only if they are in some way bounded and all sub-processes contribute to the purpose of the whole in a balanced way. They take in inputs, process them, and export outputs (products), thus they are systems by that definition (Chap. 4). In fact, as argued previously, system ↔ process; all systems are also processes! Thus, we talk about the system for understanding complex systems, the process of analyzing systems (and their environments), capturing the structured knowledge, and using that knowledge for multiple purposes in exploiting the knowledge.

Below we present the transparent box system for understanding as a model (Fig. 5.1). We will briefly analyze the whole system in this chapter in more detail as compared to Chap. 2 in order to provide an overview of the whole. Parts II and III will be elaborations of the components presented below and examples from multiple disciplines to show how they apply.

[1] The classic population model is given by: $P(t) = \frac{KP_0e^{rt}}{K + P_0\left(e^{rt}\right) - 1}$, P_0 the initial population size, K an upper limit on P, r is the rate constant, and t is the time.

5.1.3 A Generalized Process Applied to Any Domain

What follows is effectively a generalized model for the process of understanding complex systems. This is possible because the process is based on the general principles of systemness and those apply to any kind of system from any domain of science (physical and social). This will be demonstrated in Chaps. 7 and 9 specifically, but also with examples sprinkled throughout the other chapters.

There is something philosophically satisfying about the idea that a conceptual system can be devised, and then implemented in a physical embodiment, that can then be used to expose the systemness of real objects in the world. Closure under systemness? Of course, the implementation must make provisions for particularizations, such as terminologies associated with the lexicon developed in Chap. 4, the language of system. The generality of the language (due to the generality of the ontology in Chap. 3) and the methods for particularization make the process extremely powerful for explicating specific kinds of systems. But at an even deeper level, the generality and capacity for particularization makes the process a viable tool for cross-disciplinary communications. It is hoped that this will permit the re-integration of the sciences in general. This is the ideal goal of transdisciplinarity—the ability to work seamlessly across what now seem to be isolated scientific (perhaps even humanities) silos of knowledge. Indeed, in Chap. 8 we will introduce the possibility of a universal knowledgebase that could integrate knowledge from across all of the sciences based on the proposition, claimed in Chap. 3, that the Universe is a system, in which all else are subsystems. Who knows, we might actually be able to eventually answer Lewis Carroll's quintessential question: "Why is a raven like a writing desk?" For starters, our solution is that they are both systems!

5.2 The *System* for Understanding Complex Systems

The system for understanding is based on the archetype model of a complex, adaptive, and evolvable system (CAES) to be explicated in Chap. 10, as a transparent box, consists of seven operational level subsystems including a governing process. Figure 5.1 provides an abstract overview of this system. Below we explain each of these components and its contributions to understand.

Much of what will be presented in this chapter will easily be seen to relate to human-designed artifactual systems, which will be more fully developed in Part IV. However, it is important to point out that the process we are developing is as useful in coming to understand complex natural systems. That is, natural systems, such as complex ecologies, even whole planets, involving multiple disciplinary sciences can benefit from the system of understanding. In this situation, the process works to provide a means for integrating knowledge domains which is often necessary in order to grasp the nature of emergent phenomena. Each of the scientific disciplines is applied to their domain piece of the system but the use of system science as meta-science in the framework, so far disclosed, acts to integrate and synthesize whole systems structures/functions.

This is actually very similar to how the process works for systems engineering where the job of the systems engineers is to provide an integrative synthesis of subsystems developed by disciplinary engineers. The same process applies to both scientific and engineering understanding of the complexity with which they are faced in discovery or invention.

5.2.1 *The General Flow Through*

As described in the Introduction, the process we describe here differs from traditional approaches in that information flows through the process generally starting with more rigorous systems analysis and knowledge capture to provide the basis for constructing various kinds of models and simulations. This general flow is actually reverse of how most understanding processes, for example, the scientific method or classical systems analysis, have been conducted in which some initial understanding (i.e., opaque-box analysis such as system identification) leads to model constructions followed by testing (i.e., solving or simulating and predicting then comparing to the real system behavior). In the more traditional approach, the model of a system is a form of hypothesis about how a system works. The model captures what are hoped to be relevant variables and relations, when solved or simulated the results should match what the real system does. If it doesn't, the conclusion is that the hypothesis is wrong (disconfirmed) and a new version of the model is constructed and tested. Hypotheses are informed guesses as to how something works or what it will do in the future given certain inputs (environmental conditions). But what we hope to show is that this approach, while historically necessary and scientifically valid, is highly inefficient and becoming increasingly unnecessary since better deconstruction tools are coming on the scene almost daily. Along with the development of incredibly sensitive sensor technology and "Big Data" analytical methods it is becoming possible to design deconstruction methods for every scale of system we find interesting from planets to economies to organizations to people to ecological systems…well, to just about everything.[2]

The emphasis in this new process is on deep analysis first and model building second. Modeling can still provide useful feedback to the analysis process (through the governance subsystem in Fig. 5.1). There can still be iterations when discrepancies occur. No analytic technique will always be foolproof so some confirmation process is still required. But the number of iterations should be greatly minimized when the original analysis is done well and the knowledgebase captured is more complete than an opaque-box analysis can provide.

[2] For example, plans have been put forth to "instrument" whole cities, for example, place remote temperature, activity, and other kinds of sensors in key locations to collect real-time data about conditions in those locations throughout the day as part of the city planning process.

The principal advantage of putting more emphasis on deep analysis and knowledgebase production, over modeling first, is a gain in efficiency of the process and a tremendous saving of time in coming to understand what is going on. The model-and-see approach is logically valid (in the same sense that inductive reasoning is valid), but generally takes much more time to work through. And there will always be a lingering doubt about the validity of the model—have we thought of every possible set of inputs? The deep analysis first (and foremost) approach should save considerable time and increase confidence in model outputs (the models are based on real knowledge and not guessed).

It will, most likely, be difficult for many scientists to adopt this viewpoint since there is a long tradition of exploring phenomena by proposing models (hypotheses) first, or as soon as some opaque-box analysis has produced seemingly reasonable results. The traditional approach has generally worked in the long run, and this new approach might, at first, appear to be too much work at the front end, so why break tradition. We human beings can adopt fairly conservative positions at times.

On the other hand, there is already some growing evidence that attitudes toward deeper analysis are already evolving. In the social sciences, in particular, but also in the life sciences and medicine where complexities are substantial, there is movement away from reliance on ad hoc models (especially from the opaque-box perspective) toward a greater effort to understand what is going on "under the hood." For example, in the field of economics much more attention is being focused on actual microeconomic mechanisms in attempts to understand "real" macroeconomic processes (see Chap. 9). Coupled with the tremendous advances in computing power of the last decade (including the potentials of quantum computing in the not-too-distant future), along with the advent of "Big Data" analytics, machine learning, and other artificial intelligence advances, our technical capacity to do deep analysis is not lost on many scientists. It is hoped that the methodologies presented in this book will provide a framework by which these advancing tools can be used in a principled fashion.

5.2.2 Component 1: Deep Analysis

Here "analysis" means transparent-box methods. This is what we will refer to as "Deep Systems Analysis" or DSA. The system is deconstructed (decomposed) carefully to reveal the major inner workings, the internal components and their interactions, with the capture of this knowledge in a structured database (the knowledgebase—Component 2 below). Analytical methods revolve around instrumentation of systems so as to capture real data from "inside" the system, from each of its subsystems, rather than from an opaque-box analysis where only the inputs and outputs to/from a system are monitored.

5.2.2.1 State of the Art

The word *Art* is emphasized when referring to what has passed as analysis for good reason. Current approaches to systems analysis derive from projects in logistics and weapons design during World War II (WWII) when the concept of systems started to take hold. With the advent of computerized procedures used in business and government, as work systems got more complex, the term was applied to finding out what kind of information system would be needed for operations. The systems approach was still quite a new way of thinking in the late 1950s and 1960s as information technology was rapidly invading commerce and government. So, the notion of systems analysis was still very naïve. Most engineers and systems analysts (in job title) considered the system as either a "product" (i.e., the airplane) or the IT operation in support of the business (the management information system—MIS). They did not recognize that *the* system was the whole operation of the business or agency with the MIS component being just that, a component of the larger work process. Their focus tended to be on the subsystem. An airplane is a product, but the whole airplane-making business along with its market environment is **the** system that needs to be deeply understood in order to make the best product possible. The MIS is only a support system for the management decision processes that are collectively **the** governance subsystem. Not really understanding this distinction has been the source of many failures in engineering and information systems projects over the last many decades since the ideas developed in WWII were (partially) transferred to the business world.

It is an unfortunate fact that systems analysis has got a bad reputation in various contexts. We think the reason is actually pretty clear. Most systems analysis procedures are not actually analyzing the whole system in terms of gaining deep understanding of what it does or is supposed to do. Recall that deep understanding means the application of all of the principal ideas in systems science to finding out what is going on inside and out. Most often, for cost, or time, constraints, the kind of deep understanding promoted here is foregone under the assumptions that the analysts involved already know what needs to be done and there is a leap to the specification writing phase before anyone is quite sure they have gathered all of the details needed to succeed. Most often the analysts have not got a clear idea of what to do and end up making assumptions that turn out to be wrong.

For example, in critiquing systems analysis of management decision systems where the objective has been defined according to systems engineering practices ("hard" systems). Checkland (1999) observed the problem with jumping to the conclusion that an existing management decision system (say, paper based) would be a sufficient model/specification for the new, improved system. Referring to the analysis method used to develop a state-level information system, he gets right to the heart of the problem:

> All of this, of course, eliminates from the study many of the most interesting questions, especially those concerning the purposes served by the existing flows of information and the desirability of others. Ruled out of the study from the start was any consideration of the

meaning of the information flows in relation to decision-making at the state level. (Checkland 1999) [italics in the original]

For Checkland (1999, pp. 143–144), the main problem with using systems analysis as developed for engineering and operations research problem-solving was that once an objective has been identified the engineer assumes that her job is to simply find the best mechanism (meaning, perhaps, the most cost-effective) for accomplishing that objective. In working with complex social organizations involving management decision processes, the objectives are not necessarily that clear, what an engineer would say are "under-specified." Moreover, such systems have a tendency to change even while the engineering phase is underway. Checkland identified five arguments for why the hard engineering approach to systems analysis would not work. And he thus concluded that another approach, equally systemic, which he called "soft" systems, would be needed.

Thus, there are today various versions of systems analysis that range from traditional engineering (taking requirements as given) to agile methods (fast prototyping and including users in the process until they get it right) to very soft (drawing pictures and mind maps). All of them incorporate some level of systemness in their approaches and thus deserve to be called systems analysis. However, with what we have learned from project successes and failures using these methods we can now synthesize a much more holistic approach to the analysis of systems of all types. We are tempted to call it "true systems analysis" in an attempt to emphasize that it includes the analysis of all aspects of a system, not just the component pieces or the information infrastructure but the whole or true system. However, we will resist that temptation, if for no other reason than to avoid having to type too many words. Henceforth, when we say systems analysis (SA) we mean whole systems will be analyzed and not just parts in isolation, and it will be deep systems analysis, not just opaque-box analysis.

The natural or physical sciences (physics, chemistry, biology, astronomy, etc.) have fared much better in terms of developing deep understanding, at least in terms of phenomena of interest (as opposed to systems of interest). The reliance on reductionist approaches has paid off in terms of exposing underlying mechanisms that account for higher level phenomena. It turns out that such methodological approaches are actually quite close to systems analysis. The difference is that reductionist methods are directed at deconstruction specifically. Systems analysis, on the other hand, explicitly seeks to maintain knowledge about structural and functional connections between the parts exposed in deconstruction. It seeks to maintain a holistic perspective of the system even as it deconstructs the parts.

5.2.2.2 A Three-Phase Process

The approach to systems analysis to be demonstrated in this book is actually a three-phase process, as mentioned in the Introduction. The first phase involves system identification or determining the boundary of the system of interest (SOI). As mentioned previously this is not a trivial task. It involves observation of the real system (or careful delineation of the boundary in a to-be-designed system). The analysis is of the various inputs and outputs (shown in Fig. 5.1), including observation of any disturbances not directly associated with inputs from noted sources.

Once the boundary and boundary conditions have been identified, the analysis turns to the various sources and sinks in the environment that interact directly with the SOI and over its entire life cycle to the extent possible—by definition it is impossible to identify non-stationary changes that could come into play in some future time. For the noted sources and sinks, the analysis looks to characterize as fully as possible the dynamics of the flows/interactions the SOI has with the sources and sinks. As mentioned previously, the analysis does not attempt to go into the internal workings of these sources and sinks as part of the analysis of the SOI. Later we will introduce the exception to this rule when it may be discovered that the dynamics of the environmental element is not adequately described by mere modeling and when the coupling strength between the SOI and that element is such that it becomes advisable to reconsider the boundary, treating the element as part of the SOI and analyzing its internals as any other subsystem in the SOI.

That leads to the third phase which is the recursive deconstruction of the system into its subsystems and those into their sub-subsystems and so on until a stopping condition is met. The overview of this was covered in Chap. 4; see Fig. 4.5 showing "atomic" work processes that constitute the stopping conditions for recursive deconstruction. In Chap. 6, we will examine several specific methods for deconstructing real systems based on their kind and composition (e.g., how to deconstruct a business enterprise or a living brain).

5.2.2.3 Reversing the Direction of Analysis: Enlarging the Scope

The language described in Chap. 4, based on the formal structure given in Eq. 4.1, allows for an interesting approach to analysis. The basic intent is to analyze by deconstruction the SOI, which is identified as level 0, the root of the structural tree. In this way, analysis is similar to the reductionist program used in the other sciences. However, there is no reason that the analysis need only go "downward." Since the environment has already been analyzed and logged into the knowledge base, the elements, the sources and sinks, are available. The environment, recall, is actually a supra-system on a larger than SOI scale by virtue of being connected to the SOI. That is, the original SOI is only a single subsystem of the supra-system. It is possible to reverse the direction of deconstruction by considering the boundary of a new SOI, the supra-system, pulling all of the environmental elements into the new supra-SOI.

In essence, the analyst is taking a step back to grasp the larger picture. As an example, suppose a major aircraft manufacturer wants to design a new more fuel-efficient airplane for intermediate haul flights. They have within their company all of the plane design engineers needed to accomplish this task. They believe there is a market for such a plane, that the demand will be high, and their sales will more than justify the investment in design and tooling that will have to be done. The company employs a systems engineer to bring all the pieces together for the project. But this particular engineer is "seasoned" by experience and realizes that while there might be a market, or the appearance of one, the new airplane will introduce some major changes in the operations of the airlines and airports with respect to flight scheduling, turn-around times, and a host of other operational considerations. These would all be considered inputs and outputs to the airplane once in service, part of the environment. So, the engineer decides to widen the scope of the systems analysis to think of the SOI not just as the airplane, but the company and the customers, the airports and their operations. He calls in experts on these elements who analyze the impact of the new design on them. What they determine is that it will be quite disruptive for a number of stakeholders unless the airplane manufacturer supplies additional support equipment designs that will facilitate the use of the new airplane with minimal disruptions to other airport services.

As it happens, new airplane designs do go through a period of trial-and-error adjustment phases, the costs mostly borne by the airlines and airports. This is an example of an evolvable system. Could these disruptive adjustments be minimized by reversing the direction of systems analysis as described above? In the sciences, this is done to some degree by the fact that there are specialists who work on problems at a higher level than where the reductionists are working. They are *integrators* who can use the knowledge gained by reductionists to map out the higher-level functions. They put the pieces discovered by the reductionists, of the larger puzzle together. For example, once biochemists understood what the adenosine triphosphate molecule was and saw how ubiquitous it was in the cytoplasm, people working at the whole cell level realized its role in distributing energy to the various work site organelles, like ribosomes, and how these molecules interacted with those sites.

5.2.2.4 The Science of Systems Analysis

Analysis is a process of deconstructing a system in order to find the subsystems that underlie the operations, behaviors, of the system. At its roots this is a reductionist approach, but with an important difference. Systems analysis makes a concerted effort to observe and maintain knowledge of the causal connections between subsystems. In the sciences, this corresponds with the attempt to understand the context of phenomena. For example, in the analysis of the structure of the DNA molecule, known at the time to be the major molecular constituent of genes, James Watson and Francis Crick almost immediately recognized the significance of the double helix construction of the molecule and the implications of the four nucleotides (adenine, guanine, thymidine, and cytosine) forming a genetic code. The deconstruction of a molecular structure was interpretable as the key to genetic knowledge because of the higher-order knowledge of what genetic information meant. Scientific deconstruction is systems analysis but not necessarily embedded in a process for recognizing the significance of a found structure or function in the context of a larger whole.

The deconstructive work of systems analysis is not much different from the reductionist program in sciences in terms of process. Both look for the subcomponents and their interactions. The reductionist paradigm seeks sub-phenomena, while the process of systems analysis is always alert to the context of those phenomena as subsystems of the larger whole. The sciences have been so spectacularly successful because they have been doing systems analysis without the formality supplied by a systems approach, until recently.

Systems analysis is the process of deconstruction of higher-level processes (phenomena) to find out what sub-processes underlie the process. The main difference is the maintenance of understanding the context of the sub-processes—the system in which they are embedded. For example, cellular biology has done an incredible job of deconstructing the internal workings of cellular organelle and metabolism. The cell is the overarching system and the organelle and metabolic processes are subsystems within. This has always been "understood" or implicit in the process of coming to understand the cell. The only difference between a systems approach and the historical approaches of cellular biology is that the latter was pursued in what we respectfully call an ad hoc framework. The mechanics and purpose of each new discovery of an organelle, like the mitochondria, was pursued primarily to see what the thing does. Only later, after working out what it did, was there a realization that what that was contributed something to the whole cell system. Eventually the systems approach was implemented as those who were able to step back and look at the integration of the bits and pieces started to put together a "bigger" picture.

The current program in systems analysis is to recognize the advancements in knowledge that came from the after-the-fact integration of reductionist-derived knowledge and to turn the perspective on its head. Systems analysis is a top-down (mostly) process in which the reductionist process is the servant of the holistic or understood-to-be-integrated framework at the outset. Rather than an ad hoc deconstruction of a system for the sake of examining a bit for its own sake, the systems approach seeks to explicate the bits for the sake of understanding the whole system of interest. We believe that this change of perspective will make the sciences even more productive in their pursuits of the bits because it provides a holistic knowledge framework into which to plug the knowledge of bits as they are discovered.

In essence, all of the methodologies of reductionist and empirical science are applicable to systems analysis. The only difference is that we are starting with the intent of keeping the whole in mind as we pursue the reductionist methods. This is why systems science is a science in its own right.

Deconstruction, from a systems approach, is an algorithm for discovery. As already pointed out, a system can be represented as a hierarchical structure—a tree of systems and subsystems. The algorithm is a recursive method for discovering the child nodes in any node in the branches of the tree. That means that we reapply the discovery process at each subsystem node, finding the sub-subsystems which then become systems in their own rights. The process can be conducted in a breadth-first approach—finding all of the subsystems at any node—or in a depth-first manner—exploring a single branch of the tree until we reach a "leaf" node or a sub-…-subsystem that is an "atomic" unit. By "atomic" we mean a system that performs some minimal process or a component whose process is already well understood. For example, in decomposing a computing system we could stop when we reach the elemental logic gates since we know quite well what they do and how they do it. There are similar "stopping rules" for other kinds of systems that will be covered in Chap. 6. Every recursive algorithm requires such stopping conditions in order to be useful!

The process of deconstruction (reductionist method) in the systems approach must be captured in a structured knowledgebase. The structure of this knowledgebase is determined by the principles of systems science and the formal definition given in Chap. 4. The structure of the knowledgebase is the key to exploiting the systems approach.

5.2.3 Component 2: The Knowledgebase

A database is a set of structured files that contain data relevant to an organized system (the systems program that runs the database engine is called a DBMS). For example, an employee database contains data that is relevant to the employees of an organization such as their personal data, pay rates, employee numbers, etc. Databases are the memories of organizations, used by various programs (not the least of which is payroll) to keep track of the employees, inventories, and customers (to name a few). A knowledgebase is a database that has a more defined purpose than just keeping track of data. The main difference is that a knowledgebase is a highly integrated set of databases that constitutes all that is known and knowable about a system.

The most dominant database model used today is called a "relational" database system. The relations between data elements are based on model schemata. For example, in an employee data base, a single file (called a table) can contain employee personal information such as address, years employed, social security number, etc. Each employee is assigned a unique employee number which acts as the "key" to any specific employee's personal data. Tables are composed of rows for each employee number (in the key column), with the rest of the data occupying columns with headings giving the type of data contained. Another table can contain payroll data for each employee. Data such as pay rate, tax rates, withholdings, etc. are kept in rows as with the personal information table. The employee number is again used to identify each employee; it is the one thing that is common between the two tables.

The reasons for keeping these data groups separate are several, including security and ease of maintenance. In using relational DBMS for different applications, the tables are "related" by, in this example, the employee number. Say, for example, an application is used to print payroll checks and route them to envelopes for mailing to the employee. The application first computes the pay for the employee from the data in the payroll database. It then cross-references the employee number in the personal information database to set up the mailing of the check, retrieving the address, etc. Relational databases (RDBMS) engines have been perfected to do these kinds of operations using a relational language called SQL (pronounced SeQueL).

Unfortunately, a pure RDBMS is not an ideal way to represent and store systems knowledge. Systems are sets of *objects*, as established in Chap. 4, each being quite complex. Objects are whole things that are not represented easily in RDBMS tables. Fortunately, there has been a growing interest in object-oriented databases (OODBMS) in which the storage element is an object and the object can have considerable complexity. At present, the field of OODBMS is not as developed as RDBMS and so the ability to readily store system object knowledge is still in a state of flux. However, the major outlines are clear. In Chap. 8, in Part II, we will see examples of the way in which we will capture the knowledge from systems analysis and store it in a way such that it is searchable and "relatable" to other objects (e.g., subsystems with interactions).

5.2.4 Component 3: Modeling

The most active area of using systems theory in the sciences and engineering is the building of system models for simulation in a computer (as indicated above). Indeed, many of the sciences have adopted sub-fields called "systems <science name goes here>" based primarily on the use of simulation modeling. Unlike earlier mathematical models involving differential equations, where higher-order or nonlinear differential equations proved intractable with respect to closed form solutions, simulations can involve arbitrary durations and numbers of internal feedbacks. Numerical methods are used to approximate within nearly arbitrary precision what a formula based on continuous (real number) methods in the calculus would produce if solvable.

All models are abstractions of the real systems they attempt to emulate. The construction of a model involves artful choices by the modeler regarding the "boundary" (conceptual in this case) and the granularity of internal components. They must also choose what they believe to be relevant variables to monitor as the simulation run progresses. This is called "instrumenting the model" and refers to the periodic recording of the variables of interest in order to collect data to be used in dynamical analysis. The reliance on the pre-existing knowledge of the modeler, plus their experience in constructing meaningful models (based usually on having constructed not-so-meaningful models earlier in their careers!) makes the part of systems science more of an art than a science. The approach works to some degree in engineered systems development, in part because there is already a rich library of system models (templates) that have been proven in the past and can be used to guide new designs. Thus, it is more knowledge-based than happens in the sciences when exploring phenomena.

We have argued that, traditionally, the use of models has been based on having a lack of transparent-box knowledge and/or an inability to observe and measure internal mechanisms of a system. The models substituted for that kind of knowledge by using informed best guesses as to how a system worked to construct the model. If the simulation (solution) of the model then produced results effectively similar to the behavior of the real concrete system, the analysts felt justified in concluding that their understanding was sufficient to use the model for its real purpose—prediction. In the process being set forth here, we assert that more analysis (deconstruction or transparent-box analysis) be done first before constructing models. That means a model is less built as a crutch to understanding as a tool for confirmation of knowledge already gleaned from the analysis. At the very least this reversal of roles means that models used for prediction purposes should have higher levels of confidence associated with their results.

Model generation is based on the fact that all of the relevant details of structure and function of a system at any arbitrary level of organization have been captured in the knowledgebase and are available for use. The model itself is a replication in, say, software code very similar to a system dynamics model in Stella or similar environment. A group of software tools can be developed that draw from the knowledgebase and generate the code. Alternatively, and in support of the code for simulations (see below) tools such as SysML might be developed to generate conceptual models (diagrams) for visualization of the system at a particular level of organization (i.e., abstraction). These kinds of tools and their uses will be covered in the relevant chapters to follow.

5.2.5 *Component 4: Simulations and Hypotheses Testing*

Once a model of a system, at whatever level of organization, is available it is available for running in simulation. The code generated in Component 3 above can be run on a computer, after determining appropriate time steps. The same visual framework used for analysis and data capture can now be used to animate the simulation. Output from the simulation, the data recorded for the variables of interest, can then be analyzed and graphed appropriately just as is the case now. Many of the tools needed for this component already exist.

5.2.6 *Component 5: Generating Designs or Policies*

In the case of engineered systems, the ability to generate a set of design and performance specifications is essential to success. In the current approach to doing this, the process is labor intensive as various specialist engineers have to analyze the requirements and apply domain-specific knowledge (e.g., electrical engineering) to generate designs for some specific functionality within a subsystem. Engineers are expensive people to employ. Moreover, the production of truly qualified engineers in many fields is lagging sadly behind the needs of modern society. Learning to be an engineer is hard work and many modern students are not inclined to pursue this line of education.[3]

Today we have a significant amount of knowledge of how to produce design specifications given that we have well-defined functional specifications. It is entirely conceivable that once a to-be-designed system is captured in the knowledgebase (i.e., we have an in-depth description of the structures and functions of systems and subsystems, etc.) the process of generating design specifications can be readily automated and easily checked.[4]

This is actually a corollary result to the generation of models discussed above. A design is, after all, an abstract version of a concrete system, containing meta-data regarding the construction of components from the lowest level of organization up to the highest level. In other words, the same knowledge that is captured in a knowledgebase that allows us to generate models is the exact knowledge needed to generate design specifications. In fact, in engineering, the typical approach is to co-generate specifications along with models used to test the designs being proposed. The two processes are Siamese twins co-joined at the hip, as it were.

5.2.7 Component 6: Monitoring Behaviors

This subsystem reflects the reality of complex adaptive and evolvable systems. Every such system must monitor its own behavior (state changes over time) relative to the desired results. In this case, a monitoring subsystem is in place to determine whether or not the process is producing reasonable results. The main method for doing so involves measuring the outputs of several other subsystems (components) such as the modeling, simulation, and design specifications/policy generation processes. These are all the "action" outputs from the entire process (like the movement outputs of brain decisions processing). It has to be determined that they are producing results consistent with the objectives of the systems understanding process as a whole or, otherwise, there have to be governance interventions in the process to direct it toward "better" outcomes. The methods of monitoring (measuring and comparing to goals—standard first-order cybernetics) will be developed further in Part III.

[3] Sadly, even those who do wish to major in an engineering degree are often not adequately prepared for the rigor of the field or they have been given a very superficial orientation to what is involved in it. Our modern education system puts emphasis on the fact that engineering professionals make good salaries and uses that as a motivation to get students interested in the fields. Then once they start taking the courses needed to obtain the degrees, they realize how hard it is. Motivating students to take difficult subjects based on future potential salary rewards is probably not the best way to increase the number of engineering graduates.

[4] For example, there is a growing body of knowledge in what is called model-based design in which tried-and-true models of designs for "paradigm" systems serve as templates for developing designs of specific systems matching the template.

5.2.8 *Component 7: Process Governance*

Chapter 12 is devoted to the architype model of sustainable CAES called "governance." This pattern is ubiquitous in all CAESs in nature and is so important to understanding why systems persist over time that it deserves a whole chapter. As with all such systems, the system of system understanding has its own governance sub-process to ensure it gets the job done right! Those involved in various kinds of systems processes today will recognize parts of this sub-process, for example as "project management." But viewed from a holistic systems perspective it is so much more.

One immediate implication of a governance process is not only to obtain and maintain stability in the system process, but also to manage the expenditures of resources (costs) as compared with the receipt of resources (revenue or benefits). We outline that function next.

5.3 Cost-Benefit of the Process[5]

The systems understanding process is undertaken within the context of how much benefit does society or a profit-oriented organization gain from it relative to the costs incurred in doing it. The process obviously involves both direct costs and risks of additional costs if something goes wrong. So, the main question everyone will ask is: Does the benefit of deep understanding gained via this process exceed the costs involved in undertaking it? It's a fair, but greatly misunderstood question.

The current prevailing concept of economics has led managers and investors to focus attention on the short-run. Quarterly performance seems to be the main decision factor in working out budgets for investments. The sentiment seems to be: "Take care of quarterly earnings and the long-run will take care of itself." There may be little that can be done at present to change this sentiment. The economic theories of neoliberal capitalism provide internal positive feedback loops that reinforce it.

However, here we present some arguments against this sentiment or, more correctly, for a different sentiment that puts more value on long-run thinking. We will return to this theme in Chap. 9, "Analysis of the Biophysical Economy," after demonstrating that a systems approach to studying the economy exposes some of the fallacies lying behind the neoclassical theories and the neoliberal capitalism that dominates the global economy today. In anticipation of many arguments, we acknowledge that this may be a quixotic quest for now. But, at some point in the not-too-distant future, we predict there will be a major disruption of the global

[5] In this section, we will be referring mostly to business and government projects, however, it should be noted that most of these factors apply equally to large science projects such as The Human Brain Project, which seeks to build an exa-scale supercomputing platform for research support of models of the human brain at multiple scales of resolution.

economy as a result of the short-term, profit-oriented focus. It is that focus that leads to taking shortcuts in systems projects (e.g., mega projects). As already pointed out, the failures of these projects are already costing billions of dollars lost to failures of various kinds. This portends future problems as projects only get larger and more complex.

In this section, we review various costs and benefits associated with using the deep understanding process just outlined and compare them with those associated with shallow understanding.

To begin with, a deep understanding process will be as costly as the complexity of the system we seek to understand. That is, the costs, roughly indexed in human labor hours, increase with the complexity of the system, as we have defined that metric in Chap. 3. Moreover, though there is no solid empirical evidence to support this, the intuition is that the relation is nonlinear, possibly exponential and almost certainly quadratic at least. The reason is the overhead needed to consider all of the subsystems and their relations for systems that have greater hierarchical depth seems to expand with that depth. The number of interactions and especially the communications channels (message flows) needed to coordinate more subsystems appears to increase in this fashion. This is supported by a simple result from graph theory when a graph is generally dense, meaning having connections between most of the nodes with each other. In a simple complete bidirectional graph, the number of edges, N, increase as $n * (n - 1)/2$, n being the number of vertices, which is order n^2. However, a system's internal subsystem connections are not simple unidirectional graphs. Most interactions are flows or directional and accompanied by bidirectional communications, which increases the number of edges at least by two. Thus, the more subsystems found in any level of the system, the longer it will take to analyze the connectivity.

The deep understanding process requires a thorough analysis of the subsystems and their interconnectivity at every level. The approach most often taken today is to make some assumptions about systems based on prior experience with similar systems. That is, analysts avoid doing a deep analysis by assuming that a system is comprised of subsystems and interconnections as seen in other (presumably) similar systems. This saves a lot of time. If the assumptions are correct, then one need merely copy the architecture/design of the model system and consider that sufficient for making claims about the nature of the current SOI. The problem is that all too often they are not the same. Even small differences can make disproportionate large differences in the target system. Nonlinearities in subsystem functions (e.g., a misplaced amplifier) can lead to unintended behaviors, among many other things that can go wrong in blindly copying what is assumed.

For one thing the model system being used to form these assumptions may not be correctly modeled in the first place, it having been, itself, the result of assumed designs (or hypotheses in the case of scientific decomposition). It simply is the case that when it comes to actually understanding a system there is no substitute for the kind of deep analysis covered in Chap. 6. Only after having performed that kind of systems decomposition should one be confident that the models and sub-models are veridical.

But time is money. The longer it takes to get the analysis right, the longer it takes before a reward is realized. Given the belief in the time value of money, in our modern capitalist enterprises,[6] this amounts to diminished returns on investments so there is little incentive to "waste" much time. Coupling this with the overconfidence of most people, scientists and engineers, in their abilities and expertise, you have a toxic formula for disaster.

So, there is a compelling reason for not doing a thorough analysis of the SOI as outlined in Chap. 6. It costs too much. Similarly, there are high costs associated with collecting the data and maintaining an adequate knowledgebase as described in Chap. 8. So-called knowledge management can be very expensive in time and labor. So, there will be a tendency to do the minimum in the realm, for example, of model building (see Chap. 15) or generating detailed specifications for designs.

The bottom line (pun intended) is that doing deep systems understanding has a high upfront cost on the analysis phase. There is also the possibility that total costs will be very much higher. Each of the phases as described in this chapter probably has additional costs associated with the level of effort required. And this alone would dissuade organizations from adopting the process in order to gain deep understanding.

But.

What about the long-term costs?

What does it cost over the life of a complex system to operate it if the design is buggy or inefficient? What does it cost if the system fails to deliver its promises?

The kinds of complex systems that are the targets for today's considerations, for example, complex social systems, or the Internet of Things, or ecosystems to be managed, entail significant costs if something goes wrong because some little part of the system was not sufficiently understood. In other words, the cost of not getting it right potentially far exceeds the costs of doing the project well or not getting to market rapidly.

Take the argument for the latter concern. If we don't get this project done quickly someone else will. We'll miss our opportunity. We'll miss getting income right away.

So what?

What is the benefit to anyone of putting a product out there that is doomed to failure due to un-understood design? Who profits? The case studies abound where organizations of all stripes pushed a project through only to find their long-term profitability (or cost minimization) over the life cycle of the project/product usage suffered tremendously.

What, then, of benefits? They have to exceed costs in order to make the project worthwhile.

That is actually part of the systems analysis. It is undertaken in the first place because the existence of a system to perform a function that produces X in benefits,

---

[6] It is understandable that this attitude arises in engineering projects within for-profit organizations or non-profit ones where limited resources means controlling costs. Sadly, however, the same kind of thinking has entered the realm of the sciences where grant monies are bid upon and "useful" products are often the goal of the research.

is perceived as solving a problem that currently costs Y and X ≫ Y is the goal. What should be invested in the project depends on how much this relation produces and over what time scale. In the case of sciences, the measures of X and Y are not as clearly defined—what is the value of new knowledge? In the case of an engineered solution, it should be possible to estimate the relative values. In engineering it is possible to continue to test the proposition that X ≫ Y as each stage of the process proceeds. If at some point it appears the proposition is not valid or, at least, in question, then it is possible to stop further developments and cut losses of to-date investments. It should be noted, however, that the ability to abandon sunk costs often takes a level of wisdom that is not particularly granted to many managers!

5.4 Conclusion

The process outlined in this chapter is a system for gaining deep understanding of any system of any complexity. It consists of seven component sub-processes all of which are required to do good job of understanding the system of interest. Details of these sub-processes will be presented in the rest of this book. Since the analytic process is easily the most important component, Chap. 6 will go into great detail as to how this process works. Chapter 7 will provide an example of following the analysis process and capturing the data relevant to the definition of system from Chap. 4. Then Chap. 8 will elaborate on the structure and functions of a knowledge-base which is built from the data collected in analysis and organized according to Eq. 4.1 and subsidiary equations in Chap. 4.

The process of gaining deep understanding of complex systems is, unfortunately, expensive as compared with how systems are analyzed and specified today. The reason has to do with the drive of profit-oriented capitalism and market competition—we rush to get things done as quickly and expediently as possible. But this leads to errors in analysis and judgments of efficacy that cannot be caught at an early stage; they must be experienced as failures to be detected.

That mode of engineering and building complex systems entails even higher costs to society and users in the long run. In truth, we don't really have a good idea of the total costs to society of these kinds of failures. But in many case studies where short-term vs. long-term costs have been tracked, invariably the long-term costs of poor understanding have far exceeded the short-term costs even after taking into account the time value of money.

As humanity seeks to push the envelope of complex systems or seeks to better understand complex systems such as the whole Earth ecology, it will find that trying to do so profitably, that is, benefits outweighing the costs, will require a rethinking of the upfront effort, admitting higher costs of deep analysis, but gaining much higher returns on investment in the long term.

Part II
Analysis for Understanding

While Chap. 3, in Part I, was mostly about system ontology, this part is devoted to system epistemology – how to gain understanding (knowledge) of systems.

In the next three chapters we lay out the method of deep systems analysis. Then in Chap. 9 we provide a cursory examination of using this method to better understand an extremely complex system – we'll examine the system of energy flow through our economies.

The method of deep analysis is based on the theoretical considerations covered in Part I. In particular, Chaps. 3 and 4 provided the basis for understanding systemness in general. Armed with that knowledge, we know what to look for as analysis proceeds, we know how to do the looking, to construct an algorithmic procedure, and how to maintain the relations between components in a structured knowledgebase that will allow us to 'reconstruct' as detailed a model of the system afterward. From that latter ability, we can conduct simulations to determine the accuracy and completeness of our system knowledge.

Chapter 6
The Process of Deep Systems Analysis

Abstract The first component discussed in Chap. 5 is the deep analysis of a system of interest using a top-down decomposition procedure. This chapter will provide the guidelines for how this procedure integrates the objectives of a reductionist analysis with retaining the holistic aspects of systemness by using the recursive system definition of Chap. 4 that preserves the interrelations of subsystems at all levels of organization in the system. The procedure is one of "deep" analysis meaning that it is an algorithm for guiding a recursive process exposing increasingly deep details of subsystems and components until we come to stopping conditions, down any leg of the hierarchy based on finding "leaf nodes" representing "atomic" processes. We end the chapter looking at "advanced" concepts of complex systems such as fuzziness, adaptability, and evolvability considerations. These will be revisited in the coming chapters.

6.1 What We Seek to Achieve

In this chapter, we focus on the analysis phase briefly described in the last section of Chap. 5. This is the part of the process where knowledge of the system is obtained through deep analysis and captured in the knowledgebase. Guidance for the procedures comes from the formal definition of a system given in Chap. 4 along with the language of systems derived from the ontology of Chap. 3 and the formal definition. What we will be describing in this chapter is essentially the procedures to be followed in performing this analysis in the abstract, that is, as they apply to any arbitrary system. In the next chapter, we provide examples of how these procedures can be used to analyze three particularly complex specific systems. In Chap. 8, we will cover the nature of the knowledgebase itself—how it stores knowledge for retrieval and use, again as briefly described in Chap. 5.

G. E. Mobus, *Systems Science: Theory, Analysis, Modeling, and Design*,
https://doi.org/10.1007/978-3-030-93482-8_6

The procedures we describe are:

- Define the System of Interest (Level 0).
- Environment, Boundary, and Flow Analysis (Level −1).
- Recursive Decomposition of Subsystems (Level 1…*m*).

Within each of these procedures, we will cover the methods for collecting data and entering it into the data structures of the knowledgebase. It is the organization of the data in these structures along with the relations between these structures that turn the data into knowledge.

The work to be described in this chapter is not without predecessor thought. Specifically, the works of George Klir (2001) and Peter Checkland (1999) have been instrumental in forming the concepts presented here. Klir was a mathematician who was interested in the abstract representation of systems (as described in Chap. 4) and who developed a rigorous framework for thinking about systems, most particularly of what we now call the "hard" sort. His work, being primarily mathematical in nature, seems to have gotten lost in the general systems literature, at least in the Western world. Our aim is to bring his insights down to earth, so to speak, by making the procedures implied in his abstractions more operational.

Checkland investigated systems that involved human actors (agents) and complex decision-making and came to the conclusion that such systems could not be characterized in the same way the so-called hard systems were done. He (with others) developed the concept of soft systems to deal with much less well-characterized systems like organizations involving social interactions (which include human emotions and motivations, biases and beliefs). His work and thoughts are pursued today in the frame of organizational systems thinking. Without losing his insights as to what makes a soft system soft, we will attempt to show how more formal constructs can be brought to bear on such systems. Our objective is to show how all systems can be understood in the same framework of systemness and that doing so can give rise equally to scientific understanding of natural phenomena as well as being the basis for generating designs and policy prescriptions. In other words, to unify hard and soft systems under a single notion of systemness.

At the time that both of these thinkers were formulating their approaches there still existed formidable hurdles with respect to characterizing very complex systems, particularly human thinking. And so systems science has taken on a bifurcated set of tracks that cater to hard problems versus soft or wicked problems. Our work will attempt to demonstrate that the deficiencies in characterizing the hard aspects of so-called soft systems are giving way to more rigorous methods (e.g., functional imaging of living brains during perception and conceptualization). Thus, we claim, a general system understanding methodology applicable to both hard and soft systems is now amenable. We briefly examine the status of Klir's approach and that of Checkland's as they represent the dichotomy and then proceed to outline the methodology that we assert will reconcile it such that there is only one concept of systemness to be employed.

6.2 Perspectives on Systems Analysis

In the next two sections, we will briefly review Klir and Checkland to provide some framework for arguing the resolution between the hard and soft systems perspectives.

6.2.1 Klir's General Systems Problem Solver

The idea that a formal approach to understanding concrete and abstract systems through a systems-based methodology was described by George Klir (2001). His approach was very abstract and covered the range of systems problems[1] very generally. He established a basic knowledge (epistemological) framework as a hierarchy of system components and categories, then proceeded to sketch out how the problem solver (GSPS) would work to capture the system knowledge from a specific domain, concrete system.

The subject of this chapter is in line with Klir's concepts. What the chapter describes is the author's view of actual methods to be employed, starting with procedures for capturing what Klir referred to as system knowledge.

6.2.1.1 Epistemological Hierarchy

Klir starts with describing a way to categorize systems based on a hierarchy of forms. For Klir a system could be the actual entity of inquiry, the "thing" that is a kind of system or what he called a "source system" or also an "experimental frame." Or it could be a higher-level abstraction. The next level up from the source system was a "data system" in which actual measurements (and their number support) of parameters were to be stored. Figure 6.4, below (Sect. 6.5.1.5), shows how a data system is obtained by measuring a time series of output flows. Figure 6.5 goes on to show a source system fully instrumented and collecting both input and output flows over time. The combination of the source system and the data collected over an appropriate length of time constitute the data system.

In Sect. 6.5.3, below we see the next level in Klir's hierarchy. This is the construction of a generative system. From the analysis of the input/output data, we can estimate what is called a transfer function. Having such a function allows us to compute a semi-unique output from the system for any combination data on the

[1] Mathematicians tend to use the word "problem" to specify whenever we want to achieve something or figure out why something works the way it does, not just when something doesn't work properly and we need to figure out why. A biologist would describe the need to find out, for example, the details of a particular metabolic pathway as a challenge, but since nothing needs fixing would not consider this a problem. On the other hand, how life got started in the first place is still problematic since we don't have sufficient sources of information to construct a model.

inputs. In other words, we have arrived at a model of the system of interest that allows us to make predictions.

At the highest level in Klir's hierarchy is the "structure system." This is the composition of source/data/generative systems that produce higher order systems (what we have been referring to as the higher levels of organization). This is Klir's version of the recursive decomposition of complex systems into sets of simpler systems.

Using this hierarchy, systems can then be classified by combinations of these characteristics. **SE, SD, SG, S^2E, S^2D, S^2G** represent structures of source, data, and generative systems, including second order structures (**S^2**), structures of structured systems.

6.2.1.2 Metasystems

Metasystems are categories of similar systems. So, for example, all living cells fit into the general category of "cell" even though there are significant differences between cell types (their underlying source, data, generative, and structure system characteristics may vary according to their "thinghood").

The **M** operator represents the categorization operator of systems of the same types as metasystems. Thus, **ME**, **MD**, and **MG** are constructions of the categories of all source systems, all categories of data systems, and all categories of generative systems, respectively. As with structure systems, higher order metasystems are permissible. Moreover, complex combinations, for example, a metasystem of structure system of source system is describable (Klir 2001, 87).

6.2.1.3 Conceptualization of Systems

Whereas Klir's description of system knowledge and a methodology for obtaining it was an abstract system for classifying kinds of systems and system components and for gaining system knowledge guided by this epistemological hierarchy, he did not provide a broad set of examples of how this was to be done. He also described the GSPS in very abstract terms, a computing system that contained systemhood[2] expertise (he imagined an automated expert system driving the internal operations). As with the epistemological framework he did not provide much detail on its operation and uses (see his block diagram, Klir 2001, 94). In this chapter, we begin to work out some important details of how a real system for system understanding, fulfilling the potential of the GSPS, might be realized and applied. Klir's examples tended to be mathematical or logical and relatively simple. While he did work with fuzzy system concepts (as described in Chap. 4), the kinds of system problems

[2]This is Klir's term which we take to be essentially what we have been calling systemness. He considers the world to be comprised of things that have specific thinghood qualities, for example, color or size, but also systemhood qualities, for example, the characteristics of being a system that transcend thingness.

called "wicked" (Checkland 1999), for example, social systems involving human decision-making, were not represented. The intent of the current work is to bridge the gap between "hard" systems knowledge and "soft" system methodologies and soft systems knowledge (Checkland 1999). The examples we will demonstrate in Chaps. 7 and 9 will demonstrate how the method of deep systems analysis can be applied to such soft systems and wicked problems.

Readers are encouraged to explore Klir's work, especially (2001) as it was extremely influential, or certainly inspirational in what follows.

6.2.2 *Checkland's Soft Systems Methodology*

One of the more influential voices in the systems thinking and its practice in the realm of "human activity systems" was that of Peter Checkland (mentioned above; see especially Checkland 1999) and his conceptualization of what he called "soft systems methodology"[3] (SSM). He recognized what he felt were some fundamental differences between engineered systems such as airplanes, space shuttles, and submarines, and social-based systems such as a corporation or non-profit charity. The former ones are characterized by "hard" requirements for behavior and engineers need only detail the specifications for performance, costs, and such, to have a basis for implementation. Human activity systems (HAS), on the other hand, have generally messy requirements for what the activity should be, even when the organization has well-articulated goals in mind. Associated with the non-hardness of HASs is the fact that human participants are the agents making decisions and taking action, and humans are notoriously not rational in the same way a mechanical system is rational (Kahneman 2011). Moreover, humans are subject to extensive noise and distortions in their perceptions as well as suffer from too much influence from ideological beliefs in their decision-making. In other words, it is generally the human factors that make decision processes in HASs messy and wicked.[4]

Soft-systems thinking, as Checkland characterizes it, derives from actual experiences of people trying to use general systems thinking to analyze wicked problem domains and finding that the engineering approach, hard-systems thinking, could not succeed in dealing with understanding such problems. On the face of it this seems like a reasonable conclusion and there are now two distinct schools of systems analysis in play, one for the so-called hard systems and the other for the soft ones. Figure 6.1 is adapted from Checkland (1999). It shows a process that is followed in SSM.

[3] By "methodology" Checkland means a set of principles applied to a family of related methods developed to address a particular kind of problem domain.

[4] The term "wicked" is applied to problems that are too complex and are greatly underspecified so that finding solutions is highly problematic.

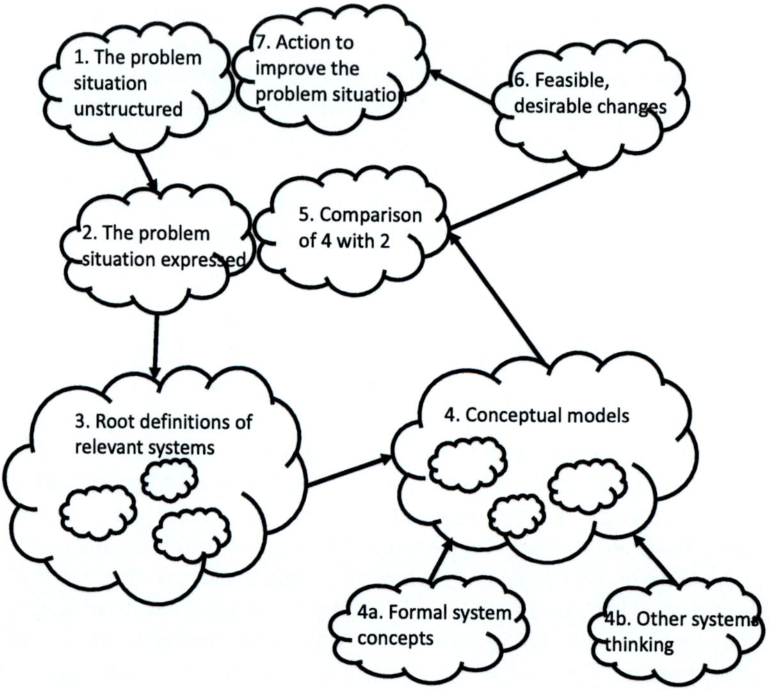

Fig. 6.1 Checkland's general outline of SSM, replicated from Checkland 1999, Fig. 6, Chap. 6, page 163

6.2.3 Synthesis

What is the actual difference between Klir's hard (and abstract) system and Checkland's soft (and concrete) systems? The framework definition presented in Chap. 4 provides a way to see that the differences might be best characterized not as a schism or dichotomy, but as a matter of degree of complexity and levels of organization. For Klir systems were abstract representations of "things"; he characterized the systemness as "thinghood." He sought to boil systems down to pure mathematics. Many other system scientists and engineers have come down on Klir's side but do not take the extreme constructivist position that Klir did. They view systems in the world as real and not just mental constructs. Still, they reserve the position that much of the systemness associated with these real systems is a matter of human consciousness. What we tried to do in Chaps. 3 and 4, culminating with Eq. 4.1 and subsequent equations, is shown that there is realness to systemhood that exists without the need for human observers. We argued, in fact, that the human brain itself is a computation system that is innately programmed to capture and encode systemness in the world. For Klir a system, $\boldsymbol{S}$, is just the tuple, $<\boldsymbol{T}, \boldsymbol{R}>$, where $\boldsymbol{T}$ is the set of things that comprises $\boldsymbol{S}$ and $\boldsymbol{R}$ is the set of relations between things. This is a very sparse definition and has been echoed by many other systems scientists.

Checkland could not reconcile the simple mathematical definition with what was for him the reality of human activity systems—messy "wicked" problems. With humans in the loop, he could not see how one could use the "hard" methods of engineering and mathematics to completely understand these systems. The vagaries of human behaviors made that impossible. The vast majority of system practitioners today still follow this line of thinking, not without cause. But is it a sufficient cause to warrant the seemingly binary schism between hard and soft systems? We do not think so.

Equation 4.1 and the recursive definition provided in Chap. 4 provide a structure that can capture (at least in principle) even the behaviors of human beings. For example, the *H* object, the history or memory capacity of a system (an example of which is Eq. 4.8), provides a very flexible mechanism for including detailed biographies, not unlike the data on user visits to websites analyzed by "big data" algorithms to learn something about that user and make predictions about where they might go next, or what ads they might like to see. The history can be used to track the pattern of a person's decisions and actions and thereby provide a model with a basis for predicting future behavior. Of course, this must be proven in the context of the system of system understanding being proposed in this book. But the existence of big data sets and their uses in modeling human behavior is already being done. There is no reason to believe it could not be expanded and refined in the context of soft system understanding.

Chapter 4 also provided an integration of several modes of communication and did not restrict itself to the mathematical definition alone. We proposed a language of system that can be expressed in, essentially, natural (verbal) language as well as graphical. We think this provides the resolution to treat these so-called soft systems in the same "formal" way one would treat the so-called hard ones. In Chap. 9 we will see this proposal in action, systems analyzing the human society economy (at least partially) which will definitely require that we take human beings into account.

With that we turn to the procedure for deep systems analysis.

6.3 Obtaining Knowledge of the General Systems

6.3.1 Formal Procedures

The point of this chapter is to develop and explain a set of formal procedures for analyzing any system, but particularly complex adaptive and evolvable systems. The complexity of these systems is so great that anything less than a formal method would easily get lost in the process. We know this is the case because it has been demonstrated time and again. Take, for example, the CAES we call "the economy" (see Chap. 13). Anyone who has given the workings of the economy any thought has conceived of it as a "system," but generally not in the way we propose. That is, they realize that there are many complex parts that interact with one another in complex ways but they rarely go much beyond this intuitive stance with respect to trying to better understand how the economy as a whole works. Formal methods as are found, for example, in the study of econometrics have been developed based on the typical modeling approach—best guess the variables and the equations that govern them. But as is becoming abundantly clear, this approach does not produce models that have any kind of valid predictive power. The problem with econometrics has been that the underlying assumptions of neoclassical economics (the academic version) are not based on any kind of reality! We will revisit this issue in Chap. 9 to show how a systems approach to economics produces a very different set of assumptions and, hence, different predictions about how the economy will behave in the future.

The procedures described in this chapter are formal in that they follow the principles of systems science and the definition of system given in Chap. 4.

6.3.2 Representations of the System

In this chapter, we will use several different representations of a system. The basic representation of a system is, of course, in the language of system, systemese as presented in Chap. 4. But there are different ways to structure the system descriptions that achieve different purposes. These are all interrelated and cover the same data, but provide different ways to perceive or use the data. The data itself constitutes the "knowledge" of the system and is the core representation. It is captured and stored in readily retrievable forms in the knowledgebase (see Chap. 8). This is a database system with the schema for relating data elements defined by the formal definition in Chap. 4.

A primary representation insofar as human perception and interpretation will be the various system "maps" or flow diagrams as we have been using. The map employs graphic icons that each represent parts of the system and show how those parts link together explicitly. See Fig. 6.11 below for a system map.

A third representation is the system "tree" diagram. This representation is, essentially, the system map viewed as if from the side with each sub-sub…system drawn at the appropriate depth in the tree. Figure 6.1, below, demonstrates this kind of representation.

The fourth form of representation is the set of equations that describe all of the subsystems' behaviors. All of the boundaries, flows, and interfaces, etc. are implicitly part of this representation even though not readily visible as such.

6.3.3 A Preview of the Most Complex Systems of Interest

In the Introduction, we introduced a type of system based on complexity and capacity to endure changes in the environment, the Complex, Adaptive, and Evolvable System (CAES). We further elucidated the nature of these systems in relation to the hierarchy of increasingly complex systems that have evolved through ontogenesis in Chap. 3. We have not, however, explicated the nature of a CAES sufficiently to make its existence relevant to the project at hand, namely the analysis of truly complex systems. In this section, we will provide a short preview of the nature of CAESs in order to rectify that shortcoming for now. This will be important because in Chap. 9 we will be delving directly into a CAES, namely the human social system's economy, using the methods described here, so a short preview of the nature of a CAES at this point will be necessary to make progress. For those who don't mind jumping around, we point out that Chap. 10 and Part 3 will provide a more complete exposition of the nature of the CAES model. What we offer here is just an appetizer!

The concepts of complexity and adaptivity in systems has been around for a fairly long time (Holland 1998, 2006; Kauffman 1993, 1995, 2000; Miller and Page 2007; Mitchell 2009; Nicolis and Prigogine 1989, just to name a few). All of these referenced authors also refer to systems that evolve to become better adapted to changed environments. But they rarely distinguish between the adaptivity of an individual entity versus the evolution of an entity to become more adaptive. In biological individuals the former capacity is fixed and the individual cannot itself evolve. Rather, the general category (species) to which individuals belong evolve through the reproduction of favored genotypes through generations. In supra-biological systems such as an organization (e.g., a corporation), however, the individual system has the ability to change itself (evolve) to become more fit. In general, the more complex systems are, the more adaptive they become, and ultra-complex systems, containing adaptive subsystems, may achieve a capacity called *evolvability*, or the ability to change themselves.

For now, put simply, a complex system is one that contains many heterogeneous parts and many levels of organization as covered in Chap. 4. At higher and higher levels of organization the complexity of subsystems includes the ability for those subsystems to obtain the capacity of adaptability or the ability to change internally in order to compensate for changes in the environment that have impact on the functions of the whole system (cf., Mobus and Kalton 2015, Chap. 10). For example, a system has the internal capacity to compensate for changes in the external temperature by increasing its internal temperature (warm-blooded animals). The underlying mechanisms are cybernetic subsystems that are involved in, for example, homeostasis and other response mechanisms.

6.3.3.1 An Adaptive System

Briefly, here, an adaptive system is one that is able to sense a change in a critical environmental parameter (e.g., temperature) and alter its internal operations in order to compensate for that change. The change, itself, must not be radical or outside of adaptable boundaries; the system is pre-designed to accommodate the range of changes but any changes outside that range will be detrimental and result in damage to the system. Homeostasis is an example of a mechanism that provides adaptivity. As long as the homeostatic range is within the preset (phenotypic) capacity of the system, the latter can adjust its internal operations to compensate. Adaptivity depends on a system having the capability to sense the change in the environment that is relevant to its functioning, make an appropriate decision to act to compensate for the change, and have the range of optional actions, what we call "requisite variety," and the necessary power in action to make the compensation effective. We will develop these ideas more fully in Part 3 of the book.

6.3.3.2 An Evolvable System

Adaptability can make a system resilient in the short run (assuming that the nature of the changes that warrant adaptive responses are within the ranges of adaptation built into the system). But in the case where changes are trending in a direction that will eventually lead outside the adaptive range that is built into the system, for example, the warming of the global atmosphere or the acidification of the oceans, then an additional mechanism for affording a greater capacity to modify the internal responses to those changes is needed. Evolvability is the ability to make or allow changes to internal mechanisms so that the system can accommodate changes beyond the typical range and it is a further method for achieving long-term adaptability to major changes. For example, in biological species, individuals may "suffer" mutations that do not immediately impact the phenotype under nominal (ordinary) within-range conditions, but under certain stressful conditions (i.e., a change in the environment pushing the limits of the range of preadaptive response) can be released so that some individuals exhibit an increased ability to adapt to the stressing changes. In a large population there will be a critical number of individuals with this particular capacity that they will be more fit than their conspecifics and survive the changed conditions thus leading to a new, more fit population. When enough such favorable mutations accumulate the individuals in this population may be so different from related (and historically ancestral populations) that they are effectively incapable (or unwilling) to mate, should they come back into contact.

The key to biological evolution, and evolvability within a species, is the fact that it is successful because of the size of a population that permits a large enough number of non-directing mutations such that at least a few of these will prove advantageous when the change comes about. When population sizes fall below a critical level, there would not be enough individuals with the "right" mutation to constitute a viable subpopulation. The population goes extinct.

Another proviso of this scheme is that the rate of changes must not exceed the rate at which potentially useful mutations can accumulate. For example, the current rate of global warming is extremely high in comparison to prehistoric events of this sort. So, it is very worrisome that many species, especially of higher multicellular organisms, may not be able to evolve at a sufficient rate to ensure viable individuals in any population, no matter how large.

Human beings are actually transitional as evolvable systems. They cannot modify their physiologies to be more adaptive, for example, being more heat tolerant. But they can modify their behaviors to achieve, effectively, the same end. This is because the human brain, with the remarkable capacities of the neocortex and, particularly, the prefrontal cortex, are able to act as an evolvable system, learning new concepts and altering behaviors to adjust to changing environments in ways that other animals cannot. Of course, not all human beings are adept at learning new knowledge and changing their concepts and behaviors (Mobus 2019). Only those with a sufficient capacity for "wisdom" are astute enough to observe the changes in their environments and intentionally alter their mental states leading to adaptive behaviors.

Human social systems, which includes societies, organizations, institutions, and government, to name a few, are the ultimate in evolvable systems in which intentional "mutations" lead to a long-term sustainable, viable system. In Part 4 of this book, we will revisit this aspect of social systems as evolvable based on the concept of intentional-organization and intentional evolution brought about by the nature of human consciousness and individual human evolvability.

6.3.3.3 CAES as an Archetype

A CAES is an archetype model, to be fully explicated in Chap. 10. An archetype model is one that specifies all of the generalized working parts of a whole system of the type. Used in analysis it guides the analyst by asserting what is to be expected to find as the analysis proceeds. It suggests questions that should be asked in the process of discovery. The patterns and sub-patterns presented in the archetype are found within the actual system being decomposed. Alternatively, used in design of a system, the archetype acts as a template for the design. Recursively applied as in analysis, that is, designing higher order systems with lower order CAESs, extremely complex systems with variations of adaptivity and evolvability can be composed, with appropriate interaction flows, to form the higher order CAES.

The actual origin of the CAES archetype model derived from an amalgamation, integration, and of the works of many previous systems thinkers. A partial list would include: Ashby (1958), Beer (1959, 1966, 1972), Boulding (1956), Checkland (1999), Churchman (1960, 1968a, b), Churchman et al. (1957), Forrester (1961), Fuller (1968, 1970, 1982), Klir (2001), Koestler (1967), Miller (1978), Morowitz (1968, 1992, 2002), Odum (1983, 1994, 2007), Prigogine and Stengers (1984), Rosen (1985, 1991), Shannon and Weaver (1949), Simon (1957, 1991, 1998), von Bertalanffy (1968), Wiener (1950, 1961). Along with the best ideas from these workers, the model incorporates more recent views of governance, agency, and agent theory, and the theory of nested economies (i.e., that the metabolism of cells is nested within the physiology of a multicellular being such as a human, and that physiology is nested within the extant social economy which supports life). Perhaps the biggest influence on the author's development of this archetype that is isomorphic across living and supra-living systems was the work of Stafford Beer (1959, 1966, 1972) who developed the Viable System Model (VSM) that includes many of the features found in the CAES model. More will be said of this in Chap. 10.

Since our main interest is in CAESs involving humans and decision processes (humans as agents) we will tend, in these pages, to focus on human social system subsystems. An extensive review of many different kinds of system with the properties of adaptivity and evolvability, including biological species, human beings as learners, human social systems, and ecological systems as subsystems of the Ecos have verified the main points of the CAES subsystem archetypes.

6.3.3.4 CAES Subsystem Archetypes

Chapter 10 will provide an overall description of the whole CAES model. But that model is composed of three sub-models that interrelate with one another, are tightly coupled. Every CAES will have these three sub-models. Chapters 11, 12 and 13 will proceed to treat each of these as a focus of discussion while pointing out how they cannot be handled as completely independent of one another. The sub-models are: agent and agency (how decisions are made and turned into praxis), economy (how work gets accomplished to provide the system with necessary goods and services for its own use but also for export), and governance (how decision types are distributed across the society and economy). The human social system is (or should be) a complete and viable CAES. But it is comprised of sub-social systems, organizations, institutions, and governments, which are themselves CAESs (or should be). In other words, larger CAESs have smaller CAESs within. Each smaller CAES has its own set of subsystems; its own agency, economy, and governance. Moreover, each of these subsystem CAESs may be found to be composed of yet smaller viable CAESs (or should be) such as departments, committees, and so forth. Finally, all of these sub-subsystem CAESs are composed of people (and increasingly AIs) who are obviously agents, but are also CAESs in their own right. Remember, the human brain is capable of evolving new thoughts and behaviors. Each human's economy is what we call its physiology. The brain is the main governance subsystem. The lowest level subsystems having humans as component parts may also have many CAS artifacts (e.g., computers) and many more simply complex or simple systems as tools for accomplishing the purpose of the CAES.

In the following chapter, we will explore how the methods described in this chapter apply to all kinds of systems, not just CAESs. But in Chap. 9 we will return to the analysis of human social system and go deep into the analysis of the social economy, in particular we will show how what most people think the economy is, is not at all what a viable CAES economy would be. A fuller explanation will need to await Chap. 13 where we reveal what a viable economy looks like from the standpoint of how a CAES works. But we think the reader will be able to see the main thrusts of the arguments given in Chap. 9.

6.4 "Deep" Analysis

As mentioned in the Introduction and discussed in the last chapter, the term "systems analysis" has been used in several different contexts since the mid-early twentieth century to the present. In engineering the design of a "hard" system (in Checkland's terminology, 1999, page A16), the term has been applied to the determination of what are called "requirements" and to an analysis of designs that would fulfill those requirements. What the "device" or system was supposed to do was a given and so the analysis phase was limited to how it should do it most efficiently and at least cost. In the field of information systems, where the term actually gave rise to a job title, "systems analyst," the work of analysis has never been to actually analyze the real system (a "soft" system in Checkland's terminology), that is the underlying organization to be served by the information SUBsystem.[5] In a vein similar to the engineering of physical artifacts, the purpose of the computational and communications subsystem was assumed a given. A "needs" analysis, in this field, amounted to little more than the same kind of requirements gathering as done in engineering.[6] The analysts asked the users/stakeholders what their needs were, assuming that the users actually knew what they needed (as distinct from what they wanted). This approach to analysis is at best partial and, depending on the experience and "wisdom" of the analyst, open to serious vacancies in the completeness of the resulting knowledgebase. Requirements are a pale image of the actual system.

What, then, is *real* analysis? The word "analysis" has several related definitions and is used in different disciplines in slightly different ways based on the medium of study in the discipline. The number one dictionary definition, however, encapsulates the broad meanings of the term in all of the various disciplines. This definition is from Dictionary.com: "1. the separating of any material or abstract entity into its constituent elements (opposed to synthesis)." The number 2 definition amplifies why separating something into its constituent parts is important. "2. this process as a method of *studying the nature of something* or of determining its essential features

[5] An information system is, in fact, just a subsystem of the larger supra-system which it serves. In Chap. 10, we elaborate the role of information systems, or actually the network of message flows and processing that is part of the governance subsystem of a CAES, like a corporation.

[6] The field of software engineering has always had difficulty being just like hardware engineering. Part of the problem stems from the nature of software, which is subject to a vast array of methods for achieving complex functions. Software development is more often like prototyping than product production. In Part 4 chapters we will return to this issue and suggest ways in which an overarching systems engineering process could be used to ensure better software development outcomes.

and their relations" [emphasis added]. In other words, the process of analysis is taking something apart (carefully) to find out what it is made of, and, hopefully, how it works.

The main process of the sciences is analysis. The philosophical and practical belief that something can be understood by taking it apart, called reductionism, and that the ultimate behavior of the whole thing is nothing more than the collective behaviors of its constituent parts has been the top guiding principle for science since the invention of empirical methodology.

In the last chapter, we introduced the analysis of a system and considered the opaque-box/transparent-box procedures. Empiricism requires that as one deconstructs a system one needs to run functional tests on the components. This is not much of a problem for reverse engineering an actual "box" piece of hardware. The parts of a computer still function properly even when they are not in a whole computer, if you set up the tests appropriately. However, hearts or livers, even when kept artificially alive, may not function at all as they do in the body of a living organism. Possible setups for testing their functions would have to be elaborate and extensive, and, in any case, assumes that the experimenter knows what effects the rest of the body has on those organs in their in vivo states.

The methods of systems analysis (as described above) have to date been only partially about taking the system apart. "Requirements gathering" for a physical device or product, if done very carefully, constitutes a kind of virtual analysis, that is, the analysis of something that does not yet exist except in the minds of the prospective "users." In the software world using the same notion of requirements gathering is barely an analytic process. For one thing, the presumptive analysis is not being carried out on the actual system (where the work is done) but on the information subsystem.[7] The relationship between what the "user" knows about the information requirements and the actual requirements of the work system depend entirely on that user's depth of knowledge of the system itself. All too often that depth is not great. Systems analysts in the software development business would do much better if they analyzed the actual work system (examples to be provided later) to find out what its information requirements are.

In the systems world, analysis has gotten something of a bad rap owing to a misconception about the issue of functionality of the whole versus the functions of the parts. The phenomenon of emergence further muddies the waters leading whole systems thinkers to statements such as: "the whole is greater than the sum of its parts." Aphorisms like this are useful to remind researchers that when engaged in analysis they cannot lose track of something very important, namely that the parts have relations to one another in the intact system and that it is those relations that

[7] Systems analysis in the computer-information systems world (or management information system, MIS, as it used to be called) tended to be strictly an exercise in analyzing the computing/communications/reporting aspects of a subsystem that was supposed to serve the work system. Only occasionally did systems analysts ever venture into a deep analysis of the work system to verify that the "user's" requirements matched the decision processes needed to manage the work processes.

give rise to the behavior of the whole (Rosen 1985, 1991). Deep analysis can include several different ways to observe this dictum. No relations need to be injured in the analysis of this system.

6.4.1 Inhibitors

This subject was touched upon in Chap. 5. Here we reemphasize two major inhibitors to gaining deep systems knowledge.

When quarterly profits are in question, these inhibitors will come to the fore to reduce the effort put into deep analysis. As mentioned in the Introduction and Chap. 5, this is ironic since short-term profit gains are all too often more than offset by longer-term losses due to the project or product failures, or due to the long-term maintenance life-cycle costs that are, in turn, due to poor design (due to inadequate knowledge about the concrete system). The inhibitors to following a principled process such as will be described in this chapter are both real threats if not carefully considered and poor excuses to minimize the analysis phase in favor of getting right into design.

The profit motive drives most of what we do, even in supposed non-profit operations, like academia and science. The desire to do the most possible with the least drain on capital and labor is just a fact of any society that runs on liberalist ideals. Novelty and innovation—being first to market, for example—and many similar motivations get in the way of doing a thorough analysis. And this is why we too often don't get things right the first time through. There is an old saying among engineers (chaffing at the insistence of marketing and management to get the thing designed quickly) is quite well understood in that field: "Why do we never have enough time to do it right the first time, but always enough time to do it over?" It captures the dynamic of modern projects in a competitive marketplace.

6.4.1.1 Complexity and Analysis

The degree of difficulty associated with deep analysis depends generally on the complexity measure of the system being analyzed. Recall that complexity in this sense means the logical depth of the hierarchy of organization coupled with the numbers and kinds of components and relations at any given level. It should be obvious that the more levels there are, the more numbers of kinds of components, etc., the longer it will take to accomplish a deep analysis. It will also require much more work in decomposition and data capture. In most current systems analysis projects, this is often called the "scope" problem. The larger the boundaries of the system, the more complex it will be and therefore take more time and resources to accomplish.

In order to determine, in advance of expending the effort to do a deep analysis, it is important to get a firm grip on the complexity of the system. But this may create a conundrum. By definition we do not yet have the details of the system including just how complex it might actually be. Fortunately, we do have considerable experience with other similar systems. For example, we have case studies of many projects in the realm of management information systems from which we can obtain a fairly good description of their complexity.

In Chap. 12—Governance, we will dedicate a section to the management of the systems understanding process. We will return to the issue of complexity and how it affects the overall success of projects, and especially how to handle the scope problem.

6.4.1.2 Cost of Analysis

Complexity plays into another aspect of analysis and that is what appears to be a nonlinear relation to costs. It seems that as complexity rises the costs involved in doing it rise more quickly; perhaps not exponentially, but a higher order quadratic for sure. Indeed some authors have measured complexity in terms of cost estimates (or results). Cost overruns are generally attributed to what some have called "runaway complexity."

Costs include manpower, capital, equipment, and time. And a substantial portion of project management is given to managing costs to stay in budget. This is the way of capitalism, but it is also the unfortunate cause of failing to do deep analysis before generating designs (as described in Chap. 14 (the unlucky chapter)). And, as described in the Introduction, the management constraints on systems analysis generally (all too often) leads to even higher costs later when the built system fails to deliver the real service needed by the stakeholders.

Deep analysis is going to be costly and that should be clearly understood at the outset. Deep analysis means taking your time to discover all of the relevant knowledge of the system so that nothing important is overlooked. From this dictum there is no escape. The problem is with financial management imposing unrealistic budgets on projects to produce products that marketing management has enthusiastically, if ill-advisedly, sold to a customer. There is no solution for this formula and enterprises will continue to suffer failures so long as they keep getting the cart in front of the horse. This subject will be taken up again in Chap. 12 on governance.

Using deep analysis is a strategic approach that is meant to save overall lifecycle costs and thus improve profit performance in the long run. More importantly, for companies, governments, and NGOs thinking about their long-term thrivability, it will ensure customer satisfaction more surely than happy-talking them.

6.4.2 *The Objective: Reductionism with Maintenance of Relations*

Deep analysis is a form of reductionism in the scientific investigation and understanding sense, but with a critical difference. Because the procedures are dependent on the a priori formal definition of system (and the language of systems) the deconstruction of a system is guaranteed to preserve the functional relations between subsystems, thereby preserving the whole system's behavior.

This is reductionism in the sense that the whole really does depend on the behaviors of the parts, but only in the context of the behaviors of all of the parts taken together. When we get to the generation and use of models of systems, based on what our analysis process produces in the knowledgebase, we will see how this is guaranteed in our analysis methodology. The preservation of relations derives from the analysis based on using the system language and on the formal definition of system from which it comes. By starting with this formal definition, we build into the process the fact that the relations are primary characteristics. In effect the analysis could not proceed if we ignored them, even a few of them.

6.4.3 *The Three Phases of Systems Analysis*

We will briefly review the three phases of analysis as described in the prior chapter and then provide complete descriptions of each phase in the next section. For the present, realize that the descriptions of objects identified in the sets, graphs, and lists comprising the data about a system, ***S***, will be collected in forms appropriate to the type of object. Actual examples of these forms will be provided in the section below. And examples of filled-out forms will be provided in the next chapter.

Recall from the last chapter that we identified three phases for systems analysis. Two phases are somewhat preliminary, but absolutely necessary in order to succeed in the third phase where the majority of the work will be done. In this section, we will briefly review and explain the significance of these three phases. In the next sections, we will expand the methods and procedures to be followed. The three phases are: system identification, environmental analysis, and recursive deconstruction.

In describing these phases, it may at first seem that they are carried out in a linear fashion, one after the other. However, this is not the case and should not be assumed. The entire process should rather be considered as potentially iterative. That is, in conducting any one phase it may become apparent that something important was missed in the prior phase. For example, in doing an environment analysis it may become evident that there is an unaccounted-for flow from a newly identified entity that mandates re-entering the system identification phase and boundary condition analysis. When doing the recursive deconstruction of a system, it is possible that discovery of a subsystem that receives a flow not accounted for in the parent system leads to re-entry of the higher-level organization analysis. The accounting system (tracking the inputs and outputs) provides a trigger that signals such a need for iteration over the higher level. This aspect of the analysis should be kept in mind as we describe what transpires in each phase. In the next chapter, we will provide a few examples of this "iteration over recursion" aspect. One way to visualize this is how the science process works. If failure to replicate results occurs in an empirical study, new experiments are designed to check the original findings. Science is thus recognized as a self-correcting process overall and over time.

6.4.3.1 Identification

This phase is tightly coupled with the next phase in the sense that the analysts may need to engage in first one then the other in a piecewise iterative fashion. The phases should not be confused, but there are times when a system boundary is not immediately identifiable until one also starts considering the environment. For example, it may be difficult to identify a boundary and the conditions of flows across that boundary until one has identified candidate sources and sinks.

Nevertheless, the system of interest (SOI) can only be understood when the boundary of the system has been established and the interfaces it has with the environmental entities have been identified. We will work with the formal definition of system and the language of systems from Chap. 4 to do this work.

The first order of business is to analyze the boundary of the system to identify the input and output flows through that boundary by way of the specialized subsystems called "interfaces" (Eqs. 4.7 and 4.8). At this stage we do not know what inside the SOI is producing the outputs or receiving the inputs, but we can identify the points at which they traverse the boundary.[8] Then, by substituting individual elements of $\boldsymbol{C'}$ for receiving interfaces ($\boldsymbol{r}_{i,0} \in \boldsymbol{I}_{0,0}$) and elements of $\boldsymbol{C''}$ for source interfaces ($\boldsymbol{r}_{j,0} \in \boldsymbol{I}_{0,0}$) from Eq. 4.8, we can then construct the tri-partite graph, $\boldsymbol{G}_{0,0}$, mapping inputs to and outputs from the SOI to environmental entities (see Fig. 4.9b).

Once the $\boldsymbol{I}_{0,0}$ set is completed (or as complete as possible for any given iteration) we are now ready to complete the analysis of the boundary object B (Eq. 4.7). This requires a thorough analysis of the boundary's physical properties to be encoded in the set $\boldsymbol{P}_{0,0}$ in Eq. 4.7.

Ultimately the objective of system identification is to specify the "grand" transfer function of the system process; to be able to say what the outputs are (quantitatively) and how they are arrived at given the inputs (quantitatively).[9] For this we will

[8] We will be explaining the nature of the protocol component of the interface in the next section.

[9] It need not always be the case that systems must be identified with quantitative relations. There are many fuzzy systems for which we are keenly interested in qualitative relations and may not need to necessarily quantify every aspect. However, we maintain that this does not mean such systems are not ultimately quantifiable. For example, psychology has been primarily interested in qualitative descriptions of the mind, such as inferring mental states based on behaviors and first-person reports. Modern psychology works hard to find quantitative measures for these aspects. And with the advent of neurobiological imaging technology, the ability to "measure" directly

be constructing the ***T*** set from Eq. 4.2 through use of the ***H*** object. We use Eq. 4.9 from Chap. 4 to start constructing (at least at this stage) candidate functions for each of the inputs to and outputs from the SOI at time intervals defined by $t_{0,0}$ in Eq. 4.2. As much as is possible from the opaque-box analysis and the instrumenting capabilities of our analytic tools we begin to characterize each output in terms of things like flow rates or volumes per unit time, substance characteristics such as composition, and other relevant factors (e.g., energy content from a fuel processing plant). This is a first pass at resolving questions about how the system processes its inputs to produce its outputs. The final resolution must wait until we get deeper into the third phase, where the details of sub-processes that actually do the work will be discovered. In the below descriptions we will introduce the use of machine learning techniques to do causal analysis of the outputs-given-inputs history that will help us characterize very complex transformation relations between inputs and outputs. Traditionally such transformation formulations have been characterized with sets of differential equations, and for simple systems this is still a legitimate approach. However, for very complex systems the transformation functions may be nonlinear and multivariate (with noise also possible) so that a clean set of functions may not be possible. Modern machine learning methods (e.g., what is called "Deep learning") have been used to capture input/output relations that cannot be characterized so neatly.

6.4.3.2 Environmental Analysis

There are a number of components and aspects of the environment of any system that have to be grasped in order to proceed. First is the enumeration and characterization of the various entities and external reservoirs that constitute the sources and sinks for the flows of inputs and outputs (and forces) that impact the system.

This characterization does not go so far as to develop models of the entities themselves. Those are outside the boundary of the SOI. But it does include determining things like the flow rates of substances and messages from/to the entities and external reservoirs. As an example of the latter, consider the carbon dioxide and other greenhouse gasses emitted by the human social system (see Chap. 9) into the atmosphere and indirectly into the hydrosphere. Both of these reservoirs have absorption characteristics that we need to know in order to consider rates of output from human activities. When atmospheric scientists are studying the atmosphere as a system, receiving the CO_2 emissions, they will be concerned with how the gasses impact their SOI (and, of course we are all concerned with the feedback of that impact on things like the climate).

Environmental analysis will involve Eq. 4.5 in Chap. 4 where the ***Src*** and ***Snk*** sets and their associated flows will be completed.

operations in the brain while subjects are behaving or reporting their thoughts is rapidly changing psychology to a quantitative systems science.

But, how do we go about discovering all of the inputs and outputs for a system when the environment is complex and there are potentially many different kinds of sources and sinks as well as many different individual entities in each kind? Further complicating things is that many entities are not purely sources or sinks. In CASs and CAESs the SOI is actively communicating with those entities and even observing other entities not involved directly in flows of materials or energies. Thus, a source of a material resource may also be in two-way communication with the SOI; we will see examples of this in Chap. 9 on the economy subsystem and Chap. 12 on governance. This means the source of material may also be a source of message and a sink of messages. Moreover, a sink for products might also be a source for some other material. The same entity may be both a source and a sink of many different flows. In the examples to be given later in the book, we will show how to handle this graphically and in terms of the programmatic treatment. All of these (and other) complications might, at first, seem to make the parsing of the environmental entities and flows extremely difficult. How might we proceed in an efficient procedure?

The answer is to start with a *generic* or *archetype* model of a CAS or CAES (to be elaborated in Chap. 10) in which we represent at least one input of energy, one of matter, one of message and outputs of heat, product/behavior, and wastes. Each of these is top ontological categories that we established in Chap. 3. Complex systems will generally have many inflows and outflows in each of these categories, with subcategories further differentiating them. For example, a system like the economy (see Chap. 9)[10] gets many different forms of energy to do its work. The economy obtains some forms of energy such as fossil fuels, real-time sunlight, hydroelectric power, and so on. The category "energy" can then give rise to (at least) three categories of subcategories of energies. When modeling the economy, it might not be necessary to be more specific (i.e., sub-subcategories). All coal mines, for example, could be lumped into one giant mine and the total production of coal (of all kinds) could be aggregated in some appropriate flow rate measure (e.g., tons per day).

Or a complex system might produce many different kinds of products (low entropy material things) or services (performing work on other systems). In Sect. 6.5.1.2 below, we recommend the identification of the system of interest starts with analysis of the products (and other outputs). Using the generic model one can then begin to formulate analytic questions about what subcategories of products or wastes there might be, as with the handling of energy above.

This same strategy can be used for all of the generic inputs and outputs. However, a special case might exist for message flows. It turns out, in the general cases for CASs and CAESs, that there will be messages exchanged between all sources and all sinks as mentioned above. Rather than just represent, for example, a flow of material from source A to the SOI, we might as well acknowledge a priori the existence of message channels between A and the SOI even at this stage of analysis. As noted above we will provide examples of doing this in both Chaps. 9 and 12.

[10] See, in particular, Figures 8.3 and 8.4 to get a sense of how this decomposition of environmental entities and flows is accomplished by moving down from general categories to more specific ones.

6.4.3.3 Recursive Deconstruction

Once the SOI has been identified and characterized along with its environment it is time to start the recursive procedure of deconstructing the opaque-box to find out what is inside. The variety of techniques for exposing the internals of any system will depend on the material medium of the system. We will provide several examples in the next chapter. For now, we simply refer to the act of discovering subsystems and the flows between them (or, more generally, the interactions between them).

The procedure may be tackled either using a breadth-first or a depth-first approach. In practice, some combination of these will be appropriate. In general, however, a breadth-first approach is the more preferred since going deeper into one subsystem will create a set of unanswered questions about sources and sinks that cannot be resolved until all of the legs of a depth-first approach are completed. At best, depth-first will only expose candidate sub-subsystems. This can be useful as a "look-ahead" procedure to find what potential sub-subsystems will be coming up in the analysis but will not yield a complete mapping of flows and influences that would be needed to provide consistency checks. For our present purposes, we will focus on the depth-first approach since it is the most efficient one in terms of keeping track of all relevant flows at each level of organization.

The breadth-first approach means that all of the subsystems of any given level of organization will be determined along with the map of flows between them. Once the subsystems and the flows are determined each subsystem will be similarly deconstructed as if it were the SOI. Assuming a complete analysis has exposed all of the subsystem inputs and outputs (with their boundaries and interfaces) it becomes possible to switch to a depth-first approach local to a specific subsystem. For example, if the real questions to be answered only involve a single subsystem, the analysis of all the other systems will act to substantiate the environment of that subsystem as it becomes the focal SOI. We will see an example of an analysis that starts at a very large scope (the whole Earth!) and narrows the analysis down to a sub-subsystem, the economic system of the human social system, in Chap. 9.

6.4.3.4 The Final Outcome

The product of a deep systems analysis is a knowledgebase containing all of the relevant data describing the objects (subsystems and components), flows and influences, and transformation functions (relations). Figure 6.2 shows a hierarchy tree with the relations made explicit. Information captured in the knowledgebase allows for the reconstruction of such a tree. The tree on the left shows a macroscopic view (also called a "collapsed" view) of just the major subsystems hierarchy, down to the level of components. Not shown are additional details of objects that are defined in the formal definition of Chap. 4 such as flows and interfaces. The expansion shown on the right side of the figure provides a more microscopic view that includes some of these objects. A tree object such as this can be visually "browsed" with selected objects expanded for view. The actual data structure containing the detailed data for any object could be shown at a mouse click. Indeed, those same visual access tools could be used to construct the tree in the first place.

Chapter 8 will provide additional understanding of the knowledgebase and how it can be used to manage various other functions of the knowledge understanding process.

Given this brief review/preview it is now time to provide details for how the analysis process is to be accomplished.

6.5 System (of Interest) Identification

The general process of analysis starts with identification of the system of interest, treated usually as an opaque box (see Fig. 6.2 below). At level 0 this involves a complete characterization of the system boundary (especially the list of interfaces), the inputs and outputs, and some preliminary information regarding the sources and sinks. Of ultimate interest is the nature of the transformation of inputs to produce outputs; of raw material and high potential energy going in and products and wastes coming out taking into account appropriate time delays from input to output. The transformation function(s) is characterized as the dynamic behavior of the SOI, meaning that it is characterized as the temporal variations in values of output

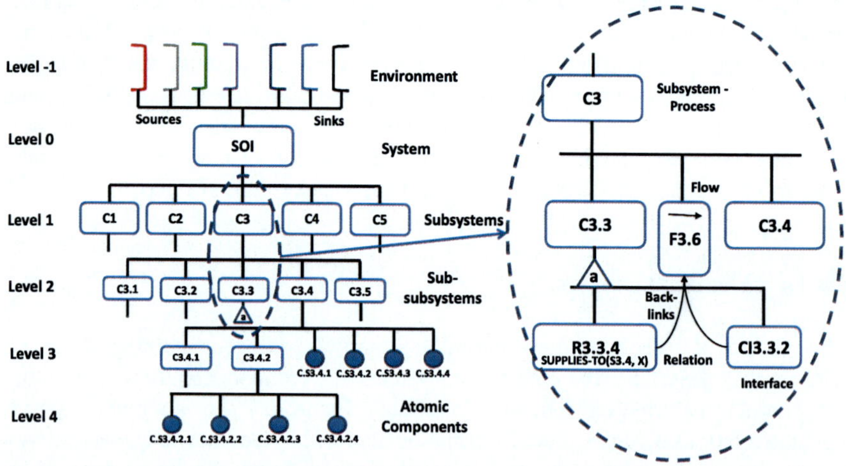

Fig. 6.2 One view of the final outcome of systems analysis is a complete hierarchy tree showing all of the environmental entities/reservoirs at level −1, the SOI at level 0, and the deconstruction tree of subsystems, sub-subsystems…, and components. Subsystem C3 is shown exploded into additional elements such as a flow between C3.3 and C3.4. C3.3 is further decomposed to show details of relations and interfaces. Note that the coding system for subsystems is based on the *C* set in Eq. 4.1. Alternatively, by Eq. 4.3, the designator C, standing for component, can be replaced by S, standing for system

parameters relative to those of inputs.[11] In this section we walk through the process of system identification as the first phase of the analysis process. The objective will be to populate the formal definition of a system, given in Chap. 4, with actual data objects, such as a flow (of a substance from a source to a sink) with real or design required values of units per time step, frequencies, or other relevant parameters needed to complete the characterization. Here we will be focusing on the general principles and methods for any arbitrary system (the fact of systemness means these will be universal). In the next chapter we provide examples of how these procedures work in practice.

6.5.1 Initialization

How do we get started? In theory, we could start with an analysis of the boundary or any arbitrary flow (in or out). This is because we have an a priori model of systemness that is the actual beginning point for all analysis. That is, we have a template of what a system is in general terms; what we are going to call a *presumptive SOI*. We don't necessarily know what the products (or services) are, but we know that the system *must* have them or it would not exist (see discussion in the next section on the nature of "purpose"). In the engineering sense, the products (or services) are givens.

In practice we have developed a structured procedure to be described below. In any case all analyses begin with the consideration of several key questions. By consideration we don't just mean a providing cursory answer. We mean applying hard-nosed reasoning and the use of evidence to back up the answers. By "key" we mean that these are literally the questions that must be considered for every systems analysis, no matter the complexity or what we "think" we know in advance.

6.5.1.1 The Questions to Be Answered

The most frequent trigger for doing a deep systems analysis is one or more questions coming to the fore regarding a phenomenon that is not understood. Even when the intent is to develop a product or design an organization, the driving force is often a question about how a new product or new organization can perform its functions better. The first key question, then, is: "What is the *purpose* of the system?" This is tantamount to asking: "What does the system do?" And this leads to the follow-on questions: "How does it do what it does?" and "Does it do it well enough?"

[11] A form of identification involves only the use of output data, partly described in the Introduction. These methods consider only the output data over time and attempt to derive behavior on this basis alone. Since most real systems are sufficiently complex (and adaptive, if not evolvable) and involve complex internal functions, this approach is reserved for mostly simple physical systems.

In the sciences, we often start with the last questions because it may seem as if we have already determined what a system does by historical observations of its behavior. But this can be an illusion! In physics it is easy to identify an interesting system, say a ball rolling down an inclined plain or a pendulum swinging (and undergoing precession). In biology, the task is much less simple. In the case of the ball rolling down an inclined plain, that is not really a system. It does not need to have a "purpose." Rather the system is likely to be an experimental setup with measuring tools in place and an observer (the physicist) ready to characterize the action, capture some data, and analyze the later in order to answer some fundamental questions about motion or gravity (forces), etc. So, the purpose of the experiment—the real system—is still to provide information to an outside observer.

Biological systems have their own motive forces and dynamics. The purpose they fulfill is dual. They must stay alive and they must procreate; all living systems are endowed with this programming, even if specific individuals in a population fail to follow through (e.g., worker bees do not reproduce, but the queen passes on their shared genes in their stead). So, with biological systems the questions to be answered are complicated.

Adding to that complexity is the evolutionary problem that in order to carry out the biological mandates, the average organism in a species has to be fit in its environment. That means, it has to be capable of fulfilling its mandates in the average set of situations it finds itself in. It must be generally capable of capturing and using energy to maintain itself (or grow when it is immature) and to either replicate itself (asexual reproduction) or find a mate and procreate. In a sense, the fact of the biological mandate makes the "purpose" question simple. It is the follow-on questions that are much harder to answer.

When we start talking about supra-biological systems, ecological, social, organizational, and so on, the questions of fitness become much more difficult to answer. But this is exactly why we are turning to deep analysis.

In the case of artifact, procedure, or policy designs—of human engineering—the question of fitness is no less important. It is one thing to say that a particular machine should perform this-or-that function. It is quite another to specify the function such that the performance meets the requirements of the environment in which it will operate. Poor fitness in this arena may result in customer dissatisfaction leading to loss of sales in the future. Here, not only may the machine be unfit for duty but the organization that produced it is unfit as well. In this sense, fitness means knowing how to understand the requirements and then doing a superior design job to produce the artifact.

Because the overall objective of a deep analysis is to answer these questions, purpose, function, and performance, the logical place to start the analysis is at the output end of the system.

6.5.1.2 Start with the Output

With respect to artifact systems, too often analysts will assume they already know everything about a system's output or behavior and so will tend to focus on the inputs or even the internal processes.

And just as often the assumptions about outputs will be wrong, especially with regard to the questions to be answered. For all systems, natural and artifact alike, a good rule-of-thumb in initializing an analysis is to do a thorough study of the outputs relative to the environment. Initially the analyst may have a presumptive model (Fig. 6.3) and have some preliminary ideas about inputs, boundaries, and outputs. This is strictly a conceptual model. It may be a seed for Eq. 4.1 at level 0, but the rule to focus first on the output(s) is still in effect.

6.5.1.3 The Presumptive SOI—Level 0

The time has come to begin filling in the details of Eq. 4.1, repeated here.

$$S_i^l = C, N, G, B, T, H, \Delta t_i^l \tag{4.1}$$

Recall, also, from Chap. 4 the elements of Eq. 4.1 at a designated subsystem index, i, and level index, l, are:

C set of components (or subsystems) of the parent system.
N is a graph of components in ***C*** and their interconnections, ***L***.

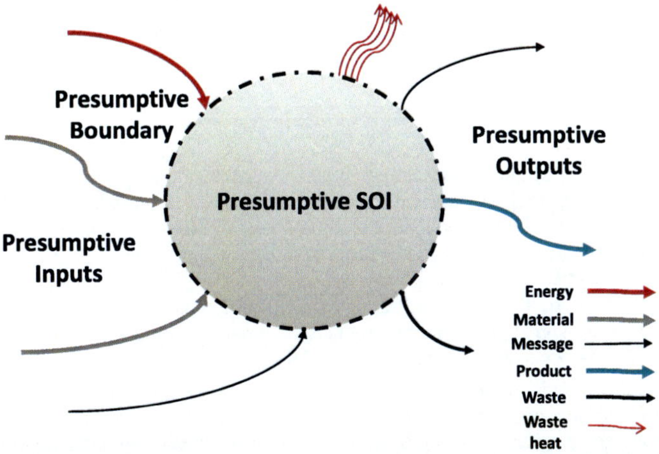

Fig. 6.3 The initial condition for starting a deep analysis may include some presumptions about the SOI, its boundary, its inputs, and its output. Since all concrete systems follow this basic form, this conceptual model is a reasonable place to start

G a tripartite graph with ***Src*** (set of sources), ***Snk*** (set of sinks), nodes in the environment and ***F***, the set of links between sources and sinks and the components of the system.

B the boundary object contains a list of interfaces and boundary physical characteristics.

T the set of transfer function for each component.

H history of components' states.

We can now begin defining S_0^0, the system of interest at level $l = 0$. At this point the sets ***C***, ***N***, ***T***, etc. are empty. The time constant, Δt, however, must be estimated now as this will guide the selection of sampling rates to be used in measuring input and output flows. To obtain a reasonable estimate for Δt we can use a variety of signal processing methods after collecting the time series data (see more discussion on this below) on the product output.[12] For example, we can use Fourier analysis[13] on the time series to identify the range and power distribution of fluctuations (frequencies). At this stage we are still working with rough estimates based on judgments, say for example, what high frequencies represent mere noise and can be ignored (or filtered appropriately). The highest meaningful frequency can then be used with the Nyquist-Shannon Sampling Theorem to derive an estimate of Δt. However, Δt will remain an estimate only until we have many data streams to analyze.

6.5.1.4 Creating the Product Flow and Terminals Objects

A first order of business is to identify the product of the system—the output that gives the system its purpose. For complex systems, of course, there may be a number of products and customers. For example, a corporation has a large number of what we typically call "stakeholders," ranging from shareholders, to nominal product customers, to employees who receive wages. All of these stakeholders are

[12] As strange as it may sound, even material outputs can be treated as signals. For example, the shipments of products from a manufacturing plant often happen in bursts or pulses (volume or weight) rather than a steady flow stream. The same is true of biological systems. Real concrete systems, unless specifically designed to produce absolutely steady flows, have these fluctuating outputs. This is significant because these signals convey a lot of information about what is going on inside the system itself, the channel through which the flow occurs, and even aspects of the receiving sink. Thus, signal processing methodologies, developed for communications engineering, often have relevance to all kinds of material and energy flow situations.

[13] Throughout this description the emphasis will be on the overall procedure and not on specific analytic tools. The latter will be mentioned to provide guidance as to when they may be appropriate to use. The overriding assumption here is that readers who are familiar with the use of these tools will recognize what is meant and need no further guidance as to what to do. More general readers are invited to investigate these tools as they are mentioned, for example in Wikipedia. The author does not presume to tell readers what tools they should use, only what kinds of tools would be appropriate at various points in the general procedures.

technically customers and receive products (dividends, physical products or services, wages, respectively).

The interface of the SOI is the subsystem on the boundary that exports the product to the customer(s). The simplest case is a customer that accepts all possible output from the SOI and is, therefore, in open-loop relation with the SOI. This situation is probably rare in most concrete systems. As a result, the normal situation is depicted in Fig. 6.4b where communications between the customer and the SOI are established. This is the situation in which the function of the interface is a bit more complex than that of Fig. 6.4a. We will address this situation in the analysis of the boundary and when it becomes clear that interfaces are special kinds of subsystems. A good example of an interface with communications would be a shipping and receiving department in a manufacturing plant. Shipping products out to customers involves paper (or increasingly digital) flows for invoices, shipping manifests, etc. that work to coordinate the actual product flow.

A product flow object requires the creation of three different objects in the knowledgebase. Recalling Eq. 4.6 from Chap. 4,

$$\boldsymbol{B}_{i,l} = P_{i,l}, I_{i,l} \tag{4.6}$$

we will be creating an element of the $\boldsymbol{I}$ set. We will also be creating an element of the $\boldsymbol{G}$ set,

$$G_{i,l} = \left(C'_{i,l}, Src_{i,l}\right), \left(C''_{i,l}, Snk_{i,l}\right), F_{i,l} \tag{4.5}$$

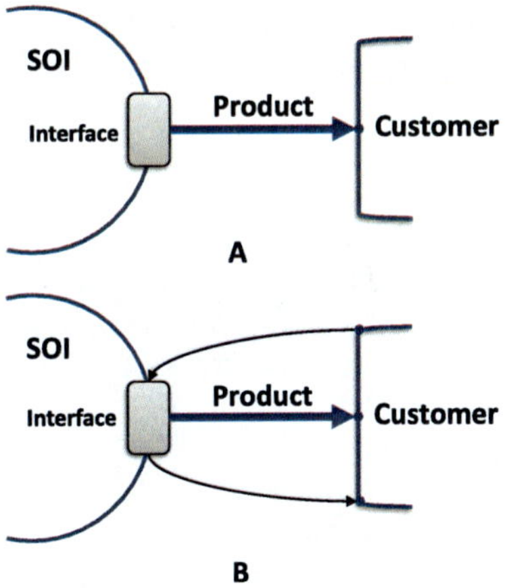

Fig. 6.4 A single product output is identified. A product is any output that is actively accepted or sought by a "customer." (**a**) A customer accepts a product flow in an open loop scenario. (**b**) A customer has established a protocol relation with the SOI wherein the interface (on the boundary) also includes the exchange of messages (thin black arrows) in order to coordinate the flow of the product from the SOI to the customer

namely, the element ($C''_{i,l}$, $Snk_{i,l}$) and the element $F_{i,l}$. Since we are starting at level 0, the index l will be zero and all of the i indexes will be 1.

In other words, we will create the interface, the flow (of product), and the customer (sink) objects, which will be captured in the knowledgebase as skeleton objects (for the moment). These will then be fleshed out with relevant data. Once completed we will be ready to proceed to a full SOI identification.

6.5.1.4.1 An Interface Object

Because interfaces are special subsystems, all $i_{i,j,l} \in I_{i,l}$ are really $c_{i,j,l} \in C_{i,l}$. The $\boldsymbol{I}$ set of interfaces are represented in the $\boldsymbol{C}$ set of components for system $\boldsymbol{S}$. The difference between $\boldsymbol{I}$ objects and other $\boldsymbol{C}$ objects is that the former include provisions for a special sub-subsystem called a "protocol." Otherwise, the two notations are interchangeable.

The interface in question here is the subsystem of the SOI which is responsible for the "management" of the transport of product across the boundary and transfer of such to the appropriate customer. The latter is achieved by maintaining an existing channel for the flow, on the SOI end.

At this point in analysis, we will not be decomposing the interface object. Rather we are just establishing the object in the knowledgebase, reserving its recursive version of Eq. 4.1, giving it a coding and inserting it into the $\boldsymbol{B}$ object (Eq. 4.7). A product interface for the SOI would be item $i_{0,1,0}$ in the $\boldsymbol{I}_{0,0}$ set (if there are a number of products, the items would be sequentially numbered i = {2, 3, … n}). All additional slots in the description form will be left "TBD" awaiting the recursive decomposition phase of the boundary analysis.

So, the first interface object $i_{,0,1,0}$ is a data structure that looks like Table 6.1:

The table can, of course, contain additional data items. However, let us consider what is shown here.

The first element item, "Interface" establishes this item as belonging to an interface list that is part of a boundary object. The data code, "I 0.1.0" indicates that this is an interface in the 0th list of interfaces (I_i) at level 0 ($I_{i,l}$), and it is number $j = 1$ in the list ($i_{i,j,l}$). For the SOI we might generally expect there to be only one boundary with one interface list. But we maintain the indexing scheme for consistency. At levels greater than 0 we will have to keep track of multiple boundaries and interface lists.

Table 6.1 An interface object on the boundary of the SOI

Element	Data
Interface:	I 0.1.0
Name:	Main product output Interface
Description:	Exports the main product X from the SOI
Type:	Material output
Output flow code:	F 0.1.0

The final element in Table 6.1 indicates a cross-reference to a flow object (see below). This item will allow the linking of the flow of product ***X*** to the output from the SOI through interface $i_{0,1,0} \in I_{1,0}$. Specifically, flow dynamics are codified in that object.

In Chap. 7, we will revisit these database objects and show more details for specific kinds of systems.

6.5.1.4.2 A (Product) Flow Object

Each flow of material, energy, or messages is represented by a flow object, $f_{i,j,l} \in F_{i,l}$. We create a product flow object. At this point, it is possible to identify the nature of the product (i.e., material, energy, or message, what it is specifically), its relevant parameters (e.g., volume or weight, flow characteristics such as pulses or continuous, etc.).

In the section to follow, we explore the methods for measuring flows through instrumentation, data collection, and data processing. This is the key to determining the dynamic properties of the flows, and ultimately determining the transformations from inputs to outputs. Here we will indicate the results of such measurement (or specifications) as part of the flow object in the database. Consider the data in Table 6.2.

A number of other characteristics of the flow would be encoded in this element since flows are the key dynamic elements of any system (the Etc.) Here we are only interested in providing an example of the kind of data element that would be stored in the knowledgebase.

However, the final two items are critical links in that they identify both the source of the flow and the sink for the flow.

It is possible that a single output flow n-furcates between multiple customers. If this is the case n-separate flow objects must be created for each flow. It will then be the responsibility of the software to keep track of the mass flows (i.e., ensure the

Table 6.2 A flow object in the database

Element	Data
Flow:	F 0.1.0
Name:	Main product
Description:	Product X exported to customer Y
Type:	Material, low entropy/high organization
Average wt. per unit time:	15 kilos per Δt
Peak wt.	22 kilos per Δt
Etc.	
Source:	I 0.1.0
Sink:	Snk 0.1.0

Table 6.3 A customer is a sink object in the database

Sink:	Snk 0.1.0
Name:	Customer Y
Description:	Receives outputs from SOI product
Type:	Material—Product X
Ave. wt. per unit time	8 kilos per Δt
Product received:	F 0.1.0

conservation of mass principle) so that the sum of all the flows equals the total outflow of the product from the interface.[14]

6.5.1.4.3 A (Customer) Sink Object

At the initial state of system identification, it is only necessary to identify that a sink (customer in this case) exists; create an object in the knowledgebase, and identify it as the sink for the product flow. Table 6.3 shows a customer/sink object.

The important link here is the last item, the Product Received. This establishes, in the knowledgebase, that the flow of F 0.1.0, the product, is received by the sink, Snk 0.1.0 and completes the flow path from source (the interface from the SOI) to the customer (sink). When we reflect on the importance of network theory (Principle 3, Chap. 2) we can see that this embodiment in the database (to be used as the knowledgebase) is a key to our understanding of the system we are studying.

Undeniably, this is a lot of detail to keep track of. However, it is essential to do so if we are to make any sense of what is going on in the system as a whole. Fortunately, with the right software we can readily capture this detail using the visual frontend described earlier. Simply drawing a flow link between the SOI and the presumptive customer calls up the need to fill in the details of what the flow is and its characteristics. Granted the analyst may need to do a great deal of research to fill in the forms, but that is exactly what an analyst is supposed to do!

6.5.1.5 Measuring and Using the Variables, Instrumenting the Flows

This step is fundamental to system identification. We will never be able to characterize a system until we have real data about how it works. Ultimately, we will see that system identification involves instrumenting all flows into and out of the system boundary. In Fig. 6.5, we see the basic scheme for determining how to characterize the behaviors of the initial elements of the product flow.

[14] In practice, the author has found it easiest to create separate interface objects for each outflow of the same product. It is an accounting problem, not a systemic problem!

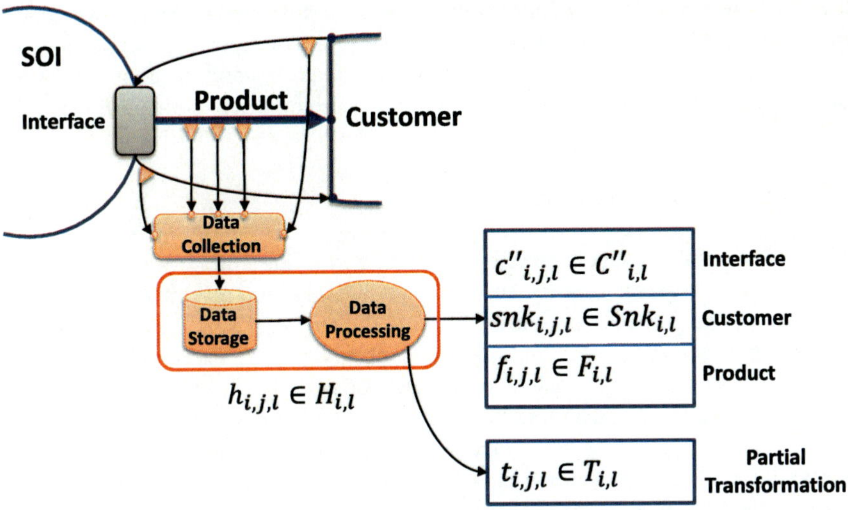

Fig. 6.5 Real-time data are gathered from measuring relevant parameters of the product flow (e.g., volume per unit time). Data on the communications between the interface and the customer are also collected (e.g., number of messages per unit time). The data storage and processing (becoming part of the SOI history object) are used to characterize the interface, customer sink, product flow, and provide partial information regarding the SOI transformation object

What is needed is a set of "sensors" (as defined in Chap. 4) that measure the various important parameters of the flow. For example, if the product is a physical object, we need to measure the number of units through the channel per unit time. There are other possible parameters that might be important to capture. For example, in the case of a physical product, measures of "quality" are also important to know. Such measures are generally couched in statistical terms. These parameters are going to become more important when we start deconstructing the SOI as they will also be subject to internal measuring as part of the quality control, management, subsystems.

Sensors appropriate to each relevant parameter of the flow are installed to capture data using the time constant determined previously (Δt), the sampling rate.[15] In Fig. 6.4, sensors are placed on the channel in which the flow of product is contained.[16] The figure also shows sensors monitoring the message channels. The selection and

[15]The underlying assumption we have been working on is that all data and functions are expressed in discrete form (i.e., digitally). Arguments have already been made that there is an equivalence between discrete and continuous dynamics when the time constants (used for the former) are sufficiently small that the approximation to continuous functions (for example) are sufficiently accurate and precise. Using a discrete value for a time interval, Δt, is thus justified.

[16]Non-channelized flows, for example broadcast messages or diffusion of materials, can also be readily sensed.

use of a sensor is a highly technical subject that is best covered in books devoted to, for example, cyber-physical systems.[17] There are several aspects to measurements of flows for the purposes of system identification. In today's world, with the advent of digitization of just about everything, we describe these for discrete-time methods.

6.5.1.5.1 Parameter Selection

There are several relevant parameters associated with flows that will come into the analysis as it progresses. At the current stage, we might be most interested in the flow rate of the product from the interface to the customer sink. Later we will be concerned with quality issues as well. Material products are measured in terms of weight of a unit measure, such as decigrams per unit object, like a manufactured electronic circuit unit. Total mass, however, is the sum of mass measures for each of the individual components in the final unit. These are important since they plus the masses of waste products are used to determine that the full accounting for flows through the system are made ultimately.

Flow rates for something like this kind of product can be measured in units of product exported per unit of time rather than total mass per unit of time—the sensor counts the number of units that pass it and the data collection process in Fig. 6.4 keeps this count for each unit of time.

Flows of energy are also measured in a similar fashion. Here the question is how much energy of a certain quality (e.g., amps, current, per second at 120 volts) flows from a source (the interface) to a sink (the customer) per unit of time. The potential difference between source and sink (voltage in the case of electricity) is an important parameter just as the pressure difference between two water reservoirs determines the rate of flow of water in a pipe of given internal diameter.

In both matter and energy flows, the key concern is for how much substance flows in the time units used.

In message flows the situation is analogous but the measurements are different. What we would be most interested in messages is the amount of information that is contained in the message. But this is not something that can be measured directly from the message itself. Information is somewhat analogous to the power factor of energy flow[18]—it is related to the amount of change (work) that will be accomplished

[17] The term "cyber-physical" is relatively new but it applies to systems that employ computers and lots of sensors (both of the external world and the machine's own self) and emulate natural living systems that interact with the world in more complex ways. A dramatic example in the news as of this writing is the self-driving vehicle (Alur 2015; Wolf 2012).

[18] A somewhat more useful measure of energy flow is actually something called "exergy" or the amount of energy in the flow that will accomplish work in the receiving system. But this measure depends on characteristics of the work processes in the receiver and not on the amount of total energy exported by the sender. The same thing is true for receivers of messages. Since information amount in a message is based on what the receiver is expecting the message to be (Mobus and Kalton 2015, Chap. 7, Quant Box 7.2, page 281) one cannot measure information directly from the flow of the message.

at the receiving end. If the receiver's work process is inefficient (wastes more energy than is technically necessary) then the amount of useful energy flowing between the interface and the customer is actually less for the same amount of total energy flowing. For example, an SUV gets fewer miles per gallon of gas. The same number of joules are available in the gallon burned in a compact as in an SUV, but the former will get more work done (moving the same driver and passengers more miles driven per gallon) than the latter. Similarly, the same message may be sent to different receivers, each having different a priori knowledge, resulting in different changes in the behaviors of each.

However, messages are exchanged between systems for the purpose of sending and receiving information that will make changes in the communicating parties. Therefore, measures of message flow are still relevant at this stage of the analysis. The traditional measure of messages is something like encoded symbols per unit time (e.g., bits per second), which is readily sensed with electronic messaging (both broadcast RF and digital wires).[19] The actual information flow through message channels can only be determined after we know more about the influence that these messages have on their recipients. In the case of the interface-flow-customer case under current review, this means we will need to analyze the "behavior" of the interface upon receipt of messages from the customer. Since, by definition, we will not be analyzing the customer per se, we can only post hoc estimate the information content of messages from the interface to the customer from changes in customer overt behavior—for example in accepting fewer products per unit time or from the subsequent messages sent back to the interface. We will take a closer look at this problem below in the environment analysis.

The ultimate selection of parameters and the instrumentation of the flows for measurement results in the design of the data collection and the history object shown in Fig. 6.4. Given a selection of Δt, emplacement of the sensors and a data capture/storage mechanism in place it is time to start measuring the behavior of this one specific element of the system.

6.5.1.5.2 Time Scales

The time step, Δt, chosen represents the smallest unit of time in which a "noticeable" or relevant change in the measurement can occur. The data capture mechanism, shown in Fig. 6.4, is set up to take a measurement on all of its sensors for each step. However, patterns of important behavior may actually not show up except in very different time scales. The data collected at the resolution of Δt may contain noise. But more importantly, real-world, concrete systems of any complexity at all demonstrate patterns that vary according to different time scales.

[19] Sensing photonic signals from the outside is problematic (as eavesdroppers fear) but here we are assuming that all parties agree to the instrumentation of all flows so that photonic detectors (sensors) can be inserted into the channel directly.

The world is fundamentally comprised of systems that operate in multiple time scales simultaneously. Their behaviors during short time intervals (some small discrete multiple of Δt's) can be, and often are different when viewed over longer time intervals. The world is full of non-homogeneous, non-stationary processes. It is influenced by nonlinear processes characterized by deterministic chaos with inexplicable switches to new strange attractor basins, self-organized critical systems, and catastrophe. None of these phenomena operate on particular time scales. Therefore, as discussed below, the data collection required for very complex systems, for their product outputs, must cover very long time periods in order to capture patterns that would not show up in short samples. Practically speaking, some very complex systems, like ecosystems or human organizations, must be observed for years, perhaps hundreds of years, in order to be certain that all of the relevant behaviors are found.

Consider that the human brain is in the business of observing the world (or its corner of it) continuously for its whole life. It is designed to change its internal representations over time as the various components of the world change on these different time scales. We don't perceive changes in long-time-scale phenomena like sea levels or mountain building. But we do perceive changes in the courses of rivers over years, or the flooding of those rivers over days. The brain is the epitome of an observing system able to operate and modify its representations in multiple time scales.

6.5.1.5.3 Time Series Data

Not yet said explicitly is the notion of what this data stream looks like. What is being collected is a set of time series data points or measurements made in each Δt for a very large number of Δt's. If the selection of the Δt has been done well, the variances in measures across some interval of time (x number of Δt's) will represent real differences and not noise or measurement errors. In that case, the analysis of the data series should yield true dynamics. However, the analysis based on a single Δt is not sufficient for real-world systems.

6.5.1.5.4 Multi-Time Scales and Non-stationary Series

In the real world, nothing stays the same forever. Typically, time series data are analyzed based on some established time window, some x number of Δt's. This, however, is not necessarily adequate to characterize the dynamics of a flow. To see this, consider the following. Suppose we collect flow rate statistics over $x\ \Delta t$. We compute the average value and the standard deviation or variance. We now have a measure of what the flow dynamics are over that time window. Suppose, then, at $1000x$ time units later we collect the same data stream and do the computation again. In most real-life situations, we will find that the newly computed mean and standard deviations will be different (slightly) from the previously determined values. As a rule, most analysts will accept this difference as due to measurement error between the two samples. They may be tempted to average the two statistical values and believe they have a more "realistic" estimate of the values. Just to be safe, they may come back and take a third sample set at another $1000x$ time units. What if they find yet another set of values? Of course, they have recourse in the Law of Large Numbers, that is, if they took enough windowed samples over very long time scales, say $100{,}000x$, they would feel justified in claiming whatever overall statistics they calculate must be approaching the processes true statistical values.

There are two problems that present themselves. The first is what if the mean values of each window sample are slightly higher each time window? This can indicate a trend in a longer-term change that greatly complicates any analysis they do. What are the characteristics of the trend? Will it go on forever? The second problem occurs when the statistical properties of the sample vary up and down with each window. At first this might look like just a problem with window error, especially if there is any kind of alternating behavior, first up, then down, etc. But a careful further statistical analysis of the window data might reveal large variations in the window means and variance in the within window variance. This could be trouble.

Non-stationarity is a general fact of complex adaptive and evolvable systems. Such systems can demonstrate stable statistical characteristics for long periods of time, and then, suddenly, deviate (sometimes drastically) from their long-term statistical properties.

6.5.1.5.5 Data Analysis

If a time series represents a simple stochastic process, then ordinary statistical properties such as mean and variance (first and second statistical moments) will adequately characterize the flow for modeling purposes. Unfortunately, most real-world systems, and particularly CAESs, do not have simple stochastic behaviors over longer time scales (see the discussion below regarding environment analysis and identification of sources and sinks for more clarification re: signal processing).

In general, these behaviors can be characterized as sporadic (that is recurring but not necessarily periodic), episodic (that is lasting for variable amounts of time), and erratic (that is the variance in amplitude within an episode is, itself, variable). In general, the statistical properties of these occurrences are non-homogeneous, non-stationary. As such, the analysis of the data must take these possibilities into account. This is no small requirement. Statistical methods for detecting these properties are still in their infancy. But as more work is being done on real-world phenomena from a systems perspective, we suspect that more attention will be paid to the characteristics of these phenomena. We have seen this develop in fields such as multivariate statistics where co-correlations and co-variance methods have been developed to reveal causal relations. The same is likely to occur in non-stationary analysis. It will be an interesting area of active research in the near future.

One interesting area where the impact of non-homogeneous non-stationary properties has been felt is in machine learning. It turns out that real neural systems in real living brains deal directly with the problems associated with non-homogeneous non-stationary processes. Synaptic plasticity shows a dynamic that seems to incorporate this phenomenon directly (Mobus 1999).

Even as we develop more advances in analytic methods such as non-stationary process analysis, it is still incumbent on analysts to understand that these long time-scale issues affect their picture of what systems are doing. We will address how this affects analysis of systems for the present in terms of how systems evolve (Principle 6 in Mobus and Kalton 2015). Those who take on the analysis of complex systems would do very well to keep in mind that "things change" and that whatever data they collect and analyze currently are likely to change in some future time frame. All solutions are provisional!

6.5.2 *Proceed Around the Boundary*

Once the primary product flow, its output interface, and its customer sink have been at least partially characterized, it is time to analyze the entire boundary along with the actual inputs and outputs that traverse it. This entails doing the same kind of analyses for each and every input as well as non-product outputs (e.g., waste products or heat). The process will be to traverse the boundary, finding outflows and inflows, and repeating the process of identifying the interfaces. Flow, interface, and source or sink objects can be created and their data forms at least partially completed.

For each flow discovered in the traversal the analysis requires instrumentation and data collection as described above. Eventually, data will be collected on the inflow and outflow parameters setting up history objects for each (Fig. 6.5 above).

6.5.2.1 Completeness and Sufficiency of Flow Discovery

The more complex a system is the more difficult it will be to complete a survey and analysis of interfaces and flows around the boundary. Once an initial analysis of the boundary has been done the analysis turns to the second phase, environmental analysis, in which it is possible that new, unaccounted-for flows will be discovered (uncovered) and require reconciliation with existing analysis. The nature of the knowledgebase that is being built up allows additions of new objects as they are discovered. It is always possible to return to an identification analysis when new flows are discovered as a result of analysis of the environment. We will return to this issue below.

Always, the construction of the system identification phase must be considered tentative. It is even possible that once decomposition of the SOI is started we will find flows that had previously been unaccounted for. For example, we might, in analyzing the inputs to an internal subsystem, discover a flow that had not been previously encountered. This is the nature of a discovery process.

Fortunately, the way we have constructed the system definitions of Chap. 4 allows us to recover and add to the structures we have already created. There is never a time when we have to close off the addition of new structures—they are simply added to the sets or lists we have already created. That is, the structures (objects and their relations) are extensible such that they may be easily amended as new information emerges. We will provide some examples of this in the next chapter.

6.5.2.2 Identification of Outputs and Inputs

Once we have identified the major "product" outputs, their flows, interfaces, and sinks we are ready to proceed to analyze the rest of the outputs and then the inputs.

6.5.2.2.1 Wastes

Non-product outputs may be classified generally as wastes or waste products. There are two major categories. Energy used up in work processes always exits the system as waste heat, that is energy at a very low potential that cannot be used to do any additional useful work. It might be noted that waste heat from one system (process) might still be useful to another process where the nature of the work does not require high temperature differentials. For example, space heating (to keep occupants comfortable) requires very low-grade heat. Technically this is not accomplishing work per se. However, it aids in letting the occupants of the space to accomplish other work in comfort. Even so, it will eventually radiate into the environment as completely unusable energy. The total energy of this low-grade heat is the same, but it simply is too diffuse to be usable in driving useful work process.

Material wastes may also "diffuse" into the environment but not in the same way as waste heat. The latter will eventually radiate to deep space from the Ecos. Waste materials, on the other hand, will simply diffuse into the ambient space. For example, waste chemicals will end up in diffuse concentrations in the hydrosphere where they may still have chemical or biochemical effects on other aspects of the biosphere.

In natural systems, the typical pattern for material wastes is that there are low-order organisms that can recycle them. Or they may be incorporated into sediments to be recycled by geophysical processes. In both cases the final waste products can be inputs to regenerative processes, that is, they can be useful resources for more complex synthesizing processes.

6.5.2.2.2 Inputs

We next proceed around the boundary and consider the inputs to the SOI. These are classified as either resources or disturbances. As with the identification process for product outputs the nature of the analysis is to identify the flows and their dynamic characteristics as well as the input interfaces and their protocols.

6.5.2.2.2.1 Resources

A resource is defined as any input that is used in a work process to accomplish the purpose of the SOI and becomes a component in the products (with some residuals in the waste materials and waste heat).

6.5.2.2.2.2 Disturbances

Disturbances are any impactful encounter with the system that is not a routine flow of a resource. These are the hardest aspects of a system to analyze yet are crucially important in terms of how disruptive to a system's normal function they can be. The analysis of disturbances can be done using traditional risk assessment methods. Risk is defined as the impact (cost of responding and repairing damage, see also Sect. 10.3.2.1 in Chap. 10) times the probability of occurrence.

Some kinds of disturbances are readily identified. For example, fluctuations in a flow that exceed the capacity of the system to have an adaptive response. What happens to a car manufacturer plant when the shipments of engines stop, even for a short time? Then other kinds of disturbances are more difficult to predict in order to anticipate their occurrence (and take precautions). Tornados can be very disruptive. Finally, still other kinds of disturbances can seem to come out of left-field, as the saying goes. They are completely unanticipated. Most people have now heard of the "black swan" phenomenon.

Another type of disturbance is more easily understood and analyzed, that is the wear-and-tear or entropic decay of components that lead to subsystem dysfunction. A great deal is known now about how physical things wear down and go into senescence or need repair.

There is just so much an analyst can do in terms of identifying the type and quantifying the risk. But some attempt at doing so is necessary, especially for the first kind that are endemic and just the result of some cause that is not necessarily perceived.

6.5.2.3 Creating the Data Objects

6.5.2.3.1 Flows, Sources, Sinks, and Interfaces

Figure 6.6 shows a schematic representation of an SOI that has been completely instrumented so as to collect data on the inputs and outputs over time (instrumentation of the interfaces has been omitted for simplicity). The SOI is now considered as

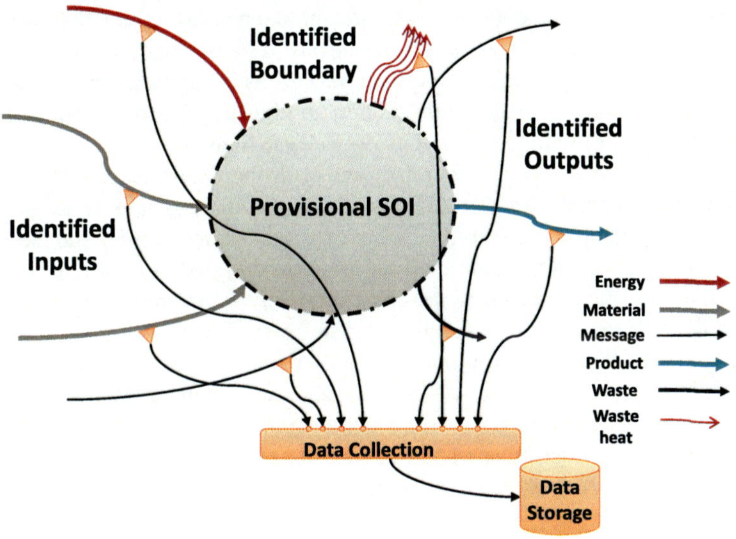

Fig. 6.6 Instrumentation of the entire SOI is used to collect data on the inputs and outputs over time

a "provisional" system in that the identified inputs and outputs and their instrumentation are only assumed to be complete in a first pass around the boundary. It is possible that the environmental analysis (phase two) will reveal additional details or exposure of flows that had not yet been considered and must be incorporated into the model before further analysis can proceed. However, absent such an exposure, we can consider the situation depicted in Fig. 6.6 as a complete condition for the opaque-box analysis of the SOI. We are now in a position to produce the first pass identification of the system of interest.

This will entail computing a transformation formula in which the product output(s) are time delayed functions of the resource inputs. Or:

$$O_t = \mathcal{T}\left(i_1, i_2, i_3, \ldots i_n\right)_{t-n} \tag{6.1}$$

the output vector at time t is a transformation of the inputs at time t-n, n, an integer multiple of Δt.

The outputs of a system at time t are the result of transformations that occurred to inputs that arrived at a past time. Equation 6.1 oversimplifies this relation since inputs that arrived at different actual discrete times may be involved in a discrete unit of transformation, that is, a unit of $\boldsymbol{i}_2$ may arrive before a unit of $\boldsymbol{i}_3$. The discrepancies are usually accounted for by the fact that subsystems include internal stocks that mediate the flows of substances. More on this aspect later.

At this point, we are now ready to complete the system identification in that we can estimate the transformation function of the provisional SOI. We have collected time-based data (time series) on all of the flows into and out of the SOI. It remains to analyze these data to estimate the transformations that occur within the system to complete an opaque-box analysis. It is important to reiterate that a mere transformation function is not the end goal of a deep analysis. A transformation function would be adequate if we didn't care how a system transforms its inputs to produce outputs. But we have already claimed that that is not our objective. We will want to dissect the system in order to find out how it manages these transformations.

Many system identification exercises end with the determination of a system transformation function because the purpose of the analysis is only to grasp the functions. For example, a controls engineer may only be interested in the identification of a physical plant transformation in order to design a controller to keep the plant within operating specifications. This is perfectly acceptable as far as it goes. It is possible to "control" a large number of systems processes without a deep understanding of what the system is or how it does what it does. And as a general rule, this is perfectly adequate. But in cases where we have need to understand a system that has broader implications (e.g., how an economic paradigm affects the well-being of the average citizen) mere grasp of how to control a system is not sufficient. We need to dig deeper.

6.5.3 Estimate the System Transformation Function

The output vector O_t in Eq. 6.1 is given by the transformation function $T()$. This is, in fact, a composite of multiple output functions for all of the output interfaces around the boundary. How these functions are derived is, again, a matter for multiple textbooks, and there are many of those. Here we describe the main points that analysts must consider.

Recall that we are still working on an opaque-box model of the SOI. Recall also that for simple systems this may be completely adequate for practical purposes. But for CASs and CAESs, deep knowledge is essential because the system function estimated at this stage cannot anticipate internal adaptations or evolution that will change how the system processes inputs to produce outputs.

6.6 Environmental Analysis

Phase 2 of the analysis involves discovering and characterizing the entities and sources of influence on the SOI.

6.6.1 Limited Models of Entities and Influence Sources

Sources and sinks do not need to be modeled in any substantial form. At most all that is needed is a minimal model of them as "naked interfaces," that is, the points at which flows leave the sources and flows from the SOI enter the sinks. And even there the most that is needed is an analysis of the protocols used. In very many cases the SOI communicates with sources and sinks in order to coordinate the flows of materials and energies. In such cases, it is necessary to identify the message flows, as in Fig. 6.4 above, and identify the specific message/information relations that affect the flows of substance. The methods for doing so are the same as identifying input and output interfaces for the SOI itself. These will be covered below so this section is a placeholder for how that is to be accomplished with respect to source and sink interfaces.

The relevant information for sources and sinks involves the characteristics of the flows (in or out) without specifying the causal aspects of those flows. Note that if it is essential to capture a causal model of a source or sink, then it is probably the result of having not quite understood the boundary of the SOI properly. The likelihood is that what had been counted as an environmental entity should have been considered as part of the SOI itself. As an example, consider the study of a lake system in a mountainous area. Originally the boundary might have been considered as the obvious shores of the lake, with inputs coming from streams arriving from the watershed area. In assessing the input of water through these streams it becomes clear to the analyst that the flow volumes are unaccountable under these assumptions. They decide that the flow volume variations have to be better understood and so expand the SOI definition to include the geography of the watershed. The environmental analysis now also expands so that the input and output entities may become the microclimate conditions around the watershed.

With respect to the internals of the sources and sinks, once the analyst has characterized the long-term flows (i.e., generation from sources and absorption into sinks) it is unnecessary to have further knowledge about how these are obtained.

That said, we will look at the situation when the characterization of the chosen environmental entities still leaves questions unanswered, as in the example of the watershed. It turns out that the beauty of this recursive procedure is that it can be reversed in direction so that what had previously been external entities can be brought into the SOI definition—the boundary can be expanded—invoking the very same analytic methods but in an "outward" direction.

This raises a difficult question. In the recursive decomposition process, we have a stopping rule regarding the recognition of when to no longer decompose. When we have the atomic processes (Chap. 4, Fig. 4.7) as leaf nodes we know to stop any further decomposition. Going outward has a similar stopping condition. What triggered the inclusion of formerly environmental entities as now part of the SOI was the existence of unreconciled questions about the dynamics of the flows from/to such entities. The stopping condition for reverse recursive analysis is when the analyst cannot detect any such unresolved questions.[20]

[20] Here we have a possible conundrum that analysts have to be aware of. How far "out" do we continue to include entities and treat them as inside the SOI if we continue to detect unresolved questions? In one real sense such an outward progression could be infinite—there are always questions generated about complex systems! This is more of a "practical" matter than a theoretical one. This is where the judgment of the analyst comes to play. In the real world there will always be more

6.6.2 *Identifying the Sources and Sinks*

The completion of the environmental analysis will be the characterization of source and sink entities, including any presumptive sources of disturbances.

Recall from above that the identification of inflows and outflows proceeds by circumnavigating the boundary looking for flow entry and exit points and their interface objects. In that process, we "named" and coded the source and sink objects but limited our analysis to the metrics of the flows themselves. We now return to the list of sources and sinks, the ***Src*** and ***Snk*** sets in Eq. 4.5. Table 6.3, above, provided a limited data object for the product sink or customer. This was a placeholder until we undertook a more thorough environment analysis. The only "measurement" contained in the object was a replication of the flow average obtained from the flow analysis. In the identification of environmental entities, we need to go one step further and produce a limited model of the behaviors of the sources and sinks.

6.6.2.1 Measuring the Behavior of Sources and Sinks

In the section, Data Analysis, above, we observed that long-time-scale behaviors of real-world CAS and CAES entities have several levels on uncertainty. These systems are in the general class of processes called "stochastic" or probabilistic in terms of their behaviors. There are many sources of randomness that affect the workings of CASs and CAESs. These can be from external sources (and sinks), from noisy disturbances from the environment, from internally generated noise, or from nonlinear dynamics and chaos. This gives rise to flows that can be found to be sporadic, episodic, erratic, as well as non-stationary. Figure 6.7 provides a sense of what this might mean.

These discrete-time measurements are based on neural firing rates in sensory cortical neurons; each bar is a measurement of a neuron's rate of firing action potentials at a time instance. The time intervals between measurements is a constant, Δt, usually measured in milliseconds (the vertical scale is in a number of pulses fired during the sample period). Each neuron is exposed to varying stimuli at varying intervals and for varying amounts of time. This is what is going on in the brain but it reflects the actual encounter of the brain—a system—with actual real-world sources of the signals that stimulate their firing.

Even simpler entities that are sources or sinks for matter or energy flows may have seemingly indeterminate behaviors. Parts suppliers, for example, ship varying quantities of parts to a manufacturer, the packages arriving at varying intervals, containing, possibly, varying quantities. The variations may be due to varying needs by the manufacturer or to disruptions in the production facilities of the supplier (unknowable in general). Or a package could have gotten lost in shipment.

questions that cannot be answered without expanding the boundary of analysis! We will attempt to address this "open question" below.

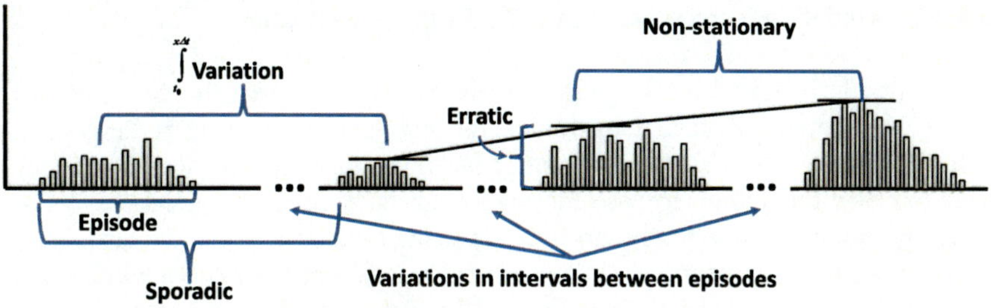

Fig. 6.7 This time series represents examples of neurons in sensory cortex firing whenever they are excited by stimuli in the sensory field. Firing comes in bursts or episodes that generally ramp up and then decline. Episodes are of varying intensity, varying duration, and over the long-timescale can vary in total firing. Some signals can contain noise and be erratic. Finally, firing properties may vary in an indeterminate way over very long-time (and varying) intervals. The vertical scale represents action potentials per 1/tenth second. Ten samples on the horizontal scale represent one second of firing. There are usually long periods of quiescence during which very low frequency and random firing may occur. These are represented by the ellipses between bursts

As mentioned before, any time series of measurements can be treated as signals and many of the methods of signal processing may be used to analyze the data. It is beyond the scope of this book to go into depth in various approaches to, or the theories of, signal processing. There are a number of very good resources for this, including online free texts (Prandoni and Vetterli 2008; Smith 1997). The important point to recognize for the systems analyst is that in order to adequately characterize the inflows and outflows, and hence the sufficient identification of sources and sinks it is essential that considerations for the kinds of variability and stochasticity illustrated above to be taken into account.

Having an adequate characterization of outflows from an SOI (via corresponding interfaces) will provide additional and essential information about the inner workings of the SOI, particularly the subsystems that export the outputs. However, such information can also be used to characterize the sinks for the outflows. That is, if the variations in the outflows are due not to SOI internal issues, they can be assumed to be caused by something in the sink's internal processes (which remain unknown). Some of this will not be known until the subsystems of the SOI are analyzed, but some of it can be identified by virtue of the analysis of the communications channels (if any) between the SOI interface and the sink (as depicted in Fig. 6.5). For example, if a sink is an active agent (customer) that signals a halt to receiving product until further notice, then some portion of the sporadic-ness of the flow may be attributed to the sink rather than the SOI. Communications signaling to modulate flows are found throughout biological examples as well as social systems. The analysis of both input and output interfaces includes an analysis of messages between the interfaces and their source/sink entities. The analysis should identify causal correlations between signals received or sent and subsequent actions by the source/sink entity and the interface. This will be demonstrated below.

6.6.2.2 Modeling Sources and Sinks for Engineered Systems

Thus far we have been discussing the analysis of concrete and existing SOI environments. The strategy has been to instrument the real flows into and out of the system, collect very long-timescale data streams and estimate functions that would generate such to refine the knowledgebase for environmental entities and interfaces. What about, however, the case for systems we are going to design? We could also adopt the methods of instrumenting the environment into which this system will be put, and this is probably the best starting place. We identify with which entities in that environment the new system will interact and collect data as before.

There is, however, a fundamental weakness with this approach taken alone that has to be addressed. Namely, the environment will behave differently simply because of the behavior of the new system.[21] It is possible to derive functions for sources and sinks as above, however, in order for them to be useful later when modeling the new system, they will need to be modified to respond to that new system. There is no simple way to approach this problem.

Consider as an example a company wants to introduce a brand-new product—a new technology—into the marketplace. It expects that this new product may compete to some extent with existing products (like Blu-ray competes with DVD) and, at the same time, introduce all kinds of new possibilities (like new features in Blu-ray not possible in DVD). Companies produce "pro forma" statements, which are essentially models of what their revenues, costs, profits, volumes, and other metrics are expected to be into the future. They do this to assure themselves that their profitability profile will be positively affected by the move. Too often companies that are driven by the emotions of what might be called "wowy-zowy" technology make unwarranted assumptions about things like market penetration rates (volume absorption by customer sinks), pricing options, and so on, because they have a wishful bias and want the new introduction to work out. They produce a favorable sounding pro forma so as to attract investors or obtain financing from banks.[22]

More thoughtful (and hardnosed business) decision makers will insist on building as accurate a pro forma model as possible. They know that being eager for something to work out doesn't mean it will. How to accomplish this is no easy task and it is never guaranteed to be correct. There will always be guesswork involved in simulating the environmental entities. Even so it behooves managers to do as thorough a job as possible in accurately measuring whatever environmental flows they can (e.g., it may be possible to retroactively look at the market dynamics of prior new products of similar kinds).

For engineered systems such as information systems to support operations of an organization, the situation is somewhat in between measuring actual environmental entity behaviors and supposing those behaviors in pro forma models. It depends on

[21] An analogous situation would be to consider what will happen in a particular ecosystem when an invasive species is introduced.

[22] Unless, of course, you are Apple, Inc. in which case you have your own financing in the form of reserved cash on hand!

whether the new system is simply replacing an old system in the same environment, or will be serving a completely new organization function. In the prior case, the information flow channels already exist so it should be possible to observe the existing work processes and use their behaviors as models. In the latter case, the organization is actually in control of the design of the sources and sinks (of message flows) and so the larger supra-system is already known (or better understood) as the environment of the to-be-designed information system.

6.6.2.3 Refining the Knowledgebase for Sources, and Sinks

At this stage, we are ready to refine the data objects for sources and sinks (e.g., in Table 6.3). In most instances, this will simply involve deriving an equation of function estimated from the transfer functions of the flows completed above. Such functions will take the forms:

$$o_{Src_{i,0},t} = f(q,t) \tag{6.2}$$

where o is the output of a source, q is a measure, and t is time; and,

$$i_{Snk_{i,0},t} = f(q,t) \tag{6.3}$$

with i being the input to a sink.

These functions should match those in the corresponding interfaces in the boundary object, but with reverse flows, for example, outputs from sources are inputs to the receiving interface.

6.7 Deconstructing the System

With flows identification and environmental analysis essentially complete[23] it is time to begin converting the opaque-box of the SOI into, first, a gray-box and then a transparent-box. We will decompose the internals of the SOI at the next level down of subsystems. This, recall, is level 1. We will identify all of the subsystems at this level and map the flows of the inputs to the SOI and outputs from it to internal subsystem. We will also map the flows between subsystems such that all of the inputs and outputs are completely accounted for.

[23] As already indicated, we must always consider these analyses as "provisionally" complete. As we enter the next phase of analysis, we may yet find unaccounted for flows that will require a return to the analysis of the environment.

6.7.1 From an Opaque Object to a Transparent System

The final phase of the initial analysis is to convert the opaque-box into a transparent-box, to elucidate all of the subsystems and internal flows of the SOI. The outcome will be a map of these objects and flows at level 1.

6.7.2 Finding Subsystems

The logical place to start this analysis is the processes that are associated with the various interfaces that we discovered as part of the boundary analysis. Since our interest is primarily in CASs and CAESs we will assert a rule for their identification. Such processes are generally active. That is they are work processes that use energy to accomplish their functions. Processes (or subsystems, remember) that receive inputs from the environment, through their respective interfaces, are called "import" processes. The work they do involves not only regulating the interface but also, generally, distributing the inputs through appropriate flow channels to internal processes that do work on the inputs to produce intermediary products and outputs (finally).

Processes that "push" products and other outputs out into the environment are "exporters." They generally consume energy doing work to move the outputs from the interior of the SOI to the environment, to the various sinks. Flows, after all, require a pressure differential between source (in this case the SOI) and the sink (the environmental entities). For example, cells have to expend energy pushing molecules out against a gradient; companies have to expend energy shipping products to customers.

Figure 6.8 depicts the general situation with an SOI interfacing with its environment.[24] It imports resources and exports products and wastes. This is where we start the analysis of the level 1 (recall Fig. 6.2 above) subsystems. Import and export processes are the subsystems that link the environment to the internals of the SOI.

[24] A general, but not absolute, convention for drawing these maps is to put the sources on the left and sinks on the right, that is, flows go from left to right for environmental elements. However, there are situations (many times in fact) when an entity is both a source and a sink. For example, when there is a communications protocol involved in controlling the flow of material or energy, say from a source, that source will also be a sink for messages sent from the system's interface (see Fig. 6.4 above). If a source of, say, substance A is also a sink for substance B, one method for representing this is to have two objects in the map, one on the source side and one on the sink side. The two objects can be given the same "name" and a logical linkage in the database, but are acting as different entities in their behaviors as sources and sinks.

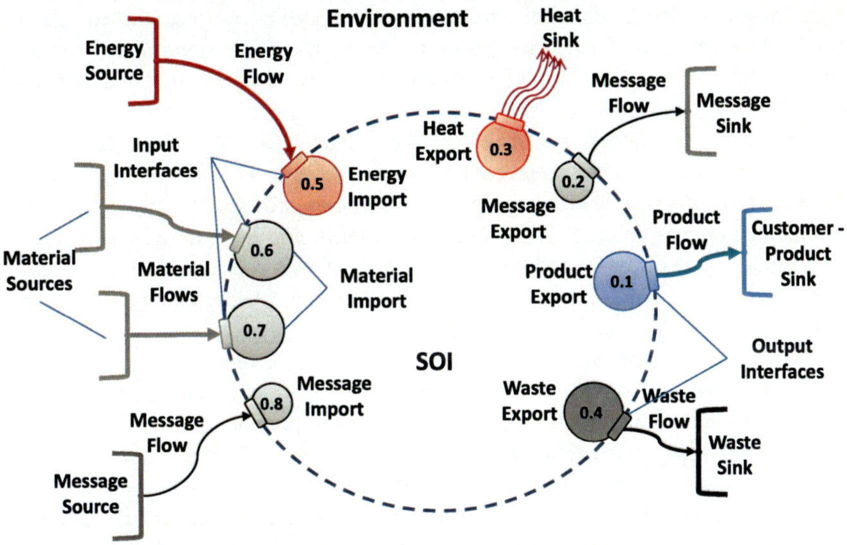

Fig. 6.8 The first step in deconstructing the SOI is to find all of the subsystems associated with the input and output interfaces. For inputs these are import subsystems. For outputs these are export subsystems. Importation processes may also function as distribution subsystems while export processes may function as collection subsystems. Note the process identification numbers. Since the analysis started with the output of the product flow, we have assigned it as level 0, subsystem 1 (0.1). Since all subsystems of level 0 would have a prefix of 0, we will drop this numbering in the future. The prefix, 0, is understood. See below for the actual treatment of the interfaces and subsystems

6.7.2.1 Start with Interfaces and Their Associated Processes

These are the first subsystems that we identify. This is because each of these will have outputs (for importers) that get distributed to other internal work processes (as shown below) or collect outputs (e.g., products and wastes) from internal work processes and prepare them for export.

A reasonable way to envision the import and export processes/subsystems is that they "own" the interfaces, or the interfaces with the external world are essentially subsystems of the import/export processes. It is these latter processes that are responsible for controlling the interface protocols. The interfaces will first be identified in the boundary object (see below) and cross-referenced in the component set of the subsystems as they are further analyzed in decomposition.

As an example of the role of an import process, consider the energy importation process. In a plant cell, this is the job of the chloroplast. This organelle is responsible for importing light energy at visible wavelengths and exporting carbohydrate molecules that can be used, later, as fuel for respiration (oxidation of the sugar molecules to produce adenosine triphosphate molecules, the energy distribution mechanism for all cells). The interface that accomplishes the importation of light energy is the molecule chlorophyll; it absorbs photons and transfers electrons to other molecular processes within the chloroplast that then complete the process. A similar function is performed by the electrical interface (through circuit breakers and an electric meter) of a manufacturing company with the electric grid that supplies electrical energy for the machines.

The reason we counted interfaces as part of the boundary of the SOI (originally) is that they occupy a very special role in a whole system. They are both part of the boundary, as previously treated, and part of the import and export processes that are, technically, internal to the SOI. This can cause some confusion for new analysts. Is the interface part of the boundary or part of the internal processes? The answer is "both." This will be reflected in the way import and export subsystems are handled in Eq. 4.1.

Recall that a boundary object, $\boldsymbol{B}_{i,l} = \langle P_{i,l}, I_{i,l} \rangle$ is a tuple of sets, P, called properties, and I, called interfaces. As each interface is identified, starting as suggested with the product output, it is given an ID code and included in the I set.[25] The actual data object holding the interface information was given in Table 6.1 above. In database terms, the ID code is the primary key for the object in a table of interfaces; the prefix, 0, is required here so as to distinguish the object within the interfaces table. The on-screen identification would be coded I 0.1 (as in Table 6.1), where the I stands for interface. In Fig. 6.8 interfaces are assumed to take the same code as their owning subsystem (importers and exporters).

Once the interfaces have been identified we can turn attention to the "owning" subsystems. These constitute the first items that will be populating the $\boldsymbol{C}_{0,0}$ set from Eqs. 4.1 and 4.2. At this point, we can only identify the existence of the component subsystem and create a data object for it as in Table 6.4 below.

Equation 4.2 provided for components, $(c_{i.j.l}, m_{i.j.l})$, to include a membership function in the event that a component of the set had fuzzy properties. In Table 6.4 this is shown as a value of 1, meaning that the component subsystem is always as

Table 6.4 The first component subsystem identified in the SOI

Component:	C 0.1
Name:	Product exporter
Description:	Exports the SOI product (X) through interface I 0.1
Type:	Process
Transfer function:	T(p, t)
Output(s):	F 0.1.0
Input(s):	TBD
Boundary	P = TBD, I 0.1
Membership:	1

[25] By this we mean the new interface is given a unique code, for example, 0.1 for the product interface, and is added to the set $I_{0,0}$. As the analysis progresses around the boundary, each new interface is similarly added.

member of the set. More generally this field could contain a function that returns a 1 only *when* the component is a member of the set $\boldsymbol{C}$ in any Δt or $x\Delta t$ window.

Note that the data object in Table 6.4 indicates a Boundary object with a P set yet to be determined but an I set containing a single member, the interface cross index of the interface identified in the boundary of the whole SOI. This is how the interface to the external environment is handled between the SOI and, in this case, the export subsystem, $\boldsymbol{C}$ **0.1**, for consistency. The interface needs only be identified initially in the SOI boundary analysis and then cross-referenced in the export (or import) component that is discovered.

Once again, the analysis proceeds around the boundary, identifying all of the importer and exporter subsystems associated with the interfaces already identified in the boundary $\boldsymbol{I}$ set. In the next section, we will conduct analysis of these component subsystems as a first step in delineating all of the internal subsystems.

At this point, the inputs to the exporter subsystems and the outputs from the importer subsystems (into the interior of the level 0 SOI) are not known per se. We will identify these flows and discover their sources (inflows to the exporters) and sinks (outflows from the importers).

6.7.2.2 Identifying Internal Work Processes and Flows

The final stage for completing the knowledgebase for level 0 of the SOI is to identify the internal subsystems/components, finally filling in the $\boldsymbol{C}$ and $\boldsymbol{N}$ objects in Eq. 4.1. Starting with the internal processes that were identified as "owners" of the interfaces with the environment we will treat each as a new problem in system identification. Each import/export process will be given identification codes at level 1 (Fig. 6.9).

In order to accomplish this task, we will "follow the money" so to speak. That is, we will start with the discovery of the outflows from importers and inflows to

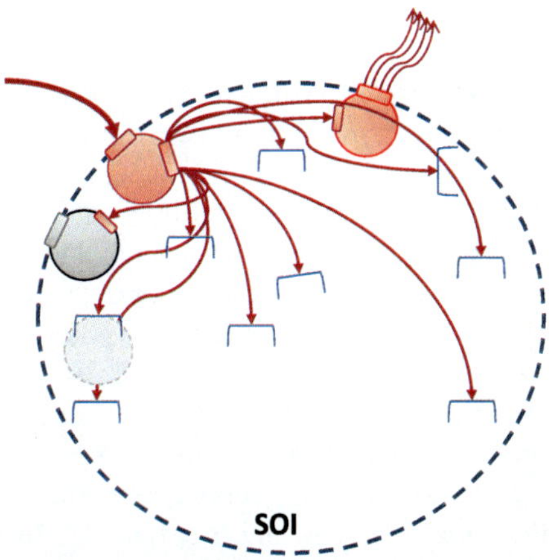

Fig. 6.9 The energy input importer process has been identified and a boundary analysis of its outputs reveals multiple channels of flow to internal sinks. These will be identified later, but at this point the energy importer is treated as an SOI for the purposes of analysis. See text for explanations of the other subsystems

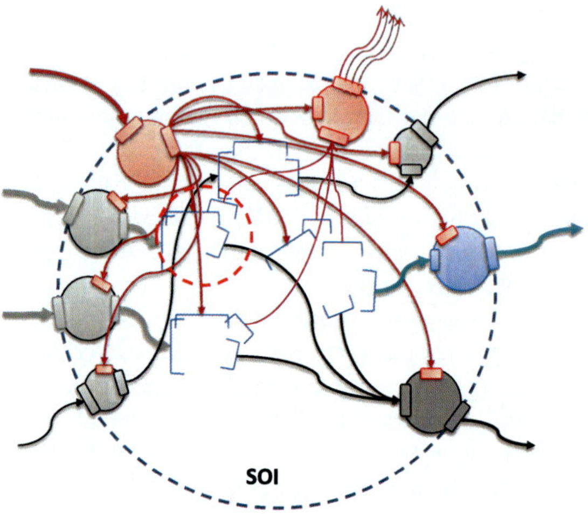

Fig. 6.10 Mapping internal work processes begins by mapping flows from inputs to candidate processes and outputs from those to the export processes already identified. Here the import of energy, matter, and messages is shown distributed to various internal work processes, including the importer/exporter processes, that convert material inputs to intermediate and then final products for export. Note the clusters of internal sources and sinks established through the environment analysis of each importer and exporter. Also note the red dashed oval encompassing one cluster of sinks and sources. That is the starting point for discovery and opaque-box analysis of the internal work processes

exporters through the same kind of boundary analysis we did with the level 0 SOI (see Figs. 6.9 and 6.10 which show a sequence of analysis states leading to Fig. 6.8 above). In the section below, *Recursive Deconstruction*, we will discuss the formal treatment of deconstruction. Here we describe the first steps in performing this deconstruction more informally.

6.7.2.2.1 Importers

A reasonable place to start is with import processes. Taking each in turn and treating each, now, as an SOI in its own right, we replicate the system identification process, this time looking specifically at its output flows to what we will find to be the internal subsystems. In Fig. 6.8 the red oval importing energy (upper left quadrant) through the red interface has a number of output flows of that energy to numerous customers. It is both an importer of the resource and a distributor of that resource to other subsystems. Therefore, the energy importer has another interface associated with its boundary (shown as another interface on the red oval). This one is only associated with the importer. In the case shown in the figure, the interface actually provides for multiple channels through which the energy flows to other internal processes. Energy is generally consumed by every other subsystem, including all of the other importers and exporters, so the figure shows red arrows going from the importer to all of the other processes. In anticipation of the next steps in this stage of analysis, the figure also shows two other components, one a material importer, the other the waste heat exporter. Initially, when looking strictly at the energy importer, these would be treated the same as all of the other internal receivers of energy, as sinks (open rectangles as in the system language icons presented in Chap. 4). Below the material importer component shown (with an energy receiving interface) we show the second material importer (as in Fig. 6.8) as a "ghost" outline underlaying a sink icon. This represents the fact that as we move to a more careful analysis of this object, we will associate the sink with an actual importer subsystem that requires energy input to do its work.

Once all of the outputs from the energy importer are accounted for (and entered in the knowledgebase) we will have another system (a level 1 system) identified as an opaque-box, with its own Eq. 4.1, but a level, $l = 1$ and index, $i = 5$. This is the first occurrence of the relation in Eq. 4.3, $S_{i.j,l+1} = c_{i.j,l}$. Recall from Chap. 4 description of this equation, the compound index, $i.j$, is the coding for the tree relation of the new SOI to its all of the other internal nodes in the tree structure. In this case, $i = 0$, and $j = 5$.

Figure 6.9 shows the results of a boundary analysis of subsystem, $\boldsymbol{c}_{0,5}$ in which outflows from the energy importer are distributed to multiple internal subsystems within the SOI. This is a preview of the recursive decomposition process that will be covered below.

Raw material resources are imported by the two larger gray ovals in Fig. 6.10 (as in Fig. 6.9). Presumably one is importing material A and the other material B. Each of these outputs their material types unchanged. Their job was to actively obtain their respective materials and send them on to other processes (the green ovals).

By tracing these various flow channels, we discover the internal subsystems that receive them and, presumably, process them. We will return to these entities shortly. For now, they can be treated as sinks for the output flows of the importers.

6.7.2.2.2 Exporters

Similarly, in the product exporter interface, we can identify the exporter subsystems associated with the interfaces already discovered in the level 0 SOI identification analyses. Once discovered, each of these is also treated as a subsystem SOI and the inputs to it are identified. In Fig. 6.10, all of the exporter processes use energy and so receive energy inputs. In this simple model, of course, all of the energy inputs must come from the energy outputs of the energy importer. In more complex real systems, we do not necessarily know where the inputs to the exporters come from—yet.

The darker blue oval in the figure represents the product exporter. What we find when examining its boundary, aside from the energy it needs, is the input to it of the product flow. At this point, we do not know the nature of the source entity (the large light blue oval). But we can provisionally code it and reserve data slots for it in the knowledgebase. In the figure, all of the importer and exporter subsystems have been put through the system identification process so that for each of them, their interfaces and input/output flows have been identified.

While at this point, we can treat the receiving (from the importers) and sending (from the sources to the exporters) entities as sinks and sources, noted in Fig. 6.9, they will not remain un-modeled entities for long.

6.7.2.2.3 Internal Flows

Indeed, by following the flows already established, that is, outputs from importers and inputs to exporters to their respective sinks and sources, we can account for the mass and energy balances (though we have left most waste heat flows out of the analysis to avoid more clutter, these would also be identified in analyzing subsystem 0.3). At this point, we can identify all of the interfaces for all of the importer/exporter subsystems, their respective flows of energy, matter, and messages, and their respective sinks/sources (respectively) as shown in Fig. 6.9.

6.7.2.2.4 Internal Subsystems

The clusters of sinks and sources in the interior of the SOI in Fig. 6.9 represent internal subsystems that do the work that transforms the inputs to the level 0 SOI into the outputs therefrom. It should be the case that with the identification of outputs from importers and inputs to exporters we have discovered what we might call the "next layer" of subsystems inside the SOI boundary (the first layer being the importers and exporters themselves). These clusters strongly suggest the existence of such subsystems even though we do not at this point know what they are. If objects in our system include, in their descriptions, locations in space, we do have a clue to the linkage among these source/sink objects. When we followed the known flow from/to each importer/exporter subsystem, doing the boundary analysis and identifying each one's environment, we located those objects and the knowledge that there must be internal work processes in the interior of the level 0 SOI tells us that that is what these clusters represent.

The general procedure, at this point, is to choose a likely candidate cluster of sinks and sources. In Fig. 6.10 we have encompassed such a cluster in a red, dashed oval to highlight how the sources and sinks seem to align and suggest the existence of a subsystem.

The role of the analyst, at this point, becomes more detective-like. This is where it is necessary to actually go to that presumptive object and start looking for clues about its boundary and whether or not the various source and sink objects in the cluster are, in fact, part of a contiguous boundary. For the scientist-analyst, this may involve setting up observations (instrumenting). For the design-analyst, this will involve querying the relevant users/stakeholders or engineers who are directly involved in that piece of the whole.[26]

Otherwise, the analyst might provisionally assume that the cluster is a subsystem, assign a boundary to it (as an opaque-box), and insert the various source/sink objects into it as interfaces. That is, we create a trial subsystem and treat what had been a set of sources and sinks (for other previously identified subsystems) as interfaces to the new object.

What ensues is analyzing each new interface in exactly the same manner as we did with the original level 0 SOI. This time, however, we start with the inputs, since their characteristics are already identified. Generally speaking, a good place to start is with the energy inputs, since all work processes use energy and we already have identified the energy flow from the energy importer, we know what the interface should be like. This can be verified through instrumentation and analysis of the flow dynamics, generally correlated with heat output from the process—Fig. 6.10 shows a few heat flows from the yet-to-be identified process shown, to the heat exporter process as a single thin red arrow with open head. The correlated flows of energy in and low-grade energy out will provide important clues later regarding the amount of work being done by this presumptive process.

In a similar fashion, we can consider material inputs. One good rule of thumb in choosing which material input to consider next is to look at the embedded energy in the material. For example, a simple ingot of steel has only the embedded energy of making the steel from ores and pouring it into molds. A shaped steel part, like a sheet or beam, on the other hand, has more embedded energy since more work was done on the original ingot to form it. We can think of the highest embedded energy input as the "main" input in many cases. Another example is the importing of food to a biological entity. Animals eat plants and other animals, the tissues of which represent very high amounts of embedded energy.

When it comes to messages as input the strategy is similar but very hard to implement directly. For message quality the relevant measure is, of course, information content. But this can only be assessed by the effect it has on the SOI receiving it. This is not at all easy to assess, especially when so many other variables (inputs) may be changing dynamically. For this reason, it is best to leave message analysis to last. Interfaces for messages in and messages out can certainly be identified as best possible and given a place in the boundary object. Indeed, sometimes it is possible, as in the case of human/machine communications systems, to analyze the interface to see what kinds of messages transfer through and how their protocols are done. From such analysis it is possible to make educated guesses about the quality of the messages, but these must be revisited after a more thorough understanding of the other parties to the communications are known, and one has an opportunity to observe behavior changes that result from the receipt of information.

[26] This is essentially what systems engineers do. They pose questions to the domain engineers who know what that particular piece of the overall design is.

Next, we perform an opaque-box analysis on the red, dashed oval cluster assuming the oval will be replaced by an actual subsystem boundary. That result is shown in Fig. 6.11 below. Taking all of the previously identified sources and sinks in the cluster, analyzing them as interfaces, and inserting them into a presumptive boundary we create a new internal subsystem. And we create a provisional knowledge base object (Eq. 4.1) for it. Further analysis of this new object identifies a new interface and flow, an intermediate product (green arrow in the figure), and creation of yet another sink object. The latter appears to be part of the cluster of sources and sinks in the center of the SOI. As we did with the opaque-box analysis of all of the other subsystems, the importers and exporters, we complete it for this new object and confirm that inputs and outputs correlate through the transformation function.

We now do the same operation for the other clusters.

The result of following this discovery and mapping process is shown in Fig. 6.12. A careful bookkeeping analysis of all of the flows will verify the total mass and energy balances for the system. With this information, we can refine the transformation function for the entire level 0 SOI because we have more detailed knowledge about what goes on inside it!

The new processes discovered are assigned Eq. 4.1 objects with the number codes as shown in the figure. At this point we have a level 0/level 1 tree with all the components, flows and relations determined and mapped.

However, we still need to refine our understanding of the message flows in the system. In Fig. 6.12 you will see an orange oval labeled "Governance Process." The only message flow in the figure is shown coming into the system through import process C 0.8. The output from this process goes to the governance process, now numbered C 0.12. Finally, we see a message output from that object to an export process, C 0.2. We don't know much about the sink that receives that message but may have to reiterate the environment analysis of the level 0 SOI to have a better idea of it.

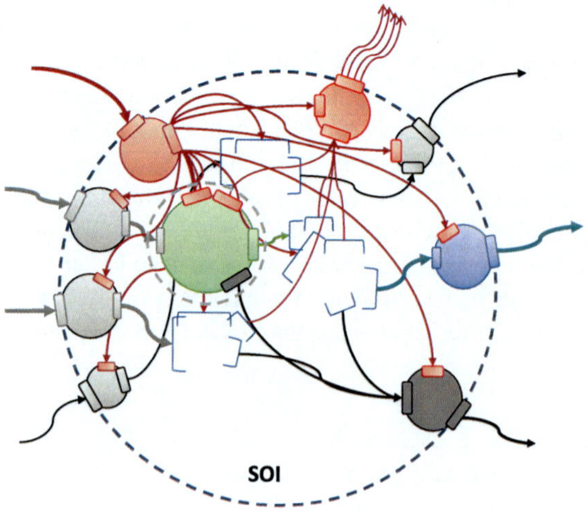

Fig. 6.11 The analysis of the first cluster of sources and sinks shown in Fig. 6.9 has identified the interfaces and its environment. All of the input flows and the heat and waste flows have been resolved because the sources and sinks previously noted had already been identified as import and export objects. A new output (the green arrow) and its sink were identified. Note how the sink identifies with the cluster of sources and sinks already noted

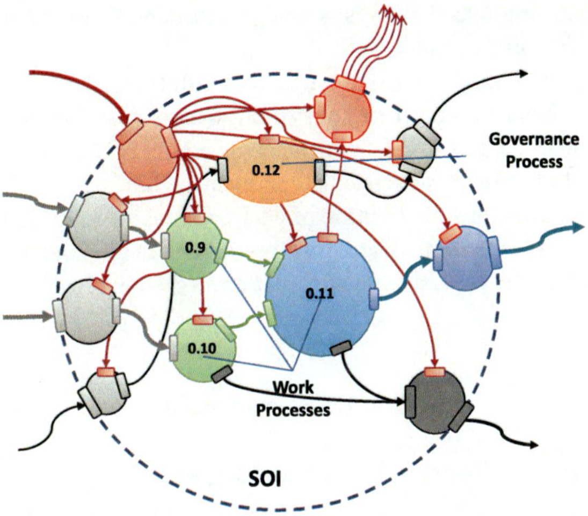

Fig. 6.12 The analysis of the internal flows and processes results in a complete decomposition of level 0, the SOI

What, in fact, is the case in all CASs and CAESs is that they all have some kind of internal regulation and management functions that process messages for information and make decisions that impact the operations of the other internal processes. As will be argued and demonstrated in Chap. 12 on Governance, this is a universal capacity of all such systems. Simple systems, such as heating and air conditioning systems, may have very simple cybernetic-based control systems, like thermostats. But all adaptive systems need multiple levels of regulation and management in order to maintain their overall behavior and fulfill the system's global purpose (see Mobus 2017 for a discussion regarding the requirement for a governance system within a CAS/CAES in order to achieve and maintain sustainability).

The vast majority of communications within and between systems are for the purpose of governance. In Mobus and Kalton (2015), Chaps. 8, 9 and 10 we give considerable coverage to the subjects of information, knowledge, how the latter is produced by computation in special information processes, and how it is all used in cybernetic frameworks (including a brief introduction to the hierarchical cybernetic governance system to be covered in Chap. 12 of this volume).

Accordingly, we will not expand much on the flow of messages in this chapter. The governance process shown in the figure is just a kind of placeholder. In reality, there will be multiple communication channels between all of the processes and with a "central" governance process (if one exists). For example, the flows between interfaces are generally modulated (regulated) by message flows between protocols in the interfaces (recall Fig. 6.3b). If process ***C*** 0.11 in Fig. 6.11 were a final assembly manufacturing process the flows of intermediate products (e.g., subassemblies) from ***C*** 0.9 and ***C*** 0.10 would be controlled by messages from the former to each of the latter requesting parts to be supplied for the assembly process (sometimes called a "parts kit"). If one of the subassembly processes is running behind, they would send a message back to the final assembly process telling it so. Of course, then, a decision maker who is actually embedded in the final assembly process might not take kindly to that fact!

Which introduces the idea that will be further elaborated in Chap. 12, that each sufficiently complex subsystem will have its own internal governance process. Governance, in most real (and successful) CASs and CAESs, is distributed among all of various subsystems and even within their next-level down subsystems. The nature of this distribution of decision-making and authority will be covered in Chap. 12. It is mentioned here just so you will understand what you see next when we now begin the recursive deconstruction to the next level down and beyond.

6.7.3 *Recursive Decomposition*

6.7.3.1 Going to Level 2 and beyond

There are two basic strategies for doing the recursive decomposition of a system. Both of these strategies mirror computer science "tree search" algorithms. A tree is a kind of data structure, essentially like that in Fig. 6.1a but without the level −1 objects. The algorithms for traversing a tree structure such as this start at the "root" node, level 0, and progress down the tree structure, visiting nodes at deeper levels (depth in CS lingo). One method is to traverse the tree down a path all the way to the terminal (called leaf) nodes and then backtrack up to the last parent node, finding the next path to deeper nodes, until all nodes at the deepest level have been visited. It then backtracks back up a level and repeats the traversal on the next pathway (if it exists). This is called "depth-first" traversal.

The other strategy for traversal of a tree is called "breadth-first." In this strategy, each node in a tree at the next lower level is visited (discovered and characterized). After all of the nodes at that level have been visited, the algorithm goes back to the first discovered node and repeats the process, going down to its children nodes, then exploring each node at the parent level in turn.

The system-subsystem-sub-subsystem organization is, effectively, a tree form of a graph. Therefore, we apply the same strategic exploration strategies as guides to the decomposition process.

6.7.3.1.1 Depth-First Deconstruction

One strategy is called "depth-first" decomposition. Figure 6.10 (above) suggests this possibility. Once any subsystem at any level has been identified and characterized as an opaque-box system (flows, interfaces, etc.) it is possible to go to the next level down inside that subsystem. For example, in Fig. 6.10 subsystem ***C*** 0.9 has been identified and characterized while the other internal subsystems at level 0 remain "mysterious." It is possible, even before determining the other subsystems at this level, to focus on ***C*** 0.9 (in this example) as the new SOI immediately and begin decomposition of this system as shown in Fig. 6.13.

The completion of the opaque-box analysis of ***C*** 0.9 allows for at least the identification of the in and out flows (along with placeholders for the relevant interfaces). This means that we have all the information needed to proceed with the deconstruction of ***C*** 0.9 in the same way we did with the original SOI. We can do a transparent-box analysis of C 0.9 to discover all of the components at level 1 (***C*** 9.1, ***C*** 9.2, … ***C*** 9.5 in the figure; recall that the prefix, "0.," has been dropped). But alternatively, we can choose one interface on C 0.9 and proceed to discover and characterize the importer subsystem, in this example, ***C*** 9.1.

In other words, we do not need to deconstruct all of the SOI, necessarily, in order to delve deeper into the system's sub and sub-subsystems. Once ***C*** 0.9 was discovered and characterized we could ignore the other subsystems in level 1 and focus on decomposition of ***C*** 0.9 alone. Of course, once we identified ***C*** 9.1 as the first sub-subsystem (presumably the product output from ***C*** 0.9) we could similarly ignore the other sub-subsystems at level 2 and proceed to deconstruct ***C*** 9.1, and so on down the levels as indicated by the red path lines in Fig. 6.13.

The reasons for doing this involve the control of the scope of an analysis, and that involves considerable foresight with respect to the nature of the level 0 SOI. For example, suppose in the analysis of an organization it has been a priori determined that a "problem" exists in a particular department deep in the organization (i.e., component ***C*** 9.1.2.4 in the figure). A depth-first decomposition analysis could be

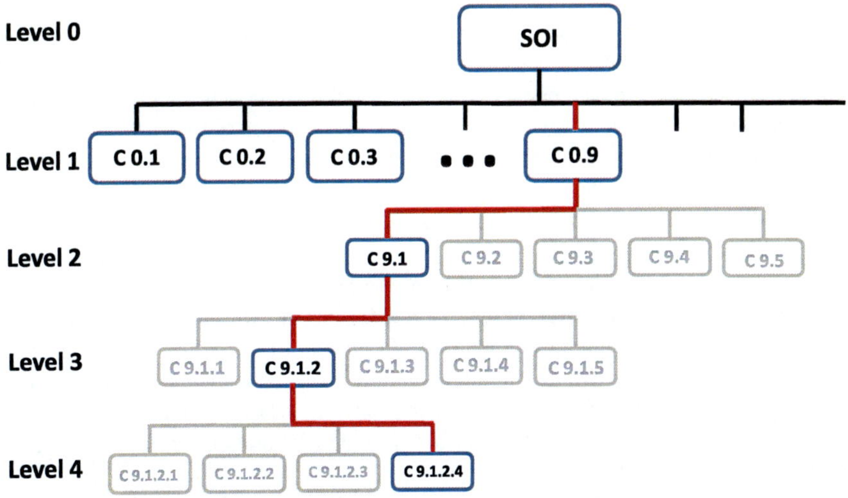

Fig. 6.13 The red path represents a potential depth-first decomposition of a system. The analyst may choose to pursue the decomposition of subsystems (and sub-subsystems) based on criteria mentioned in the text. The pursuit leads, in this example, to a sub-sub-sub-subsystem (level 4) before any other sub-…systems are identified and decomposed. The red path identifies a depth-first decomposition. The grayed components at each level exist but have not yet been identified or characterized and thus are not included in the analysis

used to rapidly focus in on the workings of that department while "ignoring" the rest of the organization. However, that does not mean the beginning of the analysis is the department itself. There is no shortcut to rapidly circumscribing the problem area which is what a depth-first decomposition accomplishes. A principled and structured analysis of the system still requires starting at a sufficiently high level in the organization in order for the analysis to ascertain the complete context of the sub-[−...]system where the problem exists. The number one mistake made by systems analysts (especially in organizational contexts) is to too rapidly vector in on the subsystem without adequate consideration for the context of how that subsystem works. A depth-first decomposition from the top level of an organization is used to establish that context and helps assure that all of the relevant factors (that is flows in and out of the system of interest) have been accounted for.

6.7.3.1.2 Breadth-First Deconstruction

The second approach is to complete the analysis at each level before going down to the next level. For example, in Fig. 6.12 all of the importers and exporters as well as all of the level 0 internal subsystems and flow channels have been completely identified. Then, choosing those subsystems, one after another, the process of transparent-box analysis is repeated for each. In other words, we do a breadth-first analysis, exposing all of the sub-subsystems at the next level.

6.7.3.1.3 Choosing an Approach

There is no simple rule for choosing a decomposition approach. Much depends on the needs of the analysis process. For example, in the desire to understand as completely as possible a "new" system—to gain deep understanding—the breadth-first method would be most appropriate. On the other hand, as mentioned above, if the analysis is of an existing system such as an organization, in order to identify a "problem" area, the depth-first method might be more appropriate. This is a decision that needs to be made by the analyst in charge (Fig. 6.13).

If we observe the big picture of the process of Science (with a capital S), we will see that it has been a general process that alternates in efforts between these two approaches. Sometimes we delve deeply into the mechanisms of a system—the reductive approach or depth-first. Sometimes the sciences proceed in a breadth-first manner trying to resolve questions of how various mechanisms work at a singular level before delving deeper.

As examples, consider the works done in natural biology versus molecular biology. In the former, the science of ecology seeks relations between species, life histories, food webs, etc. seeking to obtain a broad vision of how different subsystems relate to one another, before delving into the details of, for example, trophic energy flows. In the latter, the emphasis has been on discovering, for example, how

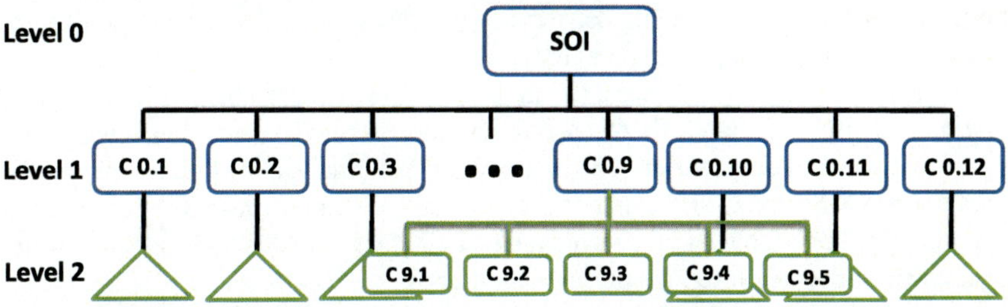

Fig. 6.14 A breadth-first decomposition completes the deconstruction of each of the subsystems at a given level (here level 0) before proceeding to the next level. All of the subsystems of the SOI constitute level 1. All are identified before any one of them is subsequently deconstructed. The green triangles represent the decomposition of each subsystem at level 1, with *C* 0.9 decomposition shown explicitly as a set of sub-components, as in Fig. 6.12

mitochondria work and then discovering how mitochondrial DNA contributes to this mechanism.

The sciences have been unguided by any coordinated decision as to which approach would work best. And this has worked quite well so far. Science has been spectacular at doing systems analysis without benefit of a structured, principled approach because it has been guided by the scientific process that is the search for objective truth. Quite fortunately the methodologies and overall science process has guided the accumulation of both deep and broad knowledge. But it has also revealed the problem of not having a more disciplined approach. We are now on the verge of seeing how the real complexity of reality cannot be managed by ad hoc searches for truth. We have come full face with why the systems approach is mandatory for future progress in understanding reality.

From now on, it will be necessary to consciously pursue breadth-first and depth-first pursuits of understanding according to the needs of that pursuit. In the next chapters in this section, we will provide some examples of this pursuit based on conscious decisions grounded in systems science rather than blind search.

6.7.4 Reverse Decomposition

Up to this point we have probably left the impression that analysis is strictly a top-down process. This is generally the case, but not exclusively so. Suppose during the decomposition process we find anomalies that cannot be resolved by further decomposition. We will, from time to time, encounter new questions that cannot be answered with the information already available. What do we do then?

For the analysis of any SOI it can happen that the environmental analysis did not adequately capture a complete understanding of the sources and sinks originally identified. This generally is the result of leaving something out of the original SOI boundary analysis. In other words, the opaque-box analysis failed, originally, to identify an important input (or sometimes an output).

A frequent symptom of this is a failure of the mass or energy balance of transformation functions. An analysis of an internal subsystem indicates that there is an input that wasn't caught in the original opaque-box analysis of the system. Let's face it, real CASs and CAESs are not easy to analyze! Almost invariably something is going to get overlooked in the initial analysis. This is why it is necessary to iterate back to earlier stages when we run into anomalies.

Quite often the problem is that there was an input to the SOI (at any level of decomposition) that was not accounted for in the initial analysis. In these cases, we are faced with several approaches to re-analysis. We can simply add a new source and input, looking for the relevant interfaces on the SOI boundary and the importer subsystem. Or we can consider the possibility that we have overlooked an essential component of the SOI that should actually be included. That is, we may need to "refactor" the system to include previously unrecognized inputs, or we may need to expand the original boundary such that an external source becomes a subsystem of the SOI. More often we do both.

Figure 6.15 provides a view of a system in which the analysis exposed an energy importer and distributer subsystem. The interface analysis along with the environment analysis produced a mapping of the flows of energy and messages that allowed the importer to obtain needed energy from the energy source as shown. This model indicates that a single kind of energy is used internally (let's say electricity).

However, let us say that during subsequent analysis of energy uses within the SOI we discover that there exists a second kind of energy usage (e.g., gasoline for company cars) that had not shown up in the original analysis. This means that something was overlooked in the original boundary and environment analyses.

From the principles of systemness, in particular network structures in hierarchies, we can reverse the direction of analysis. We start by re-analyzing the importer subsystem and discover a second energy flow and interface. We thereby discover a second energy flow/interface and then follow the second input flow back to its

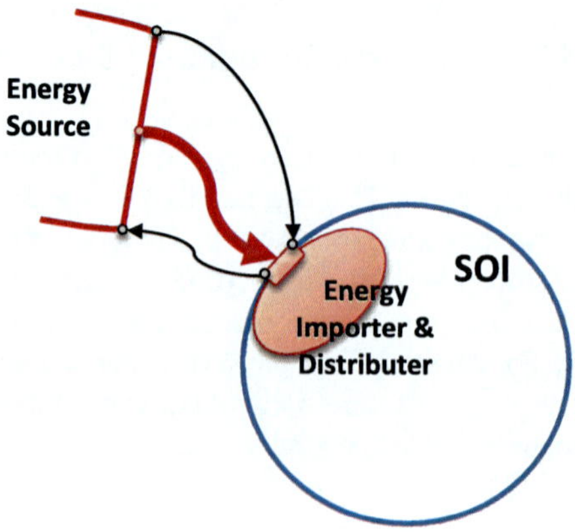

Fig. 6.15 The initial analysis of a system resolved an energy importer and its interface (with protocols) with an energy source in the environment

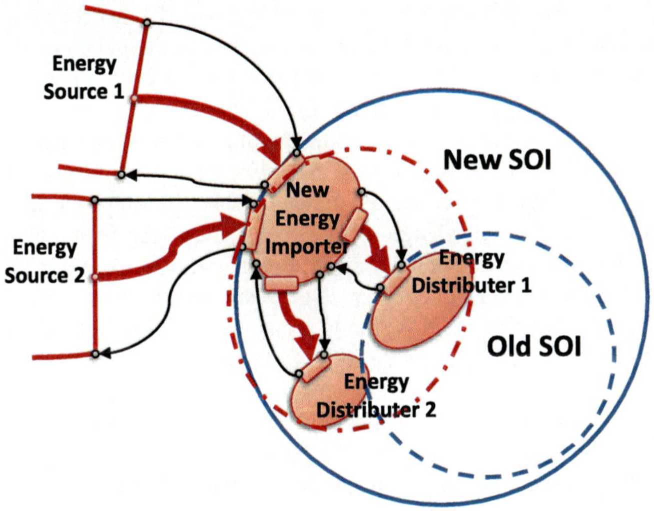

Fig. 6.16 The new SOI results from two operations that recognize a new importer subsystem and refactor the work of importation and distribution into three separate subsystems recognizing the distinctions in the two different energy types

interface. A reanalysis of the energy sources and flows reveals that the source of energy inputs to the original SOI was actually a previously unmentioned (unrevealed) subsystem that properly belongs to the SOI as the "ultimate" importer. The restructured situation is shown in Fig. 6.16.

Note that the revised boundary analysis requires the addition of a new interface for importing the second kind of energy. The environment analysis also reveals the second source.

The red dashed oval in the figure indicates how the original analysis gave a single energy importer as in Fig. 6.15. Since by the definition of a system this single subsystem is already considered to be decomposable into yet finer sub-subsystems, the discovery of another energy flow simply means leaving the original importer/distributer as coded and occupying the knowledgebase (tree) and inserting three new subsystems below it. In Fig. 6.8 the energy importer was coded ***C*** 0.5. These three new subsystems would thus be ***C*** 5.1, ***C*** 5.2, and ***C*** 5.3, respectively. The new energy source can simply be added to the environment structure just as the interfaces can be added to the boundary structure.

6.7.4.1 Expanding Boundaries

There are many reasons for this operation. Generally, it is because an external source or sink were originally thought to be outside the "control" of the SOI beyond the protocols of requests and notifications that accompany any regulated flow. Those protocols (e.g., purchase orders and receipts) are in the domain of cooperative cybernetics. This will be more fully explained in Chap. 12, but for now note that cooperation is a volunteer action. Neither agent/actor in a mutual procedure is compelled to follow through except to the extent there is some kind of reward for doing so (like the sender will get paid for doing so).

In other cases, the external source or sink might actually be under more compelling forces emanating from the original SOI. For example, many big-box retail chains exert more than just cooperative influence on their suppliers, dictating, for example, prices that they can charge. At first glance, the chain and the supplier seem like independent systems in their own rights. But in reality, the supplier has become a captive of the chain and is, in effect, a de facto subsidiary. Thus, the true analysis of the SOI (chain)-supplier relation should probably include the supplier within the boundary and also include a decomposition of the supplier as an essential component of the "real" system.

In very typical system dynamics modeling it is not uncommon to realize that an important variable that had a non-trivial impact on the dynamical results had been left out of the original model and needs to be included. This is an example of expanding the boundary of an abstract system, which seems to many as an arbitrary exercise. However, it is not much different from the above big-box chain-supplier example.

6.7.4.2 Refactoring

At times the expansion of boundaries, or even just the discovery of a previously left-out subsystem, will alter some aspects of the component coding that had been done to that point. In the example above of the second energy flow and interface, the boundary list of interfaces can be expanded but the ordering of the interfaces already discovered and coded makes it awkward to track the new objects. For example, the original interface for energy input was coded ***I***5, corresponding with the single importer subsystem ***C*** 0.5. The next interface was to be found in the importer ***C*** 0.6 and numbered I6 accordingly. Now it turns out that there are actually two interfaces associated with ***C*** 0.5. Since these interfaces are not only subsystems of the importers but also associated with the boundary of the SOI this can create a bit of confusion. We could simply add the new interface to the end of the boundary list, numbering it ***I***9, and make a special note that indicates it really belongs to ***C*** 0.5.

A more elegant approach is to refactor the coding scheme so that ***I***9 is numbered ***I***6 and all thereafter are renumbered (incremented). There is no necessary condition that would require the interfaces in the importers' boundary lists to have the same number as the subsystem itself, especially if, as in this case, there are more than one interface associated with a more complex importer. At the same time, it would be useful to keep the interfaces in the SOI boundary list in sequence order of discovery.

The same conditions can be invoked in the case of discovering a previously undiscovered subsystem. This generally comes about as in the energy flow example above by finding anomalies in the internal mapping of subsystems and their flows. If, for example, the sum of the inflows to subsystems does not match the sum of the outflows of other subsystems then there is a clear indication of a gap in the analysis. Upon reanalysis the missing pieces can be added to the map and refactoring of codes as needed can be done.

Refactoring a complex system by hand would be a monumental task. But with the appropriate software tools for doing the analysis (to be discussed in the next chapter) and filling in the slots in the knowledgebase, the chore becomes trivial.

6.8 Iterative Analysis

Real systems analysis of extremely complex systems is in no way easy! But, for deep understanding it is essential. With the process being guided and done by human decision makers it would likely be impossible not to make mistakes and leave essential details out inadvertently. Fortunately, the procedures of analysis we have outlined in this chapter provide principled ways to recover from such mistakes. Or, alternatively, when the problem is a lack of analytical tools with which to ferret out answers, a portion of the system must necessarily remain an opaque-box with only a best guess system identification as a placeholder for more definitive knowledge. The analysis must await the invention of the appropriate tools.[27]

By providing both pathways down from the top toward the revealing of the smallest components and from the components up to the very top (or as shown above to expand the scope of the original SOI) we have a way to iterate as needed over the space of the subsystems to be discovered and characterized. Whenever an anomaly appears at any level of the system model, we can backtrack upward to discover what we missed at the parent level, or reiterate the downward recursive decomposition using a depth-first approach to get to the problem.

But for the lack of specific deconstruction tools, we have no excuse to not understand complex systems.

6.9 Considering Advanced Concepts

In the following four sections, we will briefly discuss how to deal with the more advanced concepts of complex systems that involve issues of fuzziness, adaptivity (CAS), evolvability (CAES), and memory recorded in the ***H*** object. All of these are factors found in every CAS or CAES and so must be considered when analyzing or modeling such systems.

[27] Not unlike how the better analysis of the brain had to await higher resolution EEG, single neuron unit recording methods, and fMRI tools in order to turn the opaque-box into at least a gray-box.

6.9.1 Dealing with Fuzzy Systems

A fuzzy system is one in which components may be in the system during some period of time and outside that system (and in another system) at other times. A complete specification of the component in the set of objects that comprise the system (recall Eq. 4.2 repeated below) requires a temporal membership function, the $m_{i,j,l}$ items in the equation.

$$C_{i,l} = \{(c_{i.1.l}, m_{i.1.l}), (c_{i.2.l}, m_{i.2.l}), (c_{i.3,l}, m_{i.3.l}), \cdots (c_{i.k,l}, m_{i.k.l}), \cdots (c_{i.n,l}, m_{i.n.l})\}_l \quad (4.2)$$

Equation 6.1 provides form for the membership function.

$$m_{i,j,l} = \begin{cases} 1, \text{ifcomponentisalwaysamember} \\ f_{i,j,l}(\tau_i, p), \text{otherwise} \end{cases} \quad (6.4)$$

where τ_i represents a time period in which the component can be found as a member of system i, and p is a probability factor as when a component is "likely" to be a member of system i during period τ_i. Recall from Chap. 4 that most systems have periodic or quasiperiodic behavior.

The idea that a component is a member of multiple systems but at different times and with different probabilities means that the component must be accounted for in those multiple systems. For example, let us consider a component $c_{i,j} \in C_i = \text{A}$, (the level index has been dropped for simplicity). Suppose the membership function associated with this component in A is,

$$f_{A,j} = \begin{cases} 1, if t \leq \tau_A \text{withprobability} p_A \\ 0, otherwise \end{cases} \quad (6.5)$$

The problem is how to represent the very same component in a second or more other systems. Figure 6.17 depicts the situation in which a single component has membership in three different systems at different times and with different probabilities. The figure shows a period of $\tau\Delta t$ for the supra-system containing subsystems A, B,

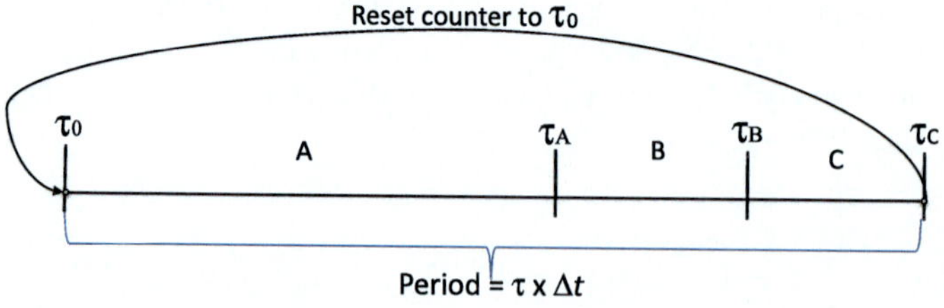

Fig. 6.17 This depicts a periodic cycle during which a component may have membership in any of three systems depending on the value of the counter, t

and C (Fig. 6.17). The period is divided into three sub-periods in which a single component, $c_{i,j}$, will be found based on the count of t. If $t \leq \tau_A$, a constant, for example, then with probability p_A the component will have the identity $c_{A,j}$.

Figure 6.18 shows the situation when the counter t is less than or equal to the constant τ_A, the component is located within system A.

Consider a human being who has a day job (system A), an evening gig playing guitar at a local pub (system B), and goes home afterward to eat and sleep (system C). This is not a hard scenario to imagine. The human is the component in question. Membership functions for the human component's presence in any of these three systems would be relatively easy to specify. The total period is a day and the sub-periods are the portions during which the human moves from one system to the next. This is a routine schedule, so is easily represented. As another example, one a little more exotic, consider the membership of an electron in a valence orbital in one or the other atom sharing that orbital in a covalent bond. The membership function would probably look suspiciously like the Schrödinger equation!

There are probably many ways in which a system can be fuzzy beyond the membership of a component subsystem. For example, a variation on the fuzziness described above would consist of membership conditions on interfaces, which are, after all, just special kinds of subsystems associated with the boundary of the embedding system. We call this a "fuzzy boundary" in which, for example, it is difficult to locate a "solid" boundary in space and time. Fuzzy boundaries have been a difficult subject conceptually and have even led some theorists to reject the idea that there are real boundaries at all, asserting that boundaries are always an arbitrary choice of the analysts. Indeed, there are arguments to justify skepticism with respect to pinpointing a physical boundary. We already saw how it is sometimes necessary to "expand" the boundary to include inside the system an environmental entity that had been previously (inadvertently) not included. This kind of exercise can lead one to conclude that boundaries are arbitrary and subject to the analyst's whims.

But just because a boundary is fuzzy does not mean it is arbitrary. It simply means that it will take more information to specify it.

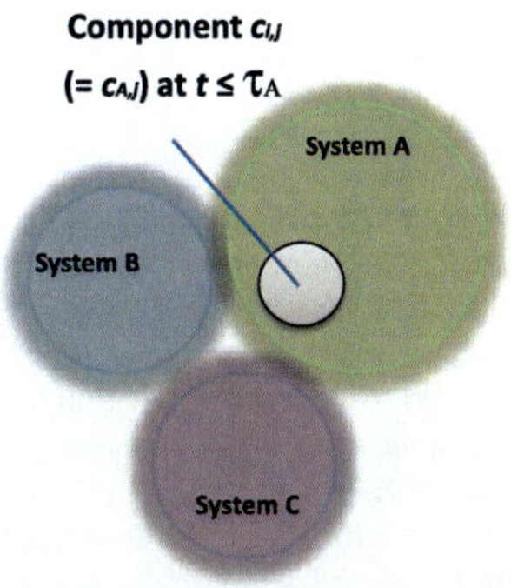

Fig. 6.18 A component can be a member of any of three systems (subsystems of a larger supra-system) but at times specified by the sub-period as in Fig. 6.17

6.9.2 *Dealing with Adaptivity*

Adaptivity refers to the capacity of a system to alter or modulate an internal function in response to external changes in the environment. All CASs have to adapt because they operate in variable conditions. Simple or merely complex systems may be capable of operating normally in a wide domain of environmental variables without altering internal parameters. Adaptive systems, however, include mechanisms that can change their operations in response to changes in demands made on them. An airplane autopilot can respond to a wide domain of factors like wind speed and direction to stay on course. But there is a hard limit (not counting on over-engineering) to the domain size for these variables. So, from this perspective, we could count an airplane with an autopilot as being in the category of CAS.

CASs have built-in capacities in the form of transformation functions of various (generally of specific) subsystems that allow them to continue operations in light of changes in normal environmental conditions. For example, if the input flow of a vital resource is reduced below the optimal level for proper functioning of the system's overall performance. CASs include backup subsystems or ways to substitute other resources for those inputs being restricted.

An adaptive transformation function involves having a limited range of outputs related to a larger domain of input variations (many-to-few, see Fig. 6.19). The

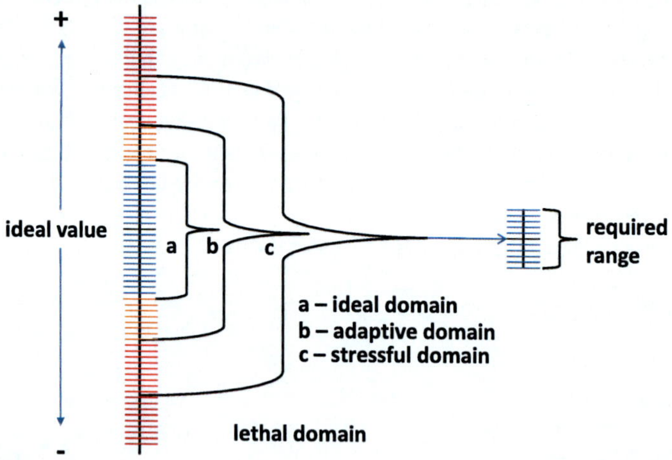

Fig. 6.19 An adaptive function has to be able to map, generally, a larger domain of input values to a limited range of output values (neither domains nor ranges need necessarily be symmetric about the ideal as shown here). The domain is divided into a set of subdomains around a central point or ideal value. Subdomain **a** is the small domain of ideal values or values which can readily be compensated for into the range. Subdomain **b** are values plus or minus the limits of subdomain **a** that require the system to modify its internal functioning in order to continue mapping the actual input value to the required range. Subdomain **c** is the values above and below subdomain **b** that require more work to accomplish the mapping than can be sustained by the system for longer periods; the effort is stressful. Finally, values outside of domain **c** are lethal and there is no mechanism that can accomplish the mapping

underlying process that produces the transformation undergoes internal changes or shifts in operations to handle input conditions that are outside an "ideal" domain.

Below we discuss several examples from the biological world.

The need for adaptivity arises from operating in a stochastic world in which the relevant variables cover a wide domain. Fortunately, in our world the dynamics of these variations are not random or dominated by noise. They are generally smooth functions over time, although occasionally a catastrophe or step-function-like variation can occur. But since the world isn't in constant turmoil life has been able to evolve limited capacities (in individual entities) to react to changes and compensate for conditions that get out of the "friendly" domain.

6.9.2.1 Homeostasis

Homeostasis is a well-understood feedback control function that is used to maintain a critical variable such as internal temperature or blood pH. The mapping function employs a mechanism that counters the input variable when it tends to get further away from the ideal value of the domain. For example, with body temperature, the input domain is the external temperature, which can vary considerably diurnally or seasonally. To counter the heating (+) or cooling (−) effects of ambient temperature on an animal's core, a mammal (a warm-blooded animal) may shiver, its muscles thereby generating extra heat, when its core temperature starts to drop (the range for core temperature is extremely small). Or it may pant or sweat when the temperature starts to rise.

Homeostasis is found employed widely in every biological process, including what are called demand-driven processes, where a depletion of a critical component signals the restocking operation to return levels to nominal. It can operate normally within the ideal domain. It can operate with some additional work and consumption of more than usual amounts of resources such as materials and energy reserves when the input value is in the adaptive domain, but not for very long (without the next described mechanism kicking in). And if the domain is in the stressful values the mechanism may break down physically. Both hyper- and hypothermia occur in the stressful regions of the domain and can cause irreparable damage to tissues if the temperatures remain above or below the merely adaptive domain respectively for too long.

Of course, a wall thermostat that controls heating and cooling in a building is an example of homeostatic device, even if not substantially complex.

6.9.2.2 Autopoiesis

A slightly more complex form of adaptive function is found in autopoiesis or self-constructing/maintaining mechanisms. As noted above, demand-driven processes are used to signal the need not just for components, but for the construction and/or repair of more complex subsystems, generally that supply those components. For example, when mitochondria, the energy supply subsystem in cells, are decommissioned other mitochondria are put under a stress load and respond by reproducing. Both the internal genes and those nuclear genes associated with mitochondrial construction are put into play to produce whole new mitochondria to relieve the load. The same mechanism kicks into play when long-term demand for muscular work (as when someone does strength training with weights). Not only does each muscle fiber require more energy reserve, but all of the other support systems need to be "bulked" up as well. More fibers are needed to adapt to the stresses as well as more blood supply and lung capacity.

Autopoiesis is called into play to repair damage to other subsystems. For example, when homeostasis alone is not sufficient to counter a domain value in the stressful region and damage to that mechanism itself occurs, then the repair (once conditions return to the ideal region) of the mechanism can be accomplished. But it uses materials and energy out of reserves and incurs a cost above normal operating costs in those substances.

6.9.2.3 Adaptive Response

In general, biological systems (at the single individual level) have various capacities for adaptability in response to changes in relevant parameters in their ambient supra-systems. These capacities evolved over time when the parameter (or condition) tended to vary from a previous norm (the ideal domain) slowly enough that genetic mutations giving rise to changes in the mapping function could pursue those external trends. We will return to this below in dealing with evolvability.

The basic adaptive capacity follows a pattern. Minor variations within the ideal domain can be handled through the existing mechanism (homeostasis). Short-term variations outside of this domain may be handled by recruiting extra resources for a period corresponding to the demand variation. A general rule of biology and the condition of mechanisms and tissues is, "only maintain readiness to the degree that the average demand requires." If for short periods of time, the conditions go outside of that mapping, be prepared to recruit help, but once the situation normalizes, go back to the average state of preparedness.

If, however, conditions persist overextended time frames, cells, and tissues have the ability to build their homeostatic capabilities up through autopoietic mechanisms as described above. That is, the capacity is adapting to higher levels of demand and the average level of preparedness may rise accordingly. The weight training, muscle building example above is in this category. If a person just starts weight training the strength capacity does not adapt immediately. It takes time and repetition to reinforce the signals that there will be higher demand occurring as a "new" norm. Over time, the muscles respond by building bulk and strength capacity. The person's body has adapted to new conditions and the level of preparedness for dealing with heavier lifting has risen.

Of course, if that person, after adapting to heavier weights, loses interest in looking buff and stops training, over time the muscles will lose strength and preparedness, returning to their pre-training status, or at least nearly so. The loss of capacity seems to take longer than the gaining. This phenomenon is called memory and it fades at a rate much longer than the initial learning.

Mobus (1999) provides a detailed analysis of the phenomenon as the basis for neural synaptic plasticity and learning in biological neuronal networks. The paper includes analysis of various "costs" associated with ordinary operations (homeostasis), construction of new capacity (autopoiesis), and repair of damage by not having an increased capacity for response to stressing stimuli. The purpose of adaptive response is to minimize these latter costs by possessing a greater capacity of preparedness.[28]

6.9.3 *Dealing with Evolvability*

This is the most difficult aspect of a system (a CAES) to analyze and characterize. Evolvability is a capacity for a system to go beyond mere adaptivity and change the adaptive function entirely. Usually this happens in a speculative fashion, as in genetic mutations in populations that give rise to new capabilities that are then tested for efficacy by the selection function of the environment. This strategy works well when there is a large population of evolvable elements as is the case in genes in a population of a species. To be clear, the evolvable system in that case is the species, not the individuals even though the mutation affects genes in various individuals (as, for example, a cause of cancer). For genetic mutations to affect the population or species it has to be in the germ line cells (i.e., sperm or ova). They are then heritable.

The general principle may be called "copy error with inheritance." A mutation in a gene in a germ cell is such an instance in the biological world.

Human organizations are not a population of similar entities in the same sense that a species is. Even so organizations operate on codified knowledge like genes and chromosomes in biological entities. For example, most organizations have bylaws, policies, and procedures. The latter are like genes in prescribing how some subsystem is to behave. And organizations work at persisting in time. The encoded knowledge has to be actualized (performed) by humans. They are analogous to the enzymes that encode the knowledge in the gene's DNA into messenger RNA and, in turn, like ribosomes that produce the proteins from the mRNA messages.

From time to time a human can misinterpret a procedure (or a policy) and implement something different from what was intended in the original code. If the changed implementation is effective it can be repeated over time leading to what amounts to a genetic mutation. At some later time, someone may notice that the original code is not being followed, but the new "practice" is working better than the

[28] In Mobus (1999), the exploitation of adaptive response in neurons produces a capacity for anticipatory behavior that is shown to reduce overall costs of responding to higher demands since an animal, if it can predict the onset of an excessive demand, can act to avoid it.

original would have. The procedure may be rewritten to reflect the newer practice and so the mutation is, in this sense, heritable.

Yet another example of an evolvable system is the mammalian brain with its neocortex. In the case of humans, anyway, it seems to be able to encode an infinite variety of concepts, both from actual experience and from a creative process we call imagination.[29] The latter are the interesting case because it resembles mutation-like change in concept space that can lead to altered or new behavior. Once again, if that change leads to success for the actor, then the originating concept is reinforced and inherited into the future time frames.

Evolvability is sometimes described as a species having the ability to allow increased mutation in certain genes when under environmental stresses. However, other writers simply require enhanced variability potential. All agree that genetic mutations that occur to create that variability must have a relatively high frequency of viability (most random genetic mutations would presumably be detrimental or neutral).

For other kinds of systems, for example, ecosystems, economies, or organizations, evolvability might be a combination of chance change (like mutation) in a particular subsystem and intentional modification in anticipation of future conditions. Organizations that do strategic planning (successfully) and human beings who "think ahead" are examples of intentional modification of subsystems in anticipation of the future.

The formal definition of system given in Chap. 4, especially Eq. 4.1, does not directly make provisions for handling evolvability in a system. The concept implicitly involves considering very long histories of the system (or class of systems). Thus, we will briefly outline an approach to dealing with evolvability as part of dealing with the system history. Evolvable systems, especially in the realm of model building, is, itself, an evolving area. In other words, more research is needed!

6.9.4 Dealing with Memory—And the H *Object*

By definition an adaptive system must retain a "memory" of its past experiences. It behaves differently at a later time than it did at an earlier time for the same set of conditions because the past conditions were recorded in some manner that influences the condition/action mapping. Thus, an adaptive system's history impinges on its future.

As described in Chap. 4, the history object is, or can be, fairly complex and varies from system to system. For example, a computer-based system, or subsystem of a larger work system, maintains exact recordings of states and events in digital coded form. These recordings are stored in databases or, by extension, on archive

---

[29] There is a growing body of evidence that many other species, including birds, generate some novel representations that lead to adaptive behavior.

media. If records are ever deleted, they are gone. Alternatively, the human brain (or any animal brain for that matter) does not store explicit representations of states and events in the same way a computer does. Memory in the human brain works on completely different principles that are only now beginning to be understood (Squire and Kandel 2009). However, some broad comments will help in understanding why memories are difficult subjects.

The word itself has different meanings based on context and field of study. In general, however, it refers to the incorporation into the system of a lasting impression that has impact on a system's future behavior. As covered in Chap. 3, this what we generally mean by "knowledge," except we differentiate between knowledge that is modifiable based on on-going experience versus the knowledge that is a priori built into the structure of the system that allows it to interact with other systems. The first kind is "learned" knowledge and is a characteristic of an adaptive system.

A system that retains a memory of experiences that it has had behaves according to both its current situation and the memory of prior situations for which the memory is relevant.

There are fundamentally two kinds of memories resulting from experiences.[30] The first is the exact recording of states of the world (conditions) and of the system itself. In human psychology these are called episodic memories, as when you remember specific incidents in your life. Such memories, in the case of humans, are not accomplished the same way a digital camera captures every pixel and fixes them in its memory. And human episodic memories are notoriously prone to modifications.

Episodic memory is one example of a more general ability to record specific concepts (e.g., facts) and have ready recall of those memories for use in declarations. This is called explicit or declarative memory. We do not have a clear idea as to how much explicit memory capabilities other animals possess. We do know that many mammals and birds are able to recognize other individuals, for example, but we do not know if their internal memory of, for example, faces can be brought into consciousness in the same way we can.

The second major category of memory is called "implicit" or "tacit." One commonly experienced form of implicit memory is called "procedural" memory, as when you just know how to ride a bicycle without having to consciously recall each action you need to take to do so. Implicit memories appear to be formed from repeated experiences of successful actions in response to contextualized situations. A major form of this kind of memory for humans is what we call intuition.

Whereas in episodic memory the concepts are explicitly recorded and recalled, even if only fuzzily, in tacit memory experiences appear to be used to develop and reinforce a situation/action mapping that works for all similar contexts.

[30] The following discussion as applied to human memory in particular is based largely on the works of Squire and Kandel (2009).

This is a mechanism that seems to apply to all adaptive systems (Alkon 1987; Mobus 1999; Squire and Kandel 2009; Sutton and Barto 2017).[31] Basically, the memory trace is a kind of time weighted average of all of the experiences accumulated to date. The Adaptrode, which models the dynamics of biological synapses (Mobus 1994) uses the following formula for updating a synaptic weight:

$$w_t^0 = w_{t-1}^0 + \alpha x_t^0 - \delta w_{t-1}^1 \tag{6.6}$$

$\boldsymbol{w}$ is the synaptic efficacy weight. The superscript refers to a time domain; 0 is the "real-time" domain, whereas 1 is the time average window. The variable $\boldsymbol{w}^1$ is a slowly moving "basement" value. α and δ are rate constants, generally much less than 1. And $\boldsymbol{x}^0$ is the real-time input (an action potential or rate of action potential arrivals at the synapse). The Adaptrode is maintaining a time averaged value that ranges between 0 and 1 that represents the history of activation. Long periods of quiescence results in a weight near zero. On the other hand, a high frequency of stimulations pushes the value toward 1. In (Mobus 1994) it is shown how this mechanism operates to form a memory trace in a neural network (see also, Alkon 1987 for the biological basis for this model).

Thus, memory traces in the brain are based on time-averaged experiences and are the basis for tacit memory formation. This is thought to be the basis for all memory encoding in the brain. Explicit memories appear to be based on a new trick that the neocortex can do, using implicit encoding but putting the memories in effectively isolated parts of the cortex (especially more frontal areas).

The ***H*** object in a system knowledgebase may be instantiated by some form of network-based encoding scheme such as this. In Chap. 12, on governance, we will discuss in greater depth the meta-system called a "decision agent." We will see that this is a special case of an adaptive (and evolvable) system that relies on having a storehouse of experientially based, implicit knowledge. That knowledge is acquired by learning. Thus, it is probably necessary, in order to capture knowledge of such systems, that the ***H*** object be, in effect, a brain-like mechanism (or, in other words a human emulating AI!) Clearly, a great deal of research is needed in this arena.

6.10 Conclusion

In this chapter, we have given a rough outline of the process of systems analysis that leads to deep understanding, or the complete deconstruction of complex systems, turning all opaque-boxes at all levels in the decomposition hierarchy into transparent-boxes, and complete with transformation functions for each.

[31] This is certainly the case for living systems. But it is also the case for a number of manmade systems that are considered adaptive, for example, Kalman filters.

Systems analysis is the heart of the entire system understanding process. It is the process that extracts information about the system and collects that information into a systemic data structure (the knowledgebase to be described in Chap. 8) given by the formal definition of a system given in Chap. 4.

The procedures described in this chapter provide maximal flexibility in divining the details of systems and how they work. The analyst has a maximum amount of flexibility in pursuing system details in either a top-down or bottom-up fashion (iterating as necessary). Assuming the necessary instrumentation is available to turn an opaque-box into a transparent one, the procedures provide an algorithmic approach to obtaining knowledge of the system.

In the next chapter, we will describe a set of specific tools that would be needed to support the activities of a full systems analysis and then a series of examples of how the process works with a few representative CASs and CAESs.

Chapter 7
Analyzing Various Kinds of Example Systems

Abstract This chapter examines the application of the methods of Chap. 6 to analyze a variety of systems at increasing scales of complexity and hierarchical organization. We start with an example of a "merely" complex system and then proceed to complex adaptive and then to complex adaptive and evolvable systems. We end this chapter with an examination of the human social system (HSS) as a CAES subsystem of the Earth and the Ecos. The purpose of these examples is to show how the methods of Chap. 6 can be applied to all varieties of system types and complexities. These examples provide, essentially, starting points for the top-down deconstruction process and do not attempt an in-depth analysis. However, they should provide directions for others to endeavor more in-depth analyses in the future.

7.1 Analysis of Example Complex, Complex Adaptive, and Complex Adaptive and Evolvable Systems

While the methods of analysis described so far are applicable to all systems, no matter how simple, the real benefits of using this principled methodology will be in coming to understand complex systems, and in particular complex adaptive and evolvable systems.[1] Examples of a complex adaptive system come directly from the biology of cells. These systems are extraordinarily complex when compared with most designed systems. This is because they have a large number of levels in their organization tree as well as a breadth of component types. Thus, cells have a Simonian complexity exceeding any machines (even computer networks).

Ecosystems and human organizations have yet higher Simonian complexity simply because the biological components of each includes cells (of plants and animals or human beings) as levels in their organizations. The human brain represents an interesting intermediate system in the CAS/CAES spectrum. It comprises ordinary

[1] These types of systems were outlined in Chap. 1. Unfortunately, we will not describe their specific attributes until Chap. 10, which is devoted to the explication of what we call archetype models, large-scale patterns of phenomena that interoperate to produce CASs and CAESs.

G. E. Mobus, *Systems Science: Theory, Analysis, Modeling, and Design*,
https://doi.org/10.1007/978-3-030-93482-8_7

adaptive cells (neurons) but organized in a way to allow the networks of those cells to have evolutionary properties similar to those of species/genus systems. Human organizations, subsequently, have an uncommon degree of Simonian complexity since they are composed of multifarious components, but most importantly among those components—human brains.

In the next several sections, we will explore a series of increasingly complex, adaptive, and evolvable systems. The normal process of scientific investigations along lines of specialization and the reductionist methodologies employed have gone a long way toward producing a tremendous amount of knowledge about phenomena at all levels of organization. The disciplinary sciences, practicing the general science process, have accomplished the decomposition of systems within their domains such that we already have a general body of knowledge about what things are and how they work at many different levels of complexity and the organization. What has been missing in the scientific methods is the knowledgebase construction that allows everything discovered and documented within the disciplines from being connected. The difference between a knowledgebase and a set of separate, and sometimes isolated facts will be demonstrated in the next chapter.

So much of the knowledge from the disciplines remain as islands of understanding bits and pieces, rather than a comprehensive view of how the world works. In this chapter, we seek to demonstrate how the same kind of knowledge as developed by normal science could be better discovered and organized in a more transdisciplinary way. We claim that the use of systems analysis as described in the last chapter would lead to this result. Specifically, in this chapter, we will reexamine knowledge acquisition from several different levels of organization involving the most complex systems we know of. In the following chapter, we shall complete the vision of how knowledge gained through analysis can be organized in a universal knowledgebase based on the structure of system knowledge.

A single chapter in a book could not begin to do justice to anything like a full systems analysis of these systems. We will examine several real-world systems that have been analyzed by the method of Chap. 6 and show some samples of the end products, maps, and trees, for example, to show how each demonstrates the properties of systems even though each is in a different domain. What we can do, in this single chapter, is to show how the analysis work begins and take the process sufficiently deep to give a good accounting of the first few levels of organization of the example systems. The analysis will be taken far enough to establish how the procedures given in Chap. 6 produce the systemic knowledge needed to gain a deep understanding. What we seek is a global view of knowledge. We want to see how any of the islands that currently exist fit with one another, how they relate in both a hierarchical fashion (levels of the organization) and also as co-relating subsystems of the whole world. The vision for how to achieve this latter will have to wait for our examination of the knowledgebase structure in the next chapter, but a clue for how this is realizable has already been given in the way in which a systems analysis of a particular SOI can be expanded in scope and even reversed such that the suprasystem (environment of the current SOI) may become the new SOI.

The five systems demonstrated in this chapter represent increasing scales of size, complexity, adaptivity, and evolvability. The examples chosen are all, with the exception of the last one, systems that have proven stable and productive. At this writing, the jury is still out on the long-term prospects of the lase example.

We start with an analysis of a basic computing system, a machine that is capable of some rather impressive processing. The hardware, coupled with modifiable software, is an example of a complex system (as defined in the Chap. 1). If we were to include the role of software engineers and the projects seeking to modify the software components in response to changes in the environment of the computer product, then we would have to classify that expanded SOI as a CAES. We will, however, stick to an analysis of just the machine with a (temporarily) fixed program.

Next, we will examine a living cell, a very special one in the evolution of animal species, the neuron. Our analysis was motivated by a desire to emulate the biological neuron in the course of emulating natural intelligence, as it was understood in the mid-1980s. The hypothesis was that it would be necessary to build artificial (that is simulated) neural networks out of much more biologically realistic models of neurons to obtain the same kinds of dynamics in processing and learning. For deep analysis of neurons, we turned to the then understood neurological and physiological science but integrated it with cybernetics and network theory as discussed below. Those efforts produced the Adaptrode mechanism, a synapse emulator, and an artificial neuron that produced some important neurodynamics in simulation. It also, eventually, led to the construction of an artificial agent (a robot) that behaved in many ways like a simple gastropod (a snail) in foraging for rewards and learning to avoid threats.

The third example, also a biological system, is the human brain. This system is clearly a subsystem embedded within not only the organism (a person), where it is directly responsible for the full range of hierarchical cybernetic governance of the body and behavior but also the cultural/social larger supra-system, which has a substantial impact on the adaptability and evolution of the brain structure (and the knowledge encoded within). This is a great example of a system that has a seemingly clear physical boundary but in fact is so highly porous and fuzzy that identification of real boundaries is a major challenge. It is also an example of a system that is just now becoming amenable to transparent-box analysis through advanced imaging technology and increasingly clever probes of the correspondence between brain structures and thoughts and behaviors.

The fourth example steps up to the social system level in which multiple brains and cultural artifacts constitute the main components, the social organization, such as a corporate enterprise. Here the scales of size, complexity (e.g., more than just an aggregation of brains), adaptability, and evolvability range dramatically. For example, in size and complexity alone the scales may reach global levels (i.e., international corporations). If the brain represented a challenge in analyzing boundaries, that challenge is multiplied many times over for social organizations.

Finally, we will take a quick look at the systems analysis of the entire human social system (HSS). Although we speak casually about the human species and its cultures as social *systems*, so far there has not been a systems analytic attempt to understand ourselves comprehensively. The social sciences, like all of the physical and biological sciences, are an aggregate of loosely coupled sub-sciences that investigate very different topics regarding different aspects of being human and living in a social world.

There will be no serious attempt to delve into an actual analysis of the entire HSS in this chapter. The point of that section will be to demonstrate how the process might be used to actually tackle such an undertaking. A key issue discussed here, however, will be how being inside the system and still analyzing it objectively is another kind of challenge, but how it might be achieved as a result of the systematic process. The reason for its inclusion here is that it strikes this author that understanding ourselves, our particular kind of system, is extremely important if we are to grasp the nature of the biophysical problems the HSS is clearly causing in the whole Earth system. A grasp of the HSS is also a prelude to Chap. 9 in which we take a specific subsystem of the HSS, the economy, and demonstrate a much more thorough systems analysis process that will show a better understanding of economic phenomena than is currently accomplished with neoclassical economics. We introduce the forbearer of a *systems economics*.

7.2 The Digital Computing System

We will start by showing the results of an analysis of a very simple computer by the methods of Chap. 6. In engineering terms, we are doing a "reverse engineering" process to find out what the computer is made of and what pieces are connected to what other pieces. This system borders on the line between mere machines and adaptive systems because of the nature of software. Similarly, it approaches a kind of evolvability in that computers are generally designed for expandability, of memory and input/output (I/O) capabilities. But both adaptivity and evolvability[2] are totally dependent on the machinations of engineers, hardware, and software. Still, the capacity to reconfigure some aspects of a computer, even when deployed in the field, makes it a reasonable candidate for what we might call "primitive" CAES status.

In truth, the analysis of a computer like this is not terribly useful since computer designs already exist and the design processes already in place can readily produce any new functions without going to the trouble of a full, deep analysis. The purpose for putting this example in the book is just to provide a starting place using something fairly simple, but complex enough to demonstrate how the procedure works.

[2]There is a very active area of research in reconfigurable hardware such as the wiring between components and self-modifying programs.

7.2.1 *A Simple Computer*

Many beginning computer architecture textbooks include a design of a very simple (minimal) computer along with a simulator that students can practice writing assembly language programs and run them. The design is simple enough to be covered in a single semester but nevertheless includes the major principles and design techniques used in real computers. Here we borrow a variant on one of these computers, developed by the author to teach computer architecture to first- and second-year computer science students. The original design is owed to Yale N. Patt (University of Texas at Austin) and Sanjay J. Patel (University of Illinois at Urbana-Champaign) (2004). It is a minimal Von Neumann computer that demonstrates most of the features you will find in any computer, just scaled down to the size and complexity that students can handle. In this section, we will not be getting deep into details of the computer itself (called the LC-3) since our objective is simply to demonstrate the kinds of results that are obtained from a deep systems analysis of a machine. The author's revised design (called the SC-2) is shown in Fig. 7.1 It borrows many of the features of the LC-3 but has expanded the instruction set and I/O capabilities. For

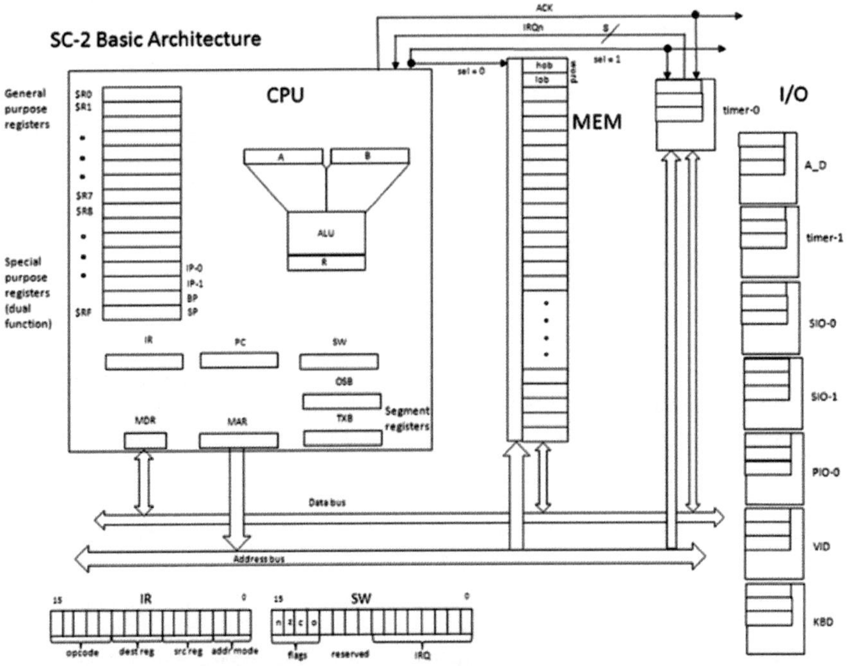

Fig. 7.1 The SC-2 computer architecture is a 16-bit (register sizes and addressability) computer inspired by the design presented by Patt and Patel (2004). This machine doubles the number of instructions but uses basically the same general design as the LC-3

those interested in the design of one of these simple computers, we recommend you take a look at Patt and Patel (2004, Chaps. 4 and 5).

Here we apply the procedures for decomposition to the computer as a system. Engineers should perhaps take care in what follows. They are used to something called "block diagrams" (as in Fig. 7.1) to show the components of a machine. But recall that the diagramming representations in deep systems analysis is meant to convey much more information than just the relative geometric relations between parts. It is also used to show functional relations through input/output flows. So, for those used to block diagramming methods, please bear with us as we approach the problem of understanding a computing system from a systems perspective. In the next major section, we will be looking at the neuron as another kind of computing system, so the methods used here, while perhaps strange looking, will hopefully make more sense once we see how they apply to other domains.

7.2.2 Decomposition

The decomposition of a computer system is aided by the fact that computers are designed in a modular fashion. That is, modules embedded in larger modules! We take advantage of this fact to show how decomposition can be applied. In what follows we have elected to leave out the numbering identification scheme discussed in Chap. 6 so as to not clutter up the figures unnecessarily. The reader may want to review the procedures in that chapter for assigning the dotted number identifiers to components in the course of decomposition. Below we only show verbal identifiers (names) of components as we suspect this will be sufficiently clear.

7.2.3 Environment and SOI Identification

The first steps involve determining the boundary of the system, identification of the system of interest (SOI), and doing an environmental or context analysis. In Fig. 7.2, we show the results of these analyses for the computing system as SOI. Remember this is a very simple machine, essentially similar to a pre-Internet CP/M or DOS computer.

Here we see that energy is supplied via a power source (A/C wall socket!). An external file device contains programs and data files. The computer will read programs/data from the files and write new programs/data back to the files (or create new files as needed). Users can be either programmers who write software (to be stored in files) or so-called end users who work directly with the data (e.g., accountants). Note that all computers generate some heat as they do the work of data processing, which must be removed from the device either by radiation or in larger systems by convection (e.g., a cooling system). All of the shown interfaces are handled through specific cable attachments that, essentially, keep the wires straight.

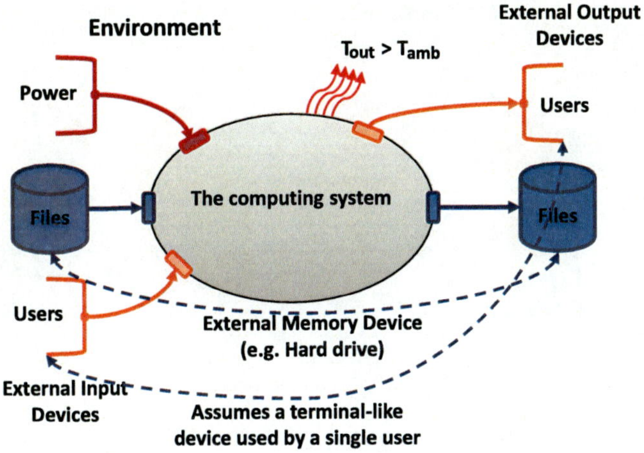

Fig. 7.2 A computing system (simplified) can be analyzed through decomposition

7.2.3.1 SOI Internals

An initial decomposition of the computer itself reveals several important details. The flows are all conducted by wires carrying electricity modulated between a maximum voltage (e.g., 3.5 volts D/C) and ground (e.g., 0 volt). A computer consists of the hardware that makes the computations possible, the software or programs that guide the hardware in doing so, and an intermediate hardware/software hybrid component called "firmware" that couples the software to the hardware. Within the software is a program that has primary responsibility for interfacing all other programs with the hardware through the firmware component, called the operating system. The details of how this works will not concern us since we are going to pursue a decomposition of the hardware.

7.2.3.2 Hardware Decomposition

We now pursue a depth-first decomposition of the hardware. The main results are shown in Fig. 7.4. For a sense of completeness, similar pursuits of either the firmware or software components would reveal, first, the existence of an operating system (OS) in software interfaces with the hardware control programs embedded in the firmware. This is a complex relation, beyond the scope of our current exercise. However, we should note that a similar decomposition of the OS and firmware components (e.g., what used to be called the "basic input-output system [BIOS]") would demonstrate the same modularity and hierarchical structure.

Rather, we will focus on the hardware decomposition as it demonstrates the overall scheme as well as any aspect of a computing system. Figure 7.3 illustrates this for the SC-2.

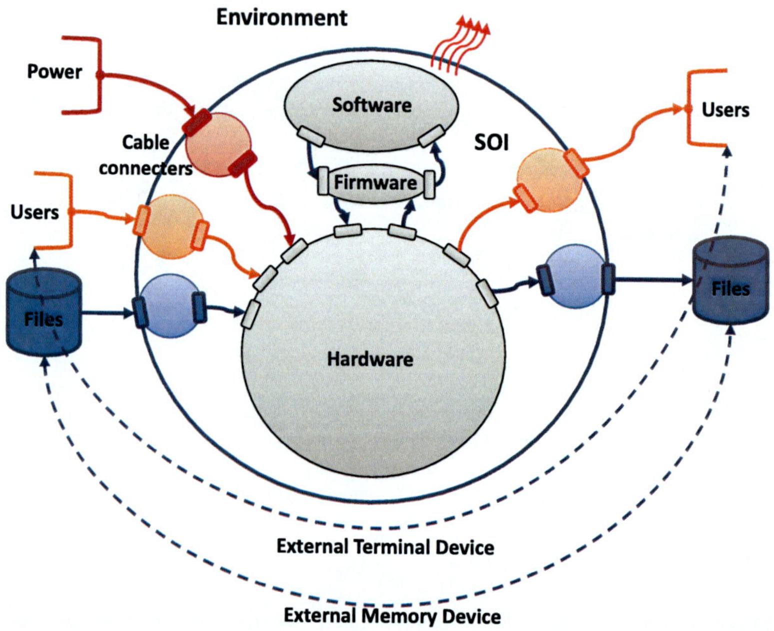

Fig. 7.3 The first level decomposition reveals the major components of a computer includes the hardware, software (programs), and an intermediate kind of hardware/software hybrid that interfaces the software with the hardware – firmware. The interfaces with the outside world are through cable connectors

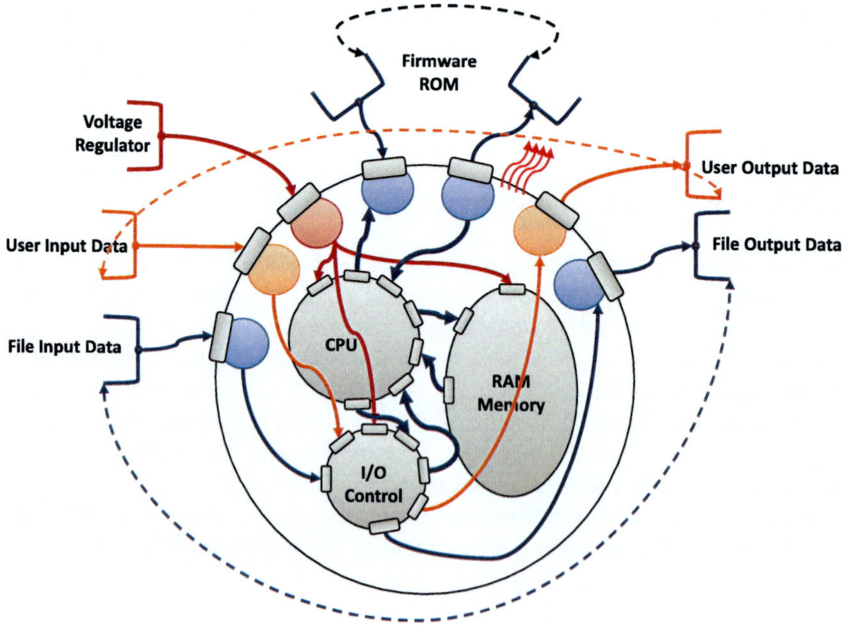

Fig. 7.4 Decomposition of the hardware component from level 1 reveals the main component subsystems of hardware and the flow paths of energy and data/messages

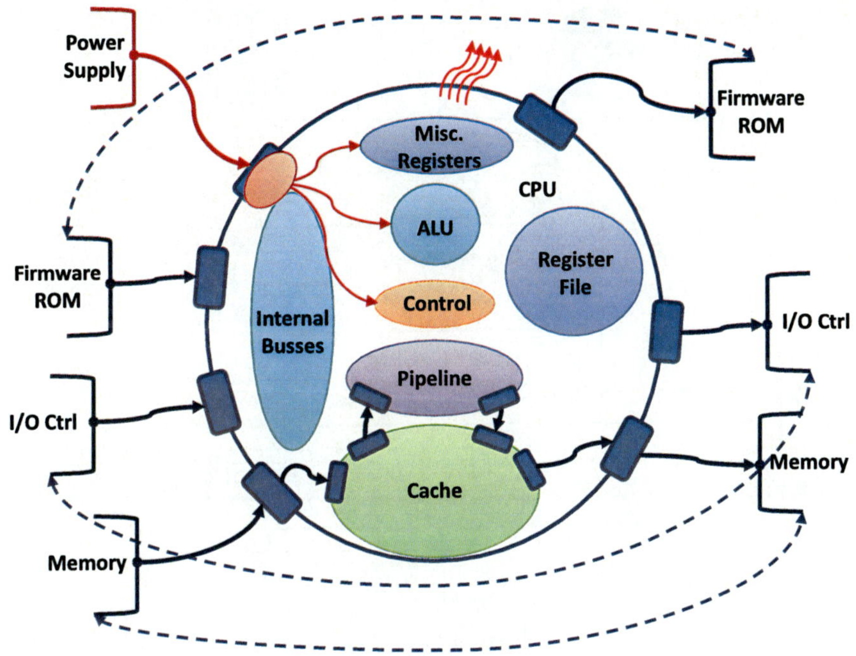

Fig. 7.5 Further decomposition of the CPU subsystem, now treating the other computer subsystems as sources and sinks. Only a few power and flows of data are shown

As the decomposition proceeds, we select the CPU as the next module to be analyzed. The CPU becomes the SOI at level 2 and is shown in Fig. 7.5. In this figure, we identify the main internal components that make up the CPU function, identifying a few of the interfaces that mediate the flows of data through the CPU.

The CPU comprises a number of subsystems, some simple, like the register file, others complex like the arithmetic logic unit (ALU). The figure identifies the major component subsystems at level 3 in the decomposition. It also shows some limited flows of power and data, for example, between the cache memory and the pipeline processor. The analysis would proceed by identifying the flows from CPU interfaces on its boundary to or from all of these elements and then between them as was done in Chap. 6.

That which flows within the computer is messages and energy. The latter is relatively easy to account for. All work performed in the various components is simply managing energy flows (switching of transistors) and results in heat. No energy is actually stored except in capacitors for very brief periods. Messages are coded in binary streams and are synchronous, that is, all flow controlled by an on/off clocked signal. So, the various metrics used to characterize the message flows boil down to bit rates. For example, the flow of data from these components to one another is handled via the internal busses (there may be several), which operate at characteristic speeds. Thus, once characterized at data rates, there isn't much else to specifying the flows (no material flows!)

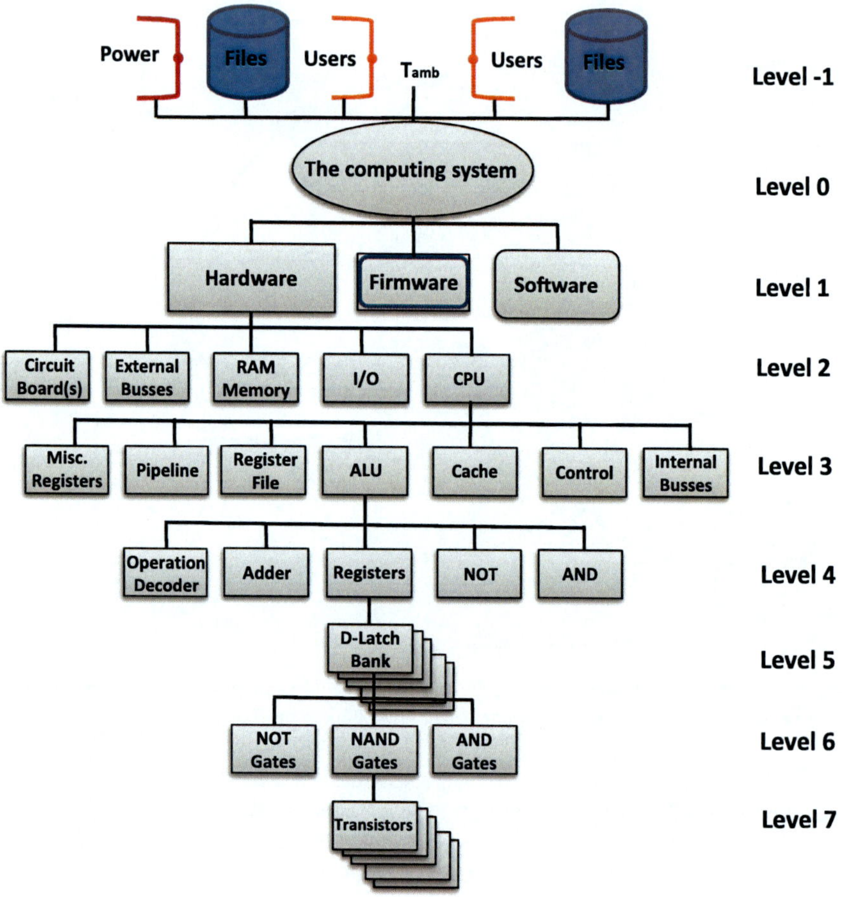

Fig. 7.6 A decomposition tree representation of several subsystems of a basic computer system down to the atomic component level (7 for the register implementation within the ALU). This tree shows the decomposition through the CPU branch and the ALU branch after tha

If we continue the decomposition, we might choose the ALU component for the next level starting place. We would find it composed of a number of registers and computational operator circuits such as full adders, parallel OR gates. Registers could be further decomposed to find D-latches lined up and connected to even more internal busses in the ALU. Finally, proceeding along this branch of the decomposition tree, we find that the latches are composed of transistors forming storage circuits. We can consider transistors as atomic components since they all work similarly and it is how they are wired up in logic circuits that determine functions beyond switching on and off (binary 1 s and 0 s).

Figure 7.6 presents a partial decomposition tree from the root (level 0) and environment (level 1) through the Hardware, CPU, ALU, Registers, D-latches to transistors branches resulting from the depth-first analysis.

7.2.4 *Knowledge and Generating Models*

In the next chapter, we will consider the architecture of the knowledgebase used to capture the components and relations during the decomposition of a system. We have alluded to this capture in Chap. 6. All of the elements shown in Figs. 7.2, 7.3, 7.4 and 7.5 and summarized in Fig. 7.6 are thusly captured in a knowledgebase.

In anticipation of those chapters, consider that if the design of a computer or a computer network were captured as a systems decomposition in the fashion just presented it would be possible to produce a simulation of the computer from this knowledgebase automatically. In fact, computer architects and designers, today, build elaborate models of their systems in advance of actual fabrication. They do this to test design alternatives and to gather metrics on performance before committing resources to building actual systems so as to make sure their design ideas are going to provide a net benefit. If an explicit systems knowledgebase existed, then modifications to specific modules at any given level could be undertaken and tested via model expression and simulation in a way that should be far more cost-effective than building such models manually. If Intel or other chip manufacturers were to formally adopt this approach, we suspect their costs of a design to production would be significantly reduced more than offsetting the costs of capturing their current designs in a systems knowledgebase through deep systems analysis. What most companies (like Intel) don't realize is that they are already incurring those costs in what they currently do anyway. What the methodologies presented in this book do is organize the activities into a system of knowledge capture and deep understanding that permit significant savings in efforts in future design efforts.

7.2.5 *Transitioning the Complexity Boundary*

A computer is one of the most complex machine examples of a system. It sits at the boundary of being a mere complicated machine and becoming a true CAS because of its reliance on software to control what it actually does. Software is mutable but only if a human programmer executes the changes. Self-modifying programs are on the horizon, however, with the advent of hardware that is reconfigurable as described earlier. We expect that within the next decade, especially with the understanding of hierarchical cybernetic management systems as described in Chap. 12, adaptive control algorithms and modifiable hardware configurations will elevate cyber-physical systems into the realm of truly adaptive machines.

For now, however, the best representative of a true CAS is the biological cell. And the ultimate representative of a highly adaptive system is the neuron cell in the brains of animals. Neurons are extremely adaptive systems that deal with computational problems, not unlike the computer described here. The nature of the computation that neurons perform is quite unlike that done by computing machines (today). Neurons operate in a highly stochastic environment and, yet, provide a reliable computation translating input signals into stable output signals that contribute to the overall computation of large-scale patterns. Neurons are somewhat analogous to the transistor-based logic circuits of the computer example, but the analysis of how they work goes much deeper. We turn to the neuron now to see how analysis can go much deeper into the biophysical phenomena that support the capacity for thought in the human brain.

7.3 The Biological Neuron—A Complex, Adaptive System

The author's primary area of research involved building a computer model of a biological neuron with the thought that such a model would be useful in then constructing more biologically realistic neural networks (Mobus 2000). The field of artificial neural networks (ANNs) was gaining increasing interest in the general area of artificial intelligence, but the standard notion of a "neuron" among engineers and computer scientists was extremely simplified and, while useful in certain limited ways, in no way emulated biological-like learning.

At the time this research began, neurobiologists were starting to acquire considerable knowledge about the overall dynamics of neural plasticity and its role in learning. The field of brain science, which included work on neurons as the primary units of function in the brain, pursued the work in the classical reductionist vein. Starting with noting how synaptic junctions changed their ability to generate a wave of depolarization of the local cell membrane, a synapse's "strength" was found to increase under certain conditions of prior excitation by incoming signals. So neuroscientists naturally went looking for the biochemical details that accounted for this phenomenon. This was the reductionist approach of the sciences.

In this section, we describe work on understanding the neuron and synaptic behaviors from a systems perspective, which, we assert, afforded some different insights than were coming from the neurobiological framework. Using the guidance of deep analysis given in Chap. 6, we will produce a system map and tree similar to what we did for the computing system above. As was true for the computer, we can only demonstrate a small portion of a neuron map and tree.

7.3.1 *Identification of the SOI—The "Typical" Neuron*

There are over 100 known types of neurons in the human brain. They vary in form and some aspects of their functions but share a core of structural and functional features. Namely, the cell body has a set of inputs, either electrical or chemical *synapses*, usually arrayed on special protrusions called *dendrites*. They have a single output *axon* that may, nevertheless, branch so that the output signal reaches many other neurons through their synapses. We will not be concerned particularly with brain structure and the wiring of neurons into networks or circuits. Rather we will focus just on the common features of neurons representing a "typical" neuron. This neuron is based on the most abundant form in the neocortex of mammals, the *pyramidal cell*, so called because its basic shape resembles a pyramid with dendritic branches on each of the "corners" and the axon emerging from the center of the base. The geometry of cell types is thought to be a controlling factor in the temporal and spatial integration of input signals, typically pulses of currents.

7.3.2 *Analysis of the Boundary and Environment*

We will not attempt a full biological description of the environment in which neurons operate, the brain tissues that include many other elements besides other neurons, such as glial cells and the fluid milieu. In short, all of the factors that are involved in any living cell such as nutrient obtaining and waste disposal will be left out of our current interest. Here we will focus exclusively on the sources and sinks of a typical neuron that are other neurons. The interest here is communications between neurons and the consequences of those communications in terms of encoding memory traces (c.f. Alkon 1987). Figure 7.7 depicts the nature of the communications channels between neurons. Signals that travel along axons are typically pulses of excitation called action potentials. They travel in one direction—away from the source cell body—toward other cells.

As shown in the figure, axons can branch to send the same signal to multiple other neurons. The axon forms a connection with the receiving cell through a synapse, which we will analyze as an interface below.

Synapses may be formed directly on the cell body membrane, but more generally they terminate on dendritic protrusions, or spines, which act as integrators for signals from multiple neuron sources (only one source is shown in the figure). Each synapse is an interface between the incoming signal and the receiving neuron system. In the next section, we will be describing the processing of a synapse in more detail. But, briefly, the incoming signal, called an action potential, causes the release of molecular neurotransmitters that cross the small gap between the presynaptic and the postsynaptic membranes. The neurotransmitter then elicits an excitation in the receiving membrane. There are many different kinds of synapses even in the same cell. Again, we narrow our focus onto synapses that have modifiable reactivity to

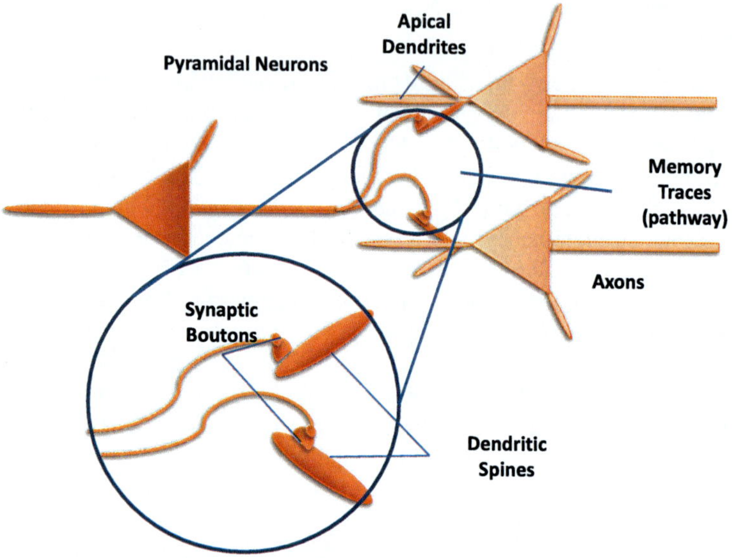

Fig. 7.7 Neurons communicate with one another

incoming signals. These are called "plastic" in that they can be modified (strengthened or weakened) and then retain their level of reactivity for a time. This is the basis for encoding memories.

7.3.3 Internal Decomposition

Figure 7.8 shows a neuron using our systemese language representations from Chap. 4. Other neurons are depicted as source and sink entities according to the boundary and environment analyses. Note that the "rest" of the cell's internal subsystems have been subsumed under a single subsystem oval but comprise all the other living cell functions that keep the neuron alive and able to perform its primary function—processing and transmitting signals.

Input processors are the synaptic junctions receiving afferent axonal terminals (left side of the figure.) These processes involve a complex set of chemical reactions that change the level of output from each one; the channels shown internally from each synaptic compartment to the central processor are labeled, "Depolarization state." The interface protocols are found to be comprised of various neurotransmitter-activated channels through the cell membrane that, when opened, allow various ions

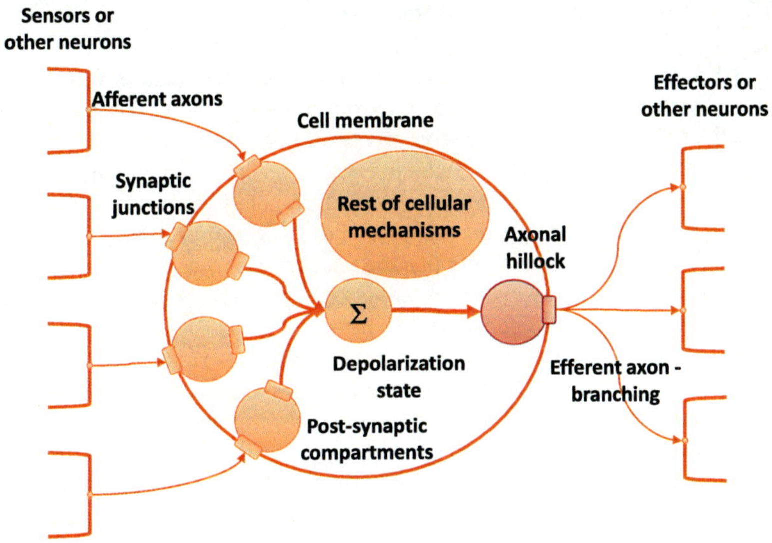

Fig. 7.8 A model neuron subsystem map

to flow into the compartment.[3] In the resting state, there is a voltage difference between the outside of the membrane and the inside. It is polarized with a net negative voltage outside. When the channels open and ions flow inward the local membrane is depolarized. The output from the synaptic compartments is actually states of local depolarization that spread out from the synaptic compartment on the membrane. The figure treats this spread of activation (depolarization) as directed channels to the central processor, which is the summation of the contributing depolarization events within a short time window. The small oval labeled with "Σ" represents this process and sends the result to the neuron output processor, the axonal hillock.

The degree or strength of the depolarization events at each synapse depends on the history of activations at each one. In general, if a synapse has been depolarized frequently in a short period of time and certain other time-correlated events have occurred, the degree of depolarization with each subsequent incoming action potential will be greater (or in the case of inhibition, lower) and that synapse will contribute more to the summation.

[3] In truth this is a little oversimplified. Some channels are activated by the currents of sodium ions flowing inward. These channels allow for the passage of other ions to flow inward (calcium) and others outward (potassium). See Alkon 1987; Squire and Kandel 2009, for details

7.3.3.1 Level 1 Decomposition—Message Processing

At level 1 of the decomposition, we are primarily concerned with three basic subsystems. These are the set of synapses (perhaps arrayed in subsets by dendritic spines), the polarization/depolarization processing of the cell membrane, and the axonal hillock which triggers the outgoing action potential signal when the total depolarization state of the membrane reaches a threshold value.

We will consider the level 2 decomposition of the synapses below. For our purposes in this level 1, we simply note that the membrane acts as a temporal and spatial integrator of all of the synaptic depolarization states. What is important to understand is that the total depolarization of the membrane at any given instant is dependent on the temporal correlation of incoming signals at the synapses. This is how the neuron determines the correspondence between those incoming signals and provides a kind of internal feedback to the synapses that figures into the strength encoding of those that contributed (see below).

The hillock acts as a binary filter. If and only if the depolarization state of the membrane is sufficient will it pass through (and possibly amplify) that state to generate an outgoing action potential. This is how the neuron acts to coordinate its output with its inputs so that only meaningful signals are sent along to other neurons.

7.3.3.2 Level 2 Decomposition—Synapse Internals

The workings of synaptic compartments, how they receive and act on incoming action potentials from source neurons, and how they change their efficacy weighting—their response strength—is quite complex and we can only give a brief outline of it here as it pertains to the decomposition of neuron processing. In (Mobus 2000), section 3 "Biological Synapses" the reader can find a more complete description of the internals of a synapse and a discussion of its dynamics with respect to its short- and long-term responsiveness.

In brief, what are called "learning" synapses are influenced by the frequency of incoming signals. If a burst of high frequency incoming signals is received their degree of depolarization is increased exponentially fast. Each incoming action potential boosts the depolarization state of the local membrane. If the frequency is low, or during quiescent periods, the membrane is repolarized, as if starting from scratch. Now, if during a high frequency burst, while the membrane is maximally depolarized, the synaptic compartment receives a secondary signal (chemical) signifying that the current burst is "meaningful," then a chain of internal chemical reactions takes place which results in the synapse being "potentiated," or in other words, it does not completely return to the polarized state after the burst is over. That is, the synaptic membrane retains a very short-term memory trace of a meaningful input signal.

If over the course of a short time (but longer than a burst period) additional bursts and secondary signals are received, the potentiation of the synapse is, itself, strengthened. And, the rate at which this new, secondary potentiation decays is much slower. In fact, much, much slower.

This is the beginning of the encoding of a memory trace. With many repetitions of the pairing of an incoming burst and a reinforcing secondary signal (which must come slightly after the primary burst starts) the potentiation of the synapse, that is its readiness to respond to future such signals, is greatly enhanced. The interpretation of this phenomenon is that the secondary signal came from a meaningful feedback signal, either from the cell itself or from external sources such as neuromodulators. That means that the primary signal must be associated with the secondary and so its receipt is important. In fact, the author has shown that the primary input signal comes to represent an anticipation of whatever causes the secondary signal and so can be used to encode a causal relation and to be interpreted as a prediction of the latter.

The overall aspects of synapses and their dynamical behavior are discovered in level 2 of the decomposition. We have given a very brief glimpse of what would be found in level 3, decomposing the internals of a synapse.

For a more complete description of these functions and their consequences for learning and behaving, see Mobus (1994, 1999).

7.3.3.3 The Decomposition Tree

Figure 7.9 shows the neuron decomposition tree down to level 2. Further decomposition of synapses (level 3) can be found in the abovementioned references.

As depicted in Fig. 7.9, neurons interconnect in a complex network in the brain. In animals that have cortical structures, especially the neocortex in mammals, brains are able to encode memories of complex perceptions. The brain of animals able to do this kind of experiential learning are, by our definition of evolvable systems, CAESs. Just as with a single neuron, even a cursory demonstration of the use of the deep understanding process on more evolved brains (such as a human's) would

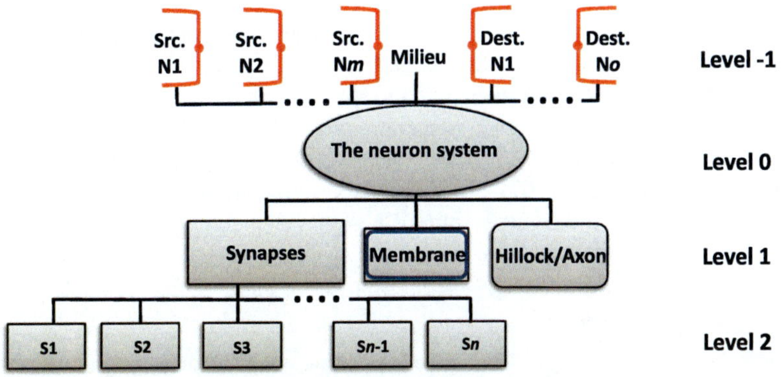

Fig. 7.9 The neuron system depicted in Fig. 7.8 shown as a decomposition tree

require many volumes. These are actually very exciting times for neurosciences, including neuropsychology. New methods for exploring living brains as they process information are making unprecedented progress in the analysis of brain functions. Improved methods of histology and neuroanatomy combined with functional imaging (e.g., magnetic resonance imaging—fMRI) data are helping neuroscientists get a much better picture of how brains work. In the following section, we will briefly describe some of these methods with respect to understanding the brain as a CAES.

7.4 The Brain—A Complex, Adaptive, and Evolvable System

In this section, we want to show how the deep systems analysis methods described in the last chapter apply even to an advanced field of science that has already discovered many aspects of its topic, and how that application can act to organize the already discovered knowledge into a deep understanding of the whole system. The idea is to use the analytic technique not to discover facts in science but to take facts that are already known from the science and insert/organize them according to the structure of Eq. 4.1 and as embodied in the deep system knowledgebase (to be described in the next chapter). This is a kind of post-reduction reconstruction of the systemic links between possibly isolated facts in any level of analysis and between the levels themselves. It attempts to reconstruct the systemness of the SOI from the current understanding of the pieces derived from normal scientific inquiry.

All of the sciences, particularly the natural sciences, have proceeded through often disparate reductionist methods to obtain bits and pieces of knowledge in specific realms of their disciplines. What is being suggested here is that the methods of deep analysis and knowledgebase filling using the already known facts of the subjects can help reconstitute the subjects as systems by showing the pathways to reintegration.

The example of the way in which the *Periodic Table* of chemistry has filled this role is instructive.[4] As chemistry became increasingly analytical (alchemy becoming an empirical science) chemists were discovering new atomic entities that had different properties. Thinkers who noticed relationships between element types, such as a progression of unit weights and repetition of forms of reactivity, began trying to organize these elements in some rational tabular form that took into account these properties. This led to the table form we see today, originally credited to Dmitri Mendeleev (1834–1907). Since chemists were working on specific elements that did not represent the full spectrum of possible elements, gaps appeared in the table. The table itself had a regular format so these gaps were recognized as indicating that some elements had not yet been discovered. That regular format helped to

[4]This example has been used by others in a similar vein but was first introduced to this author by Gary Smith, a systems architect at Airbus Defense and an active member of the INCOSE Systems Science Working Group

"predict" the properties of the undiscovered elements, especially, for example, what sorts of molecules or crystals the elements might form, which then helped chemists design assays that could isolate the to-be-discovered elements. In other words, the periodic table provided a way to predict the discovery of unknown elements and a structured way to go about their discovery. It was the organization of the periodic table based on the properties of elements that formed a meta-chemical model of all of the elements.

This is another example of how systems science should be considered as metascience. This method of obtaining systemness in any scientific field, starting with what is already known, may provide the kind of roadmap to discover the gaps, where they exist, and suggest what to look for in those gaps. All of the subjects of the sciences are either systems in their own right or phenomena that involve systemic behaviors[5] so should be amenable to this approach.

The choice of example, the brain, is made because this organ is lately the subject of intense study, is the mediator of human activity, and because it is representative of an archetype model system, the hierarchical cybernetic governance system (HCGS) that will be the subject of Chap. 12. The reader may want to peruse the subject of that chapter for background, but it isn't required reading to understand the treatment here. Our interest is in how to accomplish the task of deep analysis of a CAES representative system.

7.4.1 Brain Science

We will use the term *brain science* to denote the totality of various subfields, such as brain anatomy (the gross structures such as hemispheres and lobes), neuroanatomy, and neurocytology (the distribution of various neuron types in modular clusters), and connectomics (how neurons and modules communicate). All of these subfields have attacked the analysis of different levels of complexity, more or less in parallel and often without reference to the discoveries in other areas. An exception to this would be the remarkable work of mapping cytoarchitectonic regions to the cortex as accomplished by Korbinian Brodmann (1868–1918), a German neuroanatomist. Brodmann mapped out areas of the whole cortex based on cell types and microstructures of connections found there, leading to a set of "areas" having differences that appear consistent across individuals and, to some degree, across species. Even with this kind of integration across spatial dimensions (from gross anatomy to microanatomy), there are still many holes in our understanding of relations.

Other neuroscientists focus on the functions of regions from whole lobes (e.g., the visual processing of the occipital lobe, the most posterior part of the cerebral

[5] By phenomena, we mean the subsystem processes that are a part of a larger system process. For example, we can study the flow dynamics of a fluid, the characteristics of which may include turbulence regardless of the fluid type and physical arrangement of the source and sink. Turbulent flow is an isomorphic phenomenon that is found in many different kinds of systems

cortex) to patches that correspond with Brodmann areas (e.g., rational decision processing associated with Brodmann area 10). There are still many gaps in our grasp of what smaller processing modules are doing in relation to larger areas, but the science of connectomics is starting to fill some of these in.[6]

The brain is easily seen as a complex adaptive and (with the capacity to encode arbitrary patterns in the neocortex[7]) evolvable system. It is the governance system for an individual, for logistical, tactical, and strategic decisions and actions.[8] So, it will provide a good test of the applicability of deep analysis. In the next section, below, we address the issue of brain complexity and how it may be resolved via taking the data already grasped by the various subfields and putting it into the knowledgebase format. There are a number of projects currently underway attempting to build models of the brain. We suggest that applying the procedures of deep analysis and capture of data in the knowledgebase, described in the next chapter, would greatly advance this effort.

One of the most exciting developments in neuroscience and neuropsychology is the application of systems modeling to the human brain and correlating its workings with human behavior. This is an enormous field of study that we could only feebly present in this volume. Rather we will remark on a few recent developments that demonstrate how the brain and its activities are representative of a CAES.

A key development in brain science has been the development of new instrumentation capabilities, including the ability to visualize living brains as they work and produce behavior. Recall the discussion of instrumenting the inputs and outputs of a system in Sect. 6.5.2.3.1 *Flows, Sources, Sinks, and Interfaces* in Chap. 6. There have been major breakthroughs in the instrumentation of the whole brain as well as subsystems (i.e., neurons and networks of neurons). By now most people are familiar with functional magnetic resonance imaging (fMRI) pictures of activity zones in areas of the brain "lighting up" when the subject is exposed to various kinds of sensory and emotional stimuli. The field of correlating, for example, patient "thoughts" with brain activities has produced some remarkable results, helping to explain many different mechanisms involved in "kinds of thoughts."

But there are many additional imaging techniques that can be used at various scales of space and of time.[9] Some of these allow unprecedented looks at individual

---

[6] Connectomics, roughly speaking, is the application of network theory as discussed in earlier chapters and in Mobus and Kalton (2015), Chap. 4, to neural connectivity between regions of varying scale, both structural (which neurons are talking to which other neurons) and functional (what are the neurons saying to one another).

[7] In this discussion, we are primarily interested in the mammalian brain with the obvious intent to apply this approach to the human brain. However, this approach is applicable to all animals.

[8] In what follows, we will be treating the brain proper as the SOI. The rest of the central nervous system and the peripheral nervous system will be treated as parts of the input-output sources and sinks. Under some circumstances, it would be prudent to use the reverse analysis methods of Sect. 6.7.4 *Reverse Deconstruction* of Chap. 6, for example, when analyzing the visual system and discovering that "eyes" are actually more complex sensory organs than simple light detectors.

[9] Both temporal and spatial resolutions are still not co-extensive that is one instrument's scale does not necessarily shade into another's that is used to explore a different scale

neurons (both visually and electrical activity. Others are being employed to map the connectivity between both short- and long-range modules and regions. The technologies being employed allow non-invasive and non-destructive observation of brain components at multiple levels of the organization. This is the ideal situation for systems analysis where we seek to preserve functional relations even as we decompose the structure.

7.4.2 *Brain Complexity*

There are actually several ways to view the complexity of the brain. Figure 7.10 provides a summary of structural and functional complexities that need to be kept in mind when analyzing brain systems. In the figure, we show the structural nesting of increasingly smaller componential elements, down to the synapse.

The brain is an information processing system—a biological computation engine (Mobus and Kalton 2015, Chap. 8, Sect. 8.2.6). In fact, it is a set of computational modules, each using the same basic processing elements (neurons) but processing different inputs to produce different outputs relevant to the module's purpose. More primitive modules of the brain (see diagram below) are in the form of clusters, often called "nuclei." They receive inputs from various sources, including other nuclei

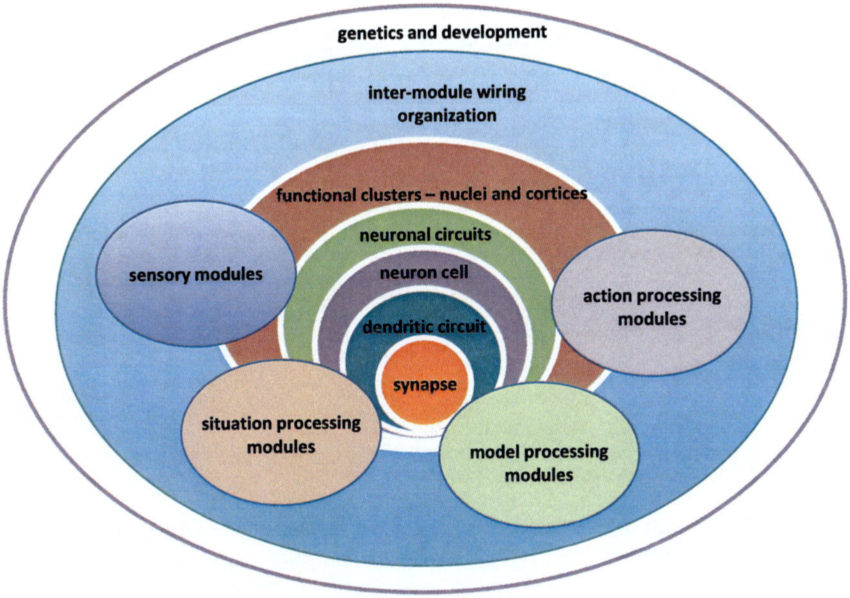

Fig. 7.10 Brain complexity can be viewed from the structural hierarchy (nested organization) or/ functional processing modularity

and sensory organs. Their internal "wiring" determines what they compute.[10] More lately evolved cortical structures are more plastic in terms of their wiring. That is, cortical modules (such as the Brodmann areas mentioned above) are able to encode variations in patterns of connections based on changing experiences. Cortical modules still have specific processing jobs to do, but many internal options with respect to how.

7.4.3 *Overall Organization—Mammalian Brain*

Figure 7.11 provides a diagrammatic layout of the architecture/organization of the mammalian brain. All animals (with brains) have the basic processing capacities to compute appropriate motor/glandular responses to the sensory inputs (both exteroception and enteroception). These are the core capabilities of brains. Over evolutionary time, as environments and body forms increased in complexity these basic

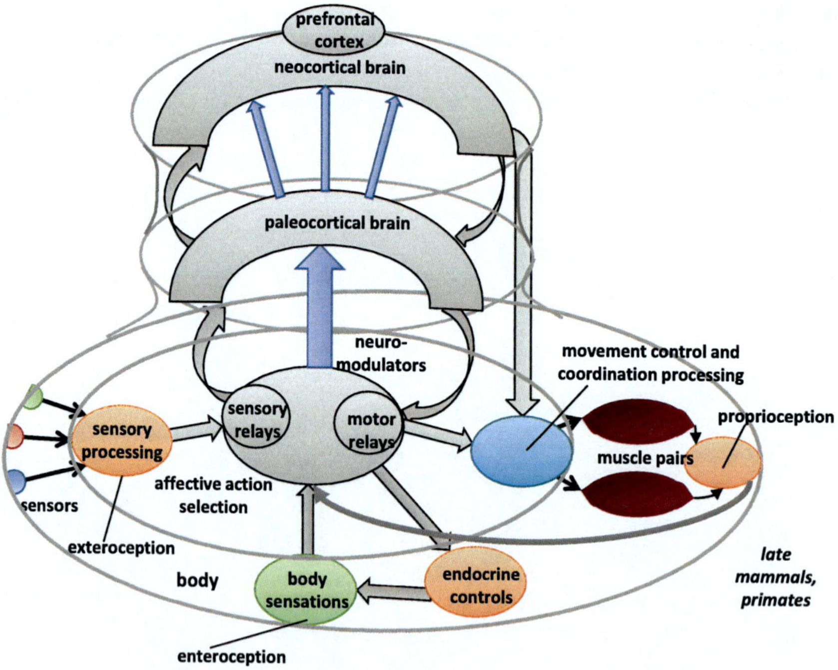

Fig. 7.11 The mammalian brain architecture is organized as a series of more lately evolved layers of cortical structures overlaying the basic, limbic, or primitive brain

[10] In cognitive science, the term "algorithm" is often used, but, in fact, these biological computations are heuristic; the term algorithm is reserved for procedures that are guaranteed to produce a specific result

processing capacities needed to expand to handle the expansion of messages coming from the outside world and the internal body (Striedter 2005). In addition, as the environment became more complex it also became more stochastic, that is associations between input signals became far less deterministic. Cortical structures, essentially sheets of grey and white matter, using repeating low-level modules (e.g., the cortical columns discussed below) provided a mechanism for encoding variations in associations. Animals could "learn" associations that might vary over time and place.

7.4.4 Levels 1 and 0 (Preliminary Look)

In this section, we take a look at a partial decomposition map after doing a levels 1 and 0 analysis followed by the start of an internal discovery of subsystems at level 1. Figure 7.12 shows an outline of the analysis at this stage. What is shown is not following the suggestion in Chap. 6 to start with the outputs (i.e., motor control and glandular control), though that rule of thumb still would apply. For illustration purposes, however, it is easier to show the internal message flow mapping from a sensor (in the retina), through a relay nucleus (thalamus), to the primary vision processing cortex in the occipital lobe.

7.4.5 The Brain Structural Decomposition Tree

Figure 7.13 should not be taken as complete by any means. Again, it is just representative of the structural decomposition tree that would be produced by the deep analysis. Note that at the lowest level shown (labeled "Level?"), many different

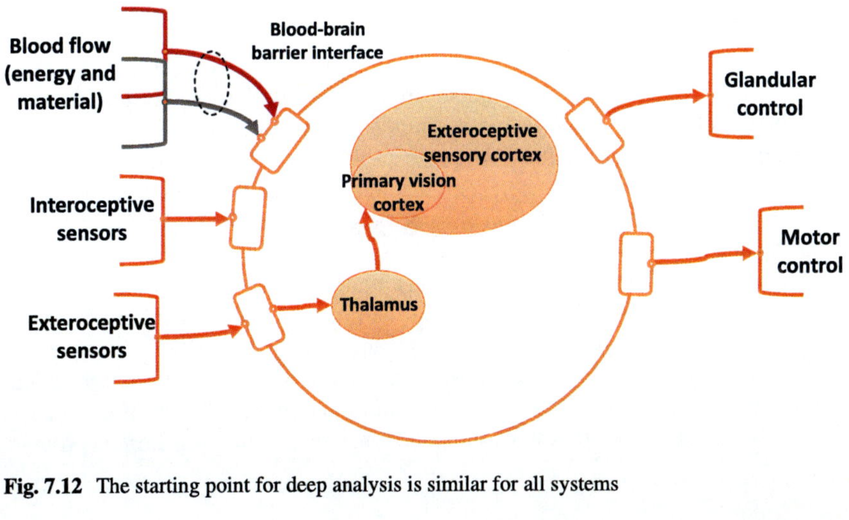

Fig. 7.12 The starting point for deep analysis is similar for all systems

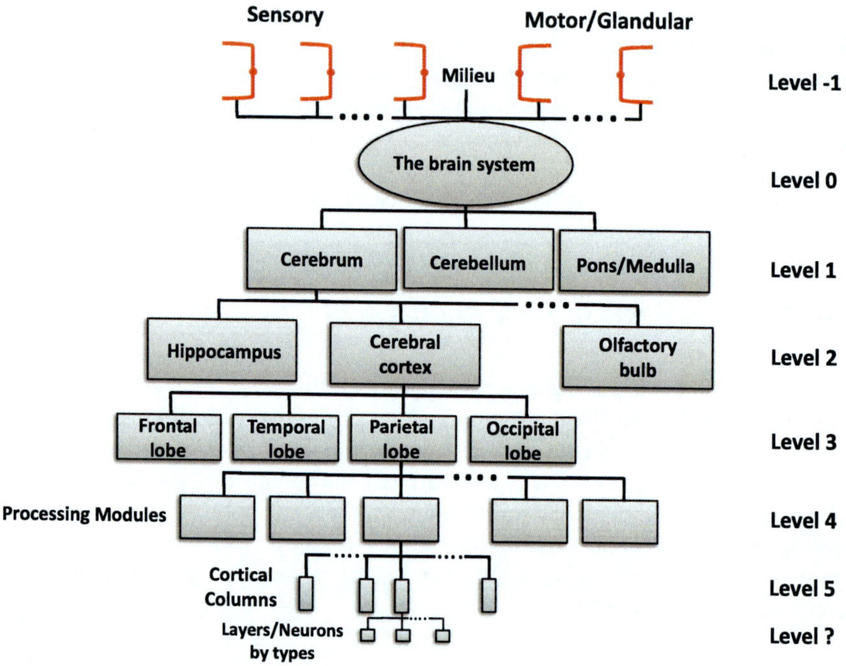

Fig. 7.13 A brain system decomposition tree (showing components only) shows the overall organization of the system

neuron types are present in the cortical column modules. Every neocortical area (the various lobes) comprises very similar columns but the various processing modules, such as Brodmann areas, may have different neuron types or mixes. The columns are organized in several layers that are continuous across the neocortex (think of the cortex as a layered sheet that is composed of side-by-side columns).

The general way these columns are internally wired is thought to provide the mechanism by which memory engrams are encoded and activated (Hawkins 2004). Columns in the sensory cortices encode features relevant to the various sensory modalities. Columns in the initial association cortices receive inputs from the sensory columns that are activated by the presence of those features during perception and encode percepts based on correlated features, for example, roundness and redness are associated in the perception of something that will eventually be conceived as an "apple."

The tree is not complete in that it is lacking the details of neurons, dendrites, synapses, etc. from Fig. 7.9. And it should be noted that that figure is not really complete in that each synapse will have a subsequent decomposition tree showing the biochemical components that make it up. This could, of course, be extended down to the atomic level (or even the elementary particle level!) but if the reader recalls the argument from Chap. 6 about what constitutes an atomic process (ultimate leaf node in the tree), then it should be clear that only the specific molecular components of synapses (as with all other aspects of the neuron) need be considered.

Even in its incompleteness, it should be clear that something as complex as a human brain can be amenable to a complete deep analysis following the methods given in Chap. 6. The singular advantage of undertaking such a project would be a reintegration of all of the functional aspects of brains with a very likely discovery of insights into how the operations of the brain give rise to the mind and cognition.

7.4.6 *Behavior*

The brain is the mediator of the behavior of the organism. For highly evolved organisms like great apes, and especially human beings, this especially includes social behavior—the ability to work in groups to achieve mutually beneficial objectives and support the lives of all in the group. The range and scope of behaviors arising from the human brain in interactions with its environment (which includes the body) are far beyond the scope of this volume. See Robert M. Sapolsky's excellent "Behave: The Biology of Humans at Our Best and Worst" (2017; Mobus 2019). For introductions into the neural substrates of overt behavior see (Mobus 1999, 2000).

7.5 The Organization System

Human beings have the ability to be social in a variety of situations. One person can be a member of multiple "tribes" simultaneously, as long as each is different from the others, have different purposes and cultures. A person can be a family member, a church member, a sports team member, and a working member of an organization.

One could reasonably argue that we have more experience with the analysis of complex systems in the realm of various kinds of organizations like businesses. This has been the result of the computerization revolution in which business processes, previously conducted using paper-based data processing, were automated starting back in the 1960s. We will not belabor the points of deep analysis for business processes. We simply claim that organizations like businesses and other institutions can be analyzed in the same way that any other system would be. Here we will only reflect on how the analysis of business processes has evolved and then on where things stand today. The reader is invited to apply the methods of Chap. 6 to a business (or other kinds of organization) as an exercise.

The form of organizations had evolved gradually over the history of commerce. They have always been hierarchical in nature, with an upper "management" and a set of work processes. In the course of growth, successful organizations needed to expand the breadth and depth of this hierarchy. The best-understood subsystems in organizations, and thus least questioned by analysis, were the accounting systems, financial and managerial. These systems had a long history of successful mechanisms for keeping track of assets, liabilities, costs, and benefits of operations. So, naturally, they were among the first to be automated by computing systems. Unfortunately, because these systems were so well understood a form of analysis emerged that sought not a deep understanding of the systems (or the firm's operations they were meant to subserve) but rather on the approach to mechanizing the paper-based methods. The efficacy of the systems was presumed and the only task at hand was to convert the methods to algorithms and reports. The question of the goals of management of the work processes was never questioned.

This approach worked reasonably well for accounting systems, which, as stated, had undergone a long evolution that had already been selected for best practices, so the assumptions made when automating them were not unwarranted. The problems started to emerge when other management information structures were automated. By the time computers were infiltrating the corporate world, businesses had already become very complex with deep and broad hierarchies. Moreover, many then-current business practices, such as finance and marketing, had not undergone the same kind of evolution and selection of best practices that accounting had. Thus, they were not as well understood. Nevertheless, in seeking increased efficiencies in other parts of their operations, firms sought to automate these other functions or at least parts of them that would provide managers with the information they needed to make decisions.

Case studies in MIS and other computerization/communications projects attempted to replace paper-based data/information processing. Today, computerization (including informatics, big data analytics) is the basis of innovative organizational activities (e.g., new services). The perennial problem that computerization projects have encountered as automation projects were undertaken is that the analysis of the organization (or proposed organization) was not deep in any meaningful sense. The analysts hardly looked into the actual operations or functional organization of the firm. They relied almost entirely on the judgments of so-called users or stakeholders to create conceptual maps of the operations and then attempt to overlay information systems on top of whatever the users told them. And all too often the users were mid-level managers who might not have actually understood some important details of the operations they managed (managers notoriously leave the details to workers). Moreover, managers too often get confused about the nature of the decisions they are supposed to make, and as will be shown in Chap. 12, do not even realize the cybernetic purpose those decisions are supposed to serve. Over the history of business automation, there has been a gradual evolution driven as much by project failures as anything. If one analyzes the way the projects were conducted of both failures and successes one finds that the latter are marked by up-front analysis that delved much deeper into the actual business processes being managed.

The state of systems analysis for business processes has advanced considerably. This is due in part to advancements in process analysis and a much clearer understanding of how these processes actually work, what sort of management is needed, and what information requirements are to support decision making. Today there exists a rich repertoire of process "patterns," such as supply chain or flexible manufacturing cells that analysts can use to guide their designs with reasonable assurance of producing a good one. It is not necessary, in most cases, to start from scratch.

Even so, as businesses evolve and grow or develop new processes (or new products) there will always be a need to do a deep analysis with the capture of knowledge in a knowledgebase to ensure the success of the enterprise. There are two basic reasons that the processes themselves and not just the information needs should be analyzed. First, the design of the process itself may be improved by such an analysis. Second, the actual regulation/control points in the process will be more readily identified and, thus, the decision agents and the "requisite variety" of the agency will become visible (see Chap. 11 for an explanation of agent/agency archetype models). That will lead to a management structure appropriate to the process.

Humans associate and organize work processes—become a system—in order to accomplish a goal, they have a purpose. Throughout the history the mechanisms employed for the organization have been more experimental than guided by systems theory. Local-scale organizations (which includes multi-national businesses and institutions) exist to fulfill their (socially agreed upon) purpose). We cannot say the same for the whole of the human species as a global-scale society/organization. The purpose that humanity serves is either very cryptic or non-existent. In the next section, we will start an analysis of the global human social system, the human species along with its accouterments of complex cultures, as preparation for considering a new organization of humanity and a society that might serve an important purpose.

7.6 The Human Social System (HSS)

You and I are human beings. That means we are part of the system that is being analyzed. How do we handle the conceptual framework for modeling a system when the observer/modeler is actually part of the system? The methods of scientific inquiry come to our aid. The ideal of scientific investigation is objectivity in observations and analysis of data. Over the last 200 years or so we have learned quite a lot about achieving scientific objectivity even when the observer is part of the system being observed. There are no guarantees per se. But there are safeguards that have been built into the scientific process to allow evaluation of the level of objectivity achieved in any scientific inquiry. Figure 7.14 provides a view of the problem and its solution.

There exist now numerous tests of objectivity when making observations of systems in which the observer is a regular participant. Recall, too, from principle 10, (Chap. 2, Sect. 2.3.10) that a sufficiently complex system can contain (or produce) a model of itself. In other words, there is nothing, in principle, that would prevent an "as-if" external observer from being able to construct a model of itself as still being in the system of interest. This, we contend, is the basis for why we human beings, applying the tools and techniques of systems science, can step outside of our own system while maintaining the fidelity of the composition and behavior of that

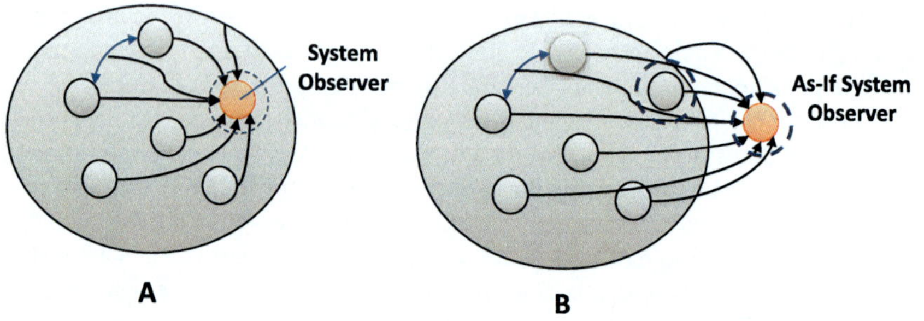

Fig. 7.14 The system observer is embedded in the system being observed. (**a**) The observer may well be able to gather data (black arrows) on the other entities and relations in the system, but does that exclude influences of those other entities on the objectivity of the observer? (**b**) Can the observer extract itself from the system in a meaningful way so as to pretend to be observing a system as-if from the outside? The figure shows the retention of a surrogate entity representing the observer as they would participate in the system

system. This is why we can have social sciences and especially a science of economics.

The idea that the whole of the human species, taken along with all of our cultural accouterments, could be analyzed for deep understanding might at first seem daunting, if not completely foolish. On the other hand, the search for understanding of the human condition has been, if not the principal effort, then certainly a substantial effort in human history. While the natural sciences, based on empirical methodologies and mathematical modeling, have sought knowledge of how the rest of the Universe works, many of the social sciences have tackled a more qualitative approach to gaining an understanding of ourselves and how we conduct our affairs.

Biology crosses the boundary between our animal characteristics and our behavioral characteristics. Many feel this is the case for economics as well. Psychology has been pursued as a natural science with hints of hermeneutics when we had so little understanding of how the brain works. But now we have very rigorous neuropsychology with appropriate analysis tools. So, we are beginning to bridge the seeming gap between biology and psychology. And with that bridging, we are also beginning to see the possibility of a sounder basis for economic science (see Chap. 9 for a better glimpse).

The logic of deep systems analysis tells us that the context or environment of a system of interest is as important to understanding the SOI as the decomposition of subsystems making it up. If we want to understand any subsystem of the HSS, it suggests we must understand the other subsystems as well. We need to understand the larger world system (i.e., the planet) and, indeed, the solar system and galactic system. Fortunately, we can start our analysis at the level of the planet because it is effectively closed to significant material flows (dust and the occasional meteor coming in and some loss of lighter atmospheric gasses). It is open to energy flows, principally from the sun with heat radiated back into space. Thus, a great deal of knowledge about the structures and dynamics of subsystems of the Earth has already

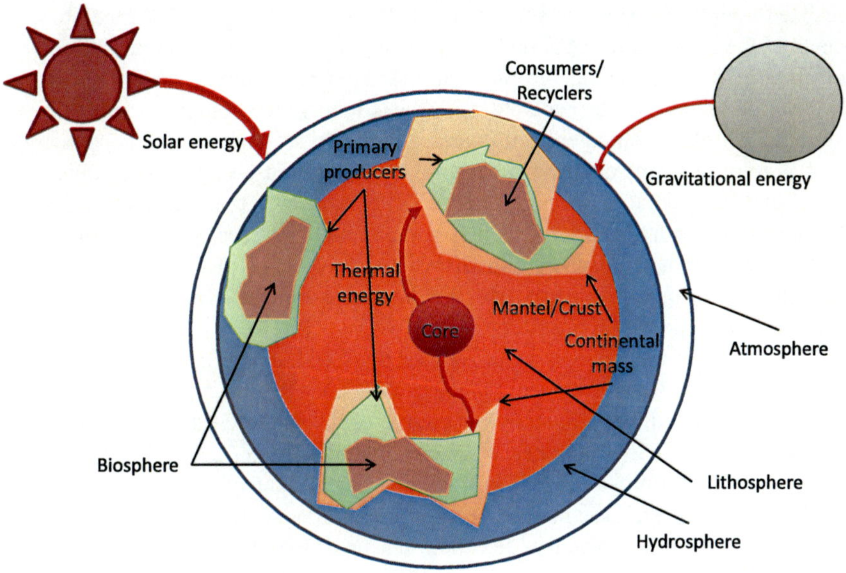

Fig. 7.15 The whole Earth system

been explicated (see Fig. 7.15). We also have significant knowledge about the biosphere and how human beings evolved from earlier hominins.

With this in mind, we argue that we can situate the HSS as the SOI within the whole Earth system—the Ecos.

Other authors, notably Immanuel Wallerstein (2004) have already launched programs of analysis of the whole "world-system," defined to be the human social system and its embedding in the physical world. They treat the HSS as a system, considering its inputs and outputs from the Earth, and then decomposing its internals deriving the interactions between major subsystems.

7.6.1 Looking Down from the International Space Station—A Brief Examination of the Earth Supra-System

In a thought experiment following the logic of extracting the observer from the system observed, let's imagine ourselves as astronauts aboard the International Space Station (ISS) looking down on our planet.

The Earth's planetary system receives solar energy from the sun (Sol). Only a portion of the total solar flux makes its way to the surface of the planet owing to the filtering effects of the atmosphere. The energy that does make it to the surface drives a number of biophysical processes, from the hydrological cycle to the production of biomass. Because the potential complexity of the Earth has not yet reached fully realized complexity (Mobus and Kalton 2015, Sect. 5.4.2, page 203) the components of the surface are continually rearranging through the auto-organization and emergence cycle (the ontogenic cycle discussed in Chap. 3), bolstered by biological and cultural evolution to generate new levels of organization and complexity of systems. All of this emergence of organization is driven primarily by solar energy, both current real-time and historic or fossil sunlight in fossil fuels, with some contributions from tidal and tectonic cycles. The surface of the planet from several kilometers beneath the continental plates, the oceanic abysses, and up through several kilometers of atmosphere constitutes the active zone of life that through its activities and evolution work to dissipate the energy influx in the form of heat radiated back into space (see Fig. 7.15).

Imagine that you are an astronaut aboard the ISS, looking out a viewport on the Earth below. You see the really big picture. And, since the ISS orbits the Earth, over a few hours you will see the really big picture of the whole Earth. From that perch, you will see continents and oceans, maybe mountain ranges. At night on Earth (in the shadow side of the Sun), you will see clusters of lights, the cities of the world. However, you will not see people. There are no borders to delineate countries. It is possible, however, that you can detect the season of the year by noticing the coloration of areas on the continents. The Northern Hemisphere may seem greener during the summer months and browner in the winter. You can detect changes even with the naked eye.

This is actually a very good perspective to start a systems analysis of the systems that are of prime importance to humanity. The whole human enterprise, viewed as a system, is embedded in the Earth as its supra-system. We need to adopt a perspective that will ensure that we do not miss anything when we analyze the human system as a whole. Systems analysis is a process of digging into the details of a whole system from a preliminary perspective looking down on the whole. We have to start there in order to understand how the system is affected by its environment. We will do this in stages from the satellite view down to the subsystem view within the human system.

In Chap. 9, we will examine a very important subsystem for humanity, the economy, and demonstrate how the use of systems analysis as was covered in Chap. 6 will produce some important insights into how the world works not previously shown by classical economics.[11] In essence, our approach is to start afresh in constructing a science of economics. We will not be repackaging classical concepts from economics, such as Adam Smith's description of the "invisible hand" or supply–demand curves. We will be tackling the problem of gaining an understanding of this subsystem by starting from the high-level view of the whole human + culture system and deconstructing it without any preconceived notions of what we will find.

In this section, we will provide the environmental analysis for the economic system, the human social system. Our approach will be slightly different from the kinds of analyses previously discussed. We already know that the economy is

[11] Today, economics is called "neoclassical" economics in an attempt to ameliorate classical economics (from the nineteenth century) with modern findings relevant to modernity

embedded within the HSS and has extensive interactions with the other Earth subsystems. What we will do here is take a quick look at the whole Earth system, and then consider the HSS as a subsystem within that. This will set us up for the analysis of the economic system in Chap. 9.

7.6.2 *The View from a U2 Spy Plane—The HSS*

Having seen from the ISS that the human system can be seen from space, we now come down in altitude a bit to see some more details of what that means. Humanity is unlike any other species on the planet. It covers the planet and it has an exosomatic (outside the body) conglomeration of artifacts that dominate the scenery—it has a culture, defined as the set of artifacts, institutions, and behaviors, that has a major impact on the whole Earth. From this height, we still see large aspects of the human system but are starting to see some interesting details as well. We see that the human + culture system comprised many various subsystems.

7.6.3 *The Human Social System in the Earth Supra-System*

The seeds for obtaining a modern, systems approach perspective already exist in the work of Immanuel Wallerstein (2004, Chap. 2), which he calls "world-systems analysis." By "world-system" Wallerstein is referring to what we have called the HSS, the human social system. His approach isn't quite the same thing as a deep systems analysis as promoted by this book, but it is guided by systemic thinking and he treats the world-economy as a whole, breaking down various components, such as markets and other institutions, as subsystems in a way very similar to what is being explored here. Thus, our efforts can be foreshortened by this previous work and we refer readers to it for clarification.

The human species is a monolithic genus (only one extant species) that occupies every continent and habitable island on the planet. In modern times, with the travel and communications technology that we have, almost all human beings are in potential contact with all others, even if indirectly (since you are reading this you are likely a mere six degrees of separation from the film actor Kevin Bacon!) Though there are still groups of humans living not too differently from our Stone Age ancestors in various pockets of the world, even they are affected from time to time and in various ways by modern technological societies. It is reasonable to argue that the human population of Earth constitutes a single social unit over the long term. Figure 7.16 shows a simplified model of the HSS[12] situated in the Earth system, which we have called the "Ecos," a term derived from the Greek for 'home.'

[12]Another term we will use for naming the HSS is the Anthroposphere. This term refers to the human species, but also our cultures, artifacts, and impacts on other systems in the Ecos (e.g., the atmosphere and hydrosphere). We will use these two terms interchangeably but more often go with HSS since it involves fewer characters!

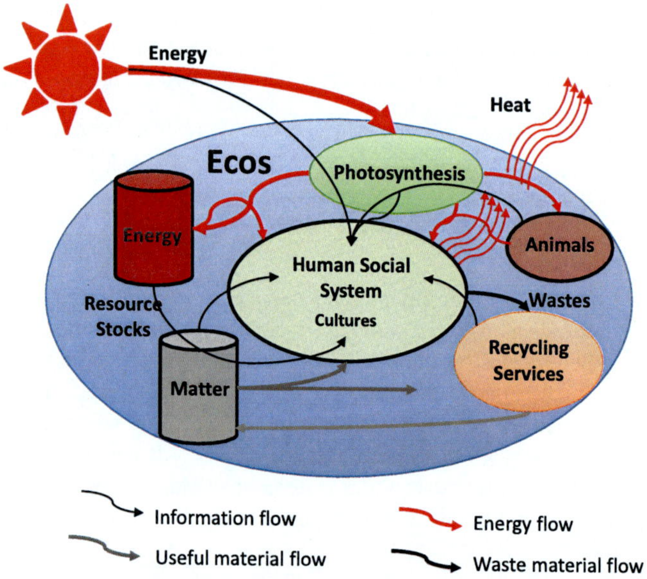

Fig. 7.16 Human society is a subsystem within the Earth supra-system, which we have called the "Ecos." Shown within the HSS is the very important sub-subsystem of "Governance." This will be the subject of Chap. 12 and will be discussed in overview in this chapter. See the text for details

The HSS is, so to speak, a fully owned subsidiary of the Ecos.[13] Our entire existence depends on the physical nature of the Ecos, and its relative stability over the life of our species. That latter point cannot be too fine. Relative stability means that the major attributes of the Ecos, the climate variations, the pH of the oceans, the tectonic activity of the mantel, the major gas and hydrological cycles, and many more, must operate within narrow ranges that are conducive to life. The planet Earth is, perhaps, uncommonly fortunate to be the right mass and composition, situated in the "Goldilocks" zone in the solar system, and orbiting a star that has been relatively stable itself over the four and a half billion years since its early condensation from the debris ring circling the sun.[14]

The Earth is effectively materially closed as a system. However, it is open, within the spectral window of radiation supplied by the sun and filtered by the atmosphere, to energy flows. That spectrum turns out to be optimal for driving photosynthesis as the primary basis of all life.[15] The flow of energy through the Earth system is the

[13] A phrase (paraphrased) attributed to the economist Herman Daly.

[14] At the time of this writing a number of Earth-like planets have been discovered in orbits around their stars that put them in their Goldilocks zones. By the time you read this, we may know if any of those planets harbor some form of life!

[15] Living systems derive primary energy from multiple sources. At present we have discovered that deep ocean hydrothermal vents provide a source of energy for a whole ecosystem that extracts energy from hydrogen sulfide spewing from the vents.

source of organizational motivation (doing work) that drives biological evolution (Morowitz 1968, 1992, 2002; Schneider and Sagan 2005).

The HSS depends entirely on the material and energy resources afforded by the Ecos. All of the biological resources (food, wood, fiber, etc.) are renewable from the flows of solar energy being transformed into biomass through photosynthesis. Our current reliance on hydrocarbon and carbonaceous fuels is based on stocks that were stored in the Earth's crust in the distant past. They too came originally from biological sources that were buried in sediments and "cooked" into their rich carbon and hydrogen compacted energy sources (c.f., Crosby 2007, esp. Part II). The atmosphere and hydrosphere, along with soils, provide the gasses and water necessary for life. The tectonic activities of the Earth's crust provide a long-term recycling system that has produced the minerals that humans have come to rely on for metals and other elements that allow us to develop modern technologies.

The Ecos is all we get for the time being. We have to find out what it needs from us (the HSS) in terms of a purposeful product.[16] We have to find out how to live in balance with the various cycles of resources and waste absorption that act over many different time scales.

Our sciences have made significant progress toward acquiring this knowledge. We have the broad outlines of what we should be doing to live successfully on the planet. In some cases, we have very definitive knowledge. However, our knowledge of the real workings of economics as the fabric of our HSS contains many "beliefs" that are in contradiction with scientific knowledge. For example, the neoclassical belief that an infinitely growing economy, as measured, for example, by year-over-year percentage increase in the Gross Domestic Product (GDP), is the ideal case.[17] This stands in stark contrast to the scientific fact that no system can grow indefinitely in a finite world. Every known biological system reaches a maximum size consistent with its access to flows of resources and its interactions with the rest of its environment.

Figure 7.17 translates the depiction in Fig. 7.16 into the graphic systems language. The sources of our resources are shown as environmental entities along with the waste and heat dumps used. Resources that are technically renewable, for example, plant and animal inputs, are dependent on solar energy and as living systems, they are continually renewing their biomasses as energy flows through the ecosystems that support them (including farms). They are technically renewable but only if the draw-down rates are no greater than the production rates. Otherwise, these resources become "flow-limited"; they cannot provide more resources than the rate of energy flow and the recycling rates for their material inputs.

---

[16] Recall the discussion of purpose in Chap. 3. The long-term stability of a system depends on the fact that subsystems fulfill a function that contributes to the whole system.

[17] This is the case of exponential growth or compound growth. The latter sounds great when you are talking about money in the bank, but a very simple calculation would quickly show that the concept is absurd in the limit.

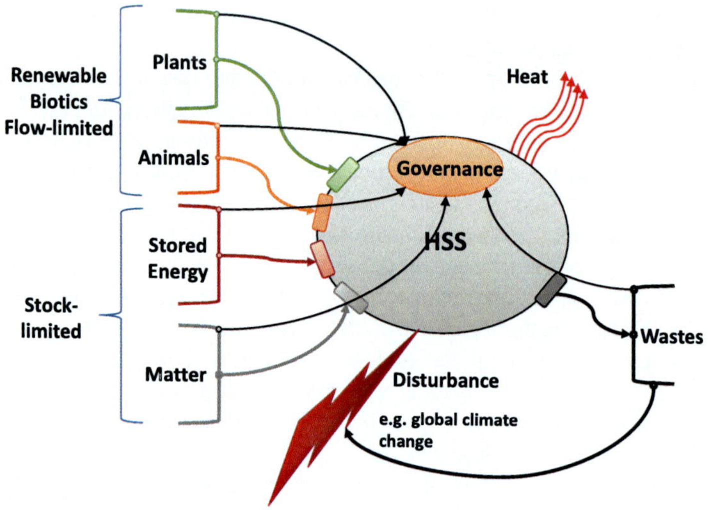

Fig. 7.17 The system language (graphic form) view of the HSS as the SOI. The Ecos has been reduced to those resource inputs and waste outputs in the environment of the HSS. As in the previous figure, we show the governance subsystem receiving messages from the external systems. The governance subsystem will be using the information in these messages to make decisions on what the various internal work processes in the HSS should be doing. Also shown is a representative disturbance (global climate change) resulting from HSS waste disposal (CO_2) into the atmosphere

Also shown in the figure is the character of the stored resources of energy (hydrocarbon fuels) and materials, such as ores and minerals. These are what are called "stock-limited"; they can be used up.

The figure is a cartoon representation of an environment and boundary analysis aggregating all inputs and outputs of each type into single arrows as well as sources and sinks being aggregated. The next step would be to begin deconstructing the environmental entities and the flows into individuated entities and flows. For example, the stored energy source would be divided into, say, the three major fossil fuels. The interfaces on the boundary of the HSS have likewise aggregated representatives of the means by which the flows pass the boundary.

In the case of fossil fuels, the supply is considered essentially fixed and definitely finite. Material resources are limited by the quality of the matter. For example, ore quality has to do with the concentration of the desired resource in the rocks from which it is extracted. The physical processes of the lithosphere that tend to produce concentrations that are valuable operate over geological time scales. Therefore, if the rate of extraction exceeds the rate of replacement, then these materials are also flow-limited. This does not consider the possibility of recycling the material, which we will consider later.

Notice what is not shown in the figure. There is no "product" or "service" output from the HSS returning something of value back to entities in the environment. The reason it is not shown is that there would seem not to be such an output, at least not one that is sufficiently significant. Recall from Chap. 3, the section on Purpose, that every subsystem within a supra-system has a function to perform that provides product or service to other subsystems such that they are fit for their environment (the environment will provide useful resources through feedback loops or select against the subsystem that fails to meet its obligations). The HSS, both in its extraction of raw resources from the Ecos and depositing its wastes (many of which are completely foreign) in the Ecos, is changing the world in ways nobody intended when they started doing so. The current scale of the HSS and the rate of its extraction/deposition processes are so great that geophysicists are seriously considering naming the current period the Anthropocene.[18] The HSS is producing effects in the Ecos, but as of the present, such effects seem largely negative.

What the HSS does for entities in the Ecos, such as the plants and animals (the biosphere) that might pass as some kind of service is to attempt the protection of some "hot spot" ecosystems in order to preserve species diversity. But this is really just mitigation of damage to ecosystems already done; a reaction to perceived damage. The "management" of certain natural resources, as in various national parks and forests, is another attempt to provide a service, but it is currently based on fairly weak models of what that management should entail. For example, forest management, in the recent past, has included preventing fires that would burn out the understory fuels on a fairly frequent basis on the fear that any fire would be bad for the forest. The result was a buildup of those fuels such that when a fire did get started, it would burn the trees worse than had they allowed many small fires to burn. When fires ran through forests under nature's management, the understory was kept low and thin so that major fires could not destroy square miles of forest trees. The good news is that human foresters are learning this about forests and there is a shift in fire management practices.

What product or service *should* the HSS produce that would benefit the entire Ecos? Do wastes from the HSS constitute some kind of "product" of value to the Ecos?

Figure 7.17 also shows the flows of messages from entities in the environment to the governance subsystem of the HSS. The latter is considered as the agent that makes the decisions about how the HSS should behave relative to the resources and waste dumps. The messages received provide some information regarding the quality and capacity of the resources and their flows. In theory (see Chap. 12), the agent will regulate the internal operations of the system in order to not cause stresses in the supra-system. As we will discuss in Chap. 12, we humans have not done a stellar job of this function so far.

In Chap. 9 (see in particular Fig. 9.1), we will decompose the HSS SOI into a few fuzzy subsystems. One of those systems is the economic system. In that chapter, we will then further decompose the economic system as we find it today (though we will consider its history as a combined intentional and evolutionarily designed

[18] This proposal is in serious consideration and may even have been adopted by the time you read this. We will be discussing this subject in later sections.

system—see Chap. 13 for consideration of human design processes). Since our objective is to demonstrate the analysis of complex systems and not to necessarily explicate all aspects of the HSS economy, we will only begin the decomposition process in that chapter. We will, however, return to the idea of designing a well-working system (a better social system as a sustainable complex adaptive and evolvable system) in Chap. 16 where we will use concepts discussed here and in Chap. 9.

The HSS has already left its not-so-positive mark on the Ecos. The proposal to rename this era the "Anthropocene" is based on the deposits of human artifacts throughout the world. Currently, the distribution of plastic particles provides a telltale marker of non-biological processes that may prove to have biological consequences. The fact that we humans are recognizing the possible consequences is heartening. If we can come to deeply understand the governance subsystem for the HSS (Chap. 12), we may find a way to moderate our collective behaviors if not actually find a purpose for our existence in the Ecos.

This section has presented a preliminary approach to the analysis of the HSS just to show that such a deep systems analysis is, at least in principle, feasible. Already, the social sciences have made inroads in their siloed analyses (see: Bourke 2011; Miller and Page 2007; Mobus 2018; Polanyi 2001; Rothschild 1990; Sawyer 2005; Scott 2017; Simon 1957; Tainter 1988; von Neumann and Morgenstern 1944; Wallerstein 2004 for a sweeping overview of findings in a number of these silos).

7.7 Summation

Everything in the Universe that can be described using the formulation given in Chap. 4 (Eqs. 4.1, 4.2, 4.3, and subsequent) is a system and can be analyzed accordingly. The methods of deep analysis, meant to lead to a deep understanding of real systems can be applied to any kind of system, though one can argue quite reasonably that the ordinary sciences have done and are doing adequate analysis of the phenomena they investigate, but this is based on the fact that most such phenomena are relatively speaking, simple or merely complex. The deep analysis procedures given in the last chapter are not significantly different from ordinary scientific analysis (e.g., dissecting a formerly living system and probing the physiology of a still-living system). Ecologists have been practicing a form of systems analysis almost from the beginning of their discipline. What makes the Chap. 6 methods different are three things. First, the procedures are based on systems principles and are formally algorithmic (as opposed to the scientific method which is more heuristic). Second, the discovery process is associated with the capture of relational knowledge as it unfolds (see next chapter). And third, because of the isomorphic quality of the formulation mentioned above across all systems, these methods can be applied as well to complex adaptive and evolvable systems regardless of the level of complexity.

Chapter 8
Capturing Knowledge of the System

Abstract This chapter will take a close look at the knowledgebase that is built up from the analysis process and how it maintains the important relations between multiple components in very complex systems. The knowledgebase structure derives from the mathematical descriptions given in Chap. 4. A sample of knowledgebase tables and forms will be used to show how to capture the information about the system from the analysis. We build a basic knowledgebase model using an existing database management tool to show how the indexing scheme works to maintain relations of components.

The knowledgebase contains all knowledge about systems, their subsystems, and all structural relations found in the system of interest. From this knowledgebase can be derived models and simulations as well as designs for artifactual systems (Chap. 14).

8.1 Why We Call It a Systems Knowledgebase

A database is an organized collection of data from which users make queries in order to get information. The organization of the data is generally based on specific uses. For example, an accounting database contains a variety of revenues and costs[1] by time-date stamped transactions (debits and credits). Routinely the accounting system produces profit-and-loss statements, among other summaries, that tell the management how the organization is doing with respect to its operating and financial goals. This counts as information derived from data by data processing programs. The further advantage of a database management system is that one can formulate special or ad hoc queries that are not necessarily tied to the overall organizational goals. For example, the Accounts Receivable department would like to know which of their customers are in the arrears in payments by more than 3 months (or any time range) in order to take appropriate actions to get the deadbeats to pay up. Marketing may pose an ad hoc question about customers that had purchased the

[1] In the form of things like Accounts Receivable and Accounts Payable, among others.

G. E. Mobus, *Systems Science: Theory, Analysis, Modeling, and Design*,
https://doi.org/10.1007/978-3-030-93482-8_8

most of a product, or how much of a product had been sold in the last year. A database management system (DBMS) allows such questions to be formulated and processed.

The example of the use of a DBMS to hold accounting data shows how the accumulation of data can be a priori organized to produce information. But should we call data knowledge?

By itself, the data cannot be considered knowledge. It is the organization of the "schema" of the database that makes it useful in the generation of information. The structure supports the efficient use of the data in a rule-based way that serves decision-making. The Greek-derived word, *episteme*, conveys this concept. According to Dictionary.com, the definition of episteme is: "a system of understanding or a body of ideas which give shape to the knowledge of that time." The key concept here is the "giving shape" to the knowledge.[2] Episteme, in our thinking, refers to the structure (or the "system") of the body knowledge and not just the facts or data. That structure has to capture the actual, functional relations between the facts and data. So, in the somewhat limited world of accounting, the database management system and the schema of the data as the "books" of the organization make the contents knowledge of the current state of the financial situation.

This will be the argument for systems knowledge. We organize system facts into a schema for system structure. In fact, that organization is based on the set of equations in Chap. 4 that collectively define a system and its components, especially in relation to its environment.

8.1.1 The Ontological Status of Knowledge

In Chap. 3, we made a bold claim that something we called "Knowledge" was a first-class ontological substance. We equated knowledge with the structure/organization of material forms. And we argued for how knowledge could change as a result of the flow of information. To recap,[3] when a system receives a message, the encoding of message states is accomplished in the modulation of energy levels per unit time and messages are conveyed serially (i.e., as a time series) through channels or broadcast media. The encoding is done by the sender. Upon receipt those energies may be amplified by virtue of the receiving system being a priori prepared to interpret the meaning[4] of the message; that is, all of the machinery for reacting to

[2] The addition of "of that time" seems redundant in that all we have at any point in history is the knowledge of that time.

[3] For an extended explanation see Mobus and Kalton (2015), Sect. 7.2.1.9 Codes, pages 274–275. Chapter 7 is devoted to explaining Information and Knowledge and their relations.

[4] As explained in Mobus and Kalton (2015, Chap. 7), meaning is based on how a receiver interprets the information in the message.

the causal influence of the message is already in the receiver,[5] The amplified energy flows contribute to additional work on the receiving system's structure, it is altered in such a way that future messages of the same kind do not result in the same amount of work; the receiver becomes more dissipative as a result of the changes in structure resulting from the work done on it. That change constitutes a change in knowledge, or the preparedness of the receiver to deal with future energy flows.

Thus, knowledge, like information, is not separate from matter and energy. Rather, it is encoded into the systemic structures of matter that allow energies to flow from high potentials to low potentials with a minimum amount of disruption to the structure of the system. Knowledge, as structure, makes a system more resilient, stable, and sustainable over time and fluctuations in the ambient conditions.

Informational messages report changes that have occurred in the receiver's environment and result in the receiver altering its own structure such that future messages reporting the exact same changes are less informational. In other words, the first time a specific message is received reporting (via its encoding) a change, the degree of that change corresponds to the amount of information in the message. If the message is repeated (after the receiver has, in fact, changed its own structure accordingly) then no information is conveyed and no additional changes in structure are done in the receiver.

The claim, thus, is that the actual structure of any system at a given point in time constitutes its knowledge of what to expect from its environment. This is why both information and knowledge are characterized mathematically in terms of probabilities of states of systems (either the sender or the receiver). A receiver expects that communications from its environment (linked through message interfaces across its boundary) will confirm its own encodings of likely states of the entities from which it receives messages, as in "I'm still in state X." That kind of message has a high likelihood of being received by virtue of the fact that it has been the most frequent message previously received.[6] Receipt, in the next time frame, of "I'm now in state Y," had a lower expectancy and so conveys more information. The receiver has to alter its own state to reflect this situation, that is, to increase its expectancy that future messages will more likely be about state Y than state X; hence, we say that the receiver has learned something about the sender, namely its new state.

In a dynamic world, where systems are continually changing their states and sending messages to other systems, the flow of information and the generation of knowledge is forever operating. Indeed, since systems of sufficient complexity can adapt

---

[5] Senders and receivers co-evolve over time owing to the history of messages sent and received. As such the receiver is a priori prepared for the kinds of messages (i.e., the average symbols in a message) that the sender might send. When the sender sends a message that has a low probability, this triggers a reaction (work) in the receiver so as to modify its own structure (Mobus and Kalton 2015, Chap. 7).

[6] More explanation for the process of representing expectations and changing them as a result of informational messages is given in Mobus and Kalton (2015), Chap. 7, Sect. 7.3.3, pages 289–296.

and evolve (particularly the latter) we assert that knowledge and information are that which grows in amount as the Universe evolves. That conjecture begs for proof.

8.1.2 Knowledge Accumulation

Systems that have mechanisms for combating the effects of entropic decay in their structures (i.e., are autopoietic) are able to encode and maintain knowledge as it is gained from the work generated by the receipt of information. They actively work to not forget what they have learned. The human brain is exemplary in the animal domain. Our memories encode not only the current states of the world, but our capacity for episodic memories constitute a record of the time series of events that led from prior states to the current state. We remember the stories of how things got to be the way they are. We remember a history even if imperfectly.[7] Human long-term memory results from the way in which neurons maintain synaptic strengths along with circuit designs that periodically refresh the stimulating inputs that promote excitability in specific pathways (speculated to be one of the functions of dream sleep) of memory traces that have proven useful in mental activities (i.e., have resulted in rewarding feedback).[8]

Computing systems employ a hierarchy of memory devices that range from highly volatile memory cells (CPU registers and main memory) to recordings of state information in magnetic domains on a quasi-permanent substrate (hard disks and tapes). While the computer is powered, the memory system does considerable work in maintaining the states of memory cells. When the power is turned off, these dissipate and the overall state configuration is lost. The genius of modern computer design, however, is that the state of the volatile components can be preserved in the magnetic substrates between power states so that the long-term memory preserves the last state of the working memories. Even these preserved recordings are subject to entropic decay effects, but over extremely long time scales relative to human interests. And, with the right sort of maintenance mechanisms in place (periodic

[7] It must be noted that while it is true our memory systems allow us to capture and maintain a trace of history as we perceived it happening, those same systems are notoriously faulty with respect to the fidelity of details, even, often, actual sequences of events. For example, we now understand that eyewitnesses to crimes all too often make faulty reports about perpetrators (e.g., identifications in police line-ups) or events (as in sworn testimony in court). The source of these faults can best be understood when we realize that our memories are not recordings in the same way that, for example, a digital recording of a movie is done. For a very excellent overview of the nature of memory in the human brain by two of the premier investigators of the phenomena see, Squire and Kandel (2009).

[8] A simple example is the procedural memory capacity for riding a bicycle. Though no one knows yet quite how this is accomplished the human brain retains the motor memory for doing so even when long gaps in riding occur. Contrast this with memory aspects in much more primitive brains where conditioned responses decay over time (see: Alkon 1987 for seminal work on nudibranch memory).

refreshing), as with autopoietic mechanisms in living systems like the brain, knowledge encoded can be preserved indefinitely, at least in principle.

The pace of scientific discovery is higher today than at any time in history. The accumulation of knowledge is staggering. But accumulated knowledge is only useful knowledge if it can be readily retrieved (see below). The rate of accumulation, especially with the traditional means of storage raises severe issues.

In 1956, Kenneth Boulding recognized the need to establish a more holistic systemic way to manage knowledge, then obtained by the sciences using reductionist methodologies, for the most part, and maintained externally from human memory, that is, the library. Speaking to the architecture of knowledge storage and retrieval mechanics, he observed:

> … the problem of the adequate descriptions of complex structures is still far from solved. The theory of indexing and cataloging, for instance, is only in its infancy. Librarians are fairly good at cataloguing books, chemists have begun to catalogue structural formulae, and anthropologists have begun to catalogue culture trails. The cataloguing of events, ideas, theories, statistics, and empirical data has hardly begun. The very multiplication of records however as time goes on will force us into much more adequate cataloguing and reference systems than we now have. This is perhaps the major unsolved theoretical problem at the level of the static structure. (Boulding 1956, 296–297)

Boulding (ibid) noted further that progress toward a solution was being made and also noted that this applied to the kinds of instruments scientists were employed to obtain knowledge as well.

In 1956, our methods for cataloging and indexing knowledge were indeed clumsy, as toddlers are clumsy when they take their first steps. The modern era of computer-based systems and massive capabilities as offered by Google and other masters of storage and retrieval offers the opportunity to establish a much more systematic approach to knowledge management. Several factors need to be addressed.

8.1.2.1 Knowledge Encodings

Knowledge encoding refers to the mechanics of changing system structures to reflect the prior receipt of information. There is a significant difference between the way the brain and a computer encode knowledge regardless of the storage mechanisms. Even so, early research on the differences in the methods points to promising ways to translate knowledge between the two. In this writing, there are several efforts to emulate human knowledge encoding (as in neural circuits) in digital forms. In principle, this is not unexpected. But it is not a trivial problem either. Meanwhile, several research groups are tackling the problem of how to do in computers what brains do in real life (Hawkins 2004).

The basic problems involve the representation of knowledge and its storage/retrieval. In computer science, the latter aspects have been dominated by algorithmic approaches to efficient methods for doing these tasks. In the history of computer science, the methods have been motivated by a straightforward notion of representation. Namely, a data item is simply encoded in bits stored in memory at any level in the persistence hierarchy. For example, an employee's number, name, pay rate, and other relevant data are encoded using binary representation codes like ASCII[9] to directly represent the data. The data are then organized into a relation called a record that can be efficiently written to an external medium (hard disk file). Retrieval involves finding the data given a key, like an employee number, which might involve a search through a long list or, more sophisticatedly, using something called a hash code for the employee number to go directly to that record in the file.

Even though the brain doesn't operate in this manner, the speed of computation has reached a point where the search methods of the algorithms compete favorably with the capabilities of the brain (which performs a massive parallel search, but at the slow rate of neural signaling!)[10] There is, today, a good reason to believe that even though the computer operates on different principles with respect to storage and retrieval of data, that it has the capacity to emulate the brain such that it becomes a powerful mechanism for capturing, storing, and retrieving information beyond the capacity of any one human brain. This is the basis for claiming that a computer-based knowledgebase can be an important tool in systems understanding.

8.1.2.2 Knowledge Representation

A fundamental problem in using computers to represent system knowledge in a way that allows ready use of that knowledge (i.e., recall and processing) has been the methods of representation, or how knowledge is organized in a memory structure in a digital system.

In Chap. 4, we presented a formal definition of the system. We also claimed that this definition is based on an ontological primacy of the concept of systemness that through the evolution of intelligence and consciousness has resulted in our (humans') capacity to think in terms of systemness. The language of systems was claimed to be based on this fundamental idea.

What we assert here is that the representation of knowledge, which is at the base about the systemness qualities of objects in the real world, must be based on those qualities. In other words, knowledge of anything should be based on the definition given in Eq. 4.1. In this chapter, we derive a knowledgebase structure, based on that definition, to be captured in a computational structure, emulating system knowledge. We present, here, a computational method for encoding and representing system knowledge as it is derived from the analysis procedures presented in the prior two chapters. The importance of this representation will become much clearer when we tackle the problem of generating models (Chap. 10) and how those models impact our grasp of policies (Chap. 12) and system designs (Chap. 14). In this

[9] American Standard Code for Information Interchange.

[10] Additionally, the speed of modern computer hardware is so great that it would seem to more than compensate for the slowness of biological neurons. The jury is still out on that issue, however.

chapter, we will use Eq. 4.1 (along with its derivatives in Chap. 4) to create a database schema that will be a usable representation of system knowledge about any kind of system. The result of this representation will be ready access to system knowledge and especially, the generation of system models for understanding, design, and policy generation.

8.1.2.3 Knowledge Retrieval

The structure of how knowledge is stored in a persistent medium plays an important role in how accessible it is for retrieval and use. Think of the library.

There are many forms of stored knowledge, different formats, and media (e.g., books, journals, microfiche). All of these are referenced through an index system (as noted above) that records key aspects of the knowledge (e.g., titles, authors, dates) and a location code so that one can go to the location in the library where the material is stored. Assuming that it hasn't been checked out, the researcher can then physically access the knowledge.

The library stores, in an organized way, what we should call potential knowledge. By our definition of knowledge in Chap. 3, knowledge is a structure but it is also relevant to an effective process (causal). The knowledge structures in a library are non-effective until they are transmitted into the minds of readers and cause a change in those minds. The researcher has to retrieve and read the resource in order for it to produce knowledge in their minds as actors and agents. The retrieval process is actually informational with respect to the researcher. The knowledge potential stored in the resources and organized in the library does not actively suggest itself to the researcher.

The methods of indexing the resources were primitive. Once, not long ago, the constrained information about the resource, title, etc., which could be called "meta-knowledge" was available on a cardboard index card in a file system. An indexing scheme, like the Dewey Decimal system, was used to aggregate collections according to subjects (then sorted in authors' last name order). One needed to first understand how the index structure was segregated by subjects and which leading index codes related to which subjects.[11] Then one had to guess as to what the contents of a resource might be, unless one were pointed to the resource by name from another resource (the bibliographies and references, for example, in journal articles or books). In the former case, one had to be willing to pay the cost of effort in retrieving the resource, reading it, and then determining whether or not it was relevant (produced knowledge in one's brain). In the latter case one could retrieve the work

[11] As much as the division of knowledge by academic departments and their rules for junior professors to gain tenure by narrowing their research to very specific topics (and the complicity of journal publishing) the Dewey indexing scheme is a culprit in creating knowledge silos. No one is to blame for any of this, of course. It was a natural consequence of trying to come up with a means to manage knowledge storage and retrieval in a "systemic" way.

but probably would be using it more to confirm already gotten knowledge, that is, it would not be informational.

Today we still basically use the same system but now our index is managed in a computerized system. This newer system permits a bit more information to be supplied within the index itself, such as an abstract of the work. An even better method for finding relevant work is supplied by online journal access. One still needs to know which journals (by name) contain articles most likely relevant to their interests. But the search for, and preliminary information about the work has been streamlined, as well as has the retrieval process itself. Today a disciplinary researcher, in many fields, need not actually visit a library except to pick up a book they ordered from the online access platform.

With the advent of the technology of search engines, the retrieval of resources has become more automated. The researcher can now aggregate a set of what are keywords to them but might not necessarily have been identified as such in the source document. Rather it is the words themselves from the body of the text that is searched and indexed in a database of resource addresses. Using a web browser, the researcher can google[12] their words of interest and get back a list of resources that contain those words (in logical relations such as AND or specific order).

Even so, with all the technological improvements in search, what have we actually gained? The fundamental problem of canalization, keeping knowledge segregated in disciplinary silos, still persists. Knowledge has become compartmentalized in a way that makes interdisciplinary inquiry increasingly problematic. While a search engine might bring us documents from a wide variety of subjects based on the keywords we put together, they can also bring a flood of documents that do not necessarily report on transdisciplinary work.[13] On the other hand, the rise of interdisciplinary research, driven by the need to understand systems more broadly, has begun to generate resources that address this need. There are now journals, for example, that cut across disciplinary topics to reflect the real needs for understanding from a more generalized perspective. This is a positive trend, so long as the quality assurance mechanisms, like peer review, are upheld strenuously. But it does not address the larger problem of how to organize knowledge more globally so that it is readily available from any number of perspectives. The usefulness of knowledge depends, ultimately, on how it can be retrieved not just from one perspective, no matter how interdisciplinary it might be, but from all perspectives.

The knowledgebase aggregates knowledge, structured by the definition of the system, such that it is available from any desired perspective. As an example, consider the new field of bioinformatics. As the name implies it is an attempt to organize biological data such that it can be retrieved and used from multiple perspectives (e.g., biological, engineering, statistical, and others). This is a step in the right

[12] It is interesting how words can morph in the language. What started out as the name of a search engine company, Google, has morphed into a verb describing the action of making a search.

[13] It is incumbent on the researcher to be very thoughtful in choosing keywords in order to minimize the flood. Fortunately, most search engines use some form of AI content analysis to rank returns according to what the AI "perceives" as the user's interest.

direction. Data have to be organized by a semantics that is general in order to be accessible thus. On the other hand, if it becomes too general, that is simply accumulated such that anyone can access it from anywhere the burden is on the intelligence of software to make it accessible and impose some particular perspectives semantics on it.[14] This is also the problem with the World Wide Web and why search engines are only a partial aid in making sense of the distributed data. Bioinformatics starts with a more structured database schema based on general biological understanding. For example, genetic sequence data can be readily linked with cellular and tissue development, and that with biological functions such as protein activity. Even though there are specific uses (research questions), say for example, searching for specific gene sequences in a species' genome. The use can come from a variety of perspectives that have in common the concept of genetic inheritance, for example, proteomics (proteins derived from genes), genealogy (tracing heritage), or evolutionary biology.

8.1.3 Knowledge Use

Knowledge is not knowledge unless it is useful! And, ultimately, that use comes from having a perspective and asking a question. For example, in the case of bioinformatics just described above, an evolutionary biologist may be able to discern the difference between convergent evolution from a case of dispersed species (Losos 2017). Significant new insights into the dynamics of evolution that indicate it is not strictly a historical happenstance have been obtained through examining the cladistics of various species that resemble one another but turn out to be derived from very different last common ancestors (convergent evolution).

Data become useful when it is transformed into knowledge by (1) being stored in structures that are amenable to multiple perspectives, (2) having mechanisms that are able to generate constructive views based on those perspectives, and (3) providing answers to questions posed by the multiplicity of perspectives interested in what it all means. In other words, a knowledgebase is only called that if the underlying data are readily transformable, first into informational messages, and then into mental representations that affect agent actions. They support decisions among researchers about what to explore next.

The knowledgebase that we accumulate represents our best understanding of any system we analyze. This can be tested. From our knowledgebase we can generate models of the system at any desired level of abstraction, including models of subsystems (at lesser levels of abstraction and more detail—see Chaps. 13 and 14).

---

[14] Data mining using statistical methods to discover patterns in large data sets is an example. The data schema for these sets is based on particular uses such as a customer database for keeping track of, say, accounts receivable. Marketing can use mining on historical data to look for trends a posteriori.

It is important to understand that this is what we are already doing in the disciplinary sciences. What we are not doing is generating these models from a system-based knowledgebase. Rather, we generate models of systems from our ad hoc and fragmented knowledge of systems of interest based on disciplinary details. This process has served us well enough in the disciplines when what we were interested in was how specific systems worked. But as we are increasingly concerned with larger and more complex systems of interest, say how human social systems interact with the environment, the traditional approach breaks down.

What we seek now is to grasp large-scale, complex adaptive, and evolvable systems like the whole human social system in the supra-system Ecos. And the normal science process is not geared to tackle that kind of problem.

Alternatively, if we start to collect and organize knowledge of the world in a true (i.e., systemic) knowledgebase, then that knowledge will become more useful for our purposes. Our claim is that by recognizing the universal patterns of systemness across disciplinary knowledge domains, we can more effectively formulate questions relevant to any discipline but more importantly any transdisciplinary inquiry.

8.1.4 The Knowledgebase Structure Fulfills These Needs

The argument advanced here is that a database schema based on the structure represented in Eq. 4.1 (and the subsequent equations in Chap. 4) provides a way to meet all of these requirements. It provides a mechanism for converting information gained from systems analysis into a knowledge structure (accumulation). It provides the template for how knowledge is to be encoded in a digital medium. It shows how to represent knowledge in a consistent, systemic manner. It provides a ready mechanism for retrieval of system knowledge in the form of maps, trees, and other mathematical representations that lend themselves to building models and generating designs. It is, in other words, imminently useful. Recognizing how the phenomena (things doing stuff) we examine are systems and capturing their parts and relations based on our definition of the system provides us with a universal basis for understanding.

As importantly, the technology of structured knowledge storage and efficient retrieval is at hand. Consider two examples of massive, online methods in popular use today.

8.1.4.1 Google™ Products, Especially Efficient Search on a Dynamic Graph

The enterprise that has caused us to make a verb out of a noun that didn't even make sense when its first product was introduced as a whole constellation of products based on the Internet, the client–server platform now called the "cloud.[15]" Its

[15] Cloud computing involves constructing massive server farms (banks and banks of powerful server-grade computers) with high bandwidth communications to the Internet. Not only are data

original product was an online search engine that canvased the World Wide Web (WWW) for linkable files, indexing every meaningful word and phrase,[16] and then providing extremely clever algorithms that let users formulate searches for documents based on those keywords and their combinations. Over the years, with massive amounts of research into tools, such as artificial intelligence, that product has achieved a tremendous capability for finding many different kinds of resources in what for most seems like the blink of an eye.

Today the Google search algorithms work hard to report back the most relevant resources to a user. Not that long ago, a user might have to scroll through hundreds, if not thousands, of links trying to establish the relevance to themselves just from the titles. But by maintaining an internal knowledgebase of relations between resources, using historical data on past retrievals that users actually retrieved, and other hidden but systemic mechanisms (including extensive data on the users themselves), the likelihood that the top five links reported in a search are going to be exactly what a user was looking for is quite high.

One major problem with resources stored in the WWW, as well as social media platforms, is the lack of verification or critical review of the contents. Librarians spent no small amount of their time filtering resources and verifying legitimate content for non-fiction work. Google has tried to mitigate this problem in the electronic medium to some degree by classifying some content by categories that have various levels of reliance. For example, Google Scholar™ handles scholarly articles linked to reputable journals and researchers. They have imposed more useful organization on the mass of knowledge that helps to make it truly knowledgeable.

Another very popular product for Google is Google Maps™. The company has linked cartography, GPS, and a massive database of locations such as restaurants and gas stations so that users of smartphones, for example, can retrieve information about where they are, where they are wanting to go, and what they will find when they get there. The product is truly a wonder of computing technology usage. At the time of this writing, according to the website, Mashable, "Combining satellite, aerial and street level imagery, Google Maps has over 20 petabytes of data, which is equal to approximately 21 million gigabytes, or around 20,500 terabytes." That is some serious data storage! Recall from the last chapter that we introduced the idea of applying systems analysis to the human social system and started by thinking about how it is a subsystem of the whole Earth. Google Maps seems to have started laying the groundwork for that project.

---

stored, but various applications such as word processing are also available making it unnecessary for individuals, or even companies, to own their own copies of software.

[16]Meaningful in the sense that they are not common or helper words, like "the" or "and." Meaningful words are cues to the meaning of the content of a document. They are often singled out as "key" words, for example, in scholarly texts.

8.1.4.2 Wikis: Hyperlinking and Collaborative Documentation

A wiki is a special kind of WWW application that allows multiple users to edit pages in order to support a community of practice (and knowledge) to capture their expertise for access by all. Wikis are in widespread use for collaborative projects, but Wikipedia and its sister projects, such as WikiBooks and WikiQuotes, use the same wiki platform to accumulate and organize the contents of a wide variety of knowledge domains.

One of the most powerful aspects of wiki technology is the use of hyperlinks embedded in the pages, which enable researchers to rapidly look at related information. A wiki page can be edited by a collaborative group. The Wikimedia platform has developed an extremely sophisticated set of tools for making this efficient for distributed contributions. The structure of a wiki page ensures that the "right" information is easily found on the pages.

What a Wikipedia-like structure represents is a mechanism for organizing knowledge in a way that makes it efficient to retrieve and useful to knowledge seekers. And it does this on a worldwide scale.

8.1.4.3 A Google-Wiki Mashup

What Google products and Wikipedia demonstrate is the feasibility of capturing and organizing massive amounts of systems knowledge. What we are suggesting here is that using the definition of a system from Chap. 4, designing a knowledgebase schema, and applying technologies similar to these two approaches provides a technical solution to how to capture, store, and retrieve the kind of knowledge needed to understand complex systems, specifically CAESs. Both organizations have aggregated experience in this problem space which could be applied to solving the problem of developing a system knowledgebase technology. They have proven that the volume of data is not really the problem. It is the organization of that data that is the key to retrieval and usefulness. In what follows, we present a preliminary vision of how system knowledge could be organized for the purposes of better understanding the systems of interest. That organization, along with the mechanisms for retrieval, provides us with the power to generate, for example, system simulation models and, where appropriate, system policy recommendations. Not only will we use aspects of distributed storage ala a Google-Wiki mashup, but we will structure that storage around a core technology, the relational database model.

8.1.4.4 A Database Backbone

Google's approach to knowledge storage is to provide a massive index mechanism while leaving files and documents distributed. The company does provide cloud-based services for the storage of specialized data. But its main talent is in retrieving files stored in Web servers, creating content-based index keys, and then retaining the

URL in its own databases so that users distributed anywhere in the world can retrieve the files of interest based on these linkages. Because web pages and other linkable files are constantly being created, modified, and destroyed (or lost) the indexes have to be continually updated. Google has developed extremely efficient algorithms for conducting regular reviews of previously indexed documents as well as discovering new documents within days (sometimes hours). Thus, the searches for highly desired kinds of documents are usually up-to-date within a reasonable time frame.

The core of all of its technologies is sophisticated database storage. We will suggest a global knowledgebase system using a database to do something very similar, store content-based index information while allowing the relevant data to be stored in any appropriate form, such as active web pages. We will also use a wiki architecture to provide a rich hyperlink structure that can be edited by knowledge analysts as they capture system knowledge.

The major aspect of our database is a schema based on the definition of a system from Chap. 4. We will organize the data in terms of the various characteristics of systemness. Recall from section KNOWLEDGE_USE above that what makes knowledge just that is its organization and retrieval. By organizing data about any system in the system schema, we make it useful and retrievable from multiple perspectives. Thus, we ensure that what we might have formerly called mere data is, in fact, knowledge that we can use.

8.1.4.5 And a Wild Speculation

Is it feasible to design a global system knowledgebase?

Recall in the last chapter that we sought to establish the context of considering the human social system, the HSS, as a subsystem of the supra-system Earth. Consider a possibility. If we treat the Earth as the "master" SOI (with the sun, planets, stars, and other space stuff as the environment) then the master index of that system would be 0. Each of the major subsystems, the lithosphere, atmosphere, hydrosphere, etc., would be 0.1, 0.2, etc. The HSS, as a unique animal-based subsystem, might be considered equivalent to any of these and be given the index, for example, 0.9. The economy, as a subsystem of the HSS would be indexed as 0.9.5, say. We will actually consider this in greater detail in the next chapter.

There is reason to believe that all of the system knowledge, which means human knowledge of the world, could be relegated to a truly global knowledgebase. The Internet provides a basic communication fabric, the WWW a basic communication protocol, Wikipedia a basic architecture, and a Google-like database index to tie it all together and provide the core structure based on systemness.

Imagine wondering what the term for the power plant in cell metabolism is (mitochondrion) and doing a search for a standard system concept—power capture and conversion, a concept applicable to all systems—along with the words 'living cell'. A Google search of "power capture and conversion" + "living cell" produced, among several, the website: Molecular Biology of the Cell. fourth edition: https://www.ncbi.nlm.nih.gov/books/NBK26882/, accessed 11/3/2017. That site provides very good information regarding power capture and conversion for animal cells (metabolism). Another site: https://www.scientificamerican.com/article/plants-versus-photovoltaics-at-capturing-sunlight/, looks at a comparison of photosynthesis versus photovoltaics, an interesting topic perhaps, but not cogent to the subject of how living cells capture and convert energy from sunlight per se.

Imagine further that you are curious about all the different kinds of such devices found in living organisms.

The designers of the Internet sought an architecture that was extensible but structured so that a message sent from one computer could arrive at a destination computer anywhere in geographical space by virtue of a routing system and an address system that gave a unique identity to both devices and a way to get the messages from one to the other efficiently (as well as robustly able to follow alternate pathways when needed).

8.2 Capture and Storage of the Relevant Aspects of a System

Recall the forms we filled out in Chap. 6 as we went through the top-down decomposition of the model system? This knowledgebase is where that information got stored. In Chap. 6, we used the definition of a system (from Chap. 4) along with the language of the system (also from Chap. 4) to guide the procedure of investigation of the system's internals, turning the opaque box into a transparent box all the way down to the nitty-gritty details of the lowest levels of components—the atomic[17] components. As we did so, we captured the various kinds of system elements at each level of organization, assigned them codes and names that would give them unique identities and locations in time and space relative to all other elements. As we entered the data into the forms, it was cross-checked for duplication, syntax, lexical correctness, and completeness. And when issued an all-clear that data were entered into the knowledgebase.

In this chapter, we will expose the internals of the knowledgebase in terms of how it is implemented in a relational database structure, provide some insights into the ways in which the entry process leads to the cross-checking, and provide some additional insights into how the knowledge, once captured thus, can be accessed to answer queries relevant to the particular system having been analyzed. This includes using the knowledgebase to generate model systems, design specifications, or policy recommendations where appropriate. We will only be providing a peek at how these are accomplished from the inside of the knowledgebase structure in this chapter. In Chaps. 10, 13, and 14, we will use the mechanics of generation to extract models, specifications, and policies to show how the whole system works.

[17] Remember, the term "atomic" in this context does not mean element atoms. It means the smallest component that does not require further decomposition.

Figure 8.1, below, shows the basic architecture of a knowledgebase (KB) system. The RDBMS contains the basic data derived from systems analysis according to Eq. 4.1 (and subsequent equations in Chap. 4). Wiki pages that expand, augment, or otherwise enhance the data stored in the database are created by users during analysis (and other operations). Links (index links in the figure) to relevant pages are stored in the database for rapid retrieval. For example, when an interface protocol is specified (a protocol wiki page) it is put in the wiki structure and a direct link to it is stored with the protocol slot in the RDBMS (see Fig. 8.1). A core engine interfaces with the users (analysts and modelers) as well as the search engine and the wiki pages.

The wiki pages contain internal links that allow rapid traversal of relations in the page views. The search engine augments these links by creating a second form of indexing of other significant words in the wiki pages, such as domain-specific names of entities and processes. The core engine can also retrieve and format views such as maps and trees.

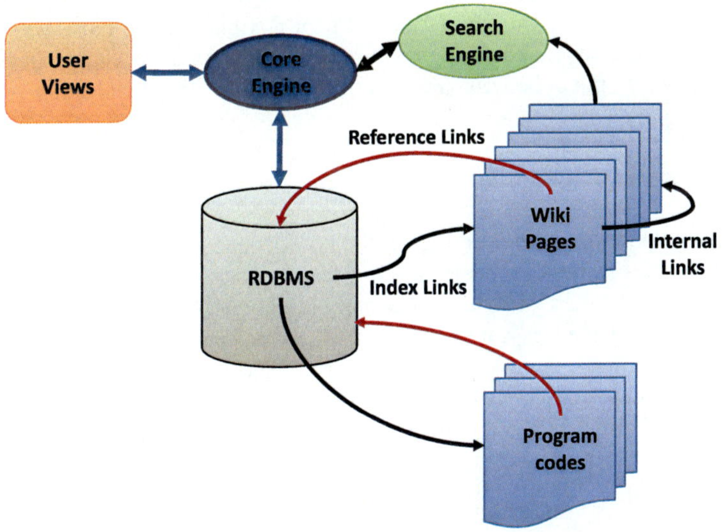

Fig. 8.1 The knowledgebase architecture includes an RDBMS with a schema based on Eq. 4.1, multiple sets of wiki pages containing descriptive and relational texts (and graphics, etc.), and a search engine supplementing the wiki pages. The "core engine" ties everything together and produces the user views (DBMS tables, wiki pages) that guide the analysis and control the retrieval of system knowledge. The KB core engine connects with the modeling core engine—see Chap. 10

8.2.1 A Relational Database Schema for the Knowledgebase

Given the system definition in Eq. 4.1, we will develop a set of relations (tables) in an RDBMS (relational database management system) that are based on the set of equations from Chap. 4.[18] The RDBMS schema captures the formal definition that gives structure to the system knowledge, which, in turn, organizes the disciplinary knowledge for a particular system. In Chap. 7, we saw how the use of the language of the system guides the collection of data for particular systems. In Chap. 9, we will demonstrate in even greater detail how that collection results in a knowledgebase of said particular system (the economic system). Now we will demonstrate how the knowledgebase architecture is used to accomplish this, from the insides.

This chapter, in essence, may be viewed as a kind of specification for the architecture of a systems knowledgebase.

You have already seen the forms that are used during analysis so the table structures, called schema, presented here should not be too surprising. In a sense, the forms simply reflect the schemas. What this chapter will do beyond repeating the data capture process of Chap. 6 is show some of the internal operations used in an RDBMS to store, retrieve, and process the data captured as part of verifying the different quality measures we introduced in Chap. 6. Chief among those was data consistency. We used that property to ensure that we did not leave 'holes' in terms of connections within or between levels in the hierarchy of the organization.

8.2.1.1 Properties to Consider

Before we expose the inner workings of the knowledgebase, we need to establish some general properties of memory systems that are important for such systems to be useful. Recent research into human memory deviations from these properties has shown how unreliable it is, for example, in witness testimony in court. We seek a memory system that is more reliable than human memory. We pay attention to these properties and check our implementations in digital forms to ensure that our knowledge of systems is reliable and useful.

[18] As often happens in this fast-paced world of technology, shortly after this chapter was written, the author was introduced to a new(ish) kind of DBMS called a "network database system," which represents data models as networks of nodes and relations. This, of course, would be ideal for representing the knowledgebase of a system since it is exactly a network of components (and a subnetwork of sub-components, etc.). However, after a survey of network database systems and their capacity to represent the hierarchical nature of networks, we decided to stick with the RDBMS structure already worked out. The use of a network database system will be the subject of future research into implementing a knowledgebase.

8.2.1.1.1 Completeness

A knowledgebase is only useful when the elements of the system are complete. What exactly does this mean? We have already seen that deep knowledge depends on a process of decomposition that reveals the lowest level of structural and functional details needed to model any system (or subsystem). This property cannot be guaranteed unless the transparent-box analysis of any subsystem, at any level of organization, is, itself, complete. This means discovering of all components in a system and identifying all of the relevant interactions (i.e., flows) that constitute the interactions between those components. The recursive decomposition of a system, as we have seen, does not lead to an infinite recursion since we can always find components that need no further decomposition. However, failure to push the analysis further down the organization tree when decomposition is warranted will lead to the incompleteness of analysis and thus errors in a sufficient description of the system of interest.

8.2.1.1.2 Correctness

This should be obvious but unfortunately is sometimes taken for granted in many current forms of "systems" analysis. The claim made here is that following the procedures of system analysis prescribed in Chap. 6 essentially forces the correctness of the data captured and recorded in the knowledgebase. The main tool for checking correctness is very similar to how a compiler checks syntax in a program. There are several mechanisms that can be built into the data capture and knowledgebase mechanisms to assure this.

8.2.1.1.3 Consistency

One such mechanism is to continually check the consistency of data entered. This can be enforced in several ways. For example, flows from sources to sinks can be checked for consistency. The output of a source has to equal the inputs to all receiving subsystems interfaces (or the sum of all receipts has to match the outputs of all sources). This is easily verified as the analysis proceeds (for example using flow graph theory). The basic rules of consistency for real systems are based on the mass and energy balance laws (e.g., conservation laws) in physics. If a subsystem is asserted to receive more input than the sum of source outputs permits, there is clearly something wrong. The data capture system can alert the analyst that a mass or energy balance equation has been violated and force a reevaluation of the flow.

8.2.1.1.4 Non-duplication

A problem in the decomposition of extremely complex systems is that entities identified at one level of organization might play multiple roles. For example, a source entity, with respect to a focus SOI, might also be a sink entity for other flows. The knowledgebase has to be able to handle such dual roles without confusion. A single entity that is, say, the source of a material flow may also be a sink for a message flow. The knowledgebase system has to be able to distinguish these differences while still recognizing the unity of the entity involved. We will provide examples of how this is accomplished.

8.2.1.1.5 Well-Formedness

A well-known problem in computer languages (and mathematics) in the form of statements or sentences has to conform to rules of form in order to be computable. Every computer language is based on a strict syntax to assure these rules are followed. Program compilers are designed to enforce said rules by throwing exceptions when the rules are violated.

Our system language is no different in that it is a formal language even though it is meant to address "natural" systems descriptions. We establish a "minimal" set of rules for how various elements of the language are construed so as to be computable (i.e., be supported by algorithmic representations). At the same time, we allow some more "free-form" constructions that provide flexibility and expressivity, making the language relate to natural languages.

Identification tags, names, type codes, and formulas must conform to a set of standards so that cross-checking (e.g., for non-duplication) can be computationally efficient. After all, the knowledgebase is contained within a computational framework. These items are "keys" that allow us to access additional information, such as descriptions and histories that are not necessarily as constrained. Examples of well-formed expressions will be provided below. And their linkage to less constrained data will also be shown.

8.2.1.2 The Schema

We use the formal definition of a system from Chap. 4 into a database design called a "schema." In relational database terms, this constitutes the set of tables and data elements. Fig. 8.2 provides an overview of the basic DBMS schema that we use to implement the intent of Eq. 4.1 in Chap. 4.

This figure varies somewhat from the way an actual DBMS schema would be implemented but this is in order to show the relations between database tables. Terms shown in bold typeface represent links to other tables. For example, the data element **Components** in the System table represent the logical linkage between the System record and a corresponding record in the Components table following the

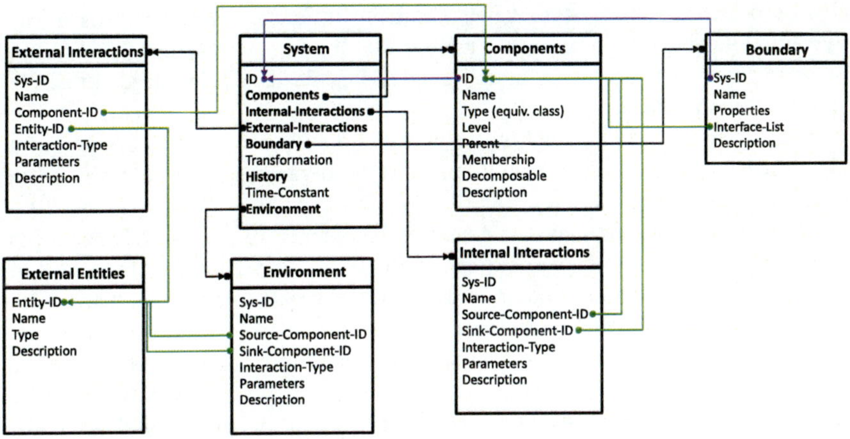

Fig. 8.2 The RDBMS tables associated with the storage of system knowledge are shown with some of the relevant links. Black arrows show table links, green arrows show internal data links, and purple arrows show identification links (secondary indexes)

results of Eq. 4.3, components at one level can be subsystems themselves and therefore treated as systems as the analysis proceeds. Various RDBMS products have some variations on their syntax as to how these linkages are made. What we are attempting to show here is a syntax-free, logical diagram of such.

This schema assumes a table for each of the components (sets and graphs) defined in Eq. 4.1. What we will now do is describe how the relations are distributed across these tables. Note that in order to make practical implementations of Eq. 4.1, we will need to add some additional relations. For example, the inclusion of environmental entities has to be addressed through a set of tables, **Environment** and **External Entities** that will provide the linkages needed to implement Eq. 4.5, regarding the graph labeled ***G***.

Additionally, note that the database schema is only a "backbone" structure for capturing system knowledge. In the examples below, we will run into examples of other data objects that are better represented in forms other than RDBMS data elements. These will be links to external objects such as pages in a wiki-like form.

8.2.1.2.1 The System-Component-System Relation

Recall from Eq. 4.2 that every further decomposable component of a system, $c_{i,j} \in C_i$, is considered a subsystem by Eq. 4.3, meaning it is also a system and therefore a member of the System table, as mentioned above. The **System-Component** tables provide this recursive relation. Any component system has an ID that reflects its membership as a subsystem of a parent system. So, for example, a component with an ID of 0.1.2 is also system 0.1.2, and since it is considered decomposable, it will be found in the **System** table with that ID. The parent systems, 0.1, will, of course, also be present in the **System** table. The main difference between items in the **System** table and the **Components** table is that the former only includes links to items in Eq. 4.1, whereas the latter includes specific data items needed to fully specify the component.

In this scheme, one can look up an entity as a system or as a component of a higher-level system. This relation establishes the organizational tree described in Chap. 3 (see Fig. 3.5).

The **System** table contains data elements that directly reflect Eq. 4.1. The fields shown in bold (in Fig. 8.3) are links to specific tables containing data from each of the elements of Eq. 4.1. The other fields contain either simple data elements, such as ID, or links to non-database documents such as wiki-like web pages. For example, the *Transformation* data element could be a link to a computer program. The *History* data element, while being an element of Eq. 4.1, is different. It could be a link to a file of time series data or a set of wiki-like documents containing the historical records of the system. We will have more to say about this element.

The **Components** table contains the set of components, both subsystems and atomic components captured in the system. The two tables are linked by the identification field (ID). An SOI is given the ID integer 0. Its relevant record in the **Components** table is also 0 and provides top-level information about the whole system. However, in addition, the **Components** table includes the set of subsystems (0.1, 0.2, 0.3,…0.*n*) as well as all of their children. Since every decomposable

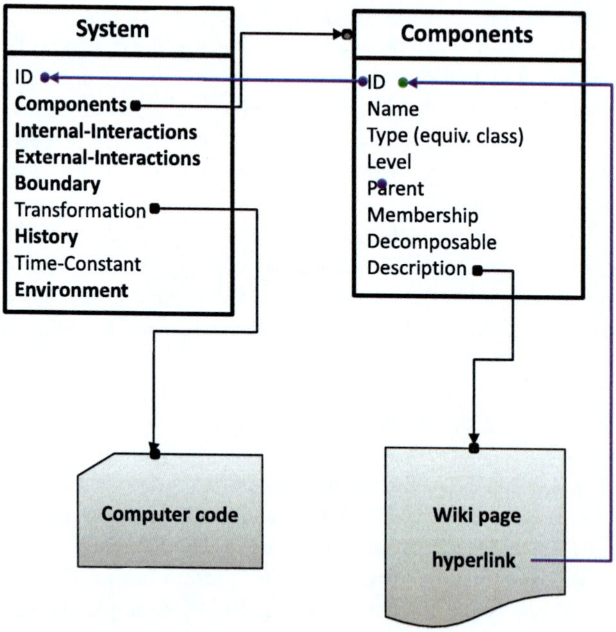

Fig. 8.3 The relation between a system and its component subsystems (Eq. 4.2) is captured in these two tables. See the text for details of the relations

subsystem is, itself, a system, it will have a corresponding record in the **System** table providing the linkages to the children nodes. Fig. 8.4 shows an example of the subsystems of an SOI as they would be entered into the **Components** and **System** tables. The SOI, ID 0, the root of the system tree, appears as a first component in the **Components** table in order to capture its relevant descriptive data. Then, the **Components** table contains all of the subsystems (those indicated as decomposable). However, the same subsystem appears in the System table so that its Eq. 4.1 information can be established. Components that are non-decomposable (leaf nodes in the system tree) are not represented in the **System** table since they are not considered as subsystems. They are only represented in records in the **Components** table.

The implementation of Eq. 4.6 in a database schema is shown in Fig. 8.5. Every system in the **System** table has a defined boundary object. The **Boundary** table contains the data associated with the boundary from the equation, namely the set of properties, such as fuzziness and permeability, and a list of **Interfaces** that transfer flows into and out of the system.

All interfaces are components (of a special kind) and so the majority of details are found in their component record in the **Components** table. In turn, the component record will be found in the **System** table (purple arrow in the figure) as in Fig. 8.5. Note, however, that an interface, as described in Chap. 4, has associated with it a **Protocol** or a special kind of function that is given by a set of operations and timing specifications. These specifications are not necessarily describable within an ordinary database data element and so the **Protocol** data in the **Interface**

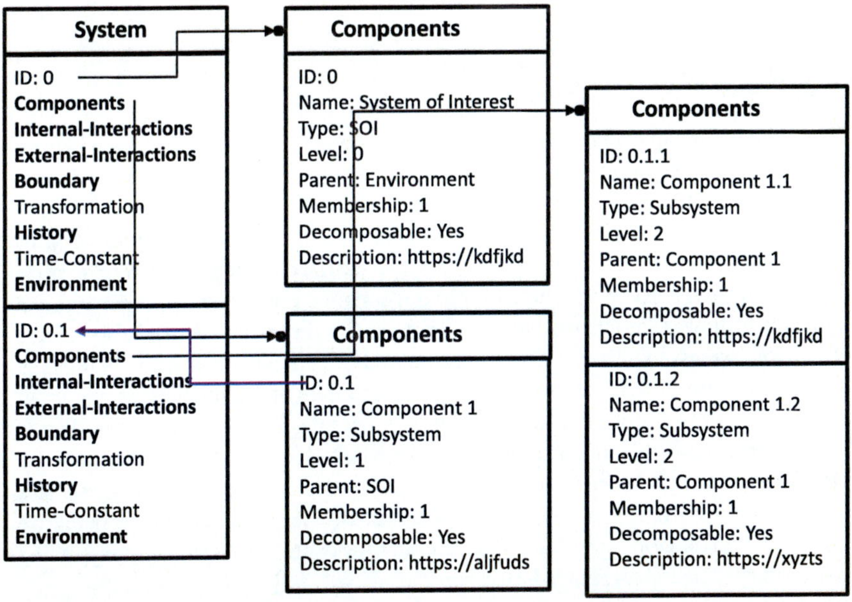

Fig. 8.4 An example of the relation between the System table and Components shows how an SOI is represented along with a subset of its first level of component subsystems

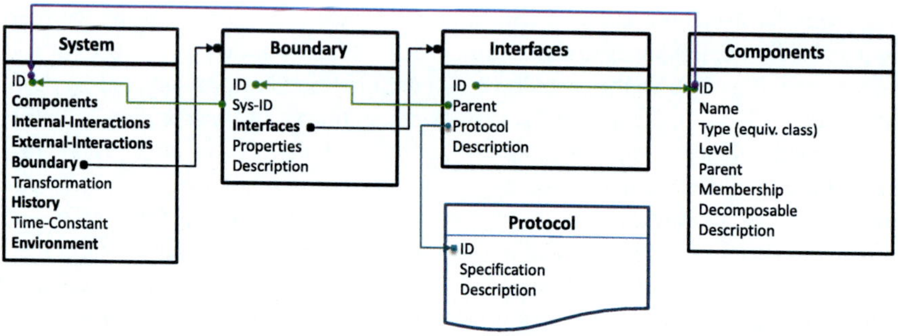

Fig. 8.5 Every system has a boundary object that is comprised of a set of properties and a list of interfaces (Eq. 4.6). Every interface is also a component of the system. And since every component subsystem is also a system the linkage returns from the component back to the system

table will generally point to an external object that is more flexible (e.g., a wiki page or a computer algorithm). Also, note that protocols are highly reusable objects. That is, the same protocol may be utilized by many different interfaces in a boundary or even among many different boundaries of other systems. Therefore, these objects may best be kept in the wiki-like part of the system knowledgebase. That is why the shape of the **Protocol** object is a "document" rather than a table. A great deal of research is needed to assess the most appropriate method for representing and storing protocols, but it is clear that this is an achievable goal within the knowledgebase framework.

8.2.1.2.2 History

The methods for capturing and storing the history of a system will vary from system type to type. For example, the history element of an elemental atom (e.g., a hydrogen atom) would most likely be NULL; that is no history is recorded. Histories only start to become important for systems complex enough to have memories of prior states. This is the case for living systems and artifacts designed to record prior states, so must at least be accounted for in our schema. Fig. 8.6 shows a generalized approach to how a knowledgebase can include the provision for a history of the system of interest. The recording of the records of history may be contained in another database or may be represented in other kinds of documents (as shown in the figure). In the simplest case, the history is a single file containing state data at discrete time intervals (similar to what has been described for the input-output data records used to analyze the opaque-box scenarios). But the issue of what constitutes the "history" record of a system is still a very much open question. The schema we have proposed here only provides a placeholder for capturing the history object. We suspect that efficient computer-based approaches will follow that model.

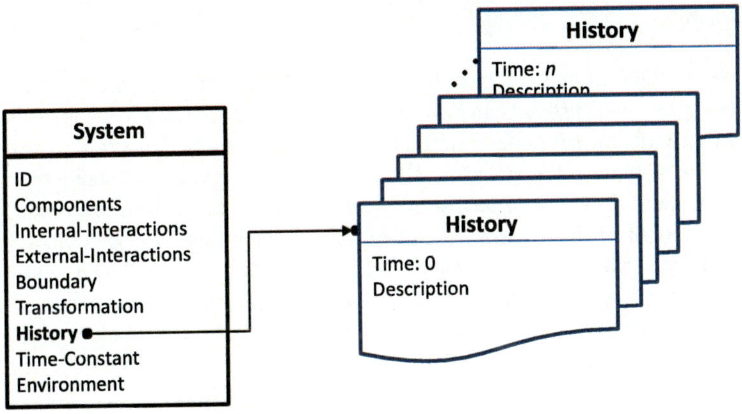

Fig. 8.6 The history element of Eq. 4.1 is captured in a series of documents

8.3 Details of Relations/Tables

In this section, we describe some of the tables envisioned to implement the structure of the system definition framework of Eqs. 4.1 and 4.2. Additional tables implement the other equations (structures) defined in Chap. 4, Sect. 4.3.3. Below, an asterisk next to a table field means the contents are actually pointers to other tables or fields implementing the network structure of the system and as shown in Fig. 8.2. Other field contents are generally things like the unique IDs of other components or the parent that is the supra-system entity. Here we will only focus on the System Table and the Component Table (Fig. 8.3 above in Sect. 8.2.1.2.1) to demonstrate more details of what can be found in the tables. With these examples, it should not be hard for a database systems designer to organize the other tables and set up the necessary linkages. The System and Component tables implement Eqs. 4.1 and 4.2.

8.3.1 System Table

The core table in the knowledgebase schema is the **System** Table. This table contains the embodiment of Eq. 4.1 with references to **Components**, **Boundary**, **Internal-Relations**, **External-Relations** (graphs), **History**, and **Environment** tables. These latter tables contain further references and actual data regarding the types of data as specified in Chap. 4.

8.3.1.1 ID

All objects in the knowledgebase have unique identification numbers. This is the relation key, no two systems can have the same **ID**. This applies to all subsystems as well. We use the dotted numeral notation to indicate this key. For example, the SOI is given the singular integer 0 as its **ID**. See below, however, for considerations of how a current SOI may become a subsystem in a larger supra-system structure.

A component of the system ($c_{i,j}$ in Eq. 4.2) shows up again in this table with a dotted **ID** such as 0.1 or 0.8.13. That is, this one table will contain the basic information of every subsystem. This is the consequence of Eq. 4.2 defining the recursive decomposition structure of a system and the treatment of each subsystem as either an atomic component or a further decomposable system in its own right. See the next section, **Component** table.

8.3.1.2 Name

The name of a system will depend on the disciplinary domain. The name can be in any spoken language, of course, so the range of possibilities is essentially endless. As currently envisioned the name chosen by the lead analyst should be as generic as possible within the domain. Note that if the system is within a type category, for example, a dog being a type (or kind) of a mammal, then that relationship will be handled in the **Component** table.

Naming conventions will likely evolve over time; we see this already in the other sciences and engineering practices. We are not, at this stage, recommending any particular convention.

8.3.1.3 Components

This field points to a table containing the components of the system. In turn, every decomposable component points back to an element in the **System** table. Again, this is the consequence of Eq. 4.2. The **Components** table will contain the data relative to the component while the **System** table treats the components as if they were systems. The fields of the **Components** table are described below.

8.3.1.4 Internal-Interactions

The **Internal-Interactions** table represents the graph of relations, in Eq. 4.4, replicated here: $N_{i,l} = \langle C_{i,l}, L_{i,l} \rangle$. $C_{i,l}$ is really just the set of components identified in Eqs. 4.1 and 4.2. $L_{i,l}$ are the linkages between internal components with other internal components. The schema for this table (Fig. 8.2) contains the two components that are interacting along with the type of interaction data. For example, if the interaction type is a flow of material one component will be the source and the other the sink. And the **Parameters** element will point to another table (not shown) containing the attributes of the substance that flows. This latter table contains the "augmentation" data that describes things like flow rates and measurement parameters, such as mass or current, as appropriate to the type of flow (or force).

8.3.1.5 External-Interactions

As shown in Fig. 8.2 external interactions involve one component in the system with an entity in the environment. As with internal interactions we need to identify either the entity or the component as source or sink, when the interaction is a flow or force. The entities are stored in the **External-Entities** table, which serves the same function for keeping track of the entities in the environment as the **Components** table serves for the internals of the SOI. The external interactions implement the tripartite graph in Eq. 4.5, replicated here:

$$G_{i,l} = \left(C'_{i,l}, Src_{i,l}\right), \left(C''_{i,l}, Snk_{i,l}\right), F_{i,l}. \tag{4.5}$$

8.3.1.6 Boundary

Equation 4.6 provided for the description of a boundary in terms of a set of properties and a list of interfaces embedded in the boundary: $\boldsymbol{B}_{i,l} = \langle P_{i,l}, I_{i,l} \rangle$. The properties of a boundary, in terms of its permeability, fuzziness, etc. can be contained in an augmented wiki page. Its exact format is still the subject of research. The list of interfaces, however, is just a list of pointers back to the **Components** table elements since every interface object is a component of the system, see Fig. 8.5.

8.3.1.7 Transformation

The transformation object is one of several types such as a set of differential equations or a computer program that embodies the way in which a set of inputs to the system (or component subsystem) is transformed into the outputs. A **Transformations** table provides an entry for each component within the system (from subsystems down to atomic components). A pointer in the table entry then points to an appropriate form for computing the transformation. Shown in Fig. 8.7 is a typical method for storing a computer program (probably the most common form today) that accomplishes the transformation. This program is written in a language that is suitable for incorporation into an overall simulation, say as a function call. The figure depicts the program for an SOI (e.g., a main() function in a C language program). The structure makes no necessary commitment to which programming language or computational environment (e.g., Mathematica or similar problem solver environment). That will be determined by the simulation environment chosen for running models (see Chap. 10).

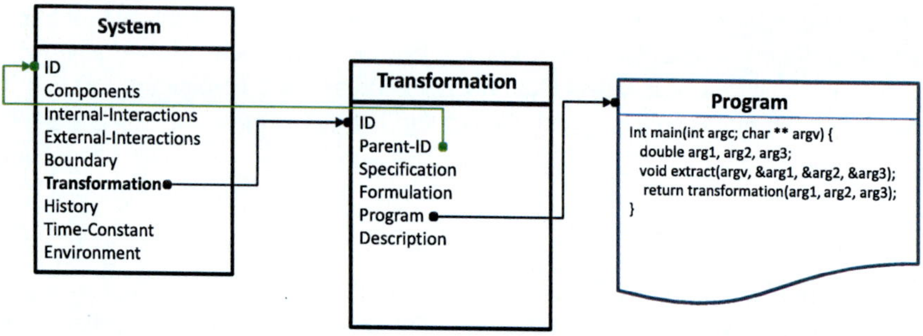

Fig. 8.7 Transformation may be captured in various forms depending on the kind of model/simulation to be derived. Here we show the relation for a computer program that produces the output values based on current inputs

8.3.1.8 History

As indicated previously, the history "object" is the most complicated aspect of the knowledgebase. This is partly because different kinds of systems employ different kinds of history recordings but also because this is a very open research area. Fig. 8.6 above suggests a structure in which we consider a time-series recording of states of the system. This can take many forms and so there should not be a rush to standardize what that should be. For relatively simple systems we can specify a time-series recording of states, for example, a living cell's input/output recordings in an opaque-box analysis. What we should record for a social system of humans is much more open to consideration. In the next chapter, we will examine the human social system subsystem—the economy—and provide some examples of what should be considered as relevant to historical documents that can be analyzed for trends and making predictions.

In general, however, it is envisioned that the history objects will be recorded in one or more of the various auxiliary data structures (e.g., Wiki or time series files) as appropriate to the system.

8.3.1.9 Time-Constant

Time is perhaps the most contentious of all characteristics to be applied to system dynamics. What is the best time constant to be used in describing the behavior of a system? There are many factors that need to be taken into consideration to set an appropriate Δt value. From the standpoint of the knowledgebase, the value selected will apply to all components identified in the current level of organization. The selected value has to be small enough in order to capture the minimal change that can be measured in the relevant parameter.

8.3.1.10 Environment

The environment not only consists of all of the entities with which the system interacts, that is, sources and sinks, but also may include unknowns that create a "milieu." The pointer contained in this field points to the **Environment Table**.

8.3.2 Components Table

The Component Table contains the detailed data pertaining to a component of a system, including the system itself which has **ID** C0. That is, this **ID** points back to the top-level system in the **System** table. The latter contains a list of pointers to components in this table and the ID points back to the same entity treated as a system. This combination of forward and backward references is what implements the relation between Eqs. 4.1 and 4.2. These references allow for queries to start in either table and quickly resolve to the proper relation.

8.3.2.1 ID

This is basically the same as in the System table in terms of format and content.

8.3.2.2 Name

This field is redundant for all elements where the component is actually a subsystem of the parent system. If the element is an atomic component, however, the entry will not be found in the **System** table so will have a unique name not found there.

8.3.2.3 Type

The typing of components as subsystems is still an area of much debate. A typology for systems, in general, was presented in Chap. 3. As envisioned in the current context type categories would be standardized but kept generic. For example, using the Simple, Complex, Complex Adaptive, and Complex Adaptive and Evolvable as the overarching category would be useful in guiding further analysis because those categories tell the analyst what sorts of subsystems should be found at the next level of decomposition. However, it might then be necessary to introduce something like sub-types to differentiate more specific KIND-OF relations. Again, we expect this standardization to evolve with practice.

8.3.2.4 Level

This is, of course, the level in the organization hierarchy. The SOI being level 0 and subsequent subsystems being at level 1 and so on.

8.3.2.5 Parent

This is a pointer to the supra-system in which this component participates. For example, if the ID of this component were C0.1.3, then the level would be 2 and the parent pointer would be to C0.1.

8.3.2.6 Membership

This entry contains the specific membership function of the component vis-à-vis the participation degree in the parent system. Recall that membership functions relate to fuzzy systems in which the component may be a member-only part of the time and each component may have its own such function.

In the case of a component that has multiple memberships in different parent systems, for example, a human being can be a member of their workplace, their home, their church, or any number of other systems (admittedly quite complex, but specifiable for averages) but when resident in one system are not resident in the others simultaneously, then the same component may have as many copies as there are memberships, but only one function per copy.

8.3.2.7 Decomposable

This is a simple binary switch indicating if the component is a subsystem (YES) or an atomic component (NO).

8.3.2.8 Description

This field could contain a hyperlink to a Wiki page containing as extensive a text description as is thought necessary to provide a useful description of the entity. What is useful will depend on the context and the considerations of the analysts. The advantage of using a Wiki here is not unlike Wikipedia usefulness.

8.4 Generating Dynamic Models

As stated at the beginning of the chapter, knowledge is useful only when it can be used to consider the future of the world and the system. The static structure of the knowledgebase captures only the elements of the system definition (as applied to a specific physical system). It contains knowledge of the **Transformation** object applicable to each element, but these are just equations or computer code. The real value of the knowledgebase is the ability to generate a dynamic model of the system for simulation. Knowledge is used to anticipate future states of the world (environment + system).

In this section, we consider how dynamic models (for computer simulation) may be generated directly out of the knowledgebase and then used to test alternative future behaviors of the system under varying future scenarios of the environment. The knowledgebase contains all of the functional and structural knowledge of the system. Each subsystem (component) includes its own transformation function along with a complete specification of the inputs and outputs. Subsystems have already been linked in terms of the relevant flows of matter, energy, and messages that couple them as structures. Thus, it is completely feasible to extract all of this knowledge in a way to construct a simulation model at (at least in principle) any scale in the system organization hierarchy. One could generate a simulation model of a single component at a very low level in the hierarchy, even a leaf node in the system tree representation. Or, at a higher level, all of the components that make up a subsystem.

It is likely that all readers are, by now, familiar with Google Maps™. When you enter a location, the software finds the coordinate information and projects a map of the surrounding territory, pinning the location into the map. The scale of the map is chosen at a resolution that allows the user to get a sense of the surrounding terrain, major landmarks/cities, and major highways. If the user wants more detail, they scroll in which reduces the scale of the map and at the same time expands the amount of detail about surrounding landmarks and roads. Scroll in enough and you start to see buildings, names of businesses, restaurants, etc., and fine details of roads. Click on any one of the pinned details and up comes all kinds of information about it, name, address, directions to get there from your current location, and much more. This is the static information about the structure of the map.

If you use the way-finding feature the software will guide you from your location, telling you where and when to turn. It plots your progress through GPS (on your smartphone). It dynamically updates the map to show you getting to where you want to go. One very useful feature of the mapping is keeping track of delays on roadways (again using GPS to determine how rapidly, or not, traffic is flowing). This is a kind of dynamic/structural modeling that can use real-time data input to project states of the world as time ticks by.

Now imagine all of that applying to a system map (and note that a city is just another kind of system) that has been developed through the deep systems analysis in Chap. 6. Suppose we treat a city as a system, set the resolution of the map so that the city occupies the screen and we have enough detail to allow us to identify the roads leading in and out of the city. Further, imagine that we can click on any of these roads and a dialogue box comes up asking us to specify the traffic load at time t_0 for each one. Click on "start" and a dynamic modeling engine begins to compute the state of the city, insofar as traffic is concerned, for time steps into the future. What you might see is the buildup of congestion in the city if the traffic inputs exceed the traffic outputs.

And now imagine that you could specify how each of the inputs would vary over time. You could specify a realistic ebb and flow of incoming traffic that would emulate data collected from the real city (recall Sect. 6.5.2.3.1 Flows, Sources, Sinks, and Interfaces, especially, Fig. 6.6) and the resulting dynamic model provides a simulation of the city traffic over time.

One further step: imagine hooking the roads into and out of the city with real-time sensors, maybe like the GPS data Google Maps uses and feeding that data stream into the model. You would get a real-time assessment of traffic conditions that might be used to control stoplight signals within the city to smooth the flow of traffic.

Dynamic models of systems can be used to simulate those systems and allow one to compare the outputs of the simulation with the behavior of the real systems. That exercise could be useful in testing the accuracy and reliability of the models. For example, if it is found that the behavior of the simulation is not in accord with that of the real system, it is evidence that the systems analysis missed some vital aspects of the system. But dynamic models are also useful in the context of agent decision-making in real-time (Rosen 1985, and see Chap. 11). The key to the efficacy of models is the quality of the knowledge obtained during analysis and its ready access from storage. This is what the knowledgebase seeks to address.

The quality of a model also depends on the depth of resolution of the sub-models that make up the whole. This is the point of deep analysis and a deep understanding of systems. We strive for greater explanatory power and greater control (or guidance). The higher the resolution of our models the greater will be those two aspects.

All models are abstractions, of course. The map is not the territory. The model is not the system (though, of course, it is ***a*** system). What we seek is a model that, when run dynamically, behaves so much like the system of interest that we can manipulate the inputs and derive anticipatory (or predictive) statements about the future state of the system under those conditions. And, as pointed out, if the model is really quite close in its predictions it has a role to play in the agency. Just as with the Google Map example, the goodness of the model depends on the resolution of scales. Deep systems analysis unveils the details of specific subsystems within the system and the knowledgebase preserves those details in a way that allows us to reconstruct the map. With the addition of a simulation engine, our models of systems can be as close to the real systems as we desire.

8.5 Conclusion

The knowledgebase is the complement of the systems analysis process. It is used to capture the details of a system's structures and functions as defined by the nature of systemness. Together, these components provide analysis and synthesis which is the objective of a whole systems approach to understanding. Neither purely reductionist nor purely integrative, the combination provides the necessary bases for understanding all complex systems.

It is true that obtaining complete (or near so) understanding is not a particularly "cheap" endeavor. It may also be true that on occasion a "quick-and-dirty" approximation of a system model may be good enough for certain purposes ("close enough for government work!") But the days in which we were forced to rely on over-abstraction due to the cost of computing power or time-to-market (profit-motive) constraints are rapidly fading into the past. With today's computational capabilities, existing knowledge deriving from science and engineering, and a growing realization that quick-and-dirty has led to some substantial unintended consequences, we should consider the notion of taking the time to deeply and completely understand systems of interest (which includes understanding their environments). Deriving the information through analysis, and integrating it to become knowledge should be our main goals moving forward.

Chapter 9
The Human Social System Economy as Subsystem: A Preliminary Analysis

Abstract The *economy* is a subsystem of the human social system, itself a subsystem of the whole Earth Ecos. Using the methods considered in Chaps. 6 and 8 we apply them to a preliminary examination of the global economy to demonstrate (1) how those methods can be used to gain a deeper understanding of an extremely complex system and (2) how insights from this analysis lead to some very different conclusions regarding how economies work in reality vs. the beliefs of classical economic thought and (3) provide a launching pad for future work in a much deeper analysis. The approach we will take in this chapter is to use a depth-first decomposition to look at arguably one of the most important aspects of the economy, the energy sector. Energy flow through the HSS is what supports the work done in all of the various subsystems we find in the HSS. One of the main tasks of the economy is to obtain energy and supply it to those other subsystems. Our analysis will include systemic considerations of the implications for the future of an economy drawing down a non-renewable energy resource, fossil fuels. These implications are found to apply more broadly to material non-renewable resources and what that means for the human social system's viability and long-term sustainability.

9.1 Purpose of the Chapter

This will necessarily be a long chapter. We can only apologize in advance. But the subject matter is extraordinarily complex. And huge. The strategy we have adopted is to do what amounts to a depth-first kind of analysis and we will focus specifically on what we think is probably the most important aspect of the economy—the one that is already becoming problematic for the HSS and will become increasingly problematic in the near future. We will focus on a deeper understanding of the energy "sector" of the economic system since a simple physical reality is that wealth is produced by processes that use energy to do physical work. Even thinking is physical (chemical) work. So, energy flows are fundamental to all economic

G. E. Mobus, *Systems Science: Theory, Analysis, Modeling, and Design*,
https://doi.org/10.1007/978-3-030-93482-8_9

activity. Classical economists have only just started understanding this and still have difficulty grasping the consequences of this fundamental reality.[1]

First a side note and something of an apology. In this book we have a difficult problem in that we need to make use of complex, adaptive, and possibly evolvable systems prior to providing a complete explanation of what those systems are. We have introduced the concepts of CAESs in previous chapters (including Chap. 1), but kept the explanations lightweight in order to not bog down the subjects of those chapters. The deeper explanations of CAES archetypes have been put off till Part III because we want to expound the models more thoroughly in one section. Nevertheless, in this chapter we are dealing with a high-order CAES and need to have some facility with using the CAES concepts in order to guide our analytic process. We will try to keep the details of CAES-based analysis to a minimum, not necessarily requiring the descriptions from Part III. But if confusion does intrude, the reader may want to skip ahead to Chap. 10 for a summary of the gory details.

Systems (and systemness) are naturally recursive structures (and concepts). The purpose of this chapter is to demonstrate how the methods derived in Chaps. 6 and 8 can be used to tease apart a super-fuzzy complex system like the economy of the HSS introduced in Chap. 7. But this introduces an interesting methodological complication. We will be using the archetype models as guides to doing so. And those archetype models require a priori explication!

Our approach is to proceed forward as if the reader is already somewhat familiar with the details of some examples of the various archetype models that, in combination, constitute whole complex systems. In Chap. 10, we will introduce several archetype models that are being applied in the analysis to be accomplished in this chapter. These will be the "master" archetypes of complex adaptive and evolvable systems (CAS/CAES) and three sub-archetypes that constitute the major subsystems of any CAS/CAES. These are: agent (with agency), economy, and governance and were introduced cursorily in Chap. 6, Sect. 6.3.3.

For example, in order to talk about an economic system, we need to have a basic concept of what an archetype economy is, and therein lies the rub. One of our premises is that using deep analysis will lead us to an understanding of what an HSS economic process should be and that is, potentially, different from what is currently considered (for example, a market-based, capitalist economy). A systems economics will not necessarily be based on classical approaches to economics. This stems from a realization that an economy, as a subsystem of a larger complex adaptive (and possibly evolvable) system, is based on an archetype model of economics that should apply across the spectrum of CAS and CAES as we have already considered. That is to say, there is a general, universal, model of an economy that can be used as the basis for analyzing the specific HSS economy, as we are about to do.

In this chapter, we will reference these archetype models without a great deal of explanation. For example, we frequently refer to "decision agents" in several

[1] As will be much more thoroughly explained in Chap. 13, energy flow is the basis for all work that reduces the entropy of materials and messages (information into knowledge) for use in CAESs.

different contexts. We assume for the moment that the general notion of an agent is sufficiently well understood that we need not belabor the concept in this chapter. We suggest, however, the reader be ready to page forward to Chap. 10 on occasion to consider more details of the archetype models. The intent of this chapter is to use the analytic techniques thus far described to tackle a beginning understanding that might be used to grapple with the CAES that is the HSS and its economy. We will restrict our analysis to a specific sub-subsystem, the energy acquisition, and distribution sector, to show how the methods work and reveal previously hidden aspects that will turn out to be immensely important to the whole economy.

In Chap. 7, Sect. 7.6, we treated the whole Earth Ecos as a system of interest and began a decomposition that identified, among other important subsystems in the biosphere subsystem, the human social system. That is, we treated the human species, along with its unique cultural attendants (for example, artifacts and institutions) as a subsystem of interest.[2] It is doubtful that anyone would contest a claim that looking at the HSS as a system of interest might be a fool's errand (except that branch of sociology called "World-Systems Analysis", Wallerstein 2004). Even though we might be able to put a boundary around the HSS with respect to the rest of the Ecos environment, say, for example, if we had a complete inventory of the full repertoire of genes that make us human, the complexity of human affairs would make the notion of doing a deep analysis seem unimaginable. But that is really the point of this chapter, to show that not only is it imaginable, it is necessary. Already the various social sciences are attempting to ferret out the mechanisms and meanings of human ideas, behaviors, and cultures. They are already attempting to come to a deep understanding of the HSS (ibid). They are, however, going about it in a fractured, uncoordinated way which leaves gaps in our understanding between various sub-disciplines. The approach is essentially historical (not a bad thing since according to Eq. 4.1, there is a "history" object associated with a complex system). But the better approach needs to include analytical techniques borrowed from the physical sciences (cf. Hodder 2012 for a review of how archeology and systems methods produce a much deeper understanding of human cultures and co-evolution).

For a long time, economics and psychology were drilling down on two seemingly different phenomena in such a way that economists knew nothing about how people actually make decisions and psychologists knew nothing about how money, as one example, shaped and distorted people's attitudes, let alone decisions. Today that is changing. Today we have a field called behavioral economics in which psychologists and economists (along with several other disciplines) work together to find out what people really do and how they really think when it comes to economic decision-making. So there has been significant process on the front of understanding humans as agents. Thus, when we look at their roles as agents in the economy,

---

[2] In Part IV, we call this an "Artifactual System" to denote the fact that the system is comprised of humans and the totality of their human-constructed (artifactual) world.

we will have a much better idea of how decisions actually get made and what the consequences are.

Still, while things are changing toward more integration across disciplines, this is a reaction to an increasingly recognized need more than a principled approach. What we have been arguing is that using the systems approach through deep analysis would provide that principled approach to understanding, providing an overall framework to all of the sciences. It could provide guidance toward integration.

In this chapter, we will begin a preliminary deep analysis of the HSS as an example of how the processes covered in Chaps. 5, 6, and 8 can illuminate extremely complex systems internals that have been a struggle to comprehend by traditional social and natural sciences, with their knowledge silos. After an elementary decomposition of the HSS to establish the internal elements that will form some major aspects of the environment of our new SOI, we proceed to analyze the economic subsystem.

It goes without saying, though we mention it anyway to make sure the reader knows we are not caught in unbelievable hubris, that the analysis of the economy—doing the work of traditional approaches to economics in all its various schools – would require much more than a single chapter in a single book. Therefore, the approach we shall take is to use the method of depth-first analysis explained in Chap. 6, Sect. 6.4.3.3 Recursive Deconstruction, on what we think is an extremely important subsystem of the economy, the energy sector.

The importance of the sector cannot be overstated in our view. Everything that happens in the HSS does so because of the flow of an appropriate form and power factor of energy. Everything we humans do, including and especially living, depends on the flow of energy through the system (Morowitz 1968; Odum 2007). It takes energy to do work. And everything that happens involves work, mechanical, chemical, or electrical, that changes or moves material. This is so fundamental that it is, on the one hand, amazing that it has not been fully grasped by the social sciences in general and, on the other hand, somewhat understandable that it has been taken for granted as it is what everyone experiences moment to moment.

What makes it essential to turn all eyes toward the energy sector today is that the HSS sources of suitable energy to run an advanced technological society are under threat of depletion. Over 80% of our industrial-grade energy comes from the burning of fossil fuels (Hall and Klitgaard 2012), which are a fixed and finite resource. Since energy used to do work follows the Second Law of Thermodynamics (2nd law),[3] the process doing the work degrades an equivalent amount of the energy flow to waste heat—energy that can no longer do useful work. That heat, as has been pointed out in Part I, is dissipated to the environment and unrecoverable. Thus, unlike matter that can to some extent be recycled, energy gets effectively used up. Thus, burning the fuels means their energy contents are on a one-way trip to dissipation. Today, a large and growing number of people are hopeful that

[3] We refer to the Second Law of Thermodynamics as a proper noun owing to its singular importance in the way the Universe works. Subsequently we will refer back to it using the shorthand form the 2nd law. Not all writers use this convention.

alternative energies such as solar photovoltaics and wind-generated electricity can replace the use of fossil fuels while continuing with our advanced living-standard societies (and while continuing the growth of developing economies as they seek to achieve the same standard enjoyed by the developed economies).[4] Whether this is feasible is, unfortunately, still not known. There are scientifically based arguments on all sides of this very complex issue. However, there is very little truly scientific (or systemic) evidence to support any particular conclusions. Below we will outline some of the challenges to obtaining that evidence and provide a systems approach suggestion for how the deep analysis could help policymakers make decisions about where it would be best to invest our dwindling resources.

Even though our focus will be primarily on the energy sector analysis, we will, nevertheless, be touching on other important economic issues such as the nature of markets and money. The economy, as will be seen in our cursory analysis, is still an incredibly complex system in which many sub-subsystems are linked quite strongly. Some of these will be pointed out in this chapter.

What we will cover in this chapter is still relatively cursory; the energy sector and its importance to the rest of the HSS (indeed the whole Ecos) as well as its own complexity require a substantial research literature on its own. Nevertheless, we think that this demonstration of the process of deep understanding will be better realized by tackling a portion of the economic subsystem of this importance and complexity.

Note that what will be presented in this chapter is only a portion of work that has been done on what is called the "biophysical economy"[5] by the author and a growing number of researchers. The author's contribution to the field has been based on foregoing the use of classical economics approaches and theories (for example, use of GDP as a measure of income) and applying strict systems principles to derive a better understanding of the economy unbiased by classical and neoclassical theories, for example, of the nature of money. The author and others have been analyzing the many other subsystems in the economy. And some significant, though unfortunately minor, progress has been made. The author contemplates a future work devoted to developing the basic ideas in this chapter in the future.

---

[4] A second motivation, maybe equally important, is that burning fossil fuels produces CO_2 that is now definitively linked to global climate change.

[5] The name of a new kind of economics may have had several origins but is most identified with a systems ecologist and student of Howard T. Odum, Charles A. S. Hall, Professor Emeritus, State University of New York, College of Environmental Studies and Forestry, at Syracuse NY (cf. Hall and Klitgaard 2012). The author had the privilege of spending a sabbatical leave with Dr. Hall in Syracuse in 2009, while developing some of the ideas presented in this chapter. Dr. Hall's work was extremely influential on that of the author's and has been equally so for the work of many other researchers in the nascent field of biophysical economics.

9.2 Analytic Approaches

Before getting into the systems analysis of the economy it would be useful to consider several relevant perspectives to take. These will be expanded as the analysis proceeds.

9.2.1 *From the Current Subsystem As We Find It*

The first perspective is to take the current global economy as it is presently understood. That is to say, we could accept the various positions of schools of economics as they have come to understand the system. This would presumably save us a great deal of time and energy. To some degree we will end up accepting some of the knowledge garnered by one or more of these schools, but only after establishing it as consistent with the analysis from a systems perspective.

A great deal of motivation for insisting on doing an independent analysis and possibly deriving some different knowledge is that the science of economics has been struggling with its topic for quite a while (see Keen 2011, Part 1). We gather from the growing criticisms of neoclassical economics that there are some internal flaws with models and assumptions in the field that have caused it to fail in the construction of scenarios and predictions. Economists have become notoriously incapable of predicting recessions, for example.[6] For this reason, we will not take at face value those models and assumptions. Our goal is to derive a systems model of economies or the global economy from systems principles and the methodologies (at least the core ones) developed in this book.

In anticipation of what we will find, it is probably no surprise to state that the current economic system is quite dysfunctional with respect to its supposed purpose in supporting the livelihood of the HSS.[7] This will be seen in the energy sector analysis in particular but applies to other subsystems as well. And this raises the question of how the system could be made functional. The answer to that question involves systems engineering or designing and constructing a functional economic system that fulfills its function with respect to human beings and also makes the HSS a compatible subsystem of the Ecos, that is, does not pollute or over-extract. Part IV of the book will be devoted to demonstrating how this can be accomplished.

---

[6] Keen (2011) and other "heterodox" economists are quick to point out that almost no one in the neoclassical school had predicted the meltdown of 2009, now known as the Great Recession (just short of a depression).

[7] That the economy of the HSS is not functioning properly with respect to its purpose is actually not really surprising. There are many kinds of complex systems that have been found to suffer various "pathologies" that cause the system to fail at some point. For example, a positive feedback loop not countered by a negative one will lead to exponential blow-up, destabilization, and ultimate destruction of the system.

We will use the systems principles and methodologies to design and engineer a workable economic system.

In this chapter, we will employ two other perspectives to assist in the analysis. The first will also provide a way to see how the history object, introduced in Chap. 4 and described in previous chapters, is essential in determining the systemness of complex adaptive and *evolvable* systems.

9.2.2 From the Historical Evolution of the Subsystem[8]

This keeps us referring to the humanness of the enterprise…

Over the last several decades the historical record of the extraction of natural resources and the creation of wealth and its distribution through various forms of trade interactions has been greatly improved through the work of anthropologists, archeologists, and historians with the aid of other disciplines such as neuropsychology. It is now possible to paint a basic picture of how cultures themselves came to be, their ontogeny, and how practices evolved from the earliest times in the late Pleistocene era right up to today. The historical perspective will help us analyze a number of economic phenomena that seem puzzling at times, such as the nature of "ownership" and "profit-taking."

Such a perspective will necessitate a background in human neuropsychology and behaviors. Since the human species is the reason that something like a society and an economy exists in the first place, this cannot be avoided. We will, therefore, touch on some of the more direct factors in the nature of human beings that directly affect the analysis of the various subsystem institutions in the economy and the energy sector thereof.

Human beings are eusocial[9] beings; some have characterized it as hypersocial (Bourke 2011; Buller 2005; Harari 2011; Sober and Wilson 1998; Suddendorf 2013; Tomasello 2014). We humans are strongly motivated to cooperate on shared objectives. The basic set of objectives involve means of living and thriving, or the biological support. The evolutionary fitness of human beings comes from the ability to organize behaviors for this purpose. Part of our historical perspective will include what has come to be called the Agricultural Revolution as it relates to biological support. The invention of agriculture and animal husbandry is part of a larger mental capacity of humans to invent new ways of doing things. Specifically, humans find ways, and invent tools, to leverage their biological support. They learn to exploit new sources of energy and new materials, along with new ways to accomplish work using less of their own physiological energy. They have technology.

[8] We are following a line of reasoning put forth by Karl Polanyi (2001) regarding the necessity to look as far back in the prehistory of *Homo sapiens* to understand social and psychological issues that then have relations to economic systems. See, especially, Polanyi (2001), Chap. 4, Societies and Economic Systems and also Morris (2010, 2013, 2015).

[9] Eusociality is the tendency for individuals to form social units through cooperation.

9.2.3 *From the Ideal of a CAES*

In Part IV, we will explore the concepts of systems design approaches to CAESs. We will use the archetype models as template models telling us what to put into our designs. We will, in fact, go through the design of an ideal HSS and its ideal economy. By "ideal" we don't mean some kind of Socratic "Form" as such. What we mean is that by using the economy archetype we can identify elements of the human economy that should be in place in order to achieve long-term viability. Clearly, exhausting our main source of high-grade energy, fossil fuels, is not a way to achieve long-term viability. So, we must consider alternatives. And, rather than automatically assuming that that means solar voltaic and wind sources, we will apply the systems analysis approach to see if these are, indeed, viable alternatives (we will ignore the hype and politics in doing so).

9.3 The HSS Decomposed

Given the HSS as a subsystem of the Ecos we now proceed to investigate it systemically using the methods of deep analysis. This will serve two purposes. It is prelude to the environmental analysis phase when the Economy becomes the focus of investigation—the new SOI at a deeper level. And it will offer an opportunity to further explore the methods of dealing with a very fuzzy system! The human components of the HSS are extremely variable and subject to evolution in the long run. In the short run, humans are continuously inventing new artifacts and new institutions which become part of the culture. Thus, the HSS, exemplar par excellence of a CAES, is overall fuzzy. But so are its component subsystems as we will soon see.

The first task is to find the boundary of the SOI. Immediately we run into difficulties. What shall we call the boundary of the economic subsystem? Figure 9.1[10] suggests the difficulty. In this figure, we show the economy as a large (purple) oval intersecting many other entities we might treat as subsystems in their own rights. It might be interpreted that the economy is co-extensive with the whole of the HSS (largest purple dashed oval). However, this is not the case. Nor is the population of humans necessarily coextensive with the economy. Put another way, while the economy is clearly a major part of the social fabric, it isn't all there is to human life or even the institutions that form part of that fabric.[11] As we will observe later, the HSS can be resolved into a set of fuzzy subsystems that organize around some major activities that are not, strictly speaking, economic. All of these subsystems interact with each other, primarily through human biological and psychological processes.

[10]The figure shows eight ovals contained within the HSS boundary. Clearly this cannot be an exhaustive set of subsystems. We have chosen these as perhaps some of the most obvious, certainly among the most important, subsystems comprising the human social system.

[11]This is an observation made by Karl Polanyi (2001) in 1944.

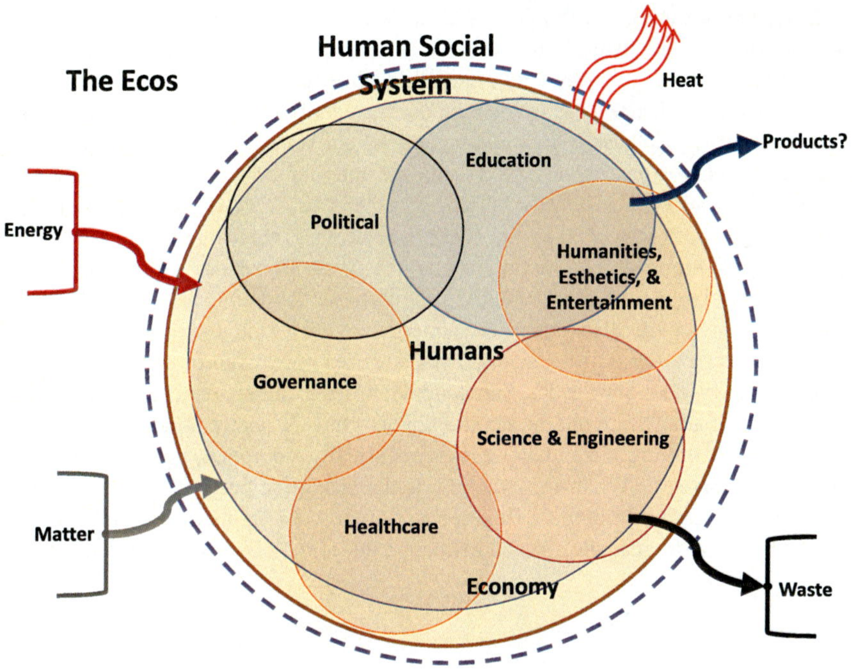

Fig. 9.1 The HSS seen as a fuzzy set of fuzzy subsystems. This is not really a Venn diagram since it does not show the complete overlapping of ovals. Nor is it meant to represent an exhaustive set. The Economy contains the main interfaces between the Ecos and the HSS

For example, a human can read a novel for pleasure (and perhaps some education) but the novel had to be written, published, and distributed. The reading is not necessarily an act of economic consequence, at least not directly or likely strongly. The writing is generally not motivated by direct economic considerations—J.K. Rowling didn't write Harry Potter books as a "job," at least not the original one—even though writing a best-seller has obvious economic consequences. But the publishing and distribution of books is a set of work processes that are very much part of the economic engine. The firms that accomplish this work are expressly set up to accomplish this function. They consume energy and materials, produce product and deliver it, and generate waste heat and waste materials in the process. Peoples' desires to read novels along with a similarly psychologically motivated desire to be expressive and tell a story (see the Sect. 9.3.2.1.2 below: Psychological Supports) created a need for an explicit set of work processes to be set up in order to couple these two desires. The activities associated with writing and reading are best concerned with the institutions we think of as the Humanities or Esthetics. The patterns of those activities may involve the doing of mental work, but nobody really pays themselves a salary for writing or reading a book. They may receive payment in royalties or they pay for the book in order to support the existence of an economic entity capable of making them possible.

Below, as we tease out the details of an economic subsystem, we will be concerned with another institution important for the development and growth of the economy, so we will include it in our analysis while ignoring to a greater extent the other subsystems. Namely, as we decompose the energy sector, we will have to take note of the role of the Science and Engineering subsystem (see Figs. 9.1 and 9.2) since the access to energy reservoirs and the rates of flows of energies into the economy has been enabled by the activities in this subsystem. It is worth noting that the activity patterns we find in the other subsystems, other than the Economy or Humans, are primarily information processing work. In science, for example, the activities involve mental model building (hypotheses) and empirical verification or refuting. That activity also generates economic activity such as the construction of labs and libraries. In addition, scientific discoveries may result in findings that can be exploited by economic work, for example, a better understanding of thermodynamics led to improvements in steam engines (and all heat engines) resulting in more efficient mining. Scientists, at least historically, do not engage in science for economic purposes. (What, for example, is the economic purpose of knowing the mating habits of wild rabbits?) Engineers, similarly, tackle their work of creating new things, not because they intend to change the economic subsystem. They do so

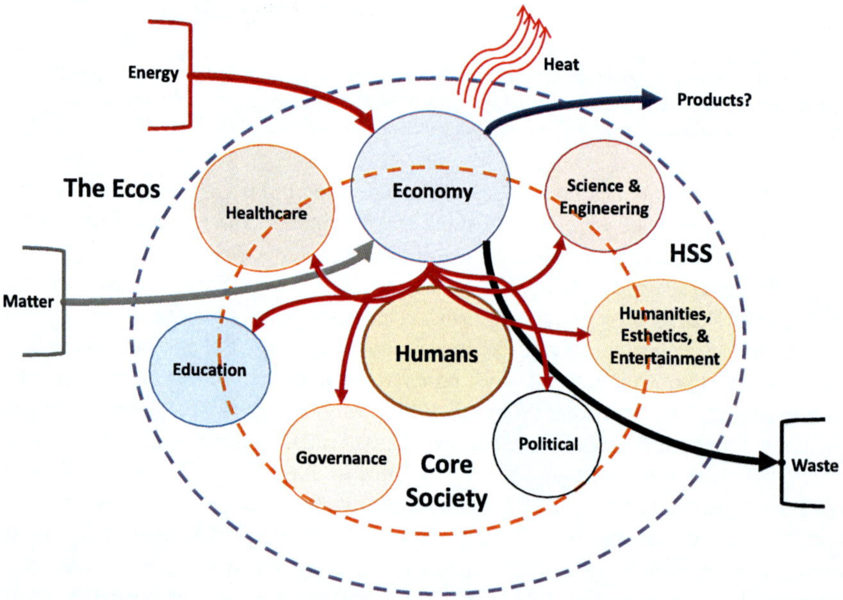

Fig. 9.2 The HSS defuzzified by firming boundaries and introducing flows (energy acquired by the economic subsystem distributed to other subsystems). Humans, their organization of governance, and political processes constitute a "core" of society (orange dashed oval). The economy and other subsystems are composed of many artifacts that sustain the core. Note the upper right arrow pointing to products (?). This poses the question: What products of use to the Ecos does the HSS produce?

to solve localized problems set to them by economic concerns for economic entities (i.e., the owners of the firms in capitalist systems). But what they do is process data, and produce information (for example, specifications) and knowledge (designs).[12]

As you can see all of these institutions/subsystems of the HSS are intermingled in extremely complex ways. The only common thread here is the nature of human beings themselves and the kinds of mental and physical work they do as living creatures. In one sense, deconstructing the HSS might seem impossible given this level of complexity. But that is precisely the job we have to do if we are to manage our affairs on this planet. How, then, do we tackle the fuzziness of the HSS and its subsystems?

9.3.1 *The Fuzziness of Systems*

Figure 9.1 seeks to depict the way in which the HSS is constituted by a number of fuzzy subsystems. These subsystems overlap to some extent; the figure is similar to a Venn diagram in set theory, but cannot really depict the nature of the overlap in two dimensions, which is actually based on temporal factors as much as locations in space. Also, we would need to show overlaps (i.e., intersections) among all of the subsystems.

To see this, consider the oval labeled Education. It overlaps with the Human (meaning the population of living beings), the Political, the Economy, the Humanities, etc., ovals in the figure, but should also overlap with Governance, Healthcare, and Science and Engineering. A two-dimensional representation does not do justice to the relations actually had between these subsystems. The figure only seeks to suggest the fact that all of these subsystems of the HSS are fuzzy. Education means that system of schools and other institutions that are responsible for preparing human young (students) to enter society as, hopefully, productive and thoughtful adults. But education plays a role in the economy just as the latter plays a role in the former. It plays a role in human life and vice versa. Clearly all of these systems interact in intimate ways that make it daunting, to say the least, to assign any kind of boundary conditions to any of them.

Nevertheless, that is exactly what we need to do in order to better understand what is really going on in each subsystem. And this circumscription must be done in such a way as to avoid the subject silos that have encumbered academia. First, we should consider what we mean by claiming these examples as subsystems in the first place. We could possibly rely on common sense to realize that the names we have assigned to them are those in general use by everyone. Most people will know

[12]We hasten to note that this claim may need to be modified for the current situation. Historically scientists and engineers enjoyed the intrinsic values of the work. Today, in a climate where innovations are supposed to lead to the creation of new companies and eventual IPOs where the engineers or scientists should expect to get rich, it is not purely the case of doing the work for the love of the work.

what we mean when we say "the education system" even if they do not exactly know where the boundary might be.

One approach could be to start by considering what elements (components) of the subsystem of interest are not overlapping with any other subsystem. In Fig. 9.2, for example, there are regions of several of the ovals that lie outside the economy. The implication is that there are components in those systems that do not directly interact with economic activities. This, of course, means that we have to clearly identify what we mean by economic activities so as to determine whether any component in a subsystem is not engaged.

As an example, consider the idea that an economic activity such as "earning the highest wage one can" (what a rational utility maximizer would most likely seek) is generally not directly involved in the decision many people in the education field make to participate in it. There might be many arguments forwarded as to why teachers would choose their profession, which is generally lower pay than working in the "private sector" that reconsider what is meant by "rewards," for example, intrinsic versus extrinsic. But that would change it from an economic decision to an esthetic one.

9.3.2 *Defuzzification of the HSS*

In Chap. 6, we devoted a section (6.9.1) to "Dealing with Fuzzy Systems" in analysis. Here we take a look at how that might proceed with respect to the very fuzzy set of subsystems of the HSS. The key here is to look at individual components within any subsystem and determine their membership function with respect to each subsystem. Many components might have membership functions that simply return unity (1) meaning they are always present with probability 1. But many components can effectively multiplex or serve roles in multiple subsystems at different times and with different probabilities. Our paradigm example is the way in which individual people move in and out of the HSS subsystems (except, of course, the Human biological subsystem) at different times and with different likelihoods.

No one will claim that identification of when, where, and at what time a component will be in what subsystem will be easy. On the other hand, we need not necessarily do so for each individual (or component). This is not too different from the problem raised in thermodynamics where temperature of a system is defined as a measure of the average translational kinetic energy of particles. Rather the analysis may, in many instances, rely on average component behaviors.

One way to handle defuzzification is to treat all subsystems as if they have concrete boundaries and interfaces through which components flow between subsystems. For example, human individuals, while certainly remaining humans, move between places and roles with some predictability. People become economic decision agents when they play the role of a consumer or that of a producer (in their job, for instance). But at other times they are playing other roles in other subsystems. For example, when a doctor is diagnosing a patient (in the healthcare system), she is not making explicit economic decisions. It may be true that these other roles may have economic consequences in the aggregate and over time, but we end up making things more complicated if we try to treat every action as being in the economic system.[13] The reasons will become clearer as we proceed.

Figure 9.2 shows the same HSS in Fig. 9.1 but defuzzified by adopting the method of boundary and flow construction. The economy has been separated from the other subsystems and will become the new SOI for this chapter. We will examine the rest of the HSS as the environment of the economy as per the procedure in Chap. 6.

The nine subsystems identified in Figs. 9.1 and 9.2 will be delineated. Note that in Fig. 9.2 we have included another dashed oval labeled "Core Society." Humans, the Political, the Governances, and most of the Economy are contained within this sphere. The term "political economy" used to be the name of the field of study of economics as it was seen to be deeply intertwined with politics and governance. Indeed, we could argue that the core society represents those close associations. However, our reason for separating these from one another is that when human beings participate as components in each of these processes, they are making different kinds of decisions as agents. We argue that a political decision is not the same as an economic one, or a governance one, even though they are strongly coupled in human minds. Below we will make the distinctions that should work to defuzzify these boundaries.

One other complication for all intentional CAESs exists that is not shown in the figures. The governance subsystem shown is not all there is to the governance of any such CAES (all of these subsystems are, themselves, intentional CAESs). The governance shown in the figures applies to the whole HSS but is only the government of the whole. Each subsystem has its own internal governance sub-subsystem that is for its own self-regulation. In Chap. 12, we will describe the architecture of a holistic governance system for any intentional CAES (and some applications for non-intentional CAESs such as an ecosystem). The situation with respect to the self-similar aspects of governance, that is governance of the whole and governance of the parts, will become much clearer there. Interested readers may wish to read that chapter first since so much of what goes on in the economy involves different governance mechanisms.

What we have done thus far is to do a functional/structural decomposition of the HSS considering the Ecos as the environment. Below we will consider the various entities in Fig. 9.2, briefly, as if we had done a further analysis of them. This is especially relevant for the Human subsystem since humans are the main agents working in all of the other subsystems.

[13] It seems that an abundance of enthusiasm for market-based mechanisms applied to all other HSS subsystems has muddied the waters further in recent decades. As of this writing, for example, the Secretary of Education in the US Whitehouse is a strong believer in market solutions to education, that is, the use of vouchers that allow parents to treat the choices of public schools as a "consumer choice."

9.3.2.1 The Human Subsystem and the Role of Humans

This is still more a question than a statement! In some sense, the whole HSS is all about humans, or at least that is the classical humanist point of view. Everything we have invented and done has been about making more humans and living comfortably while doing so. Thus, humans provide the motivation for developing all of the other subsystems and especially the economy. Since so many books and journal articles, not to mention fiction stories in all genre, magazines, movies, etc., have been devoted to answering the basic question, it would be the height of hubris to suggest the small section of this one work could even begin to provide an answer. Rather we will mention a few of the sub-subsystem aspects of human beings that seem most relevant to further examination of the economy itself.

The Human subsystem is comprised of all biomass and produced assets that are under the control of human beings—that is their physical bodies, their clothes, their shelters, their pets, their lawns (!), etc. Basically, the family possessions are considered in this subsystem. Humans are, by their own designation, consumers. In fact, in our later modeling of the economy, we see humans as the ultimate consumers. Whatever the economic engine produces it is meant to support humanity's consumption of stuff. Human possession and consumption provide the basic motive drive for everything else that goes on in the HSS.

Humans play another role besides consumption, however. They also supply their bodies and brains to do work within the other subsystems. That is, they temporarily become components of the other subsystems during the work day. The fuzzy membership function of an employee is approximated by a time-in-service equation. But there should be no confusion about the fact that the fuzziness of employment is not an insurmountable analysis hurdle. In fact, we can measure the actual amount of time-in-service for any worker. We may not, it's true, be able to discount time spent daydreaming or playing solitaire on the computer when no one is watching. But most organizations have the ability to record the amount of time their workers are on the job. That is how we calculate productivity values.

So, when humans are buying goods and services to consume, they are operating in the economy (even if shopping online). When humans are at work, they are participating in the subsystem in which they are employed even when that work may have direct economic consequences.

Human characteristics that interact with the rest of the HSS and especially between humans are described briefly.

9.3.2.1.1 Biological Mandate

The first fact about human beings is that they are animals, living creatures that answer ultimately to the biological mandates to obtain physical resources, convert those into biomass, especially during development and growth, but also in reproduction, and, importantly, stay alive even in the face of competition, disease, and predation. This mandate is programmed deep in the human brain even when we are not consciously aware of it being the fundamental driving force behind our behavior.[14] We do not claim that the mandate causes us to choose as we do in every decision we make – one could not explain altruism easily if this were true. But we do claim that it provides an incredibly strong influence on our decision processes. What needs to be explained, especially in relation to economic decisions, are the basis of selfishness, say as it manifests itself in motives to "maximize profits" even at the expense of fellow humans. (See: Geary 2005; Harari 2011; Marcus 2008; Maslow 1943; Mithen 1996; Sober and Wilson 1998; Striedter 2005 for more background.)

9.3.2.1.2 Psychological Supports

The biological mandate might be thought of as the core of motivation for human behavior. But since the species strategy for achieving fitness is based on sociality and technology (instruments used as extensions and amplifications of our biological capabilities), humans rely heavily on a hierarchy of psychological supports that involve the generation of complex behaviors. Sometimes those behaviors do not seem directly connected to or even have anything to do with the biological mandate. But why a man would want to buy a massive (muscular) automobile can easily be linked to his desires to win at the reproduction race even if he is unaware of the influence of advertisements involving glamourous people.

Abraham Maslow, the father of "positive psychology," developed a model of a hierarchy of needs, each higher need based on fulfilling the lower needs (Maslow 1943). In Maslow's hierarchy Physiological, Safety, and Love/Belonging needs formed the lowest level; these correspond with the biological mandate with the need for food and water forming the base. Safety means having shelter and protection from the elements, predation, and disease. Love is somewhat transitional in that meeting this need is where human behavior starts to look a lot less like other animals. Courtship rituals and general inter-sex behaviors are much more complex in humans. Above love and belonging, Maslow saw humans needing Esteem, or positive relations with other humans in the social system. People work hard at being thought well of by other people; reputation can be everything in influencing how successful an individual is at obtaining the needed resources and potentials for mating.

The highest level in Maslow's hierarchy he called "Self-actualization" or finding contentment in having succeeded meeting the lower-level needs. Happiness.

Another way to look at this hierarchy is that the behaviors one generates at the higher levels can actually promote success at the lower levels. For example, people who are self-confident and seem happy with their situation in life seem also to be more capable of acquiring the esteem of others and succeed in all the other levels of needs meeting. In other words, the causal arrow might just as well go downward as

[14] Sigmund Freud may have intuited this deep, fundamental, drive behind human behavior at least insofar as reproduction was concerned.

upward. Maslow envisioned the way it worked was that if someone were meeting their personal physiological needs—had enough food and water—they would be able to then take care of their safety needs. And accomplishing that, they would be able to turn attention to their love needs, and so on. For Maslow, the arrow of causality pointed from the lowest level upward. But in societies, the causality may go both ways, up and down. Certain personalities, which may be largely influenced by genetic predispositions, just seem to be more successful at meeting needs.

In any case, human behavior is vastly more complex than other animals because the degrees of freedom in choices they can make as they seek to meet their needs are so great. And this is a source of complexity in considering the economic subsystem. Ultimately human behavior, from the top of the hierarchy (seeking happiness) downward, is directed at making economic moves that fulfill all of these needs. In the modern, developed world economy, there is perhaps a greater emphasis on higher needs than the lower ones. The economic engines of modern societies have produced so much material wealth and have virtually guaranteed the meeting of lower-level needs—it is pretty hard to starve to death in a developed society—so that most individuals concentrate on the acquisition of material goods and services that they believe will make them happy. We now know that this is not necessarily the case (Gilbert 2006; Schwartz 2004).

In the modern world, in the developed nations, we have used the economic engine to generate an array of products and services that go far beyond the mere needs of human beings. Having transcended the lower-level needs via agriculture and technology, we have come to a point of focus on higher-level needs but transformed now into unconstrained wants. We are so thoroughly supported in terms of our baser needs that we have literally forgotten what it means to work on fulfilling those lower needs.

Something extraordinary happened in the evolution of *Homo sapiens* that made this possible. Below we survey a few of these aspects.

9.3.2.1.3 Creative Intelligence

There is probably little in the way of a rational argument against the idea that humans are a very special kind of animal. Humans assess the potentials of things in their environment and determine the usefulness of those things in helping them to carry out an objective. Humans are both unusually creative and unusually intelligent in terms of the pantheon of sentient creatures on the planet. The combination of creativity and intelligence (here meaning an ability to solve problems) is cleverness. Human beings see objectives (to fulfilling the biological mandate, for example), obstacles to reaching them, and then look for things in their environment and methods of using them that will allow them to overcome the obstacles and achieve the goals. Two capacities of cleverness are particularly important to economics.

9.3.2.1.3.1 Affordance

Affordance is the capacity to see a potential for use of a naturally occurring object to achieve an objective. Many animals, mammals, and birds in particular but also some cephalopods like octopi have an ability to reason about the use of natural (and artificial) objects to use as tools. The analysis of tool-ness will be handled below. But for now, consider that any object that allows a sentient agent to obtain its goal (which is conditioned by the biological mandate) is to be called a "tool." A chimpanzee finds a way to arrange some branches and leaves into a nest in a naturally occurring cleft in a large branch of a tree. A hiker sees that a fallen tree makes a good seat to rest while climbing a steep trail.

9.3.2.1.3.2 Invention

As in the chimpanzee bed example, the capacity to see how to put several different things together for a new purpose is the essence of invention. In humans this capacity has far exceeded anything in the biological pantheon. Humans actively shape materials so as to create usability but much beyond a crow's ability to shape a stick to poke into tree bark or a chimpanzee's ability to strip leaves off of a twig to use the latter as a fishing instrument for termites. We humans create "compound tools." That is, we bring together various elements to construct a *system*! In a very real sense, everything we humans do in the realm of invention is a form of systems engineering.

9.3.2.1.4 Knowledge and Communications

What most sets humans apart from the other sentient creatures on this planet is their ability to acquire and use complex knowledge. This ability rests on several component capabilities that set us apart from other species.

9.3.2.1.4.1 Beliefs

Knowledge isn't really knowledge in any absolute sense. It is just the best approximation of reality that we have induced through information receipt and structuring in our neural networks. But humans have an internal sense of certainty that is a kind of measure of how well that construction has been done. Unfortunately, the sense of certainty may not really correspond with the veracity of the knowledge (Burton 2008). This is problematic in many ways in our modern world. Below we examine the role of ideologies in organizing our HSS.

9.3.2.1.4.2 Curiosity

Anyone who has owned a pet dog, cat, or even a fish will know that curiosity is deeply ingrained in the animal brain. Animals, those that can learn from experience, depend heavily on a drive to examine any novel situation. In humans, this has reached something of a crescendo. All one need do is observe a two-year-old child for a small amount of time to verify this conclusion.

Curiosity drives us to exploration. We need to know. We have been described as informavores! Recall from Chap. 3 that information is what drives the formation of knowledge. Therefore, curiosity, the drive to gain information, is the critical process that leads to increasing knowledge of how things work.

We are intensely curious about what the things in our world and what other human beings can do for us to fulfill our biological mandates. We want to know how can we survive and thrive in whatever environment we find ourselves.

9.3.2.1.4.3 Science and Understanding the Natural World and Ourselves

Science is the embodiment of curiosity driving us to understand our world and our own selves. Science, as described in Chap. 1, is our attempt to gain understanding of our world, what we are, and what our place in this world is. This is unlike anything we find in the rest of the sentient beings on this planet. What other species asks: What does it all mean? There is something very different going on here for human beings. We need to grasp it.

Our deep desire to exist, to extend ourselves (through reproduction), drives us to understand so that we can anticipate future contingencies. Our capacity to understand through a structured methodology (science) gives us a potential advantage over mere sentience. If we can anticipate the future of the Ecos, we might be able to avoid the fates of former life forms (for example, extinction events).

9.3.2.1.4.4 Language

In Chap. 4, we saw that the key to human interactions and sharing of knowledge is communications through the use of language. No other sentient creature has the extent of communications of abstract ideas that we have achieved. Language is the means for communicating concepts between brains, for establishing meaning of phenomena, and constructing shared beliefs and intentions. Language is the fabric of social structures and institutions.

9.3.2.1.4.5 Shared Intentionality

One of the mental attributes that makes us different from our closest cousins in the phylogenetic tree is our mental ability to share intentions with our fellow (social) beings (Tomasello 2014). We, unlike, say chimpanzees, are able to communicate with our fellows regarding our intentions to accomplish some particular tasks that cannot be accomplished by a single individual and elicit their help to do so. The evolution of language may have involved the deepening of our capacity to communicate our intentions and proposals for how to accomplish them. Today, of course, we take this ability for granted. But the need to cooperate to accomplish ends by mutually considered means was a deep part of human evolution and the reason that we developed the level of eusociality that we exhibit. We humans are first and foremost social beings. Out entire existence is predicated on working in social units to accomplish work that cannot be done by individuals.

9.3.2.1.4.6 Talents and Specialization

A peculiarity of human nature is the tendency for individuals to have variable talents for performing tasks. Some people are more capable of performing specific kinds of work than others.[15] Adam Smith (1778) recognized this as an important factor in the organization of work in his seminal work, *An Inquiry into the Nature and Causes of the Wealth of Nations*. He noted that individuals could become adept at specific tasks so that their "productivity" would increase and contribute to the productivity of the whole operation. Later we will consider this aspect of human nature and the role of human natures in organizing an economic system.

9.3.2.1.5 Role of the Economy

The Nobel Prize in Economics winner Paul Krugman and co-author Robin Wells define economics as "the social science that studies the production, distribution, and consumption of goods and services" (Krugman and Wells 2012, p. 2).

Economics is widely considered a social science since it involves so much of what society does and is formed by social factors such as the propensities of humans noted above. Nevertheless, the economy, studied as subsystem, is found to have internal structures and functions that are the result of factors that are not, strictly speaking, social.

The economy depicted in the figures above is a fiction for the present. There is no global systemic economy (see comment below re: no global governance either). Instead, there are, at present, a diverse set of economic processes that are pursued by different nation-states. These economic processes encompass the Krugman and Wells list of economy sub-processes. They vary based on how the economic processes are governed. Originally, economics was called "political economy" since the political philosophies and beliefs (ideologies) held in different states factor strongly into how the economy operates in actuality. However, there is a generic pattern of economic systems that we need to consider. This pattern transcends the nature of specific political ideologies; it is valid for all economic processes. In fact,

[15] Admittedly this is a near taboo subject in the world today. On the face of it, it appears to be a throwback to the nature vs. nurture arguments and is in direct conflict with popular theories of education. Nevertheless, the scientific evidence for variance in skills and talents is quite solid. So, we will persist in requiring this as a factor.

as we will explore later in this chapter, the ideologies that drive the political process and result in the kinds of governance we see (for example, capitalism vs. socialism) are all too often the source of systemic pathologies and result in economic failures of various kinds. All economies appear to share certain repeating patterns of organization and function regardless of the organization of either the political or governance corresponding subsystems. This not only is the case for human economies but can be found true of all CAS/CAESs.

Chapter 10 provides a general economy model archetype. The claim, developed there, is that the human economy is just the expression of a deeper dynamic in all CAS/CAESs. The living molecular dynamics within a single cell—metabolism—is an economy, as is the physiology of a multicellular organism. The latter is a suprasystem that extends the metabolic processes across cells, tissues, and organs. Physiology supports the metabolism of individual cells that have become specialized for specific tasks and would otherwise not be able to survive on their own. All of the cells/tissues are mutually supporting. So too, is the case of the human economy. We will see the pattern of a social group (or society) in which individuals have become specialized and need the rest of the society in order to provide for the supports that they themselves are not able to provide.

The economic process is what keeps a system *alive*. It imports energy, material, and information. It transforms materials such that they become more "useful"; what we will call lower entropy wealth. It uses energy to accomplish these transformations and to generate forces for motion (for example, animals). It stockpiles wealth for purposes of backup against times when resources might not be available. It distributes wealth appropriately in support of all member sub-processes. The living biomass in the system (all of us and our pets and ornamental plants) consumes some portion of the wealth to stay alive and reproduce next generations according to the biological mandate.

9.3.2.1.6 Role of Governance

The governance subsystem depicted in Figs. 9.1 and 9.2 is the whole societal governance, responsible for the regulation of all of the other subsystems for the purpose of fulfilling the HSS purpose (which you may recall we could not discover in Chap. 7!). The classical humanist view is that the HSS purpose ought to be to make every human being happy or some species-centric ultimate state. If that were true, then our species has failed miserably.

Governance as a process is the information and knowledge processing needed to regulate the dynamics of the economic system and the behaviors of the components of that system, particularly some individual humans who might have a disinclination to cooperate with others for mutual benefit. What a good governance system should do is keep the whole CAS/CAES operating as close to optimal as possible. For the HSS this would mean a global equitable distribution of wealth so that all humans would benefit.

One possible reason that this is not the case in our modern world is that throughout history we have never had a global governance function. The present United Nations does not even begin to qualify. However, it is interesting to note that there has been some tendency toward unifications of nations from principalities, for example. Nations becoming empires was a further attempt at some kind of "scaling up," though one that failed for good reasons. The creation of the European Union and the Eurozone was perhaps the most recent attempt to unify a diverse body into one governance structure, but it too has had its problems (at this writing, Great Britain is going through withdrawal from the EU). And it is not quite a full governance structure.

To date, all examples of governance, at least at large scales, have been dysfunctional. As with the case of the economy, there is a universal pattern of governance as a process. This should not be surprising since the two subsystems are so tightly coupled. Chapter 10 provides the basic model template of a CAS/CAES governance subsystem. Governance is a recursively implemented hierarchical cybernetic system. That is, the basic pattern is applied, fractal like, on many scales, from individual agents, through small social groups, through organizations, through large social groups, to whole societies such as nation-states. It should be equally applicable to the whole of the HSS.

Chapter 10 shows how the governance process is applicable in several natural CAS/CAES. These complex systems, from the single cell up through the whole HSS, are composed of myriad special sub-processes (specialists) who need to work together in a coordinated fashion in order for the whole system to be sustainable. Failures of the governance system in place result in dysfunction of the economy and other non-economic processes that involve working together (for example, the science process to produce new knowledge). If we examine the existing governance systems in our world today, we can easily find many deviations from the template model, especially at the largest scales. Much of the failures of the governance systems can be traced to faulty understanding of the model and the various ideological beliefs that have dominated the political process and hence shaping of the governance system.

As compared with natural governance subsystems, say for the governance of cellular metabolism or the body's physiology, the HSS governance subsystem is far from evolved to be structurally and functionally patterned after the template model. This subsystem along with the economic subsystem and the political subsystem (below) are undergoing coevolution to produce the kinds of cultures we witness. The human subsystem (the biological component) and the cultures are themselves locked in a kind of arms race.

At the same time another kind of unification process seems to be underway, going under the name of "globalization," meaning the organization of a global economic system. The governance of this economic system is based on the neoclassical economics belief in "free markets," coupled with multi-party agreements on trade policies and procedures. Free market solutions to all social problems are the ideology of libertarianism, and the mechanism for attaining the rewards under a free market society is held to be some form of capitalism, of which there are several flavors.

The diversity of governance architectures among nation-states is reminiscent of the trend toward specialization that led to economies in the first place. Globalization and the adoption of some form of capitalism may very well represent the setup for the kind of grand transition from local-scale disparate individuals a large-scale social cohesion. Time will tell.

9.3.2.1.7 Role of the Political System

The evolution of the governance subsystems within different groups in the HSS is a truly evolutionary process involving auto-organization, emergence, and selection (Mobus and Kalton 2015, Chaps. 10 and 11) leading to variations on structures of hierarchical governance and coordination in economic processes. Throughout history human groups have been trying out different economic and governance processes that they "invented" based on desires and intentions, but without the kinds of template models we get from systems science (as described in Chap. 10). It has been one grand experiment with ideas, one after another. As humans spread out over the world and as human groups continuously consolidated into larger economic units, many different "theories" regarding how the social system should be managed have emerged and been tested, sometimes through competition, one with another, but as often by their fitness in the physical environment in which they developed (Diamond 2005).

Perhaps the least understood subsystem of the HSS is the process by which humans make decisions regarding how governance should be performed and who will be responsible. We will claim that this problem exists because of a lack of deep understanding of the governance process itself—even when an exemplar appears to be successful. Even so, decisions about how to proceed must be made, with or without systems understanding.

The variations on political decision-making are beyond the scope of this book. Human history provides the examples. From tribal councils to Devine Rights, to democracy, humans have been experimenting around with leadership authority models, probably since we became humans. These form the basis of a set of beliefs or ideologies about what is the right approach to setting up and running a governance subsystem. The political process is basically one of deciding for the HSS how that will be done.

In this sense, the product of the political subsystem is the set of instructions (policies) that will be put into positions of power, which is the government. As with the economy and the governance system, there has never been a single overarching political process (at least not one that didn't involve war).

Humans become participants in the political subsystem whenever they are concerned with decisions about governance or economic issues.

9.3.2.2 A Note on Ideologies

What makes the Core Society so fuzzy is the role of ideology in guiding human decision-making. Ideological beliefs about human nature, how the economy should work, how the governance should work, and how governance decisions should be made have, so far, been the main source of influence on decision makers. The problem with that is that ideologies are historical artifacts—concepts that were someone's best guess at some time in the past that seemed to make sense in the context of the time but have generally been found insufficient in terms of scaling up toward global levels.

Contrast ideological beliefs with scientific theories and models. The latter are based on empirical research and mathematical rigor. The former are just guesses, sometimes informed by somewhat empirical evidence, but more often based on wishes. An ideology that results in a political or economic experiment is somewhat like a hypothesis. The problem is that an ideology is not thrown out when the experiment proves it invalid (falsified). Humans are bad at observing real dynamics of the very systems in which they are participant components and very good at inventing excuses. Ideologies are cherished because they work to help people define themselves along with their belonging to some group and so become part of our personas. Once held, it is extremely difficult to overthrow their influence on thinking.

Ideologies may be related to superstition, the development of a belief that is not grounded in reality. Humans notoriously form such beliefs when they think they apprehend a correlation as a causality and especially when they perceive some particular consequence that either serves their purposes or poses a threat. For example, most people strongly believe that economic growth is good and that an economy should go on growing forever. They believe that an end to growth portends economic collapse.

An ideology of particular significance in our assessment of the economy is the so-called neoliberal belief in "free markets"[16] and their ability to solve all economic problems (which, we must say, are to most economists all problems). This ideology has invaded the organizing processes for the other subsystems as previously noted. The subsystems of healthcare and education have been colonized in modern culture by attempts to model them as economic problems, but with the caveat that they should also operate as economic processes with at least cost minimization in mind, if not profit making.

[16] The scare quotes used here serve two purposes. One is to remind us that a free market is just an idea and the other to recognize that no real instance of a truly free market has ever existed so far as we know. Thus, the idea has never really been tested and empirically confirmed or rejected.

9.3.2.3 Roles of Healthcare, Education, Science and Engineering, and Humanities, Esthetics and Entertainment

The other subsystems shown in Figs. 9.1 and 9.2 are still part of and influence the core society functions. But each has its own purpose within the HSS. Whether they are fulfilling those purposes in the modern world may be coming under increasingly pressing question due to the colonization of neoclassical capitalistic economics. Even so, we have several centuries of experience with them and can fairly easily identify their purposes.

The Healthcare subsystem is responsible for keeping humans reasonably well and making sure that our reproductive functions are kept in working order. Humans, in the form of healthcare providers, enter the subsystem in the form of jobs (for which they receive remuneration). But they also enter as patients, dare we say, consumers of healthcare services.

This might be an opportune moment to point out something about our modern economic system (at least in the developed world) that has very much contributed to the fuzziness of the HSS, and that is the invasion of a market-based approach to distribution of almost all social subsystems. It is the case that a market ideology has supplanted the social purposes and mechanisms for most, if not all, of our subsystems. While there are still some countries that maintain a healthcare system that does not depend (at least entirely) on market mechanisms of supply and demand discovering prices for services, in the United States, the invasion has been complete. A discussion of the market mechanism will be covered below, and in Chap. 10, as it pertains to one aspect of governance.

The education system is another process that is being invaded by market-style mechanisms.[17] The role of the education system has, historically, been the enculturation of children and preparing them for useful lives. That had meant developing their minds to be able to think about many kinds of social issues as well as be able to master skills deemed necessary for gainful employment. As schools are pushed (by well-meaning politicians and social thought leaders) to become more business-like, the emphasis has shifted to the latter aspect. Today, what it means to be educated is very different from what it had meant through history. We should point out that there is no evidence that market mechanisms and capitalistic approaches have served the education subsystem well.

Science and engineering involve all of the sub-processes that expand human knowledge and its application to practical use. Part IV is devoted to "Artifice" and will explore the design and engineering processes more deeply. As noted in Chap. 1, scientists discover new knowledge and engineers determine ways to exploit it. Engineers have generally been more closely associated with the commercialization of knowledge and so have a very tight relation with the economy. But, if we consider that when an engineer is doing engineering (i.e., designing a new widget) they

[17] At this writing, in the United States, the push for vouchers and charter schools is strong at the national leadership level. But the notions of efficiency and productivity, etc. are overrunning the higher education world as well. Cf. Washburn (2005).

are performing a function that is not, strictly speaking, an economic decision.[18] Engineers, as a kind of human specialist, are more motivated to solving technical problems without necessarily considering costs as a primary driver. However, they are also consumers who need to eat and clothe, shelter, and otherwise provision their families. They need to do the job according to the constraints of the employer in order to keep their jobs.

Again, we must note the invasion of market and capitalistic ideology into this arena as well. After World War II, especially in the United States, agencies were created to fund research in major universities as part of a push to enhance innovation in technology and healthcare. Projects were (and are) chosen on a competitive basis and money flows accordingly. This is actually one of the factors that is "businessifying" higher education—money must be attracted and managed.

The Humanities, Esthetics, and Entertainment, lumped together because you would think these would be very un-economic-like, are thought to have the purpose of maintaining human mental health. We study history, literature, and art for the sake of understanding ourselves. We produce plays and operas and concerts because these bring pleasure to us and soothe our demons. Humans enter these activities, as the others, as both producers and recipients of the works. Not all that long ago, these activities were supported financially by patrons (some still are to some extent). But today, they are, as with the other subsystems, being taken over by market and capital ideologies. Art is collected as much for profit as esthetic pleasure.

All of these activities have predated the market mentality and so our interest in tracing their histories should play an important role in decomposing them as subsystems of the HSS. They are among the most ancient activities of the HSS and one might easily argue that before they are given over entirely to market logic, we might wonder how they were so successful at helping human beings be fit before there was capitalism.

9.4 Deep Systems Analysis Applied to the Economic System

Now that we have tackled the problem of defuzzification of the HSS and shown how various subsystems might be delineated (or circumscribed) so that further decomposition (analysis) might be started, we can proceed. The HSS is, to be certain, an extremely complex adaptive (and evolvable) system. But that does not mean it is unamenable to analysis. If it were, then the whole program of the social sciences would be meaningless. There would be no point to doing social science if the HSS were inscrutable. Nor do we believe that is the case. The science might be very hard, indeed seem impossible at times without an overarching framework to guide the

[18] There is a practice in engineering called "Value engineering" in which the design objective is to produce a product that will have least cost while not sacrificing functionality. Of late another branch of engineering practice, called "Life Cycle engineering," attempts to minimize up-front and end-of-life costs, including externalities such as disposal costs.

work, as we are proposing. Rather, we proceed to analyze the HSS as a system and hope to show that it is as understandable as any system in the Universe.

9.4.1 Analysis of the Environment

The first step is to situate the Economy as the SOI and delineate its Environment. From Fig. 9.2, we can see that the Environment of the Economy will include the other subsystems of the HSS along with entities in the Ecos (summarized as energy, matter, heat, and wastes along with a question mark with respect to any products). Figure 9.3 shows the first step with the Economy treated as an opaque object. We proceed with analysis using Sects. 6.5 and 6.6 of Chap. 6, System Identification and Environment Analysis, respectively. Here we will collapse these two showing the final results of the level-1 analysis. In the following figures, we have also aggregated a number of flows of the same types rather than attempt to show all of the details. The purpose of this exercise is to illustrate the process of analysis of a CAES but not to produce a fully detailed one.

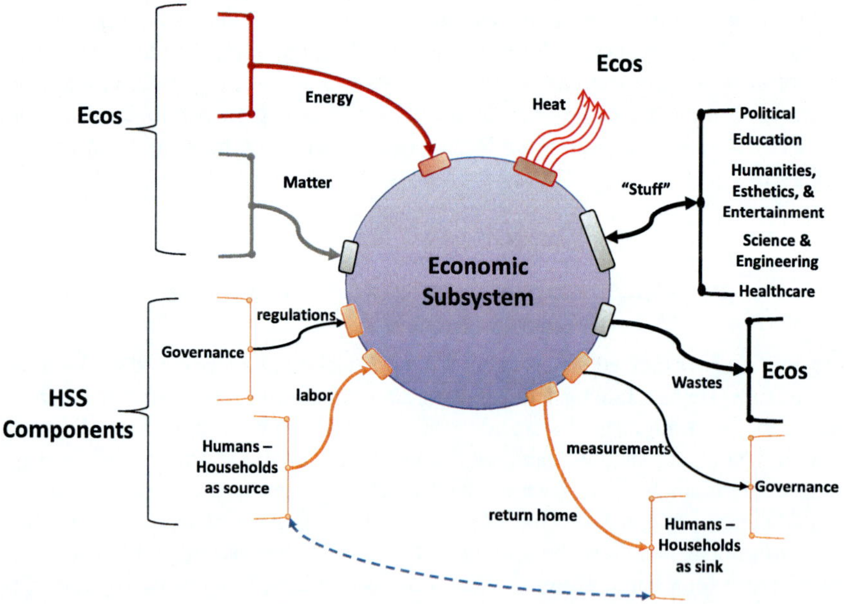

Fig. 9.3 An initial environmental analysis of the Economic subsystem reveals the major kinds of inputs and outputs, environmental entities, and relations. Note the handling of the Humans—Households as both sources and sinks. Following the convention of inflows coming from the left and outflows going to the right, the dashed blue arrow indicates these two entities are one and the same

Figure 9.4 shows a similar drawing for the Core Society. We will be increasingly focusing on the economy from the standpoint of the energy sector and how it affects interactions with the political, human, and governance subsystems along with considerations for the role that science and engineering play in the evolution of the economy. In particular, the reader should attend to the red arrows, labeled "Energy" or "Energies" (since there may be several different forms) and also to the radiation of "Heat" from various subsystems as we proceed. The economy of a CAS/CAES is the engine that keeps the system alive and thriving. That is why we will focus on it in this chapter.

Figure 9.3 is admittedly a cartoon version of what a system level-1 map would look like. We've lumped like-type entities and flows together under the generic labels that categorize them, for example, Energy constitutes all types of high potential energies that will enter the Economy. There isn't just one energy source (although one might argue that the sun almost is) or one flow of energy into the system. In Fig. 9.5 we will begin decomposing the energy flow concepts into particulars. Similarly, Humans are not only sinks for consumer goods and services but

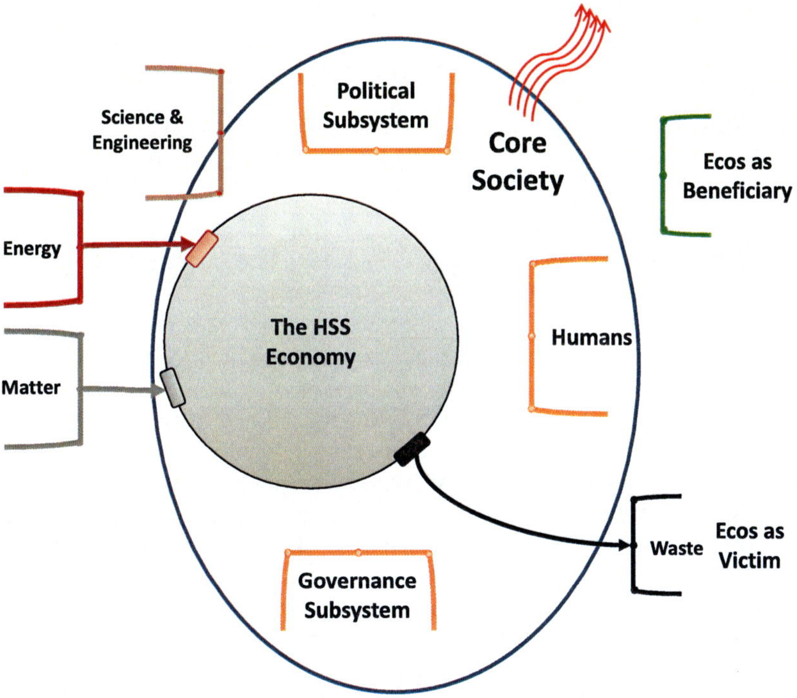

Fig. 9.4 The Economy is treated as the SOI. Sources and sinks include the other entities in the "core society" as well as in the Ecos. Also included is the Science and Engineering subsystem of the HSS. The "Ecos as Beneficiary" label replaces the question mark in Fig. 9.2 with the provision that the HSS Economy might produce some product that is beneficial to the Ecos. Wastes are otherwise

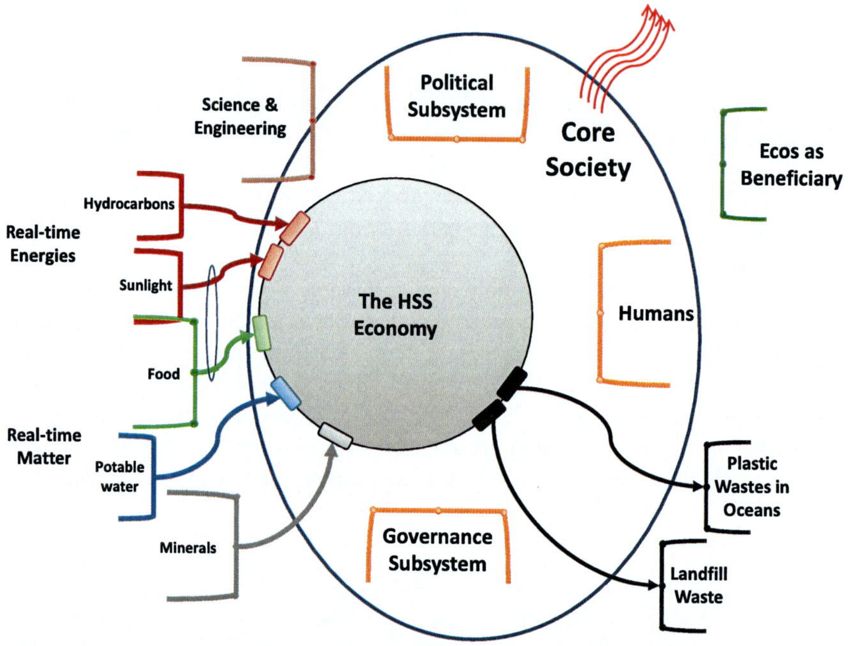

Fig. 9.5 The process of environment analysis and discovery of particular flows from or to environmental entities shows how categories can be broken into these particulars as described in Chap. 6. Energy sources have been broken into two straightforward real-time entities. Food is a general category that encompasses both energy and materials. The two flows are shown linked. Just two real-time matter flows have also been shown as sub-categories of material flows. The same is true of wastes. Interfaces for all of these kinds of flows have been preliminarily identified

also sources of labor and agency. As we proceed, we will show more details of the various roles of the entities (discussed above) and break out specific flows as they pertain to the analysis of the Economy subsystem.

Figure 9.5 begins the process of discovering inflows and outflows (as described in Chap. 6, Sect. 6.4.3.2). We take the major categories and start to identify particular and specific flows from/to the Ecos. The few examples demonstrate how ontological general categories begin to be parsed into more specific categories until we get down to very specific ones. For example, in the figure, the source called "hydrocarbons" would yield two (at least) sources and flows: oil and natural gas. Those, in turn, would be broken down into grades and reservoirs where found. We'll see this in greater detail as we drill down into the energy sector sub-subsystem below.

The next step in analysis of the environment of the Economy is to similarly find the flows to/from the other entities, the ones that constitute the Core Society, of course, but also all the other entities discovered in the HSS decomposition shown in Fig. 9.2. The Economy interacts strongly with every other entity (we could use the term institution almost synonymously) in the HSS. For example, the Economy is what delivers energy to the other entities. Another way of saying this is that the extraction and conversion of free energy to do work are an economic activity. In this example deconstruction, we will focus on the energy sector primarily.

Figure 9.6 shows the analysis starting at the major outputs of the economy to the human households. Refer to Sect. 6.5.1.2.

In the figure, we also show the human households both as sinks for the products and as sources, at least for labor (actual humans going to work) and messages that request purchases (in other words, money). One of the outputs of the economy to the households is a similar message, in form, and is labeled "Wages." Workers receive wages for the work they do while members of the economy. They reuse their stock of money to make purchases of goods and services.

In Fig. 9.7, we shift our focus of interest to the interactions between the economy and the governance subsystems. We show only a few of those interactions. The primary flows from and to the governance subsystem are messages. Going to the governance system we find, for example, messages containing "Measurements" that the governing process takes of activities and results from the economy. But we also note how the economy transfers some of its labor force to the governing process—it needs people to be the decision agents at all levels. Major inputs to the economy from the governance system are the various constraints, constitutional, laws, rules, and regulations that shape how the economy can work by determining what it

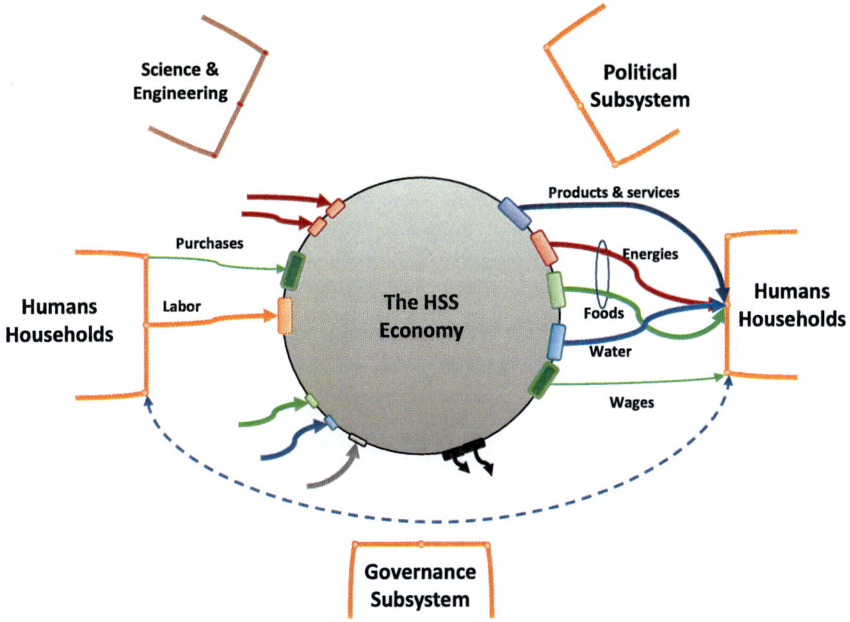

Fig. 9.6 Starting with the major outputs from the Economy to the Human Households, the sketch shows households receiving goods, services, energies, etc. Human Households show up twice, connected by the dashed blue arrow, because households act as sources for some things that enter the Economy and as sinks for some things that the Economy provides (see text for the explanation). Also shown are the other environment entities we will briefly consider in the analysis

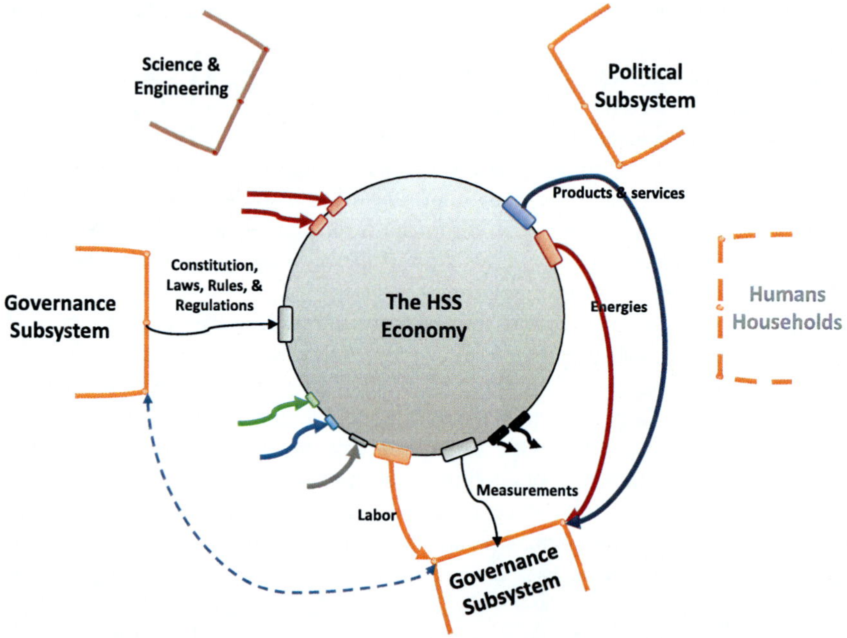

Fig. 9.7 Continuing the environment analysis, now focusing on the inputs and outputs from the governance subsystem, we find more inputs and outputs along with their interfaces

cannot do. Note that for the purposes of this analysis we are counting monetary policies as part of the governance process. In the United States, monetary policy is largely left to a supposedly independent (from the official government) institution, the Federal Reserve. While this body is supposed to operate outside the influence of politics, it is still fulfilling a governance function, so we lump it in in our analysis. Besides, different nation-states have different means of setting monetary policies and they are all still part of governance.

9.4.2 *Approximation of the System Function (See Sect. 6.5.3)*

Figure Fig. 9.19 provides us with a view of an initial approximation of the economy's overall function. That function is to convert raw materials and energy into human assets of various kinds and durations, including human and human-associated biomass. The graph shows how net energy, the energy obtained to do useful work after subtracting out the energy required to obtain the total, is related to the growth in all assets. This is the ***T*** object in Eq. 4.1 and explained in Sect. 4.3.3.4 at level 0, when the economic subsystem becomes the new system of interest (noting that the economic subsystem is at level 1 within the HSS and level 2 within the Ecos). Of course, assets have lifetimes. Food is consumed shortly after being produced. Appliances may last for several decades. A house for many decades. Heavy equipment and factories have variable lifetimes but are generally considered "fixed" assets. And land that has been developed, while still available indefinitely afterward, usually declines in value due to soil erosion or becoming a waste dump. This means that the transformation function needs to include a provision for the output of waste material—garbage and effluents—as well as waste heat.

9.5 Identification of the Economic Subsystem

In the following section, we shall outline the system identification process for the HSS economic subsystem, which will become the system of interest. We start with the emergence of the HSS economic subsystem from the social context which constitutes the main strategy of the species. We work together in order to make a living.

9.5.1 Resolving the Boundary of the Economy as a Subsystem

Section 9.3.2 provided an example of dealing with a fuzzy system by treating the boundary as if it were crisp and resolving the flows into and out of the system. The economy is so intertwined with all other subsystems that it is exceedingly hard to put your finger on a boundary. In part this is due to the fact that money is the universal currency for all institutions. In part it is because all other subsystems involve people as "workers" who receive monetary incomes so that they can participate as consumers in the market economy. And in part it is because all of the other subsystems' activities are strongly affected by and themselves strongly affect the economic activities. Nevertheless, we will try to identify the boundaries of the economic subsystem on the basis of the work processes within it having roles that wind up supporting the consumption of the members of society. We tackled the differentiation of the economic subsystem from other subsystems above in Sect. 9.4.1 Analysis of the Environment.

Even so, there are going to remain difficult boundary resolving decisions to be made. Recall in Fig. 9.3 we showed a number of these other subsystems as being in the "environment" of the economy. For example, Science and Engineering were treated as separate from the economy for the reason that the focus of the work processes in these fields is not so much on consumption products as on knowledge production.[19] The products of Science and Engineering certainly return to the

[19] It is difficult to differentiate things like corporate R&D as compared, for example, to what transpires in university and national laboratories. The Science and Engineering depicted in Fig. 9.3 more readily applies to the latter two.

economy in the form of improvements in tools and processes or just general knowledge, which may or may not have a direct impact on economic activities.

One way to develop a sense of what should and should not be considered as true economic activity versus, for example, repair and maintenance or adaptivity, which are clearly delineated in the CAES archetype model (see Chaps. 10 and 13), is to take an historical perspective or, in other words, look at what economies started out as in more primitive societies.

9.5.2 The Historical Perspective

Karl Polanyi (2001)[20] asserted that to understand the modern economy (or the one extant in the 1940s) and its supposed basis in "the market," it is necessary to tease out the most fundamental elements of economic process and that this could be accomplished by observing the economies of long ago that were not based on any notion of a market.[21] To this he turned to social anthropologists' descriptions of a number of contemporaneous "primitive" peoples. He identified a number of fundamental patterns among the various groups, regardless of the specific social organizations, that constituted economies.

Polanyi restricts the concept of an economy to situations of trade that involves the *profit motive*. We will examine this motive later in Sect. 9.5.3.1 A Note on the Evolution. For him the more salient aspects of human transactions are based on reciprocity, redistribution, and householding (production for one's own benefit). The first two are based on a group phenomenon among social beings.

Homo sapiens evolved a fitness strategy based on cleverness and an ability to see how naturally occurring objects could be fashioned into tools that compensated for the species' lack of physique (for example, lack of canine teeth). Tools are any object that assists the human pursue their biological mandate. They provide leverage allowing a relatively weak human to accomplish work that they would otherwise not be able to do. Figure 9.6 depicts a single individual as a unit of biological and "cultural" work. The human gathers resources (like a rock), consumes food for energy, and does work on the rock, shaping it so that it can be used to do additional

---

[20] We are impressed by much of what Polanyi wrote regarding the evolution of the modern market-based economy. The world of the mid-twentieth century was already heavily industrial and the beliefs in market mechanisms had become entrenched. For him to have gained the many insights he did was quite an achievement.

[21] Polanyi argued that the economy is not the same as the market(s). Economy, for him, is the natural result of social conditions and need not involve the kind of market mechanisms envisioned by the nineteenth-century economists, one that is guided, as if by an invisible hand—or, in other words, a self-regulating mechanism.

work, for example, as an axe used to chop wood for the fire. Products of the work (like clothing and shelter) can accumulate as assets.[22]

9.5.2.1 A Note on the Meaning of Tools

A class of assets that are very important is the *tool*. Any instrument or procedure that increases the efficiency of a work process is a tool. Tools provide leverage in the work process, thereby increasing the efficiency of the work with respect to the amount of product relative to the inputs of matter and energy. Reducing scrap or heat waste in a process translates into increased productivity of the process. In thermodynamic terms, the same amount of output can be produced with reduced energy (for example, labor) inputs to the process. This is the same thing as increasing the availability of total free energy to the system. If any given process requires less energy for the same output, then the saved free energy is available to either do more of the same work, or do different work.

Tools are the key to understanding productivity, the ratio of outputs to inputs of energy and materials, and differentiation or the expansion of an economic system into different kinds of work. It is also the key to understanding economic growth and the accumulation of assets. If the use of a plow increases the agricultural yield of a farmer (per unit time), then the farmer can and will produce excess crops. In the case that the crops are grains like wheat or barely, which can be stored for long periods of time, the farmer is generally motivated to do so, rather than simply work less to produce the same output volume as before the plow. This propensity, which is driven by the biological mandate and the wisdom of storing excess food for emergency exigencies, will show itself later in the evolution of economic systems in the form of *savings*.

9.5.2.2 Tools and Proto-Capitalism

Tools come into existence by work done on some lesser organized material making it suitable to be used in other work. For example, a farmer may devote some time and resources to gathering the materials needed to make a plow, shape them to the parts of the plow, and assemble the device. The resources came at a cost, as did the energy expended in the work. The farmer had to take time out from other work to make the plow. But all of this can be seen as investments since the result is more efficient production of food and potentially more excess. Generally, tools will have long life spans relative to the overall amount of work they are put to. The plow should last the farmer many growing seasons before he might need to build another

[22] For a more comprehensive introduction to early humans and the evolution of living arrangements, a particularly good story is told by Yuval Noah Harari (2011, Chap. 5).

one. Thus, the payoff extends over the long haul rather than just being a one-off event.

The fundamental notion that investment in tool creation has a payoff is at the root of the concept of capitalism as a way to make life better. The farmer had to save up energy and resources over some span of time until he could craft the plow, but afterward he reaped (pun intended) great rewards. He started with some capital (raw material assets and energy), made the investments to build the tool, and was rewarded doing so. He may even have been taking a risk in not doing more of his usual farming work that if the growing season turned bad, he would have lost capital. These notions were surely held by early farming communities. They were certainly institutionalized as those communities coalesced into hierarchically organized civilizations (Nissen et al. 1993).

Figure 9.8 graphically summarizes the situation of the work of an individual hunter or agriculturist in the late Paleolithic to early Neolithic ages. Tools were primitive; the concepts of inventing tools were brand new but so also was the amount of effort that could be afforded investing in them. The figure captures the earliest relation between the Ecos and the human being and the origin of the capital-labor economy.

Of course, it is not really the case that only individuals did work and benefited from it as individuals. Humans were always social creatures, a trait inherited from their ancestral species and shared with many of their cousin species today. To really understand the ontogenesis of the capital-labor economy we have to look at early group economies.

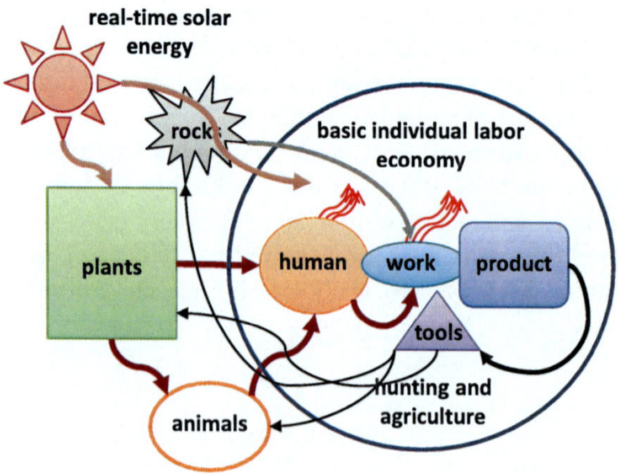

Fig. 9.8 A single individual unit of production doing work to fulfill part of the biological mandate

9.5.2.3 Societies and Economies

An individual worker as shown in Fig. 9.6 is largely a fictitious entity. The circumscription of a boundary around the individual is more an exercise in defuzzification. The main intent of the depiction is to recognize the self-sufficiency aspects of a biological unit. Humans, early in their evolution, adopted the strategy of working in groups or societies in order to greatly enhance their survivability, given their lack of individualistic "equipment" for exploiting their African environment. Working together in coordination they were able to do things that no individual could have done alone.[23]

The key to how this could work lay in the production of tools that could be shared. And that may have been enhanced by the propensity of individuals to exploit talents, to specialize. As Polanyi did, we can turn to cultural anthropologists and their studies of contemporaneous "primitive"[24] peoples to see this phenomenon in action. What we see in these societies is the division of labor along several dimensions, for example, gender or age, and especially along lines of talent. Figure 9.9 depicts this situation in a simple four-person society. Each individual does

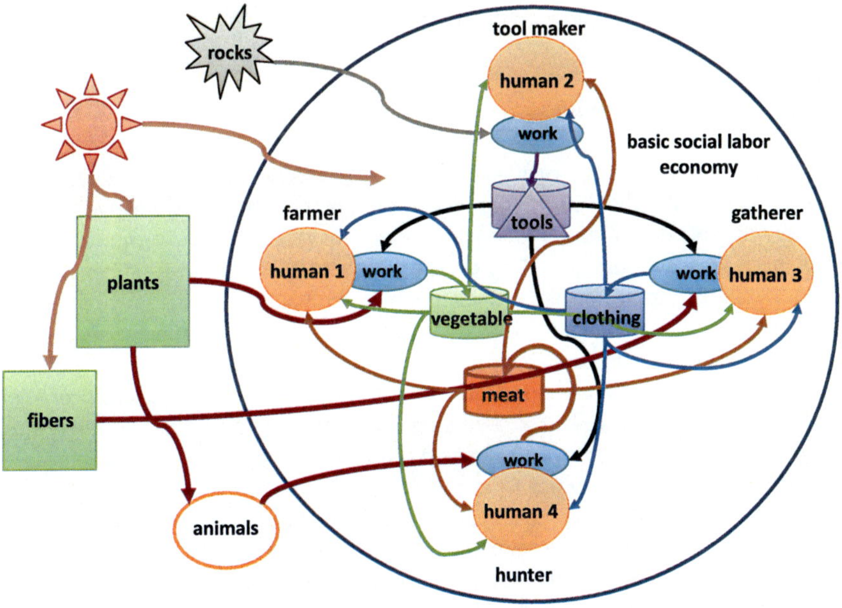

Fig. 9.9 A group of individuals specialize (to some degree) and trade their products with one another in the interest of the whole group

[23] This is yet another example of the auto-organization—emergence process in the ontogenic cycle.

[24] The use of the term primitive here is conventional and not derogatory. Primitive tribes in many, especially, equatorial and tropical areas of the world are generally recognized to be pre-literate and their societies are based on technologies that date back to prehistoric times.

necessary work and shares their output with others in the group, who, in turn, share their products with the others. As long as the basic needs of the biological mandate are filled, a society can exist in a stable group. Polanyi called this kind of transactional relation an economy but did not allow that this was a market.

While we are generally in alignment with Polanyi's requirements for the existence of a true market (i.e., the existence of barter trade and profit motives), we still note that from a systems perspective, the essence of the transactions that took place in such societies (for example, reciprocity and redistribution as described by Polanyi) still forms a complex network of relations that are mediated by judgements of the individuals who are monitoring the equitability of the transactions.

In this kind of primitive society each individual has complete access to understanding the nature of the work that others needed to do to produce their products. Under this kind of complete transparency, it is relatively easy to judge the fairness of trades. Human #3 has a pretty good idea of what it took for human #4 to bring home the meat. They are in a position to compare their efforts in gathering to #4's efforts in hunting. It doesn't have to be precise, of course. They can use proxy measures to make the assessment, such as how long were the hunters gone versus how long it took to collect all of the roots the gatherers obtained. People knew how much of various foods they needed to eat to survive, so subconsciously they could make these comparisons. Their implicit preferences would show up in their intuitions.

9.5.2.4 Tribes and beyond

Clearly the human strategy of working in groups was wildly successful. Human beings became so adept at exploiting their environments that they were able to successfully migrate out of Africa (on several occasions it seems) and learn to live in a wide variety of environments. They also were successful in expanding their group sizes. Extended families became tribes in which the magic of cooperative production provided the species with a capacity to deepen its ability to exploit many different nuances of environments.

Eventually human beings discovered the ability to plant and reap, particularly grains. Once this threshold of exploitation was achieved, we were seemingly no longer constrained in any practical sense. Grains are eminently preservable and techniques for preserving other food stuffs were developed, such as drying meats and pickling vegetables. The concepts of long-term storage of food stuffs, of producing long-lasting tools and shelters, of maintaining a geographical space for seasonal planting led quickly to the notion of permanence of location and the growth of groups into many-family tribes. Humans have a natural desire to identify with their group and so for tribal groups larger than, say several hundred, it became the custom to share practices and beliefs about the nature of the world in order to bond the tribe into a cohesive "cultural" unit with a sense of identity and place.

The key economic aspects of the growing size of tribes were two-dimensional. In one dimension, the tendency for individuals to exploit variations in talents to become specialized in the kinds of work processes they would excel in continued and refined, as technologies gradually improved. In the other dimension, the overall management of the tribe also benefited from specialization in decision-type space (recall the separation of decision types discussed in section [Management (Agent) Decision Types in the HCGS]. In this dimension, the separation of roles falls largely along the categories of ages. Adult members of a tribe might be categorized as young (just having gone through some kind of transition ritual), middle-aged (child-rearing and general work processes for the tribe), and the elders (usually characterized as wizened). Figure 9.8 depicts these two dimensions of tribal organization along with communications lines needed to cooperate among specialty work processes and among those who organize and manage work and those who make long-term decisions for the tribe.[25]

9.5.2.5 Settlements, Communities, Trade, and Early Markets

The Agricultural Revolution, with its resultant emphasis on larger-scale societies, settlement (and consequent magnification of territorial senses), and continued specialization of work led ultimately to social organizations that Harari (2011) called "Kingdoms." These were city-state operations (best examples coming from the Middle East, Babylonia, etc.). We know for a fact that peoples from many regions were traveling and in contact with other kingdoms. We know that they traded specialty goods such as fabrics and spices. So very early on, human beings were developing trading relations that involved transactions based on perceived value of those goods (Fig. 9.10).

Even within a given kingdom a vibrant place for trading goods produced within the kingdom by specialists was developing. Producers of goods (and services) gathered in a specific geographical area of cities, bringing their wares and produce to be offered for trade. It is likely that the first markets were based on barter, but we know from records (cuneiform and later more sophisticated numerical representations etched in clay and marked on papyrus) that humans in these communities were already sophisticated enough, 6000–8000 years ago to count and measure quantities of commodities as a means of keeping track of transactions. Some forms of these records constituted IOUs that, while not yet money, did allow people to trade symbols representing value. As kingdoms evolved in sophistication of writing and numeracy, eventually these symbols were incorporated into stylized tokens that represented quantities of commodities of value to the general population. The idea of token trade for value emerged on the scene of human economy not too long after the whole concept of controlling the production of food and the subsequent support of the biological mandate arose.

Very many modern notions of what an economy is emerged in that early time. Economic ideas have evolved since then in much the same way that biological

[25] Again, the reader will need to refer to Chap. 12 in order to get background on the management aspects of this hierarchical arrangement.

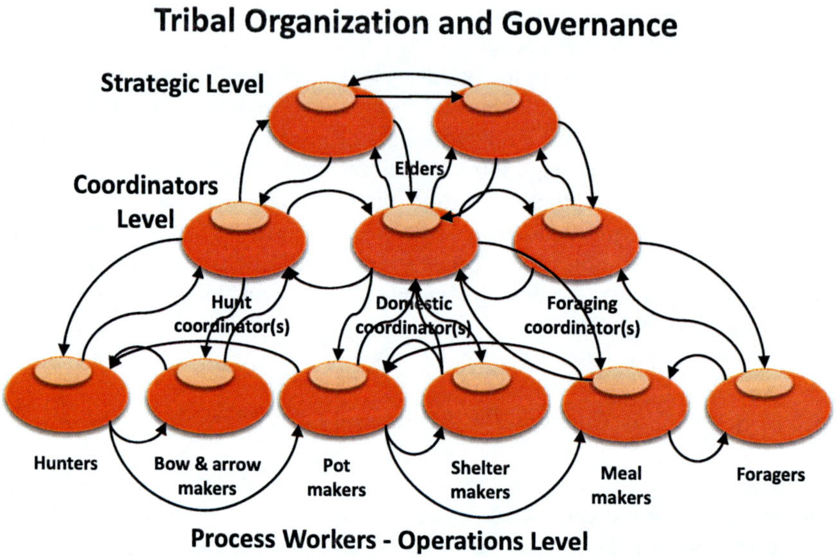

Fig. 9.10 Human communities grew and individuals within a tribe could become specialists (at least to some degree) to gain efficiencies in work. The tribal organization, still found in communities in various parts of the world today, shows the beginnings of both the distribution of work and of organizational governance. In this figure, each process is seen to have its own internal governance subsystem reflecting the fractal-like structure of the HCGS covered in Chap. 12

species have evolved once life actually emerged some 3.5 billion years ago and much the same way languages have evolved among humans as they spread out over the globe. When we take a careful look at economic activity in the modern world, we need to keep this in mind. There is a reason why some particular economic processes, like capitalistic or socialistic operations, are the way they are today based on these origins. We need to turn now to the modern perspective with this in mind.

9.5.3 A Modern Perspective

We switch now to a modern perspective on the HSS Economy as a subsystem. The objective of this section is to do a top-down decomposition of the modern economy. We will not be guided in following this procedure by what the current science of economics says we'll find. Rather we are going to be guided by systems principles and the methodologies covered in previous chapters. In doing so we will see most of the same sorts of subsystems, such as corporations and markets, but we will also see new relations and functions that have been masked in standard economics by assumptions and premises which turn out to be invalid from the systems view. We will also uncover a few systemic pathologies that may not be readily recognized as such because they have been a part of societies for so long that their manifestations are considered normal.

The HSS began much as described in Sect. 9.5.2 The Historical Perspective, above, and has evolved into what we find today. We should take a look at a few aspects of that evolution to better understand what the pathologies are.

9.5.3.1 A Note on the Evolution of the HSS Economic System

In the historical perspective above, we presented a model of the earliest small human group economic system in a tribe. This economy was most likely based on sharing and considerable cooperation among the members of the group. It depended on some degree of specialization; however, it was also likely that most members of the tribal unit were capable of doing a variety of tasks if needed—call it weak specialization. In these earliest societies, the individual agents probably had a pretty good idea of the value of products and services since they might have had to produce and serve themselves at times.

The question about the history of the HSS economy is how did it evolve from this simple set of relations to become the complex network of highly specialized producers and general consumers that we see today? The answer to that question involves the realization that the ontogenesis of the HSS economy and its continuing evolution was a result of (1) residual (from the biological mandate) motivation to possess,[26] (2) intentional creations meant to solve some local problem at a given time, and (3) ignorance on the part of intentional agents with respect to how those solutions would create new and additional problems for the social units (Tainter 1988). Human agents invented not just technologies but also institutions, procedures, and transient organizations in order to meet an immediate need or solve problems seen from a limited perspective and time scale. The number of possible interrelations between human groups and these inventions exploded exponentially even while there was no particular set of constraints on what could be tried out. As a result, almost everything was tried and so-called solutions were not necessarily tested by some encompassing environment, for it was the HSS itself that constituted the environment of the evolving economy and that enwrapping environment has acted as a buffer for the economic experiments. The testing agents, being ignorant, could not truly test and select against anything. So, the kinds of combinations of pre-existing solutions were never adequately tested until, of course, the society in which they were tried finally collapsed as a result of failures of the economic system to provide the same kind of stable, resilient, and sustainable metabolism-like support for the society.

[26] Under conditions of cultural pressures that seem to sanctify greed, this residual motivation is amplified far beyond mere biological needs. The current neoliberal capitalist system seems to have provided the epitome of such a culture. See, for example, Piketty 2014, for a very deep and wide exposition of the disposition of wealth in such a culture.

Over the millennia and centuries, a number of socio-economic model systems have been tried out more or less by happenstance. That is, "it seemed like a good idea at the time." Few have survived to the modern period. Under the influence of the agricultural revolution, which actually developed over several millennia, small groups evolved into larger groups, villages into towns, towns into city-states; eventually, amalgamations of near neighbor trading city-states gave rise to nations or federations and, eventually, empires that learned to foster internal cooperation among significantly many more individuals than the tribal Dunbar's number. But, in addition to cooperation, the larger the scale of societies, the more they incorporated coercion to foster harmony and economic stability (Scott 2017[27]; Morris 2010, 2013, 2015).

Throughout the history of state economic systems in various parts of the world, the interplay between the economic system and the governance system (and both with the tightly coupled political system) has always been authoritarian and hierarchical with those at the top of the pecking order holding considerable wealth compared with the common workers, free or slave. The idea of a hierarchical arrangement in the governance process reflects the needs for instantiating a hierarchical cybernetic governance architecture, as covered in Chap. 10. A complex system cannot operate without such as arrangement. However, the non-egalitarian distribution of wealth as well as the use of top-down coercion to achieve the objectives of governance, a command-and-control method of management, are due to the three factors mentioned above. So, some parts of the design and implementation of the socio-economic system within large-scale states are a natural evolution of the necessary architecture of complex adaptive and evolvable systems (CAS/CAESs; see Chap. 10 for a full description of these systems). But the amplification of the motive to possess and to power, along with the unconstrained inventiveness of the HSS's evolvability, produced many unsustainable subsystems, most of which unfairly favored some members of the society. Slavery is possibly the most blatant example of institutions that arose to accomplish the promotion and support of a few at the terrible expense of many. Modern humans (hopefully, the majority anyway) have finally come to recognize this horrible mistake, both from a moral perspective and from the social consequences with which societies still deal. Unfortunately, in a seemingly milder form, the class system, the same kinds of inequities and systemic dysfunctions appear to be in full form in most, supposedly enlightened societies (Piketty 2014).

The current global human economic system is replete with examples of dysfunctional subsystems that are imposed upon otherwise natural CAS/CAES organization. It would be very difficult to even mention these let alone do an analysis that

[27] Scott points out that the evolution from small gatherings to empires was not a smooth transition in scale. Rather the up-scaling seems to have developed in fits and starts with many communities forming and growing but within as few as three or four generations collapsing. The rise of Bronze Age states was filled with attempts here and there, trying different formulas for governance (mostly coercive), and always fragile against climate, disease, and wars. Even so, it is clear that the trend toward larger coordinated units of populations was the rule globally.

would supply satisfactory guidance in how to design a truly functional system (even so we have undertaken the project elsewhere and will provide a preliminary look at a socio-political-economic designed based on CAS/CAES principles in the near future).

We will draw upon three examples of sub-subsystems of the HSS economy to demonstrate how ignorance, excessive biological mandate, and intentional invention have conspired to create the institutions we have today, all so complex as to be nearly incomprehensible in terms of the roles they were first designed to play in a reasonable economic architecture. The first is the mechanism used in the market economy to signal supply and demand—money. The second is closely related but supported by a logic all its own that does not correspond with anything recognizable in a CAS/CAES economy—debt financing. The third, deriving directly from the biological mandate, is the concept of profit making.

9.5.3.1.1 Money

In Chap. 13, we will show how signals (messages sent along defined channels) are used to control the distribution of materials and energies amongst a network of work processes. In that chapter, we provide an overall model of how those messages are determined by agents governing economic processes. In the case of single-cell metabolism, this signaling is based on relative concentrations of specific molecules such as ADP and ATP. In the early HSS economy after the advent of the rise of agriculture-based communities, the role of a currency (flow of energy) fell to commodity food products, specifically grains.[28] According to scholars such as James Scott (2017), the advent of sedentary pre- and true state communities was made possible primarily because of the nature of various grains (grasses such as barley and wheat) that could be harvested and stored for long periods before being used. Grains, unlike tubers, fruits, or cows, are nearly fluid, that is, each element is small, and, in bulk, are easily partitioned, stored, and transported. They could be contained in unit vessels and units equated to work effort (the calories in a unit equating roughly to those needed to supply the next unit of labor). They supported life, though with some nutritional deficits compared with a wide variety of meats and vegetables on the menu for hunter-gatherers. Nevertheless, the convenience of grains in terms of storage, transport, and nutrition seems to have won out.[29]

[28] In hunter-gatherer societies, the human being was embedded in the ecological food web. Humans are extreme omnivores and used many strategies for food-getting. With agriculture, humans extracted themselves from the natural ecosystem, creating their own system, which many have noted is highly impoverished from a nutritional point of view. However, what humans gave up in the nature of a broad array of nutrient sources, they made up for in terms of consistency of supply.

[29] Very telling in this account is the fact that civilizational states arose in geographic locations where grains could easily be grown such as the Tigris-Euphrates alluvium (the Fertile Crescent) and the Nile flood plains.

Over the course of several millennia communities grew, and with the increasing emphasis on additional tools, housing, and other forms of wealth produced by specialists, communities developed trade markets in which, originally, barter was used to distribute goods (and services) to all members of the community. Grains, being easily handled, became a common denominator trade. One jar of grain for one chicken! Moreover, as communities evolved in terms of governance, with specialists taking over the roles of logistics and tactical managers (as described in Chap. 10), there was a need to support these non-farming agents by collecting a "tax" of some portion of grains from each farmer belonging to the community. Community and government granaries collected these taxes and the wealth was used to support the governance functions.[30]

Jars of grains were somewhat convenient for lubricating trade and collecting taxes but not completely so. They were still bulky. We now know from archeological evidence that sometime relatively early in the Agricultural Revolution and the advent of state-like communities the practice of making marks on the outside of wet clay jars that were to hold a commodity, like barely, allowed an abstract representational form of that wealth—the first form of writing. Cuneiform marks could designate contents, who owned the contents, and other accounting information needed to dispose of the contents. It was not long after that scribes took to accounting for the wealth using marks in clay tablets, divorced from the actual jars holding the wealth (Nissen et al. 1993).[31] And from there trade could be carried out in an abstract conceptual space! Clay tablets could be delivered, for example, to the temples indicating the value of taxes collected.

It was relatively easy to develop more convenient and transferable representations of wealth such as grains, cattle, and slaves in cuneiform tablets. These could be traded instead of bringing all of these wealth objects to a market and bartering them. Essentially these were "IOUs." Whoever held the tablet could claim the grains (or other commodities in storage). At some point in the late Neolithic or early Bronze Age, the idea of embossing a value mark equivalent to some standard quantity of a commodity (again usually barley) on small clay balls or metal coins to represent a more generic representation of wealth took hold. The advantage of these early forms of money as an abstract representation of wealth is that the markers

[30] On a somewhat darker note, it appears that as communities of this sort grew and became more "state"-like, they acquired slaves and non-remunerated (corvée) labor that also had to be fed (and clothed and sheltered). Thus the taxation process developed so that not only the governance agents were supported, but so too were the workers who performed "extra" duties that were not, strictly speaking, governance activities. Those filling governance agent roles apparently started seeing themselves as "important" and deserving of certain "frills" as additional compensation. This is the weak link in the governance structure that the decision agents have always been human individuals who seem to be susceptible to the allure of "power."

[31] The trick worked for more than just jars of grains. These agriculturists invented counting and number systems that let them keep track of how many oxen or goats belonged to whomever. Those number systems would then be applied to the accounting of ownership in general, especially of the possession of land units. Arithmetic and even geometry emerged relatively early in this period and provided the basis (though not the motivation) for statehood to develop.

were no longer tied to specific wealth forms, such as grains. Rather they represented amounts of abstract value and could be used to replace the clay tablets; the actual forms of wealth could be denominated in these markers—five markers might buy a chicken or a small chunk of beef.[32]

The key point here is that these markers or "chits" came to represent an amount of work that had been done, as in either the fashioning of a tool, or the amount of work that could be done, as in the caloric content of a food commodity. These markers represented energy, kinetic or potential, and that is what gave them value. At the time of their invention as abstract representations no one had any notion of energy or work as we know it now from physics. But every person certainly had a clear idea of "effort" needed to sustain their own and their family's lives. So, the evaluation of value was pretty clear. The chits, the new invention, money, was linked to real wealth, which for Neolithic peoples meant sustenance.

The problem was who would have the authority to validate the claims of value? Who would say that x number of these markers were worth some quantity of grain or a chicken? At the same time that people were experimenting with various kinds of IOUs like this, communities had evolved into more complex social hierarchical states with bureaucracies and state heads. These states represented the evolution of governance structures needed to coordinate the activities of increasingly complex social systems and especially the complex activities of maintaining agricultural production. Most importantly, the heads of states, and their underlings, had "authority." They commanded respect and fealty. They could authorize the value marks on the markers. They could determine how many such markers should be in circulation (the accounting systems being used to manage agricultural production and peripheral work had grown quite sophisticated by this time). Fiat money was born as a tool to measure the amount of wealth. Prices could be established in the marketplaces to establish relative values among wealth types. In the Bronze Age, the use of metals, perceived as valuable in their own right (intrinsic value), the development of coinage was accepted as a direct way to produce markers for wealth. Gold and silver coins eventually became the norm for quantifying purchasing power even though the metals themselves weren't much good for anything other than making jewelry. Somehow people came to equate rareness and beauty with worth. And thus began a long and unfortunate history of the stuff we called money.

Money, in the form of coinage, became a facilitating tool for promoting trade of goods and services, which, in turn, allowed an expansion of marketplace transactions. In systems terminology, money provided the signal path for a positive feedback loop. Easier trade led to expansion of trade, which, in turn, led to the need for more money in circulation.

The role of money, at this stage, in facilitating trade, was acting as a regulator on the activity of work processes. Or, rather, it allowed consumption agents (households) to signal demand for specific kinds of work to be done. If a family got into

[32] This might be the origin of the concept of "price." A single marker could still stand for a unit of grain so that the price of anything else traded was tied to the grain by some integer number.

the habit of eating a lot of bread, then they would be spending more money on bread, or wheat, or flour, and, thus, the wheat farmer, miller, and baker would receive more coins and conclude that they should produce even more of their products.

This was all well and good for marketplace transactions, but there was a problem with the use of fiat money in that the governments of these forming states did not really know how to determine the actual volume of monetary units relative to the total wealth that could be traded. If they produced too little money, the prices of goods and services tended downward. If they produced too much, prices tended upward. In other words, the price structure of markets was not stable given the ignorance of the governing agents.

Today the role of money is an extremely complex proposition. Many, many authors and academics have tried to tease out the complexities of what the various roles of a monetary system might play. The most obvious role of money is as a means to purchase goods or services in a market economy.[33] But money is also considered to be a "storehouse of value," or of wealth. As it turns out, in analyzing the situation in modern economies, this is really not the case. The existence of phenomena such as inflation or deflation clearly shows that units of a money do not hold stable with respect to purchasing power. What this means is that a given unit of money has a time-sensitive measure of value.

This, however, is a relatively recent monetary phenomenon. For most of the history of money, it actually did provide a stable representation of purchasing power (value).

The modern problem with the role of money in the economy comes not from the nature of money itself, not even from the commodity backing of monetary value. It is the result of how money is created (i.e., "printed"). Fiat money, created by states, could be a storehouse of value if the state accounting system took a realistic measure of actual wealth that exists in the community.[34] For example, if the state took care to measure the actual production of grains and cattle (say in ancient Egypt) and then produced only enough coins to represent that amount of wealth, then the money would have had a realistic backing. Unfortunately, this was not and is not today the case.

The creation of money in much of modern history has been turned over to the banking system.[35] The governments still print fiat cash, which is circulated, of

[33]The term "market" as used here is very broad and means that buyers can find and buy goods and service among many competing sellers. This model is true for a wide variety of ideological socio-economic versions such as communism or capitalism. In all such systems people use money to buy from sellers. How prices are set is another matter. In a so-called free market, the prices are presumably set by supply and demand pressures. But that is another story.

[34]The gold standard was actually an attempt to accomplish this even though misguided in terms of what actually amounted to real wealth that could back the currency. Gold is an artificial value base. But it was convenient, and in the view of many rulers of the time, it was "valuable."

[35]And, as we will see directly, this is an illusion that completely distorts the role of money in signaling work and consumption processes regarding what work gets done in the economy.

course. But the real action, the real volume of monetary production, is due to how banks treat the primary cash. As with so many other economic phenomena, this started out innocently enough. But the decisions agents have made regarding how to proceed have turned out to be devastating.

9.5.3.1.2 Debt-Based Financing

Closely related to money is the use of promissory notes to acquire assets in the short term.[36] The notes signify a promise to pay back the value of the asset, usually along with a premium paid for usage, in the future as soon as an income from the use of the asset is realized.[37] The wealth to be realized is an expectation of future work processes that would generate sufficiently more profit that the debt could be paid back with interest. This is very different from the method of financing the acquisition of assets via savings. Historically, that was the "normal" approach.

In the early days of agricultural societies, grains could be stored in communal or state-controlled granaries. In good years of harvest, there would be excess over what would be needed in the current year to feed the population. The accumulation of excess overtime was savings. During times of expansion (say of the population), it was possible to support the creation of new farmlands by loaning grains directly to, for example, young new farmers starting their own operations. The new farmers promised to pay back the loan of grain seeds from their harvests because those savings could very well be needed in the future when harvests were not as good. These savings acted as insurance against bad times, which were almost certain to happen. The granary operators acted somewhat like bankers. The grains in their care belonged to the collective or state controllers, but were not needed just then. If the loan was paid back, everybody would be restored to their rightful shares and a new unit of production would have been created in the meanwhile. This is financing growth from savings.

As societies evolved and money became the major representation of wealth (for example, amount of grain in the granary), it was a relatively easy step to go from granaries to monetary banks once money forms took on the responsibility for representing that wealth. Someone took on the responsibility for holding on to excess cash and protecting it as a service. But bankers soon realized they too could loan some fraction of their deposits to new ventures, say to finance a ship sailing to the Spice Islands to bring back cargos of desired spices. The sale of the cargo would produce something we now call a "profit" which would then be used to pay back the bank (with interest) while giving the entrepreneur an income. This was, of course, possible as long as the recipients of the spices held them in high value relative to the efforts paid for by the original loan. History is replete with adventures in trade

[36] Indeed, there are authors who insist that all money is actually based on debt. As argued here, the pattern of borrowing assets did not start as borrowing from the future, however. It is this latter pattern of behavior that has gotten us into deep trouble.

[37] In other words, this is the classical idea of investment.

driven by the thrill of customers to have something really new and exclusive, thus highly valuable and worth giving over more money than perhaps the goods were worth in terms of efforts to get. Herein lay the central fallacy of commercialism—a matter of ignorance of value.

Fractional reserve banking became a widespread practice along with the development of capitalism (another form of loaning money to ventures for profit). All of these mechanisms had the effect of creating new money that was not directly related to the fiat cash produced by the states. Money had morphed (evolved?) into something new, partly a promise that the representations (coins or bills, later) actually were worth something, for example, some weight in a precious metal like gold, and partly a relative value based solely on perceptions of buyers and sellers. The creation of money ad hoc, with fractional reserve banking for example, complicated the whole notion of valuation via denominations. The true value of a monetary unit (like a dollar) no longer had any real meaning. In massive markets the value of anything was no longer based on some intrinsic quantity, such as the amount of energy needed to grow and harvest a bushel of grain. Rather it became so abstract that it was deemed to be worth whatever a buyer was willing to pay for it (in the nominal currency).

The evolution of the economy has produced a world in which the value of anything is now arbitrary and not directly accessible. Mostly this is a result of ignorance but also the result of intentional manipulations of buyers, sellers, bankers, and governments who have all tried to take advantage of the decreasing visibility of real value to obtain a higher profit on their sides of transactions.

Eventually, in the more modern world, money was deemed a mere commodity much as the grain it originally was meant to represent. Its value was to be set by market forces (whatever those are) and currencies from different nations could be traded in international markets.

In today's world the idea of financing operations using promissory notes is de rigueur practice. Corporations think nothing of paying employees with borrowed money on the premise that someday in the future they will be able to pay back that loan because future profits will more than cover the principal plus interest, but also contribute to net profits. As experiences in the last half of the twentieth-century and the first decades of the twenty-first-century attest, that isn't really working out in most industries. Corporate debt is building apace.

9.5.3.1.3 Profits and the Profit Motive

The concept of profit seems so completely natural to most people that they could not imagine an economic system not driven by the desire to "make money." There are several reasons, however, that this naturalistic concept is invalid from a systems perspective. First, as already noted one does not "make" money except, to be discussed below, in relation to the amount of free energy that has become available to the system. If the flow of energy is going up (and as we will see shortly it is starting to do just the opposite), then the monetary authority can create additional amounts to represent the increasing capacity to do useful work. But if it remains the same rate of flow, then there should not be an increase in the supply.

When stocks in a corporation are sold or bonds are issued, this is a form of debt creation of money in the sense that the corporation promises to repay the principal as well as, perhaps, a dividend so the buyer treats the stock certificate or bond as an asset. Nevertheless, the corporation can now spend the money as they choose (supposedly on capital goods). The theory is that they will use the money for productive investment, meaning that their operations will produce excess money above what they spent which will allow for the retirement of bonds and paying of dividends to the investors. There is a fly in the ointment, however. There is no upper limit placed on how much profit one can make. If one can manage to hide one's costs and the market is still willing to pay high prices for the good or service, then the seller can pocket the excess, pay higher dividends, perhaps, or, as has been happening in the capitalist world, paying incredible salaries and bonuses to the geniuses who engineered this marvel.

Profits are probably not inherently evil, but when they become the single largest motivating factor in operating businesses then a systemic pathology enters the dynamics. The pursuit of higher profits, especially when those at the top stand to reap larger rewards for doing so, can lead to behaviors such as cutting corners, externalities (like dumping pollutants in the river without having to pay the cleanup costs), outright fraud (anyone remember Enron), and other bad behaviors that lead to lesser actual value of the product or service and exporting the true costs to other subsystems.[38]

In the section above, 9.5.2 The Historical Perspective, we referred to Karl Polanyi's differentiation between a profit-based market system and an actual economy. In his view, an economy is more like what we have outlined in the next chapter. The addition of profit seeking beyond what is needed to create and maintain a buffer against down-turns or simply to pay off debt is not really part of an economy and, in fact, distorts our perceptions of what an economy is. What parts of the modern economy are, in fact, economic and what parts are something else?

The profit motive is a derivative of the biological mandate to grow during development. The HSS has been and is still developing, albeit more slowly (Gordon 2016), and so the impetus to grow has motivated much of the collective drive to acquire more and more. This translates from the collective drive to the individual drive. CEOs believe they are owed larger bonuses. Workers are ticked off because they believe they should be getting bigger annual raises.[39] The whole zeitgeist of the society becomes one of belief that growth entitles individuals to have more. The perception is that the "pie" is growing so everybody should enjoy a bigger slice, but

[38] For example, paying low wages to workers increasing the disparity between the haves and have-nots.

[39] Which, coincidentally, contribute to inflationary pressures, which then motivate the US Federal Reserve Committee to push up interest rates, which then drives up the cost of money, which then results in, and the chain of consequences goes on. This loop is very complicated but the bottom line is dampening the growth of the economy and future raises.

the reality is quite different. As long as the population grows the effective pie is actually shrinking (see the section below on the Biophysical Economy, 9.6). Each slice of the pie, the per capita wealth, for the vast majority of humans is also shrinking (Piketty 2014).

9.5.3.2 Decomposing the Modern Economic System

We start in Fig. 9.11.

This is a reconstruction of Fig. 9.3 with the Economic subsystem fully established as the SOI. The dashed oval in the figure represents the starting place for the part of the economy that most interests us at present—the energy sector.

In Fig. 9.6, we establish level 0 (SOI) of the economic subsystem along with its level-1, its environment. We will drop the "sub" prefix since this becomes our focus of interest.

The environment has been divided into two parts. The Ecos, external to the HSS itself, is the source of energy and material resources, shown lumped in the figure. It is also the sink for wastes that are not explicitly recycled within the economic

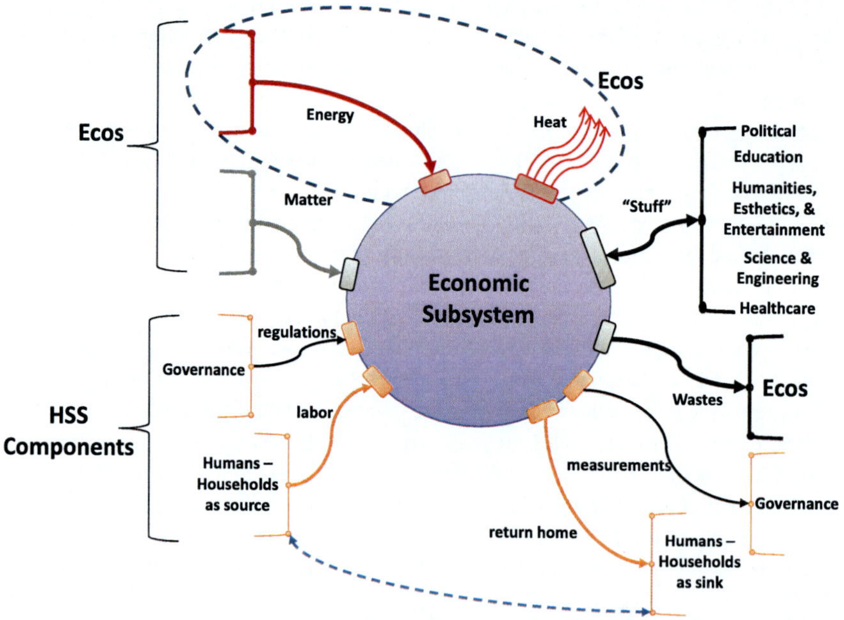

Fig. 9.11 The economic subsystem is now the system of interest. The analysis of the environment shows the relation to entities in the Ecos (outside the boundary of the HSS) and other subsystems within the HSS. These flows are merely illustrative. For example, the flow of humans into the economic system, as labor, and its return to home (for example, in the evening) would be accounted for with an appropriate membership function and flow descriptors. The dashed blue oval focuses attention on the energy "sector" of the economic subsystem

system. The other part of the environment of the economy is the collection of other subsystems in the HSS. In the figure, the Human (biological) and the Governance subsystems are shown explicitly with the relevant flows (people and messages). All of the other subsystems are shown lumped together with "stuff" being produced by the economic system flowing to them. Stuff means the products and services that support those systems.

The energy sector of the economy is circled in Fig. 9.11; that is what we will be decomposing in particular. It consists of energy capture and conversion interfaces able to channel appropriate forms of energy into the economic system. From the level 0 perspective, the energy sector also has to provide for the removal of waste heat when necessary. Often waste heat will simply radiate into the atmosphere (or hydrosphere) and eventually into space (otherwise the Earth's temperature would rise, not because of greenhouse gasses but due to thermal pollution).

In Fig. 9.12, we show a schematic view of the economy as a transparent box. In this figure, we have not specified the boundary. The energy resources (the sun symbol) are captured, converted, and distributed by the red oval. Energy flows to every component in the economy, the various other extraction and work processes as well as to the human subsystem for consumption (for example, home heating). Human beings provide what we have labeled as "labor" flowing to all of the internal work processes. Labor involves the capacity not just to do physical work but to have a self-controlling worker, with a mind sufficient to provide work guidance. Human workers bring both physical work capacity and mental models of how that work should be done to the other work processes, for example, to the extraction and various manufacturing processes.

But humans are also ultimate consumers of all wealth that the economy produces, directly or indirectly. Note that the flow of material and energy is

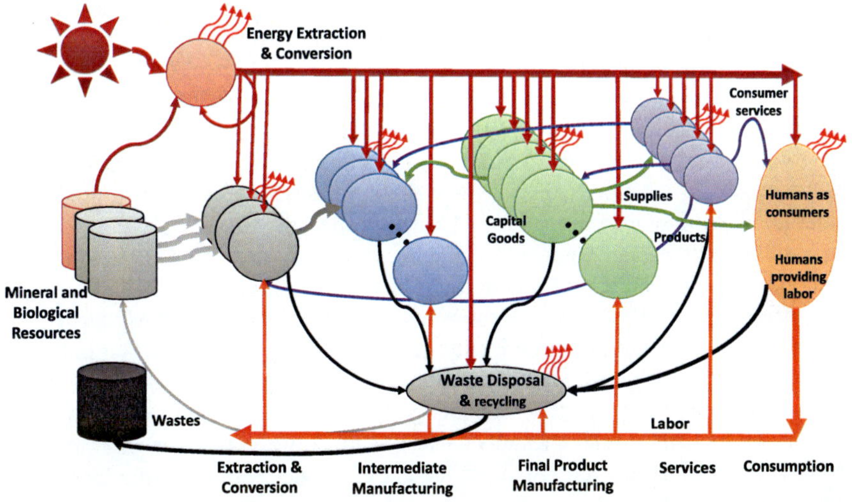

Fig. 9.12 A schematic of the modern economy

consistently, on net, from the extraction (left side) to the consumption end with wastes being removed and dumped into the Ecos.

The schematic representation in the figure shows that products and services that are ultimately consumed (the black arrows represent the resulting wastes) are produced in a staged process. Raw materials are extracted from the environment (for example, metal ore mining), processed initially (for example, metal smelting), then progressively those lower-entropy (more organized) materials are assembled into usable products and various classes of assets (for example, fixed assets like buildings needed in which to do the manufacturing). Even services ultimately depend on the manufacture of physical goods like computers or paper.

The multiple ovals in each of these pass-through stages represent the fact that there are both many kinds of intermediate and final products being produced and that even within any given kind of product (or service) there will be multiple producers who, as a rule, compete with one another for a share of the market, a subject to which we will return below. This latter aspect is modeled in a manner similar to the Leontief Input-Output model of an economy, which is the closest thing to a systems representation of an economy that had heretofore been developed (Leontief 1986).

9.6 The Energy Sector and Biophysical Economic Factors

We will now focus on what is arguably the most significant sub-subsystem of the economic subsystem, identified in Fig. 9.11, the energy extraction, distribution, and removal process. The reason energy is so important in considerations of the economic subsystem is actually very simple. It takes a flow of the right kind and potential of energy through a material system to drive the work processes that result in the construction of usable materials—the materials that go to make our "products," "services," and what we generally think of as wealth. According to Harold Morowitz (1968, p2): "…the flow of energy through a system acts to organize that system." Of course, it takes more than just energy flow. It takes controlled and regulated channels for both energy and material flows as well, and that takes decision agents sending messages to actuators, etc., to guide the application of energies to the right materials at the right time. Recall from Chap. 3 (especially Figs. 3.9, 3.10 and 3.11) the basic concept of a component work process that can use energy flow to transform materials, other energies, or messages (data) into useful outputs. The HSS can be considered as a huge and extremely complex work process converting Ecos resources into human biomass and material wealth, along with waste by-products. The economy is composed of all of the internal work sub-processes and all of them depend on the flows and proper channeling of energy. The HSS economy, as with cellular metabolism, is fundamentally a machine for doing work to lower the overall entropy of the system, even at the production of greater entropy in the Universe, including parts of the Ecos. Thus, economics is a study in non-equilibrium thermodynamics.

9.6.1 *Basics of Thermodynamics, Energy Flows, Work, and the Production of Wealth*

We start with a basic definition of wealth—that is *any object or substance that provides support for human life*. This includes material goods, natural resources, and accessible knowledge (in peoples' heads or in documented form). We also call these things assets. The *value* of any piece of wealth depends on two things, the low level of entropy in its arrangement of atoms and the degree to which it helps reduce entropy in the human person and human environment (recall the section above on the significance of tools, A Note on the Meaning of Tools 9.5.2.1). This is a rather different view of both the concept of wealth and that of value, but the reader is invited to check these aspects against anything they personally consider wealth, with one caveat.

What we leave out of the definition is *money*, and that is for a reason that will become clearer later. For now, we simply say that money *represents* wealth, but is not itself wealth.[40] This is one of the first real differences between classical economic thought and understanding from a systems perspective.

Let's consider a few examples of what we consider wealth and see how the notion of reducing entropy in human affairs plays out. From a biological perspective, the first form of wealth is the food that a family has controlling access to. Food is, after all, biological materials that are very low-entropy substances. Moreover, the best foods are those that allow the body to most efficiently maintain the low-entropy state of one's own body.

Other forms of wealth come as "hard" assets, for example, a house (or shelter). This object is not some random arrangement of materials (even a cave exhibits a low entropy compared with, say, crevices in stone), but a highly organized arrangement of materials that provide shelter from the elements and spaces in which particular activities take place (for example, the hearth). A shelter reduces the loss of heat from the bodies of the occupants, thus increasing the free energy available for useful work, among other things. The degree to which a shelter provides these services determines how valuable it is perceived to be by potential possessors. For example, a shelter that provides different spaces for different various activities, for example, cooking versus sleeping, versus workshop, allows the accouterments of those activities to be organized appropriately and reduces the effort needed to transition from one activity to another; such a shelter might be considered more valuable than a simple hut with one room used for all of those purposes.

Clothing might be considered as assets that are intermediate in terms of value in increasing useful free energy (keeping the body warm outside of the shelter). They are not as transient as foodstuffs, which are consumed in the short term; if they are well made, they will last many years.

[40] Money is sometimes called a "storehouse of value" in the sense that if you have a lot of money in the bank, you might feel wealthy. But in reality, you have simply accumulated a storehouse of purchasing power that generally should be convertible into actual wealth.

What about land? Certainly, subsequent to the sedentary settlements wrought by the Agricultural Revolution, the possession of land became particularly important. Land is an asset insofar as it can support agriculture (or support an ecosystem that provides hunter-gatherers, in which case it is a territory). And land generally needs to be improved (for example, planted, irrigated, or built upon). The value of land is also based on its relative entropic condition. A rubble field is not a great place to plant or build a home. Fertile soils are extremely complex environments that must be carefully husbanded in order to yield maximum food, forest, or fiber products.

Then there is the assets that are buried in some particular lands, such as metal or other mineral deposits that can be mined. The concentrations of specific metal oxides, for example, in ore veins are actually low-entropy products of geochemical/thermal processes that did the work of concentration in the fluxes of heat through the mantel (for example, volcanism). These resources are already in a form that makes them relatively easy to refine and use. Thus, land that harbors such resources is also valuable. The products that can be made from the refined products, for example, tin or iron, for a small investment in work/energy are, themselves, tools for improving life. The iron blade of an oxen-pulled plow is, perhaps, the quintessential example.

Wealth is produced by work processes that have been designed or have evolved so that material objects and, especially, tools have been organized to facilitate the work flow. Raw materials (or subassemblies) are brought into the work process at one end. Energy of the right type and quality, that is, high potential, is also brought in to drive the tools under the supervision of operations decision agents. We should note that work can be done on not just materials but on messages (data) and other energies. Again, we refer back to Figs. 3.10 and 3.11, which show some of these kinds of work processes at atomic levels. But all work requires the flow of high-grade energy in a form that is coupled with the moving parts of the tool(s). When work is done, a significant amount of the incoming energy is transformed into non-recoverable heat; that is, it is now energy from which no further work can be gotten. It is only good for exciting air molecules (raising the ambient temperature a bit) and eventually radiating into deep space.

All activity in the HSS is based on this universal pattern. The economy is now seen as the systems organization of all such activities in support of human life (excepting the various dysfunctions that take human life, such as using guns to kill people instead of wild game—these are believed to be anomalies, the results of failures of some peoples' brains).

Figure 9.13 provides a view of the economic activities that support human life as a form of extra-somatic (outside the body) metabolism. These are the activities organized within the economy framework that keep people alive and thriving. The diagram is more representative of a developed country's capacity to support life, but some form of each of these activities can be found in even developing societies, even if in a reduced or primitive form.

Another view of a social system economy is given schematically in Fig. 9.14. Here only the major categories of activities are shown lumped into ovals with flows from left to right representing the value-added chain that eventually leads to

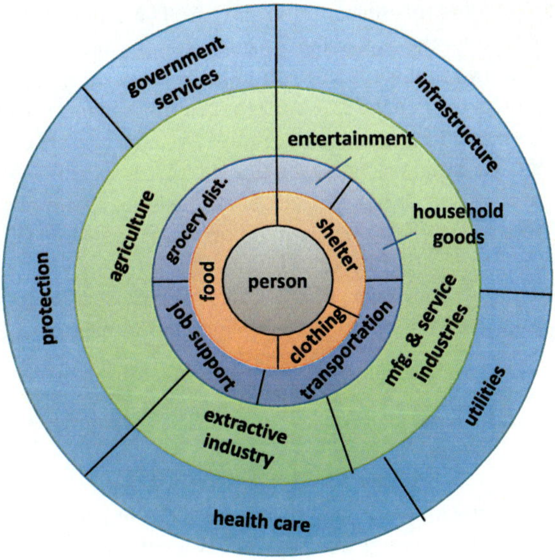

Fig. 9.13 Every person in a social system exists within a cocoon of extra-somatic metabolic-like activities we call the economy

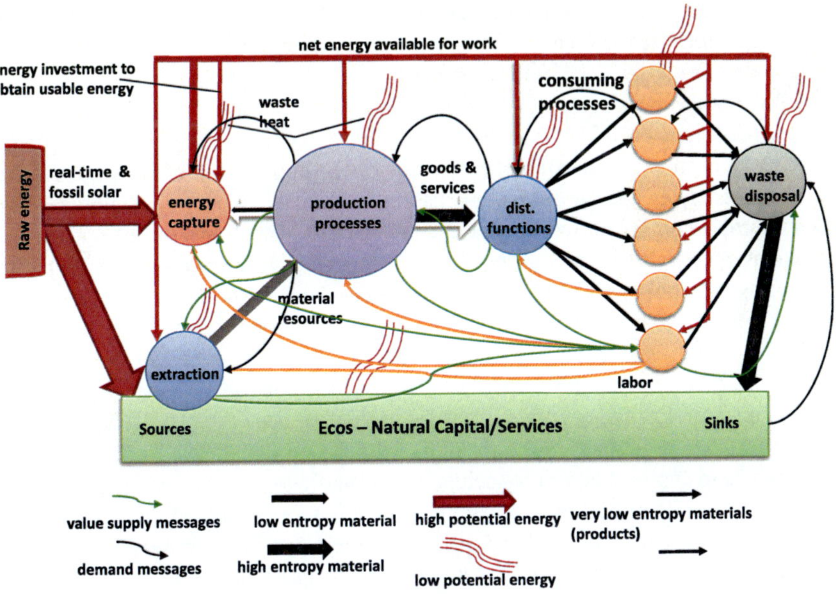

Fig. 9.14 A further abstraction of the human economy. The orange ovals on the right side represent households that both consume the products of the economy, the low-entropy objects and services, and provide the human labor needed to operate the economic production and distribution functions. Money flows in the direction of energy or intelligence sources, that is, from consumers to production processes and from those processes toward labor

households for consumption and use. The economy is shown in relation to the natural world—the rest of the Ecos wherefrom material resources are extracted.

Shown in this representation is the flows of messages that help to coordinate and regulate the general flows (described more below). There are basically two kinds of messages that can be found in this system. The black (skinny) arrows represent the sorts of complex messages that supply processes with detailed information about the situation. Examples are purchase orders, shipping documents, contracts. The green arrows represent a more abstract, and ultimately less informational kind of message that is mediated by the flow of money. These are used to transfer value from, say, consumers to producers for purchases of products and services, and from producers to households to purchase labor (and intelligence).

9.6.1.1 Energy Sources for the HSS Economy: Dynamics and Economic Effects

Over the course of human history there has been an evolution of our capacity to extract free energy from our environment to power the production of human biomass and cultural artifacts or wealth. From the origin of the genus *Homo*, our mammalian-kind have depended first on the availability of suitable organic sources of energy, otherwise known as food. At some point in the evolution of the genus, presumably *Homo erectus*, we sentient beings discovered the taming of fire. This was momentous, possibly the most momentous event in human history[41] if not in the history of all life. Fire, for example, the burning of wood, is the essence of extracting energy to increase the free energy available to humans for other work. At the very least fire keeps people warm against a cold climate, thus letting them use their own internal energies for more "productive" work. It also allows for the cooking of food materials, releasing additional calories that would not have been available to predecessors. Cooking is generally recognized as the condition that gave rise to the more cognitive versions of *Homo*, us in particular. It is important because the expansion of availability of free energies in foods is directly responsible for the ability of humans to occupy many new ecosystems beyond the African savanna.

Fire is the essence of energy sources for the economics of the HSS. Fuels, first in the form of wood but later in the form of hydrocarbon fuels, have made it possible for humans to evolve a complex society and technology. Fuels embody free energy in chemical bonds of hydrogen and carbon, which, when oxidized, release substantial amounts of energy in the form of high-temperature flows. When coupled with the phase transition capacities of water (liquid to gas phases), and the explosive expansions resulting, gave rise to the steam engine and the transformation of heat at high temperatures into mechanical work. Thus, was born the Industrial Revolution!

[41] Or, actually prehistory. *Homo erectus* is considered the most likely predecessor of *Homo sapiens*, as well as *Homo neanderthalis*, and other species (or sub-species) of humans some 400,000 years BCE. It is believed that *erectus* had realized the ability to manage fire and that capability was then inherited by the subsequent species such as our own.

Today, the economic subsystem of the HSS is still largely dependent on the phase transitions of material systems due to thermal processes. The internal combustion engine (and its variants) is another version. The turbine engine is another version of dependence on the explosive oxidation of hydrocarbons. All of these forms of fire power modern society.[42]

And that leads us to a huge problem. The amount of hydrocarbon fuels buried in the Earth's mantle is finite. To be sure, there is a tremendous amount of such fuels, coal, oil, and natural gas, still in the Earth. But it is still finite in quantity.

There is another problem with these fuels. We have to use energy doing work to extract them. In other words, it takes energy to get energy. The energy net of the work needed to extract it is what counts. We have to drill and extract these fuels at a price. And, as we use up the easy-to-extract sources, we have to increase the energy expended to obtain the harder-to-extract sources. For example, a conventional oil well on land produces 20 times (at least) more energy than it takes to drill the well, set up the pump, extract it, and move the oil from the well to a refinery. But when we have to go out to the ocean platforms to extract the same volume of oil, we get only five to ten times as much energy. When we have to extract oil from bitumen in Canada, the ratio goes down to two to five times.

There might still be a net energy return on the energy invested (EROI). But it is declining over time toward zero net. We are doing more work, extracting energy resources, than we had to in the past. And the problem is this: The civilization we built over the last several hundred years was based on an energy return of between 20 and 100 to 1. The infrastructure we have built probably requires a minimum of 10 to 1 just to maintain. Even with better technology, new infrastructure would need the older 20 to 1 or more.[43] That is a monumental problem.

What about alternative energy sources? Why can't we switch to solar electric or wind?[44] Well, we might be able to do so to some extent. But the conversion of raw energies in solar inputs requires substantial technologically dependent tools. It is one thing to simply burn a hydrocarbon fuel to produce a huge temperature difference that causes a phase transition and to transform a low potential energy form into

[42] Total 86.4% of primary energy in the world comes from fossil fuels; the rest from hydro and nuclear with a tiny percent coming from alternatives and combustible fuels like wood (Hall and Klitgaard 2012).

[43] It might even be worse. The newer technologies that make the drilling/pumping/refining more efficient actually represent a greater expenditure of energy in inventing/building manufacturing capacity/delivery, etc. The costs are amortized over all of the drilling operations. They are diffused in the energy sector and only show up in terms of capital equipment on various oil company balance sheets. Many advanced technologies rely on, for example, exotic metals which are very energy intensive to mine and refine. So, the real energy used to get one unit of energy is basically unknown.

[44] Actually, wind is a solar energy! It is the climate—the effects of the Sun on the Earth—that drives wind patterns and wind forces. The same is true for hydroelectric power. The Sun drives most energy flows on the surface of the planet. The Moon drives tidal cycles and core thermal processes drive hydrothermal cycles, but these are relatively minor compared with the influence of direct solar energy.

a high potential form. Solar energy, for example, is actually extremely weak; you can walk around in sunlight without being cooked. You can walk around in a high wind without being terribly disturbed. But consider walking around in a coal-fired furnace in contrast! No way. The thermodynamics of producing free energy is not favorable for just collecting solar energy. It takes a huge land area of collectors, even if they have efficiencies of greater than 15–20%, and a lot of technology to derive enough free energy to drive an economic system such as is possessed by the developed economies.

For example, consider a simple *thought experiment*. Can a solar photovoltaic plant provide enough energy to support an ordinary society and enough additional energy to reproduce itself, that is, produce its replacement over its lifecycle? At current levels of the technology there is no evidence that this is even feasible let alone settled. And yet this is exactly what will be necessary in order that society, as it currently works, could achieve stable existence based on solar energy.

Ultimately, the economy depends entirely on the flow of high potential energy through it. It is a system that does work by virtue of the availablility of free energy as we have covered in this book. Yet currently two factors are at work in the mainstream energy sector of fossil fuels. First, the fuels are finite and diminishing. The peak of production of so-called conventional, land-based oil has come and gone. Every year, globally on average, conventional oil production declines.[45] This same phenomenon will follow in non-conventional sources such as deep-water, shale, and bitumen mining, which are the sources that have kept total oil production up (actually growing slightly). There is already evidence that the phenomenon of peak oil is approaching for certain shale fields; the early production was extremely high but the production rate seems to fall off much more rapidly compared with conventional wells. The inference from this is that these wells will ultimately not produce as much total volume as conventional ones. Moreover, the number of highly productive geographical sweet spots where drilling is successful in a field is limited. So the methods of horizontal drilling and use of water and sand to fracture the shale rocks, initially looking promising, now appear to be nearing its peak as well, after which production will also decline year after year. No new technology is on the horizon that might significantly change that.

The second factor is the EROI decline mentioned above. As with financial investments that fail to make an acceptable return, as the EROI declines, as it must with the growing reliance on non-conventional wells and bitumen mining, at some point the returns of energy on energy invested fall below what is needed to do the physical work of society. At that point, even with lots of oil still in the ground, we have to give up on energy from oil (and gas and even coal). How will we replace the over 80% share of primary energy? How soon can we? And this isn't even about getting off carbon to save the environment. This is just the plain thermodynamics of the energy sector itself.

---

[45] There are regions in which oil production is still rising due to geopolitical consequences. However, total production for the globe is in decline.

Figure Fig. 9.15 shows the dynamics of the extraction of a finite energy reserve. The structure of the model is shown in Figs. 9.11 and 9.13. What is happening in this model is that the reserve of, say oil, is fixed and finite (or "stock-limited" in the language of system dynamics), represented by the barrel (stock) of fuel, labeled "Gross Energy." As oil is extracted, there is a kind of "back pressure" that increases. That is, it is harder and harder to do the extraction, that is, it takes more energy to do the same job. The blue curve in Fig. Fig. 9.15 represents the dynamical behavior of obtaining raw oil, say.[46] Note that in Fig. 9.11, the barrel is large and in the next figure it is shrunk, representing a diminished supply over time. Also note the red arrows that are reentrant from "Extraction" to itself. In Fig. 9.11, the thickness of the arrow represents the fact that not that much energy is required to do the work of extraction, whereas in Fig. 9.13, that arrow's thickness is greater representing the fact that the amount of work to do the same extraction has increased significantly.

Then compare the red arrows labeled "Net Produced" in both figures. In Fig. 9.11, the arrow is thick representing a significant amount of net energy being provided to the social system, "Society," primarily to support the reproduction of biomass—the population of human beings.

What Fig. Fig. 9.15 demonstrates is the relations between the growth of energy production, energy costs of production, and net energy available to society to support growth. The dynamics are dramatic. As the economic system extracts more and

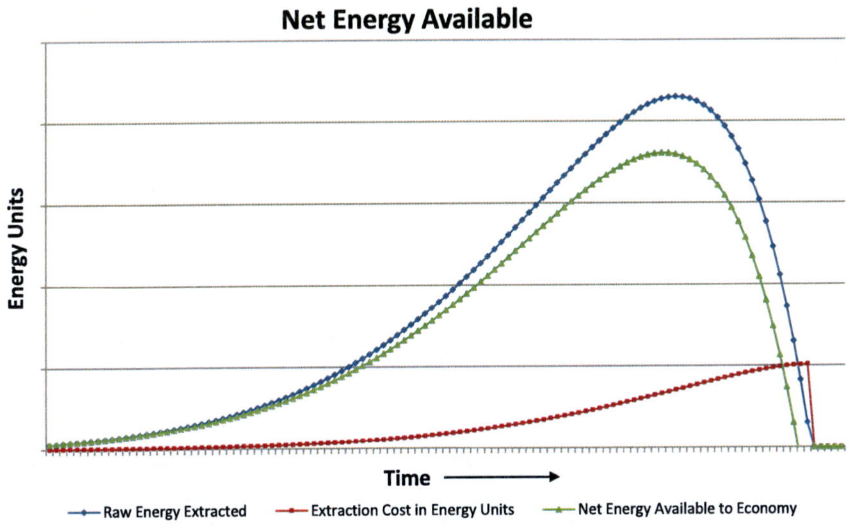

Fig. 9.15 The dynamics of extracting a fixed, finite energy source (like oil) over time, and the effect of increasing costs of extraction lead to more rapidly diminishing net free energy for society, that is, economic work

[46]The same basic arguments apply to a number of mineral resources such as coal and metals, but we are limiting our argument to energy resources. Thus, the same dynamics are applicable to natural gas as well.

more energy from its finite reservoir, the curve initially rises exponentially—the more energy we extract in excess of what we use, the more energy we have to do more extraction. Also, the demand for energy in the society puts a positive pressure on the extraction process. The green curve tracks the rise of net energy which is going to support the economy and increasing the biomass of humanity.

But as time proceeds the energetic costs of extraction (red curve), rising due to the increasing difficulty of extraction, lessens the net energy relative to the gross energy resulting in an increasing gap between the blue and green curves. And then the finitude of the resource comes to bear. The interactions between diminishing reserves and increasing costs result in a diminishing return on investment (the green curve tails off before the blue curve shows diminishment). This is subtle but important. Due to that finitude, the blue curve reaches a maximum, a peak, after which it begins to decline. And according to the model parameters it declines far faster than it rose. Yet because of the declining EROI (rising cost curve) the net energy actually peaks before the gross energy curve. The free energy to society stops growing and starts declining even before the gross energy supply curve starts to decline. We reach peak net energy earlier in time than we reach peak total energy production.

This is why that phenomenon has gone undetected. Our economic sector economists track gross energy, that is, total primary energy produced. They do not consider net energy available for useful work because they do not grasp that it is the latter that actually counts toward producing real wealth. Economists are not, typically, physicists, and they definitely are not systems scientists.

It gets worse. Recall that one of the main outcomes of increasing wealth is the support of an expanding population (biomass of humanity[47]). Thus, we need to consider in our economic modeling the effects of net energy per capita. The curve in Fig. Fig. 9.15 is scary because it suggests that in the not-too-distant future we are going to run out of energy and at a dizzying rate. But given that over the last several millennia we have increased our populations dependent on that energy, it should not be surprising that the amount of free energy available to each individual has been declining precipitously.

Between Figs. 9.16 and 9.17 we see the operational changes that occur in this system over time as a result of the twin effects of declining EROI and depletion of reservoirs (peak production). Society gets bigger while the energy source diminishes. Meanwhile, the cost of extraction of gross energy increases causing a decline in net energy available. And, meanwhile, the population has increased so that the net effect is a diminishing of net energy per person.

The gross energy source has to be extracted (and converted to usable form). The extraction work takes energy (cost) to produce a net energy reservoir (for example, tanks of gasoline). It needs to be transported to points of use, where the work gets

[47]We should probably note that it isn't just the biomass of humanity that constitutes societies. It also includes the biomass of all of our pets and lawns. We ought to also include domesticated animals of all types, for example, farm animals, since these are all biomass under the ultimate control of human society.

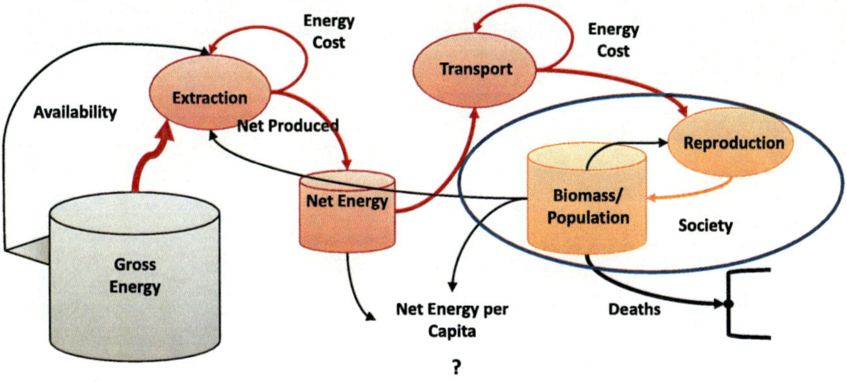

Fig. 9.16 The situation with the extraction of a fixed, finite source of energy (for example, oil)

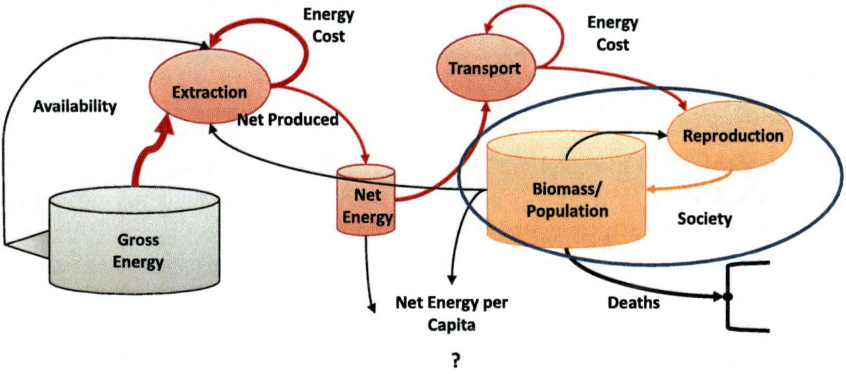

Fig. 9.17 As the source of gross energy depletes (a smaller reservoir as compared to that in Fig. 9.16), extraction work gets harder—more effort is required—and the energy cost needed to power that work goes up. This means less Net Energy available for society. If the biomass of society grows at a rate greater than the growth of energy production, then the net energy per capita (or unit of biomass) declines

done. From a societal perspective, the use of energy helps support the biological function of reproduction leading to an increase in population.

Figure 9.18 now looks at the effects of the dynamics of Fig. Fig. 9.15, but on a per capita basis. Though we use the more familiar term "per capita" which means per person generally, the term "capita" can actually refer to physical capital, in this case something like average biomass per individual. There is a direct relation between free energy and biomass production. However, people will be concerned with how much energy is available per person, so we will stick with that meaning.

The graph comes from a model that spans the period prior to the beginning of oil age, through the oil boom and peak oil, projecting into the future. We did not include a feedback loop from the effects of reduction in net energy per capita to the

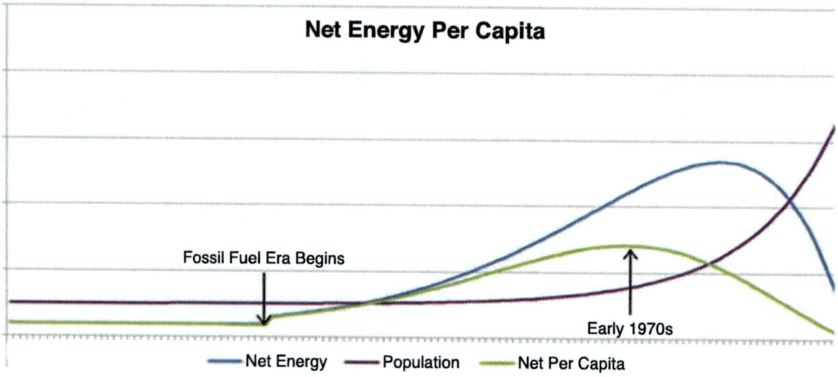

Fig. 9.18 This model suggests that net energy per capita (global average) declines earlier than total net energy due to the affects of continued population increase. The model does not include feedback from that decline to affect the population growth. That assumption is unrealistic, of course, and actual experience has shown that population growth rates in many parts of the world have actually declined over the last several decades (the purple line should actually form a logistic curve, bending toward a more flattened form shortly after crossing the green line). The position of the presumptive calendar date of the 1970s is explained in the text

population growth so the graph makes it appear that the population keeps growing in spite of the falloff in net energy per capita. The important dynamic to note in this graph is that net energy per capita (green line) peaks and then declines much earlier than total net energy (blue line) and at a much lower level than total net energy (under these assumptions about population increases).

The marking of the "Early 1970s" indicates that in this scenario the peaking of net energy per capita happens in the early part of that decade. What might be the consequences of the peaking in that timeframe and might it correspond to our actual experiences?

Clearly if there is less actual net energy to produce assets (as in Fig. Fig. 9.15), then fewer assets (goods and services) can be produced and translated into the model of Fig. 9.18; this would mean fewer assets per person on average globally. The effects of this might not be readily recognized by economists who do not yet grasp the relationship between free energy and asset production capacity. But it would show up as a positive influence on inflation, positive in the sense of promoting it. With more people clamoring for more stuff and a production system less capable of producing apace, prices on things in general will be pressured in an upward direction. Of course, declining net energy per capita is not the only cause of inflation. But it provides a steady upward nudge. Then other positive feedback loops, the ones, for example, that the Federal Reserve Bank Committee pays attention to, begin to act to amplify the effects. As inflation begins to diminish purchasing power of the average household, the clamor for wage increases to compensate eventually drives wages upward. But that just ends up eating into profits so companies eventually have to raise their prices for the goods and services they produce.

And so a difficult-to-control spiral upward is triggered, buoyed always by gradually increasing energy costs (decreasing EROI and peaking of reservoir production).

If we reflect on the actual economic evolution from the end of the nineteenth century to the present, we note that there was a significant growth in terms of increasing wealth production from the onset of the industrial revolution and a sharp upturn with the advent of the oil age just at the end of the nineteenth century. After that, combined with the developments of technologies such as transportation, communications, metallurgy, and others (more lately computation), growth of wealth, particularly in the Western world, increased exponentially. Even factoring in the effects of two world wars and a "cold war" Western affluence increased as measured by both GDP and productivity. That growth exceeded the growth in population so that per capita wealth grew providing for a more affluent middle-class lifestyle.

Then something happened in the early mid-1970s. Growth rates in GDP and productivity began to fall (Gordon 2016). The curves began to flatten out. Then, as the 1980s were coming to a close, there was a downturn overall, the bursting of the Dot-Com bubble at the end of the millennium. The economies of the world suffered a number of economic shocks, usually attributed to normal business cycle declines but also to the effects of financial bubbles bursting and dragging down the rest of the economy. Economists were often taken by surprise, that is, they did not predict the timing or depth of recessions, for example. Post facto they looked for classical economics explanations for these episodes, every economist having a favored phenomenal cause. None of them thought to look at either the overall trends (the big picture) or the role of energy costs and net energy per capita because energy, in general, has been taken for granted in all economic theories. The physics of wealth production was not considered in any of these theories. As of this writing a number of economists and historians are still uncertain as to what the long-term reduction in growth rates and productivity mean, let alone what might be causing them. This is because they simply do not understand the role of energy in driving work processes and the production of low-entropy products and services—wealth.

However, we can anticipate what that future might look like in Fig. Fig. 9.19. This is based on a model that combines the effects of fossil fuel depletion, ensuing energy cost increases, and the amount of free energy that can be applied to producing wealth or assets.

This model starts after the onset of the oil age and was based primarily on the dynamics of oil extraction. It does not take into account the seeming extension of the oil age through non-conventional drilling and bitumen mining. However, these will not change the overall shape of the curves in the graph; they will only extend further in time the events depicted. Fossil fuels are still finite stocks so the situation depicted in Figs. 9.16 and 9.17 will still play out.

In Fig. 9.1, the blue curve, labeled "Gross Energy," represents the extraction rate of oil, pushed to its maximum and constrained only by economic factors, such as financing and technology. It rises logistically (early exponential with an inflection point—just below the Y in Energy—and then decelerates) reaching a peak. This was presumed to be the phenomenon of peak conventional oil that data indicated was

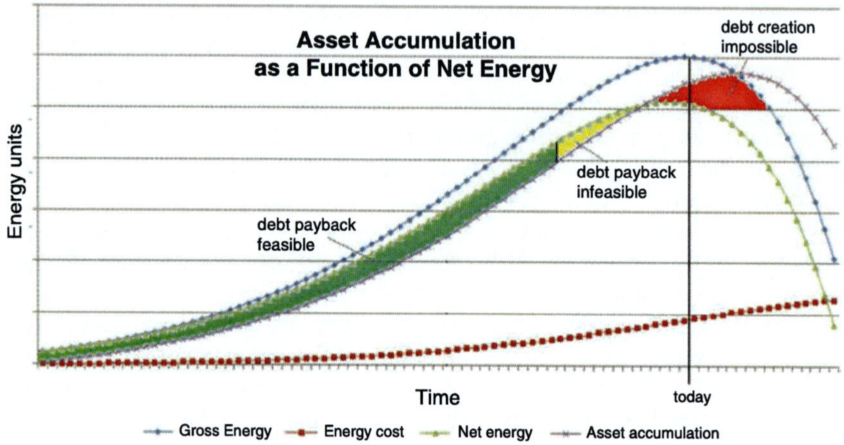

Fig. 9.19 Effects of "peak fossil fuel energy," decreasing EROI (increasing energy cost), and the production/accumulation of wealth into the future

taking place around 2012.[48] The red curve at the bottom of the graph represents the increasing energy cost of extraction and refining. As oil reserves became more expensive to get at, the costs of extraction increased. Thus, the net energy available to do useful (economic) work (green curve) as a fraction of gross energy also declined. The graph indicates the net energy reached its peak before gross energy did. That is, net energy started to decline a number of years before 2012 and is also reflected in Fig. Fig. 9.15.

The purple curve in the graph represents the use of free energy to do work and accumulate assets or total wealth. The wealth is measured in units of embodied energy in material goods of all kinds—that is, we assume that all of the free energy was turned into assets; the gap between the net free energy curve and the accumulating asset curve represents the less-than-perfect efficiency. The colored zones are the real story here. The green area represents the period in which the availability of more free energy in the future would always seem to increase. Thus, in the future one could expect to accumulate even more assets. This is the situation that supports the concept of economic growth for society. It is a time in which debt-financing makes some sense because the expectation that wealth will increase and debts can be paid back appears reasonable. But after the inflection point in the net free energy curve occurs, the gap between asset accumulation and free energy starts to narrow. From that point on (the yellow zone) there is increasingly less energy to do work with and so the overall production of assets begins to tail off as well. More importantly, the ability to pay back loans becomes increasingly infeasible.

[48] No precise time window could be declared but the projection models based on historical extraction rate growth coupled with various projection techniques such as Hubbert's method showed the peak to be somewhere in that timeframe. The then nascent fracking of tight oil had not yet started showing up in the data.

There continues to be some increasing accumulation of assets, owing mostly to the fact that many kinds of assets are long-lived. But the consequences of the peaking of oil extraction should be clear. The red zone represents an area where debt will increase without the hope of paying it back in the future. Debt, after all, is just a promise written on paper to pay back the amount borrowed (backed by current assets, not future ones) and that promise assumes that the borrower will, in fact, own more assets in the future (liquidating some will provide the cash needed).

The red zone is where debt bubbles of all sorts will develop. Unable to pay back current debt, for example, some countries will "restructure" their loans from some banking institution and, in effect, increase their long-term debt even more. The normal expectation is that by becoming fiscally prudent and somehow attracting investors in the future, those countries will be able to pay back their debts one day. But as Fig. Fig. 9.19 indicates, without some new source of energy to compensate for the decline of oil (and other fossil fuels) the loss of net free energy will cause a relatively sudden collapse of the economy's capacity to do useful work.

This model is not meant to provide an accurate or precise prediction of the future. All of the units of measure are relative to a presumed value of raw energy available. This is an attempt to explore the dynamical behavior of the boundary conditions of the economic system depending on oil as its main (over 80% at present) source of free energy. The rules apply to any fixed, finite supply of energy. So, we expect the same behavior from an economy dependent on all of the fossil fuels. The inclusion of alternative energies (but see the section below, 9.6.2.1, for an examination of this feasibility), assuming their EROIs are around 20:1 and do not decline over time, might change the maximum of the peak, but unless they took over the whole supply of free energy, the shape of the curves would remain the same.

When one combines these curves with the per capita curves (recall due to growing populations), the prospects do appear dim. At the same time, the dynamics demonstrated in Fig. 9.19 represent the worst-case scenario, that humanity will drill, drill, drill and pump, pump, pump at the maximum rate possible until it becomes uneconomic to do so. And that is because humanity is also committed to accumulating wealth at the fastest possible rate until it is no longer possible to do so. It assumes that humanity cares nothing for their grandchildren and just wants to consume and have it all now. It doesn't have to be that way.

9.6.1.2 A Short Note on So-Called Decoupling

Most modern economists will argue that there is a tendency for an economy to decouple from carbon naturally as that economy becomes increasingly a so-called service economy. Supposedly service work doesn't buy as much energy and so they are not significant users of primary energy in the way, say, manufacturers are. But this misses the fact that service companies are buying products that have embedded energy, that is, energy that has already been used to produce them (think paper and computers). So energy was consumed to support the work of services.[49] It just doesn't show up in the books of the service company as such. Those books reflect direct costs of energy, electricity, and gas, which are relatively minor. But they fail to take into account the embedded energy in the products and services they buy. In theory, those energy costs are reflected in the prices they paid for those products and services, so these never really show up in so-called carbon footprint analyses as contributions to carbon emissions.

The belief that the economy, by becoming more service oriented, is decoupling from energy usage is not borne out in fact. The use (and cost) of energy is just more diffuse among multiple other services and manufacturing. Off-shoring of production is also another way in which energy usage has been hidden from ordinary bookkeeping. Energy, and in many cases the energy source is particularly "dirty," that is coal-powered electricity, is still very much being used. In fact, given that the total global CO_2 emissions have continued to increase in spite of the supposed decoupling, the world is using more fossil fuels in toto even while the free energy per capita is in decline.

9.6.1.3 The Control of Energy Flows

Consider what work gets done in the economy. How does a manufacturing subsystem decide to build another unit of product? The building of such a unit requires the availability of free energy. But why should the subsystem take in that free energy, do the work, and provide the unit to a purchaser?

In a market-based system, the answer is that the purchaser, an economic agent, offers to acquire the product and give up something that represents an amount of free energy that they control, an amount that is equivalent to the free energy needed to build the next unit. The customer signals the value of the product by virtue of offering something considered to be valuable in the sense that it could be used to purchase other products of equal value. The production subsystem is responding to the demand signaled by the willingness of a customer agent, or, rather it is responding to an aggregate of such demands by a population of customers. It turns that aggregate of signals into a motive to produce more of the product. What is the nature of these signals that they cause the producer to do more work in the future?

A market economy is the equivalent of the cytoplasmic matrix of a cell or the circulatory/lymphatic system of a body, in which signals of demand and supply diffuse through the medium (cytosol, blood, and price). Producers respond to signals

[49] Cloud computing server farms, in fact, are huge consumers of electricity, so much so that some of the larger ones have set up their operations next to major dams in order to have access to the amount they need. Since they are getting electricity from hydroelectric, they are decoupled from carbon, technically. But hydro isn't everywhere available so server farms in the Midwest, for example, have to rely on coal or natural gas-fired generators (until really huge wind farms can provide).

that indicate demand (higher prices paid) and consumers respond to signals of availability (lowering prices).

Work processes do what they do when what they do has some value to receiving processes (consumers). They can only know what that value is when the customers provide signals that correspond to their desires (needs) for the product. Once those signals are received, the work process has a motive to continue producing its product under the assumption that the product will be desired in the future and that there are more customers in the marketplace. Car manufacturers keep making cars because people keep buying cars. And, in a replacement population, with new customers coming into the market continuously, there will always be customers in the future so it makes sense to produce the next units in anticipation of those transactions. Thus, the producers will further signal the suppliers of materials and energy that they need more of these resources currently.

Consumer agents decide that the possession of a particular product will enhance their existence (considering hedonistic aspects!). And assuming said agents had acquired the necessary accumulation of signal tokens (money), they will send said signals to the purveyor of the product, thus demonstrating a "demand" for that product. Figure 9.14 shows a few of the kinds of signals that are communicated between producers and consumers to regulate the flow of energy. The black arrows represent specific information conveyance, things like purchase orders. The green arrows represent a more generalized kind of message, one in which the information content is determined by a simple metric—price—conveyed by flows of money. This one kind of signal is simple to implement and understand but depends on the agents' grasp of what the unit measure of the currency is "worth." That is, each unit of currency should have a constant value as compared, for example, to some standard. Historically, the currency was valued in terms of units of some convenient commodity, such as barley (valued as food, but especially its use in brewing beer!)

Any agent that holds currency is in a position to direct the future flows of energy by virtue of how they spend that currency. Spending is the signal that prompts directing energy flows into the work processes that produced the good or service.

There is a major flaw in the logic of market-based producing and consuming using money to provide the signals. Producers receiving money (a price paid) from a customer assume that there will be more customers wanting the same product or service, so they direct the money to the purchase of supplies, raw materials, labor, etc., to produce more for future sales. However, consider the case of long-lasting products like cars and houses. People don't go out every other day and buy these things. Since they last a long time with some maintenance, there would be no continuing signal once the market had been saturated. This actually happens, cyclically, all the time. There is only one thing that keeps, say, car manufactures or home builders in steady business and that is growth of the population.[50] New humans have to keep coming into the market at more or less a steady pace to keep the energy flowing

[50] Growth of a population in a given area can be due to either positive birth rates or immigration or generally a combination of both.

through these work processes (firms). Some products can be engineered to break down—planned obsolescence—so that customers will have to keep coming back. But in markets where maintenance services are available (like the automobile market), this strategy doesn't work too well.

So, there is a built-in grand positive feedback loop that reinforces the growth of the population just to keep new customers coming into the market. Indeed, countries like Japan, where reproduction rates have plummeted and the population is not increasing, are quite worried that markets will not be buoyed up by growth.[51]

The plastic-based products and throw-away economy are essential to maintain the illusion of endless growth.

Overall, the flow of monetary value (money) is counter to the flow of energy, both embodied energy in products, or free energy flowing into work processes in order to produce lower-entropy products and provide services (again, shown in Fig. 9.14). Originally this is exactly what the role of currency was, to signal the producer members of the economy how they should manage their resources and efforts. Today, however, there has been a massive distortion in the signaling value of money. The overuse of debt has increased the amount of apparent money being routed and thus confounding the actual information value of monetary flows. Money itself has become a commodity and less a measure of true physical (utility) value of the units.

9.6.1.4 Entropy Reduction in the System at the Expense of Entropy Increases outside the System

Cells, bodies, societies, and ecosystems do work to decrease or maintain low entropy internally. They do this by extracting material resources from the environment and high potential energy. As living systems minimize entropy internally, they contribute to increasing the entropy of the environment in which they are embedded. Overall, the total entropy of the Universe increases even while pockets of low entropy persist—on planets like Earth. This is only possible because of the continual energy influx from the sun to the Ecos. That energy allows the Ecos as a whole to maintain or even reduce its entropic state even while it is degrading the high-value visible light photons into low-value ones, infrared.

Starting from the extraction of material resources and the capture and conversion of energy sources, the role of the economy is to reduce the entropy of bits and pieces of the system, ultimately in support of maintaining human life. Material resources are generally already in a low entropic form. For example, the metal content in usable ores is already somewhat concentrated due to the work that the geothermal processes in the Earth's mantle. The degree of concentration is sufficient to make the ore valuable since the amount of additional work to purify the metal is

[51] They are also worried that with the demographic distribution continually moving toward an older population they will have fewer workers to supply labor!

energetically feasible. The pure metal is further valuable because it can be used to produce more complex products. The same is true of petrochemicals and fuels such as gasoline. At each stage in the economic process value is added to the bits and pieces as they become more directly useful to consumers (users).

Ironically, the reduction of entropy within the system, the HSS, is accomplished by increasing the entropy of the environment.

Solar energy—rate of energy flow that restores the natural resources of the biosphere—is not sufficient to keep balance. Extraction of the mineral resources far exceeds the rate of replenishment (that is why we face peak oil). Recycling is essential—in the same manner that ecosystems involve recycling of biochemicals (nutrients and minerals).

The current structure/function of the HSS economy is such that the system is increasing the entropy of the Ecos much faster than the latter can use solar or geothermal energy to repair its pre-human state far from equilibrium.

9.6.2 Energy Capture, Conversion, and Delivery to Consumers

Over 80% of primary energy used by the global economy comes from fossil fuels. The extractive industries include coal mining, taking various forms such as deep crustal mines, and quarrying after "mountain top removal," oil and gas drilling of wells to find pockets of these raw fuels which are then pumped to the surface. As mentioned above, fossil fuels are a finite reserve that humanity has been drawing down at increasing rates. It will become increasingly expensive, in both financial and energy terms, to extract as time goes on because the industry has to go deeper into the earth, go to drill areas like deep water wells on the Continental Shelf,[52] and require increasingly sophisticated (thus expensive) technology in order to keep the production rates up.

Coal requires a minimum of further processing other than grinding into small chunks optimal for burning. Similarly, natural gas (methane) requires minimal processing, perhaps filtering out other gases like CO_2. Oil, on the other hand, requires considerable processing, the refining of oil-derived products that contain much denser energy per weight (or volume) unit such as gasoline and diesel.

Refining of oil produces a number of power delivery fuels and other useful products. These fuels are rated according to their energy density (for example, Joules or BTUs per unit weight). Jet fuel, for example, is richer in hydrogen molecules which make it burn at higher temperatures than gasoline. It is the temperature of combustion that produces the power or useful energy per unit time. Coal burns at a much lower temperature than jet fuel, but sufficient to boil water and, through the

[52] Many will remember the Deepwater Horizon tragedy in which a well "blow-out" led to one of the worst oil spill accidents in history.

expansion of steam, drive mechanical devices, particularly steam turbines which, in turn, drive electrical generators.

Electricity is one of the most efficient forms of energy for both transportation to points of use and in terms of mechanical conversion. However, there are limitations to the use of electricity especially in the transportation sector. A big hurdle in this regard is the state of battery technology. There have been some impressive developments in increasing the power density of the storage device, the duration of charges, and the lifetimes. All electric ground transportation is under development and its uses could help reduce the emissions of carbon dioxide, especially in areas like the Pacific Northwest of the United States where most electricity comes from hydro-electric sources. The same may be true in areas that can produce significant contributions to the electric grid from wind turbines. But in areas where the electricity is coming from coal, or even gas-fired power plants, electric vehicles have a substantial carbon footprint even though they are not burning fossil fuels directly.

As mentioned above, the hope of many is for alternative energy capture to take over from burning fossil fuels.

9.6.2.1 An Exercise in Evaluating Solar Photovoltaic Capture and Conversion

One of the alternative energy sources being considered to replace fossil fuels is the capture of solar radiation (light) through the photo-voltaic mechanism to produce electricity. We mentioned earlier the thought experiment in which sufficient energy needed to be collected to both supply society with electricity for consumption and provide for the ongoing replacement and repair of solar collection subsystems.

In this section, we provide an analysis framework for considering if the goal of producing a self-replicating energy source is feasible. This analysis is meant to supply guidelines for the collection of data to test that feasibility. It is an example of where a top-down deep systems analysis would lead as we look for solutions to societal problems.

Figure 9.20 provides a schematic model of a photovoltaic solar system. This system includes not only the solar panels needed to supply electricity to society for their consumption, but also a set of panels that supply the manufacturing plant (of the panels and associated equipment).[53] The manufacturing process shown in the figure should also include construction and repair activities on the manufacturing facilities themselves. In other words, we have to supply energy to the acts of constructing the facilities and energy to the maintenance of said facilities. It would be possible to adopt a wider boundary condition by including the manufacturing of vehicles (i.e., in the blue deployment ovals), which would, presumably, be electric, and even some amortized cost of building and maintaining roads for delivering

[53] We are ignoring a lot of infrastructure requirements such as power converters (from DC to AC), and the electric grid itself. The point of this exercise is to demonstrate the general methodology, not to produce a definitive answer.

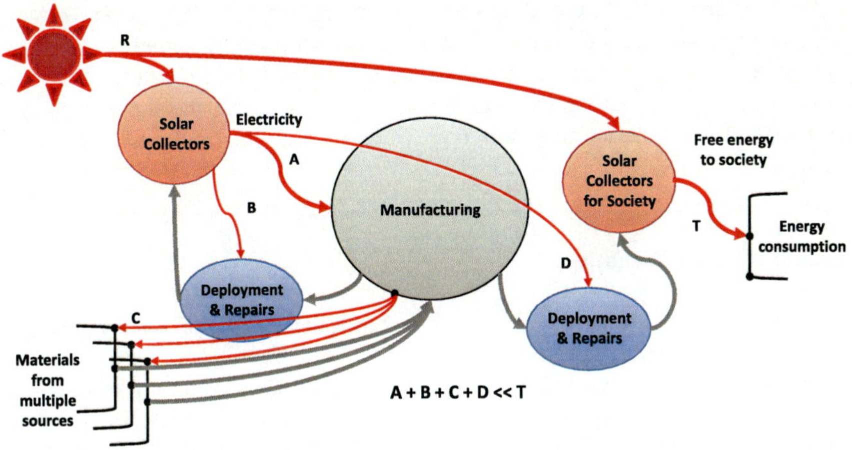

Fig. 9.20 A solar energy capture and conversion system includes the capacity to power the replacement of solar panels (including deployment and repairs). This means providing energy from solar capture to the manufacturing process but also should include the construction of the manufacturing capability as well as its repair

panels, etc. However, this model should suffice to point out the difficulty in deriving a "solution" to the power supply problem. A more realistic requirement for a system like this would be that it is autopoietic,[54] that is, it needs to be completely self-regenerating, self-maintaining, as well as producing the free energy for societal consumption.

Figure 9.20 does include the supply of power to various material extraction resources, the materials that are needed in the construction of solar panels and accouterments, which must be accomplished by electrical power. The letters A, B, C, and D represent the power supplied to all of the processes that produce solar panels, deploy, and maintain them. The letter "T," standing for Total free energy to society, is the payoff. But the necessary condition for this scheme to work is shown as the sum of A, B, C, and D must be considerably less than the total energy supply to society.

This is another version of the EROI problem that we have discussed with fossil fuels. Even though solar energy won't be depleted by collection, the ultimate economic viability of this technology depends on whether $T/(A + B + C + D) > 20$ or 30 (Hall and Klitgaard 2012). The EROI of solar must be capable of supplying enough energy to society to support social uses such as household consumption and transportation as well as replacement of the solar energy infrastructure. As of this writing there has been neither a convincing model simulation, nor a comprehensive experiment that would demonstrate this feasibility.

[54]The concept of autopoiesis will resurface in the next chapter in Part III in the context of a complex, adaptive, and evolvable system.

For the foreseeable future, it appears that the majority of society's power will come from fossil fuels. This may be increasingly supplied by natural gas (many electric power plants are converting to natural gas as this is being written). But that still involves burning carbon-based fuels and pumping CO_2 into the atmosphere. As long as humanity insists on power (and convenience), this will continue to be the case.

9.6.2.2 Other Alternatives

The same logic regarding the EROI of solar applies to all other alternative energy system proposals, wind, tidal, etc.

Famously during the G.W. Bush administration, the idea arose that perhaps corn-derived ethanol might supplement the gasoline supply. In part the thought was that corn-based ethanol would be carbon-neutral, that is, corn plants would absorb CO_2 from the atmosphere to grow, and so when the ethanol was burned, it would not add net carbon to the atmosphere! Congress in the United States acted and passed a law requiring up to 10% corn-based ethanol to gasoline. It didn't hurt that Midwest corn farmers would recognize a healthy new market for their products.

Subsequent analysis of the EROI of corn-based ethanol has shown that it is far too low to provide a positive economic advantage (Hall and Klitgaard 2012, page 336). Furthermore, taking some of the activities that are needed to subsidize corn ethanol, the carbon footprint may also be net negative.

The lessons from solar and corn ethanol in terms of systemic effects should be instructive. But political choices are often made for other than scientific reality reasons.

Wind energy appears to have a better economic advantage (ibid). A major caveat, however, is that wind turbines, to achieve very high efficiencies, require using rare-earth metals that are quite expensive and in limited supply (hence the name "rare-earth"). The same kind of analysis of solar energy self-sustainability shown above should be applied to wind as well, taking these issues into account.

Overall, the jury is still out on the overall feasibility of alternative energy source technologies. Since they are all producers of electricity, we also have to consider the various costs of converting most of our work processes to be driven by electric motors (and heaters). Since electric work is done more efficiently than heat engines powered by fire, this would be a good thing. But how fast it can be accomplished, and at what energetic cost, is still a very open question.

9.6.3 *Energy Consumption: Work Processes, Efficiencies, and Waste*

For most of the eighteenth, nineteenth, and twentieth centuries, energy has been taken for granted. Wind and water drove the nascent industries of the eighteenth century. Wood charcoal and then coal was the source of power for the nineteenth and early twentieth centuries. Then oil and natural gas began to take center stage. All of these energies either came, fundamentally from the sun, in real time, for example, photosynthesis, or accumulated over time but released in real time, for example, water behind a dam and fossil fuels (cf. Crosby 2007). In all of these cases, energy seemed abundant and easily obtained. The idea that we should be overly concerned about efficiency did not start to take hold until the age of air travel when the means of conveyance needed to carry its own fuel source and be light enough to get airborne. Engineers were generally concerned with efficiency but it was not the pressing issue for an economic system most concerned with profit-making, especially one in which sellers could increase their prices as they felt necessary if they could not reduce costs.

That is, until the availability and cost of fuels began to become more visible as a factor in all work processes. The oil shocks of the 1970s (1973 and 1979) due to an OPEC embargo brought the Western world's attention to the fundamental role of energy. In the early 2000s the price of oil climbed rapidly, peaking at $147+ in 2008 before declining again. It never fell back to its price levels prior to that event. It is believed that the price shock helped trigger the Great Recession, which followed closely on the heels of the shock.[55]

The cost of energy as a major factor in the economy finally became apparent in the years that followed. However, most economists did not really know how to interpret the meaning of the phenomenon. There were so many concurrent factors that were better understood by economists so that they tended to focus on them as causing the recession, like the drunk who lost his keys at night and went looking for them under the street light because he could see better there. One example was the mortgage debt bubble created by lenders providing sub-prime loans to people who had no prospects of paying them off. A financial bubble was something that economists had seen many times and their association with recession triggers was well recognized. However, correlation is not causation. Were the bubble bursts causes or effects? Debt financing, as discussed previously, completely obfuscates the relationships.

[55] Oil prices did tumble back to pre-spike levels in a rebound effect. However, that was a short-lived condition and oil prices have gradually climbed back up. At the time of this writing a barrel of West Texas crude is running around $50. It is estimated that oil produced from Canadian tar sands (bitumen) needs to capture around $70 per barrel in order to just break even. Note, though, that the oil market is extraordinarily volatile, so anything said at this moment is almost certain to be false at any other time!

There are starting to be some recognition among some economists that the relationship between wealth production and energy, as well as the negative effects of decreasing EROI, are primary forces in explaining economic phenomena (Hall and Klitgaard 2012). But the full extent of the relationship has yet to become completely understood. Most economists are not as schooled in systems science as they probably should be.

9.7 Systemic Dysfunctions

In truth, there are many dysfunctions that can be found in modern economies throughout the world. We cannot present a comprehensive list (though we do try to cover some of the issues in Chap. 13). Here we will just discuss a few of the most egregious. These dysfunctional aspects of modern economies have a common root cause. We human beings are biological entities and are motivated by the biological mandate to grow our populations. In all other biological species, there are natural limits to population sizes imposed by the dynamics of the environment—the ecosystem in which they live. Food availability and predation, for example, operate to prevent populations from growing without bounds. But for humans, with the capacity to produce artifacts that seemingly transcend those ordinary boundaries and, indeed, permit humans to escape from any particular environment, the biological mandate has nothing to keep it tamed. The expression of greed that is ravaging modern humans is a cultural construct but it is based on the biological mandate giving it impetus.

There are several other factors unique to humans that lie at the root of many dysfunctions. For example, we humans have a penchant for novelty. If this were left to causing us only to explore new territories or look for a better way to accomplish a task, for example, then it should be counted as a strength insofar as increasing biological fitness. But taken to extremes, where we seem to get bored with the last new thing and demand more new things, it is a driver less for innovations that truly help as much as for anything at all that feels new.

And that is coupled with another human-unique attribute that has gone to extremes in our modern cultures, and that is the ravenous appetite for entertainment. As the fossil fuel economy expanded and made available more conveniences for ordinary citizens, they turned their increased leisure time to demand more and more entertainment. Humans, unlike the vast majority of other animals that engage in play as juveniles, continue wanting to have fun throughout their lifecycle. They have been willing and able to direct more of their discretionary income toward a panoply of entertainment options. And then, especially in the era of motion pictures, this desire for entertainment, coupled with the above-mentioned demand for novelty, has driven the nature of these forms of entertainment toward their own kind of expansion and outdoing last year's versions. Think of the budget and box office take for a blockbuster movie. Some kinds of entertainment are useful in terms of allowing human mental rejuvenation and release from tensions. But in looking at the current markets for entertainment as compared with say half a century ago we see a staggering growth in capital and revenue resulting from the drive to make things bigger and better (IMAX™), more grandiose (the Oscars ceremony), more exciting (the roller coaster arms race). Granted this is enabled by technologies, one needs to ask where does this end? Of course, it ends when the energy sector we've been discussing burns itself out. But the urge to have more, bigger, better, more elaborate new stuff and entertainment will continue to direct vital resources to non-sustaining activities.

There are several major factors that operate to keep humans in general and economists in particular from seeing and understanding these things as dysfunctions.

9.7.1 Values and Value Signaling

The initial role of money was to provide a convenient means to measure various values of assets and work.[56] It fulfilled this function because it was tied directly to the use of energy (human labor) to produce those assets as a substitute for bartering. People did not consciously recognize this association except through the biological needs they experienced, for food, shelter, clothing, etc. But once grain-states emerged with their hierarchical governance and class structures (bureaucracies and kings, etc.), money began to take on additional roles as symbols of wealth and, similar to the granaries that preceded them, storehouses of value. We will have much more to say on the subject of money losing its ability to signal true values in Chap. 13 where we will have a better description of just what value means. Here we will only point out that since we humans did not really understand the need for a true signaling device related directly to the availability of free energy to do the work of creating assets, we have allowed money to become nearly worthless for that role. Nowadays we assume that the value placed on a thing or service is based on market forces finding a price—cost + profit equilibrium. That was actually the case in an era when people understood how much work they had to do to accumulate assets. What tokens of money represented was closer to the actual energy required to do work. In today's distorted monetary and market systems no one has any real clue as to the real values of things or services. This is one of the main reasons so little interest was put into recognizing the central role of energy in every aspect of the economy when there was an opportunity we could have established systemic controls on its uses in the economy.

[56] Cf., Nail Ferguson's "story" of money (Ferguson 2008) for a more comprehensive telling of a complicated and complex history.

9.7.2 Emphasis on Growth

When CAS/CAESs are nascent and until they reach a state of maturity, growth in size and complexity are natural and good. Maturity means the system is capable of self-sustaining and persistence for the rest of its natural lifespan, discounting accidents or overly rapid change in the environment. But once a system does reach maturity, the need for growth ceases and the economy switches into maintenance mode at the maximum size, a size determined what we call the carrying capacity of the environment.

Of course, what it means to call a society mature is much less certain, though there are many sociologists who have considered the problem and offer some insights (Harari 2017, esp. pp. 202–221; Polanyi 2001; Qualman 2019; Quinn 2000). The problem for the HSS is that with a continuing capacity to invent new technologies, which in the capitalistic systems require the accumulation of capital for investment in R&D and production, there is a concomitant drive to do more work and produce more stuff. The need for novelty, convenience, and entertainments mentioned above contributes to this drive as well.

And then there is the fact of growing populations worldwide. A growing population puts pressure on an economy based on labor income to create more jobs. But then those new workers buy more stuff which grows demand and the cycle is perpetuated. As long as populations continue to grow there will be more income and more consumption. The dominant measure of national income, the GDP, is now hailed by politicians and the news media alike as being "healthy" when the number grows year on year, which is exponential growth (like compound interest). The growth that everyone thinks is healthy, is nonlinear, and is a positive feedback loop. People who understand systems consuming finite stocks of resources know this is a totally unrealistic scenario (cf. Meadows et al. 1972; Meadows, et al., 2002). Systems that grow continually either blow up or implode, that is collapse.

Growth in the GDP gives the illusion that the economy is doing well and by inference, the people in the economy are doing well. But in point of fact both the aggregate GDP and the per capita GDP numbers are not what they seem. The former grows as much reflecting the inclusion of negative factors that destroy real wealth while making it seem that income is going up. For example, every time people rebuild their homes after a hurricane has damaged or destroyed them, the sales of materials and contracting work goes up. It is important to remember that income is not wealth. The growing evidence is that per capita GDP has actually been trending downward at least since the mid-1970s but probably before that (Gordon 2016).

Moreover, what growth has occurred in much of the world has gone disproportionally to those who actually do the least amount of actual work, the rentiers and senior corporate management. This growth in income is also illusory. It comes in the form of profits on existing capital or from bonuses based on debt-based financial profits. The effect is to grow capital (on paper anyway) for those who already have accumulated capital while those who have very little net worth pay out more of their income to rents and interest on loans; their capital shrinks. Today it has become common knowledge that the dynamics of the past several decades has returned societies to the days of the robber barons and tycoons (Piketty 2014).

9.7.3 *Neoliberal Capitalism*

In examining the mechanisms that have led to these dysfunctions of the social economy, we come to one prominent culprit. An increasing number of heterodox economists and many social commentators have come to the conclusion that neoliberal capitalism, indeed most forms of capitalism embodying the prior two issues, has been the major contributor to the destructive forces that look to cause the collapse of the HSS economy and the HSS itself, since the latter, in its current form, cannot exist without the former, but in a sustainable form. Naomi Klein (2015) makes direct connections between the engine of capitalism, burning fossil fuels, and the rise of CO_2 that is driving global warming and climate change.

There was a period of time when mechanisms for aggregating capital to fund new projects were much less about maximizing shareholder value and more about boosting social value which was seen as also providing an acceptable profit, a premium paid for taking the risk of putting money in. But today's various stock and bond markets are more like casinos where gamblers are betting on short-term and maximum payoff rewards. Worse yet, the casinos have been rigged by central banks to always pay the gamblers.

In the natural world, in the living CAESs there is no analogue to capitalistic economies except for the conditions of cancers, where the internal regulations on growth have gone amuck and the cells are out of control. Many an author has likened capitalism to cancer.

9.8 Conclusion

In this chapter, we have taken a cursory look at how to use the methods of Chap. 6 (and a few other concepts) to analyze a truly complex system, the HSS economic subsystem and its particular sub-subsystem, the energy capture, conversion, and distribution system. We've looked at it from a number of different views and perspectives and while we were only able to cover just a small part of the system, we did try to go into depth for what we argued are the most essential aspects for keeping the economy working. And along the way in this analysis, we ran into a number of aspects of the commonly accepted paradigm of neoclassical economics that suggested the current economic subsystem is not truly healthy.

In particular, we found that the reality of the nature of an economic system is substantially different from the observations of Adam Smith and other eighteenth-century political-economists upon whose thoughts much of the neoclassical view is based. Using knowledge of modern physics (thermodynamics is central) and guided by systems principles, the picture of what an economy is and how it actually works, as considered in this chapter, turn out to be very different from that developed by neoclassical economics academia. We can see, now, that an economic system is not closed in any sense. It requires the continuous input of medium entropy materials (ores, wood, etc.) and high-quality energy to further reduce the entropy of materials and increase their usefulness in maintaining the "metabolism" of society.

The current HSS economic system does not really conform to the natural economic systems of the Ecos. It is based on fallacies such as continuous growth and maximizing profits which are actually counterproductive with respect to the maintenance of a viable system, one that is stable, resilient to changes, and sustainable in the long run.

The systems science archetype model of an economy offers a foundational approach to the conceptualization of what an HSS economy ought to look like. The deep analysis of our current economic systems (across the spectrum of ideological models) based on systems principles shows that we are far off the mark when it comes to having an economy that is life-supporting in the same way that ecosystems, physiologies, and metabolisms are. There are some aspects which attempt to emulate the archetype, for example, using money flows to signal energy channeling. But human hubris along with a zeal to invent too often muck up the systemic workings of these mechanisms. One can only hope that as we develop a more systemic knowledgebase of economics along with a better grasp of human neuropsychology (and how to tame it), we will start to fashion an economic system that truly supports all humans' bodies as their physiologies support their living cellular metabolisms. Then, the HSS might actually become a fit component of the Ecos.

Part III
Archetype Models: Complex, Adaptive, and Evolvable Systems

In the following four chapters we turn attention to a generic or archetype model of all complex, adaptive, and evolvable systems as have been mentioned in prior chapters. In the CAS/CAES models we have assembled all of the relevant sub-models that comprise any real-world instantiation of complex systems. Chapter 10 starts with an overall description of the whole CAS/CAES model and situates the three sub-models, agents/agency, economy, and governance within the whole.

Chapters 11, 12, and 13 then provide details of these sub-models in their own rights but maintains the linkages between all three.

These models are the basis for conducting analysis of complex systems with a priori understanding of what *should* be found in those systems, The models provide guidance to the analysts in conducting their deconstruction, ala Chap. 6. And the process is the same whether analyzing an existing system (as in Chap. 9) or analyzing the requirements for a system to be designed (see Part IV).

These archetypes provide an architectural framework. They are general in that they apply to all complex systems. And, yet, they are detailed enough to supply a prototype framework for developing understanding of any system. These archetypes provide the same kind of "head start" for understanding as innate concepts in human development provide for babies learning to deal with the world (Carey 2009).

Chapter 10
Model Archetypes for CAS, CAES

Abstract There are attributes and functions required of all viable complex adaptive and evolvable systems (CAS/CAES). These are brought together in a general archetype model, which is then seen to be composed of several also general archetype sub-models. Every CAS/CAES will be shown to contain a set of work (material/energy transformation) processes that collectively comprise its internal economy. This economy and several peripheral processes involved in system maintenance (autopoiesis) and, in evolvable systems, making new arrangements internally to address major changes in the environment, require a specific organization of governance modeled by a hierarchical cybernetic governance system archetype. All CAS/CAESs can be described by their internal network of decision nodes responsible for managing the economic, peripheral, and the decision agents themselves. The agent/agency archetype describes a generic decision processor situated at each node in the network. The CAS/CAES is the result of an ongoing effort to integrate a large number of important ideas from numerous workers who have developed their own versions of models in this arena. We compare and contrast a few of these to argue for the case that the CAS/CAES archetype is a more holistic and more broadly relevant model suitable for guiding analysis as per Chap. 6 and design of artifactual systems to be covered in Part 4.

10.1 Models of Complex Systems and Model Archetypes

In Chap. 6, we provided a brief introduction to the nature of a complex, adaptive, and evolvable system (CAES) and its precursor lacking evolvability of the complex, adaptive system (CAS). This was necessary in order to proceed with the use of the methods of Chap. 6 in several complex system examples in Chap. 7—and again, useful in the more elaborate example of those methods working on the example of the human social system (HSS) economy in Chap. 9. We now turn, in this Part 3, to a fuller explanation of the CAS/CAES models and their sub-models because, in Part 4 we turn our attention to the design of CAESs. As will be explained in Part 4, our design approach is based on using the CAS/CAES archetype model as a design template.

G. E. Mobus, *Systems Science: Theory, Analysis, Modeling, and Design*,
https://doi.org/10.1007/978-3-030-93482-8_10

An important caveat regarding the subjects in this and the next three chapters: What we will cover is somewhat like the part of an iceberg that can be seen above the water; there is a vast body of knowledge regarding these subjects below this surficial treatment. What we will attempt is to address some of the most salient issues in trying to make sense of the big picture. Throughout these chapters we provide "hooks" or anchor points that attach to the deeper knowledge areas. For example, in the next chapter, on Agents and Agency, we consider the nature of decision-making and action from simple systems, for example, reactive controllers like thermostats, to adaptive controllers, like homeostatic processes in living systems, to complex and evolvable (that is learning) governance systems, like the human brain. Clearly each of these would require volumes just to outline their main points. Knowledge of the nature of human psychology in decision-making alone is massive and growing rapidly. What we point to, however, is the need to consider that mass when thinking of human agents, especially problems of motivations and corruption. We will consider the nature of a so-called ideal agent and then quickly point out some of the ways in which humans are far from ideal. The point of an agent archetype model is not to go into all of the details of how the agent works (or fails to work properly) but to show how it integrates with the other archetypes that constitute a CAS/CAES model.

In this chapter, we will explain the different model archetypes of subsystems of a CAS/CAES that are needed to produce a viable, long-enduring system. But in order to grasp the significance of these models, the reader needed to first be exposed to the whole theoretical construct of systemness and how the process of ontogenesis led inexorably to complex adaptive systems and then to complex adaptive and evolvable systems (CAS/CAESs). Chapter 9 was an anomaly in that we sought to show how the understanding of that construct and the analytical method derived from it would lead to sensible knowledge. We also wanted to demonstrate this for a very complex system to convince the reader that the concepts and methods work.

Now we have to revisit the inclusion of all of those references to this chapter by examining the larger picture of how they all relate in general to the concepts of CAS/CAES. The method to be used in this chapter is to bit-by-bit unpack the details of the archetype models as if we were doing a systems analysis (without formally doing the procedures of Chap. 6). We will first provide a summary view of the models and their relations. Then we will start unpacking the overall concept of a CAS/CAES archetype and how the other archetypes (agents, economies, and governance) interrelate to form the whole. Finally, we go into the details of each of these three models in the next three chapters.

All CAS/CAESs share a fundamental organization of subsystems and functions that allow them to persist over the long haul and continue to serve a useful purpose in their environments even when those environments, themselves, evolve over time. We will start with the general model of a CAS/CAES and then examine the major component subsystems that make these systems possible. Then, in subsequent sections, we will provide more details of the subsystem archetype models so that you, the reader, will be able to use these to analyze and, if tasked with creating a new system, design real systems.

The word "model" has many meanings but in the context of this book we restrict our concerns for the nature of models to the scientific and mathematical modeling of real phenomena and systems. In short, a model can be any abstract representation of a real phenomenon or system (hereafter we will use the single word, system, to mean both general physical phenomena and systems as we have been describing throughout the book) that captures the "essence" of that reality but does not attempt to represent all of the details of the system that are not cogent to the system's behavior in terms of understanding it. We are interested in the system's behavior in terms of how it affects the rest of the world (or us), so minute details may not be relevant and can safely be left out. Leaving out unnecessary details while still accomplishing our goal of understanding is also pertinent to computational tractability.[1] The nature of the abstraction allows for various kinds of manipulations of the model parts such that the whole model behaves in a general way just like the real system.

The purpose of a model is to provide that abstract representation in a way that a human mind can comprehend all of the working parts without getting bogged down in superfluous details that do not affect the final results. For example, if our major interest is in predicting some future state of the system, any model that captures the main variables and transition functions may serve. But, at the same time, if our intent is to deeply understand how these transitions and behaviors are accomplished, then it might be necessary to include more details than are necessary for mere prediction. Science might be viewed as a process of discovering the details to the minutest level for just this purpose. On the other hand, when engineering a system, we might be just as happy with major behavioral outcomes from more abstract models.

The "natural" sciences have considerable experience with dynamical systems models in which the main features of behavior are captured in a set of ordinary differential equations. When such equations are solvable, then a future state of a system is a simple matter of solving the equations for a given set of initial conditions and specifying a time interval. As it turns out, the vast majority of interesting complex systems models cannot be handled in this way. It works very well for "simple" physical systems (like a few gravitationally coupled bodies) but does not work for intricately complex systems like human societies.

With the advent of the digital computer, we have developed a whole new way of building dynamical models and simulating the "systems" in order to determine what their states will be in some future time. We still need to specify starting conditions. But it is unnecessary to derive a set of formulas that compactly capture the dynamical behavior of the system. System dynamics modeling and computer simulation allow us to iterate over time rapidly (at the speed of digital computation) and arrive at an end state of even extremely complex systems with multiple kinds of internal feedback.

[1] For example, we do not need to include all of the details of quantum mechanics when building a model of a chemical reaction, though that reaction might be reduced to the quantum level. They are de facto included in the overall behavior of the atoms, molecules, and their interactions.

10.1.1 Representing Models

Equation 4.1 provides a basis for representing a model of any system. It uses the language of graph theory, or more to the point, the language of flow networks as mentioned in Chap. 2. However, it expands that basic language in order to capture the semantics of processes and sub-systemness. It is asserted, without formal proof, that any system, no matter how complex, can be represented by Eq. 4.1, which captures both structural and functional information down to some level of detail. In other words, a model of any system (having the requisite systemness properties) can be represented by a flow network where the nodes are work processes that observe the laws of conservation (mass and energy) along with the second law of thermodynamics.

10.1.2 Systems Dynamics Models—Simulations

Many systems investigators have been concerned with system behaviors or dynamics. Models built from principles of systems dynamics (SD, Forrester 1961) are the basis of simulations run on computers that can trace the behavior of systems over time and generate end states of the system.

System models are often developed from ad hoc understanding of systems. Analysis is used to decipher interactions between system components, but often boundary conditions are determined by the modeler rather than emerging from a deep analysis (Meadows 2008, p. 97). Modelers are frequently faced with making somewhat arbitrary choices regarding boundaries because how a system interacts with its environment is extremely complex and absent a principled method for analysis, they are forced to select the boundaries. Most interesting systems are, recall, fuzzy and therefore obscure as to where a boundary exists. As we showed in Chaps. 6 and 7, dealing with fuzzy boundaries is not trivial and is definitely necessary in order to get a firm grasp on the nature of very complex systems. In SD projects the problem is somewhat avoided by making those arbitrary decisions about where a boundary exists and what is inside a system and what is external.

SD models/simulations have been hugely helpful, even given the problems with boundary selections, in showing how complex systems behave in nonlinear ways even when the inputs ramp up (or down) in apparently linear schedules. These models are highly successful in capturing the feedbacks (negative and positive) that give rise to nonlinearities in behavioral parameters.

SD has had several implementations since the first SD language, DYNAMO, was developed at MIT. The language(s) however do not provide representations for all of the terms derived in Chap. 3, nor does the language provide a process semantics or a hierarchical structural organization.[2]

10.1.3 A General Systems Model

In Chap. 4, we presented a specific mathematical description (or definition) of a general system in Eq. 4.1 (and subsequent explanatory equations). In Chap. 8, we demonstrated how the analysis of a real system, based on Eq. 4.1 (Chap. 4), could be captured in a knowledgebase for further analysis but ultimately for constructing models. In the next chapter, we will show how this is to be done. We will demonstrate how the knowledgebase contents can be used to generate models at various levels of abstraction. And by models, here, we mean models suitable for computer simulation. The model can be as detailed as the analysis produced. If the intent was for deep scientific understanding, then the model might similarly be extremely detailed and require extensive computing resources to simulate. On the other hand, because of the way Eq. 4.1 through 4.3 are structured recursively it is possible to generate system models that are more abstract and, hence, useful for engineering or management purposes.

A user of the modeling interface with the system knowledgebase need only indicate what level of abstraction is needed for a simulation. Since the transfer functions for any given module subsystem have been captured, it should be possible for the software to construct a simulation using those functions indicated in the level of the simulation requested.

10.2 What Are Model Archetypes?

A model archetype is a generic version of a specific kind of system model. Some authors have used the term "meta-model" in a synonymous sense. In general, we will stick to the term "archetype" since "meta" can take on several senses that might not always work.

The models to be presented in this chapter are archetypes of three complexly interrelated work processes, already presented in Chap. 3 as archetypes, that constitute the "workings" of a functional and stable complex system that is able to maintain itself in a fluctuating and possibly non-stationary environment. The three

[2]Actually, there are some newer SD modeling environments that are moving in that direction. Some support modular or object-oriented models that can be combined or re-used. Some also support a hierarchy of models, that is, models within larger time domain models. But these have been developed in what looks like an ad hoc, needs-based manner rather than from a theoretical basis as provided in Part 1 of this book.

models now being presented should be seen to be at much higher levels of organization. They are, nonetheless, isomorphic across all levels from the origin of life upward. Each of the three plays a necessary role in the whole and are highly integrated with one another to achieve completeness.

The concept of an archetype comes from observations of a large number of complex systems across the spectrum of levels of organization, but particularly from the first levels of living systems (bacteria and archaea) through the highest levels of human societies. We have devised the categories of complex adaptive (CAS) and complex, adaptive and evolvable (CAES) systems to hold all of the living systems representative categories (cf. Miller 1978, for the array of these systems). All individual organisms (i.e., single cells and single multicellular organisms) are complex and adaptive. Adaptivity to changing environmental conditions is a basic attribute of living systems. Evolvability is also seen in living systems from cells (bacteria able to allow mutations in critical functional genes under certain environmental stresses) to more complex individuals possessing brains able to learn. The details of what these designations mean will be discussed below.

These archetypes are effective guides to analysis and design since they tell us what subsystems and components are to be found in all instances of CAS/CAES.[3] In this chapter, we develop an archetype model of a CAS/CAES, identifying all of the subsystems that must be found in any instance of such a system. Whether we are decomposing an existing system or determining what is needed for a to-be-designed system, these archetypes act as guides to analysis and design. In the scientific reductionist decomposition of a particular system, such guides tell the scientist what structures/functions to look for at the next level down. In engineering, they tell the engineer what structures/functions need to be incorporated into the designs.

As used here a model archetype is a pre-defined model of the general architecture of a particular kind of subsystem that is common to all CAS/CAES systems.[4] It is so general that it can be used in any number of different contexts or specific CAS/CAES models. The governance model archetype, for example, and introduced below, is one such model that describes a general governance process for any CAS or CAES. In this chapter, we introduce four such archetype models. We expand the description of the CAS and CAES and then describe the three major subsystems that are vital to all CAS/CAES, agents, economies (as previously examined in Chap. 9), and governance.

[3] We have found some aspects of these archetype models effective in non-adaptive systems as well but will refrain from trying to make the case for it in this book.

[4] The three model archetypes of subsystems within a CAS/CAES are necessary, but may not be sufficient (Mobus 2017).

10.2.1 The General CAS/CAES Architecture

The concept of complex adaptive and a complex adaptive and evolvable system (CAS/CAES) is an amalgamation and integration of concepts that have come from many different writers (see below, Sect. 10.4). At least on Earth, the concept starts to take shape with the first living cells emerging perhaps 3.8 billion years ago, just shortly after the condensation of the planet out of the solar debris (Smith and Morowitz 2016). All living systems taken as individuals, which include single cellular organism (both prokaryotes and eukaryotes), colonies of cells, multicellular organisms (comprised of eukaryotic cells, and in more complex forms, tissues, and organs), and groups of multicellular organisms (populations), constitute the CASs, all able to adapt to some extent to variations in environmental conditions. Starting with species as a Darwinian evolutionary system, but including animals with modifiable cortices (i.e., capable of learning), especially human individuals, and then moving into groups of humans or societies with cultures we recognize systems that are not only adaptable but are also evolvable. That is, they can undergo modifications that permanently change their structures and behaviors to meet the demands of a longer-term change in their environment. Alternatively, evolution may involve, as it often does in Darwinian biological evolution, fortuitous alterations that when tested by the environment, changed or not, are found to imbue increased fitness on those individuals that possess the alteration.[5] They out-compete and out-reproduce their conspecifics.

Another kind of evolvability enters the picture with human cultures, organizations, and institutions. We have termed this "intentional modification" as opposed to "chance modification," that is, the hallmark of Darwinian evolution. Intentional means that some human brain recognized that a change in a subsystem of the culture (i.e., artifacts) or organization or institution would either improve on the existing structure/function or make them preadapted to perceived future conditions in the environment. A marketing manager sees a new market opportunity if the company would slightly alter the characteristics of an existing product that they manufacture with enhanced features. A young human being chooses a major in college and pursues changing their own minds with new and, to them, useful concepts and skills.

CAS/CAESs constitute a new phase of organization of systems. Life was such a new phase of matter when it emerged from prebiotic chemistry on the ancient Earth.[6] The CAES is yet a further phase transition to higher organization and

[5] The alterations referred to here are genetic mutations that lead to phenotypic structural and/or behavioral modifications. This mode of evolution depends on a large population of individuals in which the various blind experimental trials (mutations) can be tried out without jeopardizing the entire species.

[6] While there are still some debates regarding source of life, whether seeded on Earth from somewhere else in the Milky Way galaxy (a theory called panspermia) or actually started on Earth, we will assume the latter case as being the most likely given what is understood today about prebiotic chemistry and metabolism. Cf. Smith and Morowitz (2016).

potential for yet newer and more complex emergences (Smith and Morowitz 2016; Morowitz 2002).

Russell Ackoff (1971) provided an early attempt to produce what he called a "System of Systems" approach, by which he meant a systemic framework for understanding systems concepts. The CAS/CAES scheme presented here seeks to further that approach by incorporating a broader set of concepts more recently understood as well as updating some older, other frameworks. The CAS/CAES archetype model is a synthesis that incorporates concepts arrived at by many researches in systems science but especially in the biological sciences. It primarily maps onto the general schema of living and supra-living (e.g., human societies) systems developed by James Miller (1978) as outlined in Chap. 3. It incorporates the concepts developed by Stafford Beer (1959, 1966, 1972) and expanded by Eric Schwarz (1992, 1997) involving the *governance and management* of complex adaptive and evolvable systems based on principles from *cybernetics and hierarchical organization* (see Sect. 10.2.2 below). It provides a detailed model of Varela et al. (1974) autopoietic system.[7] It incorporates concepts developed by Howard T. Odum (1983, 2007) regarding the trophic exchanges that occur in complex systems, the *flows of materials, energies, and information*. In other words, their *economies*. It also involves concepts of the modeling relation and anticipatory systems of Robert Rosen (2002) in the treatment of *decision agents*. These are all conditioned by concepts developed by Harold Morowitz (1968, 1992, 2002) regarding the organizing influence of energy flows through systems (and Smith and Morowitz 2016, already cited) and providing a grounding for the *ontogenic cycle* from Chap. 3. These authors were central shapers of a universal picture of complex adaptive and evolvable systems as we find them on the present Earth, including the cultural addenda for the human social system. What we offer here is a new way to categorize these systems and a synthesis, an approach to unifying the many various views, if not the terminologies employed.

Figure 10.1 provides an overall architectural scheme for the CAS/CAES archetype model. It shows that the model is comprised of three generic archetype submodels: the "Agency," "Governance," and "Economic"[8] system archetypes. Below we will describe each of these and how the whole CAS/CAES archetype is composed. The agency (or agent) model is shown in the center and overlapping both the governance and economic models since agents are the key decision-makers in both subsystems (as represented in Fig. 10.2 below). These subsystem models are shown in a Venn-like diagram to emphasize the fact that all three are necessary to comprise

[7] Autopoiesis has been criticized for its incompleteness in explaining life as a phenomenon. Though it may have some shortcomings it does take into account major aspects of a living system and is more complete than the traditional lists of attributes normally associated with a definition of life. See Capra and Luisi (2014, Chap. 8).

[8] We were tempted to call this the "Metabolism" system archetype since cell metabolism is an instance of the economy of a cell and as we show physiology, the HSS economy, and the chemistry, energy flows of the ecosystem are just larger-scale extensions of basic metabolism. But we decided that the term "Economic" was the more general and would cover all of the subsystems of interest.

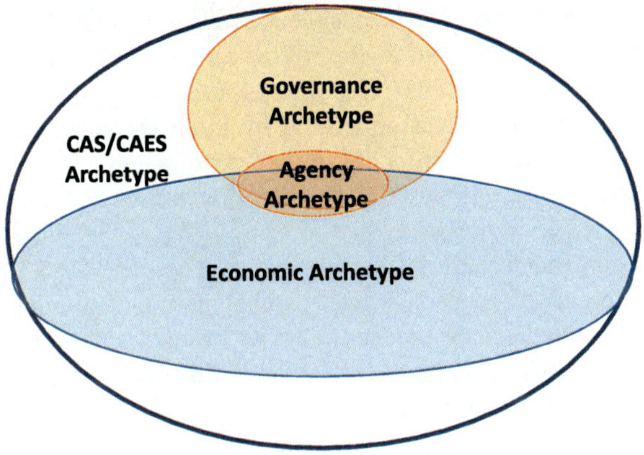

Fig. 10.1 An architectural overview of a CAS/CAES. The model archetypes for Agency," "Governance,", and "Economic" superstructures can be shown to be isomorphic across all such systems. These archetypes are fuzzy and yet can be delineated in accordance with Chap. 6 and demonstrated in Chap. 9. The relative sizes of the ovals are merely meant to be suggestive of the amount of work (for example, the amount of energy devoted) that the process involves. Note too that the actual distribution of these processes in a concrete CAS/CAES is not represented here

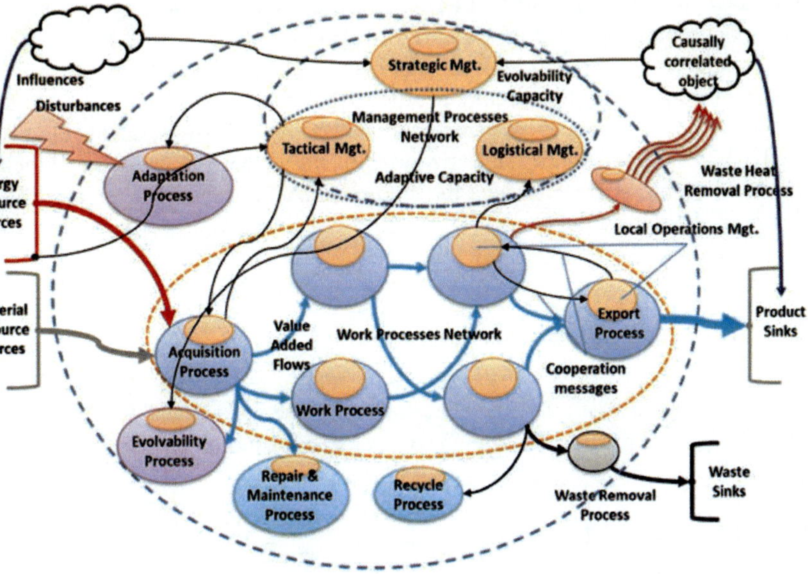

Fig. 10.2 The most complex kinds of systems we know are not only adaptive but also evolvable. This figure represents the several main component models (archetypes) for a CAES. Basically, a CAS is most of what is in this archetype except for the "Strategic Mgt" component in the "Management Process Network" and the "Evolvability Process" (lower left sector). Shown are just a few of the flows of matter, energy, and messages into the system and between components of the system. Not shown in this figure, but not to be left out of the analysis, are all of the interfaces

a complete CAS/CAES model. They are extremely integrated yet can be focal points of systems decomposition, as the following chapters will show. All CAS/CAESs from the earliest protocells to the whole Earth Ecos are governed economic systems that require constant decision-making by competent agents, as managers of economic processes and as governors of the society of agents. In terms of the HSS we have long recognized this by talking about the "political economy!" We will explore this last viewpoint in Part 4.

The claim being made here is that all known exemplar CAS/CAESs can be explained by reference to these subsystem archetype models, though for specific application domains there may be some number of auxiliary subsystems that are part of the whole particular system. For example, in Chap. 7 we identified a few subsystems in the HSS that were specific to the human society system, such as the Science and Technology subsystem. Any such particular subsystems will interface and be integrated with the specific version of the archetypes discussed here. They can be treated as specialized processing modules but are still based on the three archetypes presented here.

Below we examine the roles of each of the three archetypes and explicate the most general features of each. The central claim we make is that in order for any complex system to be long-term stable in a changing environment it must have this architecture with the details worked out pertinent to the environment in which it is embedded and the interchanges it has with those conditions. When faced with the task of understanding complex systems, the a priori knowledge that this is the relevant architecture we should look for will make the analysis (and subsequently the design if that is the objective) far more efficient compared with casting about blindly trying to figure out what is happening and why (Chap. 13 will be devoted to a new approach to systems design and engineering based on these archetypes and including an ontogenic approach where the auto-organization process is replaced by an "intentional-organization").

10.2.2 Archetype Models Overview

Here we present a brief overview of each of the archetype models. In the next three chapters, we will provide descriptions of each of the sub-models in more detail.

10.2.2.1 Agents and Agency

Agents are specialized decision-making information processes the outputs of which generate control activities. Agents have "agency" in this respect. Any decision-making mechanism that affects the state of the system and/or its environment is an agent. To be a successful agent its agency must cover the possible actions that would be needed to adjust itself and/or its environment so as to maintain a suitable relation (Ashby 1958). The agent archetype includes simple mechanical devices such as the old bi-metal coil thermostat as well as more complex devices such as the auto-pilot on jetliners and all instances of homeostatic processes in living systems.

The effectiveness of an agent depends on it making veridical decisions. That is, it decides to take the action that will satisfactorily maintain that suitable relation (cf. Simon 1998, p. 27). Faulty decisions will make the agent and its embedding system subject to selection processes (as described in Chap. 3). The decision selection process is based on an internal model of the relations between the incoming signals and the appropriate output control signal(s).

Another aspect of agency is the degree of autonomy or flexibility the agent has in making the decision. The degrees of freedom an agent possesses in arriving at a suitable decision will, in general, match the complexity of the decision environment. Agents operating in highly complex environments cannot have access to all of the state information that they would need to arrive at a certain decision. The decision model the agent uses to compute a solution will tend to be more probabilistically based and use auxiliary heuristics to make a choice. In other words, the more autonomy the agent needs, the more it will need to rely on learning to modify or adapt its model in order to continue making reasonably good decisions.

Chapter 11 will delve into agents and agency in greater depth, especially as pertains to humans as decision-makers where biological and psychological factors complicate the decision process.

10.2.2.2 Economy

CAS/CAESs may be viewed as a whole coordinated group of work processes that produce the products and services needed internally to maintain the system (autopoiesis). The system, as a whole, needs to have processes for obtaining resources and expelling wastes. It may also produce exported products or provides services that are of value to other entities in its environment and part of the supra-system in which the CAS/CAES is a subsystem.

An example of an early evolved economy in nature was the metabolism of early cells (Morowitz 1992). But we may also consider the various cycles of materials in the whole Earth system as a kind of economy (Volk 2017). Below we will briefly look at some aspects of the metabolism in cells as representative of a living economy. Multicellular organisms expand on this notion of an economy of the body through the multiple interactions between cells in tissues. We will generically refer to this as "physiology," but we should point out that it is really a coordinated system of extra-cellular metabolism. Ecosystems are yet further expansions of this basic notion in the trophic food webs and waste recycling processes. Finally, in human societies we come to the economy of society. This, too, is an expansion of the basic cellular-body metabolism. That is, the activities of the human social economy (recall Chap. 9) are all part of the most complex CAES we know of and all intended to support human life (even when some humans turn destructive toward other humans, it is with the intent of preserving one society over another).

Below we will provide the archetype model of a generic economy. The model can be seen to be applicable to metabolism, physiology, trophic flows, and the human economy.[9]

10.2.2.3 Governance

The ongoing maintenance of a CAS/CAES requires management and regulation of internal economic processes but also of agents.

The hierarchical cybernetic governance system (HCGS) model is the basic architecture of how a CAS/CAES managed and regulated to achieve stability in the face of an uncertain and changing environment (Mobus 2015, 2017; Mobus and Kalton 2015, Chap. 10). The hierarchy aspect refers to the fact that the governance subsystem is organized in layers (and possibly sub-layers), of which there are two (in CASs) or three (in CAESs). The bottom layer is devoted to real-time operational controls and between-processes cooperation (feedforward) at the level of the economic work processes mentioned above. The second layer contains agents that are responsible for coordination of the lower layer when cooperation between processes is not reliable or even feasible, in some instances. There are actually two kinds of coordination managed in this layer. Not only must the lower layer work processes be coordinated amongst themselves, but the whole system has to achieve coordination with external entities, resources, and sinks. The internal coordination is termed "logistic" and the external coordination is termed "tactical." Examples will be provided below.

The top layer, when present, applies to the most complex and evolvable systems that can learn from experience and modify themselves or their behaviors accordingly—CAESs. This is termed the "strategic" layer and it is involved in, among other forms of governance, assessing the trajectory of changes in the external environment and planning what internal changes need to be made to continue to succeed (exist) in that future situation.

Each of these layers operates in different time domains. The operational layer, as mentioned, operates in real time, meaning that it responds to the changing conditions in the same time frame in which those changes take place. The coordination layer operates in, possibly several longer time scale, working on changes in the time-averaged changes in overall operations or accounting for time lags in changes in upstream processes. The strategic layer operates over very long time scales relative to the operations and coordination layers.

These three archetype models interoperate to produce a viable system, stable and survivable, over long periods of time, effectively what we would think of as the lifetime of the type of system of interest. After describing and explicating the nature of CAS/CAESs, and then comparing the basic CAS/CAES framework to several

[9] All but the last listed economies work properly. As we saw in Chap. 9, the HSS economy has some serious design flaws; it does not fully conform with the archetype model.

former models, we will take each of the three sub-models in turn for description and explanation in detail.

10.3 Complex Adaptive and Evolvable Systems—The Archetype Model

The most complex systems that we know of are living and supra-living entities. Cells and individual people are living systems. Species, ecosystems, and societies are supra-living systems. So is the whole planet geosystem, which we call the "Ecos."[10] Any identifiable system that includes living components, like a society or a corporation, is an example of this category. Table 10.1 provides some exemplars of CAS and CAES along with notes on the archetype models to be found in each. The first column names exemplar systems ranging from the protocell thought to exist at the emergence of life through examples of living and supra-living systems all the way up to the Ecos. In the latter case, we are speculating about what an Earth system with humans and ecosystems living in balance might be like. Humans and human societies are situated between the extremes of economies and governance of single cells and of ecosystems. In this hypothetical world of "harmony," human societies might take on the role of the *brain* of the Earth (cf. Grinspoon 2016). One way to state this as a reasonable expectation is that, by the CAS/CAES archetype assertion, unless the Earth does achieve a kind of "consciousness," it cannot be long-term viable. Perhaps it would be wise to recognize what it would take to manage the planet wisely.

As discussed in the Introduction, complex adaptive systems (CASs) are able to respond to changes in their nominal environmental conditions and operate to counteract any ill effects of those changes, thereby extending their viability. The main subsystem of the CAS that accomplishes this is the homeostatic process. As further discussed, a CAS that is also evolvable (CAES) can make internal permanent restructuring in response to long-term non-stationary alterations in the environment,[11]

[10] James Lovelock and Lynn Margulis (1974) have called the planetary ecology "Gaia," the name of the Greek goddess of Earth. The Gaia hypothesis stipulates that the interactions between the biosphere and all other "spheres" constitute a self-regulating entity that maintains balanced flows (over geological time scales) of the major substances in a kind of planetary physiology (Volk 1998). The use of a goddess's name to invoke the Earth system has been somewhat controversial in the past, but has become more acceptable in recent times. This author chose to use the term "Ecos" (from the root of ecology, meaning "home") to be somewhat less invocative of a mysterious process. But, as demonstrated by Tyler Volk's 1998 book, *Gaia's Body*, the mysteriousness is much less an issue.

[11] Such restructuring may be due to either accidental (trial and error) or serendipitous discovery, as in Darwinian evolution, with subsequent testing for efficacy by the environment (selection) or intentional as in the case of a human social organization that invents new mechanisms and new behaviors. In the latter case, the environment will still be the final arbiter of what is efficacious or not.

Table 10.1 Examples of CAS/CAES and their instantiations of agents, governance, and economy models

Exemplar	Agent model	Governance model	Economy model
Protocell—emergence of life from geochemistry	Operational primarily with auto-organization and selection	Coupled organic-mineral reaction networks and auto-catalytic cycles	Energy supplied by various reducing reactions in an anoxic environment
Single cell	Operational, logistical, and tactical phenomena embodied primarily in enzymatic control of production of structural and enzymatic proteins	DNA knowledgebase, RNA-mediated transcription and translation into proteins and various active messaging and transfer RNAs	Metabolism and cellular physiology Mitochondria energy supply, ribosomes as factories, consumption to grow and reproduce
Multicellular Organism	Tissues and organs processing chemical information to decide on the responses, release of hormones, the brains of animals with specialized modules	Brains, endocrine, and immune system in animals	Physiology in which multiple organs participate in maintaining the body, blood system for main transport
Human being	Prefrontal cortex orchestrating activities in conscious mind, localized special modules for processing sensory input, perceptual processing, and conceptual manipulations	Same as for animals but with the addition of strategic level processing, thinking about the past, the present, and the future, imagination and affordance recognition, language	Same as for animals but with the addition of creating an external physiology (tools, shelters, etc.) to allow incursions into non-physiological environments
Organization (e.g., family, tribe, enterprise)	Individual human beings, motivated by biological mandate but able to hold beliefs	Hierarchy of authority, power relations	Obtain resources, increase order, consume, expel wastes
Society	Officials, representatives, bureaucrats, and other human beings	Hierarchies of authority, various models of authoritarianism and seat of power, still very much experimental	Evolving, not yet stable, possibly self-destructive
Ecosystems	Primarily operations level	Mutual constraints, distributed competition	Trophic flows, producers, consumers, recyclers
Ecos	Fully sapient beings	Human society as strategic and highest levels of tactical/logistical combined with cooperation with global ecosystems	Energy (power) applied judiciously, only necessary consumption for maintenance, recycling of all material

thereby keeping the basic system viable in the long run. Below, in Sect. 10.4.1, we will expand on what this means. Evolution can be essentially happenstance (Darwinian evolution due to mutations in a large population of individual representatives followed by selection of favorable phenotypes), a blind search through design space. Or it can be intentional/directed as when a corporate CEO decides to open a new line of product or an engineer decides to alter a component of an artifact to get expanded or different performance. This is evolvability as we have come to understand it. The internal restructuring that takes place in response to changes in the environment will also be tested (selected) but, in our example of corporation, by market forces rather than climate or predation. In Sect. 10.3.3.3, we will expand on this.

Complex adaptive systems as individual entities evolved on Earth through the same process of ontogenesis described in Chap. 3. First, the complex chemistry of carbon, hydrogen, nitrogen, oxygen, phosphorous, and sulfur (CHNOPS) and other trace elements and minerals produced the first living systems, bacteria and archaea. Then, Darwinian evolution took over producing myriad species of those first living cells and later multicellular lifeforms. Darwinian evolution introduced the species as a system (along with populations as subsystems and genus as supra-systems). These latter systems, aggregates of individuals, became evolvable, thanks to the fact that so many different (and thanks to mutation) diverse individuals were produced for environmental selection (testing).

Evolvability describes both a system's capacity to change internal subsystems to attain better adaptation and a capacity to regulate the *change in internal subsystem changes* on a needs basis. In other words, some evolvable systems have the capacity to either allow themselves to undergo mutation of subsystems (some bacteria are able to "permit" mutations when food resources are non-ideal) or, as in the case of a human organization, cause an intentional modification expected to provide a pre-adaptation to an environmental change (as when starting a new product line in anticipation of market expansion). In the biological (Darwinian) case, this is accomplished through massive reproduction with variation in the genotype due to chance mutation. In the human-made artifact or institution case (Hughes 2004), it is due to anticipatory modification (strategic decision-making).

Natural ecosystems are a bit difficult to characterize easily. They can evolve over time in terms of their component species until they reach (given enough time and stability of the climate) their climax state. This is the state in which the balance of species interactions leads to a quasi-stable dynamic. Once in such a state, an ecosystem resembles a CAS more than a CAES. A rain forest is an example of a climax system. Their evolution is in concert with biological evolution, that is, it relies on opportunity and chance, without intentional mechanics.

Intentional evolvability requires foresight and planning and apparently applies only to human constructs.[12] The whole capacity to design future configurations of material, energies, and messages—that is, invention of machines, procedures, and

[12] This is not *entirely* the case, though it is obviously the case. Brains of non-human animals possessing cortical structures, for example, the neocortex in great apes, are also considered CAESs in that they have the capacity to process concepts that deal with the future possible states of the world. Or, in other words, they are anticipatory up to a point.

policies—seems to be a uniquely human capacity, though it also seems to have roots in human forbearer species (De Waal 2016). Nothing is new under the sun, it seems.

Figure 10.2 is a general schema for a CAES in our systemese iconic language. It applies to all such systems we find on planet Earth. A CAS is very similar except that it lacks a strategic management capability and an evolvability process capability. These differences will be covered below. The figure shows a management process network, which constitutes a governance structure for the whole system. It also depicts the basic economy of the system, or the work process network where material and energy inputs are transformed into useful (low entropy) goods and services. Each of the process ovals in these two major networks have internal agents (smaller orange ovals) that are concerned with managing the processes. Further discussion of the difference between governance and economic process agents will be presented below in Sect. 10.4.1.2 Governance Architecture.

Also shown in the figure are a number of auxiliary processes that are hard to qualify as strictly economic or governance. In extremely complex systems these are themselves often complex enough to be treated as CAESs having their own internal economic subsystems, etc. These include processes that handle adaptation, that is, they respond to disturbances to restore balance in the system; repair and maintenance of existing work processes, often working in concert with the adaptation processes; evolvability processes that are able to construct new work processes under directions from the strategic layer of the governance system; and recycling processes that are able to take waste material outputs and rework them into usable components.

In Chap. 11, we will explore the nature of the agents and their agency (the orange ovals within each process oval, including the management ovals). Chapter 12 will consider how these agents are distributed through both the governance processes themselves and the economic processes which are being coordinated by the higher-level governance layers. Then in Chap. 13 we will go deeper into the economy sub-model. All of these, to be explored, are archetype models, meaning that they are generic patterns applicable to all CAS/CAESs.

In terms of Eqs. (4.1 and 4.2), we could set the SOI level 0 as consisting of three major subsystems: the economy, the governance structure, and a population of agents. That is,

$$C_0 = \{c_{0,1}, \text{economy}, 1\}, \{c_{0,2}, \text{governance}, 1\}, \{c_{0,3}, \text{agents}\}$$

However, this might not be a good way to detail an actual CAS/CAES or one to be designed as it is too abstract to support a decomposition process. Rather, in the next few chapters we will develop these "components" as originally intended, to act as archetypes that can be referred to whenever instances of them are encountered in a concrete deconstruction.

What Fig. 10.2 shows is how the three sub-models (agents/agency, governance, and economy) interrelate. The Work Processes Network constitutes the basic value-adding economy with peripheral processes for repair and maintenance, recycling, and waste removal as necessary attendants even though they don't directly contribute to product production. The Evolvability and Adaptation processes, likewise, are separate from the main economy but are closely related. They are considered work processes—those that transform material using high-potential energy—to accomplish the work of modifying other work processes to handle changes.

The figure also shows the relation to the global governance of the economy and peripheral processes. The organization of governance is a hierarchy of decision types based on necessary roles. Operational decisions for managing each work process are local to that process. Coordination-level decisions are embodied in two kinds: logistical decisions which work toward coordinating and balancing the internal work processes to keep the value-adding production flowing, and tactical decisions which are about how to best interact with entities in the environment. Finally, for CAESs the highest level of decision-making is at the strategic level where information from both logistical and tactical management along with information about the general environment (not just the sources and sinks) is used to build a model of the CAES and potentially relevant aspects of its environment to be used to anticipate future states of the world that might have a significant impact on the system. With these anticipatory scenarios, the strategic agent makes decisions about how the system should reconfigure itself to take advantage of possible new resources or avoid possible new threats.

This model of governance is called the hierarchical cybernetic governance system (HCGS, Mobus 2017, 2018). It shows how decision types are organized in this three-level structure and addresses the kinds of communications that are required between decision processes. This will be the subject of Chap. 12.

10.3.1 Complexity

Put in the context of ontogenesis we see a progression of systems that start out relatively simple or, perhaps, not terribly complex. Here complexity refers to the Simonian complexity measure characterized by the heterogeneity of component subsystems at each level of organization in the system, the number of interactions between components within and entities in the environment, and the depth of the organization tree (see Sect. 3.3.3.7.2 Simonian Complexity). Systems become more complex, historically, by accreting additional components (sometime freestanding systems in their own rights), as when some large prokaryotic cells acquired other smaller prokaryotic cells that had well-established citric acid cycles that would later evolve into mitochondria via a process of endosymbiosis (Margulis 2011). Or multiple systems at a given level would form tight associations and develop into multi-unit organizations, such as the early forms of multicellular organisms (Bourke 2011; Morowitz 2002; Volk 2017). Volk (op cit.) describes these processes with the term "combogenesis" meant to evoke the vision of heterogeneous things coming together, somehow, to create a new larger, more complex thing.[13] In Chap. 3, we used the concept of "auto-organization" to cover the same ideas but cast it in light of the actual kinds of mechanisms that enable the "coming together" of disparate components.

Complexity of systems has increased over geological time on Earth as larger organisms evolved. As this has progressed the need for internal regulation of cooperation, that is, coordination among the components in their interactions, has also developed. But so has the ability for the systems to become more flexible and resilient against external disturbances through the emergence of adaptive capacity. Adaptivity comes from the emergence of response choices in components coupled with external influences (a consequence of increasing complexity!). In the origin of life, for example, proto-metabolic processes may have included an ability to switch between several mineral sources of redox coupling for energy (Smith and Morowitz 2016). One of the most fundamental forms of adaptivity is in an ability to buffer inputs against a fluctuation in flows. Systems that could store surges in energy would be better able to withstand lags in energy inputs.

Adaptivity evolved as a result of increasing complexity in systems along with the selection imposed by the environment—the systems that evolved adaptive mechanisms survived.[14]

10.3.2 CAS

Complex adaptive systems (not capable of evolving in their own right) are systems that have degrees of freedom in operations that allow them to adapt to a corresponding range of changes in environmental conditions.[15] Living systems such as individual entities, as opposed to populations which are evolvable, possess subsystems that sense key parameters both externally and internally (for particularly complex multicellular forms) and respond to changes in those parameters that drive the parameter back to the ideal level for that parameter. For internal parameters, such as blood glucose levels in animals, they possess subsystems that either increase the level when it gets below an optimal level or decrease the level when it gets above optimal. For external parameters, such as temperature, an entity may be able to correct an imbalance (too hot or too cold) by means of several different mechanisms, including moving along a gradient toward a more ideal condition.

[13] Of course, this is what we described as auto-organization in the ontogenic cycle in Chap. 3.

[14] The other aspect of fitness is reproduction. Little is understood about when and how reproduction of systems emerged from proto-living systems. The current favorite hypothesis is that RNA autocatalytic cycles coupled with protein synthesis "somehow" managed to encode patterns of organization of proto-metabolic processes. As always, more research is needed.

[15] This was noted by W. Ross Ashby (1958) as the "Law of Requisite Variety." Our inclination is to avoid the word "law" in considerations of matters in systems science given the still nascent state of the science. The concept is, however, well recognized in systems circles.

The mechanisms involved (the subsystems) reflect the evolution of the species experiencing normal environments and normal physiologies tuned ultimately to those environments. The main thing to recognize is that all biological responses to changes that stress the organism have their limits and exact costs (in terms of energy and material) associated with both the response and any failure or inadequacy to respond.

Human organizations, likewise, have built-in adaptive capabilities in several of their internal functions. When there are fluctuations, for example, with supply shipments, or customer demand, an organization can respond by shifting internal operations so as to compensate. In modern automated supply-chain management systems, for example, many of these response mechanisms are built into the software itself so that the computers are the decision agents and provide very rapid responses to such fluctuations. In other cases, human decision agents, relying on their particular agency (power to make alterations as needed), are the adaptive mechanisms. But note that either machine or human agents can only respond with their a priori built-in agency. They cannot restructure their work domain on their own (we wouldn't want software to be able to modify itself for certain!)

In this section, we examine several models of adaptivity and discuss their ramifications in terms of costs incurred in exercising their agency and how more evolved mechanisms are able to reduce total costs, clearly an evolutionary advantage.

10.3.2.1 Adaptivity

There are three levels of adaptivity that living systems and organizations use to adjust to fluctuations in their various environments. Mobus (1999) analyzed these three based on the notion of cost reductions (in energy requirements terms) in dealing with the changes. The objective is to minimize the costs of responding to change so that the energy resources can be conserved and thus reduce the need to acquire additional resources. In both living systems and organizations, there are four basic costs associated with responding to the environment. These are costs above the direct costs of nominal operations (e.g., basal metabolic rate in organisms or overhead in organization operations).

The first is the direct cost of responding to a change, the homeostatic cost. The increased cost in energy for responding to a critical parameter is essentially proportional to how far from nominal the excursion of the parameter (both negative and positive value excursions are the general case) and how long the excursion lasts. This is called the response cost, $C_{response}$. In the social world of humans, this is the cost of responding to a crisis situation.

The second cost involves the drain on autopoietic support systems, the stores of materials, and energy along with the work needed to restore those stores. For example, after a prey animal exerts extra energy to escape a predator it must eat a bit extra to replace the energy used up in the chase. This cost may be paid back over a longer time frame, but it is still a real cost, $C_{restore}$. In the human world, this is the cost associated with re-acquiring stores of resources that were used up in responding to the crisis, such as replenishing inventory after "Black Friday" sales stampedes.

A third cost arises when the excursion is more extreme or lasts for a long enough time that damage accrues to the system. For example, if a warm-blooded organism is subjected to very cold temperatures for a long enough exposure, it can suffer hypothermia or frostbite with possible damage to some of its tissues. These tissues will need to be repaired (assuming the creature survives) and that takes up additional energy—higher costs, C_{repair}. Suppose the Black Friday stampede at a department store resulted in physical damage to the facilities. Those would need to be rebuilt.

The fourth cost involves the autopoietic maintenance of the responding mechanism itself. This can be thought of as the cost of maintaining some emergency response equipment, like a sprinkler system in an office or plant. In biological systems, it corresponds with the maintenance of tissues above that needed for ordinary operations. For example, humans that are called upon from time to time to lift heavy weights maintain extra muscle mass for the occasional demand (see extended example and explanation below). We designate this expenditure as $C_{prepare}$, the cost of being prepared. Think of this as the premium costs for having insurance.

The levels of adaptivity provide ways to minimize these costs, thus reserving resources for the entity system giving it maximal capacity to survive (and thrive) in its environment.

The objective function for response systems would be:

$$min \sum_{t-stimulus}^{t-restored} \left[C_{response} + C_{restore} + C_{repair} \right] \quad (10.1)$$

The time of stimulus onset is *t-stimulus* whereas *t-restored* is the time at which the system is completely restored and all costs have been paid. The cost attributed to being constantly prepared to respond is amortized over the *t*-restored interval.

Minimization of the total cost depends entirely on the efficiencies with which response, restoration, and, if needed, repair can be done. Of these costs, the cost of repair can be quite large as can that of response. Clearly one strategy for reducing costs would be to avoid damage by responding to the stimulus more quickly, that is, minimizing the time lag between onset of stimulus and onset and magnitude of response. This, however, often results in higher preparedness costs and restoration costs. Another strategy involves anticipating a stimulus in time to reduce its effects or avoid it entirely. Any mechanism that would allow this would likely have a slightly higher preparedness cost due to having to maintain the extra control equipment needed (see below). But preparedness costs are the least component in total costs. Whatever additional costs might be involved are more than offset by the substantial reductions in response and repair costs. Details of this approach are given in Mobus (1999), reviewed in Mobus and Kalton (2015, Chapter 9, Sect. 9.6.3.1, and outlined below.

10.3.2.1.1 Stimulus-Response

The first level is a simple stimulus-response capability. In this capability the system responds as quickly as it can to changes in its environment (the critical parameters). The system has a subsystem dedicated to making a response to the stimulus. In the general case, the response acts to counter the effects of the stimulus signal and is reactive. In biological systems, this can be seen in what are called reflexive responses. The system maintains a minimal capacity to respond to the episodic[16] changes in stimuli that are encountered. For example, an animal's musculature is maintained at the level necessary to handle "ordinary" sorts of work effort. An organization might maintain a basic inventory of parts for manufacturing to handle normal fluctuations between supply and demand for production. The system activates its response mechanisms in proportion to the stimulus and experiences costs accordingly.

A fundamental problem arises when a stimulus situation is greater than the kind of normal that has generally prevailed over the history of the system. If the temperature fluctuation is greater than ordinary, the biological system has to draw down its reserves more than usual. Similarly, an organization might need to draw down inventory more than would ordinarily be the case. The question then is how often might such extraordinary fluctuations be expected in the future? There is no good answer to that question on the basis of a single event. But, suppose such events occur with greater frequency than had been the case in the past? That is, the external processes that are generating the stimulus are non-stationary and thus not necessarily predictable. This calls into need the second level of adaptivity, a more permanent change in the capacity to respond to such events.

The cost associated with response is just the amount of resources drawn from stores, for example, energy stores that are needed to activate and sustain the response mechanism. In addition, there is an additional cost associated with restoring the resource stores to a "safe" level. If the drawdown has been significant, then the restoration activities need to be undertaken quickly, thus also requiring more power. The time lag between stimulus onset and response onset as well as response magnitude are critical factors in determining the extent of costs accumulated over the duration of the stimulus.

10.3.2.1.2 Adaptive Response

Living systems evolved to be able to change their capacity to respond to changes in the environment based on the frequency with which extraordinary demand (stimuli) occur, at least within limits. This means that a system can increase the responsiveness of its mechanism by increasing standby resources dedicated to it and/or by

[16] Stimuli occur not only episodically, but also with varying intensities and varying durations so that the response mechanism is designed to "track" the episode and respond in kind.

growing more response tissue. A good example of this phenomenon is the case of an ordinary person who decides to take up weight lifting to gain muscle bulk (perhaps thinking this will make him/her more fit!). Starting out, the person has an "ordinary" musculature needed for ordinary life. Over time, training with weights, a person can enjoy an increase in muscle mass as the periodic but frequent demand on lifting causes the muscles to respond by becoming stronger. The muscles are adapting to a change in long-term demand on their ability to respond. If the training is sustained, then the muscles will bulk up and the person will be more able to lift heavier weights on demand.

Biological systems have this capacity to adapt on the basis of reinforced demand. They do this by increasing their efforts to obtain necessary resources, material and energy, and they commit more resources to maintain the acquired capacity to respond. If the demand for continued ability to respond declines (the weight lifter stops training), the tissues will eventually revert to their baseline capacity—a rule of biological systems is "use it or lose it." This is because the energy and material economics of life are such that a system cannot afford to maintain something it doesn't really need. It is "willing" to pay the cost if the demand is reasonably expected in the future. But if ongoing experience shows that the demand was just a short-term phenomenon, it will revert to the baseline capacity.

Human organizations basically follow this same rule, or try to. If an organization increases its inventory to handle long-term disruptions in parts receiving, it will order larger batches and keep them in inventory so as to have parts on hand to meet demands from manufacturing (meeting demands from customers ultimately). The organization keeps track of costs (in dollars) for maintaining such an inventory stock and continually matches that against the actual demand being experienced. Should either demand decrease or the parts supplies are no longer subject to disruptions they will most certainly reduce the size of their inventory to reduce operating costs.

It costs more to maintain a higher level of preparedness. But not doing so carries an even higher cost in the form of repairing damages accrued from not making an adequate or timely response. Biological tissues are damaged (degraded or disrupted) when the stimulus goes over some limit or remains high for too long a period, draining response resources. A business's reputation can be damaged if it fails to supply customers with demanded products. These kinds of damage may not result in death or going out of business, but they do need to be repaired as quickly as possible and that will take additional resources and time. The strategy of building up and maintaining additional response capabilities is meant to minimize these costs which could otherwise be quite high depending on the extent of damage.

10.3.2.1.3 Anticipation

The most cost-effective strategy for both living and organizational systems to minimize costs due to responses to fluctuations in stimuli is to find ways to anticipate the onset of a stimulus before it happens and take preemptive (pre)adaptive actions. The trick is to sense other factors in the environment that might be causally connected with the factor that will produce the stimulus that demands a response. The other factors are essentially neutral in terms of direct impact on the system but may have a causal relation with the factor that does have a direct impact. As such, if a neutral factor is found to occur just prior to the impactful factor (the stimulus), then the system may preempt the onset of the stimulus by starting to respond even before the stimulus is detected.

A capacity to be anticipatory is based on the CAS having an internal model of the environment insofar as the relations between direct stimulus events and those that are neutral but have causative impacts on the stimulus entities (Principle 9 in Chap. 2). The nature of models was developed by Rosen et al. (2012) who showed that a model could be used to anticipate future states of the world and replace the more typical response mechanisms. The model used has to contain sufficient details about how entities and forces in the environment relate such that it is reasonably veridical with respect to the real nature of those relations. It must be capable of accepting input states that are hypotheticals in time scales much shorter than real time for the SOI. Today we use computer models with hypothetical input data to project what the state of the system will be at some point in the future. For example, we use climate models with hypothetical increases in CO_2 emissions to predict various scenarios of future warming and climate disruptions.

Anticipation serves a useful role beyond just reducing reaction-mode costs. When a negative consequence is anticipated the point is to take actions that will obviate the need to respond. As when a prey animal anticipates that a predator lies in wait around the corner, it chooses to take a different path to the water hole. In this case you can't quite call the anticipation a *prediction* that continuing along this path will lead to my death. Good predictions are supposed to come true in the future. In this case we use a seeming prediction (a possible scenario) to avoid the prediction coming true, therefore making the prediction false. A seeming contradiction.

In biological systems, the notion of anticipation can be seen in the effect of Pavlovian conditioning; a bell rung before a meal can cause a dog to salivate in anticipation of the meal. The dog learned to associate a bell being rung just prior to meals with the fact of an impending meal. The ringing bell then became a stimulus for the response (called a conditioned response).

Well-managed companies have people who are monitoring the external environment and are able to associate movements in factors that might portend changes in, say, demand for their products. For example, sellers of household goods (furniture, stoves, etc.) might take note of a downtrend in new housing starts, and because they had learned from past experience, anticipate a downturn in demand for their products as well. We call these sorts of anticipatory clues, leading indicators of movements.

All of these capabilities are to be found in concrete complex adaptive systems. The long-term viability of any CAS depends on relative stability of the various critical environmental factors upon which the system depends for its existence. Supplies of resources, absorption of products (and wastes), and the general milieu of the environment must remain fluctuating within boundaries established over the evolutionary history of the system. As long as this kind of steady-state condition exists, the systems that have evolved to fit in that environment can generally deal with occasional excursions to the limits. It is when some factor(s) in the environment are themselves changing over time that the capabilities for adaptation are tested. In those circumstances, systems that are incapable of making the necessary internal adjustments to their structures and functions will eventually succumb because of the relentless accumulation of the costs discussed above. Some systems, however, have the ability to make the needed modifications and continue to be fit in the new environment. In the biological world, those systems are the genera in which speciation permits evolutionary adaptation (not individual adaptation based on physiology or behavior as described above[17]) to take place. In human organizations, the capacity to alter internal structures and function in response to changing environments makes these systems in situ evolvable.

10.3.3 CAES

A complex adaptive and evolvable system (CAES) is one that is adaptive but also capable of changing its internal structures/functions in response to a changing environment. The changes here go beyond mere degrees of freedom in responding to nominal ranges of changes in environmental parameters. The kinds of changes we are referring to here involve permanent changes in the response mechanisms themselves or even creating new response mechanisms to take advantage of some new potential resource or avoid some new kind of threat.[18] The latter involves changes to the architecture (or genome) that make a system more fit (if it works) in its environment: neo-Darwinian evolution in biology (applied to genera) and organizational change in human organizations (e.g., changes in policies and procedures).

---

[17] There is an unfortunate confusion regarding the word "adaptive" as used in much of the literature on CAS. In evolutionary biology writers often refer to the idea of "evolutionarily adaptive" meaning that a species is adapted to a particular niche. The species, in this case, is a subsystem of the genus that has adapted to the niche via evolutionary mechanisms of mutation and selection. New species evolve from existing ones. Both may survive and undergo additional microevolutionary change. But the word adaptive in this respect applies to the species level of the genus. Adaptivity, used in the context of individuals within a species, is the capacity of the individual to change its response but only within some a priori range, determined by its evolutionary adaptation. It cannot completely alter its physiology or behavior to fit wholly new conditions. Humans are both adaptive and evolvable, at least insofar as their behavior is concerned. Their expansive neocortex allows for the evolution of new concepts in response to new environments. This is why we are able to "adapt" to completely new technology or to live in extreme environments like the Arctic. Human organizations are also both adaptive and evolvable by virtue of being composed of human agents. We will make every effort to make clear when we are referring to individual adaptivity (as in a CAS) versus an evolvable system (CAES).

[18] An excellent example of this is the acquisition of antibiotic resistance in microbe that results from mutations in genes.

The significant difference between adaptivity and evolvability is that the latter involves having internal mechanisms for effecting change beyond nominal changes in the environment (lower left-hand oval in Fig. 10.2 labeled "Evolvability Process"). That is, components within the system have to have the ability to be modified or created anew. Such modifications in a living organism, for example, a mutation in the DNA of a single cell, usually lead to either cell death or potentially cancerous growth. These mutations are vigorously repaired where possible since they are generally detrimental. However, when we look at a species as a system the situation allows for evolvability. If a germ cell gene is mutated, resulting in a change in the phenotype of a descendant, and if that change is favorable—leading to increased fitness of the resulting phenotype—the species may undergo evolution assuming sufficient conditions of selection are present (cf. Losos 2017).

More complex systems, such as the neocortex of mammals, especially humans, or human organizations, are evolvable by virtue of having internal mechanisms for wholesale changes in adaptive mechanisms. Brains can learn new concepts or forget (submerge) old ones that are no longer useful. Organizations can acquire new departments or dissolve those no longer useful. Both such operations serve the interest of increasing fitness of the system in its particular environment.

10.3.3.1 Adjacent Possible

Stuart Kauffman (2000, cf. Chap. 8) describes an evolutionary process in which the current state of the system (the genome, for example) can be tweaked minimally but yet allow the system to explore nearby possible states or, in our scheme, behaviors, that will either prove fit or not. In Darwinian evolution this results from, for example, point mutations leading to changes in the phenotype that can then be tested by selection. In human organizations, we see experimentation with step-wise or incremental alterations of organization that lead to improved functions or behaviors that can similarly be tested by the economic environment. With enough of these steps, especially being taken in parallel in several adjacent dimensions, it is likely that the system is moved into an entirely new space of possibilities—new functions and behaviors.

All CAESs demonstrate the capacity to "explore" their phenotypic space in a blind search for improvements, that is, incrementally better fitness. This is a necessary capability that keeps a CAES from getting stuck in a local minimum, a fitness capacity that is perhaps sufficient but not optimal. Of course, some exploratory moves may be more than just to the nearest, adjacent possible state. Occasionally systems may be able to make large jumps out of their minima as in the case of punctuated equilibria phenomena in biological evolution (Gould and Eldredge 1977). Corporations may discover a new technology and develop a new product for a new market that is distinctly different from their ordinary products. But these events are relatively rare compared with the exploration of incremental regions of fitness space. The latter is a major source of variability across a gene pool in biological genera or between organization in societies.

10.3.3.2 Chance and Necessity[19]—Darwinian Evolution

In biological systems, specifically species, evolution proceeds by seemingly random modifications to genes (mutations, cross over, and certain epigenetic phenomena in the genotype) that then affect the phenotypical properties, structures and functions, of the individual in which the mutation occurred.[20] The mutation of any functional gene or any sequence involved in the control of expression of genes is thought to be a random event, a copy error, or a result of a radiation knock-out event and is thus in the realm of "chance." The way that chance can have a role in evolution is because the events occur within a very large population of potential opportunities—a population. Many parallel chances can be tolerated due to the fact that there are many more cases of no chance (no mutation) so that the few that do occur do not put the whole species at risk.

Mutations and changes in the phenotype resulting must then be tested in the context of a dynamic environment. In biology, this constitutes some form of selection (natural, sexual, and directed or breeding). Those that improve fitness of the system tend to be propagated into the next generation.

In human systems, chance and necessity may still play a part of the general evolution of these systems. For example, a serendipitous discovery or a new idea that just popped into someone's head can lead to modifications not directly linked to external conditions initially.

10.3.3.3 Evolvability

We have come to realize that some organisms are able to modulate the mutation of certain genes in response to changes in their environments that facilitate ordinary Darwinian evolution. This is a kind of massively parallel random search for a solution to a problem in the interest of allowing life to go forward. Though the details of mechanisms are still sketchy, it appears that there is actually a kind of favoritism applied to critical genes, allowing them to mutate at above the background rate in a kind of hope-for solution.[21] This is to say that organisms can allow themselves to produce mutant offspring that might be able to be more fit in a changed environment.

---

[19]The phrase is credited to Jacques Monod (1971) who characterized the process of biological evolution in terms of random modifications generating variability (chance) and environmental selection through constraints necessitating the viability of those individuals lucky enough to possess those variations that the environment permits.

[20]For clarity, the mutation occurs in one of the genes in one of the two gametes (in sexual reproduction) that then results in a phenotypic modification that will be subjected to environmental selection. Most mutations are believed to result in a detrimental configuration that is non-viable for the zygote. But those that are not will be subjected to selection in the Darwinian sense.

[21]When genes are being read out or copied, errors occur all the time, which could lead to a mutated form. However, these changes are generally repaired by enzymes associated with the chromosomes. One possible mechanism for achieving a higher rate of mutation would be to just inactivate those enzymes associated with specific genes.

For biological organisms this is literally a "craps shoot." It is not a calculated risk but a result of having so many parallel attempts in play that the likelihood of one of them paying off (in terms of increased fitness for the lucky individual) is sufficient to ensure the survival of the species. Living systems have evolved to be evolvable. And that may explain the success of life in dealing with the huge variability of climate shifts and general environmental variations that have defined the history of the planet. Life has endured in spite of cycles of global warming and ice ages because some species have become evolvable.

10.3.3.4 Intentional, Volitional Modification

Birds build nests in order to have a place to hatch and raise their offspring. They do this quite intentionally. Beavers build somewhat more complex dam shelters for brooding. They too do this on purpose. In these cases, we assume that the intentional actions taken needed to complete the tasks are the result of instinctive drives but even so we can see a high degree of autonomy in terms of selecting specific building materials and fashioning the developing nest. Ant and termite colonies are also the result of built-in programming to construct nests but carried out by a much more hardwired set of behaviors distributed through the colonies and coordinated by chemical signaling. Given the size of an ant brain it would be hard to attribute intentions to the individual ant as it follows the program based on the chemical signal inputs it is receiving. We can attribute much more in the way of intentional action to beavers as they have been observed to solve local problems that arise when attempting to, for example, gather building materials in a complex and challenging environment.

However, though we may see intentions in the individual behaviors and responses to challenges in some animals, we have never witnessed a beaver, say, try to build a two-story dam, or a bird decide that a multi-compartment nest would be more comfortable with a separate room for the chicks.

Biological evolution led to the emergence of human consciousness, a wholly new kind of consciousness that enjoys a seemingly unlimited capability to modify its world according to imagination. This capability appears to arise from the symbol representation and processing capacity of the human neocortex, particularly in the prefrontal cortex of the brain. Human beings are capable of changing their minds, meaning that they can modify their thoughts, their symbolic representations, and their behaviors. This ability has a number of evolution-like qualities. Changes in internal representation and relational connections can happen as a result of environmental change, that is, the brain receives informational messages that alter its conceptions. Or they can be the result of accidental (that is stochastic) processing that leads to new insights. Dreams are sometimes credited with changing someone's perspective or understanding. Or they can be intentional, the result of thinking hard about something and developing a new conceptualization.

When humans change their minds, they change their behaviors, or the actions they undertake to make something happen. We then change something in our environment to make things work differently. After which, the larger environment will apply selection pressures on the new things/behaviors and some will prove worthy while others will fail (e.g., VHS vs. Betamax).

Open-ended, that is, seemingly unbounded, intentional modification seems to be a phenomenon relevant only to mankind. We modify our instruments, tools, esthetics, and even ourselves intending to solve a problem or make something better, or more comfortable, or more efficient. Then we wait and see what happens. This is still a form of evolution. It is still an exploration of the adjacent possible. But now it is not really relegated strictly to chance as was the case for Darwinian evolution. Part 4 will be covering the various kinds of intentional modification/invention/artistic creation in which humans engage. There we will see that design and engineering of artifacts is a new kind of ontogenesis (as covered in Chap. 3) in which auto-organization is "replaced" by what we will call "intentional-organization." We will also see that the term "intentional" does not mean "deterministic." Intentional organization retains an element of chance (see Chap. 13) and so is still in the realm of an evolutionary dynamic.

10.4 The CAS/CAES Framework and Prior Models

We will briefly describe two prior models mentioned above, which together contributed much to our model, and compare them with the CAS/CAES framework to underscore why a general synthesis of many models is required. The Viable Systems Model (VSM, Beer 1972) and the Living Systems (LSM) model (Miller 1978) are among the more prominent models of whole systems decomposed to sets of subsystems that perform vital functions within the whole integrated in a complex network. Both models deal exclusively with living systems as exemplars of such systems. VSM primarily focuses on the management/governance of such systems, especially comparing the human nervous system as the "management" of the human body with the way in which enterprises (firms) are or should be managed. LSM is also concerned with the management/governance problem but also explores the underlying work processes which need governing. In the next few sections, we will compare aspects of these two closely related models and with the CAS/CAES model. In doing so we hope to show why the latter is a significant advancement over these older models for purposes of understanding the whole panoply of living systems, as a valuable tool in guiding deep analysis of such systems, and to provide a holistic template for the design of artifactual systems.

10.4.1 Viable Systems Theory/Model

Viable Systems Theory (VST) asserts a set of necessary and sufficient conditions for any complex system to be considered "viable." Viable means that a system is operating more or less optimally and stably with respect to the complex tradeoffs among an array of internal processes and with the whole system's environment. A viable system is one that is capable of adapting to changes in its environment that might disturb those operations and lead to instability or breakdown.

What we have described as the CAS/CAES archetype model is somewhat similar to the theory of viable systems advanced by Stafford Beer (1972). Beer developed a theory of a viable system that was applicable to human organizations of all types. His approach to developing a model of management of a complex system was to use the human brain and nervous system, what was known of its structures and functions at that time, as an analogue to the way the management of a "firm" should be organized. Beer identified five major management functions handled by brain modules in the cortical, limbic, brainstem, and spinal cord, along with functions performed by the peripheral nervous system, and mapped those functions to levels in a management hierarchy. In the CAS/CAES model, we have used a similar framework for approaching the nature of the hierarchical cybernetic governance system introduced in Sect. 10.2.2.3 above.

He recognized the fact that a system's viability depended on a hierarchy of management decisions, which he named "Systems 1–5" (see below). He did not differentiate between adaptability and evolvability as we have done, owing largely to his focus on human organizations which, almost by definition, are evolvable. Our purpose was to distinguish those work processes and governance decision types that are needed to achieve adaptivity and evolvability when it is present. Beer's theory, VST, was the basis of the Viable Systems Model (VSM). Given the brief description of the HCGS above, we will review Beer's VSM and its relation to the HCGS.

The CAES model presented above possesses the necessary but not-quite sufficient conditions for long-term viability. Since the completely sufficient conditions include the fact that the system is *always lucky*, such as in no asteroids or comets of consequential size will crash into the Earth ever, the concept of viability (stability, resilience, and sustainability) must remain provisional. That is, unfortunately, the best any CAES can hope for.

But even absent the luck condition, a viable system can expect to enjoy a full life cycle if it is a CAS or be potentially eternal, in an abstract sense, if it is a CAES.[22] Examples of the latter are phyla in biology, perhaps.

Here we consider a mapping from the CAES archetype to elements of the VSM. As noted above, Beer's primary concern was for modeling a human

[22] Of course, it cannot be absolutely eternal! For example, we know the Universe seems headed for an entropy death in some distant future if our surmise of dark energy is correct. It is hard to imagine how life could continue under those circumstances, but since there is no theory to tell us what to expect we should probably prepare for the worst-case scenario.

organization, for example, a commercial firm. He was most interested in how cybernetics played a role in management and management decisions. And he clearly understood that agent decisions and actions map onto a hierarchical organization.

10.4.1.1 Agents and Agency

A major concern for Beer was the regulatory capacity of the basic cybernetic model. Beer delved into autonomous agency theory noting how the lower-level "echelons"[23] of a control hierarchy, his System 1–3, corresponded with the autonomic nervous system and the lower brain stem of the human brain. He relied heavily on W. Ross Ashby's (1958) "Law of Requisite Variety" in control theory. This "law," more an observation in practice than a "law of nature," states that a control system must be able to have a number of control options (and the power to enforce the choice) as there are variations in system states. He also relied on feedforward and feedback loops to allow lower-level managers the same kind of autonomy that, for example, is demonstrated by the reflex arc in the spinal column. Systems 1, 2, and 3 were described as and related to the autonomic nervous system of the brain. This system is known to be able to automatically adjust various functions in the body (mostly dealing with internal organs) and so display a great deal of autonomy with respect to "higher" authority. On the other hand, the regulatory capacities of the decision centers in these three Systems are genetically programmed. There is no discretionary decision-making. As to whether this correctly describes how things work in actual organizations, near the operational level of work processes, is somewhat dubious in today's world. In the early part of the twentieth century, workers may have been more inclined to follow prescribed decision models (as will be described in Chap. 11). In many modern organizations, not based on the assembly line production model prominent in Beer's descriptions, workers are often given more discretion because the nature of the work is so much more complex than simple "programs" or iron-clad procedures may not be specifiable. We will revisit this below and in the next two chapters.

In Systems 4 and 5, the agents and their agency are more executive in nature. System 4 agents are a mixture of higher-order tactical and strategic decision-takers. The CAES model differentiates between these functions and decision types, the former involved in adaptive responses and the latter involved with evolutionary responses. The governance architecture, below, makes this clear.

Beer's handling of more autonomous agents, those able to modify their own decision model, relies mostly on descriptions of the agency of upper management in organizations. That is, he seems to assume the decision-taking functions based on human thinking and behavior in actual management positions and does not provide any reverse mapping from the organization framework back to the human brain in

[23] The term echelon is used by both Beer and Miller to mean tiers in a hierarchical governance structure.

any detail, leaving it that the cortex is a complex switching network that must (somehow) take the decision. This is not surprising since very little details about the functions of the prefrontal cortex of the brain were known when he was constructing his theory (as opposed to knowledge of the autonomic and limbic nervous systems which was much greater at the time). The CAES model has expanded the details of agents and agency to include or incorporate what we know today about this part of the brain.

10.4.1.2 Governance Architecture

Beer described a five-level architecture that involved operations, measurements of performance (information), agent decision processes, and coordination/strategic processes.

Beer labeled each component in this five-level architecture as Systems 1–5 (S1 through S5), as mentioned above. S1 entities map onto the CAES conception of work processes and their local control/regulation/management. This is where the fundamental work gets accomplished using error feedback (positive or negative). He does not elaborate on the actual workflows, that is the flow of transformed material from one S1 to the next in the value-added chain but assumes it to exist ("Division[24] A receives a material product from Division B," for example). Nor does he differentiate work processes in categories such as import or export processes versus internal transformation processes. He does not explicate special work processes such as recycling or adaptation processes. And, as stated earlier, he does not pay special attention to the role of energy flow as a driving force through work processes, that many operational level (S1) decisions are actually decisions about adjustments in energy flow. Beer's concern centers on the cybernetics or information and decision processes. The CAES is also concerned with these but recognizes that distinctions need to be made in assessing the properness of decision capabilities based on what the decisions are actually about.

Beer defines S2 more as communications channels or the network of communications between S1 entities (to be discussed in the next three chapters in terms of cooperation) and between S1 entities and an S3 entity. S2 may be a reasonable mapping onto the logistics coordination manager in the CAES model. Here Beer seems to be more concerned with the optimization of groups of S1 entities; his research in Operations Research (OR) would certainly account for this focus.

S3 receives information from S4 for transmission to lower echelons, providing an interface with the higher-level S4 whereby longer-term performance information is transmitted upward and command-like information is transmitted downward. S3 agents "filter" the information sent upward to the higher echelons and transmits plans and change requirements downward to the S1 units through the S2 network.

[24] Beer uses the term division generically but meaning essentially what we have defined as a work process.

It is not clear that Beer distinguished tactical from logistical coordination decisions explicitly, though he discusses tactical vs. "synergies" (presumably between S1s). In the VSM, the S1 units interface directly with external entities (sources and sinks in CAES); we presume one function of the S1 controller is tactical coordination, but he never makes this explicit. His S4 unit was responsible for monitoring the environment in general and for making plans for how the whole system should respond to changes. S4 in this author's reading actually seems a combination of some aspects of tactical control and strategic control. As with the description of S4 agents above, Beer leaves it to the autonomy of human agents to describe the workings of S4 in his model. Its relation to the human nervous system equivalent is vague and seems to involve the brain circuits responsible for alertness, which were also investigated in neurophysiology in the day. As to explanation of decision models used in the neocortex he had little to say, assuming they must be something like the business decision models of corporations where the profit motive provides the basic framework.[25] The CAES model goes into the decision models used by agents doing strategic thinking (both humans and organizations).

Beer defines a level above the S4 strategic. L5, which is tasked with establishing the "ethos" within the organization, is described as the "board." He labels this "Identity." In CAES, this is a long-term result of decisions made in the strategic management level and so subsumed in that process.

One might conclude that most of the functions covered by the CAES model are also covered in the VSM but possibly distributed differently and with some blurring of boundaries between function types. There is some truth to that, but only some. Careful comparisons of VSM with CAES show significant differences. One important similarity between CAES and VSM, however, is the recognition that there is a recursive definitional relation of entities in both. For VSM, many of the Sn entities are themselves VSMs. Certainly, for very large organizations each of the Ss would contain internal functions of the VSM sort. For example, large departments, like accounting, would have S1s like bookkeeping functions, S2s like consolidations of accounts with reporting to the controller, S3s like supervisors for various accounts, S4s and S5s are in the form of the controller who has to be cognizant of the environment of the accounting department (the rest of the organization) and make decisions regarding changes in policies and procedures within the department.

Similarly, in a CAES we can distribute functions as just described. But CAES also recognizes a further level of recursion in that every human agent is itself a CAES. And this has interesting ramifications for the actual work that gets done given that humans are quite imperfect agents.

---

[25] In fairness, Beer did offer a model of a decision process using unreliable processing elements (neurons) that he based on a cytoarchitectonic model of cortical neurons (Beer 1972, p. 204) in a network that he explained as a collegial information sharing that collectively is more likely to make a right decision than would a typical rigid hierarchical management arrangement. However, Beer's treatment of the cortex and the cellular linkages is (and had to be) superficial with respect to actual networks in, for example, cortical columns.

Finally, the VSM model situates the organization (SOI) within a larger environment that includes sources and sinks, as well as external markets with competitors, and reference to what he calls "potential futures." S4, as we said, is responsible for considering these potential futures and making strategic decisions.

10.4.1.3 Economy

VSM does not explicitly treat the internal flows as an economy in the way it is defined in CAES (e.g., in Chap. 13). It does consider value-added relations among S1 entities but does not categorize these S1 entities into necessary sub-functions such as import/export, production, and distribution functions (as will be delineated in Chap. 13). There was an explicit recognition of the supplier (source), system, and customer (sink) relations. In this, he assumed that the purpose of an organization, like a firm, was to produce a product that could be exchanged with entities in the environment for equal value in some form (money). And thus, Beer does explicitly link the internal implicit economy with the external embedding social economy.

It must be pointed out that in the late 1960s, our knowledge of the isomorphic forms of work done in, for example, a living cell or a multicellular organism was a model of a generic process we would come to call an economy. We might as well call the social economy (e.g., as we discussed in Chap. 9) the social "metabolism," indeed some authors have done so. Volk (1998) likens the whole Earth's various processes in and between the major geospheres as "Gaia's Physiology," noting how the many cycles of materials, driven by the flows of energies, produce multi-timescale dynamic equilibria (sometimes those multi-timescale dynamics superpose to produce extrema, but in the long run the Earth has maintained a range of conditions conducive to life). So, it is with intermediate systems between Earth as a whole and cells in our bodies. The internal flows and transformations form a network of value-adding or restoring processes that collectively constitute an economy. Human organizations are no different.

But, in summary, Stafford Beer provided an enormously valuable insight into, especially, the workings of an organization that would make it viable as an entity over an extended period of time. He identified numerous necessary conditions and, as far as human organizations were concerned, he felt those conditions were sufficient. Where CAES differs from VSM, aside from some distinctions in nomenclature, is the degree of refinement in both the architectural layout and the specification of particular functions. Beer alluded to how VSM was applicable to living systems but at the time of his work not as much was understood about the range and roles of mechanism isomorphic across the spectrum of complexity from cells to civilizations so this tended more toward being a metaphorical modeling than a high-resolution analogic modeling. Still, it was visionary for its time and this author owes a debt of gratitude to Beer for pointing the way.

10.4.2 Living Systems Theory—James Grier Miller

Whereas Beer came at a theory of what a viable complex adaptive system involved from the standpoint of management science (operations research and cybernetics), James Grier Miller was a biologist first and incorporated systems science into his development of a generic model of functions that he found to be isomorphic across the spectrum from cells to civilizations. We introduced Miller's work in Chap. 3, Sect. 3.2.2.1, his "Living Systems Theory" in the context of the model of an ontology of systems. We noted there that the theory only concerned living systems and so could not provide an ontology of simpler systems. Nevertheless, it provides a superb example of developing a generic model of a complex, adaptive, and evolvable system.

Miller's theory is extraordinarily detailed and complicated. We could not begin to cover it as elaborately as it deserves. So, we will only provide glimpses of his treatment as it pertains to the CAES model we are putting forth. As with Beer's VSM, there is much overlap in terms of isomorphic functions with CAES.

Miller applied his "basic concepts" as given in Chap. 3, Table 3.1 to the following levels of organization: Cell, Organ, Organism, Group, Organization, Community, Society, Supranational System. This list is similar in presentation as, for example, Tyler Volk's (2017) "Grand Sequence," covered in Sect. 3.3.3.2 of Chap. 3 in the context of ontogenesis from "prokaryotic cell" to "geopolitical states." But Miller is not really concerned with major transitions or emergences per se; he leaves that to the fact of evolution. Rather, he is interested in the functions that are isomorphic and found in all levels of his hierarchy. As mentioned in the Chap. 3 treatment, Miller lists 20 "Critical Subsystems" discretely identified for their functions. Most, if not all, listed by Miller can be found within the CAES model but many have a system–subsystem relation rather than being identified on an equal basis.

10.4.2.1 Agents and Agency

Miller lists a number of functions as separate categories, for example, systems that process material and energy versus systems that process information (and Miller's use of the term information did not recognize the distinction between information and knowledge as was covered in Chap. 3). The list of information-processing functions is generally subsumed under the agent/agency model as covered in Chap. 11. For example, he speaks of transducers (sensors in Chap. 3), channels and networks (messages in Chap. 3), memory (the H object in Chap. 4, Eq. 4.1), and the decider (the agent along with the decision model in Chap. 11).

He then proceeds through each of the living system levels identifying each of these as they are instantiated in that level. For example, here is what he says about the "decider" structure in a cell.

> Components in both nucleus and cytoplasm which are involved in control of cellular processes constitute a cell's decider. The subsystems appear to have at least two echelons.

> These are described by Eigen as "legislative" nucleic acids (the higher echelon) and "executive" proteins, like enzymes (the lower echelon). [p. 272, Sect. 3.3.7.1]

"Eigen" refers to Manfred Eigen (1927–2019), a German biophysicist, 1967 winner of the Nobel Prize in Chemistry. The term "echelon" refers to an aspect of the hierarchical cybernetic governance architecture we will cover below and as we already saw above with Beer's use of that term. Agents, in the CAES model, are situated in different levels of that hierarchy according to the "types" of decisions they make. In the above example of the decider(s) at the cellular level, the nucleic acids (DNA) in the nucleus are roughly equivalent to the logistical coordinator in CAES.[26] The enzymes are equivalent to the operations manager associated with the work processes, such as synthesis of lipids and carbohydrates.

Miller similarly goes through each level of organization delineating each critical subsystem in terms of the core concepts backed up with extensive empirical evidence. It is truly a massive amount of work. And no cursory summation like this can possibly do justice to it. We mention it simply because it was another major source of inspiration to the development of the CAES model.

10.4.2.2 Governance Architecture

In Miller's conception of governance, like Beer, he arranges "control" in a hierarchic fashion (as mentioned above) naming the levels "Echelons," by which he means steps in the "chain of command" (p. 29). That is, higher echelons have broader scopes of control or influence over lower echelons.

Within each level of organization, he then elaborates the roles played by each of the major components, giving extensive examples. For example, the "decider" (agent) in an organization is "Top executives, department heads, middle managers" (page xxiii). In the Preface, he provides a table with level of organization (cells to civilizations) in the left column, critical subsystems that process information as column heads, and specific examples such as just given in the matrix entries. One can see from this array that there is strong evidence that his list of critical subsystems for processing information is truly isomorphic in that there are exemplars for each level and type subsystem.

[26] Except it should be noted that DNA is just a stored program that can only be read out when activated by a set of proteins that copy the DNA sequence onto, say, messenger RNA (mRNA) for transmission to the ribosomes for manufacture of more proteins.

10.4.2.3 Economy

The list of critical subsystems that process matter and energy is shown in Table P-2, pages xx and xxi in the Preface of his book (1978). These correspond to many of the low-level elements in the ontology of systems in Chap. 3 (see Figs. 3.17a, b, 3.18, and 3.19) and are further elaborated in Chap. 4 (see Fig. 4.12).

These are elements that operate together to produce instantiated processors. Miller's list is actual categories of work types primarily related to biology, whereas in Chaps. 3 and 4, we focused on more generic work and function types from which specific types in Miller's list could be built (at any scale).

Thus, Miller's model of matter/energy critical subsystems corresponds more closely to the notion of an economy as will be given in Chap. 13 in this volume. As an example, his "PRODUCER" critical subsystem example for an "ORGANIZATION" level is: "Factory production unit," by which he means a manufacturing company, its physical plant, and personnel. It will be clear in Chap. 13 how this concept and many others that Miller describes map onto the processes in an economy model.

Miller is much more explicit about the role of energy flow than Beer was but this is likely because he started from a biological point of view in which energy flow is of principal concern.

10.4.3 Synthesis and Incorporating Other Work

There have been many comprehensive models of complex adaptive systems proposed over the years but the two just given have seemed to this author to be the most comprehensive. Fundamentally the CAES model presented here started with an attempt to integrate these two versions. Along the way it was discovered that certain details needed to be drawn out more than either author had done and some concepts needed reorganizing and relabeling in order to be consistent and also to incorporate some more up-to-date understandings.

Both of these models incorporated some elements from many other fields of endeavor, for example, both include basic cybernetics, communications, information, and control theories. The CAES model synthesizes the two approaches (along with this author's distinction between information and knowledge and cleaning up some ambiguities regarding the definition of information in Chap. 3). At the same time, it goes deeper into the distinction of decision types as fundamental to the structure of the management hierarchy and integrates into the model some of the latest understandings of agent theory (especially humans as agents).

Finally, a major distinction that arises from the attempt to integrate and consolidate is that between a merely adaptive system and an adaptive and evolvable system. This distinction is essential to a deeper understanding of the life course histories of the different systems. CASs have life histories that are constrained by their a priori preparation to adapt to changes in the environment. Their evolution and construction assume a more or less ergodic, that is stationary over time, environments. The real environment of Earth is definitely not stationary in geological time scales. The environment will be forever in fluxes, and when the changes go beyond what any merely adaptive system can handle, it will die. On the other hand, evolvable systems have a much better chance of persisting for much longer lifetimes if they are able to reconstruct themselves to accommodate the changes. It is important to recognize this difference and to incorporate it more explicitly in our design models.

Next, we will mention the work of some others that provided major influences on the concept of the CAS/CAES model. These are just a few that show how the model is much more extensive than either the Miller or Beer models discussed above.

10.4.3.1 Howard T. Odum's Energetics

The identification of an economy sub-model as an extremely important and major element of a CAS/CAES owes much to H.T. Odum's systems ecology and in particular the treatment of energy flows involved in transformations of matter (1994, 2007; also see Odum and Odum 2001). It is notable that Odum's work was seminal in the development of two related heterodox economics fields, ecological economics and biophysical economics (as in Chap. 9). For Odum, the energetics of systems was a generic feature of all complex systems.

10.4.3.2 Harold Morowitz's Energy Flow

Morowitz's seminal work *Energy Flow in Biology* (1968) not only augments Odum's energetics but provides a fundamental organizing principle in systems dynamics. Famously he wrote, "[T]*he flow of energy through a system act to organize that system*" (p. 2, italics in the original). Energy flow, of the right kind and power, affects physical systems in many ways such as moving molecules in convective cycles, when the geometry of the containment is right. It also provides energy to form and break bonds (recall ontogenesis described in Chap. 3). His theory actually says more. After a system has evolved under the influence of energy flow, that flow is still needed to *maintain* the system, though not necessarily at the same rate. Thus, the relevance for the CAS/CAES model is that the flow pathways through the system need special attention; both the inputs of high potential energy and the outputs of waste heat are essential.[27] And the decision processes associated with the control of energy flows are quite different from those associated with material flows.

Morowitz also had a great influence on the treatment of the ontological status of information and knowledge as presented in Chap. 3 (see his Chap. VI, especially p. 126). Whereas researchers in the field of communications theory often are

[27] Morowitz's work stands in relation with that of Ilya Prigogine (see Prigogine and Stengers 1984) with respect to what Prigogine called a "dissipative system," or one that obtains a certain organization due to energy flow-through.

concerned with the "coding of information" at the transmission end of the channel (message source), relating the number of possible messages to the "entropy" of the source, Morowitz pointed out that the real entropy related to information was that of the recipient, or, rather, the ignorance of the recipient (due to a lack of a priori knowledge) determines the "quantity" of information delivered by the message, the subject of Sect. 3.2.1.2.1 Ontological Information.

10.4.3.3 Herbert Simon's Complexity Theory

In *The Sciences of the Artificial* (1998), Herbert Simon tackled the problem of complexity, both structural and functional, in a hierarchically organized system (see 1998, Chap. 9). There are many different "theories" about or versions of complexity but Simon's approach is most applicable to the CAS/CAES model.

Simon's description of systems as hierarchies of modular units (what we called components or subsystems above) gives rise to a metric that can be used to characterize the complexity of real systems. We develop that metric below.

There are two fundamental aspects that go into defining complexity, structural complexity, and functional complexity. One deals with the hierarchy of modules and submodules and the other deals with what each module does. The former is exposed by the structural decomposition process outlined in Chap. 6. The latter is much more difficult to estimate. For our purposes here, we will use the concept of a state space to the approximate metric of functional complexity.

Imagine taking a reading on every flow (connection) and every reservoir in a system and all of its subsystems every Δt instance. The state, σ_i, of the system, where i is the index of the set of possible states, $\boldsymbol{S}$, is an element of that set. The instantaneous measure of all of these dynamical elements at time t defines the system as being in state i at time t, or $i = \sigma_t$. Since by definition a system is an organized set of parts (components and subcomponents), the number of possible states is constrained and so can be represented mathematically, at least in principle, by a Mealy finite-state machine (or automaton). For systems where multiple state transitions may occur from any one state to a successor state and where these are stochastically chosen, the representation is that of a non-deterministic finite-state machine with statistically determined transition probabilities on each node out-link. The number of transitions possible in the entire state space is $\boldsymbol{T}$, a list of the pairs of from and to states. The Mealy FSM takes into account the system's interactions with the environment or context. The state transitions are determined by a combination of inputs to the system and the system's current state at time t.

As a reasonable approximation of the *size* and complexity of the state space, we can use the number of states, $\boldsymbol{S}$, in the space (number of nodes in the finite automaton) and the number of transitions, $\boldsymbol{T}$.

$$|S| = f\left(|S| + |T|\right) \tag{10.2}$$

The function, *f*, is as yet unspecified, but assumes some kind of scaling factor.

We can then define Simonian complexity measure as a sum of the number of components and subcomponents down to and including leaf nodes (atomic components) along with the sum of interactions in the *N* networks within each component module, the size of the boundary as a list of interfaces, approximating the number of inputs and outputs, and size of the state machine.

$$\mathbb{C} = \ln\left(\left(\sum_{l=0}^{L}\sum_{i=1}^{I}\left[\left[\left|C_{i,l}\right| + \left|N_{i,l}\right|\right] + \left|B_0\right|\right]\right) + \left|S\right|\right) \tag{10.3}$$

This function sums the number of all components and relations at each level in the decomposition hierarchy and the number of affordances with external entities or number of interfaces on the boundary and takes into account the state space size. The log function compresses the numerical value of the complexity measure. Note that this formulation is related to Shannon's formulation of information. ***C*** is equivalent to the amount of Shannon information that would be needed to specify the system.

10.4.3.4 Robert Rosen's Anticipatory Systems

A major contribution to the concepts of agents that use internal models of the environment to anticipate future states that would have an impact on the agent and preemptively act to take advantage of supporting states or avoid dangerous states came from Robert Rosen's notion of an anticipatory system (Rosen 1985). This author developed an agent model (used to control a mobile autonomous robot) that learned associations between classes of stimuli encountered in a stochastic environment and favorable vs. unfavorable outcomes (rewarded or punished). The robot then used cue stimuli at a distance from those classes to either seek or avoid the source of the stimulus, that is, anticipated the results if it were to come closer to the stimulus source and take the appropriate preemptive action (Mobus 1999). Subsequently, more elaborate versions of this agent model were developed leading to the intelligent autonomous agent model used in the CAS/CAES archetype.

10.4.3.5 Other Facets

It would not be possible to delineate all of the various models that the author has employed from the brilliant works of others in a single chapter. A quick review of the Bibliography should provide some indication of other pieces of the puzzle for those familiar with the systems science literature.

10.5 Expanding the CAS/CAES

In the next several chapters, we will expand the concepts of the agent, economy, and governance model archetypes. In each chapter, we will focus on the role of the sub-model archetypes and give examples of their instantiation in real systems, somewhat in Miller's fashion, at different levels in the hierarchy of organization.

Chapter 11 will examine the basic Agent/Agency archetype, focusing on the overall model but also indicating how and when model variations occur due to what decision process is being served, that is, a governance or an economic decision. Chapter 12 is about the governance archetype and will delineate the hierarchical cybernetic governance system (HCGS) that applies to all CAES (and the parts that apply to all CAS). Finally, Chap. 13 will elaborate the Economy archetype, hopefully tying up some loose ends left over from Chap. 9's review of the HSS economy.

Chapter 11
The Agent Model

Abstract Agents are the decision makers and actors that affect the system or subsystem in which they are embedded. Their decisions have causal impact with respect to the overt behaviors of the system and they, thereby, affect the environment of their system, the supra-system. What then happens in the supra-system will eventually reflect back to the SOI and the agent itself. We look at what kind of a subsystem an agent is and develop a model archetype that applies to all levels of organization within the CAS/CAES realm.

11.1 Purpose of This Chapter

Agents make decisions and take action. In this chapter, we will describe the model archetype for a generalized decision agent. This model will be revisited in the next chapter covering the model archetype of governance. Decision agents are the actors in the governance and management of CAS/CAESs. Their agency is in the fact that they can sense/measure relevant parameters, compute via a decision model what action should be taken, and then send signals to the appropriate actuators to affect the decision. After describing the basic architecture of the agent model, we will provide several examples from biology, cyber-physical systems, and social systems.

11.2 Agents

In our investigations of CAESs, we inevitably find ourselves needing to understand agency (the power to affect other components in the environment) and the nature of agents as the entities that take decisions and actions—in other words, have agency. Within even the most complex governance architectures, agents at different levels of "authority" also have different degrees of "autonomy," or the freedom to make choices or make decisions. This correlates with the complexity of the decision domain in which they operate.

G. E. Mobus, *Systems Science: Theory, Analysis, Modeling, and Design*,
https://doi.org/10.1007/978-3-030-93482-8_11

The nature of an agent, its degree of autonomy, and its agency will be examined in more detail since these are the most critical components of a governance system.

The question of who benefits from the actions of an agent is a key element. Do the actions of an agent benefit itself, benefit other entities, or benefit the whole system through the network of agents in a governance architecture? We argue that in fact, the answer is all three will benefit.

11.2.1 The Basic Agent System

In this section, we will consider the systems model of an agent and place that agent in a larger CAES in which its decisions and actions affect the operations and sustainability of the whole larger system. It is agents that constitute the workings of governance and management processes, so understanding what they are, how they work, and particularly what happens when they make mistakes is essential.

Figure 11.1 shows a basic and universal model of an agent. The operational elements of an agent are a computational engine,[1] a decision model, and an experiential memory. The latter may have been built from learning in higher animals, organizations, and artificial agents (e.g., robots) or the "memory" may be built into the structure of the system through evolutionary modifications to brain circuits that control behavior, that is, instincts.

In cyber-physical systems (e.g., robots), the three roles are easily seen to be separated, the decision model (i.e., the program) and experience memory (variables and databases) are both contained in the computer memory; he computational engine is, of course, the hardware. In brains, however, all three components are highly integrated in the nature of neural computation involving networks of interconnections between neurons; the processing, encoding of memories, and the models are all embodied within the networks of potentiated connections.

In the human brain, functions of these components at higher levels of organization are distributed among multiple specialized modules in the various cortices (see the next chapter, Sect. 12.4.3. Human Brain as an HCGS). Each module uses the same agent model working on a very specific set of inputs (from other modules) and producing very specific sets of outputs (going to other modules). The brain has been characterized as a network of agents. Minsky (1986) described the mind as a society of simple agents that acted almost like a democracy, each agent voting, so to speak, for a proposed solution to a problem. The candidate solution with the majority of votes wins.

[1] As used here and explained in Mobus and Kalton (2015, Chap. 9), computation broadly covers any process that transforms input messages into output messages. This includes the traditional notion of the Universal Turing Machine running algorithms but also biological signal processing and activation working in the domain of chemical processes.

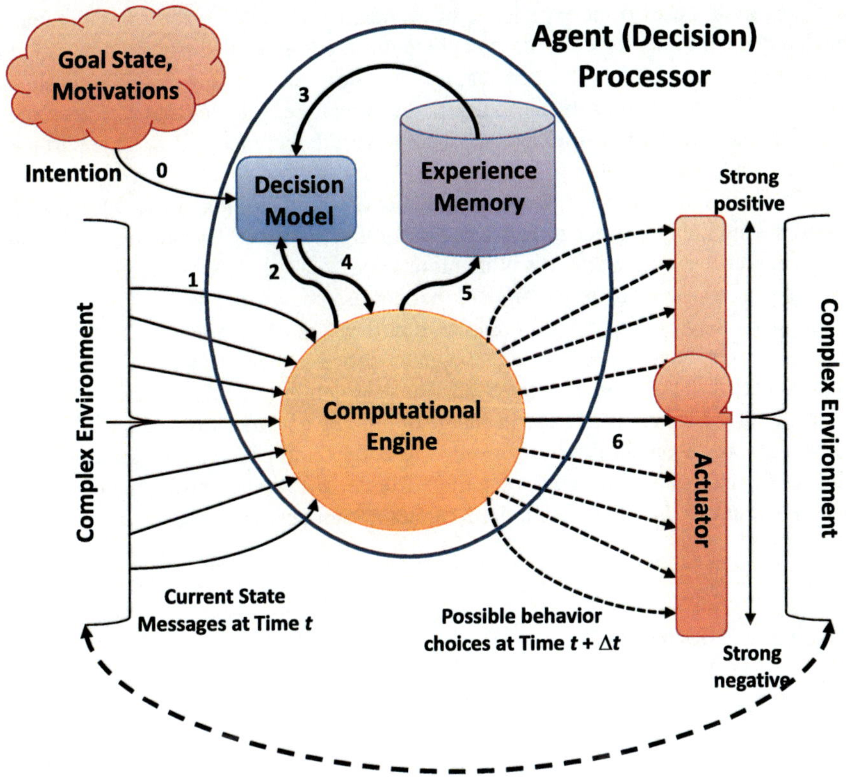

Fig. 11.1 A basic and general agent model includes a computational engine, a decision model, and an experience memory. See text for details

The agent is given agency, the power to affect the necessary response to the environmental conditions, when its output signals command an actuator having the necessary power to affect changes on those conditions (requisite variety, Ashby 1958).

Agents are supplied with motivations and goals from sources outside the basic agent model. For example, hunger is a drive arising in the hypothalamus that will drive an animal to seek and eat food. The brain centers responsible for tactical control for food seeking (e.g., foraging) are activated and the animal goes into seeking mode. Human agents often receive motivations from other humans, for example, from social norms. In our discussion below of human beings as agents, we will revisit the role of motivation and goals as they pertain to decision-making under conditions of high autonomy and intentionality as expressed in our species.

Agents are supplied with goals or motivations from external sources (represented here by a "cloud"). Sequence of events: (0) The decision model is provided with an intention (goal or motivation to attain a particular state of the world); (1) messages arrive from a complex environment providing state information at time t; (2) the computational engine uses the decision model to compute the decision; (3) the decision model uses prior experience from memory so that; (4) the computational engine can complete its work to; (5) update the experience memory (in the case of a learning agent); and (6) make a choice of action out of many possible (dashed arrows). The output controls the actuators that affect the behavior that acts upon the complex environment at time t + Δt. Not shown are the energy flow inputs to the processes and the actuator.

What the figure does not show is the situation of a complex agent, which is an agent having many degrees of freedom and multiple variables to manipulate. In the next chapter, we will tackle this issue more directly as a problem of governance of the agent as a system. In anticipation of that discussion, we mention only that agents themselves may be sufficiently complex that they are organized as a hierarchy of agents that manage the whole agent/agency process. For example, consider the office of the CEO of a large corporation. The president is not only the chief strategic manager of the company but also in charge of running a whole "department" devoted to various management functions needed to collect and process data and interpret the results in terms of strategic planning. The department is thus an agent for the whole company but it is composed of many sub-agents reporting to a "master" agent.[2]

11.2.1.1 Types of Agents

Agents can be categorized into three basic types in order of increasing complexity. Recall the discussion in Sect. 10.3.2.1 Adaptivity in the last chapter.

11.2.1.1.1 Reactive Agent

There are, broadly, two basic kinds of reactive agents, purely reactive and adaptive reactive. The first is the simplest in terms of its decision model and degree of agency. It simply takes measurements of the situation and has an algorithmic or heuristic program of response. A simple thermostat is an easy example.

The second kind of reactive agent is adaptive. That is, it has the ability to vary its response-outputs based on variations in the inputs. It tracks the input variables and responds in kind. This is the homeostatic mechanism. In extreme cases, the agent can modify its response capabilities to match longer-term changes in demand.

---

[2] The use of the term "master agent" is not to be taken literally. As will be shown in Chap. 12 (next) the idea of an agent higher up in the decision hierarchy is not that it need be a command-and-control master, but, rather, a coordinator of the activities of sub-agents.

11.2.1.1.2 Anticipatory Agent

Agents that are capable of "learning" or modifying their own decision model may incorporate new associations between stimuli and responses. They can use their new model to begin a response in advance of the onset of a situation, thus reducing the overall costs of adapting. Recall Sect. 10.3.2.1.3 Anticipation in the last chapter.

11.2.1.1.3 Intentional Agent

Agents, able to modify and manipulate internal models of the environment (i.e., of diverse objects other than mere stimuli), can form completely new models of systems and/or the supra-system's relations to those systems. For example, a human being can combine a model of a sharp rock with a model of a club using a model of sinew used to bind the two objects together, to form a new model object, an axe, to be used to chop wood. The human being has a need to chop wood, so forms an intention to construct a tool to accomplish that end. Most vertebrate animals appear to form intentions, even if only driven by instinctual or appetitive needs. We will have much more to say about humans as agents throughout the rest of the book.

11.2.2 Decision-Making

The process of making a decision and activating action is based on the model in Fig. 11.1. The nature or type of decision determines what sort of decision model will be appropriate (see Chap. 12 for a description of decision types relevant to governance). The purpose of the decision at any time $t+\Delta t$ is dependent on the data collected regarding the state of the external "complex" environment at time t along with relevant experiential memory.[3] The agent must interact through its motor outputs, collectively producing the behavior. The model that connects inputs with appropriate outputs, known in engineering cybernetics as the transfer function, and some form of context as contained in an experiential memory are used in the computation, producing the behavior. Thus, the environment is altered by the action and gives rise to a new state to be measured by the input transducers (sensors) at a later time that includes delays for the computation and action (Δt) and latency for discernable changes in the state variables of the environment, a value that is dependent on many factors and in general will itself be variable. We assume constant time for the decision process itself in simple cases, but this is not the general case as complexity of the data and memory increase.

[3] Those familiar with the concept of a finite state machine will see some resemblance. However, the general model of an agent does not require a finitude of states of the system (or of the inputs).

As we go up in the hierarchy of complexity in concrete systems, we need to take into account the role of stochastic process increasingly. This applies not only to complex environments but also to the complexities of the elements in the agent. We tend to think of digital computers as being completely deterministic, but even while using digital computers in an agency of networked agents (called system of systems, SOS, in highly complex engineered systems) or distributed computation in cyber-physical systems, the vagaries of noisy communications, time lags, and data inconsistencies in distributed experiential memories (distributed databases) can introduce stochasticity into the general system. In living systems, chemical reactions are inherently stochastic even if constrained directionally by, for example, the specificity of enzymatic reactions.

The role of experiential memory varies depending on the degree of autonomy and the capacity for learning. The diagram in Fig. 11.1 shows the various components of the decision process separately. In the case of, say, computer-based control systems this generally is the case. The decision model itself needs to include provisions for incorporating experience memory (e.g., state variables that are changed as a result of prior decisions) allowing its own structure to be modulated by experience. In using computers for control systems, this is a notoriously hard problem in machine learning. However, in the brains of mammals (and birds) the three functions are built into a single structure, the architecture of the neocortex (palladium in birds) in which adaptive memory encoding and computations are integrated. This is thought to be the nature of the unique computational capabilities of structures in the cortex called cortical columns and microcolumns.

Agents make decisions. The theory behind decision-making has been very thoroughly explored in areas such as management theory and game theory. The agents in this case are, of course, human beings. Any information processing subsystem that affects the actions of a whole system is an agent, so the principles of decision-making (and taking) are applicable in all complex systems. In this section, we examine some of the general principles of decision theory as it applies to managing system behaviors. A core set of principles are applicable to all agents from the mere mechanical controllers (e.g., thermostats) to complex cyber-physical systems such as mobile, autonomous robots, to living cells, and to human brains.

11.2.2.1 Decision-Making Abstraction

The structure of a decision process can be represented in the form of an expanding tree (in mathematics, the term tree is given to an inverted branching structure—see Fig. 11.2) in which each level represents possible states of the world that will be used to compute a decision according to the decision model employed. Each branch to a lower level of the tree represents an alternative choice based on the current state and the preferred future state that should obtain from the choice of that branch. The mechanics are that at each node in the tree, the agent evaluates the state of the world (as in Fig. 11.1) and uses the decision model to obtain a choice of the future state (branch) which should be the most desirable given the current conditions, and which

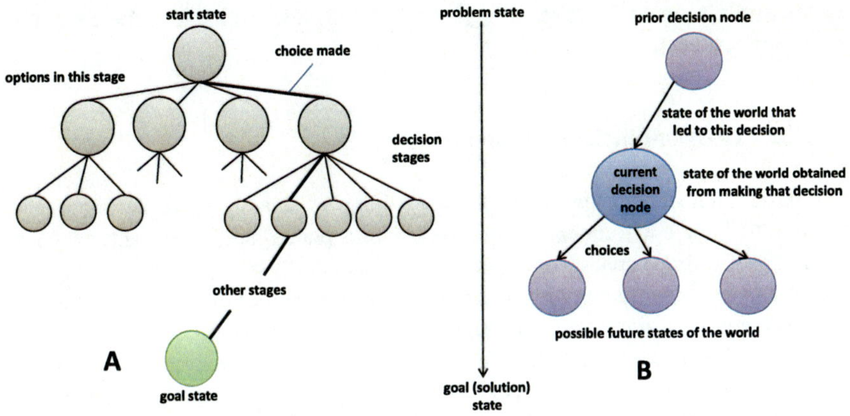

Fig. 11.2 (**a**) Decisions can be modelled as being made in a sequence of discrete stages, starting with the system being in the start state. The process is represented by a tree where time progresses downward. Each node in the tree represents a decision point. The child nodes of that node represent the choices that can be made to progress to the next stage. (**b**) The efficacy of a decision depends on how much information is available to the agent about the consequences of the next stage. Learning agents are those that can add choices at any given stage as a result of experience and reinforcement

actions of the agent should lead to that child node. Decision trees capture the overall structure if the states of the world are knowable and predictable based on the current state of the world, the degree to which the agent can ascertain it, and the amount of uncertainty from several different factors. A perfectly deterministic decision tree, such as a game of Tic-Tac-Toe played by two very alert players has perfectly predictable future states given the first move. If one of the players is less than alert, then the outcome might be surprising.

In very complex decision processes, there are many sources of uncertainty that are not just involving mistakes made by the agent. The input data may be incomplete, noisy, or ambiguous. The decision model may be sketchy and may not include all of the real possible branches (this is why learning agents evolved!). Lastly, the motor outputs from a decision might not be sufficient to actuate the desired outcome in terms of the changes made to the state of the environment.

The transition from one decision node at one level to the next decision node at the next level down in Fig. 11.2 is the result of the agent's processing the current state of the world, getting input from the environment, computing the decision based on that data and its internal model, making a choice from the options at the next level down, taking action based on the choice, and then setting itself in the new decision node at that next level down. Once action is taken, it modifies the world situation and there is a time lag before that situation stabilizes so that the sensory input about the new state of affairs can be judged reliable. Similarly, it takes time to perform real computations once the data are available. Thus, time passes before the agent can take the next step. Therefore, the agent's computational speed must be fast in comparison to the rate of changes in the world situation in order to prevent destructive lags, that is, lags resulting in actions too slowly relative to the evolution of the environment.

11.2.2.2 Decisions in Living Systems

The basic agent model applies to the way living organisms make decisions and respond to their environmental situations. While plants, and even bacteria, have been shown to have analogous agent-like mechanisms, we will focus on animals with brains sufficiently complex so as to have wider degrees of freedom in choices and relatively high autonomy. Specifically of interest, are mammals and birds.

Figure 11.3 shows the situation for two decision models, one for a simple learning agent and the other for an agent having multiple sources of background knowledge that can help modulate the decision process. These are shown from the perspective of a "current decision node" as shown in Fig. 11.2b, where "nodes" may be interpreted as neuronal clusters that represent potential future states and the decision at any one node depends on which cluster the current node will send excitatory signals. The lower nodes represent future possible states. In both figures, the outer purple oval represents a variety of a priori knowledge that might help affect a decision. It is equivalent to the decision model of Fig. 11.1. This can include genetically inherited automatic responses and instincts and, in animals that are capable of learning new behaviors, the learned relations between world states and successful responses.

We normally think of decision-making as a rational process, involving intelligence. But in animals and humans, there are a number of other cognitive processes that contribute to the making of decisions (c.f. Damasio 1994; Kahneman 2011). Antonio Damasio (1994) was the first to alert us to the idea that when we experience

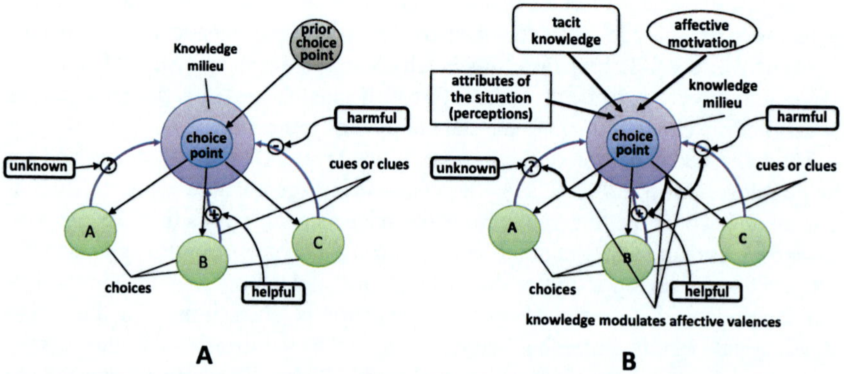

Fig. 11.3 (**a**) The agent may learn to associate choices with various cues or clues from past experience, tagging pathways with valences (+ or –). (**b**) Highly autonomous agents have enriched knowledge milieu with which to judge the best choices, even if valence tags suggest otherwise

the results of a decision made, it usually has the effect of tagging[4] that choice (linkage in the neural representation of one state to that of the new state) with a valence—it was either a positive experience or a negative one. In the figure, the small ovals overlying the links leading to the next stage contain either a plus (+), a negative (−), or an unknown valence (?). The strength of this tagging depends on the salience of the experience. Weaker valences may be altered in future or strengthened if reinforced frequently along the lines already encoded. All other things being equal, the agent in Fig. 11.3a would tend to choose the path to the choice B since at some time in the past this choice resulted in a positive outcome that tagged it with a positive valence (see Mobus 1999, for details on how this tagging actually works in neural networks).

Choices that humans make are modulated by a much more complicated knowledge milieu. This milieu is where *judgement* takes place. Seemingly, rational decisions may be highly influenced by a number of factors involving memories and feelings. Humans (and likely many mammals and birds) are able to override the simple path valences if they are able to consider states much further down the tree levels. That is, it may be that taken alone, a given path might be tagged negatively at the current stage, but there is a path out of that node to a much more positive node further down. This has to be available in the knowledge milieu at the current choice point (Fig. 11.3b). It is the reason that a chess player can sacrifice a valuable piece in a game, knowing that this will likely result in the opponent making a choice that will allow the first player to put the opponent in check (a rewarding state).

A contribution that is made in influencing decisions is the degree of confidence that the agent has regarding the choice (Burton 2009). Both affective (emotional) and tacit knowledge seem to contribute to this factor. It is believed that confidence in one's decisions is necessary in order to take action without hesitation. It helps prevent analysis paralysis or delaying making a decision until all of the relevant data have been processed. Herbert Simon's concept of satisficing may be the effect of the sources of confidence reaching a threshold that leaves the agent feeling certain that they are making a good choice. This, however, is a source of error (previous citation) when the agent believes they are correct based on faulty knowledge or emotions. A sad example of this phenomenon is when those of a certain political persuasion insist that anthropogenic climate change is not happening. They may not be lying in the conventional sense, only convinced of their own veracity because their entire knowledge of the goodness of a certain economic condition forces them to conclude that global warming is a hoax.

The classical psychological study of decision-making is converging with neuroscience at the point of recognizing that high-level decisions are made in neural circuits (notably in the prefrontal cortex) with considerable information provided by

[4] Damasio's term for this tagging was "somatic markers" in that they recorded body states or feelings, at the time the experience happened. If a person (or any animal) experienced a negative consequence vis-à-vis their body states, then the marking of the association link with a negative valence provided a kind of 'warning' that choosing that set of actions and subsequent environment state would not be immediately advantageous.

many parts of the brain, including what we might call the more primitive parts in the core brain, like the amygdala (fear processing). As this research progresses, we will be able to see more clearly the influences that emotions and biases have on our "semi-rational" decision processing. Along the same lines, as the artificial agents we invent become better decision makers with varying degrees of autonomy it will be because we have better understood the dynamics and processes.

One very important topic related to decision processing is learning and memory. In Fig. 11.3b, reference is made to "tacit" memory. This is the realm of general mental models of how things work rather than explicit memories such as episodic (specific instances) ones. These memories are learned; mental models are constructed over time and multiple occurrences of situations involving the concept, what we call inductive learning. Categorical memory or the existence of "classes" of objects, relations, and actions are in this realm. One recognizes one's own house as a "house" similar to other houses, but with particular features, including location, makes it unique among houses. There is a tacit model of house-ness that has been constructed by an inductive process of experience over time that can then be used to model specific instances of particular houses one becomes familiar with. The particular instances are held in explicit memory accessible by conscious recall to put into working memory for thinking.[5]

The amount of capacity for forming memories of mental models and the ability to learn complex causal relations and the ability to use those models in decision thinking are all factors to how much autonomy an agent might possess. As with the example of overriding a negative valence tag on a decision path in order to achieve a better "payoff" later, the more capacity an agent has for ignoring affective recommendations or confidence factors, the more autonomy they possess.

11.2.2.3 Decisions in Artificial Systems

An argument can be made that CAS artifacts should be designed to emulate animal decision-making as described here. The capacity to learn, not just affective valences, but to be able to refine models of situation-response over time and even construct new models of these after exploring new world situations would be needed in order to give our artifacts behavioral autonomy. Moreover, the ability to construct generalized models of general-situation-general-response, or what we called "tacit" memory above, would provide our artifacts with a tremendous amount of power to learn from experiences and be able to have generalized or abstract models of what is to be done in novel situations that are similar to, but not exactly the same as prior experiences have been.

[5] It is generally believed that bringing a thought (concept or model) into working memory is based on activating the kernel of the concept which, in turn, activates all of its components in associative and even perceptual memory areas. When all of the components of the concept are activated and in synchrony the conscious-processing brain is aware of the object.

There are, of course, many ethical questions to be asked about how autonomous do we want our machines to be. We would argue strenuously against producing highly autonomous combat robots given the freedom to kill if, in their judgement, they have an enemy in the crosshairs simply because we could never be certain that this was truly the case.[6]

11.2.3 Autonomy

Agents fall into categories based on degrees or levels of autonomy, that is, how much freedom the agent has in making a decision. Some agents, like thermostats and other kinds of controllers have no real autonomy; they are programmed to respond to the situation in a deterministic, or algorithmic, way. Living systems, that is individual living entities, display various levels of autonomy. Generally, we recognize less complex living entities like insects as being driven by instinctual behavior. Their decisions are hard-wired into their nervous systems in the manner of algorithms. When we reach the genera of birds and mammals, however, we begin to see more volitional behaviors and variations in responses to similar situational stimuli, suggesting that they have a range of choices they can select from. There is, obviously, more complexity involved in their decision-making processes. We also note that these creatures can suffer errors in judgements or making erroneous decisions—they can make mistakes and suffer the consequences. This is very certainly true for members of the human species who would seem to have the highest degree of autonomy and consequently make the highest number of decision errors. Our archetype model of an agent must make provisions for the degree of autonomy and must provide an explanation for why it appears that more complex systems display greater seeming autonomy. We shall come to recognize that the degree of autonomy is actually another reflection of not just the complexity of the system of interest but also the complexity (and dynamics) of the environment in which the agent has to respond to.[7]

When we come to human-level autonomy the discussion usually turns to the concept of "free-will." This is generally taken to be the ability to, in essence, regardless of the circumstances of a situation, elect, or not, to behave in a particular way. We submit, without extensive philosophical argument, that what appears to be "free" in the way of choice is no more than a plethora of optional choices to which the subject applies a subconscious procedure to make a selection and in a

[6] And in the worst-case scenario we can imagine machines having not only autonomy but motivations and sufficient intelligence to decide that humans are problematic and all are enemies, as in the Terminator movies!

[7] In other words, the complexity of the environment will be reflected in the complexity of the SOI. But that is a reflexive relation! The complexity of the SOI is reflected in the complexity of the environment in which it is embedded. Environment and SOI co-evolve.

subsequent conscious examination interprets as a conscious choice (c.f. Libet et al. 1983; Mobus 2019, Chaps. 2 and 5).

Agents may be characterized by the levels (or degrees) of autonomy, which is to say their degrees of freedom in the making of decisions. There are several aspects to the nature of autonomy, of which we will only cover two dominant ones here. One aspect has to do with the decision model used and the other is the kind of computation that is employed on the model to arrive at a decision. The range of autonomy runs from a purely deterministic (a fixed mapping from situation inputs to response outputs) to highly variable options available, the selection of which depends as much on history as on context, to experimental or possible options available to explore the territory of possibilities in order to construct a wider field of responses.

The range of autonomy can be based not only on several factors, such as the capacity for modifying the experiential memory or the decision model itself, but also on the relative *looseness* of the decision model. By this we mean how deterministic or stochastic the process of computing the decision is.

There are at least four basic approaches to computing decisions with varying degrees of this looseness.

One, the most deterministic in terms of procedure, and therefore not loose at all, is to use an *algorithm* that is guaranteed to produce an answer in finite time and number of steps. This is the classic derivative of the Turning machine model of computation embodied in modern computer languages. A solid-state thermostat embodies this approach.

The second level of looseness is called "heuristic programming" and while these can be implemented using computer languages, following a heuristic (or rule of thumb) does not guarantee the correctness of any answer obtained. Heuristics embody a probabilistic result that depends on many stochastic factors in the input. Classical examples of heuristic decision-making are so-called *expert systems* that use the IF-THEN-ELSE construct with the modification that the proposition has associated a probability factor, for example, IF (patient temperature > 99.1) THEN (disease probability > 20%) ELSE (patient is 90% likely fine) The decision being supported in these kinds of diagnostics application is simply a guide to the diagnostician as to whether to do more testing and, perhaps, what kind of testing, or send the patient home.

Heuristic programming shows up in biological systems as well. Evolution programs brains, for example, to use heuristics to achieve rapid decisions where the response has been selected through evolutionary trial-and-error testing. Even human beings are subjected automatically to relying on heuristic programming in what Daniel Kahneman (2011) calls "fast thinking." An important difference between heuristic and algorithmic (or rational) computing is speed. Heuristics can be thought of as "quick and dirty" versus the slow and precise algorithmic approach. For living systems, speed is often the winning argument as to which one is most appropriate.

A third category of looseness in decision models and their computation depends on being able to encode experiences with real-world situations and trial-and-error responses that prove successful. Whereas heuristics and algorithmic processes are "instinctive," and suitable for very fast response requirements, this third category involves a capacity to *learn* from experience. Those experiences cannot entail life-or-death outcomes; the learning agent has to be able to survive unsuccessful trials. We will have much more to say about this capacity and its consequences for agent decisions in the Sect. 11.3.4.4 below, Decision Model Refinement.

A final degree of looseness is the ability to construct a priori experimental decisions and run simulations of the subsequent consequences

Autonomy is linked with the capacity for judgement modulation on decisions as discussed above. A thermostat has no autonomy. Its decision model is fixed for all time. Similarly, homeostatic mechanisms in living systems may have slightly more variation with which to work, but are still destined to react to their sensed environment as designed by evolution with a fixed mapping from situation to responses. In the same vein, adaptive response, while much more flexible, is still dependent on actual experience, and often repetition of experience, in order to develop a reliable mapping from stimulus to response. This means that a brain exposed to a novel set of inputs may construct a novel output through "guessing"[8] the correct response and recording in its experiential memory, the results. It then takes further offline processing to modify the decision model accordingly. If the novel behavior was successful in meeting the decision objective (see below), then it will become part of the repertoire of responses used in the future. Otherwise, its model will be inhibited from further use.

It isn't until we get to very complex brain structures and correlated complex behaviors, particularly prefrontal cortical architectures, that we see a priori or counter factual "experimental" responses introduced into the mix. Effectively, this amounts to the construction and conduct of a mental simulation of, for example, "what would happen if such-and-such were true rather than thus-and-so?" Humans engage in this kind of simulation modeling in spades. But so do organizations, through the auspices of collective thinking among a group of human decision makers. This, seemingly ultimate, level of looseness, the ability to try anything but in simulation rather than risking existence on a hunch is the hallmark of a CAES.

In the current state of art, computer-based control systems are just entering the realm of autonomous decision processing, though as far as we know there is no model of artificial judgement in the way we think of artificial intelligence. The success of chess-playing programs like Deep Thought (IBM's AI entry into the field) created an artificial knowledge milieu based on the judgements exercised by human experts, collected in a large set of heuristics that acted as judgements in playing. For now, the knowledge milieu and judgement processing have to be handcrafted. Recent work on artificial neural network models (along with statistical learning methods), called "Deep Learning," are being used to automatically construct

---

[8] 'Guessing' isn't exactly correct. In fact, more complex brains may use a number of methods to make an "informed guess" about the decision. In humans we use analogies—is this situation similar to other already known situations from which we can extrapolate a decision? In animals, especially birds and mammals, there may be several "reasoning" mechanisms at work when encountering a novel situation. But in humans the number of kinds and degrees of efficacy of reasoning are substantial.

knowledge milieus for judgement-like processing in autonomous cyber-physical systems like self-driving automobiles. The methods being employed are still fairly restricted to stationary, or semi-stationary pattern learning so the nature of what is being learned is limited to minor variations on pattern sets. The major paradigm of network learning is still based on what is called fully distributed representation, in which categories of patterns are encoded in the weights of all synaptic connections through the network. This is fundamentally no different than that was being used in the late 1980s when networks were constrained in size by the available hardware of the time. All that seems to have changed, we now have much bigger, faster computers with much more memory. So, we can tackle bigger more complex problem domains. Still this kind of learning, while useful for cyber-physical systems is a far cry from the way the mammalian brain works. And so, it is not able to emulate even the animal learning necessary for true judgement and constructing complex concepts.

Computers have been used for decades in what are called embedded systems controls. The common thermostat on your wall, today, is such a system. These are relatively simple devices using deterministic algorithms to compute the transfer function converting the ambient temperature into a control signal to the furnace or air conditioner. The HVAC (heating, venting, and air conditioning) systems in whole large commercial buildings are much more complex, and today those can involve distributed computing for local zone controls, but a centralized coordinator computer working on optimization of the whole system across these zones has to monitor and communicate tuning commands to the network of computers. These are among the first examples of engineered hierarchical (layered) control systems, still simple enough to be engineered for determinacy. The distributed computers have a minimum amount of autonomy, relatively speaking, their agency is highly constrained.

More complex systems have been recently coming under the aegis of engineering, recognizing that increasing complexity of systems, reflecting the complexity of their operating environments, also may involve increasing their autonomy and agency. Complex cyber-physical systems, the Internet of Things (IoT), and other complex engineered systems are referred to as "systems of systems" (SOS) and the emerging approach to engineering such systems is becoming cognizant of the need to deal with increasing stochasticity (surprise) in terms of the systems fulfilling their intended purposes. More autonomy must be given to component subsystems so that they can continue to operate nominally in the face of disruptions in other parts of the embedding system. The design of the Internet is a very good example. Nodes, such as computers, routers, and switches (along with domain name servers, and other service providers) are designed to continue operations even when other components go down. The hierarchical control aspects of the Internet are handled by a distribution of roles to special nodes in the network (e.g., service providers and backbone providers) that issues global commands to the distributed operational nodes.

Mobile, autonomous robots are other examples (Mobus 1999). Some discretion in choosing appropriate responses to unpredicted environments have to be given to the robots so that they can avoid being thwarted by those environments. Research on how, exactly, to implement this capacity is quite active. The desire is to create robots those can act as "slaves" to human needs, recognizing that they need some flexibility, that is, autonomy, so as to negotiate complex unpredictable circumstances (which would make them useful.) At the same time, they cannot be given too much autonomy. We would not like to implement the scenarios depicted in movies like "The Terminator" or "The Matrix."

The issue of autonomy is especially cogent in the discussion of governance and management of human organizations and state governments where we are talking about the most autonomous agents we know of, human beings.[9] These are the key decision agents operating in the structures of management and governing. The environments with which they must interact are extremely complex and with full of uncertainties, requiring a high degree of autonomy and; the creativity needed to deal with those environments. On the other hand, humans are notoriously guilty of constructing and holding beliefs in decision models that are not necessarily valid in terms of how the real world actually works. Put bluntly, human beings may entertain ideologies that may seem logical but do not actually correspond to reality. At this writing, a large contingent of so-called conservatives (political) still reject the overwhelming evidence for anthropogenic climate change due to the burning of fossil fuels. It has become abundantly clear, empirically, that climate change is underway and causing increasing variations in weather patterns, the extremes of which are damaging. Their disinclination to understand what is happening is actually based on a *true* proposition, that wealth production is directly tied to energy use, which has been due to the burning of fossil fuels. As we pointed out in Chap. 9, to reduce carbon emissions means reducing the burning and consequently the energy available to do economic work, in lieu of replacement by so-called renewable sources, a possibility still being investigated scientifically. Their philosophy includes the belief that economic growth is the only good for humanity. But this flies in the face of the problem presented. Thus, their only real position has to be to deny climate change (or the human contribution) in order to protect their cherished belief in wealth production and expansion.

Such ideological positions clearly compromise the effectiveness of their agency when it comes to making legitimate and helpful decisions. So, the irony of the human position as agents in decision processes is that they enjoy the maximum degree of autonomy but also the maximum degree of choice to ignore reality's messages. Human beings appear to be free to choose what they are going to attempt to learn—a meta-decision process. And they can choose not to attend to sources of factual information if there is a hint that it would require them to learn something contrary to their beliefs.

[9] It is important to note that supra-biological entities such as corporations and states seem to exhibit degrees of autonomy themselves. However, we might not be able to easily differentiate between the autonomy of individual human agents, like the managers or lawmakers, and the autonomous behavior of the whole system. This will be an area of active research in the future.

In the realm of management, this had tended to be less of a problem prior to the emergence of the neoclassical, neoliberal capitalist ideology for the economy. With the latter, came a heavy emphasis on maximization of profit-making and shareholder wealth as opposed to customer benefits (e.g., quality, as low a price as possible to make a fair profit), employee wellbeing, or that of the environment. This was coupled with a tremendous increase in reliance on debt financing and financialization (making profits off of second and higher-order financial instruments treated as assets). Under this environment managers have been faced with different objectives in their decision models that have severely complicated and, one might say, muddied the experiential memories used to make decisions.

Thus, the current situation with respect to the degree of autonomy for human decision agents is problematic. Further complicating the matter is the kind of experiential memory constructs held by many human beings. Education, it can be argued, has not provided the majority of humans with the kind of broad grasp of how real systems actually work or of the implications of a systems perspective. The result is a poorly educated populace when it comes to insights into what constitutes good decision in running everything from families to multinational corporations and states. A significant amount of research is needed to answer some fundamental questions about the efficacy of human agents in the governance and management of organizations (which is quite an understatement.)

Implications for systems analysis vary depending on the nature of the whole system with respect to its governance architecture. In the hierarchical cybernetic system described below, agents occupy relatively clear functional positions in the hierarchy (see the next section on decision types). The key guide for analyzing a system's governance and management system is to recognize the degree of autonomy *needed* by the agent at any decision node in the structure. One approach to this would be to have a good characterization of the environment of the system with respect to the degree of stochasticity in input flows and output acceptance by entities as sources and sinks. Additionally, internal subsystems may be subject to disruption from the activities of other subsystems with which they interact. Finally, systems are always subject to their own internal components failing or degrading so as to affect their performance. Analysis of these sources of uncertainty should look for the amount of stochasticity (variance in the range of the signals, etc.) and the kind (stationary, homogeneous non-stationary, or non-homogeneous non-stationary). All of these factors need to be taken into account in order to assess the amount of autonomy needed by the agents. In situations of low variability and stationary stochastics, the agent needs very little autonomy (e.g., a situation reflected in homeostasis). At the other extreme, high variability and non-homogeneous non-stationary, where predictability is minimal, a large amount of autonomy is needed. This means the decision model needs to be not only flexible but modifiable (learning), and the experiential memory facility must be extremely large to learn the larger number of possible causal relations needed to formulate a knowledge milieu.

11.2.4 Ideal Agents

The role of agents as decision makers in governance and management can be idealized for the purpose of establishing a basic or baseline model. Idealization serves the purpose of understanding how a governance and management system can work optimally (a concept that will be further explored in the next two chapters). An ideal agent will make the best possible decision under whatever conditions of uncertainty, ambiguity, or time constraints might prevail. Note that this is different from an ideal set of information inputs which are subject to environmental variations not under the control of the agent as a receiver. Herbert Simon (1957, 1998) coined the term "satisficing" (a combination of satisfactory and sufficiency) to describe the heuristic process (as opposed to algorithmic) for making a sufficiently satisfactory (near optimal) decision given constraints on computation power and time. Simon recognized that humans, in particular, are not capable of being completely rational decision makers owing to these constraints (also see Kahneman 2011 for a more detailed discussion of human decision-making). He described the condition as "bounded rationality" (1957). Under these conditions, humans try to make the best decision they can but recognize that the decisions will not necessarily be the absolute best, hoping that they will be good enough for the situation at hand. This works only if there is some leeway in the consequences that would not seriously disrupt the overall performance of the system.

Bounded rationality also applies to the probabilistic nature of the information input. There are always uncertainties about whether the messages being received are the "right" ones and that the information content of the messages is "right," that is, free of noise. Thus, even ideal agents are subject to errors in decision-making just due to the inherent problems of computation power, time limitations, and uncertainties.

All of these factors apply as well to automated decision agents operating in CASs and CAESs. But for the simplest embedded system controls, where we have lately been using very powerful computers and very large memories (think smartphone technology), complex control systems for, say, large airplanes or nuclear power plants, are still faced with uncertainties and time constraints that can cause problems in making correct decisions at all times. Engineering real-time controls requires extreme care. Task performance in these systems (e.g., mission or life-critical systems) must be guaranteed to complete within a specified time interval in order to prevent damage to the system (not to mention passengers riding the system!)

Human decision makers, probably comes as no surprise, are far from ideal, even in terms of satisficing. With humans, there are several additional influences at play on decision-making.

In addition to the constraints and disruptions on information input, human memories can be notoriously faulty. The decision models can be tainted by ideological beliefs that are non-veridical. Rational thinking (computation) is slow (Kahneman 2011) and easily disrupted. But most of all, humans are strongly motivated by emotions and moods (Damasio 1994). Human beings evolved decision-making capabilities suited to the late Pleistocene era, for hunting and gathering, for small group social living, and for relatively less complex environmental contingencies. The built-in heuristics and biases (Kahneman 2011) served us quite well in those conditions. But now, in a technologically embedded situation and in a huge scale social environment, these are proving more often detrimental to decision-making within governance and management roles. One might reasonably argue that human beings are far from ideal as agents. Even so, it will be instructive, when examining the issues of governance for government, for example, to ask what would the ideal system look like given ideal human agents, that is, those are motivated by non-selfish sentiments, work with non-biased decision models for their particular decision types, and are versed in rational thinking. They may still be faced with having to reach satisfactory and sufficient decisions (near optimal, but not absolutely optimal), but they will play their roles in a manner that leads to success of the whole system being governed.

We will take this approach below to project an ideal government for a social system, based on the HCGS model. We will then, briefly, examine several aspects of current governments to show how they deviate from the ideal to the extent that they too often, under modern social conditions, fail to adequately provide the sort of long-term sustainable governance for their societies. In future work, the author will pursue this line of analysis in hopes of finding a pathway from the current situation to a government architecture that is more in line with the HCGS architecture and makes provisions for helping human agents become maximally effective despite the above-mentioned problems.

11.2.5 Agent Roles and Decision Types

We now embed the model of agents into organizational structures in which decisions affecting a system's behavior are to be made. This will specifically be the subject of the next two chapters on governance and economy. Agent-based systems (those relying on agency) have to operate internally to maintain themselves and produce products needed by themselves and other systems in their environment. They have to obtain the resources they need, avoid dangers or respond to threats with appropriate behaviors. Thus, we will see that decision types can be classified into three general ones and two sub-types of one of the general ones. The general types of decisions are as follows. Operational—decisions relating to the real-time work processes, maintaining stability, quality, and quantity of products required. Coordination—decisions relating to keeping distributed, semi-autonomous but effectively cooperating work processes operating as an effective whole, that is, as a system. There are two sub-types of coordination decisions. Logistical—decisions relating to maintaining optimal coordination between the internal work processes, including processes that obtain resources from the environment and those that export products and wastes to sinks. Tactical—decisions relating to coordinating the overall behavior of the system with the behaviors of the entities in the environment with which the system interacts. The tactical agents need to obtain information from those other entities in order to make decisions about obtaining of resources successfully and exporting of the system's outputs. In some CAESs, we find a third general type, strategic decisions relating to planning for future contingencies and coordinating the internal alterations of the system (learning and evolving) capacities to anticipate that future.

These decision types will be more deeply explained in the next chapter and will be further elaborated in the one following that.

11.3 Complex Environments

Complex systems evolved to be complex due to the fact that they were operating in complex environments. Complex environments generate more information that eventually incorporated into the knowledge structures of the systems embedded in them. One theory of evolution leading to increasingly intelligent creatures (that are also inherently complex) is that intelligence evolved in response to the increase in information processing loads. Of course, this is really another version of coevolution as a kind of arms race (see Mobus and Kalton 2015, Sect. 12.6, p. 568). Environmental complexity arises not only from the pioneering of new econiches during adaptive radiation (say after a major die-off) but also from new species behaving in new ways that challenge other species to evolve new responses. This is the essence of coevolution wherein there is a mutual selection pressure applied between species. In complex food webs, there may be multiple participants. There are many dimensions to consider in the increase in information in evolving environments. For example, an increase in range in animals expanding their migration behaviors extends the dimensions in space over which new behaviors are experienced by resident populations. This alone can increase the uncertainty any local species may experience as invaders penetrate an ecosystem.

11.3.1 Uncertainty in Interactions

In Chaps. 2, 3, and 4, we learned that to really understand a given system of interest one must first get a handle on understanding the environment of that system. One need not know the internal details of other entities or agents in the environment—they are modeled as sources or sinks only (which in the current context includes predators). But one does need to know what they are and what, in general, their expected behaviors will be in order to understand the inputs and outputs to/from the SOI.

Just as we are concerned with the complexity in the SOI, so should we be with the environment, but from a different perspective. There are several concerns about the complexity of the environment. For example, the entities in the environment should generally be viewed as stochastic processes. Their behavior can only generally be described statistically. But more than that the statistics of many entity behaviors need to be described themselves statistically! That is, many environmental processes are non-stationary. They can have statistical properties, such as a mean value and variance that change over longer time scales. For example, in the case of global warming, the mean surface temperature of the Earth is trending upward due to anthropogenic greenhouse gases entering the atmosphere during the industrial age. Moreover, the variances around the mean temperature appear to be increasing as well. Non-stationarity is a complication to systems that seek to stabilize their operations within an environment.

In a similar vein, the uncertainty of future behaviors of entities, many processes are now considered to display characteristics of chaotic dynamics and fractal structures. Fortunately, chaotic behavior is generally found to operate in a dynamic attractor basin that keeps it from turning into what we normally mean by the word chaos. Both chaos and fractal structures seem to be reasonably bounded in one way or another. Some philosophers think that they are at the root of beauty in the world!

We will address these kinds of uncertainty in Sect. 11.3.4.4 Decision Model Refinement below in describing how learning agents adjust their decision models under such conditions.

Additional contributions to complexity include the sheer number of entities with which the SOI interacts or could potentially interact with. On top of that, there is the fact that the range of types of entities might be substantial—both the classes of entities and the particular representatives from any such class play a part.

Finally, all of the above contributions to complexity may be affected by entities that exist in the environment of the environment.[10] Remember that systems are always subsystems in some larger supra-system, and by induction, that supra-system is a subsystem of a yet larger supra-system. And larger in space means longer in time. We typically consider the SOI as impacted by its immediate environment by a kind of "light cone-like" wrapper; that somehow our SOI is not affected by events that are happening out of view—outside of the supra-system. But this is, of course, not a reasonable assumption for any phenomena taking place in the mid-world of human-scale phenomena. As a simple (and frightening) example consider the comet collision with Earth ~65 million years ago. It may have been set on course as a result of a small perturbation in the Oort cloud that took place thousands of years before the collision! Events outside the normal sphere of our Eco system could have a significant impact on the subsystems within. Figure 11.4 shows this as a set of concentric rings around an SOI representing the realms of influence. The inner ring represents the immediate environment of the SOI with various other entities (sources and sinks) directly influencing the SOI. Out from there is a ring representing what amounts to the environment of the SOI's environment—a supra-system wherein other entities (only one shown in the figure) have causal influence over entities

[10] See Mobus and Kalton (2015), Chap. 7, Sect. 7.4.6.5, for an introduction to the concept of the propagation of influences from outside the nominal realm of a system and its immediate environment.

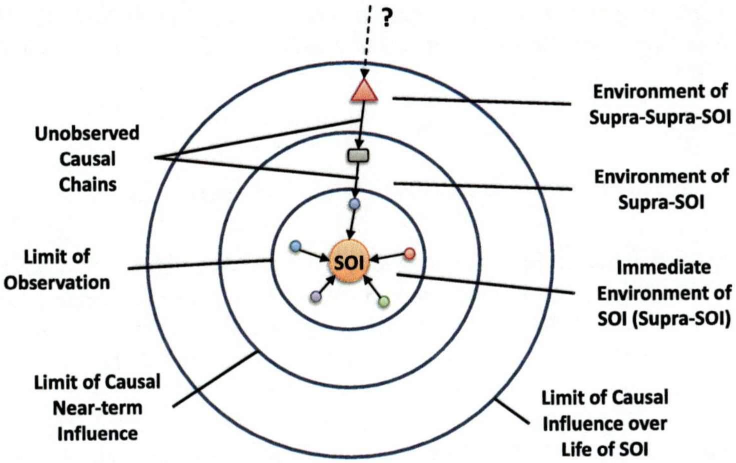

Fig. 11.4 An SOI is directly influenced by elements in its immediate environment; it is the limit of direct observation by the SOI. The immediate environment, by virtue of its stronger connections with the SOI, constitutes a supra-system that is embedded in a yet larger supra-system. Due to the temporal evolution of impacts from more distant entities, and the lifetime expectancy of the SOI, current influences from the greater supra-systems came from causal events in the past history of the SOI

within the SOI's supra-system. In other words, these entities form a supra-supra-system that may be outside of the observable world of the SOI but has indirect influence on the SOI through interactions with the latter's environmental entities.

The same argument extends outward, as per principle #1 in Chap. 2, every system is a subsystem of a larger system until we reach the extent of the Universe. However, the further out the source of influence is found, the longer distance the influence needs to travel and so events happening to the SOI that may have been the result of changes that propagated through the causal chain from far out is also far back in time. We, on Earth, are just now receiving the photons emitted by the Alpha Centauri star cluster 4.37 years ago!

The influences of distant events on the SOI could have started out being perfectly deterministic (though in general they probably are not!) But, by the time they reach the SOI in some altered form they are more a source of uncertainty than determination. This time/distance tracking issue is why we cannot make perfect predictions of what is going to happen to an SOI, what the state of the environment will be at any given instant. We rely on statistical analysis to extract patterns of behavior and estimates of future behavior resulting from unpredicted events. This applies even to simple SOIs because it is a property of environments.

As a more cogent example, consider the case of economic globalization being experienced today. Each nation might be thought of as a local SOI but having interactions with the larger regional and, ultimately, global economy. As this is being written, one of the hottest debates coursing through political rhetoric is the negative vs. positive effects of globalization on national economies, particularly labor markets. Each local economy has become subject to interactions with the larger economies that individuals or even firms cannot directly observe, yet feel the impact.[11]

11.3.2 Information Theoretical Considerations

From the perspective of the SOI, uncertainty in interactions with the environmental entities conveys information and for CAESs results in a change in knowledge in the SOI. A change in knowledge, in other words learning, is a change in structure that persists and leads to altered behavior of the SOI (Mobus and Kalton, Chap. 8).

A message conveys information when the receiver is uncertain about the message state (e.g., the encoding in the message) as discussed in Chap. 3. The amount of information in a message, in turn, causes alterations in the receiving system structure that allows it to dissipate future messages in that same state more readily. That is, the receiving system is much less surprised by the following messages and thus changes its own structures much less. Because of the second law of thermodynamics, the changes in structure, the knowledge, can never be perfect. And since the changes in structure involve embodied energies (e.g., covalent bonds) that can slowly "leak" away, thus degrading the new structure, completely dissipative knowledge is never perfect.

Knowledge is stored in memory structures in more advanced CAS/CAESs. Some of these memory structures have mechanisms for maintaining long-term histories of experiences and are stable for the duration of the agent.

The amount of information, and thus the processing load on the agent, is dependent on what the channels for message input are and their capacities. Or, in other words, it depends on what the agent can attend to. An earthworm has a very limited perception of the dirt through which it crawls. The information it receives is based on only a few limited senses (mostly of a gustatory or olfactory nature—sensing chemicals). It can undoubtedly "feel" its environment, but it does not perceive the complexity of soil. Hence, it does not need much of a computing engine or experiential memory to successfully deal with its surroundings. A human soil scientist, on the other hand, sees tremendous complexities in the nature of soils. The soil scientist receives a tremendous amount of information from observations of soils. The difference is that the scientist has many more means of perceiving and receiving messages to be processed than does the earthworm. Same environment but very different amounts of information.

[11] The nightly news notwithstanding. Individuals in advanced economies receive some information about their situations, but one might have difficulty verifying the correctness of the messages.

11.3.3 Computational Engine

The work processor of an agent is the computational engine that uses energy to process (transform) input messages about the state of the environment into output messages that will activate actuators to affect the future state of the environment (the double arrow from input environment to output environment in Fig. 11.1 is interpreted as both environments are one in the same configuration, but that the arrow from the output to the input environment implies feedback from the resulting state to the new input state after an appropriate time delay). The engine can also issue commands to change memory. In systems that can learn from experience, the model may also be altered as a result of computations.

In a cyber-physical system, like a mobile autonomous robot, the computational engine is one or more microprocessors. In a human agent, acting on its own behalf, the computational engine is the brain.[12] For an organization, the engine is a combination of human brains and computing machines.

The characteristics of the computational engine is that it follows algorithmic or heuristic rules given by the decision model to transform the inputs into appropriate outputs (Mobus and Kalton 2015, Chap. 9, esp. Sect. 9.1).

11.3.4 Decision Model

Decision science is a well-developed branch of both psychology and management science (Hogarth 1988; Montague 2006; Lehrer 2009; Schwartz 2004). It has been the subject of a branch of mathematics called game theory (see von Neumann and Morgenstern 1944 for the origin) since the mid-twentieth century. The field is vast and fairly complex, so we will not go into any detail here. We are concerned rather with the broad and general notion of a decision model used in the computation of decisions for action. These models, just as the computational engines, have many different instantiations in real systems. An online accessible introduction can be found in (Hansson 1994).

Our major concern is with decisions made in uncertain conditions, that is, in the face of imperfect information about the environment state due to the complexities discussed above. Since the whole purpose of a decision-making agent is to carry out actions that will preserve the system in the face of disturbances, decisions need to be made in a timely fashion even when complete information is unavailable. The ideal of making a decision is to find an optimal solution to a complex real-time set

[12] Interestingly in the human brain all three of the major components shown in Fig. 11.1 are combined in one fabric—the network of neuronal elements constituting primarily the cerebral cortex with some computational and communications support from the lower brain levels. In the classical Von Neumann computer model the three elements are separated as shown, though the decision model (program) is stored in a section of the general-purpose memory system. The rest of the memory is given to storing data generated during the computation.

of requirements and constraints. The computation of such a solution using an ideal decision model in time to have an effective output is the objective. However, the constraint of completing such a computation, even with an ideal model, within a reasonable time frame is always creating a less than optimal solution. One important contribution to the issue of decision-making in uncertainty and bounded by computational complexity and time was by Herbert Simon (1957), a theory he called "satisficing," in which the decision agent makes the best "estimate" of a solution that is satisfactory for the situation. It need not be optimal, but it does need to be good enough to keep the player in the game, so to speak.

11.3.4.1 Policies—Models

Decisions are to be made against the background of the goals of the system. That is, a correct decision is qualified by how well it brings a system closer to a goal. Goals are embodied in a set of policies or rules about outcomes—what should happen as a result of the agent's decision and action. A simple example is the policy of a living system to stay alive. On a seemingly more abstract level and larger scale, the policy of a for-profit organization is to take more resources (money) than it spends. The goal of the organization is to accumulate profits so as to be viable.[13]

In naturally evolved systems the policies are embedded in the structure of the system. For example, in a homeostatic mechanism, the set point represents the ideal value that a critical factor should take on. And that set point was arrived at through the process of natural selection. The policy is thus given in the system's construction and the procedures are embodied in the response mechanisms that act to restore the factor from a deviation.

As CAESs evolve new policies are either discovered or intentionally[14] determined (see the discussion of strategic management in the next chapter). There are two aspects of setting a new policy. The first is determining that a policy needs to be established as setting a (new) goal. The second is determining the policy mechanisms or procedures that are needed to enforce the policy.

This latter aspect becomes the decision model to be used. The procedures are built into the response mechanisms.

---

[13] Or, more to the point, to "seem" to be viable. We have already seen, in Chap. 9, that the goal of making profits in modern capitalistic economies is a gross distortion of the real needs of making a profit. The modern version is based on what we might call a meta-policy which is to maximize shareholder value (returns on investment). This distortion drives the agents toward less than wise decisions regarding the operation of the enterprise.

[14] Intentionality concerns a second order decision-making process in which a new goal is decided by the system, followed by the design of the mechanisms that will help the system achieve that goal. As described in Chap. 13, Design, even though a human engineer may exercise volition in choosing the specifics of the design, in many cases those choices are still basically tentative. The more novel the goal and mechanisms the more subject to tentativeness the design will be. That is, the new goal and mechanisms are exploratory initially and hence as much a product of evolution as natural mutation and natural selection are. The new goal has to be tested for efficacy.

11.3.4.2 Real-time Decisions

The decisions that need to be made regarding the operation of a work process are made in what we generally call real-time. However, we should note that there is no one absolute time scale that constitutes real-time. Rather, as per the mathematical definition of system given in Chap. 4 (see. Eq. 4.1), every system and its subsystems, have a designated Δt that defines the absolute time interval used to measure events (state changes) at that level in the system being defined. The Δt intervals for subsystems will always be smaller intervals of absolute time compared with their parent system. So, what we mean by real-time will depend on the natural time constants for the behavior of the system being defined as the SOI (at whatever level of analysis).

The agent(s) involved in controlling that behavior has to make decisions in relatively small increments of the Δt for the SOI. For example, the classical homeostasis controller (see Mobus and Kalton 2015, Chap. 7, Sect. 7.4.6.4.3 Resilience, see Fig. 7.7) must respond quickly relative to the rate of change in a critical factor in order to prevent that factor from getting beyond an acceptable range of value. Examples from biology abound (e.g., blood pH regulation), but the same principle is at work in the management-by-exception practice in organizations.

In all agent decision systems, there are a set of inherent time lags that can affect the effectiveness of the agent's actions. There are sensing and communications time lags. These should be minimized as much as possible. Real-world sensing devices take some time, called the "settling time," to respond to changes in the values they are measuring. The output of the sensory measure then takes some time to be communicated through the channel. Electrical wires and radio frequency modulation (e.g., WiFi) are very fast modes of communications. Neural signals through action potentials propagating along an axon are faster than hormonal signaling through the bloodstream. Signals reaching HSS government officials may take a very long time. Price signals propagating through the economy network (Chap. 13) may be extremely slow and subject to much noise.

Other latencies are due to the processing time required to reach a conclusion and take a decision. We have already alluded to this latency with respect to the time limits imposed on taking a decision discussed by Simon (1957). The computational latency depends on multiple factors, such as speed of the processor, the complexity of the decision model, and the adequacy of the sensory inputs. A second communications latency involves sending the final decision signal to the system actuator that will carry out the orders.

The actuator is also a source of latency that has to be accounted for. Real-world actuators, like sensors, require time to execute the orders. This latency depends on the power factor of the actuator—the more power applied, the faster the actuator can respond—but there are always practical limits to the amount of power that can be applied. It also depends on the inherent inertia of the system being changed.

Fundamentally, then, there is always a time lag between the time the sensor detects a difference in the value of the sensed critical factor and the time the response mechanism is deployed and effects a countering change in that factor. This is unavoidable and inherent. In the context of the evolution or design of agents and their sensing/actuating capacities those systems that act within the real-time values of Δt for the SOI will achieve stability and fitness.

11.3.4.3 Decision Model Construction

We look briefly at how decision models come into being as yet another instance of ontogenesis.

11.3.4.3.1 Maps

Decision models are, generally speaking, pattern recognizers. That is, they map the set of inputs at any instant to a particular pattern designation. The latter, in turn, is mapped to a specified set of outputs. The simplest such mapping pairs can be hard-coded into the decision model (the thermostat model). But as the complexity of the model increases, meaning the number of inputs and their ranges of values and the number of possible outputs and their range of values, it becomes increasingly difficult to determine the actual mapping function. Moreover, as stochasticity increases the task of providing a satisfactory mapping becomes intractable.

For such cases, nature has evolved brain structures, cortices, with flexible connectivity between input and output nodes to handle the complexities and stochasticity. We say, flexible, because these networks allow for remapping when novel or changed conditions warrant. More on this below.

Engineers, seeking to emulate nature, have turned to artificial neural networks (ANNs) for their ability to "learn" mappings from inputs to outputs usually under some form of supervision, that is where the engineer determines the correct mappings and supplies the network with samples during a training period. The beauty of using an ANN is that one does not need to know the details of the mapping relation. One merely pairs input samples with desired output designations and lets the ANN "learn" the best mapping. The word "learn" is put into single quotes because what is going on is definitively not learning in the way that a biological neural network adapts to its inputs. We will address biological learning in the next subsection.

Even so, ANNs have provided good solutions to many agent decision problems where pattern recognition and classification lead to reasonably good decisions within a restricted problem domain. ANNs are robust mappers in mildly stochastic domains. They have also been shown to be useful in some mildly non-stationary environments, such as a trending condition over a long period of time where incremental adjustments to the map can be made.

But if engineers ever hope to emulate real animal-like decision models they will need to find mechanisms that can actually learn from experience, without a teacher or supervisor, can learn associations in a single instance if needs be or at least not require hundreds to thousands of pairing instances to produce a mapping. And most importantly, the decision model must be able to modify the mapping on the fly, in real time. This remains the holy grail of learning agent research.

11.3.4.3.2 Inference

Higher-order agents not only have basic models that map from inputs to outputs for rapid decision-making, they may also be able to compensate for information input deficits and/or "holes" in the maps they possess by constructing chains of reasoning. This is a non-experiential mode, that is, not based on the model's current memory. Rather, using some set of inferencing rules the agent may produce a novel mapping for conditions not previously encountered, a conditional mapping that needs to be tested before committing to permeance. Here we will consider just three forms of inference as examples of how agents might construct a chain of reasoning and, thus, a model for use in future decision-making. The first is the most reliable in terms of the veracity of its conclusions, but also the least "practical" in ordinary agent reasoning (and here we refer primarily to human thinking). It has, however, found its place in algorithmic computational systems. The second relies heavily on experiential modes. It is generally reliable given enough experience, but is not as well-founded as the first. Finally, as an example of a very common form of human inferencing we provide an example of what has been called backward or reverse inferencing, that is from effect to cause.

Deductive reasoning is the most infallible means of inferring the case of the matter or the state of affairs. Unfortunately, it relies on sets of rather rigid rules and premises that have to be taken on face as being "self-evident," so-called axioms. Where these axioms come from and why they are considered true is not easy to explain. Too often, the set of axioms, at least in human reasoning processes, can look like ideology—there is probably a strong relation between these notions. In the fields of mathematics and logics, this approach works extremely well since the axiomatic bases of various threads of deduction seem to be quite literally self-evidentially true. In the realm of messy human decisions (wicked problems) this is rarely, if ever, the case. Sherlock Holmes notwithstanding, deductive logic seems to play little role in ordinary human decision-making, but where the premises are relatively simple, it can be useful.

Inductive reasoning is one of the most relevant modes for humans to arrive at models that support reasonably (satisficing) sound decisions. In this form of reasoning, a model is constructed based on experience, or more to the point, repeated experiences that can lead to a generalization, a rule of thumb, essentially.

Abductive reasoning can be thought of as the reverse of inductive reasoning in that given an observation of a phenomenon one works backwards through possibly several chains of cause-effect relations to identify the most likely prior cause of the current observation. The question is, given the current observation, what happened just before that would result in the observed state of affairs? And then, what prior to that would have resulted in that state of affairs, and so on. Abduction, unlike deduction, is not probably correct inference. There can be many explanations for the state of affairs, and some considerable amount of context information may be required in order to raise the likelihood that it provides a veridical inference.

This form of reasoning may involve several other inferencing techniques such as metaphorical or analogical construction of a candidate model. It may also involve imagination. These are not unlike subroutines called upon to supply tentative knowledge/model pieces.

These three methods for inferring missing knowledge about states of affairs give human decision makers considerable autonomy.

11.3.4.4 Decision Model Refinement—Learning

As suggested in the section on Inductive reasoning, above, and Sect. 11.2.2.2 Decisions in Living Systems, and Fig. 11.3b, the real essence of animal decision models is based on three kinds of memory encodings. The first is the linking of decision nodes in the tree-like structure, that is from any given node what are the options for the next node? The second is to attach some kind of hint or valence to indicate what the most favorable selection would be based on past experience. And the third is to have at hand a knowledge milieu or subconscious model of the context in which the decision is being made.

In real-world environments, as we saw in Sect. 11.3.1, Uncertainty in Interactions, it is possible that no two situations are always the same due to the influence of factors on the near environment that are further away in time and space. The real environment must be considered non-ergodic and hence non-stationary, indeed the worst kind of non-stationarity, what is called nonhomogeneous non-stationarity. Homogeneous non-stationarity is where the statistical properties of relations may change over time, but they change in a somewhat predictable fashion, like a trend line. Nonhomogeneous non-stationarity, on the other hand, is generally speaking, unpredictable. The statistical properties may be varying in one direction at one point in time and then reverse course at another point in time. The homogeneous kind may be due to some systemic bias or an external force that is stable over time. The nonhomogeneous kind may be due to chaotic dynamics. Particularly pronounced nonhomogeneous changes can come in what we call "catastrophes" in the technical sense.

The fact that much of the dynamics in the real world is non-stationary requires that agents be able to adjust their decision models to accommodate the changes.

Before humans had evolved, animals had the capacity to adapt to mildly homogeneous non-stationarity, for example gradual climate change due to natural causes through biological evolution. Human evolution produced animals that could actually deal with nonhomogeneous non-stationarity, such as, for example, intermittent draughts, with their capacity for mental evolvability. They could adopt new behaviors when conditions changed.

11.3.4.5 Role of Judgment in Human Decision-Making

For many agents in many environmental situations, the decision process can be more or less mechanical. If a mapping exists (either deterministic like a thermostat or non-deterministic like a Bayesian neural network (below)) then the computation is relatively straightforward. Given the state of the input messages you choose this particular output.

In Fig. 11.3b, we introduce the role of a generalized influence on decision-making in the form of what is called tacit knowledge present in the knowledge milieu surrounding decision nodes in the network (e.g., a decision tree). Tacit knowledge is the kind of knowledge that we accumulate with various life experiences that seem to have similar attributes and that our brains automatically integrate into our subconscious (possibly during sleep and REM dreaming). One of the more easily understood forms of tacit knowledge is procedural knowledge or "how to do something," which builds up over time and repetition (practice). Procedural knowledge is the basis for "expertise," the capacity to do a job (or ride a bike) without having to think about how to do it. In fact, most forms of tacit knowledge cannot be captured in language or even brought to conscious awareness.

Tacit knowledge is thought to be at work in conditioning decisions by providing background knowledge that can be applied to the current situation, which resembles the many historically experienced situations of the same sort. Perhaps, tacit knowledge is at work in what we call intuition. The latter is a vague feeling that one should go one way as opposed to another even when all of the other cues and valences suggest otherwise.

Some people are better at listening to their intuitions than others. And some people have effective (good) intuitions. Such intuitions have been associated with the kind of supra-intelligence we call wisdom (Mobus 1999). Wisdom is the ability to make appropriate choices in the face of uncertainty and ambiguity (both factual and moral) that will prove efficacious in the long run (a rich literature on wisdom can be found in Sternberg 1990a, b). The problem seems to be that real wisdom is rare and often relegated to elders who have had long lives to accumulate veridical tacit knowledge. Thus, it seems to be the case that human agents can not necessarily be counted on to make the wisest choices as individuals, and especially at young ages. As Surowiecki (2004) points out, the wisdom of the crowds, too frequently touted as making up for individuals' lack of wisdom, is also limited to specific kinds of problems/questions that favor a statistical estimation. Democracy (or representative forms thereof) is touted as relying on the wisdom of crowds to select the best governance agents. But experiences in recent years in a number of democratic nation-states should call this assumption into question. This issue will be raised again in the next two chapters with respect to human society governance and economy archetypes.

11.4 Agency

Once a decision has been made the agent acts. The ability for the actions taken to have an efficacious effect on the external environment (e.g., the problem to be solved) depends greatly on the agent's possession of the "power" to affect that part of the environment that needs to be changed. That is to say, the agent requires effective agency in order to make a difference. W. Ross Ashby (1958) introduced what he called the "law of requisite variety" or an agent having the degrees of freedom to act so as to produce a meaningful control action. In a cybernetic feedback loop, the controller needs to have at least as much "variety" or range of actions as that of the physical system to be controlled may vary from its ideal status. In other words, for every force, there needs to be an opposite and equal counterforce that can reestablish balance or dynamic equilibrium.

11.4.1 Affective and Effective Action and the Subsequent State of the Environment

The decision made by an agent is translated into an action that is meant to affect the state of the environment. And that change must be effective in the sense that it improves the situation for the agent or at least minimizes any negative consequences resulting from the state of the environment.

Agency begins with the capabilities of the sensory apparatus used to monitor the environment. Living systems have sensory arrays in several modalities, starting with the ability to detect specific molecules as well as determine the gradients of concentration. This capability is present even in bacterial cells. Going up the evolutionary ladder, organisms evolved additional modalities based on the natural availability of information carrying message, say in sound waves and reflected light. Evolution has produced a matching between the sensory acuity and the agent's capacity to discriminate changed states of the environment that could have an impact on the future state of the organism. In psychophysics, this discriminatory capability is known as "just noticeable difference" or JND. As Bateson famously put it: "A difference that makes a difference" (2000 [1972]).

In designed artifactual systems (as will be covered in the next part of the book), the engineer must carefully consider the sensory capabilities relative to the effectiveness of the action choice that will be made by the agent. Given the state-of-art in sensor technologies today this is generally not problematic.

Sensory systems have evolved to be quite complex in their own right. Often in living systems the sensors of a given modality are set in two-dimensional arrays wherein every sensor is capable of measuring its relevant parameter across a field that intersects with its near neighbors. This is a neat trick that improves acuity even when individual sensors are noisy or not that accurate. However, the information collected in the array needs to be processed considerably in order to extract the topological correspondence of variations in the image projected on the array.[15] Many algorithms have been developed for such image extraction in artifactual systems and the cortical structures of brains provide that capacity in animals.

The effectiveness of the agency also depends on the efficacy of the decision model itself. Once the state of the world is ascertained the model is consulted to determine what, if any, action should be taken. Maybe none is needed. The model in anticipatory agents will provide information on the expected next state of the world if no action is taken in the current time step. The model will also project forward as the consequences of the future state that will be obtained in that case. Then, if the projected future state should be acted upon preemptively, the appropriate action can be selected and the agency can provide an advantaged move in the case of a possible good outcome, or take evasive action in the case of a bad potential outcome.

Again, in living systems the models that affect these capabilities are the result of brain evolution. That is, up until humans. While many mammals and some birds are found to be able to anticipate situations that are either favorable or dangerous and take effective actions, the time range for these anticipations are generally measured in minutes or hours at best (some evidence that some great apes can demonstrate anticipatory behaviors over days has been advanced, c.f., De Waal 2016, Chap. 8). Humans go much further in being able to anticipate future states in weeks, months, and in some cases, years in advance. Moreover, the state variables that humans work with vastly exceed those of other animals.

In artifactual systems, engineers are not yet able to design anticipatory decision models except for very limited numbers of state variables and short time scales. Consider the case of the autonomous automobile. The models of road conditions, pedestrian behavior, and so on, are quite extensive. But they do not anticipate completely novel situations based on temporal correlations anywhere near the huge state variable space as available to human decision-makers.

Assuming that the agent can make a "good" decision to take an appropriate action, it only works if the agency includes an adequate range of reaction and the actuator power to affect the change in the environment needed to improve the success of the agent. Again, evolution equips living systems with the motor capabilities to fulfill this requirement, at least on average. In a non-homogeneous non-stationary world, there are always possible situations that can arise for which the agent does not possess sufficient agency to respond adequately. It is in such cases that injury or destruction to the system in which the agent is acting can occur. This is why in biological systems there are many copies of the same system (individuals in a population for example) with variability in the ranges of their agency. If some individuals have agency capacity to handle the situation they will survive and replicate their capacity in their offspring.

[15]This applies to all of the exteroceptive modalities as well as a few interoceptive ones. We use the term image generically to refer to the variations from one location on the array to another across the array.

11.4.2 *Feedback Through the Environment and the Causal Loop*

Effective agency should affect the state of the environment. But this means that the environment enters a new state in which other agents or processes also react and make choices and take what for them should be an effective action. Which then means that the state of the environment changes again for the agent of interest. There is, thus, an endless loop of state changes, which will, in complex environments with a number of other complex entities, be highly stochastic. Any agent's actions will alter the environment of other agents, each of which will respond according to their models and agencies. The environment is, thus, again changed for the agent of interest who must go through the process of decision-making again. It never really ends. The loop may be temporarily suspended when the agent sleeps at night, but even then, for advanced animals anyway, the brain goes into housekeeping mode, consolidating new information and updating the model as needed while throwing out any memory traces that do not contribute to making the model more efficacious in the future. Individuals cannot alter the efficacy of their sensory or affecter capacities, but they can improve their models to some extent, based on their experiences while in the active loop. If a choice made in some situation, based on the then current model, works to the advantage of the agent, then the strength of the pathways leading to that choice (as depicted in Fig. 11.3) will be reinforced, if not, then that pathway may be weakened for future reference.[16]

Supra-systems composed of myriad systems, what we have taken to be equivalent to a particular system of interest's environment, is a complex network of causal loops, all agents involved affected by all others to one degree or another. Natural ecosystems, as one example, evolve to a dynamic stability by virtue of the agents within being adaptive over long time scales. Human societies do not evolve to such stability because individual humans are adaptive to a degree, but they also are evolvable and motivated to change things without necessarily understanding the consequences of those changes. Human societies either grow or collapse. Indeed, all past societies that deigned to grow while they could, eventually did collapse or dissolved away (Tainter 1988). We need to understand human beings as agents in this light.

The great bulk of human decision-making depends on the nature of the mental models, both explicit and implicit, that each individual has constructed over the course of their lives. Put rather simply, two humans can have wildly different models of the same system in the world and thus different beliefs about the world and subsequent biases in their decisions regarding that decision. Perhaps the paradigm example of this is political beliefs. There seems to be a strong polarization between so-called conservative and so-called progressive views of how the governance and economics of a society should work. People from both perspectives have lived in the

[16] In animal learning, negative experience does not necessarily lead to forgetting the pathway entirely. It may be saved, but weakened, just in case the last experience was an exception. See Mobus 1994, 1999.

same world (especially the case with the advent of extensive international commerce and media) and observed the same phenomena to some degree. But both have derived different meanings from those observations. There is some growing evidence that there is a brain basis, itself based on genetic differences regulating brain development, for perceiving the world starting from these two poles. Thus, we can expect that as a person develops in life, their perceptions and consequent models of how the world works will become increasingly supportive of their innate political persuasion.

We will come back to explore the situation of society governed by human agents, and particularly their foibles, in the last chapter of the book.

11.5 Complexity and Autonomy

Complexity of systems and environments is correlated with levels of organization (using Simon's hierarchy of organization concepts, 1998, and see Chap. 4, Sect. 4.3.3.7.1., Simonian Complexity). Biochemistry is more complex than inorganic chemistry. Multicellular organisms are more complex than their cells. Societies are more complex than a population, and so on. As we proceed up the physical hierarchy from atoms to societies the degrees of freedom in choices increase tremendously.[17] Behaviors are more lawful at the simpler levels, somewhat rule-like at intermediate levels, and seemingly historical accident at the highest levels. So we find that agents gain autonomy as we rise in the hierarchy. Atoms have no autonomy whereas societies are free to choose any conceivable path to behaviors not explicitly denied by the laws of physics. Human societies can choose to pollute the environment no matter the consequences, until, of course, the physical consequences are realized.

Thus, there seems to be a relation between the complexity of a system and the autonomy that system wields in dealing with its environment. The complexity concept applied to agents involves, primarily, the complexity of the decision model it uses to compute a decision. And that complexity reflects the complexity of the environment that the agent must contend with. Simple agents, such as thermostats, have correspondingly simple decision models because they deal with a simple environment (and few variables). And such agents have no autonomy whatsoever. As the system within which an agent operates, its supra-system (that is the CAS/CAES it is part of), engages with more complex environments, the agent's decision model must also become correspondingly complex, what Ashby (1958) called "requisite variety." That agent displays a higher level of autonomy. Take, for example, an amoeba "crawling" around in a pond' bottom muddy environment looking for food.

[17]This reflects the amount of information that an agent must process. The more choices that an environment presents (another way of saying the number of states that a meta-system can be in) the more information is needed to process that choice.

Its food-seeking control mechanism needs to have the agency to extend a pseudopodium to sample some portion of the amoeba's environment. It must also have the agency to withdraw a pseudopodium in the event that some chemicals in that region are noxious. What we see in the overall behavior of the amoeba seems like trial-and-error searching, implying some role for uncertainty. It also implies a level of autonomy in that the amoeba can "choose" a direction to probe.

The author's own work on autonomous robots illuminated this aspect of autonomy (c.f., Mobus and Fisher 1999; Mobus 2000).

These papers describe the emergence of a strange attractor dynamics in a neural oscillator, known as a central pattern generator (CPG). Figure 11.5 shows an artificial neuron-based CPPG that generated motor control signals for the MAVRIC robot (ibid).

The circuit had been designed with the intent that it would produce a smooth zero-centered sinusoidal waveform to drive a foraging robot to and fro as it looked for stimuli. The positive values would drive the left wheel of the robot while negative values would drive the right wheel. Instead, the circuit produced a "noisy" sinusoid (Fig. 11.6) that caused the robot to conduct what we called a "drunken-sailor-walk," not quite random but also not quite directed. Analyzing the data generated by the CPG we discovered the strange attractor dynamics (Fig. 11.7). The pattern that had emerged was serendipitous in that the search approach that it drove

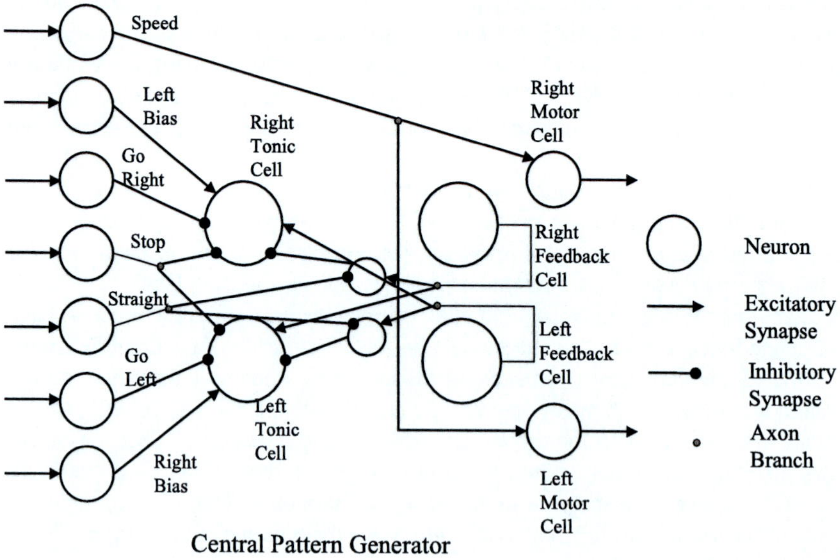

Fig. 11.5 An artificial neuron central pattern generator. From Mobus and Fisher (1999) with permission

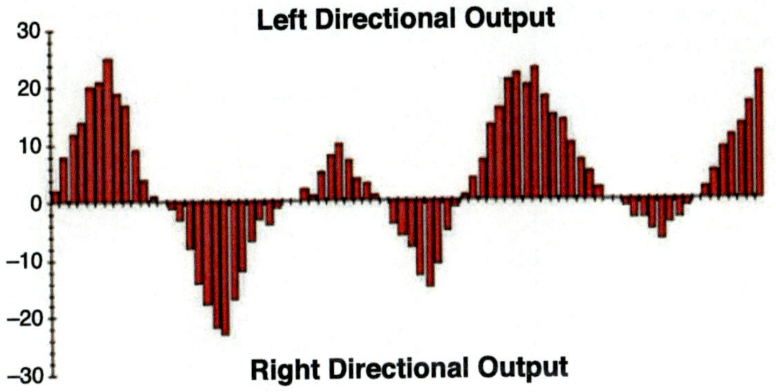

Fig. 11.6 A noisy sinusoidal-like signal produced by the CPG in Fig. 11.5. From Mobus and Fisher 1999 with permission

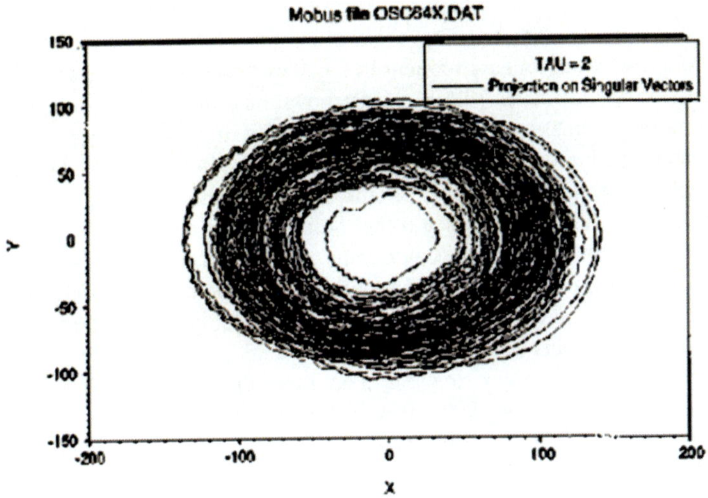

Fig. 11.7 The strange attractor basin shown in a phase space map from the extended data set of what is shown in Fig. 11.6 above. From Mobus and Fisher 1999 with permission

improved the robot's chances of finding the sought stimuli over a pure sinusoid. Additionally, we discovered that it replicated the actual kinds of search patterns found in foraging animals. This experience showed that we should expect the unexpected, but possibly better, behaviors when our system designs include adaptable components (neurons with adaptive synapses).

Figure 11.8 shows a sample of traced paths followed by the robot as the left and right drive wheels alternated in speed following the kind of signals seen in Fig. 11.6. The reader may now see why we chose to describe the robot's search pattern as a "drunken-sailor-walk."

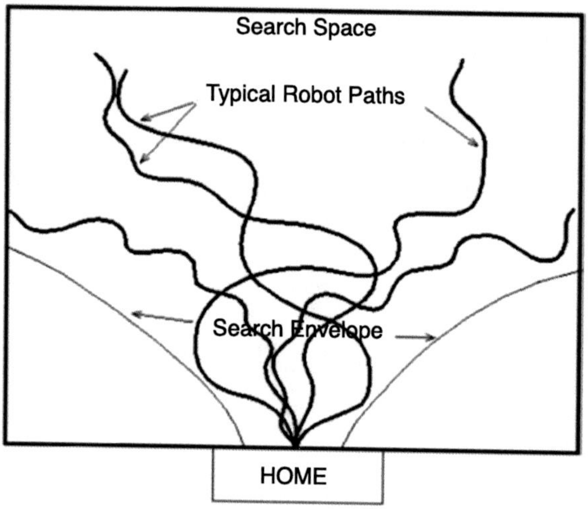

Fig. 11.8 Showing a sample of typical search paths taken by the robot. From Mobus and Fisher 1999 with permission

The robot searched its environment in a pattern that mimicked that of foraging animals (Mobus and Fisher 1999). It was shown how this kind of search pattern, versus a more systematic or a random walk, actually improved the robot's chances of encountering stimuli of interest (e.g., "food") before its batteries depleted.

This kind of search, like that of the amoeba, demonstrates the fact that a certain amount of stochasticity injected into the more complex decision model increases the agency and thus the apparent autonomy of whole system. If a simple neural network like a CPG can make a robot appear to be autonomously searching its environment, is it any wonder that the human brain can produce significant autonomy? The models that humans use are extensive, adaptable, and evolvable.

The robot searched its environment in a pattern that mimicked that of foraging animals (Mobus and Fisher 1999). It was shown how this kind of search pattern, versus a more systematic or a random walk, actually improved the robot's chances of encountering stimuli of interest (e.g., "food") before its batteries depleted.

This kind of search, like that of the amoeba, demonstrates the fact that a certain amount of stochasticity injected into the more complex decision model increases the agency and thus the apparent autonomy of whole system. If a simple neural network like a CPG can make a robot appear to be autonomously searching its environment, is it any wonder that the human brain can produce significant autonomy? The models that humans use are extensive, adaptable, and evolvable.

11.6 Conclusion

Agents are the special kind of information-processing work processes those make decisions and drive actuators to effect change. As processes they , themselves, are systems and can be as simple or complex as needed for the kinds of decisions they need to make. At the high end of complexity, they can also be adaptive and evolvable, full CAES status, as in the brains of birds and mammals.

Decisions need to be made whenever a system or a component of a system receives multiple variable inputs and it has to choose an appropriate action to produce the appropriate outputs at appropriate levels. The Agent/Agency archetype model is the basis for modeling all of the kinds of details needed by agents and media will vary from instance to instance, but all real agents will follow the same basic archetype.

And both of the other two archetype models, to be discussed in the next two chapters employ the agent archetype extensively; are, indeed, only possible because of the agency described in this chapter. Now we will see how agents are employed in governance and economic decision-making.

Chapter 12
The Governance Model

Abstract The key to the success of a CAES in continuing to be viable within its embedding supra-system is that it is capable of stable management of its economy, and has a capacity to adapt to "normal" changes in that supra-system. Further, to be viable in a non-stationary changing environment it must have an ability to modify itself when those changes are not normal, that is, it must be able to evolve strategically. The key to all of this is the efficacy of the *governance subsystem*. A governance system based on the hierarchical cybernetic governance system and employing veridical agent decision-making is critical to maintaining a stable and sustainable whole system. In this chapter, we examine the nature of a governance architecture, employing the HCGS model, that keeps the economy operating efficiently and effectively, and the activities of the decision agents (e.g., human beings in the human social system) so that the whole CAES functions in fitness with respect to the supra-system.

12.1 Purpose of This Chapter

The complex adaptive and evolvable system (CAES) framework based on principles of systems science allows us to approach the concept of governance in a fundamentally new way. The traditional methods of political science and management science attempt to understand governance by deconstructing and parsing existing and historical representative versions of governments and organizational management.[1] No doubt there is great value in approaching the understanding of governance from these approaches to analysis insofar as providing an "anthropological" view of what we humans have done over the course of our cultural evolution. But this is also a narrow, anthropocentric-oriented view. And it assumes that historical accounts will help us get to a deeper understanding of what governance is all about. It must be understood that the various forms of governance and organizational

[1] In general, we will subsume management under governance, the latter term being an umbrella for all forms of system control.

G. E. Mobus, *Systems Science: Theory, Analysis, Modeling, and Design*,
https://doi.org/10.1007/978-3-030-93482-8_12

management have all been experiments based on best guesses as societies and organizations continued to experience two related driving forces. One was the increasing scale of populations and the other was the increasing complexity brought on by cultural evolution, itself driven by growth in technological advances. These experiments and the "isms" that attended them have all fallen short in many different ways, so to study them as the source of understanding the systemic nature of governance cannot succeed. We will be looking with new eyes.

When we invoke systems science approaches, we do not start with the human experience but rather ask what kind of system is a governance subsystem. And we then look to the whole of nature to provide the insights and answers. Other kinds of CAESs include governance subsystems in order to achieve sustainable existences. We should ask: What is going on in these other kinds of systems and what can we learn from them to guide our own efforts to design a systemic and workable governance subsystem for the human social system (HSS)?

By examining the governance of CAS/CAESs we might hope to get guidance for how we should be approaching our social and economic subsystems governance. We treat the governance subsystem and its archetype model as a template for how the HSS could achieve a more systemic design for its government and organizational management. To the point, the examination of a range of CAS/CAES types from the earliest life, through multicellular organisms, through groups of organisms, and finally to early groups of humans reveals a common architecture for governance structures, that is, the structures that embed the agents described in the last chapter, and achieve the goals of stability and sustainability that make systems viable. All previous systems achieved those goals through, fundamentally, evolutionary processes. What worked survived, and what did not, did not.

Human societal governance, which means governance of the economic subsystem (to be described in the next chapter) and the interactions between human agents, has also undergone significant evolutionary changes. We describe this as cultural evolution. Cultural evolution is significantly different from biological evolution yet shares some generally similar features. Both are forms of ontogenesis but the latter relies more heavily on purely stochastically generated variation and auto-organization whereas the former involves "intentional-organization." We will explore this concept more fully in the next part of the book. For the moment we will only state that with human consciousness we recognize a major phase transition in the way the Universe evolves (recall Chap. 3, Sect. 3.3.1.1). Humans invent new ways of doing things based on their capacity to anticipate future states of the world with that new way compared to that future state without it. They then go about an exercise in problem solving to obtain the new way (or tool) and, if successful, set about implementing it. However, this process is still subject to stochasticity. It is subject to mistakes. And it is more-or-less incremental, that is, by way of improvement over the old way. We submit that the history of inventions of governance architectures without the benefit of systems science has been riddled with many blind alleys that resemble the blind alleys we see in biological evolution (tree of life depictions). What we propose here is that with the benefit of systems science we might be able to minimize the accidental design of a governance architecture and discover a design approach that is far more likely to succeed in providing the HSS with a sustainable future.

In this chapter we give a deeper description of the hierarchical cybernetic governance system (HCGS) introduced in the context of complex adaptive and evolvable systems discussed in Chap. 10. The HCGS is that governance/management subsystem that internally regulates the behaviors of subsystems within the larger complex system. We submit that this architecture will turn out to be a better approach to the governance of the HSS, and will attempt to show this in Part 4.

Agents make decisions and take actions in the context of governing the workings and behaviors of complex systems. In Chap. 10, we elaborated the model of a complex adaptive system (CAS) and a complex adaptive and evolvable system (CAES) and introduced the role of governance as that process that maintains near-optimal operations and coordination of the CAS/CAES with its environment, its sources and sinks, and mitigates the effects of disturbances. We introduced the concept of a hierarchical cybernetic governance system (HCGS, see Fig. 10.2) that partitions decision types, as described in the prior chapter, into levels corresponding to time scales of decision-making and action.

In this chapter we will elaborate the nature of the HCGS as a generic governance model archetype that can be used to guide the analysis of CAS/CAESs and/or their design. This is a further elaboration of the governance model introduced in Mobus and Kalton (2015, Chap. 9).

Our thesis is: *The structure and operations of an HCGS when enlightened agents are working at the decision nodes, is what makes a CAS/CAES stable and sustainable* (Mobus 2017). By enlightened we mean agents that make decisions based on reasonably veridical decision models as described in Chap. 11 and without biases or hidden agendas. We will show examples from natural systems that have evolved HCGSs in which the decisions made are based on constrained models that correspond with the actual environments in which they operate. Of course, of greatest interest to us is the governance of human activity systems such as municipalities, businesses, non-governmental organizations (NGOs), nation-states, and, ultimately, the whole human social system as introduced in Chap. 7.

Meuleman (2008) provides an argument for the notion of generic governance archetypes. He reviews what he takes to be three different generic architectural patterns of what he calls meta-governance structure/dynamics.[2] He categorizes these as: hierarchical (as in top-down command-and-control), networks (depending on trust and cooperation among decision makers), and markets (using transaction-centric measures of performance and exchanges of value as discussed in Chap. 9 and again in the next chapter). He refers to these as "styles." Examinations of both public and private governance subsystems reveal that organizations may tend toward one of these architectures being dominant. Militaries, for example, are usually strongly hierarchical whereas small- to medium-sized high-tech companies lean

[2] Recall from Chap. 10 that we elected to not use the meta- prefix for our generic models. Rather we put these models forth as archetypes found to be isomorphic at all levels of complexity and organization among CAS/CAESs.

more on networks. The economic system is argued to be governed by market mechanisms. More generally, and especially when we look at natural governance in living systems, we can find all three of these superimposed on the organization. Meuleman argues that this is, in fact, what should be expected as each style addresses different mechanisms for achieving cooperation and coordination. The optimum in management is achieved when all three mechanisms are used. He argues that it takes a strong form of "metagovernance" to make sure these architectures work in consonance with one another. We assert that the HCGS to be described here (and covered also in Mobus and Kalton 2015) provides exactly the kind of meta-governance architecture that naturally incorporates all three of Meuleman's styles.

There is a fundamental concept that needs to be addressed before going into the details of an HCGS structure, the core principle of cybernetics, using information feedback and feedforward to make adjustments in processes. And we will do that below.

Complex, dynamic systems interacting with a complex and generally non-stationary environment need to have means for adapting to ordinary fluctuations in that environment. But even more importantly, systems that interact with non-stationary dynamics need to be able to undergo evolutionary changes (that is, heritable modifications to phenotypes) internally in order to sustain their existence into an indefinite future. Systems not possessing mechanisms for internal self-regulation as well as coordination with environmental entities will not long persist in the real world.

The core of living systems is the metabolism of single cells.[3] Cells are CASs that have mechanisms for adapting to moderate changes in their environments. Multicellular organisms are also CASs up to the evolution of the hominids with larger prefrontal cortices. Some great apes show some capacity to form new concepts when in captivity in novel environments (De Waal 2016), but as of the present, the evidence suggests that only human beings have the capacity to think truly strategically, that is, for long time scales into the future and broad spatial scales. The human brain is thus an example of a CAES. The brain, especially the neocortex, makes individual humans evolvable systems. Prior to human existence the natural world's degree and rate of changing was relatively low over long geological periods, thus Darwinian evolution accounted for adaptive changes in species; only species and higher genera were CAESs with evolution providing the strategic component. Occasionally, cataclysmic events occurred that were fatal for some substantial portion of species, but not all. In general, the core of life has been maintained somewhere on the planet such that life has been preserved.[4] With the advent of human culture, with artifacts and technologies, the Ecos has become strongly non-stationary. The changes taking place in the Ecos due strictly to human activity are

[3] In the following chapter, we show that core metabolism is a generic economy model for life.

[4] Indeed, such cataclysmic events, such as the meteor or comet that is thought to have struck the Yucatan peninsula 65 million years ago, causing massive die offs and extinctions create opportunities for surviving species to undergo adaptive radiation subsequently. Life manages to flourish in spite of non-stationary environments.

now being considered sufficient to rename the current epoch as the "Anthropocene." Those changes are taking place rapidly compared to non-human caused changes that take place on geological time scales.

We have already used the concepts of the governance model in several previous chapters (e.g., Chap. 9) using the model archetype to guide our gaining of understanding of the HSS economy, or at least a subsystem of the economy. Now, all of the methods of systems analysis and modeling discussed up to this point will be used here to gain understanding of how very complex systems manage to maintain their existence in a constantly changing world. We present a general model of management and governance subsystems for larger complex adaptive and evolvable systems (CAES). We then will demonstrate how these concepts work in both natural and man-made systems.

As complex systems are being analyzed, the questions of how a system is self-regulated internally and how it manages to interact successfully with its environment guide the examination of its governance architecture and management subsystem. This chapter will provide the basic structures and functions that are part of every such system. What we propose is a step toward a general theory of governance.

CAESs are found to be composed of subsystems that have *agency* and degrees of *autonomy* as discussed in the previous chapter. An agent, as described in that chapter, is a special information processor that can take decisions and behave in variable ways depending on their information input along with the decision models with which they operate and a computational engine for doing the processing. Decision agents are not capricious or should not be. They follow a set of guidelines (the decision model) for arriving at decisions for what to do next. This does not mean, of course, that their model is necessarily veridical. Indeed, as we will argue below, in the case of human agents there are too many times when the models are contaminated by a number of biases and flaws that make them suboptimal—a major source of what goes wrong in societal governance. Their behavior, in turn, however, has causal impact on other components in the system including, and especially, other agents. Every work process in any CAS or CAES will have an agent decision maker with sufficient autonomy to make the decisions that should, in principle, be able to maintain optimal operations of that process, subject to constraints imposed by interactions with other work processes and agents.

Cellular metabolism is fundamentally working at the operations level (recall from Chap. 7 that metabolism is the first form of an economy model). Decisions are made in the sense of homeostatic regulation. The agent model that applies generally is the homeostat (see below, Sect. 12.3.2.1, "Feedback," for the description of a homeostat) using error feedback to signal a response. It is implemented in biochemical reactions.

However, an argument can be made that some coordination-level decisions are also involved. For example, the production of a specific protein is regulated by the relative concentration of messenger RNAs (mRNA, ribonucleic acid) for that protein in the cytosol near ribosomes. That concentration changes more slowly and the mRNAs are manufactured by enzymes reading out the code from the gene's DNA (deoxynucleic acid). And that rate of read out was triggered by demand somewhere in the cell for the protein needed. In other words, a deficit of that protein somewhere in the cell sets in motion a chain of communications that result in the increased production of the mRNAs, which results in increased production of the protein and eventually reestablishing the needed concentration. Similarly, tactical decisions are made when the cell receives messages from its environment, say a particular molecule that is an agonist activating a membrane channel allowing a current of ions to flow into the cell and set off a cascade of signaling mechanisms as well as work activities, for example, the activation of a postsynaptic membrane by the receipt of neurotransmitter molecule. Penetrating channels through the membrane open to allow the influx of sodium ions and that leads to depolarization of the membrane. If the depolarization event is sufficiently strong it gets propagated outward from the synapse and if reinforced with other similar depolarization events may lead to the neuron firing off an action potential that will propagate along the axon to signal other neurons. Thus, the neuron is a sophisticated information processor. Many more examples of cells interacting through tactical mechanisms are available.

In multicellular organisms, we can similarly point to many instances of operationallevel and coordination-level decision mechanisms. In most cases, for example, the endocrine system, these are implemented in biochemical interactions, but now at the level of tissues and organs. In some cases, that system is triggered by brain processes, that is, computation that more closely resembles what we usually think of as an agent.

In all of life up to the human species, the strategic level of agency is subsumed in the process of Darwinian evolution. Nature makes the de facto strategic decisions through selection of what works best in any particular econiche. Humans are still subject to evolution but not entirely dependent on it.

Humans are the epitome of agency and the need for management and governance of human organizations, from families up through nations, is certainly well recognized. Not only are human beings agents, but also their degrees of freedom in behavior are nothing short of spectacular. Thus, whatever a human decides and does can produce significant amounts of information in the system, that is, they can generate a great deal of surprise to other humans. Humans can also suffer information overload, or data processing jambs, that impair their decision-making capabilities. Humans rely heavily on internal tacit and explicit memory, in the form of concepts, to model what they have experienced in the world. Memories can be slow or faulty and lead to information overload as well.

Finally, humans are subject to sensory distortions and noise—they don't always see the world as it really is. Thus, humans as agents are notoriously bad when it comes to situations involving great complexity and fast dynamics. All of these factors lead to the need for an overall governance architecture for organizations and management practices carried out so that harmful mistakes in judgments can be minimized.

This chapter will provide a general model of governance and management that can be applied to any CAES. The implementation details will obviously vary from one kind of system to another. But the basic model will be found operative in all.

This chapter provides a general overview of the governance model archetype[5] and will provide some examples from various CAS/CAESs found in nature and human activity systems. This model archetype describes a hierarchical cybernetic network of distributed decision makers, agents, as described in Chap. 11. The hierarchy is a leveled structure of cybernetic processes and we refer to it as a hierarchical cybernetic governance system (HCGS).

At the base of the hierarchy is the operational level, which is comprised of all of the organized work processes that constitute the CAS/CAES. The organization of the work processes is such that the CAS/CAES carries on a basic economic process (Chap. 13), supporting processes (e.g., reproduction for all living systems), and auxiliary processes (e.g., philosophy for humans). The governance system is tasked with decision-making duties that facilitate all of these processes working in concert, meaning long-term synchrony, and continuing to be stable in the face of disturbances from the environment (e.g., temporary loss of a resource). The proper governance of a CAS/CAES should lead to its sustainability in its environment.

Operational-level subsystems—the identified work processes—are able to cooperate to some limited extent and may achieve some degree of local stability and synchrony by inter-process communications. However, this ability does not scale in very complex networks with many work processes that interact directly or indirectly.

Overall stability and synchrony are achieved by the level above, the coordination level. This level operates on a longer time scale than the real-time scale of the operations level. In fact, there may be several sublevels within it, each higher level operating on a longer time scale than the one below. Two kinds of coordination governance are observed. The first is the logistical coordination, which ensures the synchrony and stability of the internal working processes of the system. The second is the tactical coordination needed to synchronize and keep stable all operations interfacing with external entities and forces, for example, the importing of resources, exporting of products, services, and wastes, and mitigating effects of disturbance within the range of adaptive response of the system.

Very long-term sustainability of the system calls for its ability to strategically modify itself (or be adventitiously modified) so as to adapt to major changes in the environment in order to remain fit for continuation. In systems that are evolvable, they may either have a top level that engages in strategic decisions or they may be modified through Darwinian modification and selection.

This model relies heavily on cybernetic and control theories. We will describe these qualitatively. The details of various principles in controls and quantitative models were reviewed in Mobus and Kalton (2015) and where appropriate we will provide references to rigorous treatments.

---

[5] Another term that can be used is "governance framework," which is an outline of what a governance structure should look like. A major difference between what we call a governance archetype and the way governance framework is used is that the former is based on a formal model of governance that is applicable to all CAS/CAESs from living cells to the HSS as a whole, whereas the latter is often used in the context of a specific kind of organization and so may not necessarily be general.

12.2 The Concept of Governance

Every CAS/CAES relies on a hierarchical cybernetic governance subsystem to regulate its long-term behaviors such that it keeps producing products that keep it fit within its environment.[6] This is as true for biological systems as it is for man-made organizational systems like enterprises. The principles of such a system are covered in Mobus and Kalton (2015, Chap. 9). In this chapter, we will expand and extend the basic model provided there to see how it applies to the process of understanding complex systems. Governance and management systems are a consistent pattern of system archetypes that emerged, particularly, with the emergence of life (Coen 2012; Morowitz 2002, Chap. 11). They are ubiquitous throughout nature and man-made systems. We have understood the core principle of feedback loops used in operational management for some time (Wiener 1950; Beer 1966, Chap. 13). Researchers have recognized the need for a general concept of hierarchical cybernetics underlying the nature of more complex systems, especially human organizations that have to operate in a complex environment for very long periods (c.f. Beer 1972, Chap. 5).

In another domain, the author has explored how such a system, a brain, should be designed for a mobile, autonomous robot operating in a fairly complex and non-stationary world (Mobus 1999; Mobus 2000). The author simulated the brain of a simple animal (like a gastropod) to control a robot's search for positive reinforcing stimuli and avoid negative reinforcing stimuli. The robot had to learn cue stimuli that were causally associated with those it sought to go to or to avoid. The brain employed a model of an adaptive hierarchical cybernetic system for the management and control of behavior.

12.2.1 The Purpose of Governance[7]

In the following chapter, we will introduce the concept of a generic and universal model archetype of an economy. In anticipation of the next chapter's subject, the economy is the aggregate of coordinated work processes, as defined in Chap. 3, of a CAS/CAES that produces "assets" or highly organized, low entropy, material structures that are of use in the on-going sustenance of the system. We will see a

---

[6] A stronger claim is that every system needs an internal control subsystem in order to continue existence against an uncertain environment. Even atoms maintain themselves via feedback mechanisms, both in the nucleus (strong forces) and between nuclei and electrons in orbitals around the nucleus. The mutual attraction of positive protons and negative electrons keep the electrons bound unless a surge of input energy overly excites an electron (forming an ion).

[7] In this chapter, we will be primarily concerned with CAS/CAESs and not so much with CS (complex systems) or SS (simple systems). These latter are governed by built-in mechanisms (e.g., governors for dynamic systems) or constraints imposed by environmental forces. Our primary interest is with adaptive and evolvable systems.

number of examples of economies at different scales. But we will also see that, for example, the economy of the HSS is actually just an extension of, and wrapper around, the biological economy of population physiology, and that is an extension of and wrapper around the individual economy of body physiology. And, in turn, that, it is argued, is just an extension of and wrapper around the metabolic economy of individual cells, which are the fundamental units of living systems. At all of these concentric scales, an economy is just the interactions among numerous work processes, all operating to produce that which is needed to maintain the health and well-being of the whole system. As we saw in Chap. 9, the main dynamic of an economic system is the flow of high-potential energy distributed to all of the different kinds of work processes at the appropriate rate. The channeling of energy flows through the economy is a part of the architecture of an economic system and will be more fully explained in the next chapter. The regulation of the work processes and the flows of energies through the system (including the transport of materials through the system) is the primary purpose of governance and management. The agents we examined in the last chapter are the deciders with respect to the regulation process. Prior to the advent of human societies, agents in non-human systems had little to no discretion with respect to their decision models and how to compute decisions for action. But humans are substantially autonomous agents and do have the ability to make other decisions that are best for the economic system. Thus, a second role for governance in human societies is the regulation of humans as agents, including the detection of "cheating" or malfeasance and the punishment of such to encourage human agents to make better decisions.

The governance subsystem is that which ensures the smooth and efficient working of the whole economy for the benefit of all of its components. The existence of a governance system is to provide the needed constraints on the dynamics of component subsystems. Under the assumption that the supra-system is well designed (or in the biological world, evolved for fitness), the governance processes keep the system from taking excursions that would prove to be dysfunctional. These constraints need not be rigid (e.g., dictatorial). It can be advantageous to the whole system to allow some kinds of excursions in order to explore the adjacent space of possibilities. This will be discussed below.

All material/energy work processes are inherently unstable left to their own devices. They are subject to disturbances from outside as well as degradations of structures from within (entropy). At a very minimum, a work process will need maintenance and repair. They are also subject to fluctuations in the rate of inputs and the absorption capabilities of the sinks. All of this is just another way of saying there is no such thing as a perpetual motion machine! Even information work processes (computation and communications) suffer from entropic decay (parts wear out) and disruptions of the energy supply (power outages).

No complex system can enjoy a sustained existence, with a lifetime that is substantially longer than that of any of its components, unless it has the ability to detect problems with structures and functions and the ability to make changes that will restore nominal function. Sustaining nominal functions in light of disturbances or disruptions in source/sink relations is the job of homeostatic mechanisms. Repair and replacement of components is the job of autopoietic mechanisms.

But on top of these basic regulation and maintenance requirements another problem poses itself. In any truly complex system, there will be many heterogeneous work processes that need to work in concert with each other to produce final products. The environmental interface processes, the importers and exporters, need to regulate their activities in concert with the entities in the environment, the sources and sinks, as well as considerations for changes in the general milieu. Mere *cooperation*, that is, when several work processes must communicate with one another to make adjustments to flow rates to the benefit of the composite work, breaks down when the number and kinds of processes exceed some limits. When that happens the only way that the composite work can maintain efficiency, effectiveness, and stability is to have a new layer of regulation for *coordination*. This is a set of new kinds of agents using much more complex decision models than is the case for mere homeostasis. The addition of a layer of coordination has associated costs; the overhead in energy consumed is proportional to the complexity being managed. But the benefits of coordination are great. The fitness of any CAS/CAES that has evolved a comprehensive coordination level of management is substantially greater than that for systems that fail to do so or have an inadequate coordination level.

The coordination level just described is responsible for managing the operations-level work processes over a longer time scale than the real-time scale of the work processes themselves. This is because it is concerned with trends in performance deviations rather than the real-time error that an operations agent reacts to. Coordination operates to counter deviation trends rather than reacting to more-or-less instantaneous errors. This is a major factor in maintaining overall stability of the whole system.

Some CAESs with very sophisticated computing power are able to achieve a form of second-order coordination with the environment over much longer time scales. They can produce models of the environment beyond just the entities with which the system ordinarily interacts. They can use those models to anticipate longer-term changes in the environment that could have an impact on the system sometime in the future. This is the capacity to act strategically. The system must be evolvable in order to be able to modify itself to be able to handle whatever changes may ensue.

In this chapter, we examine the architecture of a governance archetype model comprised of these three levels of management. We will situate agents within the decision nodes in the hierarchical network and we will outline the basic mechanisms involved in achieving the objective of governance.

Of course, the governance system that probably we are most concerned with is that of the human social system (HSS) introduced in Chap. 7 and further examined in Chap. 9. The history (and prehistory) of human societies is one of an evolution of governance from simple systems (e.g., tribes and groups) to larger communities (city-states) to empires. That evolution has been progressively toward the architecture described in this chapter but is greatly complicated by the influence of cultural evolution, that is, the coevolution of human knowledge, cultural practices, and technologies. The HSS is viewed as a CAES that is still evolving and seeking fitness in the larger context of the evolution of the Ecos itself. Various forms of governments have been "tried" and generally found wanting. The single most important question we could ask is: Is it possible that a form of HSS governance could be instantiated that would make it truly sustainable in the Ecos? Our suspicion is that if the answer is yes, then the form of that governance will embody many of the features presented in this model archetype.

12.2.2 Definitions

There are a number of terms we should explain as they will be used in multiple contexts yet retain their basic meanings. Governance, management, and administration are used to designate different levels and scopes of responsibility and authority. Governance is the overall umbrella term covering all forms of regulation mechanisms. Management is used to describe decision types that need to be made when something isn't functioning properly within the scope of the manager. Administrators are those agents tasked with keeping normal operations going. They monitor operations and activate appropriate responses based on established policies and procedures (or naturally evolved operations in all living systems).

It will be important to first make some clear distinctions about the words that we are using. Governance (and government), management, and administration are too often conflated with the result of there being poor delineation of duties in human organizations. Natural governance systems sorted things out by virtue of the evolution of working models through natural selection. Some examples of the latter will be provided below. But in human organizations the situation is more muddied. An example of the problem might help explain.

The modern education system, especially in higher education, provides a good example of blurring the distinctions between the kinds of "work" that needs to be accomplished in a well-functioning governance system. The situation is a result of the rapid evolution of the social demands put on the education system over the last century. Universities and colleges tend to be fairly conservative institutions that found themselves in a radically changing environment without the internal mechanisms for rapid adaptation. Initially the governance of colleges and universities was shared between a board of regents (like a board of directors) composed of prominent private citizens and a few senior professors, and the representative body of the (generally senior) faculty, a senate. The regents' role was merely to provide a degree of oversight in coupling the institution with its constituencies and only in the sense of making sure that the institution was fulfilling its strategic commitments. This task was not very complicated because the strategy of higher education had been worked out centuries before and had not changed substantially until around the end of World War I. Nor were there many management tasks (as defined below) needed by the faculty senate or departments. The pedagogy and curriculum, as well as the divisions of departments, had been well established so that the primary concern of universities was simply maintaining operations of teaching courses. The main form of governance, therefore, called for administration of policies and procedures. That is why those who take on management-sounding titles/roles, for example, deans and provosts, in the education system today are called "administrators" and not managers.

Unfortunately, in the modern university, especially since the end of World War II, the mission and strategic situation have changed considerably. Universities are now expected to operate more like businesses and compete for students, faculty, and research grants and not just on the basis of providing high-quality education (especially for undergraduates) but on many fronts with many new constituencies/stakeholders (Cuban 1999; Geiger 2004; Newman et al. 2004; Rhodes 2001). Public universities have to do this in the face of diminishing state funding meaning that they are faced with the problem of finding new sources of revenues. Now the strategic situation is incredibly complex and requires a whole new notion of governance and management that goes far beyond mere administration. Yet, sticking to historical methods, most "administrators" in academia come from the ranks of professors who ended up chairing a department and getting promoted from there to a deanship. They are not trained in management decision-making but find themselves continuously having to make new sorts of tactical, logistical, and operational decisions in systems that are far from stable in their operations. University administration used to be just that, manned by administrators. Once the nature of higher education started to change from just being institutions of education to encompass research (with the incursion of monetary concerns), professional-like athletics, marketing, finance, and other commercial-like activities, those administrators were called upon to make decisions they were not particularly prepared to make. In other words, they were not effective agents for the kinds of decisions they were called upon to make.[8] The results have been very disturbing for the institution of higher education in the USA.[9]

University administrators are very smart people—most have earned PhDs. But they have not studied management science or worked in management/governance environments (e.g., a profit-oriented business such as they are supposed to emulate) where they would have thus learned the difference between administration, management, and governance.[10] Thus, we see middle managers (e.g., deans) too often

[8] We should be cautious in painting with too broad a brush. There have been many very capable and thoughtful university administrators promoted from the ranks of the professoriate over the years and they have grappled nobly with problems of the modern academy. However, if we look at the state of the academy in the USA, honestly, we might reasonably conclude that the majority of administrators never really achieve the capabilities of real strategic management.

[9] From various colleagues around the world, we hear today that the same trend is diffusing around the globe.

[10] To be fair some education administrators have studied management concepts and know the vocabulary of terms, like "strategic planning." However, few have actually had to perform management functions in any kind of organization other than academia. Understanding of the deeper meaning of terms comes from experience, not from the mere application of logic.

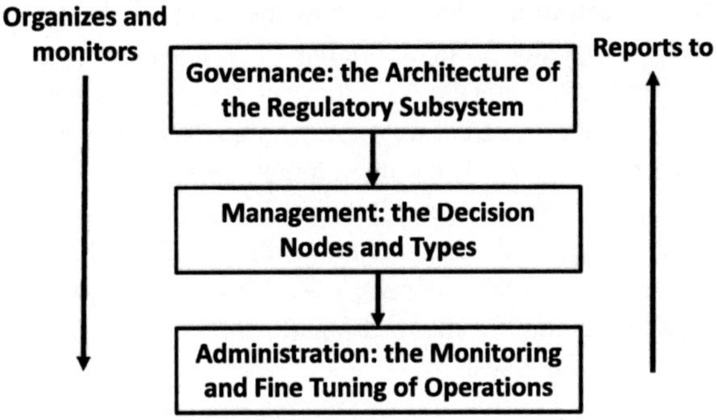

Fig. 12.1 The relations between the three primary functions involving decision agents constitute a hierarchy of responsibility

taking on operational-level decisions when they should stick to logistics or tactical decisions. This distinction will be delved into in much greater detail.

Figure 12.1 provides a graphic of the relation between governance, management, and administration as a hierarchy of responsibility.

Definitions of these should help in understanding the roles of each.

12.2.2.1 Governance: Establishing an Architecture That Will Succeed

Governance is the process (subsystem) by which a CAS or CAES maintains its internal functions such that the whole system sustains existence for extended time, what Stafford Beer called a "viable system" (Beer 1972). Governance includes the structural specification of a government (architecture, e.g., the offices and their relations to one another depicted in a classical organization chart) and functional specifications of decision processes that occupy specific nodes in a hierarchical network (management duties). The lowest level of governance is the set of policies and procedures that constitute instructions for the ongoing operations of the component work systems, those directly involved in obtaining resources and producing products. These are what need to be administered.

The governance architecture is the ultra-long-term plan for how the whole system is to be controlled, regulated, adjusted when needed, repaired when needed, and so on. It designates the structures and regular functions of the system so that it fulfills its long-term mission of remaining viable within the environment in which it operates. It is the design for the "brain" of the system in terms of the major subsystems and their functions. Within those subsystems operate decision processes—agents—that are responsible for managing those functions.

Governance needs to be established early in the evolution of systems. Those that do can expect to have long lives. Those that don't will be short-lived with a possible tortuous existence. For naturally evolved organic systems this is guided by variation through genetic mutation and natural selection in brains and physiologies. For supra-organic systems, such as organizations and states, the design of the architecture of governance is the result of a meta-governance process, such as developing a charter and bylaws for an organization or a constitution for a nation. The process can be seen to be a subsystem of the larger social system in which it is embedded. That governance architecture has to be designed such that the system is responsive to its environment (a CAS—think constitution of a nation), and in general is able to modify its own structure when the environment changes substantially (CAES as described below—think amendments to constitutions).

Once a governance architecture is in place, the ongoing operations of internal subsystems and the interactions with the environment require decision-making processes for a variety of purposes. That is, the various offices in the architecture require the emplacement of agents able to make veridical decisions and guide the activities of the system.

Additional terms that will be used in the discussions ahead are explained below.

12.2.2.1.1 Authority and Responsibility

The term "authority" shall be used to designate the agent that has the ability to effect a change in structures or functions lower in the governance hierarchy. It means the management of lower-level managers.

Responsibility is a designation of what an agent's purview should be. The basic responsibility of any manager is to monitor the activities of lower-level processes, receiving reports (below) from those activities (measuring qualitative and/or quantitative attributes), exercising the decision model, and sending commands to the lower level as needed. In other words, it is a reapplication of the basic cybernetic principle at a higher level of organization and scope (see Sect. 12.3.2.1, "Feedback").

12.2.2.1.2 Command

A command is any message sent from a higher-level manager to a lower-level manager in the governance hierarchy. The lower-level manager acts on the information content of the message to make a change in the process it controls. In natural systems, commands are simply communications that arise in the course of a higher-order recognition that a lower-order process is not behaving as needed. In human organizations commands may arise from other than operational needs, motivated by a variety of emotional states. One reason that humans so often do not come close to "ideal" agents (see discussion below) is that they can succumb to the "power" relation, an evolutionary hold-over of the status structures in primate societies.

12.2.2.1.3 Report and Monitoring

Higher-level managers receive regular reports from lower-level managers summarizing the ongoing operation of the lower-level process. The summaries are often in the form of time-averaged measures of behaviors of the process. For example, a purchasing manager may receive a report from the receiving department quality control manager regarding the number of defective parts of each kind that were received during the month. The purchasing manager uses this information to monitor the behavior of the various parts suppliers and may need to take actions (such as switching to a different supplier) if the number of defects goes above a threshold.

12.2.2.2 Management: Decision-Making Functions Within the Governance Architecture

The word "management" is used broadly to describe a number of activities involved in controlling the productive operations of an organization and its interfacing with its environment. In this chapter, we will refine the concept of management, and differentiate its activity from governance and administration, so that we can clarify a number of systems issues with respect to the overall success of a CAES. However, we should note that management is not strictly isolated from governance in CAESs that need to change some aspects of their own structures. Officers of a corporation, for example, need to act as managers of their offices, for example, the Vice President (VP) of Finance may also manage the Controller's office, who, in turn, administers the accounting system. They may also be called upon to make what are strategic decisions, for example, when a chief financial officer (CFO) decides to restructure a firm's investment portfolio for potentially better returns. Such decisions lead to possible changes in some aspects of governance architecture. For example, the CFO's decision may lead to the creation of a new portfolio management function.[11]

It is very difficult to distinguish, at times, between a management activity and its relation to the governance of an organization, but such distinctions are important in order to provide clarity to large-scale systems analysis of such organizations.

Management is a sub-activity within the framework of the governance architecture and, itself, a framework for administration (defined below). Management, in the present context, is the process of *making decisions* involving adaptations of operations when the environment changes or when internal disruptions cause dysfunctions (management by exception—making a decision only when things go wrong). Management is performed by decision agents with higher degrees of autonomy to assess local conditions and affect actions as required to restore normal operations as soon as possible. Here, "normal operations" refers to the operational processes that have been defined a priori as needed for the overall accomplishment of the whole system's mission. Their subsystem status is defined within the overall system to be governed. *Management, then, is all of the decision activities directed at keeping a system's particular subsystem operating within pre-established nominal ranges.* Managers are agents that have much more autonomy than

[11] We realize that the most common use of the term "governance" with respect to corporations or other organizations is made more narrowly to mean something like "oversight." But this is just a peculiar usage. Clearly, governance as applied to the running of a nation-state or a municipality involves government where active decision-making results in changes at the operational level. We use governance in a much broader sense as described in the text.

administrators (see below). They are problem recognizers and solvers and, so, require a great deal of more intelligence and judgment than administrators (who might be characterized as homeostats—see Sect. 12.3.2).

What is most crucial to the performance of good governance is that managers at every level and in every subsystem completely understand their function (decision types they are supposed to be making) and stick to the agenda. This is not particularly a problem in natural systems like cells and animal brains, except in the event of serious malfunctions or disease. However, it seems to be a problem in human organizations where the human agents all too often forget their function, or, as in the case of the university administrator, do not completely understand what it was in the first place. And, with human agents there is the problem of personal agendas that can run counter to the objectives of the organization.[12]

12.2.2.3 Administration: Keeping the Wheels Turning Smoothly

As long as conditions are nominal (as they are supposed to be under normal operations), administration is the process of monitoring operations. An administrator's duty is to follow procedures to ensure the fulfillment of policies, not to deviate from those procedures. Should the administrator determine that a disturbance to the process has occurred in the face of following procedures, their responsibility is to refer this fact to the manager of the operational unit for correction (which may require deviation from procedures).

Administration, in natural (living) systems, is homeostasis. An administrator is free to respond to error with the normal mechanisms available. The same principles of "same-staying" apply to human societal systems at all scales. In human governance parlance, this implies bureaucracy.

Administration is a more or less mechanical process with a minimal amount of discretionary decision processing (e.g., should I use a #2 pencil or ink?) or autonomy. An administrator is essentially a homeostat with a narrow range of reactions. It is the typical dotting of i's and crossing of t's. It is also the bane of bureaucracies

[12] In human organizations and governance processes, there is a new emergent process owing its origins to the degree of individual autonomy held by the agents (human)—the political process. This is actually a dual-level process that is a recurrent version of the governance/management process discussed in this section. The normal political process that corresponds to on-going management once a governance architecture is established involves the interpretation of policies and procedures—the various political parties, representing and holding different ideological interpretations of, say, the constitution of a nation—and the decision process for selecting which interpretation shall be used in the near term. There is a higher-order or meta-political process that involves decision-making regarding the very architecture of the governance system itself. It too may be influenced by ideological positions but it results in the evolution (or sometimes revolution) of the governance process for a given human social system. The subject of political process is, unfortunately, beyond the scope of this chapter since it applies only to human social systems. But clearly, it is an important subject to be explored. Our hope is that the methodology presented in this book might help in such an effort.

when some tasks could use a bit more autonomy but the administrator chooses to fall back on the "rules" to avoid having to take any out-of-the-ordinary actions. A major problem in human governance is bureaucratic creep, or over-reliance on rules when change is needed.

12.2.2.4 Confusion of the Roles in Human Governance Systems

Often the same human agent has to operate in all three roles, as governor, manager, and administrator, in the course of operations but under varying conditions (e.g., the VP of Finance example above). This is certainly true for human agents in social systems, since human beings can and do present different personas depending on who else they are interacting with. But this same kind of role multiplexing can be found in a few instances in natural systems as well.[13] As a result, it is easy to become confused about what role is to be played in various situations. As described above, many higher education administrators have been unwittingly cast into all three roles without a clear understanding of the requirements and constraints of situations that demand specific actions (decisions or mere monitoring with minor adjustments). Thus, we see university presidents embroiled in logistical decisions simply because they do not clearly have in mind the difference between strategic-level decisions and logistical-level ones.[14] Unfortunately, this phenomenon is not limited to higher education institutions. For human social systems in general, the complexity of modern systems has reached a point where human agents are easily confused about these matters. For another example, take the case of a representative in the national House of Representatives for a district in a particular state getting involved in the details of a pork-barrel issue (an operations-level issue) when the intent of governance at this level is determining how the citizens of that district "fit" into the national narrative of income distribution. The representative succumbs to trying to grab perceived immediate benefits for constituents rather than working on the larger national picture of economic well-being that would benefit those constituents in the long run.

---

[13] For example, our emerging understanding of the control networks in the genome gives some examples of non-protein coding regions of DNA (what used to be called junk DNA) that code for regulatory strands of RNA (e.g., interference RNAs) in response to signals from the cytoplasm. Thus, the transcription machinery can serve a dual role of making coding RNA and, as needed, non-coding RNA. The complexities of this machinery are just starting to be elucidated but the notion of dual purpose of some proteins is now established. The same kind of phenomenon is found during development of an embryo where various proteins are used for different purposes at different stages of development.

[14] In fairness we should point out that there are very high-level tactical decisions that would seem to bleed into strategic ones. For example, when developing long-term plans, say for a year out, for interacting with a specific vendor or acquiring a new vendor of the same material the tactical manager is seemingly close to doing strategic planning. But, as will be discussed shortly, there are important distinctions. For example, a strategic decision might involve switching to a new material resource entirely, which can have very long-term consequences for operations.

The cognitive confusion of roles and duties seems to be widespread. Consider the plight of a department manager in an organization. Most of the time they are just administrating the operations of that department. Occasionally, something goes awry requiring them to make corrective decisions; they might be called upon to make tactical or logistical decisions—tactical in the sense of working with other departments and logistical in terms of changing some work process to gain efficiency. And then, very occasionally they are in the position of needing to make a strategic decision, perhaps observing an opportunity for their department to take on additional responsibilities within the organization. Unless a nominal manager[15] knows which hat s/he is wearing for which particular kinds of decisions it is too often the case that s/he could apply the wrong kind of decision model, say applying logistical thinking to tactical or vice versa.

This capacity for a nominal manager occupying a mid-level operation position to make both management and strategic decisions is due to the autonomy of the human brain. Every individual has the necessary "equipment" in the brain (specifically the neocortex in the frontal and prefrontal lobes) to make all three kinds of decisions for themselves (e.g., a strategic decision about what major to take in college) and that ability can be extended into the social domain so long as the complexity of the social organization is not too great.

However, the ability to switch between decision modes is not well understood. Most managers cross lines between operational, logistical, tactical, and strategic decisions without ever realizing what they have done. The result is considerable confusion in fulfilling roles and mistakes being made.

A major objective of this chapter is to sort out these roles in decision types and levels in a CAES in order to better understand the governance model archetype and how it applies in many kinds of CAESs. We are particularly interested, of course, in how the HSS is to be governed to achieve sustainability in the rapidly changing world.

12.2.3 Decision Agents in Governance Systems

We now situate the agent archetype model within the context of a governance system. In Chap. 7, we introduced the hierarchy of decision types relevant to the governance of a CAS/CAES (see Fig. 10.3). Below we provide general explanations for the roles agents take on in a governance system. Those roles are determined by the kinds or types of decisions that are needed to be made in the various levels of the governance hierarchy.

The three levels of decision types are: (1) operational, (2) coordination, and (3) strategic. Here we briefly explain, reiterating what was covered in the last chapter.

[15]We use the term "nominal" with manager in order to indicate that the agent's role is generally given the overly broad title of manager.

Below, in Sect. 12.3, "Hierarchical Cybernetic Governance," they will be explained in detail.

12.2.3.1 Operational-Level Decisions

Operational decisions are generally made in what we call real-time. They involve making local decisions that affect the operations of the local subsystem, specifically the material, energy, and message work processes (harking back to Chap. 3, Figs. 3.9, 3.10, and 3.11). The time scale for these decisions is the same as the rate scales for flows of material, energy, and messages into and out of these subsystems. Examples include the time scale for a ribosome to process the stream of amino acids being linked into a polypeptide (microseconds), the time scale for a liver to process glycogen and output glucose into the blood (milliseconds), or the time scale for a manufacturing work cell to produce a product (minutes to hours).

Every work process has at least a minimal decision agent that adjusts the operations to match the real-time flows being processed. This is the basic cybernetic feedback principle used to maintain the quality and quantity of desired outputs feedback (see Sect. 12.3.2 below). However, in a network of processes forming supply chains (outputs of some processes are inputs to other processes), mere feedback is not enough to maintain the workflow.

Operational-level decisions employ real-time feedback but also *cooperative* feedforward messages from other operating units. Multiple operating units often form a tightly coupled work-flow network and a corresponding tight network of communications that permit cooperation to play a large role in the decision-making process. These networks operate based on well-defined, formal communications protocols, at least in systems that are well regulated at the operational level. For example, the concentration of various amino acid-tRNA (transfer ribonucleic acid) complexes in the neighborhood of a ribosome signal the latter as to the availability of those needed to fulfill the mRNA specification being processed. The liver is signaled through several channels about the need for glucose in the bloodstream. And a work cell foreman sends parts kit requisitions to the parts inventory kiting function.

Cooperation through feedforward signals and response feedback is an efficient mode to achieve smooth operations, but it has limits based on the time delays inherent in communications channels and the number of "hops" a message might need to take. Here a hop means a message or some variant on it must pass through a node (a neighboring work process) before being passed on to the next node. The regulation of flows in a supply chain is like this. One downstream node signals its supplier to hold up a shipment of parts and then that supplier has to signal its vendors that supply the materials used in making the parts to hold up their shipments, and so on. And there are processing delays inside the neighbor further delaying the signal before it is sent to the next node in the network. Propagation delays, channel bandwidth, possible noise, and other classical communications "problems" act to restrict cooperation signals to only a few nearby nodes. Thus, analyzing a complex system for the extent of cooperation and its effectiveness should take these issues into account. In complex human organizations the dysfunctions that occur when there is an over-reliance on cooperation are legion. What typically happens is the formation of informal communications networks that depend on personalities and non-standard messages with protocols essentially absent to try to compensate for the inherent problems in the formal channels. But these ad hoc networks only serve as band-aids and generally have problems of their own. And, they too have scaling issues. In large complex organizations near neighbor communications for cooperation cannot provide the needed coordination among all work processes. This same principle applies to economic markets, as will be taken up in the next chapter.[16]

When cooperation becomes impractical due to the scale of the whole operations level, that is, the number of work processes and the complexity of the network of flows, it is necessary to introduce a new level of management, the duties of which are to make decisions that help coordinate the activities of a cluster of operations-level subsystems. This principle applies equally to governments and economic systems.

12.2.3.2 Coordination-Level Decisions

Coordination decisions constitute the next higher level in the hierarchy and are needed to keep multiple operational units working in concert with one another when the cooperation mechanisms discussed above cannot provide optimal operations. This is often the case when, for example, a supply chain gets long and response times for downstream processes can be negatively affected by variations in upstream processes. Coordination is achieved by giving commands to operational-level decision agents that modify their decision models or change an operational parameter (e.g., the set point in Fig. 12.5), at least temporarily, in order to, for example, change the rate of processing of a local process. Coordinators seek to balance processing so as to achieve a larger optimum production.

There are actually two kinds of coordination decisions. Internal operational units need to be coordinated since the outputs of some of those units are inputs to other units (see Sect. 12.3.3.2 and Fig. 12.7). The other kind of coordination decisions involve operations that coordinate the activities of the whole system with the entities in its environment with which it interacts, the interface processes for importing to and exporting from the system of interest (SOI). The first kind of decision types belong to the category of *logistical* coordination. The second kind is called *tactical* decisions.

Coordination decisions involve monitoring various units in the coordinator's purview over a longer time scale relative to operational-level cooperation time scales (real-time with some small lag added). Logistical decisions generally seek to

[16] It is interesting that those who lean libertarian in their worldview seem not to understand this. Enthralled with Adam Smith's invisible hand and the idea that self-interest will still lead actors to act in ways that ultimately benefit the society, they seem to believe that cooperation will automatically keep things working correctly. This in spite of massive evidence to the contrary.

optimize overall operations on this longer time scale. Tactical decisions are also made over longer time scales. They involve monitoring the internal subsystems that are engaged in importing resources or exporting products and wastes as well as observing the behaviors of the external agents (sources and sinks) with respect to acquisitions and exports. Recall from Chap. 4 that sources and sinks are generally not modeled in the same way that the SOI is. Tactical decisions, therefore, involve considerably more uncertainty than do logistical decisions. Tactical decision agents often use more complex decision models of those entities that are essentially opaque-box. They also generally collect more information on the behavior of those entities.

12.2.3.3 Strategic-Level Decisions

Finally, at the highest level in the hierarchy, decisions may need to be made involving the very structure of the whole system itself in view of very long-term changes that might take place in the environment. These are strategic decisions that can and need to be made by evolvable systems. In some sense these are meta-decisions or decisions about what sorts of decisions need to be made in the future. Not all CAES can make explicit strategic decisions. For example, a biological species (or genera) does not make strategic decisions to achieve evolutionary change. Normal neo-Darwinian evolution makes those decisions (about fitness) through mutation/variation and selection. But human beings are, by virtue of their learning neocortices, evolvable systems that are faced with making such decisions (what career do I want? What kind of mate do I want?). These decisions are based on far more complex decision models, memories, and a more powerful computational engine than can now be accomplished with electronic computers. The time scales for strategic decisions are far greater than mere coordination decisions. The spatial scope in terms of external entities is far greater than for tactical decisions. For example, such decisions often consider possible disturbances that the system should take into account when deciding how to reorganize.

As the logistical decision agents influence operations-level decision agents when needed and tactical decision agents influence environmental interface operations when needed, strategic decision agents influence tactical and logistical agents through something we could call a "plan" for future activities. Both logistical and tactical coordinators, as part of their decision models, can be programmed, so to speak, to take certain actions (and thus influence operations) at certain points in time, in a longer time scale, given certain external conditions obtain. This will be expanded below in Sect. 12.3.5.

The HCGS model to be presented below maps these decision types onto the hierarchical control model, for example of Findeisen et al. (1980). Whereas the latter is a theoretical model of temporal domains for decisions—higher levels in the hierarchy operate over increasingly longer time scales—the former maps the kinds of decisions to be made onto the temporal domains. Each higher level works, essentially, on time-averaged values from the level below it and issues commands downward when needed.

12.2.3.4 Considering Ideal Agents

In the last chapter we introduced the notion of an ideal agent as one having a sufficiently veridical decision model, sufficient sensory inputs, and sufficient actuator outputs so that it will make satisfactory decisions the majority of the time (statistically). This notion of an ideal is not of a perfect agent, obviously. This is because the world is fundamentally stochastic and no agent can have absolute information or knowledge or processing capacity to make perfect decisions. What makes an agent ideal is that it does not inject its own non-relevant agenda into the decision-making process. We would not want our thermostat to "feel" cold and decide to turn on the furnace for its own comfort.

The governance agents of evolved systems achieve ideal-hood by virtue of natural selection. Here is where we butt into a fundamental problem for human agents. The author suspects that human evolution has not yet produced a mental capability to approach the ideal agent with sufficient autonomy and veridical decision models. Humans are largely guided by beliefs and a background of ignorance about how the world actually works. We can hope that the progress of scientifically derived knowledge about the world, and the guidance of systems science, will help inform human beliefs such that they will move closer to the role of ideal agents.

While we can hope for improvements in the decision-making capabilities of individual human minds due to the knowledge that science and systems science has to offer (i.e., the evolvability of the human brain), it may yet prove to be the case that human beings, as a species, will need to evolve further (Mobus 2019).

12.3 Hierarchical Cybernetic Governance System (HCGS)

The model of an HCGS is derived from observations of governance mechanisms both in naturally evolved systems—living systems—and in human social systems, organizations, institutions, and nations. Naturally evolved systems carry considerable weight by the fact that they have proven effective in managing living systems and making them sustainable. The observations of human social systems are more problematic in that these systems have a spotted history insofar as sustainability and stability are concerned. Organizations often dissolve after short lifespans. Institutions all too often become dysfunctional and unable to support the social functions they were designed for. And nations? When looked at through the lens of history from the emergence of nation-states to the present, nations and empires have a pretty dismal track record for providing a long-lasting, stable, and population-supporting model. When we compare naturally evolved governance, including tribal governance through the early Holocene, with human governance since the Agricultural Revolution, we begin to see the problems.

This chapter will attempt to elucidate the architecture and mechanics of naturally evolved governance systems and compare (and contrast) these with human social systems' approximations of governance systems in the realms of both civil society

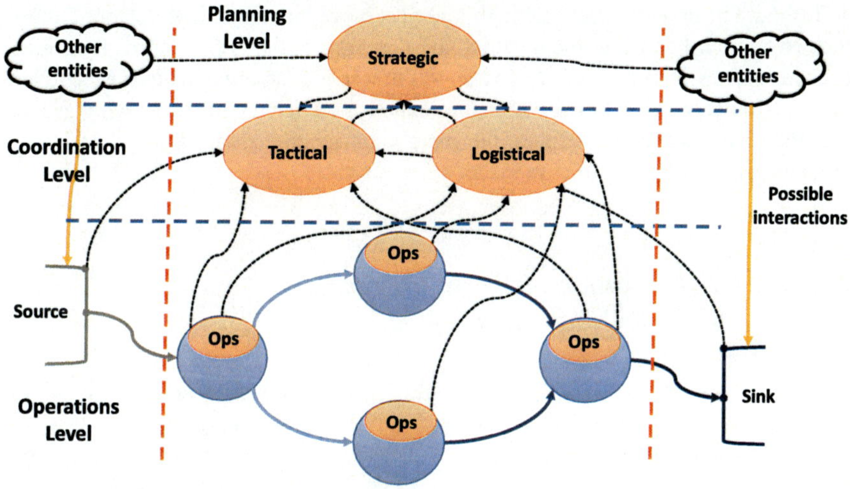

Fig. 12.2 Outline of a hierarchical cybernetic system

and economics. The main intent, however, is to keep focus on the nature of governance systems that are seen to provide the benefits of keeping the whole CAES stable, able to thrive, and internally harmonious for the normal life cycle of that system (Fig. 12.2).

12.3.1 Energy Requirements Per Level

In each section below, on the various levels in the hierarchy, we will include a short discussion regarding the energy requirements devoted to the management in each. Recall from the previous chapter that making decisions requires energy for computation, memory storage, and communications. Actuators require power. There is a proportionality between the amount of computation and other work done by an agent per unit time and the power requirements, that is, the free energy consumed per unit of time, for the various decision types and levels in the hierarchy.

The energy requirements at each level are different because the higher you go into the hierarchy, that is, from operations up to strategic, the more energy each agent requires because the decision models get more complex, hence requiring more power.

As we will discuss in the next chapter, there is a relation between money and energy (the former being a token marker for the latter). Hence, there is a simple rule that follows from the energy requirements relation: The more complex a whole system is, the more levels and sublevels required to manage, the more taxes need to be levied to accomplish the goals of governance. That isn't what many citizens want to hear, perhaps. But it is the way things work.

It should be noted, however, that this subject is pregnant with research opportunities. Not much has been documented about energy costs associated with various agent and decision types. All we do know is that in human organizations executives get paid outrageous salaries while line supervisors barely make a living wage. Whether that is a true reflection of the energy consumed making strategic decisions or an anomaly of human thinking is an open question.

12.3.2 Administration of Work Processes: Classical Cybernetics

Norbert Wiener (1961) noted that regulation of any dynamical process subject to disturbances from the environment required a periodic measurement of error between the current status of a system's path toward a goal state, and the ideal path. The amount of error would then be used to correct the path to reduce the error to zero. This is the famous negative feedback mechanism that constitutes control of a process to achieve its goal state. All CAS/CAESs are based on mechanisms that implement this principle. He termed this cybernetic. It is the basis of all management and governance processes (c.f., Mobus and Kalton 2015, Chap. 9).

12.3.2.1 Feedback

Cybernetics is the science of control and coordination in systems (von Foerster 2003; Wiener 1961). The basic theory of cybernetics is the use of feedback information to cause a system to modify its activities in order to maintain an output function in a viable or nominal value range in the face of disturbances that might otherwise cause the output to deviate from a desired value. The system is goal-maintaining in this sense. Figure 12.3 shows a basic control system and the principle of negative feedback used to counter whatever deviation from the goal state. The "Product" in this sense could be a physical product (specific chemical, specific object) or a directed motion or force. This system is described more completely in Mobus and Kalton (2015, Chap. 9). The agent in this case is an administrator as described above. The "policies and procedures" are completely embodied in the control model with the "set point" embodying the current management decision.

The principle is that a value (either quality or quantity) of the output (product) is a result of the work process operating normally. If a disturbance to the process causes the value to vary from an ideal, as represented by the "set point" constant, either higher or lower, an error signal is generated and fed back to the computational engine that uses the control model. This information is used to generate a control signal that activates an actuator (e.g., a motor) that changes the internal operations of the work process in opposition to the error.

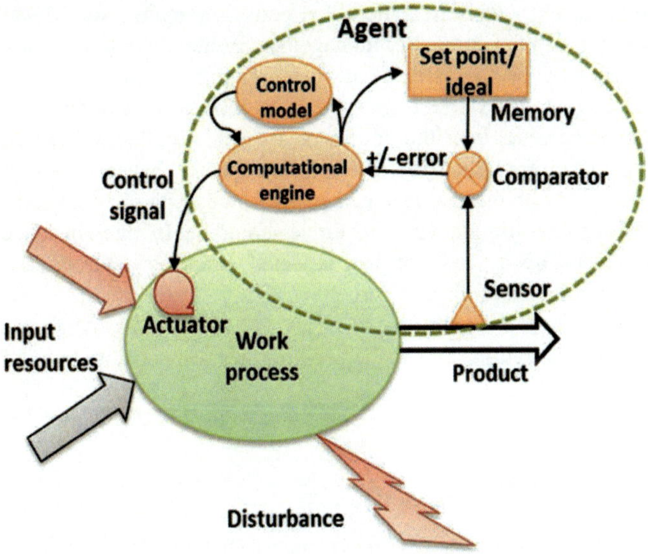

Fig. 12.3 The fundamental cybernetic system operates to restore the proper function of a work process in light of some (manageable) disturbance. The agent in this figure is essentially the same structure as depicted in Fig. 12.1. The circuit works by comparing a measure of the product (quality or quantity) with an ideal value (set point) and the consequent error (information if the error is non-zero) is used by the control model to generate a control signal to an actuator inside the work process. (Modified from Mobus and Kalton 2015, Fig. 9.5, with permission)

This is the basic principle of cybernetics, maintaining nominal operations in spite of disturbances. The actuator must be able to affect the operation in a range sufficient to counter the normal range of variation caused by the normal disturbances encountered by the system.[17] We will borrow the term "homeostat" from Ashby (1958), who designed and built a mechanical/electrical device that demonstrated the ability to counter disturbances giving it that name. We wish to appropriate the name for any mechanism, chemical, electrical, mechanical, or human-based, that fulfills this principle, and is acting as a process administrator.

12.3.2.2 Requisite Variety

Every agent must have a range of actions that can be taken to restore nominal activity to the process being controlled. The "law of requisite variety" proposed by Ashby claimed that the stability of a system's operations required that the range of actions must be equal to or greater than the deviations from normal operations that disturbances might cause. A homeostat must be able to generate countering actions

[17] This is often referred to as the "law of requisite variety" formulated by W. Ross Ashby (1958).

to the disturbance or the system could not effectively remain stable.[18] Stafford Beer also addressed this in the context of management in human organizations.

A homeostat must be possessed of sufficient range based on the "ordinary" range of disturbances that could affect the system operations. In naturally evolved systems, this is determined over the course of a species' collective experiences and selection for those representatives that have a capacity that lets them survive the marginal excursions. In human-designed systems, we try to use data about environmental conditions to estimate what that range should be. In many instances of engineered systems, we design to more than expected variances just to be safe. We have to specify the boundary conditions somehow.

Homeostasis, as described for a wide array of CAS/CAESs, by many authors, is the mode of administration. This is the pre-programmed capacity to handle normal ranges of disturbance.

12.3.2.3 Management

Autopoiesis and adaptive response mechanisms, as mentioned in Chap. 10, Sect. 10.3.2.1, "Adaptivity," go beyond simple homeostasis. They provide a higher level of adaptivity in which a homeostat might be modified or adjusted so that it can respond over a greater range of disturbances over a longer time scale. In our vocabulary, this is the role of management over administration. CASs have built-in capabilities to extend or enhance responses to changes as needed. As described in Chap. 10, when a disturbance or critical parameter is pushed out of nominal range or is kept at an extreme for extended time, the system responds with making modifications that help keep it viable even in the face of these higher stresses. Living systems such as cells and multicellular individual organisms have evolved elaborate management of resources such that extended capabilities are called upon only when absolutely needed. This is a cost savings strategy.

Variations on this basic model include the use of feedforward signals (to be explained below), for example, measuring the value of inputs as compared with ideal values and using this information in the control model. Some forms of anticipatory control are obtained from such an arrangement, but the control model is made more complex as a result. For engineered systems where the variations in conditions can be calculated in advance, it is often possible to use feedback alone, but for natural systems, anticipation is generally necessary in order to avoid damage.

Figure 12.4 is a more general model that will be used for agents in an HCGS structure.

The basic cybernetic principle discussed above is the basis for the lowest level of management and administration in complex systems. All processes within a CAS/CAES are managed in this way. Recall that all processes do some kind of work,

[18] This result should not be confused with overreaction, which can lead to increasingly wild oscillations that are also a decline of stability.

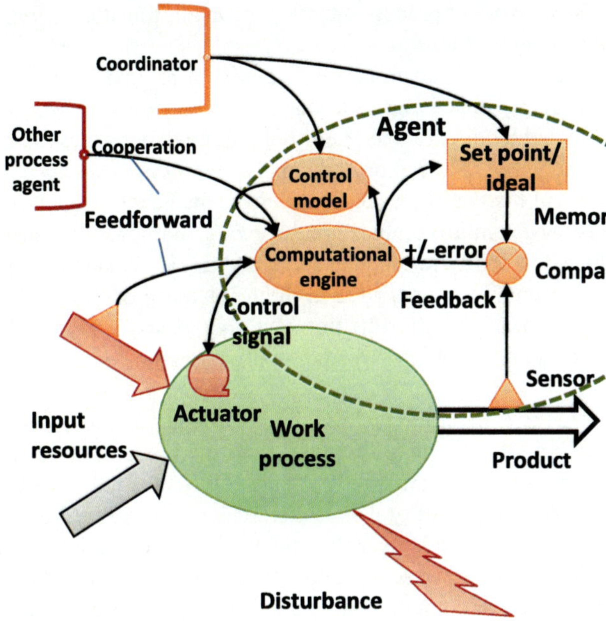

Fig. 12.4 This is the same basic management model as in Fig. 12.2, but now with feedforward and "feed-down" signals. One feedforward signal is the result of sensing the input factor. The other is sent to the agent from another process agent providing information about resource quality and/or quantity for cooperation. The feed-down signals are shown coming from a coordinator. One of these signals might direct changes to be made to the control model. The other is a simpler modification of the set point (as when someone changes the desired temperature in a thermostat)

which may be transforming materials or messages. Thus, all complex processes will involve some kind of basic management, including processes that are part of the governance system itself.

Figure 12.4 includes message feedforward and feed-downward signals. One of the feedforward signals comes from measuring an input. Together with the feedback from the output these constitute all of the local process control. The other feedforward signal is shown coming from another entity, another process that is sending information (either as a customer or a supplier) about its situation. The feed-downward signals are shown coming from a "coordinator." Both of these additional sources of information will be explained below.

12.3.3 Operations-Level Networks: Cooperation

Work processes do not exist independently in complex systems. They form a complex network of processes that generally work in concert to accomplish the goals of the whole system. The next chapter on a model archetype of the economy will drive this point home. The network of processes that do the major transformation work of the system is at the operations level.

In such a complex network of processes, we will often find local clusters of nodes that are more tightly linked together. That is, they are found to be near neighbors in space and linked by direct flows.[19] For optimal operations these nodes need to coordinate their activities. For example, a supplying process may need to notify the receiving process of fluctuations or interruptions in the flow of what it is supplying. The latter must have a more complex control model that takes this information into account and has the ability to regulate some additional actuators internally to compensate. This is usually accomplished in the receiving process by it having a stock reservoir in which it can buffer the flows, storing excess when the flows are higher than required and drawing down on the supply when the flows are lower than required.

Operational networks with clusters of this sort are directly related to what, in the economy model archetype, we call a "market," with the provisos stipulated there that market mechanisms work best in tightly coupled clusters (small worlds). As we will explain, the market is generally also a hierarchically organized small-world model with nodes in localized clusters having the strongest coupling with the most frequent interactions and clusters acting as nodes in larger (fractal-like) clusters over longer distances and lower coupling strengths. Local clusters generally operate through a node that acts as a hub and is responsible for the longer-distance interactions with other cluster hubs. See Fig. 12.9 ahead for a graphic representation.

12.3.3.1 Basic Organization

The problem can be characterized thusly: A complex system is comprised of many interacting work processes, each under local basic control, but each also subject to disturbances that might affect downstream receivers of upstream products. If each of the processes is monitored and controlled by a local controller (as in Fig. 12.3 above), then the whole system may generally be expected to operate in nominal form. Local disturbances should have minimal impact on overall system operations. Figure 12.5 shows a typical organization of basic work processes at level 1 in a complex system. These processes must work in consonance with one another, each doing what it is "supposed" to do to satisfy the needs of downstream processes or customer sinks. What is shown in the figure constitutes the "operational" level of the system and the processes are controlled by local real-time agents.

We start from the assumption that a system has evolved or was designed to have this organization. The central concern is how all of the various processes are to be kept working together in a "more-or-less" optimal fashion. Moreover, the various acquisition and export processes need to operate in consonance with the sources and sinks.

[19] An exception to nodes needing to be near neighbors in space is the case of information processors linked by high-speed communications flows.

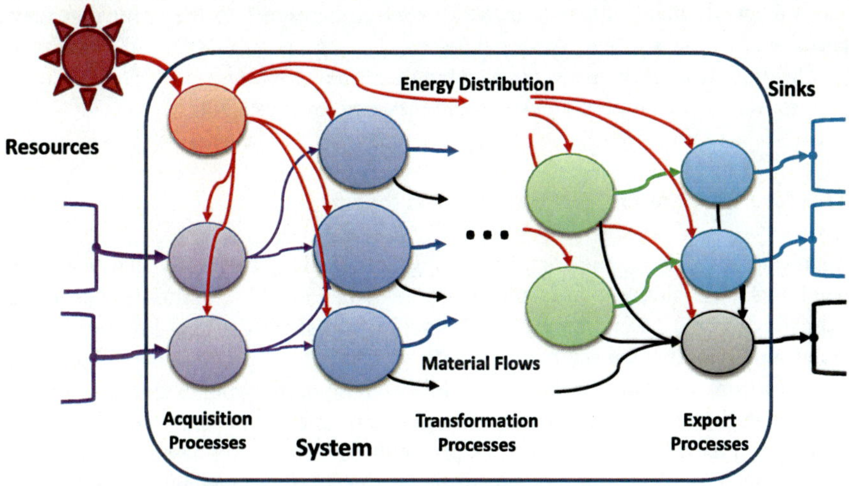

Fig. 12.5 Every CAS or CAES is composed of many internal subsystems that perform work processes, transforming materials and energies. Acquisition or import processes bring resources into the system while export processes push material out to the environment sinks (e.g., customers and garbage dumps). All processes use energy in a high-potential form (red arrows) and produce waste heat (not shown). Generally speaking, the internal flows may be very complex depending on the number and kinds of processes

The problem of maintaining system-wide optimal or near-optimal operations becomes increasingly difficult as the system increases in complexity. Both the exposure to and timing of disturbances become increasingly difficult to manage by feedback control alone. For example, should two (or more) disturbances co-occur with sufficient magnitudes, the combined impact could overwhelm a downstream process's ability to compensate.

In all CASs and CAESs that have been examined in detail, we find three fundamental control mechanisms operating at this level. The first is, as mentioned, the fundamental negative feedback mechanism of cybernetics. A local agent (e.g., as in Fig. 12.4) monitors the performance of the work process and generates control adjustment signals to the internal actuators as needed to keep the process producing at optimum (given the constraints). An example from the metabolic processes in a living cell would be the way in which the synthesis of ATP (adinosine tri-phosphate) from ADP (adinosine di-phosphate) and a free phosphate molecule with energy supplied by the oxidation of glucose (through a complex set of reactions and intermediates) is regulated by the local concentration of the various intermediate molecules within the matrix of the mitochondrion. As more ATP is needed by the cell, more of these intermediates are produced to up-regulate the production in the citric acid cycle. When less is needed, fewer of the intermediates are produced to down-regulate the process.

In human organizations, the role of the familiar "line supervisor" or "foreman" is to make sure the work being done under their watch is being done correctly. They take action to correct variations before there are significant problems. In a real sense, every human worker is a local agent with respect to the work that they perform.

The other two mechanisms are also familiar. They are ways to achieve non-local cooperation between work processes and they employ messaging.

12.3.3.2 Cooperation Between Near Neighbors

There are several options for establishing cooperation, especially between near neighbors. The simplest mechanism is shown in Fig. 12.6, assuming agents as depicted in Fig. 12.4 earlier (e.g., with feedforward signals). Here, Process **A** produces a product that is used by Process **B**, which, in turn, produces a product used by a downstream customer. Both processes contain more complex decision agents that are in direct communications. The problem here is that both agents have to share a common language as well as protocols for interacting based on their own states. Assuming that such protocols have been worked out and encoded into their control/decision models,[20] the agents inform one another of variances in their situations that allow them to take some kind of actuation response to help correct the problem.

Suppose Process **A** is disrupted by an interruption of flow of Material **A**. Agent **A** can notify Agent **B** of the situation (that is, provide feedforward messages to Agent **B**) by virtue of monitoring its own status receiving Material **A**. The agent is projecting a disruption to the flow of its intermediate product to Process **B**. Agent **B** can then use this information to make internal adjustments, for example, it might have internal stocks of the intermediate product (stored in anticipation of such a

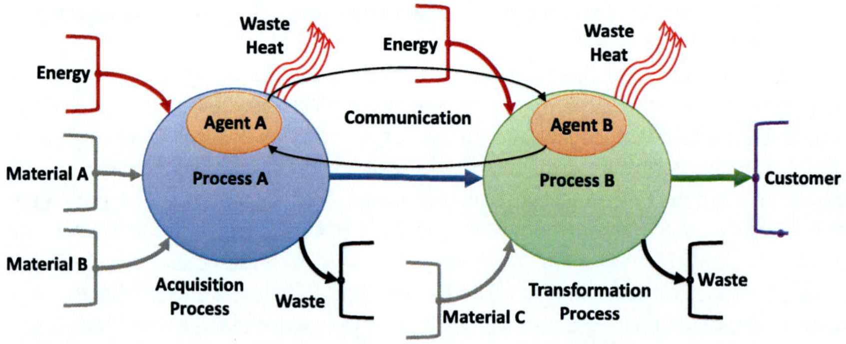

Fig. 12.6 Two systems (processes) cooperate when their management agents exchange messages that help regulate the flows from the producer to the consumer process (blue arrow)

[20] In real systems, this is actually a result of auto-organization and emergence processes as covered in Mobus and Kalton (2015). In designed systems such as organizations, the communication protocols have been a priori designed. However, it should be noted that most such designs originated in an evolutionary process of essentially trial and error.

disruption) that it can use in order that it does not experience a diminishment of its production. Of course, these kinds of compensation actions are time and duration sensitive. But this is an example how, if the two relatively independent agents are communicating with the appropriate message protocols, they can cooperate to smooth out the overall operation and the ultimate customer is not (immediately) affected.

Figure 12.7 provides a more complicated situation; three of the processes extracted from Fig. 12.5 are shown in a more complete cooperative *network*, combining elements of the basic cybernetic model in Fig. 12.4 with the communications shown in Fig. 12.5 above. This depiction shows that agents in various work processes will need to communicate with multiple other agents. In this case, an acquisition process supplies two (or more) internal work processes (the figure only shows the agent/work relation in one process). The flow rates to each of the receiving processes could be very different with different timing. There can even be priorities that must be accounted for in the case of emergency conditions.

Clearly there are going to be complexities involved in providing control models that can adequately provide correct responses in all of the agents involved in these local connections.

As should be becoming clear from these diagrams the nature of communications between cooperating agents and the local control models used by each become increasingly complex just to obtain cooperative interactions. After taking a look at two near neighbor forms of cooperation we will revisit the complexities involved in achieving cooperation in the whole network (such as in Fig. 12.5).

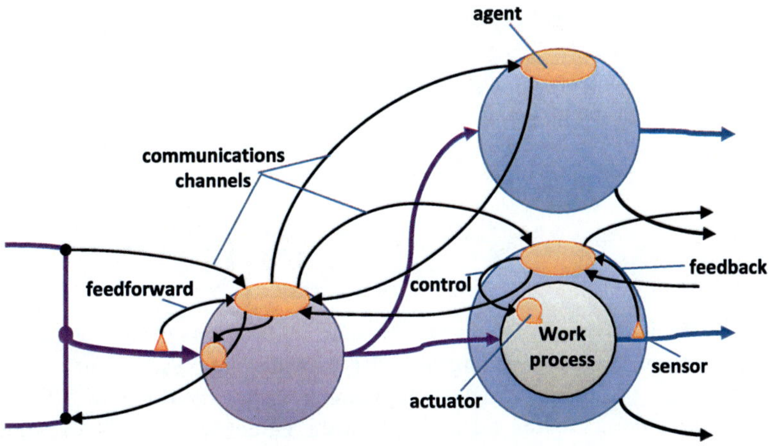

Fig. 12.7 More details of the communications channels needed to provide internal control (lower process on the right) and cooperative messages between processes. The acquisition process (purple on the left) also shows the use of sensing inputs to provide feedforward information regarding the flow of the input. Additionally, the acquisition process may communicate with the environmental source (or just be able to observe what the source is doing)

Sub-processes within systems comprised of many cooperating sub-processes do not attempt to communicate with every other process. The overhead for communications would soon overwhelm any possible advantage. Moreover, the further the distance between sub-processes would mean the greater the time delay distortions that would occur in the messages, further aggravating the problem of cooperation. Instead, successful systems employ a hierarchy of coordination controllers whose job it is to maintain a modicum of "local" cooperation while also providing a level of more global coordination.

12.3.3.2.1 Cooperation by Messages

In Figs. 12.6 and 12.7, we show messages communicated between the internal agents making operational decisions for the work process in which they are embedded. Not shown, but necessary, are the message interfaces. Recall from Chap. 4 that interfaces involve protocols for sending and receiving, and this is the case whether what is flowing is material, energy, or messages. In general, for processes that have a tight coupling as shown in those figures messages can be exchanged to help coordinate the activities of the work processes so long as there are no major systemic problems.

12.3.3.2.2 Cooperation by Exchanges

When systems are in a formative stage of development (e.g., during the origin of life), the first form of cooperation that is established is the exchange of substances, material, and/or energy. Such exchanges are the result of fortuitous auto-organization among aggregates of heterogeneous processes (Mobus and Kalton 2015, Chap. 10, Sect. 10.3). This is depicted in Fig. 12.8 below.

If one or both of the processes are evolvable (and from the first primitive autocatalytic cycles active in the origin of life are thought to be), over time the exchanges might be complemented by message flows to help coordinate the exchanges.

12.3.3.3 Feedforward

The fundamental cybernetic feedback control in Fig. 12.2 can be augmented with the inclusion of feedforward signals that provide anticipatory information. Feedforward can be obtained from simply monitoring the inputs to have advanced warning that some disruption might occur in the output (affecting the feedback error) in the near future, with a time delay based on the transit time through the process. Using an anticipatory model plus, generally, the use of internal buffers to supply the substance of the input in case of diminished flow or absorb excess flow, the operations controller can maintain optimal output in spite of these kinds of disruptions.

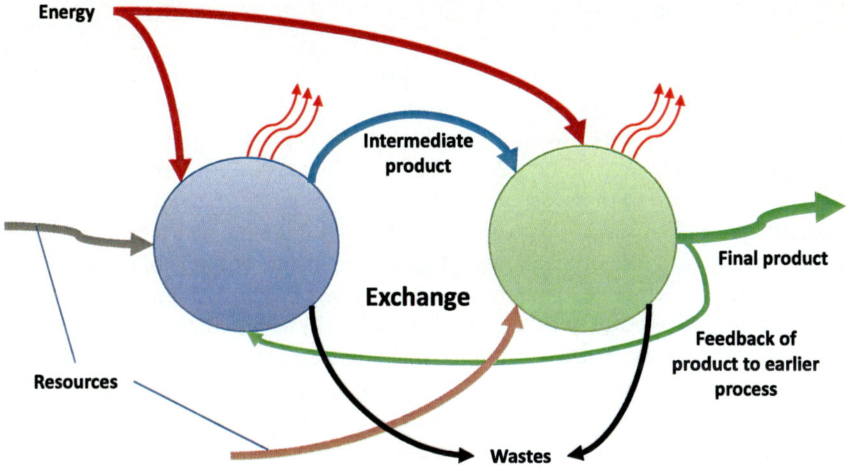

Fig. 12.8 Processes may establish cooperative relations by exchanging substances that are useful to each. In this example, the blue process produces an intermediate product that is needed by the green process. The latter produces a "final" product (so far as these processes are concerned) some of which is sent back to the blue process that might be using that product for consumption purposes. As long as the two processes cooperate, they will maintain an optimal overall production of the final product

In the context of cooperation, however, the feedforward signal comes directly from the supplying process. In Fig. 12.6, processes **A** and **B** are communicating with one another directly. **B** is supplying **A** with a form of feedback and **A** is supplying **B** with a form of feedforward.[21] There must exist a protocol for message interpretations in both processes. Both messages convey error values. The information content (value) conveys the significance of the message, while the message is about the flow state (or dynamic). For example, **A** may signal **B** that a lower rate of flow is about to be experienced. Or, another example, **B** may signal **A** to increase the flow rate if **B**'s stock of the substance is running low. In either case, neither **B** nor **A** are necessarily expecting those messages—they are thus informational. And they relate to the flow rates of the substance going from **A** to **B** at some near future time.

Feedback and feedforward are thus complementary messages that, with the right protocol in place, can act to coordinate behaviors without an explicit command from above to do so.

[21] These conventions are the result of considering the flow of the substance in question going from the left (**A**) to the right (**B**). Thus, **B** is feeding back information to **A** regarding what it has received and **A** is feeding forward information regarding what **B** is about to receive.

12.3.3.4 Decision Models Required

The decision model required by simple feedback mechanisms is relatively straightforward (once time delays are taken into consideration). For example, for general analog controls the proportional, integral, derivative (PID) control algorithm suffices nicely (Mobus and Kalton 2015, Chap. 9, Sect. 9.4.2). However, once you incorporate feedforward information the decision model becomes much less straightforward. The basic PID controller needs to be modified with additional decision functions. The Adaptrode model (Mobus 1994) is one example of such modifications that incorporate adaptive response based on anticipatory signals (feedforward). Other examples in natural systems will be provided below.

In human organizations, the use of feedforward from upstream processes is quite well known. For example, a parts inventory management system can "tell" a production kitting manager how many of a certain part will be available in stock. The kitting function can then use that information to schedule the withdrawal from stock to make up the production kit. As will be seen in the next chapter, this capability is the essence of a supply chain process.

Anticipatory feedforward is relatively easy to incorporate in the decision models that agents use in cooperative clusters. However, the more complex a cluster of cooperating processes grows, at some point a limit in the effectiveness of cooperative signaling is reached. At that point cooperation alone cannot provide solutions to optimal behavior.

12.3.3.5 Complexity and Breakdown of Cooperation

Near neighbors in a network may be able to achieve stable, coordinated operations using homeostatic and augmented decision models that take the other entities into account (Chap. 2, Principle 9). Cooperation depends on mutual respect for protocols (and no malicious agents sending false messages).

There are several additional concerns in cooperative situations. The message exchanges are reciprocal and there is danger of establishing a positive feedback loop in which one agent, responding to an abnormal condition signaled by the other agent, inadvertently amplifies the abnormality and feeds that back to the first agent with the consequences that cooperation breaks down.

Even if near neighbors can successfully cooperate over a wide range of operating conditions, both may simultaneously be neighbors to other processes, such as the situation depicted in Fig. 12.7 that will be affected by their interactions (attempts at cooperation with one of the two processes) in a way that is not readily handled by the recipient. Figure 12.9 hints at the problem. Process **B** receives a message from process A informing it that a disturbance will be felt in the flow of intermediate product from **A** to **B**. **B** must process this message and then inform process **C** that, with an appropriate time delay, its own output to **C** will be disturbed. In theory this allows **C** to prepare (preadapt) in advance.

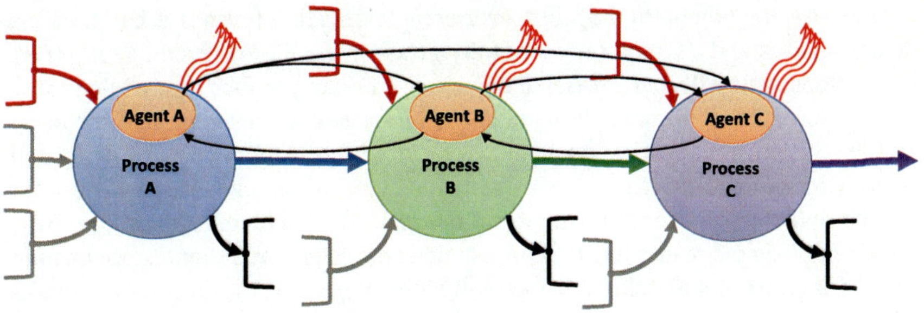

Fig. 12.9 A communication channel can be extended to nodes further distant in the network. This arrangement is meant to mitigate propagation delays in transmitting information to more distant nodes through intermediate ones that have to process their messages before they can transmit to their neighbors further downstream

However, there is a processing delay in **B** that shortens the time that **C** will have to prepare. Worse yet, upon receiving the message from **B**, **C** may start preempting just as the disturbance is actually reaching it. In a worst-case scenario, this could lead to overcompensation with ensuing instabilities or oscillations that could damage the process. What Fig. 12.9 suggests is that **A** could communicate directly with **C** so that the propagation delay could be avoided. This is certainly technically feasible, but now imagine complexifying this with the need to communicate with many more downstream recipients. Or imagine every process in Fig. 12.5 communicating with every other process, including feedback along long-distance channels (called a peer-to-peer [P2P] network architecture). In theory, this should lead to the maximum in cooperation, but at what cost?

The actual technical problems with designing a cooperation-only architecture, as this might be called, are many and beyond the scope of this chapter. Two important ones will be mentioned to give the reader an idea why this scheme cannot work in practice and even, likely, in theory. First, there are still communications delay problems even when the channels are set up in a direct peer-to-peer (P2P) network. Moreover, the longer the length of the channel the higher the probability of interference or noise injection that would possibly distort the signal and lead to false messages. The longer the channel (physically) the more complex any communications protocol would have to be, including error detection and re-send, and security concerns.

Second, the decision models used by the agents (who are doing the communicating) are suddenly much more complex in order to send the right messages to the right other agents. In human-designed systems, today, it is possible to mitigate the communications problem through the use of an Internet protocol (IP; in fact, the TCP (transmission control protocol)/IP and protocol suite could be used directly). This is exactly what is being contemplated in the design of social–technical hybrid systems under the general name "Internet of Things" (IoT) in which human–human, human–machine, and machine–machine communications are facilitated by the Internet (especially including wireless).

However, the admonition against over complexification (even if it can be done cheaply) still stands. If you look carefully at all of the P2P or even client–server applications, you will find excessive complexities owing to the "desire" that these applications take all possible decision nodes into account—made more difficult by having so many connections. As a general result of this level of complexity foisted on each agent, there are usually many software bugs lurking in the code waiting for just the right decision condition to occur to wreak havoc with the application. How many times has the reader had to reboot their home router, or a running application on the computer or smartphone? Since software engineering is still somewhat in its infancy (from the standpoint of systems theory), as more complexity is added to communications/computation systems, we can expect a multiplicative increase in such bugs. In all operations-level process networks it is neither feasible nor even desirable to use P2P communications and complex decision models to achieve cooperation.

Natural governance systems have evolved a different approach.

Figure 12.10 provides an abstract version of a set of subsystems within a larger system. This map derives from a second-level decomposition of an SOI into its subsystems and then those, into their sub-subsystems. As shown in Chap. 6, this is one of the methods for verifying selections of boundaries when no physical one is found.

Note that there is a detectable difference between any node taken at random, and groups of nodes that cluster. That is, there are more (and sometimes stronger) links between small groups of nodes than between those groups and other nodes. Intuitively we realize that the nodes with a higher density of links are more tightly

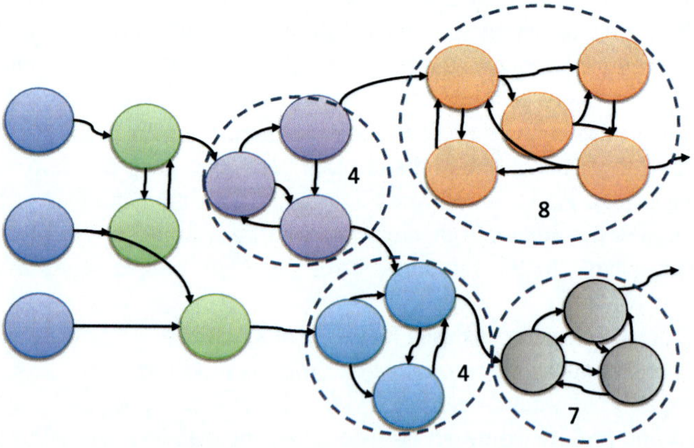

Fig. 12.10 Processes within systems tend to form clusters or groups. In fact, this is just another reflection of the notion of subsystems within the larger system. Each of the blue, dashed ovals contains a group of processes that are tightly linked (counts of the number of links between neighbors can be used to indicate clustering). The implication of this kind of clustering is that members of a group are better able to cooperate since they are close together in process space. These groups generally correspond with subsystems

coupled and must be working together more so than with other such groups or independent nodes.[22] In a concrete system, context analysis of the dynamics of work and flows into and out of as well as among the members will identify the validity of considering them as a group.

It turns out that cooperation between members of any such subgroup is more feasible than between groups or over the entire system, so long as the clusters do not extend too far. Groups of four or five processes may be able to achieve a reasonable level of cooperation within themselves. But then we reach a limit.

However, we need to keep in mind that a map such as in Fig. 12.10 only represents one level of decomposition. Each process shown in the figure could, in fact, be further decomposed if it is itself a complex system. Thus, it is reasonable to conclude that P2P schemes for communications in the interest of a "democratic"-style cooperation-only governance architecture are really not workable. Throughout the world of naturally evolved CAS/CAES (and this includes human societies), cooperative interactions are possible and desirable insofar as they are efficient for getting local groups working together near optimally. But the problems associated with long-distance cooperation presented above will defeat success for any large and very complex system.

Therefore, another approach, suggested by the kind of topology of real systems shown in Fig. 12.10, emerges. When, in a group or cluster, one of the nodes takes on the role of an agent whose task it is to coordinate the cluster, the problem of long-distance cooperation is elevated to a new level in the cybernetic hierarchy.

12.3.3.6 Energy Requirements for Operations-Level Administration/Management

The energy consumed by operations-level agents is relatively small owing to the simplicity of the decision model using error feedback to adjust the operational parameters. Of course, with each operational unit becoming more complex, the complexity of the decision model increases and the agent power requirements increase accordingly. A surrogate for energy costs in a corporate operations-level unit is the number of clerks, data collectors, etc. needed to support the supervisor. This is analogous to the number of different enzymes and intermediate products being catalyzed and the total amount of ATP consumed in anabolic metabolism. More complex pathways take more energy.

Additional energy costs come from the degree of cooperation being employed at the operations level. Cooperation increases the complexity of each agent's decision model since they now have to take into consideration the direct communications and interpretation of messages that increase the load on computation.

[22] In graph theory, groups of vertices with more edges between the members of the subgraph than with other vertices in the rest of the graph are called cliques and there are algorithms for detecting their presence. Strictly speaking this applies to undirected graphs, but variations on directed graphs are known.

12.3.4 Coordination-Level Decision Managers

Ultimately complex systems cannot rely on simple cooperation to achieve optimal performance. Moreover, they cannot always use cooperation when it comes to working with external entities in the environment. By definition a system does not have adequate access to the internal operations of an environmental entity—a source or a sink—and therefore cannot necessarily have the kind of communications channels necessary to work with such entities.[23]

Coordination is a form of inter-process control that arises when the communications situation has become sufficiently complex and the channel delays are long such that reliance on direct communications is no longer reliable. An attribute of communications between distant process agents is that the time scale over which the dynamics are affected is, itself, long. The change in an output of process **A** may not be impactful on process **Z** for an extended period. Such time delays are disruptive. And it should be clear that if process **A** is supposed to notify **Z** of its situation, the problem of complexity is deeply aggravated. **A** would have to communicate not only with **Z** but also with every process from **B** to **Z** and every one in between. The overhead of computation and communications is untenable.

Instead of trying to extend the nature of cooperation to all of the sub-processes in a complex system, the natural tendency is to introduce a higher-level cybernetic processor, a coordinator, whose job it is to intervene in the normal feedforward and feedback loops of process controllers to provide coordination. Figure 12.11 introduces the implementation of a coordinator agent whose responsibilities are to monitor the activities of a cluster of closely associated work processes and provide them with longer-term guidance so as to optimize their collective behaviors with respect to the final output of products (blue arrows) going to downstream processes. The depiction in Fig. 12.11 is of a coordinator that is working to balance the flows of materials imported by the purple process to two (or more) work processes that must share that material for their independent processes. As can be seen in this depiction, the role of communications and agent decision-making is starting to dominate the internal flows that have to be understood.

Figure 12.11 includes some communications channels between the three work processes representing the cooperation channels. To the degree that the messages sent along these channels can assist cooperative decisions on the part of the three agents, the complexity of the decision models of the coordinator is reduced (the coordinator does not have to tell each process what to do all the time!). In such a system, a hybrid of cooperation and coordination, the computational load on all agents is kept manageable.

[23]The possibility is not, of course, prohibited. But neither is it a given possibility. A corporation may have a very deep communications capability with its suppliers but will not necessarily know what is going on inside its major competitor.

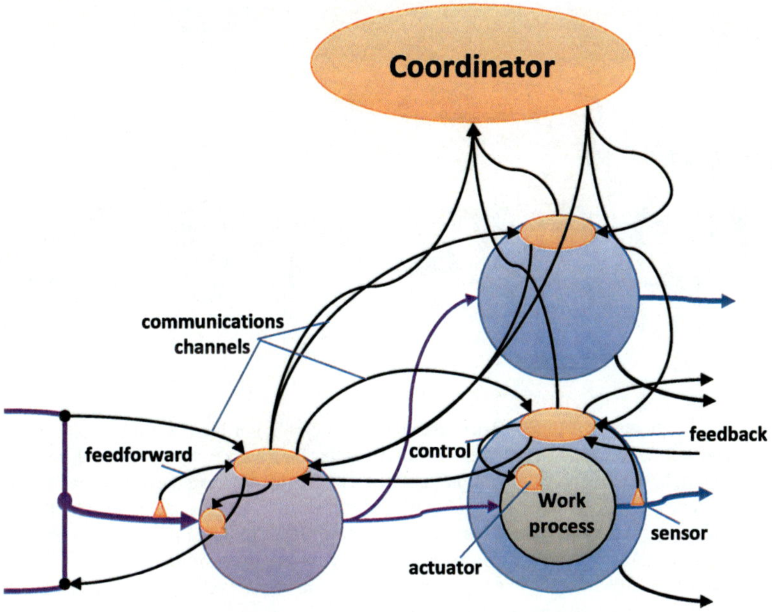

Fig. 12.11 At some level of complexity and dynamics, subsystems are no longer able to merely cooperate and need a higher-level perspective coordinator agent to keep them operating at optimal performance

12.3.4.1 Coordination at Larger Scales

The clustering of near neighbors (in Figs. 12.10 and 12.11) that allow local cooperative behaviors to facilitate the cluster behavior as a subsystem, only requiring coordination in unusual circumstances, is a prototype for a more general approach to coordination management as shown in Fig. 12.12.

Just as a sufficiently large number of neighbors in a cluster warrant a coordination level of management, so too, a large number of subsystems will require a higher-order (and longer time scale) coordination. This is achieved by a new level of coordination above the first level of coordinators at the subsystem level.

The figure depicts two (out of many) work subsystems (clusters) that are sufficiently complex as to require some form of coordination-level management. The tactical and logistical coordinators in each subsystem have models of the whole subsystems' behavior given inputs and outputs as well as the models of each work processes' time-averaged behavior (see discussion of these models below) by which they provide feed-downward messages.

However, the introduction of multiple subsystems creates a need for an additional level (super-level) of coordination, that between subsystem coordinators. In the figure, we only show two subsystems in order to convey the basic concept. In a real complex system, there will be many subsystems and all of their coordinators will need to be in communications with one another. Not unexpectedly, this

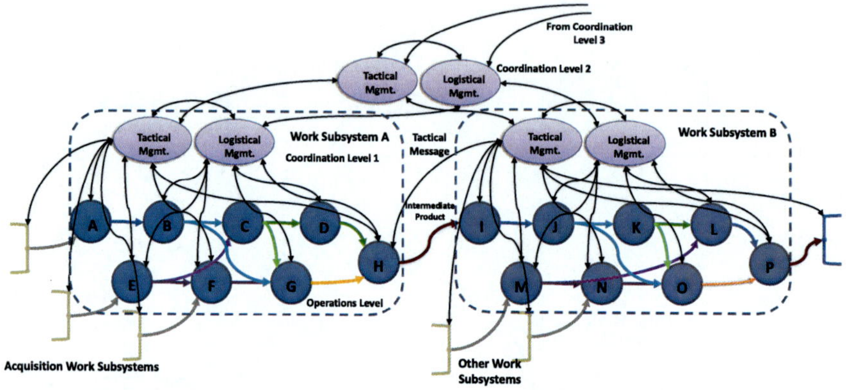

Fig. 12.12 Coordinated clusters of works that correspond with our idea of a subsystem. The output of subsystem **A** is input to subsystem **B** (along with other work subsystems in the network). Each subsystem has its own local coordination level and the two subsystems (and possibly several more) are themselves coordinated by a sublevel of coordination with the larger scope. Blue ovals are operations-level work processes. Purple ovals are coordination-level processes. Coordination level 1 is local to the subsystems. Coordination level 2 works to coordinate the coordinators in level 1. If the whole system is sufficiently complex, a coordination level 3 would be responsible for coordinating the level 2 coordinators, and so on

architecture is the same as cooperation between work processes, but now between coordination processors. We will revisit this situation below in Sect. 12.3.4.5, "Coordination Between Logistical and Tactical." Note for now, that all logistical and tactical coordinators at the same level have communications between them for cooperation purposes. The tactical coordinator at the subsystem level reports to the tactical coordinator at the next level up and the same is the case for the logistical coordinators.

This architecture forms a pyramid (hierarchy) of coordinators from the operations level upward that is as deep as is needed in order to facilitate coordination across the span of the whole system. The architecture is fundamentally that of so-called middle-management, with the number of sublevels dependent on the number of operations subsystems overall—roughly the number of levels is $log_s(m)$, where $s \geq 2$ is the "span of control" and m is the number of subsystems.

Span of control is medium dependent, that is, the kind of *thing* the system is, and is affected by transmission delays, bandwidth, computing power, and decision model complexity. A generalization can however be made. This architecture can be shown to reduce the overall overhead of management and lead to maximal stability of the system behavior if, and only if, the subsystem sizes and complexity are within the range of capabilities of the coordinators. This is a recursive feature that extends upward to however many sublevels are required. In naturally evolved systems we can see this principle in operation in examining the details of actual near neighbor work processes and their coordination controls (to be discussed below).

12.3.4.2 Work of the Coordinators

12.3.4.2.1 Monitoring Longer-Term Behavior

A coordinator is strictly an information processor.[24] In other words, it is basically just a decision agent but with a much more comprehensive decision model and more computational power than operational-level agents possess. A coordinator has to monitor the behavior of all of the operations-level processes under its ken. This is generally done using time-averaged or integral quantitative methods. The operations-level processes (their agents) can "report" data to the coordinator on intervals longer than the real-time scales on which they operate. In Sect. 12.4 below, we provide examples of this phenomenon in the coordination of synaptic compartments (operations-level work processes). The coordinator must collect data from all of its subject work processes using the same time constants so that it can efficiently compute errors in any of them in any particular time frame.

12.3.4.2.2 Decision Models

Coordination models are more complex than operations-level ones. At the base a coordination model has, as its components, the operation level subsystems' models of themselves in a coupled array of models. These are, of course, also time-averaged and/or integral terms summing up the longer-term behavior of each component subsystem and comparing those with the array of ideals for each. The algorithms employed are some forms of optimization over the entire array in which deviations from individual ideals are translated into an overall performance metric for the group. From there, the coordinator has a simpler time computing an appropriate response for the overall group and mapping that back to actions that individual units can take.

Another class of decision models that are becoming better understood and "more popular" in human-built systems is pattern recognition and learned responses. Artificial neural networks with Bayesian learning capabilities are being employed increasingly in all levels of real human governance/management systems and will be further discussed below. Of course, the premier such system is the human brain. We will also reflect on its capabilities for governance of the individual and small social groups below.

Additional modeling approaches found in both natural and man-made systems include finite state machines (e.g., models used to regulate biochemical interactions in metabolism), hidden Markov chains (used to approximate models of sources and sinks), and other statistical methods. Since all of these modeling methods have been thoroughly explored elsewhere in great detail, we will not belabor their uses here.

[24] By information processor we still mean that the process does work but now only on messages (data flows) for the purpose of extracting information.

12.3.4.2.3 Feed-Downward Messages

Coordinators have two basic ways to influence the operations-level agents. They can alter the value of set points (in Fig. 12.3) or they can alter the decision models (programs) of the operations-level agents (see Fig. 12.12). Altering a set point simply causes the operations agent to adjust its actuator commands up or down in accordance with the new set point. This is easily seen in the thermostat model. By setting the thermostat desired temperature the system's "seeking" behavior is altered and a new "optimum" is established.

Altering the operations-level agent's decision model is more complicated. There are a number of ways in which a PID model, for example, can be altered to produce a new mapping from input sets to output sets. The simplest modifications would be to change one of the formula constants. For example, consider the simple output proportionality, $o = af(i)$, where $\boldsymbol{o}$ is the output signal, $\boldsymbol{i}$ is the input signal, and $\boldsymbol{a}$ is the constant of proportionality. Simply changing the value of $\boldsymbol{a}$ will change the mapping from $\boldsymbol{i}$ to $\boldsymbol{o}$. Of course, the coordinator must have a model of the behavioral change in order to know by how much to change the value of $\boldsymbol{a}$. The realm of adaptive control theory provides the basis for analyzing and designing such systems.

We can now differentiate between the two major categories of coordination control in the HCGS, logistical and tactical coordination. Logistical coordination involves getting internal subsystems to work together so as to produce a near-optimal set of behaviors, thus a near-optimal overall system behavior. It involves getting the right stuff to the right place at the right time and at the right cost. Tactical coordination is the tricky business of coordinating the embedding system's behavior with that of the entities in the environment of the system with which it must interact.

12.3.4.2.4 Energy Requirements for Coordination Agents

The power requirements vary among agents in the coordination level (and its sublevels). But in general, the wider the scope of management, e.g., a second-order logistics coordinator managing a number of first-order logistics managers, the more complex the decision model and the more computational, communications, and memory capacities will be required. We will discuss the differences between logistical and tactical agent requirements in each discussion below.

12.3.4.3 Logistical Management

Logistics is a familiar term in human organizations and especially the military. In general, it means getting the right things to the right places at the right times and for the right (optimal) prices (costs). It is generally viewed as problem in optimization over a complex operation with requirements and constraints shaping what constitutes an optimal or near-optimal solution. But, as we shall see, even natural systems display elements of logistical coordination management with respect to coordinating internal flows of materials, intermediate products, and getting final products ready for export.

Logistics is one of the more mature management sciences with extensive mathematical models used to solve the problems of complex operations (e.g., operations research was developed in the early and mid-twentieth century coming out of the military needs during World War II to move supplies and troops around efficiently in an ever-changing war topography). Since the science of logistics is treated well in many other places, just as with control theory, we will not attempt to rehash that subject here. Rather we will mention just a few aspects of logistical management that are relevant across the spectrum of CAS/CAES models. These will be revisited in the next chapter regarding the management of economies.

12.3.4.3.1 Time Scales

Logistical decisions operate on a longer time scale than the real-time scale of the operations level as a rule. If operations-level behaviors are measured in, for example, minutes, then logistical decisions will be made using data that are time averaged over hours or even days. There are a number of different kinds of logistical decisions, which we will discuss in the next section. Some types might require quickness, say in emergency situations, or to respond to an ad hoc condition, such as in adaptive response. But by and large, logistical coordination depends on observations of the multiple operations that need to be coordinated over time to compute, for example, trends.

12.3.4.3.2 Decision Models

As stated above, logistics is an area of management science that has been studied intensively and there are well-understood models that can be adapted to particular instances. We find these models already in place in natural CAS/CAESs. Those systems achieved sustainability through the evolution of implementations. Below we will examine just a few examples of logistical processes in some of these systems.

12.3.4.3.2.1 Strategic Requirements

The whole system is supposed to do something. It is supposed to interact with its environment in such a way as to be "fit" in that environment. It is supposed to produce actions or products that serve the larger embedding supra-system to achieve this. The structure/function organization of an existing system was the product of a strategic process. In the case of systems up to humans and their organizations, strategic "decisions" were (are) the result of evolution. By definition, evolution of species leads to modifications that either do make the species fit or result in negative selection. That a species (and its populations) exists is prima facie evidence that evolution has found a solution (or set of solutions) to that species' strategic problem. We will revisit the concept of *intentional* strategic management, as it applies to human individuals and organizations, in Sect. 12.3.5 below.

The strategic requirements are those that motivate logistic behavior. In the case of CASs, they have been incorporated into the structures/functions of the system and so the highest level of strategic management is automatically doing the "right" thing. In CAESs, however, the situation is different. There can be an explicit strategic level of the HCGS that sends commands to the upper echelon of the logistics agents causing them to alter their decision models and thus causing long-term changes in internal processes. For example, an individual human can, at a young age, decide they want to go to college and they change their devotion to study in order to get good grades. The upper echelon logistics managers affected are those involved with directing the whole human's behavior with respect to extracting useful information from the study materials that the tactical agents have obtained.

The numerous homeostatic mechanisms that respond to changes in conditions have limits within which they can act. Those limits are set by strategic requirements. For example, when a species of animal evolves an ability to survive in a shifted temperature regime from that of the parent species, this is a case of a strategic move that results in setting new limits for temperature tolerance.

12.3.4.3.2.2 Moving Stuff: Coordinating Sources and Sinks

The most general form of logistical decisions involves the transport of material through the system. Once a decision is made that an intermediate product should be sent from subsystem **A** to subsystem **B**, along with volume/weight and timing considerations, the logistical manager must compute the best way to make this happen in the context of the rest of the ongoing operations.[25] The logistic manager then needs to monitor the behavior of the flow over time and analyze statistics regarding that flow to determine if it is operating correctly. If not, the manager has to, perhaps, adjust the set points of the two processes to bring the flow into the correct regime. Or the manager might need to modify the decision models of either or both of the operations agents.

12.3.4.3.2.3 Budgeting Allocations

Logistical managers are responsible for making decisions regarding how much of a resource should be allocated to each process. Energy is a general example of something that needs budgeting based on how much work any given process is expected to accomplish per unit of time.

[25] This could be done to modify the system given strategic decisions to do so, or it might be from a need to adapt to temporary conditions.

12.3.4.3.2.4 Optimizing

An ongoing problem in most logistical decision models is how to find optimal or balanced solutions subject to numerous constraints and requirements. In reality, systems do not expend the resources needed to find optimal solutions. Rather, in the vein of Herbert Simons' notion of satisficing, most CAESs (and CASs) find a near-optimal solution given the plethora of constraints and requirements. Harking back to our review of homeostasis and autopoiesis, systems are reasonably "happy" to be within some range of the actual optimum. All decision agent models involved with regulating these processes take this loose requirement into account. They must not overreact to variations from the ideal in order to prevent uncontrollable oscillations. This is true at the operational level but is also true at the logistical level. The only real difference is that the time increments used at the logistical level are longer and represent averaged increments at the operational level.

12.3.4.3.3 Commands

Logistical managers issue messages (generically called "commands") to the operations-level work processes to get them to change their behavior. There are two basic kinds of commands: one changes one or more parameter set points to nudge the process toward slightly different rates of production; the other message involves making changes to the operations-level agent's decision model itself. These are, understandably, more complex sorts of messages with a more complex interface protocol. But a relatively simple example of this is when a software update is downloaded to a computer program. The software engineers, operating as logistical managers, provide the operating software with updates to their routines to, presumably, improve performance.

12.3.4.3.4 Energy Requirements of Logistical Agents

Often, logistical agents are distributed according to the type of decision models as described above. A transportation logistics agent need not be concerned directly with budgeting issues and vice versa. Both may have optimization problems to solve, however. The division of labor across different logistics problems, however, means that any one agent has a relatively simple decision model to compute. Logistics problems are often elaborations on the basic cybernetic model in that they deal with managing errors and so their mappings from inputs to outputs are computationally tractable with modest computing power.

However, coordinating multiple logistics coordinators generally involves an increase in complexity that is at least multiplicative if not exponential. Indeed, the pressure that directs the implementation of higher-order coordination sublevels is exactly the same that has driven the creation of sub-process formations throughout hierarchical structures. At some boundary of complexity and scale, it is cheaper to abstract to a set of subsystems and introduce a new higher level of control. The energy requirements for this new level (or sublevel) are greater than the sum of energy requirements for all of the subsystems. It is overhead, the cost of increasing complexity. In natural evolution, the cost is offset by the rewards of stability, resilience, and sustainability of the whole system. In human societies, it may help sustain profits (i.e., creation of wealth).

As we will discuss below, in general, logistical energy costs are modest and manageable in comparison with tactical management.

12.3.4.4 Tactical Management

Tactical decisions are somewhat similar to logistical decisions except that, one of the two cooperating entities is outside the system and not exactly under the control of the manager the way the logistics manager has control over both (or multiple) operations entities. This means the tactical manager needs more information processing capabilities with respect to building some minimal but sufficient model of the external entity (following Principle 9 in Chap. 2).[26] They also need some sensory apparatuses set up to observe the behavior of the external entity. In some more advanced forms of entity–entity cooperation, communications channels very similar to those between cooperating work processes might be established to facilitate the interface. A great example of this from the natural world is the communications between flowering plants and pollinators in the form of flowers that signal the presence and location of nectar. The plants know what will attract a bee and the bee knows how to read the message in the flower.

Tactical managers are responsible for situating the whole system in a sustaining relation within its environment. They are responsible for the importing of all resources needed by the internal work processes and for exporting products and wastes to the environment. They manage all of the work processes that are associated with in- and out-flows but do so in cooperation with logistics managers (shown in Fig. 12.12 as two-way message arrow between logistic and tactical agents). The tactical manager is responsible for the flows coming into or out of the system through these interface processes, but the logistics managers are responsible for the flows out of the import processes into the internal work processes and the flows into the export processes from the internal work processes.

Higher-order tactical managers may be tasked with searching the environment for resources. Search methods depend on the substrate system, that is, animals forage, humans seek information (informavores), a parts-purchasing manager looks for alternate suppliers with better prices, and so on. There is, surprisingly, an underlying algorithm to all stochastic searches, that is, searches for a resource that might be found in seemingly random distributions in space and time (e.g., food patches like

[26] This is actually also the case when the system must coordinate with the milieu of the environment and not just an identifiable other entity. For example, an animal monitors the weather conditions and seeks shelter when it turns cold or rainy.

flowers for bee foragers). Non-random searches are used in more structured environments. We mention a few of these below.

12.3.4.4.1 Time and Space Scales

As shown in Fig. 12.12, tactical management agents are generally leveled based on the scales of time and space. Here, the space scale is considerably different since the tactical decisions involve ranges outside of the boundary of the whole system that vary considerably. The time scales relevant to the import or export processes must be in concert with the real-time scales on which the internal work processes operate. But the time scales relevant to the entities in the environment must also be considered. Internally, a system might enjoy a certain amount of synchronicity through logistic management, but externally the world is asynchronous and, quite often, seemingly haphazard. Even quasi-periodic phenomena such as hourly temperatures over 24 hours may be chaotic. Real-world events are, in general, "sporadic (recurring but not necessarily periodic), episodic (lasting for variable amounts of time), and erratic (variance in amplitude within an episode)" (Mobus 1999).

Thus, tactical management is generally more complex than logistic management. Moreover, tactical managers need significantly more sophisticated sensory and actuation apparatuses. They must be able to recognize the existence of a resource source or product customer against a background of the co-containing environmental background. And, of course, they must be able to do so at distances much further away than is the case for logistics managers that are close to the work processes they coordinate.

12.3.4.4.2 Decision Models

Tactical agents require decision models that are considerably more complex than those of logistical agents. The models are used to compute a "plan" of action, that is, a set of actuation signals that produce sequenced behaviors that are in concert with the entities and milieu of the environment. Such plans can be as simple as a single actuator command issued directly in response to the model being used, or as complex as a set of actions to be carried out over extended time scales. For example, an earthworm exercises a repetitive sequence of undulations as it crawls through the soil. It responds to sensory inputs that detect food concentration gradients to change course. There is no planning as such. A predator, on the other hand, needs to build a plan of attack that is based on the current conditions of the environment and the prey. We will examine the plans below.

12.3.4.4.2.1 Strategic Requirements

As with logistic agents, commands that condition tactical agents are strategic in origin. And as with logistic agents, the strategic decisions were made via Darwinian evolution of the species for natural CASs (as individuals) and speciation for natural CAESs. For humans and human societies (organizations), strategic decisions are achieved through intentional modification. That is, the decisions are volitional, a matter of choice by a strategic thinker. Whether that thinker is good at strategic thinking is another matter, of course.

Strategic commands are in the form of goals set for the system to achieve, for example, find food. The hunger signal is built into the natural system by evolution. The tactical agents involved then go into action to fulfill the need for food. For a human deciding they want to go to college, the goal can be achieved by studying hard and making good grades. The tactical agents include all of the facilities that student has for studying and finding information.

12.3.4.4.2.2 Plans

A *plan* is a sequence of steps that *should* be executed in order to achieve a goal. Plans are not perfect algorithms. Rather they represent a sequence of actions that are the best approximation of actions that, if successful, will lead to the goal designated. But as Robert Burns, the sixteenth-century poet, noted: "The best laid schemes o' mice an' men / Gang aft agley (often go awry)."[27]

Tactical agents work according to a plan of action. This plan is generated from a set of plan steps known to the tactical agent. It is assembled in response to commands from the strategic management level regarding a sequence of goals to be achieved. In a simple case, a predator feels hungry and its tactical brain modules start formulating a sequence of actions that will lead it to find prey. In a more complex situation, a strategic decision to lower costs of the manufacture of a product will lead purchasing managers to look for lower-priced parts. A plan of action is assembled (perhaps modeled and simulated to test for reasonableness) as a sequence of actions to follow in order to achieve the goal. See the discussion of strategic goals below.

12.3.4.4.2.3 Minimal Models of Environmental Entities and Milieu

Tactical agents need to have some ability to compute expectations regarding the behaviors of entities in the environment with which they interact. This is analogous to logistics managers having models of all of the operational-level processes that they manage. The problem is that, by definition, the tactical agent has no way to gain insight into the internal operations of those entities—they are outside the system. The same observation obtains for the milieu that is relevant to some tactical agents (e.g., those tasked with keeping a mobile system in a proper temperature

[27] Robert Burns, *To a Mouse* (1785).

domain). Thus, tactical agents are required to develop or otherwise have a minimal model of the entity's behavior over time in order to make predictions or anticipations regarding that behavior in the future. This is a hard problem and a computationally complex one. Thus, tactical agents have need of much more sophisticated decision models and more computational power to run them. Indeed, the decision models must include a simulation of the entity given observations of other environmental conditions.

Thus, a tactical agent is more challenged with making a decision than is a logistical agent simply because the environment is more complicated than is the internal milieu. Minimal models, that is, models sufficient to make reasonable anticipations, will depend entirely on the complexities of the environment and the entities/milieu with which the system must interact.

A minimal model for a tactical agent would be able to capture enough of the entity's behavior over time such that the agent could make reasonable predictions about what the entity might do in the near-term future (long-term behavior is the subject of strategic management covered below). Tactical agents may, of course, have more than just observational communications established with the environmental entity. A parts-purchasing manager may have the ability to query a parts supplier regarding the availability and cost of certain parts needed as inputs. But this is not quite the same as having a causal model of the entity in question. All that a tactical agent can do is recognize that an entity exists, that it has behaved in such-and-such a way in the past, correlated with what it sends in messages, and that is all. Unless we invoke industrial spying, there is no way to gain access to the details of its internal operations, or a systems model of how it works. The tactical agent has to rely on the proposition that its minimal model of the entity is reasonably predictive of that entity's future behavior.

12.3.4.4.2.4 Pattern Recognition

Tactical agents are more often required to observe and interpret patterns of state or activities of entities in the environment. Thus, their decision models are more likely to require some capacity for learning patterns to form a mapping from those patterns to appropriate actions by the interface processes they command. Pattern mapping learning is an area of very intense research today. The field of deep learning using artificial neural networks is extremely active and finding applications in many cyberphysical systems such as autonomous vehicles. Animals, of course, rely on biological neural networks that are capable of more than just forming a mapping. In humans, at least, we know that they can formulate exploratory mappings autonomously.

Once a tactical agent has learned a pattern and a mapping from that pattern to an appropriate behavior plan, it is competent to manage the system's behavior resources to accomplish the goal.

12.3.4.4.2.5 Actuation Capabilities: Working Through Import/Export Work Processes

The tactical agent commands one or more import/export work processes that receive or push out materials, energies, or messages. Since these work processes manage flows (with perhaps some minimal transformation processing), the main focus is on the operation of the interfaces they have with the external environment. These processes will tend to depend on flow rate monitoring as the basic operational control, that is, flows are at their required levels with errors derived from higher or lower than ideal flow rates (and at specified flow times). The tactical manager, in a similar fashion to the logistical manager's influence over internal work processes, can influence the import/export processes by adjusting set points for flows. However, since generally there is also a reservoir or stock buffer involved in managing flows, another variable that might be resettable is the stock level maximum. The tactical agent has to compute what the long-term ideal level of a buffer should be based on past flow activities (and overflows or underflows of the stock) and issue commands to the import/export processes to adjust accordingly (assuming they have adaptive or evolvable capabilities).

The set point adjustments for flow rates assumes that the import/export work processes have some adaptive capacity for ramping up or down the actuators associated with the flows. Pump motors might be sped up or slowed down. More pore channels might be inserted into a cell membrane to allow greater influx or outflux of molecules and ions. A household may have to increase or decrease the number of trips to the grocery store to stock the refrigerator.

In the event that the equipment built into the system for ordinary flow rates is inadequate to handle short-term fluctuations, then if the system is evolvable, the tactical manager may call upon the modification mechanisms to increase the flow actuators. Perhaps a more powerful pump is needed. The tactical manager requests the plant management department (in charge of maintenance and modifications) to install one.

12.3.4.4.2.6 Search

One of the more important decision models for higher-level tactical managers is the search for satisfactory situations relative to the environment. This includes the search for satisfactory resources, e.g., a plant having access to good soil, water, and sunlight (accomplished by seed dispersal mechanisms) or a foraging animal's search for food, the search for protection (a fund manager seeking a good hedge position), or the search for a suitable place to dispose off wastes.

There are two general classes of search algorithms that tactical agents can use depending on the behavior of the target of their search and the environment through which the search must proceed. The first, and most general, is called stochastic search. This is where the target resource (or other situation) is distributed in space and time in a quasi-random fashion (i.e., the sporadic, episodic, and erratic characterization given above). In other words, the agent does not have a priori knowledge of where to find the resource and must conduct a search that will produce results in a reasonable time, that is, before the system is depleted of its resource stores. A foraging animal needs to find food before it starves. Mobus (1999) developed a model of the kind of search that is not a random walk but a semi-directed quasi-randomized walk he termed "the drunken sailor walk." Implemented in a wandering robot it looks very similar to the path taken by searching scout ants as they walk across a relatively smooth surface. It weaves back and forth in the same sporadic, episodic, and erratic fashion, yet remains going in a general direction.

The second class of searches is ordered (i.e., algorithmic). When a search is being conducted in an organized structure, such as a list or a street (looking for a name or an address, respectively), there are rules to follow that are guaranteed to produce a result if the item is in the list or the address is on that street. Search algorithms for networks (graphs) and many other topologically ordered structures are well known. However, it should be noted that the structured environment had to be constructed in the first place, before the search can be done efficiently. For example, a list might first be sorted by lexicographical order. Then a binary search (recursively splitting the list in the middle and determining whether the searched-for item comes before or after the middle item and taking the upper or lower sub-list accordingly until the search ends in finding the item or failure) can be done in roughly $\log_2 N$ time, where N is the length of the list.

Natural environments tend to be unstructured in the way a human-made world is. Thus, the first class of search algorithms dominates the world of tactical governance.

12.3.4.4.3 Commands

Tactical agents coordinate the work processes that acquire resources, export products, and dispose off wastes. They determine the timing of activation of the interfaces with the external environment, flow rates of the inputs and outputs, and, in cooperation with the logistics agents, the flow rates from the interfaces (or local buffers) into the work processes internal to the system.

Tactical agents in mobile systems (e.g., animals or military units) order motor units (muscles or attack formations) to position the system in accordance with its objectives (eat or attack). These are coordinated through plans as previously described.

There are many more aspects to tactical management than is the case for logistical management and so this description must necessarily be abbreviated. We cover only those that are common across all CAS/CAESs.

The agents managing an interface process operation are responsible for activating an action that is coupled with actions of environmental entities such as sources or sinks, but also with milieu conditions. For example, a mobile system may be under the control of a foraging or hunting tactical agent, following a plan of attack to fell a prey. It has to find the prey, chase it, and bring it down. Once that act has taken place, it hands off control (cooperatively) to a tactical agent responsible for assimilating the prey (eating it). Afterward, the food is managed by operations under the control of a logistics coordinator that issues commands for digestion and absorption.

There is a great deal of commonality between this description of a hunter chasing and eating its prey and a purchasing department looking for the best price and quality requirements for parts to be used in its manufacturing company. And this is not unlike an army attacking and vanquishing an enemy and then turning the prisoners of war over to prisons to be prosecuted.

The actuators that are managed by tactical agents are, as a rule, more complex and more powerful than those managed by logistics agents. They are often manipulated in temporal sequences (like the running gait of a cheetah) and also need to have a fair amount of flexibility built into the execution of those sequences.[28]

12.3.4.4.4 Energy Requirements of Tactical Agents

Not surprisingly, given the huge increase in the computational complexity of tactical decisions (over logistical ones), they require much more energy to execute. Put simply, those systems that obtain benefits from a tighter coordination with their environmental entities must pay a higher cost of processing interactions. Animals that move about, both hunters and hunted, must have larger and more energy consumptive brains in order to survive.

Tactical decision models, recall, not only require the basic internal cybernetic mechanisms as logistic models but also need to have constructed and compute a partial model of the entities they are coordinating with. If those entities are cooperative, the models need only be minimal and correlative rather than complex and causal. But if the entities (or for that matter the milieu) are competitive or threatening, then the models need to be much more elaborate and hence computationally much more expensive.

12.3.4.5 Coordination Between Logistical and Tactical

Tactical decisions are often driven by factors that arise from entities in the environment. These, recall, are not necessarily modeled (at the tactical level, see below for strategic-level models of environmental entities). At the same time, the decision process of a tactical agent is constrained by logistics from the insides of the system. The tactical agents are responsible for acquisition of resources and expulsion of wastes or export of products (in a timely fashion). But those processes, in turn, are

[28] A fascinating example of this flexible programming may be exhibited by linear activation circuits in the cerebellum of mammal brains. These circuits are strongly implicated in muscle activation and coordination where, under the control of the motor cerebral cortex, a sequence of activations can be modified in mid-execution. The linear circuits are like computer programs (that have been learned) that have "if" statements embedded and the cerebral cortex, responding to real-time input, is able to change the parameters. Hence, a cheetah can run after a haphazardly escaping gazelle.

conditioned by the needs of the internal work processes that are, basically, under logistic control. This means that tactical agents and logistical agents need to be able to cooperate at any given sublevel in the coordination level.

12.3.5 Strategic

To achieve long-term viability in a forever changing world, CAESs must be able to generate internal changes to its own structure and functions that better match up with the forces in the environment. They need to be able to adapt what they do and how they do it so as to continue to be fit in a different set of circumstances.

Fortunately, the world we are in is relatively stable for long-enough periods to allow CAESs generally to take advantage of their capacity for evolvability, whether serendipitous or intentional, to become adapted to changes that do occur. In contrast, we know that this wasn't the case for the world of 65 million years ago when an asteroid crashed into the Yucatan peninsula and led to the mass extinctions that included the dinosaurs. The amount of change involved in an incident like that was too great for the vast majority of species then extant. Some, fortunately, survived to provide the seeds of a new efflorescence of life, the adaptive radiation of new species.

Darwinian evolution serves the purpose of making strategic "decisions" for natural CASs. Human engineering provides the same service for CAS artifacts. But human beings as individuals, and societies of human beings, undertake intentional evolution. The human brain[29] is a CAES and by virtue of that human organizations and societies are also intentionally strategic. That are the sorts of decisions to which we now turn.

12.3.5.1 Observing Other Entities and Forces in the Environment

A prime responsibility of the strategic agent is to observe more of the environment than just the entities with which the tactical agent is concerned. The reason is that those many other entities may have direct or indirect causal relations with the resource sources or product sinks that will change their behaviors in the future. The

[29] In reality it may be that many mammals and birds are also capable of possessing concepts that are modifiable to some extent. We know, for example, that chimpanzees organize hunting parties and war parties, indicating that they have an ability to think into the future and lay plans. Crows and parrots are also known to behave as if they were thinking ahead and assessing their environment to change their behaviors somewhat. De Waal (2016) asks if we are smart enough to know how smart animals are (from the title of his book)? And he suggests that many species demonstrate great cleverness and adaptability when confronted by changes in their circumstances. We suspect that the capacity to think strategically (which is considerably more than just being clever) is something achieved only by animals with brains with prefrontal cortices (Mobus 2019) or the equivalent in birds.

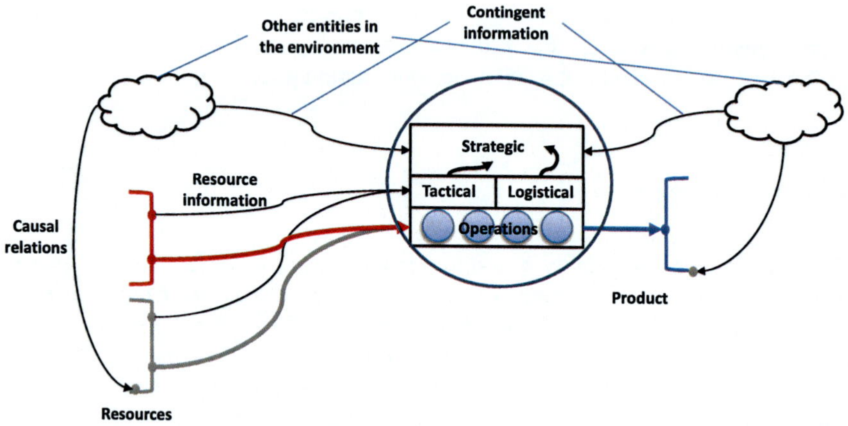

Fig. 12.13 The strategic manager observes other entities and the milieu in the environment. There are many possible "other" entities that have direct or indirect causal relations with the resource entities (sources) and/or sink entities that affect the future behavior of those entities

tactical agent may supply information regarding the long-term behaviors of the sources and sinks (as longer time-averaged behaviors) but the strategic agent must directly observe the rest of the environment. The strategic agent also receives long-time averaged performance data from the logistical agent. Figure 12.13 depicts the situation.

The clouds represent other entities (milieu observations not shown) that may or may not have causal relations with the sources and sinks. The strategic agent observes the behaviors of such entities over a longer time scale and is responsible for finding the causal correlations if they exist. The reason this is necessary will be explained below. It has to do with the need to anticipate those changes and take preemptive action to avoid damage.

12.3.5.2 Anticipatory Models of the Environment: What to Expect in the Future

The purpose of monitoring the long-term behavior of other entities in the environment in addition to the ones that have a direct relation with the whole system is to detect and exploit any causal relations that exist between those entities and the ones upon which the system depends. The strategic decision model of the environment can then be used to form expectations for what may happen in the future that would require evolutionary change in the system. For example, an automobile manufacturer that depends on a sole-source supplier of a critical part may take note that that supplier has lost another major customer that could put them in jeopardy and thus threaten the companies' future supply line. Or perhaps the supplier is known to be carrying a heavy debt load and the central bank raised interest rates that would squeeze the profits of the supplier. This is an example of a milieu condition change.

An anticipatory model is one that *learns*[30] about causal correlations between entities and milieu conditions in the environment and uses that model to anticipate scenarios that may play out in the future (out to some time horizon that is within the computational competence of the agent). Mobus (2000) has provided a working model of this learning in a robot.

Initially, the strategic agent may not have any information about which entities are relevant to its model. Clearly the agent cannot observe everything at once even if their sampling frequency is low. If the agent has background knowledge (either explicit or implicit) it can "make educated guesses." That is, it can select entities that seem most likely to be related to its sources and sinks. For example, if the sensory arrays and modalities available to the agent are sophisticated (e.g., vision, auditory, etc.) the agent may be able to discern systemic patterns in the arrays of data.

12.3.5.2.1 Causal Correlations

A causal correlation imposes a time ordering on events that must consistently not be violated. The brain's ability to encode causal relations and to modify the model in non-stationary environments is an archetype for modeling in general. Machine representations of networks of causal relations are the basis for computer-based models. These models are both structural and dynamical, expressed in Systemese from Chap. 4.

12.3.5.2.2 Anticipating New Sources or Sinks

Another aspect of building a causal model of the environment, even if some entities included are not found to be immediately related to current sources of sinks, is that the model can be used to identify new opportunities, new sources of current resources or sinks (customers) for current products. If the agent is sufficiently sophisticated and capable of recognizing affordances, it might be able to determine completely new resources that might be exploited or new products that it should produce for new customers. Human beings are quite good at this. Unfortunately, not much is known about how the brain computes affordances so we have little understanding of how this might be accomplished in a machine form. For the present, then, and probably into the near future, the ability of an organization or society to recognize affordances will rely on human insight.

---

[30] Learning causal correlations is covered in Mobus with respect to how biological neural networks appear to do it. In general, any learning mechanism that can deal with non-stationary relations should be able to be used for the purpose. In this era of Big Data and Deep Learning, we expect to see these applied increasingly to more structured strategic model building.

12.3.5.2.3 Anticipating New Threats

A causal model of the environment can be used to identify new threat conditions that may require evolving the system to counter (be preadapted to the change). For example, human societies are currently learning about the threat of global warming and climate change (e.g., sea level rises that will inundate coastal cities and island nations). The leaders of these societies, the putative strategic agents, are beginning to consider what adaptations may be needed to counter these threats. The chief executive officers (CEOs) of corporations are constantly concerned with product and financial market conditions and spot potential problems before they negatively impact the firm.

12.3.5.2.4 Observing the System Itself

Strategic agents actually have two models to work with. The model of the external environment provides anticipatory scenarios about what might happen in the environment to provide new opportunities or pose new threats. But they also have a model (usually very abstract) of the whole system itself (Principle 10 in Chap. 2). This model can arise from the same kind of learning procedure as described above. Or it can be "given," that is, provided by the structure of the system itself. For evolvable systems the model is adaptable to changing conditions.

The purpose of the self-model is to assess capabilities relative to external opportunities and threats. For example, in the case of a new opportunity, e.g., a new product to serve a new market, the strategic agent has to assess the system's ability to produce it. It may determine that significant investment in new structures/functions will be required, in which case it issues commands to the tactical and logistical agents to make this happen. The system then evolves to take advantage of (be fit in) this new environment.

12.3.5.3 Planning

Strategic agents develop long-term plans for their future behavior based on the models of the environment and of themselves. If they identified new opportunities or threats, they have to consider how prepared they are to meet the challenges, or how they might recruit the resources needed to effect a change in the structure/function of the organization/society. This is the problem of affordance processing mentioned above, for which there are no current machine-computable models.[31] Some humans are quite good at doing this as the strategic agents of organizations; they are CEOs

[31] Affordance is more than pattern recognition. Current automated pattern recognition can be trained to recognize an object for its category properties but that is not enough to suggest uses of the object in other contexts. This would be an interesting area of research to pursue—linking pattern recognition to identifying affordances.

of various kinds of organizations. Of course, not all organizations will necessarily have a CEO that is good at affordance, so not all organizations do a particularly good job of strategic management. For non-autocratic governments, it is very unclear how or even if affordance is accomplished. Other than the conduct of war (in various forms) governments do not seem to be able to think strategically (see discussion below). Autocracy might be successful if the leader is good at strategic thinking and affordance. For most of history, however, such "wise" leaders have been in the minority based on the fact that most historical civilizations, which were often autocratic or oligarchic, have collapsed, with the reasons understood to be poor strategies (Diamond 2005). As Diamond (2005) points out, many past societies were caught unawares of changes in climate (a milieu factor) and failed to adapt in time. Others chose to pursue unsustainable practices such as distant conquests or soil-depleting farming practices.

Planning is a process of using the two anticipatory models discussed above, coupled so that the output of the environment model is input to the self-model and vice versa, considering multiple possible states of inputs to the models, and then aligning the system's capabilities against the outcomes to test, in much faster than real-time, whether the system is sustainable (c.f. Rosen 1985). And if it is not under its current configuration, what would need to change to make it so?

12.3.5.3.1 Playing What-If with Model Simulations

A "what-if" game is played with environment- and system self-models. Factors in the environment, including the other entities and milieu factors described above, are varied in a range of plausible scenarios. The factors' range of changes goes from low levels of change to high levels of change in a set of model simulation runs. The outputs of the simulation are what are called best-case to worst-case scenarios. The middle case is generally thought to be the most probable outcome, but this depends entirely on assumptions made about the values chosen for the low end and the high end of the factor values.

Human beings practice this sort of simulation when they rehearse a scene they expect to play out in real life in the future. For example, when an employee anticipates asking their boss for a raise, they will imagine the scenario in their minds and with each run vary some aspect of what they do and imagine how the boss will respond. Since the brain can construct systems models of its own personality and the boss' then it is engaging in exactly this sort of what-if game playing. We can use this simulation approach much faster than real-time processing so as to project what is likely to happen into the future. We can thus anticipate which things might happen and what our more successful moves would be. Of course, the success of using this kind of planning depends on the efficacy of our models. Not everyone is sufficiently good at constructing veridical models or in eliminating emotional biases in selecting scenarios.

12.3.5.3.2 Using Scenarios to Set Up Plans of Action

Above, in Sect. 12.3.4.4.2.2, Tactical Management, Decision Models, Plans, on tactical plans we considered the assembly of action steps needed to accomplish a goal. In most cases, the plan is actually a series of plans, each set up to accomplish strategic steps toward a strategic goal. The strategic agent does not determine the details of the actions to be undertaken. Rather its job is to set up a sequence of goals that the tactical or logistical managers should accomplish. The sequence needs to be designed so as to lead ultimately to a strategic goal. For example, a human being may decide that going to college (while, say, a junior in high school) is in her best interest and so considers the tactical goals she will have to achieve to get into college. She might consider her grade point average (needs to be high), her extracurricular activities (need to be diverse), and her community involvement (need to be engaging and sacrificial). She thinks this through strategically but then hands off the goals to the part of her consciousness that deals with tactical details. She will work up tactical plans to realize these goals. The same brain is participating in strategic as well as tactical thinking. This might tend to blur the lines between the two, which is a very general problem for human agents when they try to explicitly participate in either kinds of thinking. We will examine this problem below when considering the human situation in governance explicitly.

For evolvable systems, the strategic decisions may involve modifications to the structures and functions of the operational and coordination levels of the organization. For human brains, this is limited to the act of changing one's mind about a concept. No changes in biological functions are permitted.[32] However, changes of mind can lead to changes in overt behavior. For example, a human who has had liberal political leanings might reflect on, say, the debt crises, and shift his thinking to become more conservative, after which, voting for conservative candidates.

In organizations and societies, there are no real constraints on what evolutions might take place.

12.3.5.4 Changing the Organization

Ultimately the purpose of strategic management is to evolve the organization so that it remains fit in the ever-changing environment in which it seeks to persist. In naturally evolving systems such as a species or an ecosystem, the changes are achieved through semi-random mutations (or invasive species) spread among a large population of individuals followed by natural selection for those few changes that led to phenotypes better equipped to persist, or, as in the case of an ecosystem, greater

[32] Although neuropsychology is considering forced adaptations of physiology based on held beliefs. The most straightforward such adaptations are seen in the placebo effect, now known to be a real shift in biological function due primarily to the patient's belief about their own condition. This is an area of intense research—the problem of mind over matter in the arena of physiology. Stay tuned.

stability and more efficient energy flux. The changes are chance but based on the fact that they take place in a large population of fecund individuals there is bound to be some change that is beneficial and will be propagated to the next generations.

In intentional, volitional systems such as the human brain or human organizations, the changes are made by design or adoption of and importation into the system of a structure/function observed in other systems. The change is still provisional and risky in that it is the strategic agent's best guess about what will happen in the environment in the future and that the design will be capable of countering the environmental situation or take advantage of a new opportunity. The anticipatory models that produced the guess are only probabilistic and only as good as the model-building procedure used to create it. A weak model will produce weak scenarios and coupling those with models of the "new" organization will likely not produce results in which one can place any confidence. The ability to implement a change that will be successful in boosting the future fitness of the system depends very heavily on the capability of the strategic agent to construct good models (see Chap. 11).

The strategic agent is responsible for deciding not only what changes should be implemented, but also why they should be implemented. For natural biological evolution that is a given. The genus evolves new species in order to continue into the future. For intentional and volitional systems, the situation is somewhat more complicated.

12.3.5.4.1 Considering Motivations, Values, and Beliefs (Ideology)

Humans and human organizations are motivated to continue their existence and are aware of the alternative consequences. At almost any time of day (or night for nocturnal creatures) an individual is aware of some motivating drive such as hunger until they take action to satisfy the need and it is satiated. All such drives are cyclical and may alternate in their presence in the individual's awareness.

Needless to say, human motivations for decisions and actions are much more complex than just the ordinary biological drives. Humans construct additional layers of motivations on top of those. Some of these, for example, what we call "values," are, at least in part, socially constructed or culturally derived and so are owing to historical developments (some of which were volitional change at the time, but most were likely ordinary accidental evolutionary).

The subject of the complexity of human motivations and values is still in the process of explication and is filling volumes of psychology papers/books so must necessarily be far beyond the scope of this chapter. Our purpose in pointing it out is to underscore the need for the strategic agents in human organizations and governments to take this factor into account when contemplating changes to meet a strategic need. A most common example of why this is requisite is the case of an organization implementing some change (ordering it from the top-down) without getting buy-in from the lower echelons in the organization. The cases of subtle passive resistance and outright sabotage are well outlined in, for example, past issues of the *Harvard Business Review*.

Where the effects of motivations, values, and beliefs become obviously potent is in the social subsystem of politics.[33] Political decisions are different from governance decisions. They are about what prevailing ideology is to be used to interpret the decision options given in the governance agency. Ideologies are a grouping of beliefs about how the world works, sometimes called a "world view." These beliefs need not be grounded in the reality of how the world actually does work (or should work). Neither are they necessarily internally consistent. Ideologies work to shape our concepts of governance and, as we are witnessing currently, are not based on any understanding of reality. Ideological beliefs are clear evidence of the breakdown in human agency and veridical decision-making of human agents. This is a subject we will revisit in Part 4.

12.3.5.4.2 Implementing Change

Figure 10.3 is a model of a complete intentional CAES showing auxiliary processes not directly involved in the ongoing economics. Processes such as the adaptation, repair and maintenance, and evolvability handle the exception cases due to disturbances (the first two) and long-term significant changes in the environment (last one). Figure 12.14 shows the relevant processes from Fig. 10.3 along with additional details of how a strategic decision to modify an internal work process comes about.

The strategic management process monitors the environment and detects a possible opportunity or a threat by comparing the current capabilities of the lower-level work processes to the nature of the threat or opportunity. The capabilities may have been determined by the logistics management (in this example) from a current state of time-averaged past performance. The strategic manager, exercising affordance, determines what needs to be done in the form of modifications to a work process that would then meet the future needs of the system to thwart a threat or seize an opportunity. The strategic agent sends a command to the evolvability process directing it to modify the work process. It requires knowledge of the current work process (its structures and functions) and proceeds to "engineer" the modifications that need to be made. With that specification and material/energy resources from system stores, it takes as input the old process and returns the modified version with new structures and functions.

---

[33] We are using the term, politics, in the very broad sense of human interactions directed at deciding what agents should be situated in the key decision-making roles in a governance structure. We most commonly think of politics in the context of governments but the same processes of argumentation and negotiation leading to decisions about who gets to decide occur within all human organizations. For example, it is well understood that in a non-government organization, official titles do not necessarily point to who is actually providing leadership.

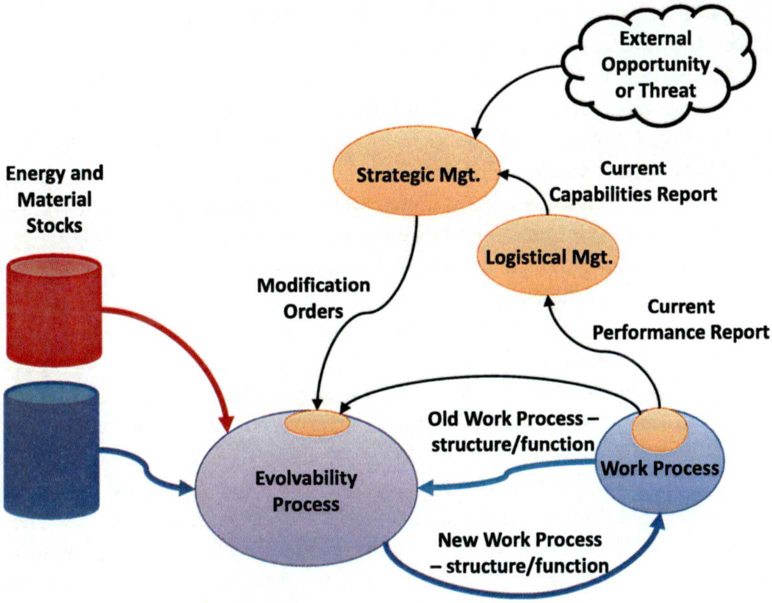

Fig. 12.14 The strategic decision to modify a work process is implemented by a special work process, the evolvability process, which is capable of turning a strategic command into the physical modification

For example, suppose marketing has identified a currently underserved customer base that would consume a modified product that a company already builds. To serve this base, the work process that manufactures the current product will have to be changed so as to produce the new version. Automobile companies famously bring out new models every few years based on this "theory."

The evolvability process should be seen as a complete manufacturing capability for tools and procedures that is a subsystem of the whole CAES. It is like any other work process in the sense of requiring energy and material inputs that it is capable of converting to the changes required by the strategic decision. But it is not a "routine" process. It is highly flexible and contains a design and engineering component that can translate the strategic command into a usable and efficient design (Part 4 covers this topic).

A very similar process takes place in the human brain as concepts are being "refined" or corrected. Though the details are not yet understood, it appears that the hippocampus is involved in orchestrating memory formations and reformations. Concepts are distinctive patterns of neural activity correlated across many brain regions. Engrams are formed when neural clusters representing percepts and concepts fire in synchrony with signals from the hippocampus that tell these clusters that they are part of a causally correlated assembly. They then strengthen their synaptic connections.

Evolvability processes may be as elaborate as needed to achieve changes in structures and functions in the system. Some organizations are able to make whole new processes in situ. Others have evolved means of incorporating smaller autonomous organizations that already do what the larger organization seeks to do. We call those operations mergers. Not too surprisingly, when corporations acquire other corporations to start serving a different market they are following a very old tradition in CAESs that started, perhaps, some two billion years ago when some larger prokaryotes formed symbiotic relations with other, smaller prokaryotes that eventually gave rise to eukaryotic cells.[34] This will be discussed below in the example of governance in cellular metabolism.

The act of making a strategic decision in government is no different from what happens in smaller organizations, except that the detailed architecture of a government may affect where and how such decisions are made. In the case of a representative democracy with a legislative branch of government, the decision to make changes may come from legislation. However, as will be discussed below in the Sect. 12.5, "The HCGS and Governance of HSS," the layout of the strategic decision-making agency is muddled. The changes affect policies and laws. The latter are determined by the legislature whereas the former fall within the jurisdiction of the executive branch. This division of power can get very confusing as well as contentious. Suffice it for now to say that however the rules for governing the society are decided, there are a set of operational agencies (mostly under the control of the executive branch) that are tasked with carrying out the change order.

12.3.5.4.3 Monitoring on-Going Performance

Once a strategic decision is made and the command executed in the actual change to the organization, the duties of the strategic agent are not completed. This is an issue that is sometimes lost on executives who issue commands and then think they have earned their paychecks. In reality, the strategic agent has to continue to monitor (through reports supplied by the tactical and logistics managers) the actual performance of the changes to make sure they are doing what the anticipatory models indicated. This is a higher-order, longer time-scale but nevertheless cybernetic activity.

There are many possible criteria for success and no simple set point error signal will tell the strategic manager how the whole system is performing. For Darwinian evolving CAESs, such as a species, the monitoring and assessment are performed by natural selection. But in intentional CAESs such as brains and organizations, the monitoring needs to be explicit and verifiable. Moreover, whereas in natural

[34] The process of symbiogenesis is thought to be the origin of eukaryotic cells and a precursor to multicellularity in animals, plants, and fungi.

systems where the law of large numbers is at play, in intentional systems there is a limited number of conditions for success. If the marketing manager's projections of sales turn out to be wrong, then the company may take a financial hit that will weaken it.

In the government of a society, the need for ongoing monitoring and attention to variances is all the more important. One of the aspects of adopting policies or legislating rules for commerce or behavior is the emergence of unintended consequences. In the governing of society these can take generations before they are recognized. And one of the crucial failures of governments is to not continue monitoring the social or economic impact of the changes they mandate and evaluate the effectiveness of their changes, or the consequences they never considered originally. A prime example is the various ways in which the fossil fuel industry has been subsidized financially in order to boost production and profits. Meanwhile, the increasing consumption of fuels has led to major releases of CO_2 into the atmosphere with consequent raising of the global mean temperature due to the greenhouse effect. That has to count as the granddaddy of unintended consequences.

12.3.5.5 Energy Requirements for Strategic Agents

Strategic management activities are information processes, which, in general, require less power than manufacturing processes. However, of the various types and levels of agents in a governance architecture, these require considerably more power. This can be seen in the energy requirements of the human brain, the latest evolved module, the prefrontal cortex, which is a power hog compared with other organs, including muscles. The whole brain consumes some 20% of the energy available to the human body at rest. Surprisingly, the prefrontal cortex (where "thinking" takes place) may only consume about 6% (Marieke et al. 2008). But this figure is somewhat misleading in that the prefrontal cortex is actually directing the firing of many different neural clusters throughout the brain. So, its power consumption alone is not indicative of the total power consumed in strategic thinking.

Another way to assess the power requirements of strategic agents is to consider the budgets for operations of executive offices in corporations. Here too, the overall budget of the localized office might not be significant with respect to the budgets of other management functions. However, just as with the situation with the brain prefrontal cortex commandeering resources from other parts of the brain, the strategic management of an organization often works by distributing chores to lower down offices that then provide the work required and consume the energy directed at strategic decisions. The corner offices on the top floors are generally responsible for a considerable amount of energy expended on information processing.

12.3.6 Comparing Stafford Beer's VSM with the CAS/CAES Governance Architecture

In Chap. 10, we introduced the generic CAS/CAES model archetype comprised of three subsidiary model archetypes (agent, HCGS, and economy). We also noted how the CAS/CAES model is very similar to Stafford Beer's viable system model (VSM) in terms of describing what a system needs to do to be a sustaining viable one. We now return to the VSM concept and describe here some key differences between the VSM and the CAS/CAES models especially with respect to how the governance subsystems are organized and operate in both. Recall that Beer was mostly concerned with applying principles of cybernetics to the management of organizations like corporations or NGOs (Table 12.1).

Table 12.1 Relationships between elements of the HCGS model and Beer's VSM

Stafford Beer's Viable System Model [mapping relation in brackets]
System 1 in a viable system contains several primary activities. Each System 1 primary activity is itself a viable system due to the recursive nature of systems as described above. These are concerned with performing a function that implements at least part of the key transformation of the organization.
[Operations-level units—Work processes]
System 2 represents the information channels and bodies that allow the primary activities in System 1 to communicate between each other and that allow System 3 to monitor and coordinate the activities within System 1. Represents the scheduling function of shared resources to be used by System 1.
[Cooperation communications network]
System 3 represents the structures and controls that are put into place to establish the rules, resources, rights, and responsibilities of System 1 and to provide an interface with Systems 4/5. Represents the big picture view of the processes inside of System 1.
[Similar to the coordination level but not differentiated between tactical and logistical]
[Systems 1–3 concerned with the "here and now"]
System 4 is made up of bodies that are responsible for looking outward to the environment to monitor how the organization needs to adapt to remain viable.
[Half of the strategic level]
[Concerned with reconciling the "here and now" with the "then and there" (below)]
System 5 is responsible for policy decisions within the organization as a whole to balance demands from different parts of the organization and steer the organization as a whole.
[Other half of the strategic level—Concerned with the "there and then"]

Source: Wikipedia Viable System Model

12.4 Examples of the HCGS in Nature

We humans commonly differentiate between what we call "nature" and the "artificial." As we hope to show in Part 4, in reality, there really is no absolute schism between those CAESs that resulted from evolutionary development (nature) and those that have resulted from human intentional organizing, that is, inventing those artifacts. It is all ontogenesis.

Nevertheless, prior to human minds evolution produced some remarkably stable, resilient, and sustainable CAS/CAESs, living and supra-living systems that achieve those qualities by virtue of the governance of an HCGS subsystem. In this section, we will take a cursory look at a few examples, enough to push the point that this architecture is found throughout naturally evolved systems. And the point is that if this has been nature's solution to governance, then we humans might benefit from paying attention and emulating nature's way in our own form of governance. We look at three examples of HCGS architectures in living systems, starting at the base scale of individual cells and the regulation of cellular metabolism, then considering the regulation of multicellular organisms' physiology. Both of these are governance structures and functions relative to what constitute the "economies" of the two scales, as will be elaborated in the next chapter. From there we will examine the governance of groups of individuals or a "society" such as herds or packs, and finally, an example from the human sphere, the human brain as an HCGS.

12.4.1 Regulating Cellular Metabolism

Enzymes regulate basic chemical reactions, both anabolism and catabolism. Protein-based enzymes are the main agents in governing metabolism in the cell. Figure 12.15 shows an example of the hierarchical cybernetic regulation mechanisms involved in the synthesis (anabolism) of a product that is active in the Krebs Cycle. It is a combiner work process mediated by a specific enzyme. Another enzyme detects the concentration of the metabolite and is stimulated by an insufficient concentration. The reduction of a concentration of the metabolite causes the detection enzyme to generate a signal that transmits into the domain of the genes; it is received by what we can call an "activation enzyme" that, in turn, causes the gene that is responsible for coding for the original enzyme to produce the mRNA signal. That, in turn, is used by the ribosomes to crank out more of the original enzyme, thus increasing the rate of metabolite production and thereby increasing the concentration of it in the Krebs Cycle.

The anabolic process is part of the basic economy of the cell, the metabolism that produces energy units (ATP molecules) that are used by all other processes, including detection and activation enzymes. The latter are part of a higher-order regulation loop that operates over a longer time scale than the anabolic process and, though not directly shown in the figure, coordinates among multiple different

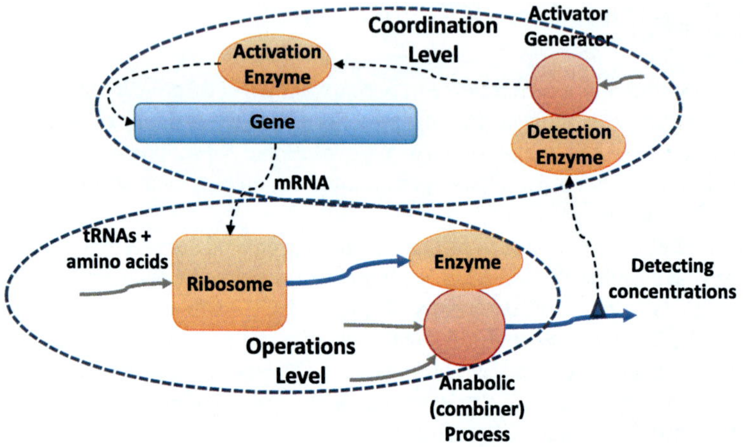

Fig. 12.15 This figure shows a hierarchy of regulations involved with managing an anabolic work process. See text for explanation

metabolite-producing processes. Thus, it is part of the logistical coordination layer of the HCGS.

Single cells are generally not considered evolvable in the way we have been using that term. Species of cells, say bacteria, are evolvable both in the narrow sense that individual cells can "permit" mutations in select genes to not be repaired in response to external stressors on the chance that the mutation might be useful in thwarting the stress, but also in the broader sense that species do evolve in the Darwinian sense. The point of this note is to recognize that single cells are CASs only and this means that the governance system—the HCGS—does not include a strategic layer.

That said, we can also supply examples of tactical-level coordination in cells, such as managing to find food and expelling wastes as needed (though cells don't particularly coordinate their exporting of wastes with the environment!). That is, unless the cells are members of a larger community, a multicellular organism, then tactical coordination between cells and cell types in tissues becomes essential.

12.4.2 Regulating Multicellular Physiology

In this example, we look at what we can call the energy sector of the body (not unlike what we saw in Chap. 9), the regulation of blood glucose, a low-weight sugar that serves as energy packets distributed via the circulatory system. All cells in the body extract glucose molecules from the blood as their fuel. The regulation of the right levels of glucose concentration in blood is a function of the endocrine system. That is a system of organs, tissues, and cells that uses chemical signals (hormones) distributed through the body, also via the circulatory system, to achieve balance in operations. In other words, it is a major part of the logistical coordination in organism physiology and it cooperates with the tactical coordination governance as will be seen later.

There are a huge number of organs, etc. that participate in the endocrine system. There are many kinds of signals and responses in other organs. We will focus on just one function in this section.[35]

Just as ATP was seen to be the "currency" in cellular metabolism, glucose is the currency in whole body physiology (glucose being the fuel in metabolism that produces ATP) and very analogous to fossil fuels in society. When the concentration of blood glucose falls below a certain value, about 90 milligrams per deciliter, the pancreas produces glucagon, a hormone that signals the liver to produce glucose by converting glycogen to glucose and injecting it into the bloodstream. The pancreas is acting as a logistics agent in that its time course of action is longer than the rate at which glucose is absorbed and used by body tissues. And it is essentially resetting the set point for liver delivery of glucose to compensate. Contrariwise, if the concentration of glucose in the blood rises above, say, 120 milligrams per deciliter, the pancreas produces insulin to tell the tissues to take in more glucose (which they can store locally if not needed) and the liver to reduce its production of glucose. Thus, between the pancreas and the liver, a fine balance in the availability and rate of use of the primary energy carrier in physiological terms is maintained.

Both organs are acting in twin roles, as operational controllers, e.g., the liver converting glycogen to glucose, and cooperating logistics coordinators, e.g., the pancreas producing either insulin or glucagon signals to mediate the liver cells producing glucose. This is because the organs harbor different cells that fulfill different decision agents in the process. Figure 12.16 shows the systems view of the organs and their inputs/outputs.

The pancreas will produce glucagon as part of the general fight-or-flight response mechanism that is initiated by the nervous system—recognition of a dangerous situation—and triggered through the brain's connection to the endocrine system via the hypothalamus. In a dangerous situation, the animal needs more energy to run or fight and so more glucose will be needed by the muscles and other tissues. This is the tactical–logistical cooperation that matches the needs of the animal's actions with respect to an environmental entity or situation, and the internal production of energy availability for the response. Without the extra energy the animal would soon fatigue and fail to survive the threat.

[35]The glucose regulation subsystem was covered in Mobus and Kalton (2015, Sect. 4.5.1) in the context of the networking nature of systems. This discussion provides more details of this function in terms of governance architectures.

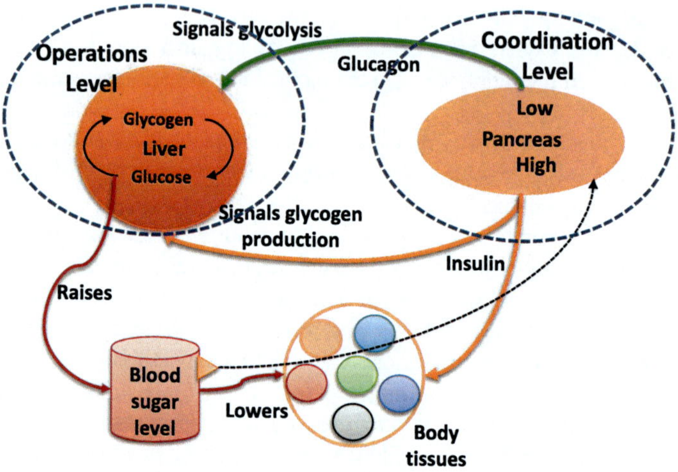

Fig. 12.16 Physiological regulation of the blood sugar level

12.4.3 The Human Brain as an HCGS

The human brain is a premier example of the hierarchical cybernetic governance system in the realm of biology. It is especially so because it is the first brain to be capable of full strategic thinking (Mobus 2012, 2019). In this section, we present a brief description of brain modules that constitute the HCGS. We will not so much focus on the details of anatomy, even though we name a few of the more important parts for orientation. For readers who want to become more acquainted with cognitive neuroscience, please see Baars and Gage (2007) and Carter (1999).

12.4.3.1 Operations Management

The economy of the human body is the physiology of the body itself. As body plans evolved from the earliest bilaterally symmetrical animals (e.g., annelid worms) through chordates and more lately birds and mammals, the parts of the brain responsible for coordinating the complex of organs and tissues have also complexified to keep everything operating in balance. This includes the limbic core, sets of "nuclei," clusters of neurons devoted to single functions such as monitoring blood oxygen levels. This core communicates through two main subsystems, the autonomic nervous system and through the endocrine system. The autonomic nervous system, as its name implies, is responsible for automatic functions with generally cyclical, short time-constant behavior such as breathing and heartbeat. The second system, the endocrine system, is part of the logistical management of the body and will be covered below.

Much of operations-level management is performed by highly distributed nuclei both in the brain stem and in the spinal cord itself. The autonomic system is a set of nerves outside of the spinal cord that run directly from or to the brain stem. These regulate non-conscious activities like digestion by innervating smooth muscles in the lining, but also stimulating the release of digestive enzymes.

Basically, for every organ and most tissue types, the hindbrain is involved in regulating the activities. The limbic core (that is, the hindbrain) is evolutionarily the oldest part of the brain and developed just to manage bodily functions.

12.4.3.2 Logistical Management

The endocrine system uses hormones circulating in the bloodstream to activate (or dampen) various tissues in response to changes in body states. Some endocrine functions may actually be better characterized as logistical since they operate to balance activities among various tissues. This system is affected by the brain stem or the limbic core via the hypothalamus and particularly the pineal gland.

But the autonomic system also has logistical functions in that it coordinates with various parts of the basic operational management to alter operations in various tissues based on variations in body states. For example, it can increase heart rate in response to adrenalin increases in the blood (that being part of the cooperative interactions between tactical and logistical coordination).

In general, logistical management functions evolved to meet the needs of increasingly complex body plans.

12.4.3.3 Tactical Management

All living systems must interact with their environments. For non-photosynthetic organisms (heterotrophs) this means acquiring food and other resources. For many organisms, some bacteria even, and all animals, this means moving in the world. Filter feeders like corals can forgo overt motion since they literally are surrounded by their food; but they do move their filter-feeding apparatuses to get that food. For most others this means moving about in the environment in search of food sources, foraging search (Mobus 1999). And that requires means for sensing the environment and a means of motivating the body through the environment. Between these two means, the brain must process the incoming data regarding the state of the world that pertains to the organism, decide what actions to take, and then direct the motor system to affect those actions, that is, the brain is the agent. The organism must negotiate a pathway through its environment. It must locate the sources of its resources and undertake actions to acquire them once they are found. All of this requires tactical management.

The brains of all higher animal organisms are largely devoted to tactical management. The human brain is no exception. The vast mass of the human brain is devoted to interacting with the world. The basic sensory processing that took place originally in the limbic system (outside of the core) has been augmented by expanded pattern recognition capacities provided by cortical structures such as the hippocampus and paleocortex in advanced vertebrates (e.g., amphibians and reptiles). These structures allowed animals much more capacity to learn aspects of their environment that could not be anticipated by pure instinctive (built-in) models. These animals were far more adaptive in the face of environmental contingency than their ancestors. And then, evolution, building on the successes of cortical structures, produced the mammals (and birds evolved from dinosaurs) with a neocortex, a structure apparently capable of constructing representations of almost any kind of concept.

Most of the neocortex is devoted to enhanced tactical management. For example, the occipital lobe of the cortex provides for extremely enhanced visual processing, far exceeding the capabilities of the amygdala (processing threatening situations) or the cingulate cortex (paleocortex) processing comparative impressions. The neocortex allows the organism to build a much more comprehensive representation of things in its experienced environment and through memory recall better construct interaction plans with those things. The vast majority of neural mass in the bird and mammal brain is devoted to interactions with conspecifics, that is, social interactions. In humans, this means being able to construct models of other human beings for purposes of anticipating those others' behaviors and to formulate cooperative interactions with those recognized as "my group."

12.4.3.4 Strategic Management

We come, at last, to the level of strategic management in the human brain. We assert, and have written extensively about, the human brain is the first kind of animal on the planet that has achieved an extensive level of strategic thinking (Mobus 2012, 2019). It is likely that many mammals, especially our closest kin, the great apes, have some capacity for strategic thinking (De Waal 2016), but it appears to be very limited in terms of time scale and capacity for generating scenarios. Humans underwent a major evolutionary transition some 180,000–200,000 years ago with the acquisition of recursive language and abstract representation of concepts (Mobus 2018) that included the ability to think much further into the future and consider consequences of current actions on future states of the world (c.f., Lombardo 2006, for a discussion of the evolution of consciousness of the "future"). This corresponds with a major expansion of the part of the neocortex called the prefrontal cortex, and particularly the region labeled Brodmann Area 10 (BA 10 in Fig. 12.17), the frontal-most part of the prefrontal cortex. This module of brain is now thought to be the seat of long-term thinking and planning, in other words, strategic thinking (Mobus 2018).

Figure 12.17 summarizes the structure and HCGS functions of the brain. Each human being is governed by the activities of the brain. Their bodies are regulated by

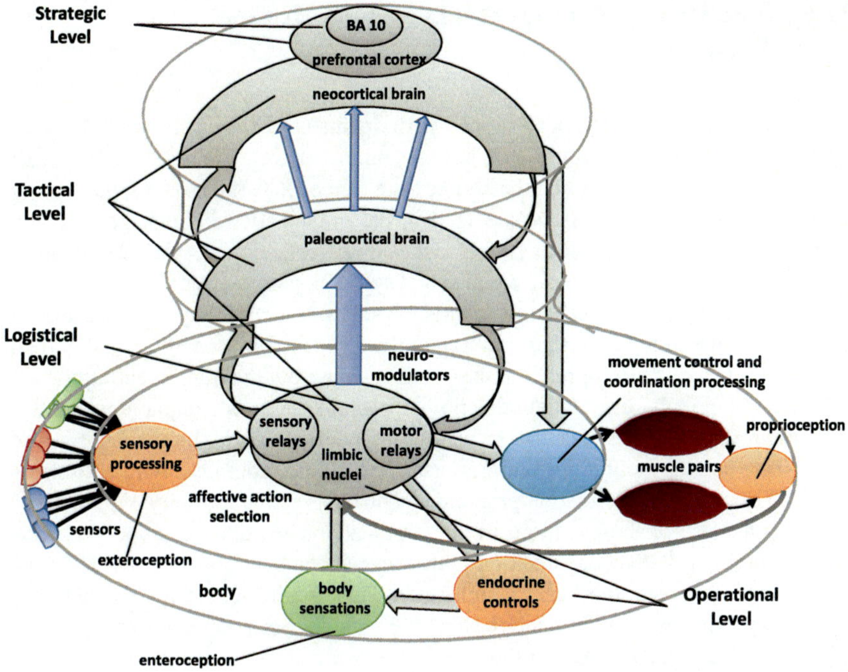

Fig. 12.17 The human brain is one of the best examples of an HCGS that governs all aspects of the human body and its interactions with the environment, including strategic interactions. Lower, older parts of the brain are associated with operational and some logistic management functions. Some newer areas of the limbic system are involved in both logistic and tactical management functions (and coordination between these two coordination managements). The two cortical structures are involved primarily with tactical management, which includes some support for the planning of tactical moves that verge on, or support, strategy behaviors. The prefrontal cortex and, specifically, Brodmann Area 10 (BA10) are involved in fully strategic management functions

evolutionarily old brain functions for direct operational control and by logistical coordination, primarily through the endocrine and autonomic system, that evolved as body plans became more complex early in the evolution of animals. And animals that needed to hunt for resources, especially food, evolved tactical management functions to ensure they could survive in changing environments.

As environments continued to become increasingly complex, brains evolved some capacity to anticipate future conditions. Some degree of strategic management emerged but it reached its culmination in the evolution of the human brain. Now we have an example of an HCGS that provides us a model of an idealized model of governance. The major question is: what can we learn from such a model that might benefit our understanding of governance systems?

12.5 The HCGS and Governance of the Human Social Systems

Social systems are aggregates of human participants who have a sense of common purpose. Whether it is a family unit, a local community, a social organization (like a club), a city, or a nation-state, the architecture of the HCGS, or something bearing some resemblance to it, can be detected in the organization of management and administration. All such social systems make strategic, tactical, logistical, and operational decisions. Sometimes, in small social systems, one or a very few individuals are responsible for these decisions. In larger systems, the roles may be distributed in an "official" hierarchy as in a corporation or the military, with specifically identified individuals responsible for making them and carrying out the implementations.

What is uniquely different about a human social system as compared with naturally evolved CAS/CAESs is that the agents are individuals with human brains. And as we saw above, the human brain is, itself, an HCGS complete with a strategic level. This sets up an interesting problem and also a great opportunity. Figure 12.17 provides a sense of the situation. In any organization we find all three levels and agents within each level capable of all three levels themselves.

The figure depicts the basic architecture. This architecture can be recursively applied as societies grow in scale (basically the number of people) to manage complexity. As noted above, as a system grows in size and heterogeneity it must divisionalize at all levels, adding sublevels where necessary in the coordination level, due to the span of control problem. But the relations shown in Fig. 12.18 remain.

One might ask why operations-level agents need to have strategic or tactical, if not logistic, decision-making capabilities. There is an advantage. A human agent

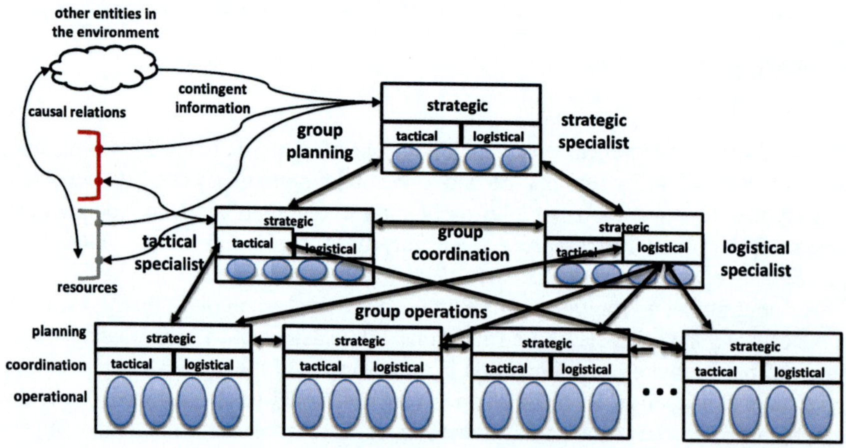

Fig. 12.18 An HCGS for human organizations is a set of nested hierarchies. Blue ovals represent work processes. At the operations level, these subsystems may be processing materials and energies as well as information. At the coordination level, these represent specific information processes that require management. The same is the case at the strategic level

can manage far more activities, greater complexity of work processes, meaning that having coordination-level and strategic-level decision-making abilities allows them to interact with the other agents (i.e., other *departments*) strategically. This results in the potential for evolvability anywhere in the social organization. For example, an operations manager, given notice about improvements in the product that they are responsible for producing, can plan and execute the necessary changes without being given detailed instructions from above. In other words, the hierarchy of human agents permits the efficient distribution of all three kinds of decisions. Top-down micromanagement is, in theory, unnecessary.

This architecture provides a tremendous amount of flexibility and "modularity" with respect to how any agent in any position in this organization can be replaced (at least in theory) with any other agent (or newly recruited agents). Notwithstanding individual talents (note in the figure that the rectangles associated with the level and decision types are shown a bit larger, representing a specialization relation), thanks to the ability to learn and adapt, many human individuals are equipped with the basic capabilities to be placed in any of the positions. This is exactly the basis for things like promotions in corporations and politicians running for higher-level offices. It is also the basis for being able to cover a critical position if an employee or member is unable to perform their duties. It is thought that one of the main reasons that humans have been so successful as a species is that working in social groups with such flexibility gave them a distinct advantage in the game of survival (see Chap. 9, Sect. 9.5.2.3 Societies and Economies, Fig. 9.7, and Sect. 9.5.2.4 Tribes and Beyond, Fig. 9.8).

The problem, however, is that the agents as depicted in the figure have more to them than ideal decision-makers. We humans have many drives, needs, wants, and a panoply of emotions to underscore them. We have ambitions and expectations that can too often supersede the rational decision process and our individual strategic thinking can lead us to less-than-optimal tactics. In other words, humans are too often not motivated to make the best decisions for the well-being of the society. This is a fundamental conundrum.

We will not belabor the human foibles that sometimes keep the governance of social systems from performing the function of maintaining the system as a whole. Psychologically, we are what we are. However, in Chaps. 15 and 16 we will address this problem of humans as unreliable agents and discuss design approaches that might mitigate the problems.

Below we will compare typical governments of nation-states with the HCGS architecture and functioning to get an idea of where the governance of human social systems stands in relation. We will consider some plausible mechanisms that could be implemented to compensate for human foibles along the way. But primarily we seek to demonstrate some major flaws in the current state of thinking about the structure and function of governments as we find them today. We need to remind ourselves that governance structures evolve over time. Those that survive do so because they prove capable of keeping their substrate system viable in the long run. Insofar as human governments are concerned, very few, at the nation-state or empire level, have proven viable for very long. In light of the rate of state failures and state disruptions currently taking place, we need to ask if this is because the various governments, we have seen to date, deviate from the model of governance that nature has found so useful.

12.5.1 Typical Government Architectures and the HCGS Model

In this section, we take a look at the differences between the ideal HCGS and the kinds of approximations that human-designed governments approach it. The governments of nation-states bear some resemblance to the HCGS, but only superficially.

12.5.1.1 Governmental Roles

Regardless of whether the political system is based on democracy (in some form) or authoritarian power,[36] three basic functions have to be performed. In a form of democracy such as the representative form of the USA, these functions are usually performed by separate branches of the government. In dictatorships, all functions may be under the control of the dictator even if performed by separate agencies. Regardless, all three are universally thought to be needed for governments to operate.

The typical governmental architecture found around the globe is based on a tripartite division of responsibilities into three distinct, and in the case of democracies generally, co-equal branches. These are the executive, the legislative, and the judicial branches. This architecture is repeated at various scales from nation-states to counties to cities of varying sizes. Names of offices and titles will vary but the functions being performed are the same at each scale.

Executive branch duties are not dissimilar from chief executive and chief operating officer duties in organizations. The executive branch sees to the administration of the laws of the land and the management of public infrastructure. As a rule, the executive is also responsible for public health and safety, as well as defense against foreign powers (in the case of nations).

The legislative branch is tasked with making the laws of the land. Laws cover a variety of social interaction arenas such as commerce and trade, citizen behavior, and maintenance of individual and group rights.

[36] Here we will be primarily concerned with a basic democratic social organization since that form has been spreading across the globe in the last century. Democracies come in many shades; some even retain a monarchy! Some are economically democratic, that is, they embrace some forms of capitalism where individuals can pursue entrepreneurial interests, even while maintaining autocracy when it comes to social behaviors.

The judiciary's primary responsibility is to administrate justice, to make findings of legality, to sanction those who are found to be breaking laws or harming others, and to interpret legislated laws in terms of higher-order laws such as constitutions, which are considered the ultimate set of laws.

These three functions, as mentioned, occur at multiple scales and would appear to form a kind of hierarchy, cities at the lowest level, counties or their equivalents next up, states or their equivalents next up, and nations at the top. However, this hierarchy is not the same as the HCGS. It is true that city governments deal in local, real-time issues, but so do counties for their rural areas. It is also true that the kinds of decisions that are made at the state levels tend to have a longer time scale relative to counties and cities. And it is true that nations operate on a much longer time scale than states, as a rule. But this resemblance with the time scale layering in the HCGS is not because governments are HCGSs. It is an artifact of the scale of size and complexity. It just takes longer to do things on a national level as compared with a state or county level.

As a challenge to the use of systems science, thinking, and analysis to the really big problem domains (as in Chap. 9 when we tackled the biophysical economy), we will examine a "typical" government architecture.

12.5.1.2 The Basic Problem: Mapping the HCGS onto the Government Architecture

A starting place for using a CAES model to guide the analysis of a society with multiple scales of government would be to inquire as to where one finds the three levels of the HCGS in the government. Clearly the economy (to be defined in the next chapter) of the society is embodied in the various industries, consumption processes (domiciles or households), infrastructure, and associated institutional organizations. What we are looking for is how are all of these *managed* for sustainability.

How do the operational, coordination, and strategic levels of the HCGS map to the typical government architecture of nations? We might be tempted to assign each branch of government some role in HCGS terms but this turns out to be quite difficult. Is the executive branch the strategic layer? Is the legislature the tactical or the logistic coordination? And where do you put the judiciary? With only a little reflection one realizes that there is no clean mapping.

Our current notions of governmental architectures come to us from history, and historical accidents. The executive branch, especially when embodied in singular heads of state such as presidents or prime ministers, derives from the evolution of early kingdoms and "headmen" autocrats who directed the activities of the early grain-states. In theory the king takes care of strategic management and the various ministers or secretaries with their clerks and other workers are the coordinators. The only real strategic decisions to be made are embodied in foreign affairs issues. But the executive is also tasked with one of the main tactical management activities as well, the military.

The notion of a singular executive continued on through many secular government models in the Enlightenment and down to modern liberal democracies such as embodied in the U.S. Constitution (Article 2). The problem comes down to the difference between an executive "function" and an executive "person." The governance architecture of an HCGS implies that there is an executive function, that of the strategic manager. But this does not imply, necessarily, that this function is, or should be, embodied in a single person. Indeed, the agent computational model, with faulty or error-prone agents, suggests that a redundant structure, several agents operating in a single function, is more appropriate. This is hardly surprising since human social systems were originally designed with a council of wise elders providing the executive function for the tribe. There are too many psychological reasons why a singular person should not be given the sole authority to make executive (and strategic) decisions, the simplest expression of this being: "Power tends to corrupt, and absolute power corrupts absolutely. Great men are almost always bad men…."[37] As pointed out in Chap. 11, human agents are far from ideal and, indeed, are far from wise (Mobus 2019). At the same time, a committee of sufficiently wise persons would be more likely to arrive at wiser decisions. We will revisit this notion in Chap. 16 when we consider the design of a more systemic human social system.

Similarly, the legislative branch of government seems ill-suited for the task of establishing rules for the operations of a CAES. Legislatures tend to be centrally organized, even in so-called democracies. In a republic such as the USA, the decision-making authority of the people is vested in representatives elected to congress. The legislators are typically distant from the constituencies and may not have a good understanding of the issues that are local to those constituencies. There are many more problems of representative legislation, too numerous to list completely. Here are just a few that have developed in representative democracies such as the USA. First, the scale of the governed system has gotten immense. The complexity of the system has gotten beyond the comprehension of ordinary legislators. Second, in the USA, the political divisions owing to a two-party system that prevails in the country create a we-versus-them mentality and gridlock in the legislative process. Third, the invasion of monetary aspects into the political process has resulted in the distraction of attention from studying the ins and outs of legislative proposals when legislators end up spending far more time trying to raise money for their reelections. Clearly, the design of the legislative branch is ill-fitted to the real issues raised by an evolving system needing to acquire new adaptations to a changing environment as well as a changing interior (e.g., the rapid developments in technologies).

[37] Attributed to John Emerich Edward Dalberg-Acton, first Baron Acton, but repeated by numerous political commentators.

12.5.2 *The HSS as a CAES*

We contend that to be a viable system for the long haul, a society must meet the operational and governance requirements of a CAES. These include a balanced functional economic subsystem that obtains resources without depleting them, produces products of value to the supra-system embedding it, exports by-products without polluting the supra-system, has an HCGS competent in making operational, logistical, tactical, and strategic decisions, and possesses work processes capable to providing repair and maintenance, adaptive responses, and generate evolved solutions to problems caused by non-homogeneous non-stationary processes in the supra-system, or even the supra-supra-system (e.g., the Earth's environment in the solar system). Extraordinarily complex systems that do not have these characteristics cannot survive for long in continually changing world. The key to success is having a governance system that embodies the HCGS at all scales.

There isn't a lot we can do about humans as agents, that is, to be ideal agents. The only solution to the problem of faulty components is redundancy and majority-rule, that is, putting the decision-making in the hands of "committees" as opposed to singular individuals. These groups of decision-makers cannot be too large as the process of arriving at a decision can get too cumbersome when too many people are involved.[38] We will return to this issue in the final chapter.

As pointed out in the prior section, the governments of most societies today are not designed in such a way that the roles of decision-making in an HCGS are clearly delineated. These government architectures are the result of humanity starting from a position of ignorance (especially of systems theory) and allowing a hit-and-miss evolution of government structures to provide the blueprints. From the Agricultural Revolution and the formation of the early grain-states to the present, societies have been grasping for a form of government that would make them a viable system into the indefinite future. None have yet succeeded. But some have worked out semi-systemic mechanisms that have gotten them closer. In Part 4, we will consider how our knowledge of systemness, systems science, and the CAES model can be brought to bear on the question of what kind of governance architecture is most appropriate for human society.

12.6 Conclusion

Every CAS/CAES embeds its own HCGS in order to regulate its internal and external behavior, to maintain stability in the face of a changing environment, to continually monitor its own viability and regenerate itself as needed. Every CAS needs at least the operational layer of governance to maintain effective and acceptable production of products and services within the system economy. Sufficiently complex

[38] If you do not believe that, try participating in a typical academic department meeting sometime.

systems also require coordination in terms of both internally balancing behaviors of the work processes and managing the extensive behavior of its import and export processes—coordinating their activities to cooperate with sources and sinks. CAESs further require some degree of strategic management in order to deal with highly complex and changing environments, beyond merely coordinating with external entities. At its highest level of strategic management, the HCGS provides anticipatory, preemptive actions that help ensure the long-term viability of the whole system even as the world around it changes in completely unpredictable ways.

Governance applies to the economy of the CAS/CAES but also, reflexively, to the agents that populate the economy and the government/management subsystems as well. Agents monitor and affect other agents when they appear to be off-track in their decision-making duties.

As an archetype model, governance of a CAS/CAES is a given structure; it is the starting place for analysis and design of any CAS/CAES. Analysis will elaborate the details of the structure for a particular instance. Designs of artifacts (see Chap. 14) will be more certain to meet desired performance requirements when using the HCGS model in designing the governance of the artifact.

Chapter 13
The Economy Model

Abstract In this final chapter of Part 3, we consider the systems nature of economies. That is, we will look at an economic system from the systems perspective. As we noted in Chap. 9, the human social economy has some serious problems, especially with respect to the energy sector and its poorly understood relation to the financial sector. In this chapter, we will provide a general archetype model of what an economy is, in fact, and what it does with respect to the whole of a CAES.

13.1 What Is *an* Economy?

In Chap. 11, we argued that the evolution of systems, generating increasingly complex ones we've labeled CAS and CAES, has resulted in the emergence of a generic agent model archetype. In Chap. 12, we argued the same with respect to the emergence of a generic governance model archetype. Now we will complete the argument for a set of generic CAS/CAES pattern archetypes with the internals of these classes of systems, the pattern of sub-processes that constitute the whole activity of such systems. All such systems demonstrate an internal and external dynamic that we will call *an* economy.

This chapter stems from noticing an interesting pattern that shows up in all CASs and CAESs. Below, in Sect. 13.2, Considering a Generic Economic System, we will explore this pattern in more detail, but fundamentally it is the way in which these systems obtain resources in energy and materials, how they do internal work to maintain themselves and grow/replicate, and how they export waste materials and waste heat back into their environments. Sometimes such systems export something that is not of use to themselves particularly but is of use to some other entities in the larger supra-system in which they are embedded. In such cases, we call the exported stuff "products."

What all of these systems do internally is to use the energy and material imports in work processes that build and maintain essential internal structures. Waste products are inevitable in all such processes in nature. The organization of the work processes and their ongoing management (which collectively we can call the governance of the system) is the pattern of which we spoke. Whether we are talking about

G. E. Mobus, *Systems Science: Theory, Analysis, Modeling, and Design*,
https://doi.org/10.1007/978-3-030-93482-8_13

the internal organization and dynamics of a living cell, of a multicellular organism, of a Neolithic tribe, a single household, a complex organization, or a modern nation-state, the patterns of organization and dynamics follow them. These patterns cluster under the title "economy."

The word "economy" derives from the Greek "oikonomíā" meaning the *management* of a "household." Eco (oiko) is the Greek word for "home." In order for a home to be successful (fit in the Greek society) the "master" and "matron" of the home had to plan and practice "economic" activities—raising adequate food to support the household, for example. This meant managing processes so that the household recognized an accumulation of wealth that would ultimately provide for the support of subsequent generations of the family. Greek philosophers such as Socrates, Plato, and Aristotle spent no small amount of ink pointing out that economy was an essential aspect of the good life, primarily so as to support the head of household being able to devote time to philosophy and politics (c.f. Leshem 2016, Introduction). In order to have time to think and participate in the governance decision processes of the state, one had to make prudent decisions about how to manage one's own affairs at home.

The term "economy" may have focused on the management of the household, but what is more important in the long run is that the various economic activities—the work processes internal to the household—are properly organized in the first place. In this chapter we will take this focus, adding in the governance mechanisms from the prior chapter to complete the integration of the whole CAS/CAES.

Fast forward 4000 years. The management of a single household is still an issue, of course. But now that household is embedded in an extremely complex milieu of sources of income and the goods and services needed as well as the costs associated with maintaining and provisioning the next generation of family. That embedding is in something we now, casually, call *the* economy, by which we mean the aggregate of buying and selling goods and services in something called a "market."

The HSS economy is, however, not the only example of a resource-extraction-production-of-wealth-consumption management system. At an abstract level, an economy, is a fabric of transactions, a network of sub-processes of resource acquisition, transformation (value added) production, and consumption, in which biological beings are sustained. Below we will introduce the notion of different kinds of economies, operating at different scales in living and supra-living systems.

In the systems view, an economy is a way of managing the flows of high potential energy and of the transformation of high entropy materials into low entropy assets via work processes that use that high potential energy to do useful work. The low entropy assets support the existence and stability of the system.

For example, we observe the nature of a living cell, the fundamental unit of life. Cells have evolved to be stable (sustainable) over relatively long periods of time relative to the time scales upon which the underlying biochemical reactions occur. Molecules come and go, but the organizing pattern of a living thing persists. What is happening within the realm of metabolism—the economic system of a cell—is management of resources to benefit the whole cell.[1] That system is not perfect in the sense that it can go on forever in any single cell. Entropy has a way of catching up with a single entity eventually. But, the lifetime of a single cell relative to its internal processes is impressive, to say the least. If the HSS were a cell with such a stable economic system, it could persist for many millions of years. And, including the fact that most cells replicate, achieving a kind of immortality through reproduction, some kind of HSS might persist indefinitely.

Similarly, at the scale of multicellular organisms, they have achieved an even longer time scale of stability by evolving cooperation between many different cell types within a single organization—the body, in the form of physiology[2]—a matrix or web of interacting processes in different kinds of cells, tissues, and organs. We human beings, as an example, can sometimes live for close to 100 years while the cells that make up our bodies live for only days or months.[3] Our physiology allows our pattern of organization to persist longer than the lifespans of the cells that make us up. Physiology is to the body its mode of economy. Our bodies are designed to process energy and material inputs to support our organization of cells, tissues, organs, and so on. Most of the time the underlying economy of the body works quite well. Of course, entropy catches up with that system as well. However, not before we have had an opportunity to reproduce ourselves in the form of children who carry our characteristics forward in time. A population, and a species of beings, can persist over indefinite time so long as its economic subsystem works to achieve that.

The human brain is a unique kind of society of information processing modules. It crosses the threshold from being a CAS to becoming a CAES by virtue of its capacity to learn and even construct new knowledge over its lifetime, and knowledge (as long as it is veridical and useful) are the assets produced. The brain is a biological entity, so includes metabolism (e.g., of neurons) and physiology, as it is one subsystem in a body. But it has a new kind of dynamic that can be described as an economy of knowledge produced from information as the inputs. We will refer to this information/knowledge economic system as neuroeconomics. This is an economy of thoughts and memories. Thinking and storing memory traces takes energy as well as material (metabolism of cells) and produces waste products. The human brain produces knowledge which it shares with other human brains through language and other forms of communication. Those products, as discussed in Chap.

[1] We adopt the term "metabolism" referring to the economic system within a cell. We will use the term "physiology" when referring to the economy of a multicellular organism. For larger-scale supra-organismic organizations such as populations, species, ecological systems, and societies we will use a variety of terms but generally refer to the complex webs of energy flow and work processes as an economy.

[2] Following footnote 1, we should point out that the term "physiology" actually includes many more biological activities, such as reproduction, that may or may not involve the metabolic processes that supply resources, matter, and energy, to the organism and then convert those resources into useful structures and macromolecules such as proteins.

[3] Until recently it was thought that brain cells could live almost as long as a human, dying only in small numbers as aging brought on various forms of cognitive deficiencies. Recent evidence for the generation of new brain cells throughout one's life suggests that this is not entirely true.

9, are extremely useful to other human brains. Which leads us to the subject of Chap. 9, the human economy of goods and services, transactions, technologies, etc. That, of course, is what most people think about when you use the word "economy." Most people do not realize that what we humans do in trade and commerce is really not that different in kind from what single cells, single bodies, and single brains do to manage their respective resources and modes of production.

At the highest end of organization, the ecosystem emerges as a system achieving a high degree of stability through many complex feedback loops that serve to govern through mutual constraints.[4] Ecosystems are collections of multiple species engaged in a food web organized into trophic levels (primary producers, primary consumers, secondary consumers, and decomposers). In all biosystems the primary product is biomass—living cells and tissues that behave in ways that contribute to the overall stability of the particular ecosystem in which they are embedded. Decomposers (bacteria, fungi, various worms, etc.) play a particularly important role in the recycling of nutrients. In dynamic balance, an ecosystem achieves a stable state called a climax state.

Of course, the Earth system itself is always undergoing change—plate tectonics shift continents around and change climates—but life itself, in the form of cells, bodies, populations, species, and genera manage to persist. They do so because their internal subsystems of resource management and allocation are evolved to produce that result.

Our thesis is that the economy model archetype is a repeating pattern at scales from the metabolism of individual cells to the whole planet.[5] There is a fractal-like quality in viewing different scale economies as *nested*. We discuss the significance of nested economies, that is, the cellular economy (metabolism) within a body economy (physiology), within a human population economy (what we call *the* economy) in Sect. 13.3, Nested Economies, below.[6] First, we need to consider why we claim that all of these processes are to be called economies. For one, they all are instantiations of a generic pattern of what we described above as the progressive transformation of high-entropy materials into low-entropy materials through work processes driven by high-potential energy flow through the system. We situate this pattern as a model archetype in the context of CAS/CAESs. Coupled with the governance archetype discussed in the previous chapter we have the basis for a

[4]The HSS should be understood as existing within the global ecosystem but with an obvious caveat. Humans create new components within their ecosystems—their cultures. They have colonized the entire planet, so act as invasive species that upset the balancing forces that had brought those ecosystems to their climax states.

[5]In Tyler Volk's 1998 book, *Gaia's Body: Toward a Physiology of Earth*, he describes large-scale processes and cycles of materials that constitutes the planetary physiology and by inference Gaia's economic system!

[6]At the time of this writing the author began reading Smith and Morowitz (2016) and was pleased to note that they make the case for the phenomenon of life as reflecting this nested quality of biological processes. They do not use the term economy but describe the biosphere in terms of an extended complex network of metabolic activities that extend up from cellular to the ecosphere.

complete understanding of these classes of systems, and thus the basis for understanding our own CAES, the human social system.

13.1.1 What Is Not in an Economy

In this chapter, we will cover the generic model of economy as it applies to a number of CAS/CAESs and especially the hierarchy of life—cells, organisms, and societies. The reader may be surprised to learn that there are a number of topics, issues, processes, and concepts with which academic economists (particularly neoclassical economists) deal regularly that turn out not to be part of a natural economic model. We will mention a few examples here to provide some scope/perspective (some of these were mentioned in Chap. 9).

In the treatment of a natural CAS/CAES economy, the archetype model, you will not find mention of many things such as profits, perpetual growth, debt-based financing, or any number of modern concerns in societal economics. The reason is simple, these elements are not found in pre-human societies, nor even early human economies. Some of them have correlate mechanisms in the archetype, but their implementations in the modern economy appear to be distorted and misbehaving.

13.1.1.1 Financing Based on Debt

For example, the notion of financing an asset in the present based on the possibility of having a future income sufficient to pay back the debt and interest to boot appears to be derived from the practice of borrowing from excess savings (e.g., a greater amount of grain in the granary due to an extra productive harvest). Only it has gradually morphed into a habit of borrowing from future earnings on the assumption that a growing economy means there will be higher income in that future time. Recent experience in many forms of debt-financing is showing that this assumption is not valid. Economic growth has slowed considerably of late (Gordon 2016) so the realized ability to pay back the loan, when what was once the future got here, has diminished considerably. The global finance system had to be bailed out with cash infusions by governments to prevent too many bankruptcies. Even so too many people lost significant assets (like homes) after the 2009 Great Recession. The current state of affairs in global finance, where both public and private entities rely increasingly on debt to continue operations and where banks have the ability to extend credit by digging into deposits through "fractional-reserve banking," essentially creating money out of nothing, looks more and more like a Ponzi scheme.

Natural systems do not borrow from the future because incomes do not go up over time, endlessly. The global economy of the Ecos is a steady state, circular economy, materials cycle driven by the energy flow from the sun, primarily, but also from tidal forces and thermal flows from deep in the mantel. The world is essentially materially closed (receiving a light dusting from left-over solar system formation and an occasional meteorite). Thus, the global economy has to make do with what it has. The biosphere, too, while participating in some of the grander geosphere cycles such as the water cycle and carbon cycle, recycles and reuses its stocks of biochemicals. Thus, the economy of Earth is in a long-term steady state flux with sometimes impressive deviations (e.g., Snowball Earth or the End Permian Event), though most of those were quite long ago when the Earth was still evolving to its current form. By comparison the swings from ice age to interglacial seem mild.

The point is that the Ecos is not growing. It is a fixed-sized pie, which means that if any one subsystem grows larger, then other subsystems must diminish. For the biosphere, this means biodiversity decline. The growth of the HSS has meant a loss of species for the world. It also means that the HSS economy is constrained because it depends ultimately on the rest of the Ecos being stable and able to provide its services.

Living systems exist in constantly changing environments in which resources may, from time to time, be in low supply. Life solved this problem by evolving the ability to maintain stores of resources against a time of need. For example, an animal's fat stores contain energy that can be converted from fat to sugars when carbohydrates are not available. The animal need not starve so long as eventually it finds the external resources. When it does it is motivated to take in more than it needs, strictly speaking for maintaining. It has to rebuild its store of energy against future downturns. Similarly, early societies developed methods for preserving some foodstuffs against hard times (even just seasonal variations like dry seasons).

The excess taken in during these times of abundance is not a profit (see below)! The stored resource is properly understood as savings. There will be a time in the future in which the excess will become a deficit and whatever was saved during good times will be used to maintain the steady state.

During the early parts of a lifecycle, in the periods of growth, an organism must take in more resource than is needed just for maintenance of their biomass at the time. They are investing those resources into new biomass as they increase in weight and develop tissues. As they reach maturity the growth rate declines and they eventually enter a routine of maintenance as suggested above.

The same is true for a social system, ordinarily. Prior to the Green Revolution agriculture was only gradually improving productivity in the western world, due mainly to the uses of tractors and other farming machinery along with better breeds of food plants and animals. This meant that the supply of food in those more developed states was adequate to abundant, but this wasn't the case in the underdeveloped areas of the globe. Subsistence farming still predominated for most of the world's populations. In both cases, however, there was not such an excess.

As the global economy evolved over the early twentieth century, people got used to the idea that tomorrow we would produce even more wealth than today. At some point this seemed to give reason to think that we could borrow from that future time today. We could spend today wealth that would be produced tomorrow. This is accomplished by finding those who had already warehoused excess wealth and borrow from them with the promise to pay back the loan in the future, along with interest (or dividends). Essentially, we believed that we would be able to do this because the future would be essentially like the immediate past in which we really had this capability because the economy was growing.

Prior to the advent of debt-based financing as described above the human socio-economic system was propelled by the ever-increasing recruitment of high-power energy in the form of what started out as cheap[7] fossil fuels, coal, oil, and natural gas. It is this growth in energy flow that powered increases in material transformations and produced the pattern of seemingly endless growth. However, as we saw in Chap. 9, fossil fuels are a finite non-renewable resource that can be depleted and that has now begun to show up in the fact that newer extraction has become costly. But so engrained in our beliefs that energy is still "cheap" we continue on building debt on the theory that sooner or later energy will get to be cheap again and we will be able to grow ourselves out of debt.

13.1.1.2 Profit

The current study of economics includes the role of profit in economic growth. But neoclassical economics also holds the premise that growth is a normal and perpetual state of affairs, which, as argued above, is not the case. CAS/CAESs grow when developing (e.g., reaching maturity) but eventually enter into a steady state where fluxes provide temporary gains or deficits but over the lifetime of the system, it remains at its maximum size. This can best be understood in terms of the amount of some crucial stock within the system. For example, energy reservoirs, such as the glycogen stored in the liver of an animal can temporarily increase, such as after a meal, but also deplete, say after a vigorous workout.

The concept of profit is so ingrained in our thinking that it will be difficult for most to realize that it is based on a mistaken account of the dynamic fluxes of materials and energies in mature systems. While the human socioeconomic-cultural system was undergoing real growth through improvements in productivity gains (technology) the excess of income over costs was real enough. The HSS could be compared with the growth and development of an animal from embryo, through neonate, through childhood, to sexually mature adult. Once that mature state is reached, a negative feedback signal turns off the production of growth factors. The body enters a quasi-stable, steady state with some occasional need for growth, say in the storage of body fat in anticipation of lower food intake during winter months but also working down of those stores during the time of low food intake.

The origins of the concept of profit in human society is thought to be in the earliest grain-states' ability to produce excess grain over the consumption (Scott 2017). Originally, as a biological function similar to body fat storage, the idea of excess

[7] Cheap, meaning here, that the amount of energy consumed in getting the fuels was quite low. For example, in the early days of oil recovery the ratio of energy out to energy in was thought to be roughly 100:1 measured in barrels of oil (i.e., it took one barrel of oil to recover 100 barrels) giving a net return of 99 barrels. And for a while this would leave 98 barrels available to generate new wealth since one of the 99 barrels became investment in future recovery.

production in any given year was to hedge against crop failures in some future year. Farmers soon learned that there were often long-term variations in climate conditions that made such a strategy prudent. Grains can be stored for long periods so afforded such a hedge (ibid). But there was no way to monitor exactly how much excess was produced in any one year nor a way to monitor what might be needed to accommodate future needs and thus strike a balance in the long run. Thus, excesses could accumulate, notwithstanding calamities owing to major climate events like floods or droughts in various locales. Unfortunately, what this allowed is a growth in population to account for the balance. And a growth in population fueled the need to plant more crops leading to the advances of the early grain-states toward becoming empires.

Having excess foodstuffs available was always a good thing, and so became a habit of thinking and working toward. The stores of grains represented real wealth and it became necessary to account for the amounts in storage and the ownership. Thus, was born writing and numeracy (Nissen et al. 1993). The combination of ownership (or controlled possession) and the motive to produce excesses above one's own consumption, coupled with the evolution of the accounting markings in clay to become the beginnings of a monetary system of trade, and the sense of power given to the owners of the wealth all underwrote the rise in the notion that profits, for their own sake, were good things. More profits equated with more power. The rest, as the saying goes, is history.

Until very recently the HSS was indeed developing and growing. Profits turned into progress and growth of human-controlled biomass. After the Industrial Revolution the pace of development and productivity increases due to technological improvements accelerated. Starting in the late mid-twentieth century, however, that acceleration slowed and by the end of the century had turned to deceleration (Gordon 2016). From a systems perspective, this looks exactly like a signal of a system approaching maturity.

Yet the notion of profit maximization, a major tenant of capitalism, remains firmly entrenched in the thinking of economists, politicians, and the common citizen.

Governments and central banks around the globe are turning to any financial sleight of hand they can think of to keep the illusion of profits for their own sake going. Production processes, too, have resorted to practices that are meant to produce the illusion of profitable operations (as much to secure their own ability to keep financial support coming in) by diverting true costs of production to the external environment (so-called externalities). For example, dumping wastes into nearby waterways to avoid the costs of recycling them.

13.1.1.3 Continued Accumulation of Low Entropy Materials (Capital and Ownership)

While it is true that living systems can accumulate energy and materials to support growth and reproduction there are limits on how much of these will be stored, just enough to act as reserves. The stores are for buffering against lean times but also for investing in new biomass or reproduction. Even then, the stores are just enough to support the new life to a point where it becomes independent and starts obtaining its own resources.

Humans have evolved a very different point of view, which, as we saw in Chap. 9's analysis of capitalism, has become pathological and is leading to wealth disparities that are unsustainable (except in the minds of those who have the bulk of the wealth).

You won't find this phenomenon anywhere in the other economy examples.

13.1.1.4 Monetary Signaling Absent the Backing of Free Energy

Every example of an economy includes a means of signaling to control the rate and timing of production processes. But in the current human social economy the signaling mechanism, money, has become extremely distorted and cannot fulfill its original purpose. Originally the price of a piece of production represented the amount of free energy (applied through labor) that was consumed to accomplish that production. That is, how much free energy should be directed at doing that particular work was a function of how much a buyer was willing to pay and that amount translated immediately into how much energy the buyer had consumed to produce an income for himself. In other words, prices more truly reflected costs of doing work and money was invented as a convenient messaging system for conveying the information about both. But not long after the first coinage was invented, corresponding to the emerging hierarchical governance structures of early grain-states, money tokens began to take on a value in their own right. As labor specialization increased such that products could be produced and sold in a market, money was readily understood to be easier to hoard than, say, keeping a granary or salting meats to keep in storage for later consumption.

It took a long time to decouple the backing of money tokens with actual wealth (produced by the consumption of free energy) but with the advent of significantly cheap energy the relation between energy flow and wealth production became opaque to all. Along with the advent of borrowing against the future and the charging of interest (rents) for the use of money the original role of money as a signal used to control the flow of energy to work was slowly strangled. The chokehold was tightened by, a leftover from mercantilism's fascination with precious metals, the adoption of a "gold standard" to establish some stability to the international monetary system. Gold is not energy; it is not food. Moreover, though it is rare, there were still mines producing gold bullion in many places in the world.

That there should be a standard measure backing money units is completely sound. But it should be something that has actual meaning and relation to the production and consumption of wealth. Free energy is a measurable quantity and it is directly related to the amount of work, that is wealth production, that can be done. Moreover, this is exactly what we see going on in our other examples of natural economies (see below). Wealth, that is, low entropy assets, comes from the doing of work. Every work process has an inherent and measurable efficiency with which it accomplishes its task with a given quantity of total energy input (free energy is that fraction of the total that directly results in the accomplishment of work—the rest is just given off as unusable heat). Thus, it is perfectly feasible to know, in advance of actual energy flowing into the economy, how much free energy is available in the energy stocks, say a barrel of oil. A unit of money, like a dollar, could be pegged to a unit of free energy and the amount of money put into circulation would be pegged to the amount of free energy that was available. This would be a rational standard by which to manage the volume of money. The gold standard leads to multiple fallacious notions and coupled with fractional-reserve banking, the debt financing discussed above, has been the source of economic upheavals, perhaps too numerous to name (c.f. Goldstein 2020, Chap. 10 for a review).

The final death knell for tying money to gold came when the United States, under President Nixon in August, 1971, effectively took the US dollar off the gold standard to let its value float free in the currency markets, turning money into just another commodity. Coupled with the concept of rents paid for use of money, for example, interest or dividends, this completed the decoupling of the signaling capacity of money from the real production of wealth (and thus from the true costs of production). As we will see, this mechanism is a bogus way to think about management of an economic system. None of the other economy representatives have anything like this because they adhere to physical reality, whereas we humans seem to think we can create any reality we desire to have.

13.1.2 What Then Is an Economy?

All complex systems (CS, CAS, and CAES) are engaged in doing work that results in a benefit to its embedding supra-system. This may involve transformation of higher-entropy material into useful products, or transportation of materials over distances. But in some way these systems consume high potential energy to accomplish their work and both the embedding supra-system and the SOI itself, are better off for that work. In Chaps. 3 and 4, we saw multiple examples of basic or atomic work processes. And in Chap. 4, we saw an example of how these processes could be coupled, that is outputs connected to inputs, to form more complex work processes (see Fig. 4.13). In all cases, work processes transform "raw" inputs into useful or useable outputs. The organization of networks of processes along with their regulation to assure the smooth ongoing transformation of higher-entropy materials into lower-entropy materials that are further usable is the nature of an economy.

Here, though, we are primarily interested in the economies of CAS/CAESs. We will consider paradigm cases across several scales of space and complexity. The smallest and least complex CAS will be a single living cell (both prokaryote and eukaryote examples). The next larger and more complex system is the multicellular organism (both plants and animals). And the most complex system we will consider is a social system (in particular human societies).

The economies we are interested in have their own domain-specific names. They are: metabolism, physiology, and what most of us think about when we use the word "economy," the social economy. We will see that the social economy is an exosomatic (outside of the body) version of the endosomatic economy (physiology) which, in turn, is an exometabolic version of cellular metabolism.

All economies do useful work. They employ networks of work processes, each work process managed by a local agent, and aggregates of interacting managing agents communicating with one another for cooperative regulation of the local cluster of work processes. When there are many such clusters a higher-order management is required so we find specialized agents acting to coordinate those clusters at a larger scale of space and time. The economy is managed so that it continues to produce value throughout the life of the CAS/CAES. This is the generic construction of an economy archetype. The particulars will vary.

13.2 Considering a Generic Economic System

We will now consider the various attributes of a generic (archetype) economy regardless of scale or level of organization. This will be followed by an expansion of examples of economic activities and processes.

13.2.1 Common Attributes of an Economy

A number of features of a generic economy are found in all economic systems. The objective of an economic system is to produce low entropy goods and services that are used internally to support and perpetuate the various subsystems that do the work, that is, are autopoietic. Some economic systems may also produce and export products (or services) that are useful to other entities in the embedding supra-system or environment (generically called "customers"). A properly functioning economy allows a CAS/CAES to persist in its environment for time scales much longer than that of its overt behavioral dynamics. We see in living systems and supra-systems the coupling of subsystems within cells, bodies, ecosystems, and the whole Ecos such that the outputs from some subsystems are inputs to other subsystems and that in toto, the flows follow network chains that provide closure for materials in a well-functioning, that is non-dysfunctional whole system. We take the analytic view of a single CAS/CAES embedded in a larger CAES where its outputs include products (or performs some work service), waste materials (which are usually by-products that provide input to recycling processes[8]), and waste heat that cannot be used to do work, but can be used to maintain a slightly elevated temperature.

13.2.1.1 Products and Services

An economy is based on the notion that a CAS/CAES, as an individual subsystem, must obtain some immediately usable resources from environmental entities (suppliers) or, more generally, construct usable resources by doing work on raw materials, energies, or messages, transforming them to a point where they can be used by the system. That is the transformation of material resource (higher entropy) inputs produce a (lower entropy) material output(s) and these are then resource inputs to a downstream CAS/CAES subsystem. At least some of the subsystems will produce some final products which are either consumed by internal subsystems or exported to the environment and receiving customers.

At a metabolic level, the economy within a single cell, the chain of decreasing entropy products is seen in the citric acid/Krebs cycle which uses O_2 and low weight metabolites and produces several usable energy-containing molecules for internal consumption, some amino acids (for building proteins), and waste CO_2. It is a cycle in that it regenerates many of the metabolites as long as fresh energy sources are available as inputs. This is known as a circular economy.

In a physiological economy (in multicellular organisms) there are two main sorts, autotrophs (e.g., plants using photosynthesis) and heterotrophs (e.g., animals that eat plants and other animals). The autotrophs produce their own food, that will feed into the Krebs cycle of the cellular metabolism, using sunlight (generally speaking). They require raw materials, CO_2, H_2O, and a host of minerals they obtain from soils, for example, to manufacture carbohydrates. The individual cells/tissues in plants have specialized in their functions such that the food manufactured in the leaves needs to be distributed to other cell/tissue types so that they can do their jobs (like root cells absorbing water and minerals). Heterotrophs eat autotrophs for their food, which are both material resources and energy resources. Heterotrophs need to first digest the food stuffs to make the energy and raw materials available for supplying cellular metabolism and the cellular mechanisms of maintenance. Products and by-products are produced within certain tissues and cell types to be taken up and used by other tissues/cell types. Waste CO_2, excess water, and unusable waste material are expelled by exporting subsystems such as lungs, kidneys, and rectum for vertebrates.

[8] Water and carbon dioxide are waste products for animals but resources for plants as an example.

13.2.1.2 Organization of Work Processes

Work processes obtain their input resources, including high-power energy, do work on those resources to modify or transform them into something of greater value to export to other work processes as input resources to them. Economies are organized in chains of work processes, each able to add value to their inputs through progressive transformations toward final products. These chains may be found embedded in more complex networks whereby products from one or more processes flow to resource inputs of one or more downstream processes (c.f. Figs. 13.1 and 13.2 below). The flow rates and directions need to be regulated for balance.

13.2.1.3 Energy Flow and Work

The fundamental purpose of an economic system is to channel the flow of free energy to work processes that reconfigure matter according the needs of the system as represented by the supply chain model above. The system must have means for extracting free energy from environmental sources. Whether this is ingesting sugars for metabolism or drilling for oil with refining to produce the fuels we use in the human economy, the key tactical work that is done is getting a reliable flow of free energy into the system.

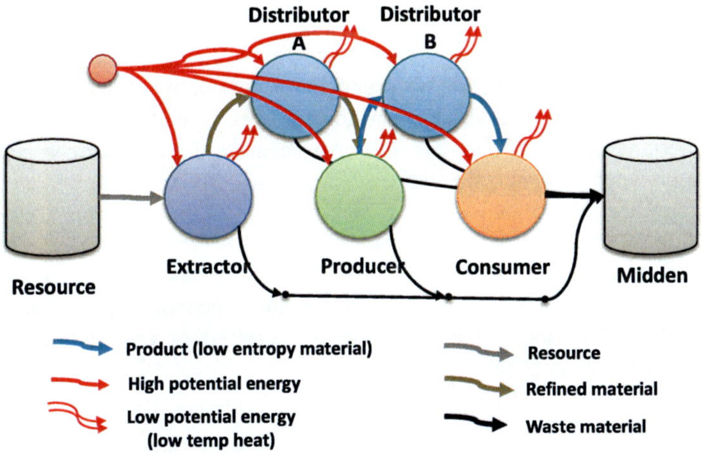

Fig. 13.1 All economies are based on a basic value-adding supply chain. Resources, such as medium, or even high entropy materials (i.e., semi-organized) are extracted by special processes and passed to producer processes, that do work to reduce the entropy of the material further (i.e., increase the organization). In turn, the product is passed to consumers. Energy is consumed by all processes as value is added. Consumers degrade the low-entropy materials through use and need to discard wastes to some environmental sink. All work processes in the supply chain produce some wastes (heat and material) because no process is 100% efficient in its uses of materials or energies

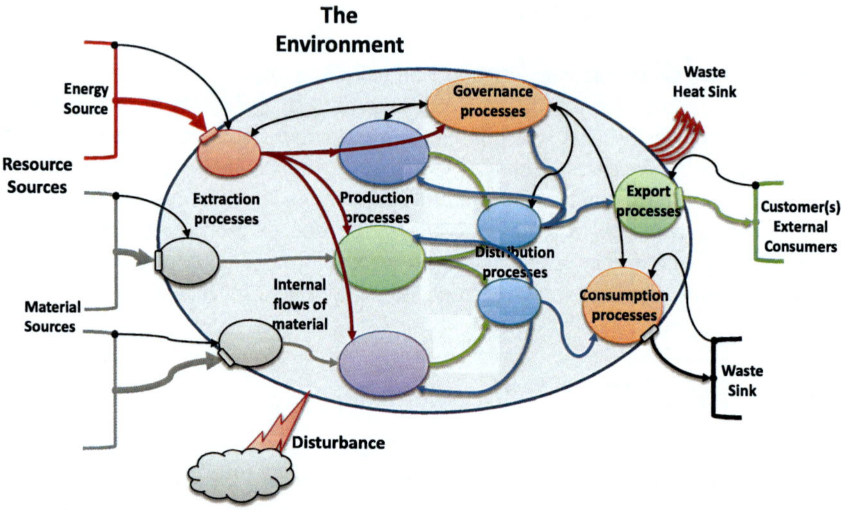

Fig. 13.2 A schematic representation of a generic economic system archetype. An economy is based on a network of interconnected supply chains. A few representative flows are shown. Thin black arrows represent message flows, red arrows represent energy flows and other colored arrows represent various material asset flows. Extraction/importing is shown on the left (as per the convention we established in Chap. 6). Flows internally generally go from left to right with exporting processes pushing products and wastes out into the environment. Internally some assets produced later in the general flow are sent to more forward processes for their use

Subsequently the energy must be channeled to the work process sites. Metabolism in eukaryotic cells achieves this by producing a ubiquitous molecule, adenosine triphosphate (ATP) that acts as a current flowing from specialized extractors called mitochondria (site of the Krebs cycle already mentioned and described below) to various sites where the energy packet each molecule carries is used. In the physiology of a multicellular organism energy packets in the form of sugars and fats are conveyed to all tissues through a network of tubes that branch fractally in order to reach every corner of every tissue (as well as other metabolites and gasses). The electric power grid does some of the job for the human economy. Below we provide more details on how these subsystems work and why they constitute a universal archetype (or sub-archetype).

13.2.1.4 Specialization of Work Processes

Materials are transformed from low organization (high entropy configurations) to useful organization (low entropy configurations) in steps. There are several primary kinds of transformation work that will be described below. For now, we want to call attention to the organization of the whole system as an ordered sequence of specialized processes. Specialization of work to modular or encapsulated processes is a core feature of all economies. Specialization translates into efficiency in doing work. Evolution shaped the designs of specialized metabolic processes, physiological tissues and organs, brains, and ecosystems (where member species are the specialists). Humans design organizations and machines to achieve the highest efficiencies possible under given constraints. Specialization allows the design of work to be particular to a given step in the value-adding chain.

However, it is important to recognize that all more complex work processes derive from the composition of simpler work processes. That is there is a nestedness of work processes as one decomposes the system (as per Chap. 6). Thus, at a fundamental level all work processes are nested compositions of what we described as "atomic" work processes in Chap. 3.

13.2.1.5 Redundancy and Multiple Pathways

In a supply chain network, there are generally many similar specialized work processes operating in parallel. There are many advantages to this kind of architecture. Chief among these is the advantage of redundancy. No work process can be guaranteed to work ceaselessly forever. They are subject to degradation and may fail at some point. Having a population of work processes of the same kind, in a reconfigurable network, means that there should always be some sufficient portion of that population fully at work. Of course, the processes that have failed need to be repaired or replaced and this is exactly what living systems do. It is what human organizations do as well.

At the same time, too much redundancy becomes pathological. Living systems regulate the amount of redundant capacity to just that needed to provide the backup ability required under nominal conditions. Living systems have an ability to recruit additional capacity on an as-needed basis for short-term pulses of demand (e.g., homeostasis), but they do not build additional capacity unless they experience long-term changes in the demand level that warrant the commitment of extra resources. See (Mobus 1999) for an analysis of adaptive response and associated costs.

13.2.1.6 Internal Competition and Evolvability

Whenever there are multiple work processes of the same kind operating in a complex network there is the potential for competition between those processes for limited or constrained resources. At the level of cellular metabolism such competition is thwarted by a tight regulatory system that ensures that there is no excess redundancy. In a multicellular body's physiology, the situation is much more complex and though there are similar regulatory mechanisms in place to maintain a balance there are also more opportunities for competition between organs and tissues for limited resources used by all, for example, glucose in the blood used by all cells as energy sources. An extreme case of competition occurs when some cells become cancerous and send signals to commandeer the blood flow; they then grow without bound and eventually consume the whole body. Under the right regulation, however, competition is limited in these kinds of systems.

At the ecosystem level, however, competition is much more prevalent and plays an important role in constraining the components (various species) from dominating the system.

Among humans as well as among human organizations, competition is viewed as the rule rather than the exception. That is the case in the current view of a neoliberal economy where all producers are free to produce their products as much as they want and to compete with each other by reducing their prices or increasing their quality. Market dynamics under capitalism and something approximating laissez-faire conditions are thought to lead to competition driving these improvements. However, it should be noted that there have been many other kinds of economic models adopted by human societies throughout history that were not based on employing competition to this extent (Polanyi 2001 edition).

13.2.1.7 Governance, Management, and Agent Decisions

Mechanisms for making sure work processes maintain near optimal balance of flows are found in all economic systems. That is, the governance of an economy, as demonstrated in the previous chapter, is essential in order for the entire economy to work for the good of the whole system. We find this is the case in the examples of metabolism in cells, physiology in bodies, and to be generally true in family dynamics that support reproduction. The situation is less certain when we examine the economic system of societies. As pointed out in the previous chapter the governance architecture of the HSS is still in flux and the decision-making capacity of the agents is often questionable. We suggest that the evolution of a stable HSS economy is still in process. In the final chapter of the book, we will provide some suggestions about the structure of the HSS that would be far better integrated with the governance and agencies involving humans.

13.2.1.8 Summary

These general attributes of a generic economic system can be found in all CAS/CAESs. All such systems examined by the author and many other investigators are found to have this general pattern (suggesting that it is universal). Evolved systems such as single cells, multicellular organisms, genera, and ecosystems are long-term stable, in other words sustainable, by virtue of having well-organized and working economic processes with well-organized and working governance. This chapter will explore the archetype of a generic economic subsystem model. We start with the basic concept of a material transformation chain, otherwise known as a supply chain, because it produces products that are low-entropy configurations of materials that can be used to support the existence of the whole CAS/CAES in which the chain operates.

13.2.2 *Economy as a Value-Adding Supply Chain*

13.2.2.1 A Basic Supply Chain

All economies in all CAS/CAESs have a fundamental subsystem model, the archetype of which we present here. Figure 13.1 shows the basic supply chain model that is found in all economic systems. The basic model includes processes that extract resources from the environment, processes that transport materials, processes that transform the materials into useful, low entropy, products, and processes that use those products and export waste products.

The basic supply chain is a fundamental unit of an economic system. In the figure, we show a simple linear chain to fix the basic ideas. An economic process exists whenever a system contains the functions performed in this figure. The subsystems that perform these functions are specialized to the functions (we will return to the concept of specialization below). Those functions are: Extraction of resources from the environment, production or configurational modification work, distribution or moving materials from point of production to point of use, and consumption or use. The whole economic activity generally moves from left to right in the figure, however, we will see later that feedback loops are important in the networked supply chain covered next.

The extraction and conversion of a material good from the environment is specialized to find and import that good. A living single-celled animalcule may obtain macromolecules through a process of phagocytosis, a human being eats food and breathes in oxygen, a corporation has a purchasing department that orders and receives shipments of parts. Value is added to the imported goods through preliminary transformations. The animalcule needs to "digest" the macromolecule and the human needs to do likewise to the food. A mining operation needs to break ores up for preliminary processing.

A resource material is relatively simple insofar as organization is concerned. Some resources such as metal ores are essentially rocks that contain a high concentration of the metal making it possible to refine the latter, throwing out the non-metal bits. Others, such as timber are structurally low entropy but not in a shape that is suitable for further use, so needing sawing into boards, etc.

Distribution processes are what move the various materials through the system. In a cell, for example, an elaborate network of microtubules provides a means to cart various cellular macromolecules from points of production to points of use. In animal bodies, the cardiovascular system transports nutrients (e.g., glucose), dissolved gasses (O_2 and CO_2), and numerous signaling molecules (e.g., hormones) around the body. In the human economy, the train and trucking industries provide ground transportation of material goods.

Producers is a general category of work processes that do major transformations of material to create usable products, also called assets, for use or consumption by other processes. In a long supply chain, end products are built up in stages of production (or value added) so that a producer situated "downstream" might also be viewed as a consumer of an "upstream" producer's product (the above figure only shows one generic producer but we will shortly introduce multiple stages of production). Products are classified as intermediate, that is, component parts, or final (before being used or consumption).

Products may also be cross-classified according to their use and longevity. For example, a ribosome, in a cell, (see below) is a stable, long-term product that is, essentially, equivalent to a physical manufacturing plant in the human economy. Producers add value to materials by virtue of increasing their organization and conformation so that they may be used subsequently, either to produce tools (as introduced in Chap. 10), capital goods[9] (which are really just long-term tools), intermediate products (parts), or final products for the ultimate consumer process.

That process, the consumer, degrades the low entropy products produced by the economy and exports[10] the low-organization material to an external sink, here labeled the "Midden."[11] The concept of a consumer is a difficult one. In an ecosystem, a consumer is generally an animal that eats plants (primary consumers) or animals that eat the primary consumers (secondary or carnivores). In a cell the consumer is the whole cell itself and its purpose is just to repair subsystems, stay alive, and be capable of reproducing. The same can be said for the multicellular organism. For the human economic system, the same rules apply but something quite different is at work. People need to stay alive and reproduce, certainly. But consumption in the human economy has taken on a whole new meaning that results from the inclusion of improving technologies. This was the subject of Chap. 9.

13.2.2.2 The Supply Chain Network

Here we present a more realistic generic pattern for an economy,[12] a network of interconnected supply chains. Figure 13.2 provides a schematic representation of the economy of a generalized CAS (we will distinguish CAESs from "ordinary" CASs later). We have identified the internal subsystems that are operative in any economy whether of primitive humans, the current HSS, or living systems such as cells, bodies, and ecosystems.

---

[9] In ordinary economics we make a distinction between capital goods and other long-term assets and consumables. The latter generally refer to things we use up and turn into wastes, for example, food, paper goods. In fact, however, taking the very long-term view, we do use up long-term assets eventually. Most such assets may require maintenance, parts replacements, etc. And eventually such assets become obsolete or just wear out.

[10] Later we will add a specialized waste export function to the chain.

[11] The term "midden" means a garbage or trash dump. It is generally applied to archeological sites where evidence of life styles and practices can be determined by what the local population threw away. We are using it in a more general sense to mean any sink into which waste materials of any sort are dumped by exporting processes.

[12] This treatment suggests that an economic subsystem is a first-class ontological concept as mentioned in Chap. 3.

All real CASs employ the major subsystems shown in the figure. All such systems acquire energy and material resources from their environment. Most acquire information from the sources and sinks in their environment. And they produce products and wastes (heat and material) that are exported to entities in the environment. This is a generic pattern that can be used as a basis for analysis of any CAS at any level of organization (cell, multicellular organism, societies, and the global HSS).

Internally, the economy consists of work processes that import and convert raw materials into progressively more usable forms; that is, organizations of matter and embodied energy that benefit other work processes downstream or, ultimately, a consumption process. Energy is captured (or extracted) and converted to usable net free energy[13] to power the work processes. Work processes transform the materials to lower entropic forms, that is, more organized and more useful to the ultimate consumption processes. Distribution processes (which are specific kinds of work processes) convey the intermediate and final products of production to internal consumption processes as well as export processes that expel products and wastes back out to the environment.

The whole system is coordinated by a governance process (see Chap. 12) that manages the internal process as well as making sure the system is coordinated with the external entities in the environment (including responding to disturbances). For more evolved CAESs governance includes some aspect of strategic management, as outlined in Chap. D. Governance of an economy involves several mechanisms that help regulate the various transactions (exchanges of material and energy). For example, market transactions between entities, mediated by direct exchanges of equivalent "value" between source and sink entities help regulate local flows of materials and energies. The signaling of value is through some kind of representation of free energy.

This generic model is what we meant before by "systemic analogy." No matter what kind of CAS or CAES we examine we will find these basic subsystems and interfaces with an environment at work.

This generic economy is, of course, only part of the whole picture for what constitutes a fully functional CAS or CAES. We have not yet addressed the issues of adaptability or evolvability. Those are yet to come.

[13] As a reminder, free energy, in physics and chemistry, is that portion of total energy which can be coupled with a work process to do "useful" work. The latter means work which results in a configurational change in a system, 12.g. capturing some of that energy in the form of molecular bonds. Depending on the stability of the configurational change, it might later decay, giving off the embodied energy as heat, as when molecular bonds give way to disruption and a complex molecule comes apart into simpler component molecules.

13.2.3 Market Dynamics

An economy is actually a plethora of producers, consumers, and distributors operating such that low-entropy goods (and service work) produced by producers get to consumers via flows of materials and energies along multiple pathways. Those flows need to be controlled by the economic agents working together, for the most part, cooperatively via established message channels and communications protocols (signals). Consumers have to signal a distributor that they have demand for a good (or a service provider for a service). The distributor, in turn, has to signal the producer to supply the good for distribution. And the producer, in turn, has to signal the extractor or refiner, to obtain and make ready for supply the resources needed. Since there are populations of consumers, distributors, and producers, in general, the signaling mechanism must allow consumers to cast multiple signals to multiple distributors and for them, in turn, to cast multiple signals to multiple suppliers. There is a network of signaling and flow relations of the many agents resulting in a general flow from extractors to producers to consumers to waste sinks.

This network constitutes a market through which goods and services move. The marketplace of an economy has several general characteristics in all instances.

13.2.3.1 Market Characteristics

From cellular metabolism to human societies, markets will all demonstrate the following characteristics.

13.2.3.1.1 Redundancy of Processes

Resilience is largely based on the redundancy of multiple kinds of processes. Multiple resource extractors, multiple producers, multiple consumers, and so on, are producing the same results, sometimes via slightly different processes. Having a multiplicity of similar processes increases the likelihood that at least some of them will continue working under varying stressful conditions. For example, in cellular metabolism there are several different cyclic pathways through which metabolites operate, all producing the same basic products, especially molecules of ATP, the cell's main power current. Tissues perform the same kind of redundancy in organs of animals. Here all of the cell types may be the same, but the massive numbers of them ensure that in spite of disease or injury, at least some of them may continue to provide services to the body. In a society of humans, several individuals may possess similar skills and abilities and, thus, be able to provide alternative workers for specific jobs. In the modern social setting, multiple firms may produce very similar products, with essentially the same functionality so that buyers have choices among various producers.

It should be noted that in metabolism and physiology redundancy does not necessarily imply competition between producers or between consumers. The network of producers and consumers operates more like the Internet packet switching protocols in which there is an attempt to balance throughput. Production and consumption processes need not be in synchrony. If a producer process is lagging at some time, the consumer may switch to a different source.

At the level of cellular metabolism, the various product molecules, both small metabolites and large macromolecular are fairly uniform or commodity-like. Thus, all of the producers produce the same basic product. Consumers of those products do not particularly care where the products came from and there is no differentiation of "brands" or consumer preferences to consider. Similarly, in body physiology products of tissues are still molecular objects with no particular distinctions to be made. When we get to ecosystems, however, there are many differences in consumable biomass forms (the products), and consumers show definite preferences for one form over another. This is what constructs the food web within an ecosystem.

In the human economy, the redundancy in production and consumption involves considerable variability and preference determination. In western societies in particular we find a multiplicity of toothpastes, tomatoes, and cake mixes. Presumably all toothpastes clean the teeth, all kinds of tomatoes can be used in salads, and all cake mixes produce excessive sugar-based treats. But for human consumption, variety is the spice of life.[14]

13.2.3.1.2 Open Channels Between Processes

For any market to operate efficiently and effectively there must be clear and open channels of communications between all of the processes. Certainly, producers and consumers must be able to communicate things like supply and demand. At the cellular level, this is accomplished by the various chemical reactions being conducted in the aqueous environment of the cytoplasm. Diffusion is the main method for propagating signals. Indeed, some very effective poisons and antibiotics work by disrupting the signaling channels.

In an animal's body signaling is both via blood circulation (hormones and immune system agents) and via the nervous system. Here, too, channels of communications must be open and clear.

But it isn't just communications channels that need to be maintained clear and open. The actual products need to have the means of ready transport between producers and consumers. As with the diffusion of signaling and energy molecules in cell metabolism there are some metabolite and macromolecules that diffuse from points of production to points of use. In addition, there is an internal network of microtubules that actively transport larger macromolecules or organelles in a guided fashion.

---

[14] How variety of products actually creates a problem for human agents is explained in Schwartz 2004.

13.2.3.1.3 Signaling Protocols for Cooperation

Communication signals (message flows) are used intensively in economies in order to achieve as much cooperation as possible between producer and consumer parties (including extractors as producers and waste exporters as a kind of end consumer). The general pattern is to have a series of communication channels between processes from the first extraction to the last waste disposal in the supply chain. In order to affect any transaction between any two processes in the chain, a very strict protocol, both for communications and for management of the substance flow, must be in place and adhered to in order to affect transactions.

A transaction is any transfer of a product or service between a producer/provider and a consumer/beneficiary. A key aspect of a market economy is the ability for these two kinds of agents to interact directly without, as much as possible, imposed coordination from a higher-order regulator (e.g., a logistics agent per Chap. D). They must signal one another of their states (status) in terms of readiness to supply or to obtain. Such a protocol must include flow rates, onset and offset timing, etc. Highly evolved mechanisms exist to provide the communications channels (message flows) and the cooperation protocols at both ends. A basic pattern is: the consumer signals an interest in obtaining and how much (demand), the producer signals a willingness to provide and availability (supply), the consumer signals the time it is ready to accept supply, the producer acknowledges and commits to supplying quantity at the specified time, the producer signals onset of the flow (shipment), the consumer may signal readiness to accept flow, the producer signals completion of pushing the flow (quantity), the consumer signals receipt and completion of the flow at their end. This general pattern of flow control protocol can be found in metabolic cycles and the HSS economic process.

13.2.3.1.3.1 Noise Suppression of Signals

Noise is any disruption of a communication signal (used to transmit information between agents). It can be induced from the outside of the channel (e.g., electromagnetic interference) or be inherent within the channel (e.g., thermal vibrations of the channel components). Either way, noise can disrupt the information quality of a signal and lead to process errors that would, in turn, disrupt production and transport of products (and services).

In the realm of communications, the problem of noise disruption is a constant problem. There are a number of methods that have been developed in communications engineering for suppressing noise (in a general sense) in order that the information in a message not be lost or garbled (e.g., the "signal-to-noise" ratio be kept sufficiently high). These include ways to enhance the bandwidth of the channel and of coding signals with redundancy that allows for error detection and correction. It also involves careful design (or evolution) of communication protocols (end protocols between senders and receivers as described above) that include providing retransmission of signals that have been irretrievably disrupted.

It turns out that there are very similar mechanisms operating within all economic systems. In cellular metabolism the main mechanism for noise suppression is the buffering of errors in signals, which are molecular transports (Smith and Morowitz 2016) via cytosolic reservoirs.[15] More generally the damping of errors or noise in signals is handled by various buffering approaches.

13.2.3.1.4 Buffering of Movements

It should be noted in general that processes will tend to operate at varying rates at different times (asynchronous, episodic, and sporadic) due to local conditions within each. Yet to keep a supply chain/network operating efficiently it is necessary to provide temporary storage—buffering—of substances within the process subsystems in order to smooth out the overall flows. This is the case for material flows (e.g., inventories), energy flows (e.g., batteries or hydroelectric reservoirs), and also messages (computer memories used to store message packets in Internet routers). Different producers produce their products at different times and at different rates. Similarly different consumers use up these products at different times and rates. Temporary storage devices (as described in Chap. 3) are ubiquitous solutions to smoothing out the rates of flows to allow for cooperative regulation between processes being able to handle the bulk of flows and transactions.

13.2.3.2 Signaling in the Control of Consumption, Production, and Flow

Every process needs resources to do its work. It must obtain these resources from suppliers, either extractors or early supply chain producers. It needs an effective means to signal to the suppliers how much it needs and when it needs the resource. Final consumers of the final products start the process of signaling demand. Their rate of consumption and, thus, their future demands, really motivate the entire supply chain. Ultimate consumers in any economy are the drivers for everything else that transpires in the economy. A ribosome generates a demand for tRNA + amino acids, mRNA from the nucleus, and ATP from the mitochondria. A muscle cell generates a demand for glucose from the liver as it contracts to do physical work. And a household generates demand for produce, meats (assuming non-vegetarian diets), and services as it rears children.

In cellular metabolism this signaling is accomplished at the molecular level with concentration levels of a particular molecule that, in higher concentrations, activates a production process. For example, a higher concentration of ADP in the vicinity of a mitochondrion activates it to produce more ATP (which simultaneously reduces

---

[15] Beyond the scope of this book, but covered in Mobus and Kalton (2015), is the theory of noise filters, in particular, the FIR filter. Think of the ability of a capacitor to buffer fluctuations in voltage changes in a circuit.

the concentration of ADP—feedback). It is an indication that the cell is actively metabolizing and needs more energy.

Signaling in multicellular organisms is still molecular but with molecules that are dispersed within the intercellular fluid (or blood). For example, when the cells in an animal take up more glucose from the blood to make more energy available within the cells, the body responds. Cells in the pancreatic islets, monitoring the blood glucose levels, and setting in motion several physiological signals that work to regulate the level. Blood glucose levels are maintained by homeostasis and it is extremely important that the signals are clear and essentially noise free. Disruption of signals (for various reasons) in this case leads to the diseases of diabetes (e.g., Type I, II).

In the HSS economy, the signaling of demand (and, indeed all signaling) has been given over to the price mechanism. Demand is signaled by what price (in monetary units) a buyer is willing to pay for a commodity, product, or service. Since buyer willingness is, itself, subject to influences that may or may not reflect actual value, we consider these signals as particularly noisy, not reliable, and the source of disruption to the long-term viability of producers. There are, of course, other disruptive factors in the use of prices to signal demand in the human economy. Some of these are mentioned in Chap. 9. Nevertheless, the price mechanism, however flawed, represents a way of signaling demand (and supply) so counts in the general archetype. Is it a good one? That remains to be seen.

Consuming processes, such as ribosomes, organs, and households are informed about supplies of resources by signals sent from the producers. Again, in metabolism and physiology these are mediated by molecular signals (i.e., concentrations in the cytosol or blood).

Cooperation signals, that is, signals that flow directly between producers and consumers to coordinate their interactions, evolved to increase the efficiency and speed with which transactions could be completed. The ATP/ADP loop mentioned above is an excellent example in cell metabolism. Similarly, the homeostatic maintenance of blood sugar levels is largely handled by cooperation between tissues directly.

In human markets, most transactions are also regulated by cooperative signaling between buyers and sellers. The price system, again, is used as the mediator. Buyers and sellers agree on a price to be paid for a product or service and the transaction is completed with the buyer taking possession of the product or benefiting from the service, and the seller receiving a quantity of an abstract marker called money. The latter is a marker representing a real physical quantity, namely the capacity to obtain the inputs to production of its product in the near future. The signal here is telling the receiving producer that they should obtain those inputs in the future and produce more product because there will be buyers and demand in that future time. The producer can use those markers to become a buyer of the inputs and keep the production process going.

All human economy markets are based on an aggregate of producers and consumers who must compete on some level to obtain transactions. The signaling levels are established by allowing market dynamics generate something like an average price such that the majority of buyers and sellers are willing to make a trade (clearing the market). However, in this scheme there are some winners and some losers, thus human markets, at least in the capitalist versions, have an inherent amount of waste above and beyond that generated in cooperative and coordinated economies, such as metabolism.

This is quite different from other natural economies (metabolism or ecosystem food webs). In metabolism nothing is produced until there is a signal of need. Ribosomes do not have to compete with one another for ATP molecules, with some losing out because there aren't enough to go around. They don't have to offer higher levels of ADP in order to win the competition. Rather the mitochondria ramp up or down production according to the needs of all of the ribosomes, something like a just-in-time delivery system.

13.2.4 Governance Processes

Chap. D provided a description and deeper explanation of a "generic" governance subsystem as depicted in Fig. 13.2. Here we provide an overview of system governance as it relates to an economic subsystem. We state categorically that every CAS or CAES involves and contains elements of a governance system to keep it doing what it needs to do to fulfill its purpose. The need for governance stems from the sheer complexity of a network of the various processes as covered above in Sect. 13.2.2.2 The Supply Chain Network. While it is conceivable that a simple, linear, and short supply chain, such as in Fig. 13.1, might be able to operate using process to process signaling (cooperation only), given sufficient capacity of the decision agents, in the more general case of a deep network of multiple work processes some form of imposed coordination is necessary. Hence, an economy needs a coordination level of governance, at least. The more complex the system is, the more coordination is needed.

By the logic of evolution (selection) and in order to be a sustainable ongoing concern a system must monitor its behaviors at every level of operations and provide control feedback whenever a subsystem is failing to do its designated job. Moreover, all of the internal subsystems must be coordinated in their behaviors so that the overall behavior is sufficiently effective.[16] Since every system is part of a larger supra-system, e.g., a corporation is a subsystem in a larger HSS economic system or a ribosome is a subsystem of a cell, it must coordinate its behaviors in acquiring resources, exporting products (or performing services), and eliminating

[16]We would ordinarily say "optimally" effective, but as anyone who has ever tried to manage a complex organization will know, optimality is a rarely achievable ideal. Fortunately, there is generally a region around an optimum that is sufficiently close to optimum so that the organization still functions well enough not to get selected out of the ecosystem. Still a well operating management/governance subsystem is required to achieve this level of operation.

wastes with other entities in that larger environment (its sources and sinks). Thus, two basic forms of coordination of subsystems is required, logistical and tactical.

The governance of an economic system rests on several interrelated mechanisms, all of which are part of cybernetic loops, hence the name, *hierarchical cybernetic governance system* (or HCGS). This structure should be thought of as the architecture of governance. Management, as explained in Chap. D, is the process of agents making appropriate decisions based on where they are in the hierarchy. Governance in this architecture can be broken into several interacting components based on the kinds of decisions and communications linkages employed.

13.2.4.1 Interprocess Communications and Cooperation

Ideally two or a small number of operational subsystems can communicate directly with one another and manage to self-coordinate on a local scale. This is possible because all of the operational decision agents share a common model and objective of what the cluster is supposed to accomplish. When this is the case, coordination is really achieved through *cooperation*. So, a primary means of managing small-scale coordination is through the sets of messages sent and received, their interpretations, and proper responses. For example, in a supply-chain model, vendors of components to be used downstream in assembly are informed by the consuming process what is needed in some short time horizon. They respond cooperatively by making sure they can fulfill that demand, or signal back that a delay may ensue. Either way the two entities establish a cooperative interchange to manage the flow of goods.

In biological systems, there are found strong hand-shaking cooperative activities, especially in metabolic processes. For example, the supply of energy in the form of ATP to various manufacturing organelle, for example, a ribosome, is based on a signaling process whereby the ribosome lets the local mitochondria (through diffusion) know there is need of more of that molecule. Assuming the mitochondria have enough raw material (e.g., glucose and oxygen) they can ramp up production to satisfy the needs.

Even larger-scale cooperation is seen in cellular functions. For example, highly stimulated post-synaptic compartments in neurons can, through an elaborate signaling mechanism, let the genes responsible for producing the mechanism for producing various membrane channel components that these are needed to increase the sensitivity of those synapses encoding memory traces. This process takes place over the spatial scale of the neuron body and over a much longer time scale (see Chap. A for more details).

13.2.4.2 Operational Decision Processes

These decisions are actually distributed throughout the whole system. Each subsystem shown in Fig. 13.1 contains its own governance sub-subsystem, just as the whole economic system has a governance subsystem. This pattern is recursive and self-similar at almost all scales. Even the subsystem labeled as the governance process contains a similar hierarchical structure. That is, it contains operational-level decisions, that is, the ongoing operation of the process of governance. For example, a US senator's office, or any legislator's office, has a staff of people who handle the day-to-day functions of the office, such as preparing documents for legislation. Decisions need to be made regarding what gets done and when at a low level. Other kinds/levels of decisions need to be made as well in the senator's office. We'll discuss those below.

The point is that all processes shown at the level (and below) of the governance process have internal governance sub-subsystems in their own right. And their internal operations need decision makers working at the appropriate level.

Generally speaking, operations-level decisions are made in real-time and are used to correct processes that have strayed from nominal operations. This is the basic cybernetic feedback mechanism covered in Chap. 12 (see Fig. 12.3).

Economically relevant operation decisions involve, for example, transactions, such as sales and purchases of goods and services. And here it is only managing the arrangements, say, of transfers (e.g., purchase order processing). As a rule, the determination of price and negotiating contracts, etc. is handled at a higher level of decision processing, namely the tactical level.[17]

In the HSS economic subsystem, the role of consumption is generally associated with households where goods and service are primarily received and used. Consumption means using something up, leaving only wastes. The day-to-day acts of consumption are operational-level decisions based exclusively on local conditions. However, work processes are also consumers in some sense. A manufacturing operation uses up a variety of "supplies" in going about its business with the waste (e.g., paper forms) being expelled.

13.2.4.3 Market-Based Decision Processes

Economic systems typically involve diffuse and distributed decision processes in which some level of flow or storage is arrived at through a kind of consensus process. That is, among a population of agents, the actions of individuals average over the population to assert a value that is then used in subsequent decisions. The price setting based on supply and demand in classical economics is an example of this mechanism. There are multiple constraints on the decision effectiveness of market-based coordination and its role in economic governance.

---

[17] Here we have another example of the cause of fuzziness in systems. In some purchasing operations the purchasing agent handles the operational-level management (of purchase orders) and the tactical-level management of negotiating prices. Thus, one person spends some time being an operations-level agent and some amount of time being a tactical-level agent. Even when it is difficult to parse the role of a single individual, it is essential to keep in mind that one person can be a member of several classes of subsystems in the course of a few hours' time.

In the limit, a position very often taken by free-market advocates (e.g., neoliberal economists), the market mechanism is self-regulating via the supply-demand relation. The market, the average of those individual decisions, "clears" at some price where supply equals demand. Everyone who can buy the commodity does so and producers/sellers are satisfied with their incomes.

A truly "free" market is considered by most, even the advocates of lassaiz-faire, to be a (at best) useful fiction. Markets do operate to shift prices based on availability of certain commodities and their desirability among buyers. And a great advantage of a market is that it can increase the flexibility of conducting transactions and trades. However, over and over again, we have seen in history massive failures in the markets underlying, especially though not exclusively, capitalistic forms of economic systems. Many books have been written about the "causes" of market failures. But in the context of this section, it should be clear that the question of governance of the market needs to be front-and-center. The market might be a useful mechanism to achieve some form of economical governance that is to say one where the overhead of governance does not unnecessarily burden the benefits of governance.

Market-based trade and distribution mechanisms ultimately depend on competition, between buyers for limited commodities and between sellers to provide those commodities at prices buyers appear ready to pay. Competition between multiple entities seeking to thrive in a constrained resource environment is, of course, a major part of evolutionary logic. Competition is seen to enhance capabilities—the most fit competitors survive by adopting "better" methods or strategies. During the formative years of economic theory in the nineteenth century, the logic of competition in evolution seemed to provide the main support for the notion of free markets. Survival of the fittest was thought to be the operative rule.

Today, we realize that evolution includes a hefty dose of cooperation as well as competition (c.f. Bourke 2011 and Stewart 2000). There has to be a balance between the two for species and individuals to succeed. And while competition may work to hone skills, it is cooperation that accounts for the major transitions from simpler to more complex organizations (e.g., from multiple kinds of prokaryotes to the origins of eukaryote cells, Morowitz 2002; Smith and Szathmáry 1995). Cooperation among individuals of a social unit became the main mechanism for success of the human species, while competition between social units (groups) provided the impetus for improvements (Sober and Wilson 1998).

Markets work best when they are of small, local scale. This is true in metabolism (local diffusion) in physiology (exchanges within a single body), and with the HSS economy. Information tends to degrade over longer distances, costs for transporting goods are higher, and opportunities for disruptions increase in large-scale markets. The degradation of information alone has been identified with market failures (in the human economy) such as asset pricing bubbles. Economic decision-makers cannot be expected to make reasonable (let alone rational) decisions when sources of valid information cannot be trusted. Even more insidious, decision-makers may not be aware of the problem and tend to treat all messages as providing valid information even when it is actually misinformation. Economic decisions in markets need to be based on full information.

13.2.4.4 Management (Agent) Decision Types in the HCGS

A more complete model of the governance of an economy is provided by the HCGS. Decision nodes in the network of economic processes are managed by agents as described in the last chapter. All economic activity is managed by the same hierarchical framework. However, decision models for these agents stress the management of energy flows and entropy-reducing work processes; they are based on cost-benefit-tradeoffs. To recount, agents are classified in the HCGS by the level in the hierarchy, the time scale over which their decisions are made, and their role in the HCGS structure. At the lowest level of module operations, the real-time manager makes decisions of how to "tweak" the local operation based on measuring process output characteristics such as units produced per unit of time, or various quality parameters. When these measurements detect deviations from ideal values, the manager takes action to reduce the errors—classical feedback or homeostasis. Operational managers are generally able to process feedforward and feed-down messages which are used for cooperation and changes in production schedules respectively.

Coordination agents, that is either logistical or tactical agents, operate over longer time scales working on aggregated and generally time-averaged data from the operational level below. All of the internal processes, the real-time managers, are under the aegis of a logistical agent, that provides feed-down messages. All of the interface processes, importers and exporters, are under the aegis of tactical agents who are cooperating with receiving or sending operational process agents.

Strategic agents, when we are looking at a CAES, operate over the longest time scale with respect to the lifecycle of the whole system. They manage the logistical and tactical agents through feed-down and receive their inputs from those coordination-level agents. Just as the coordination level agents work with aggregated and time-averaged data (intermediate time scales), the strategic agents acquire very long time-scale data summaries with which they process information on the viability of the system and from which they formulate "plans of action" for evolvability.

13.2.5 Extraction of Resources

There are three basic resources that are needed by all CAESs (and CASs). All of these systems need material resources in order to replenish their own material structures. Entropy is always at work degrading physical structures (protein denaturing or home repairs). But also, the system itself is continually consuming structures. This is the fundamental nature of life and supra-living systems. This is what we mean by dissipative systems. They are constantly needing to import materials and energy (of the proper form and power) in order to do the work of constructing structures for use. They are constantly in a battle with entropy.

Energy comes in either potential (stored but available) or kinetic (doing mechanical or chemical work) forms, for example, fossil fuels or sunlight. The generic system must have a mechanism for capturing energy and converting what it captures into a usable form—into what physicists call *free* energy or the energy available to do useful work.[18] That energy then needs to be distributed to all of the other work processes so that they can, in fact, accomplish their work.

Materials, in general, have to be acquired through active mechanisms; they need to be *imported*. Living things have to eat and digest. Society has to extract minerals from mines. The processes that do the importing work, of course, also need energy.

As energy needs to be converted to a usable form, so too, materials are generally extracted in a raw form that needs processing before it can be used in production. Digestion can convert proteins in food into amino acids that can be processed into the proteins the cell or body needs. Lumber can be sawn into planks and studs for construction. Iron ore can be smelted into iron (or steel) ingots.

The task of resource extractors (as in Fig. 13.1) is to locate, import through the boundary of the system, and possibly modify the raw resource in some manner to make it suitable for distribution to producers. Here we examine briefly a few of the agents involved in these processes.

13.2.5.1 Resource Finding and Quality Testing

Most organisms have to have a means of locating their resources before they can extract or import them. Locating the source of a resource can take many forms, but generally involves differential sensing of a physical cue in order to detect a gradient. Following the gradient then allows the organism to locate the source and commence importing (Mobus 1999). In more complex animals with brains and visual sensing, the latter work to pinpoint the resource after being drawn to its general location by sound or olfaction.

There are a number of ways to test the quality of a resource. For example, in the case of a chemical gradient, the absolute concentration of the cue signal can be sensed to determine if the quality of the resource is good enough to warrant further acquisition work.

Photosynthesizing plants are not ordinarily thought of as "hunting" for sunlight, but in fact many plants have growth control mechanisms that implement strategies for finding the ideal level of light. For example, understory plants make do with the dimmer light that filters through the forest canopy. At the same time, they are, generally speaking, passive in obtaining sunlight and they are subject to the luck of the draw in terms of where their seeds might land. Plants, photosynthetic autotrophs, possess specialize organelle in their leaf cells called chloroplasts in which photons

[18] In any flow of energy only a portion of the total energy can actually do work. Nicolas Léonard Sadi Carnot developed the formalism for an ideal heat engine (machines that are driven by a temperature difference). While Carnot's result applied for heat engines doing work, the same energy flow principle, from a high potential source to a low potential sink, applies to other forms of work.

are able to excite electrons into a chain reaction that allows the final fixation of carbon into carbohydrates, for example, sugars and starches. The sugars are then available for oxidation processes within mitochondria that produce the energy packet ATP.

Plants obtain mineral resources such as nitrates and water from soils through roots that penetrate it and actively seek the sources underground. They obtain CO_2, as their source of carbon, from the atmosphere through pores in their leaves' undersides called *stoma*.[19]

Animals, including single-celled animalcules such as *Paramecium*, generally move around in their environments to find food for ingestion.[20] They use a variety of sensory modalities to locate food sources. Many employ a prototypical foraging strategy in searching for stochastically determined food locations. For example, ant colony scouts can be seen wandering over even flat surfaces like a sidewalk in a seemingly haphazard path. The wandering is stochastic but not really purely a random walk (Mobus 1994, 1999); it was dubbed "the drunken sailor walk" in honor of the times the author did manage to get back to his submarine somehow. Hunting, foraging, browsing, and grazing are just a few modes employed by different species in finding food. In more complex animals, gills or breathing are used to obtain ubiquitous oxygen.

The agents responsible for animal behavior in finding and eating food is, of course, modules in the brain that obtain information on the environment, looking for cues regarding the presence of food and testing the quality, for example, the amount of food available for ingestion. Honey bees test the quality of a flower patch by sampling several flowers for nectar. If they find a very low quantity, they will abandon the patch to search for more promising prospects. Taste or smell (olfaction) is another quality test for foods for many animals. If it tastes bitter, it might be poisonous.

For human organizations from families up through nations the search and quality testing of potential resources is an ongoing problem. Finding food in most developed countries is generally a matter of driving to the supermarket, at least for those who have adequate incomes. The same is true for finding fuels for transportation and cooking. The whole society is arranged around providing these resources (which we will discuss later). In underdeveloped regions, the quest for food and cooking fuel is more complicated.

Modern human economies have complex acquisition processes at the front end of the chain or network. Based on the capitalist model of corporations, companies are organized to explore for raw fuels and refine them for use in the general economy. Large industrial agricultural operations grow and supply the bulk of food to the economy. The decision agents in these organizations need not only have expertise in finding and mining or growing, they need to make decisions for how to proceed based on the need to make profits in their operations.

---

[19] Also used in the regulation of H_2O.

[20] Of course, some parasitic varieties of organisms may simply embedded themselves in other animals bodies where they sap the nutrients from the host's tissues.

We might note that the profit motive for decisions and actions among human enterprises has its roots in living systems being programmed to maximize gain so as to have resilience in the face of scarcity by saving any current excess. However, in humans, at least in the developed world, the motive has outlived its purpose since the times of want are rarer and shallow. That drive seems to have morphed into a kind of pleasure-giving motive beyond practical purposes. We will return to this issue later.

13.2.5.2 Resource Importing/Extracting

Importing and extractor processes have to have agency in actually acquiring the resource needed by the system. In cells there are complex channels embedded in the cell membrane that are designed specifically to allow the one-way passage of particular molecules into the cell cytoplasm. These channels constitute the interfaces the organism has with its environment, and the selectivity of the channels with respect to what they let in is part of the interface protocol discussed in Part 1 of the book. Single-celled animalcules, as noted, may be able to ingest larger particles such as bacteria into special digestion chambers called lysosomes. There the particles are broken down into the constituent components (amino acids, sugars, and fatty acids, for example) needed as resource inputs to metabolism.

Cells, both as individual organisms and as members of a tissue in a multicellular organism, are immersed in semi-aquatic milieu in which the resources are dissolved or similarly immersed. Animalcules may actively swim (as do some bacteria) to capture their food, as noted above.

Whole organisms, in particular animals, eat. They ingest food and digest it much as is done by the animalcules discussed above. Once the nutrient molecules are broken down, they enter the blood stream for distribution to various organs, for example, the liver, for further processing.

A family goes shopping in the developed world. There are still a few hunter-gatherer societies in remote regions that continue to practice the arts of hunting game and foraging for fruit or tubers.

Farming, logging, mining, and drilling/pumping are the main resource extractors for the HSS. In Chap. 9, we explored the energy sector and the consequences of decisions about extracting finite resources like oil and coal.

13.2.5.3 Refining a Raw Resource for Use by Producers

Amino acids are the building blocks of proteins, the main structural and functional workhorses of metabolism. Many amino acids are acquired through the digestion of foodstuffs; some may need to be internally produced to have a full complement of all 20 needed to construct proteins.

Similarly, food eaten by an animal must first be digested, broken down to simpler component molecules before they can be absorbed and used.

Cooking and meal preparation is an example of refining a food resource in a domicile, or at least it was in the eras before fast-food chains became the thing. The cook/chef is the agent doing the refining.

In an organization like a manufacturing company the receiving operation, getting material parts from other companies, involves unpacking and putting the parts into inventory, ready to be drawn out when needed to go into manufacturing. The receiving clerk is the agent.

13.2.5.4 Monitoring Demand by Producers

In both plant and animal metabolism a universal signal for obtaining more resource is the relative concentration of adenosine diphosphate (ADP), a molecule similar to ATP but missing a terminal phosphate—the free energy in ATP resides in the final bond in the chain of phosphates and when it is broken, releasing the third phosphate in the chain, the ADP molecule is recycled to the mitochondria. The activities described above are regulated by this relative concentration differential. More ADP signals a need by the producers, for example, the ribosomes, for an increase in resource acquisition.

We humans and all other animals that eat periodically, as opposed to constant grazers, get hungry, signaling us to switch to the food acquisition mode of behavior. The tactical part of the limbic system and cortical areas take over and coordinate the body's activities to food-getting.

When the family refrigerator looks low on requisite food stuffs, someone in the household goes shopping.

And when an item in inventory goes below a certain level it means that production has been using it up and it's time to order more of that part so as to not hobble the assembly line.

13.2.6 Production of Useful Goods and Services

Once resources are acquired and pre-processed the system's production processes can begin constructing useful material structures for the ultimate consumer (or for export to external customers). The entire economic system is organized around the need to supply consumers with structures (and services) that they require to survive and thrive. In the living cell, this is represented by the construction of enzymes and cellular organelle in tightly closed cooperative network. Cells produce metabolites that are available to other kinds of cells for many different purposes (e.g., regulation hormones). Companies produce products and services that are used by other companies and individual consumers.

Production, as we will see later, can be divided into several categories by who receives the products and how those products are used downstream of where they are produced. Products are considered assets. Put another way, anything that is produced from a work process that cannot be used downstream is waste and needs to be eliminated from the system. Assets are useful to someone or something in the system. They are generally categorized by their "residence time" in the system. Fixed assets such as buildings and cellular nuclei have very long lifetimes relative to that of the whole system. Fixed assets still require continuing inputs of material and energy for maintenance, meaning the work of repairs. Assets such as machinery, automobiles, or ribosomes (in cells) last for a while but eventually need replacing. The shortest lasting assets are things that get consumed completely by the processes they support; food, gasoline, plastic wrapping, are a few such items.

Producer processes are the main value-adding or material entropy-reducing actors in a CAS/CAES. Depending on the level of complexity, that is, how many levels of organization are involved, the kinds of decisions that a producer process may be involved with can run the gamut from production operations (simple error feedback) to strategic, as in the case of a department chair in a university college who is trying to figure out where new funding may come from.

Producer decisions include when to start or ramp up the work processes that produce the products or ramp down as needed. This involves a tactical connection with resource suppliers and customers.

Our main example from cellular metabolism is the ribosome where proteins are manufactured. The material entropy of the collection of amino acids that make up a specific protein is reduced by the structural organizing of those amino acids into a polymer with the right ordering of acids that result in the protein assuming a folded shape that allows it to do the job it needs to do (e.g., an enzyme). The ribosome consumes ATP energy in the process of constructing the peptide bonds that accomplish the polymerization chemistry. They give off waste heat to the cytoplasm which must be exported, eventually, through the cell membrane. The entropy of the universe is increased by this exporting of waste heat—potential free energy has been converted to the work of forming chemical bonds plus the waste heat produced. The total entropy of the universe is increased even while the local entropy, the new, more useful organization of matter, has decreased. That is the meaning of "value-added" work.

The ribosome is an agent following a "programmed" decision model. It uses a program in the form of a messenger RNA molecule—a template for the construction of the protein that is a message transmitted from the genetic code in the nucleus—and decision rules built into its structure. When a three-nucleic acid code is brought into the work zone, the ribosome accepts a complimentary transfer-RNA (tRNA) that binds to the "right" amino acid and initiates the formation of a peptide bond between the end of the growing polymer and the new amino acid.[21]

[21] Naturally it's a little more complicated than that! The ribosomes actually produce polypeptides, chains that are shorter in length than the final product, proteins. These polypeptides are then stitched together to make the final protein by other enzymatic proteins (a real chicken and egg conundrum! This is because many sub-products, the polypeptides are actually used in a number of different final product proteins. Life isn't simple.

At the level of multicellular physiology economics, just about every tissue type in the body is a major producer—of itself! Cells within a tissue that are damaged or wear out are replaced through mitosis by healthy cells. So, muscle cells that degrade are replaced by new muscle cells so as to maintain the movement capacity of the organism. Muscle mass may also be increased with increasing demand on the muscles. If a person takes up exercise or weight lifting, for example, the increased long-term demand on the existing muscles results in them increasing their mitosis so there will be more muscle cells to help carry the load.

Another major producer for vertebrate organisms is the liver. The liver is thought to have several hundred functions, making it the ultimate manufacturer and service provider of the body. One of its critical main functions is the metering of glucose into the bloodstream. Glucose is the main energy carrier, absorbed by all other cells which then put these molecules into the internal metabolic production of ATP, as discussed above. The liver can remove glucose from the blood when there is a spike, say after a meal. It then converts the glucose to glycogen, a molecule that is more stable and stores the latter until the blood sugar level falls. It then reverses the process and converts glycogen to glucose and puts it back into the blood for distribution.

The agents responsible for directing the functioning of the liver are various but the major agent, being an operational decision maker, is the tissues composed of a particular cell type, the hepatocyte. These cells act collectively to sense the chemistry of the blood and decide which functions to perform to maintain homeostatic conditions in the blood.

Households are not generally considered producers, except for the fact that new human beings are usually a product of there being a household. But there are other economic productions that can be attributed to households. For example, households are commonly the core units for the production of labor for the social economy. That is, the household is the place where a worker lives and receives sustenance when not at work. But other goods and services may be associated with households. If a household is part of a farm, then the growing of food is clearly a production function. Making clothing or furniture used to be a part of what went on in households—not so much today in developed nations. Not that long ago every high school in the United States had classes in "home economics," which referred to "managing the home."

These classes were historically attended mostly by young women (in the days before feminism came awake) who would presumably one day be "housewives" and the home managers. The male role was to bring in income with which to purchase necessaries. In that model, the housewife was operational, logistical, tactical, and (unbeknownst to the husband) the major strategic agent. She made decisions on what to cook for diner, when and how to tidy up the rooms, what foods to buy from the grocer (and how to store them for later use), and looking after the children's education. Today, while this model is still found in many households, it is no longer *the* model. Many households have far more complex arrangements for effective agency—too many to go into here.

In the social economy producers of products and services are ubiquitous. A typical manufacturer takes in raw materials or medium entropy parts and through its work chain produces a low entropy product. The agents distribute according to the HCGS in what we call "management."

In natural ecosystems, plants hold the role as "primary producers" that is in respect to the food webs. Plants construct the original macromolecules that are needed for their own growth and when consumed by herbivores become "product." Similarly, the herbivores convert the organic compounds in plants into animal tissues, such as muscle, that becomes product for carnivores.

In ecosystems, the main form of agency is the complex food webs (who eats who) and the trophic levels. These kinds of systems are subject to some instabilities, for example, when an abundance of herbivores (say rabbits) triggers the reproduction of a carnivore (say coyotes) leading to sometimes wild oscillations in populations.

13.2.7 Distribution—Moving Atoms, Energies, and Bits

Since the various subsystems in an economy are distributed in space a fair amount of energy as well as conveyance vehicles are needed to move matter, energy, and messages, in all their various forms. The vehicles must be constructed in the production stage (above) and then powered and controlled are specialized routes (flows) to convey substances from their sources (e.g., extraction storage) to their destinations for use (e.g., production processes). All of these pathways may be described as flow networks, even when, for example, the flow is based on diffusion or convection. Things have to move or be moved through the system (essentially from left—inputs—to right—outputs in Fig. 13.1).

In eukaryotic cells a complex membrane called the endoplasmic reticulum, ER, is a major transporter for many products of synthesis.[22] There are multiple other mechanisms within the cell to move various products from points of origin (synthesizing micromachines like the ribosome) to points of use.

In physiology, blood, which is pumped around the circulatory system, is the main but not only means of transporting materials from points of production in various tissues, to points of use in various other tissues.

In the human economy, we have trucks, ships, and airplanes, of course, as well as multiple other means of conveyance of humans themselves. The bottom line (as they say in economics) is that all of these various forms of transportation consume energy; the heavier the load the more power is required; the longer the distance that power consumption must be sustained. Light loads and shorter distances are considerably more efficient in terms of overall energy consumption.

[22] The ER has many other duties, including some forms of product synthesis. But its role as a major highway within the cell provides us with a good model of controlled distribution.

13.2.8 Consumption

Consumption, as used here, is the act of degrading a low entropy material object or a store of potential energy, in order to do other work. In a biological system, food stuffs are broken down into component parts to be reassembled into structures that are useful to the consumer. This kind of process also consumes energy (e.g., to break bonds) and usually produces waste materials that need to be exported to sinks in the environment or risk toxic effects or clogging flows.

In the HSS humans living in their homes are consumers of products and services from the production sector. This consumption is oriented around the act of maintaining life (reflecting the biological mandate). Cells consume carbohydrates to produce energy stores in packets of ATP (see below). Businesses consume lots of paper, electricity, and paperclips but also human labor. All consumption results in some wastes. Energy is consumed doing work and produces waste heat.

Counted in with the consumption of material and energy (resulting in wastes) the use of intermediate-term, "hard" assets such as equipment and vehicles (or ribosomes and cell membranes) causes wear and tear that will eventually require repairs. In the modern human economy, another factor that produces wastes is obsolescence of equipment such as 3-year-old computers or smartphones. Owing to the complexity of these devices it is easier to replace them with newer, more up-to-date ones and scrap the old ones. We are not aware of anything analogous in cell metabolism.

13.2.9 Exporting Products and Wastes

Every economy must make provisions for the disposal of wastes and the exporting of products to the larger supra-system. Humans have routinely dumped their wastes into the nearest convenient hole or river. Cells exude their waste products into the surrounding intercellular medium. In a multicellular organism there are supra-systems responsible for collecting and removing such wastes (e.g., CO_2 or uric acid) such as the circulatory system.

Cells often export substances such as hormones or molecules that act as signals to other cells. These would be more in the category of products since they are actually usable in the larger supra-system. The HSS, on the other hand, does not appear to produce anything that might be of use to other systems, except, perhaps, our excrement used by bacteria as food.

Waste removal, for both material and heat, and product exporting are often active processes that require energy to be used.

13.2.10 A Note on Disturbances

Almost by definition there is not a lot that can be said about the sources of disturbances; they tend to be spurious. There are generally no interfaces through which disturbances are "received." Their effect is felt within the system in unpredictable ways and times. Some disturbances may have internal sources such as a breakdown of a component as a result of wear and tear unmaintained. Another, and very common, internal source of disturbances, one that has been experienced by most people in the human economy, comes when a positive feedback loop in a market goes unregulated causing what is known as a "bubble" to grow. Eventually bubbles burst (or at least deflate rapidly) causing some economic pain in the process. We will examine the market mechanism, from the systems perspective, later in this chapter.

Disturbances, regardless of their source and magnitude, are compensated through some regulatory mechanisms that act to restore a more stable dynamic process (see Chap. 12). In cellular metabolism this regulation is the well-known homeostasis and autopoiesis.

13.3 Nested Economies

Above we suggested that the various forms of natural economies could be viewed as nested like concentric rings, with the simpler CAS versions in the center and more complex CAES versions outside. One can think of this as the economy of the whole HSS is comprised of families (households and the centers for reproduction). Families are comprised of individuals' bodies (physiologies), which are made up of various kinds of tissues/cells (metabolisms). The claim is that metabolism is the economy of a cell, physiology is the economy of a body. More importantly, the physiology of a body is an extension of metabolism into the "community" of cells that make up that body.[23] In other words, physiology is seen as "exo-metabolism." In the same way, we argue that the economy of a society of bodies is an extension of physiology in the community of people.[24] Thus, we can say economies are nested with those of simpler CASs inside more complex CAESs (Fig. 13.2).

This argument follows from the ontogeny of systems presented in Chap. 3, where more complex systems arise from simpler systems forming combinations or cooperatives through the ontogenic cycle (see Sect. 3.3.1 The Ontogenic Cycle). The history of biology has witnessed "Great Transitions" and emergences of higher levels of organization (Calcott and Sterelny 2011; Morowitz 2002; Smith and Szathmáry 1995). From the origin of, relatively speaking, primitive simple cells and

[23] Of course, there is actually another two layers between cells and bodies—namely tissues and organs, so our treatment is not exhaustive but, we think, amply illustrative of the points.

[24] We make a similar argument for the nature of ecosystems. See Sect. 3.2.2.6 H.T. Odum's Energese for a brief on this relation.

early metabolism (Smith and Morowitz 2016) to the evolution of prokaryotes to the emergence of cooperation between different varieties that led to the eukaryotes, and so on, living systems have been organizing at higher levels of complexity for, perhaps, 3.5+ billion years on Earth. At every transition the new forms displayed more complex metabolic mechanisms and increasing differentiation of cell types as specialists. With the advent of multicellular life forms new intercellular communications as well as useful metabolite production produced physiological processes and an economy of the organism. The evolution of human beings and their social systems is in direct line with this ongoing ontogeny. We, therefore, have much to learn from our progenitors.

To demonstrate how these economies (systems) are nested and mutually dependent we will examine the inner two levels of Fig. 13.3 and make some cursory comments about the Family economy. We'll get back to family and society levels in the final chapters of the book.

Note that the figure is about societies being comprised of families, which are comprised of individuals (and their bodies), which is comprised of myriad cells. This is the reverse order of ontogenesis and so we are not so much concerned about single-celled organisms in this account.

13.3.1 Metabolism

The most fundamental core process of natural CAS/CAESs is cellular metabolism. It constitutes a basic economy because it is the extraction of raw materials (carbon, water, trace elements) from an aqueous environment, the extraction of free energy from sources in that environment, the production of useful low entropy configurations of materials (i.e., macromolecules and structures) for sustaining the living state of matter, and the consumption of those materials as a dissipative system.

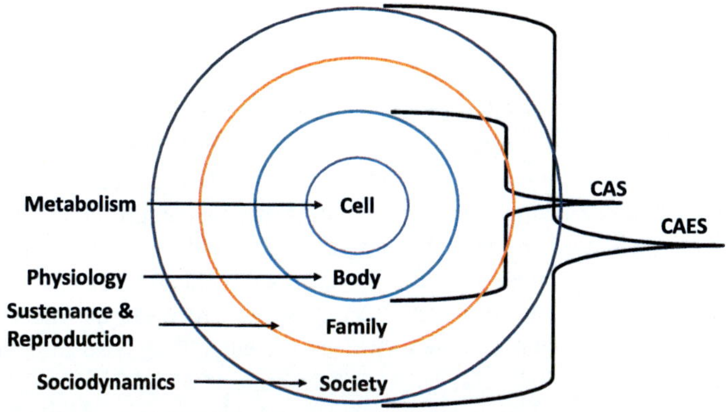

Fig. 13.3 "Simpler" economies are nested within more complex ones

Anabolic processes build structures using free energy to accomplish the work. Catabolic processes break them down extracting some free energy that can be used in anabolism and degrading non-functioning molecules to their components. Both production work processes and degradation processes produce waste heat that dissipates. There are multiple possible sources of free energy. Photosynthesis is the most well-known source of energy on the surface of the planet. However, some extremophile bacteria and archaea obtain energy from inorganic reactions involving, for example, iron and sulfur compounds. In either case, energy flows through the cell driving all of the processes.

Biosynthesis (anabolic) is used to produce biomass, either in the form of a growing cell (after mitosis) or as replacement of degraded structures. In the former case, the production of biomass leads to population growth in unicellular organisms or individual multicellular organism's growth and development.

Metabolism is an extremely complex chemical reaction network that so long as supplied appropriate resource molecules (and sunlight in the case of green plants and algae) maintains a dynamic pattern of organized processes. Everything that happens in the cell's economy is mediated at the molecular level. True, some of the molecules involved are extraordinarily complicated, such as the ribosome (below). Many molecules, such as the ubiquitous water and amino acids are relatively simple by comparison.

The workhorse molecules in metabolism are the multitude of proteins, long polymers of amino acids, just 20 of which make up the building blocks of the proteins. Proteins can act as enzymes, catalysts for various chemical reactions. They can also function as structural components and motors to move other molecules around. Many other organic molecule types as well as non-organic trace substances are used for various purposes, such as energy storage or as components in structures, for example, phospholipids used in building cell membranes.

Here we will provide a very general overview of metabolism as economy and only mention a few examples of implementations from the generic model archetype.

13.3.1.1 Governance

Since the only medium available in cells is molecular all computations are done through chemical reactions. Similarly, all signaling is through molecular flows of various kinds. Even within this organization there are finely tuned feedback and feedforward loops that maintain operational level management of things like reaction rates and gradients. Most of the governance of the cell's metabolism is handled through cooperative signaling. See Chap. D for more on this subject.

Ultimately the governance of metabolism is the provenance of the genetic code. Genes determine the amino acid sequence of proteins. They also influence the secondary structure of those proteins that determines their enzymatic properties.[25] The genetic code is the ultimate decision model for all of the decision agents in living systems.

13.3.1.2 Resource Conversion

Within eukaryotes specialized energy extractors called mitochondria are good examples of resource converters. Energy comes into the cell in the form of complex macromolecules such as polysaccharides or fats. These are digested in special vesicles and broken down into simpler compounds such as glucose. The latter are then processed by the mitochondria, which oxidize the sugar under controlled conditions to transfer electrons and drive the synthesis of adenosine triphosphate, ATP, a molecule that is ubiquitous throughout the cytosol and is the main free energy current used in all other processes, such as synthesis (see next section) and movement. Figure 13.4 provides a diagram of this process. Note that an ATP molecule releases its energy to the work process by cleaving a phosphate, converting it to adenosine diphosphate (ADP). The latter diffuses back to the vicinity of the mitochondrion where it acts as a feedback signal—its concentration rising will ramp up the production of ATP, using the ADP as the base and adding back a phosphate. This is a good example of the governance of the activity of this process through a mechanism that looks awfully like a market.

13.3.1.3 Goods Production

At the core of metabolism is the production of the various proteins that are needed throughout the system. This job is handled by a complex manufacturing unit called a ribosome (Fig. 13.5).[26]

A ribosome is a complex structure composed of subunits and made from both proteins and RNA molecules. Its job is to construct a polymer of amino acids according to a specification written in a molecule of RNA (mRNA specifically) that carries the message from the nucleus to the ribosome. Essentially this is a blueprint for manufacturing a specific protein.

Amino acids are found in the cellular medium, or cytoplasm, attached to another kind of RNA molecule (tRNA) that is said to "transfer" the amino acids from the cytoplasm stock to the ribosome site of production. The mRNA carries a code

[25] It is even more complex than this! Genes are only the beginning of the story. Epigenetic constraints are now known to have a much stronger influence on genetic expression than we had previously imagined.

[26] Cells need many other kinds of macromolecules such as polysaccharides and various lipids. But proteins, being both structural and functional (as enzymes) molecules could arguably be most important.

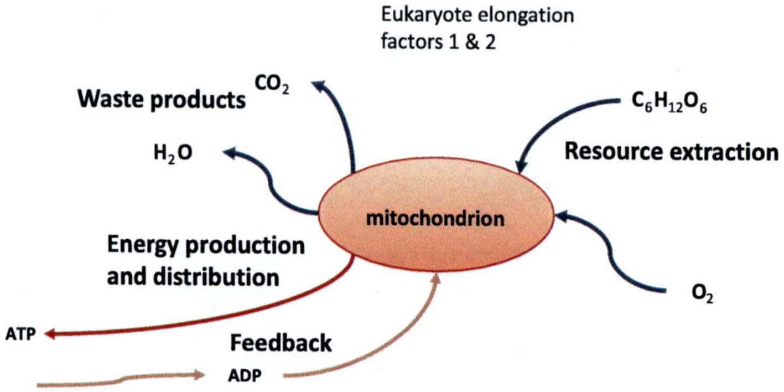

Fig. 13.4 The mitochondrion is a particular organelle that produces transportable energy packets, molecules of adenosine triphosphate or ATP, which diffuse to the work processes where the energy is consumed in doing work. See text for more explanation

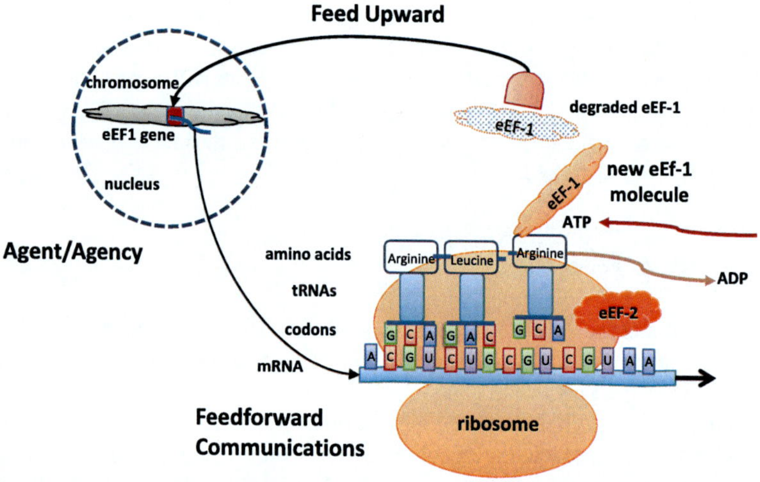

Fig. 13.5 A ribosome is a manufacturing plant to produce the proteins that can then be used for a wide array of purposes. Not shown is the input of energy from the conversion of ubiquitous ATP molecules into ADP and phosphates

written in triplets of the nucleotide bases, adenine (A), cytosine (C), guanine (G), and uracil (which is an RNA analog of thymine in the DNA of the genetic code).

13.3.1.4 Distribution

Two large complex proteins, actin and myosin, function as actuators within the cell and in cell movement. They can be found in many instances of transporting materials, especially against gradients, as mentioned above. They are ATP-dependent, working similarly to the ribosome in extracting the energy of ATP molecules which is then used to cause the molecules to contract or change conformation.

13.3.1.5 Consumption

In the context of metabolism, products that have been produced for use in the metabolic processes, macromolecules such as enzymes and the various organelle such as ribosomes tend to degrade at physiological temperatures and need continual replacement. The macromolecules, when they lose their functionality may be digested to produce some reusable metabolites and some by-products are exported through the cell membrane as waste products. However, metabolism is its own producer and consumer, as it were. Anabolism produces low-entropy structures and catabolism breaks down unneeded or unwanted structures raising the material entropy.

13.3.1.6 Exporting Products and Wastes

All cells must excrete waste products directly into the environment. Under ideal conditions, these diffuse rapidly into what amounts to an infinite reservoir, becoming too diluted to harm the cell or other cells nearby. Cells exist in a fluid or moist environment where currents and diffusion can be counted on to disperse waste products. But there are situations in which cells are confined to an approximately bounded environment that does not allow the dispersion of wastes. For example, consider the fate of yeast cells in a wine fermentation process. The yeast cells take in sugar for fuel and export alcohol (ethanol) molecules as waste products. There are actually two constraints put on the population of yeast cells. One is the availability of sugar, which is fixed and finite. The other is the toxicity of ethanol in concentration. A wine bottle or beer keg is a closed system with a finite capacity to disburse the toxin they produce. Eventually the yeast cells die of either lack of food or toxicity of their own wastes. This condition is universal for all biological systems (all CASs) including all multicellular organisms such as *Homo sapiens*!

13.3.2 Physiology in Multicellular Organisms

The economic system of a multicellular organism is, essentially, a new level of organization beyond the metabolism of single cells. Such organisms are built from a set of tissues and organs that take on the role of specialists. These tissues and organs specialize in particular physiological functions that are, largely, extensions of metabolic functions at the cellular level. They can only work in the context of the whole organism but by cooperating and being coordinated by a supervisory agent they can contribute to the stability and persistence of the whole organism. All such tissues/cell types have a high degree of adaptability and are, thus, able to operate in a highly variable range of conditions within the whole organism. And, thus, the organism is able to operate in a highly variable range of conditions within its own environment.

13.3.2.1 Governance

The governance of a multicellular organism is now a matter of coordinating the activities of a huge number of interdependent cell types in a wide range of tissue types. Three main subsystems of the organisms are involved in this governance. These are the brain (in animals), the immune system, and the endocrine system. In plants the latter two are integrated as a system that acts to counter infestations by grazing insects and higher-order animals as well as disease microbes and fungi. We will focus on animal models (brains) since these are what leads to sentience and are most applicable to the human condition.

The brains of even the most primitive animals (e.g., worms) are primarily involved in sensing both internal and external stimuli and computing an appropriate response to the situation. Brains generate behaviors that are, under the right circumstances, designed by evolution to allow the organism to survive the challenge and in so doing, procreate.

But the brain is involved in more than just the overt behavior of the organism. By monitoring all of the internal states of the physiology, the brain acts as a primary coordinator of everything that goes on in the physiology of that organism.

In organisms higher than worms, the brain is responsible for monitoring many low-level physiological conditions (e.g., oxygen levels in the blood) and initiating responses to changes according to the dictates of the homeostatic requirements of the whole organism.

The physiology of multicellular organisms is an economic system wherein various tissues produce products (biochemicals) that are circulated via several different "humors" such as blood to be picked up by target "customer" tissues/cells that then react to those products. All of this is to provide every cell in the body with what it needs to carry on its own metabolism and contribute its own production to the tissue in which it resides.

13.3.2.2 Resource Extraction

The primary resources needed by a multicellular organism are: (1) gasses such as oxygen, for respiration, and CO_2 for plants' photosynthesis; (2) organic and inorganic molecules, food for animals; (3) chemical energy; and (4) information about the state of the environment around them. Animals have lungs or gills (or skin) that conduct the exchange of gasses, bringing in oxygen and expelling the CO_2. Plants take in CO_2 during the day and produce oxygen, these gasses flowing through the stomates of leaves, reverse directions of net flows at night when the plant switches from photosynthesis to respiration.

All living systems require various organic compounds as precursor raw materials for constructing more complex biomolecules for metabolism. Animals usually take in more complex molecules, whole proteins, fats, and carbohydrates, for both the component molecules they contain and for the bound chemical energy obtained by breaking down those molecules during metabolism. Plants and animals alike require water, of course. Plants draw in the organic/inorganic material through their root systems. Animals eat plants and other animals. Foraging animals have particularly elaborate methods for locating and obtaining food, but all have mouths and means for preprocessing (like chewing) the food for digestion. The latter step, in the alimentary canal, is necessary to break the large and complex food particles into biomolecules that can be absorbed in the gut and release the minerals also to be absorbed.

Plants have chloroplast organelle containing chlorophyll (and several other) pigments that conduct photosynthesis, turning sunlight into stored energy in heavier weight molecules like sugar ($C_6H_{12}O_6$). They will then use this stored energy during their respiration phase to produce their much larger biomolecules such as enzymes, fats, and complex carbohydrates. Plants are called "primary producers" because they construct all of the needed biomass for food for animals. And animals get their energy from that food.

All living systems sense their surroundings in multiple ways. Plants are sensitive to sunlight and some can actually track the course of the sun across the sky dome with their leaves in order to maximize the light they receive. Their root systems sense and grow toward moisture and minerals, etc. They have chemical signaling mechanisms for detecting attack by animal "predators" and they can react by producing toxins to dissuade the eaters. Animals have sensory arrays that transduce energy forms such as pressure, impinging light, or modulated vibrations into internal electrochemical signals carried by nerves to the brain. Most animals higher in the phylogenetic tree of life have vision and hearing to detect distant objects and events.

In all of these cases, living systems are equipped with acquisition (import) processors of greater or lesser complexity depending on the resource and the means by which it is obtained. In all cases, input messages that convey information are used by the governance system (tactical coordination) to control the use of the acquisition processors, for example, the eyes of a predator following the fleeing prey and chomping down with jaws at just the right time.

13.3.2.3 Goods Production

The primary product of all living systems is biomass. They take low-weight organic and inorganic molecules, along with, in the case of plants, CO_2 and water, to construct higher-weight biomolecules, generally very long chains of these simpler molecules, such as proteins. Every cell in the body is involved in this process since living cytoplasm is constantly replenishing itself (autopoiesis). As older cells die and are sloughed off new cells emerge from populations of stem cells. In immature organisms, tissues grow as the rate of new cell production exceeds that of cell death—the whole organism grows to a mature state.

Growth and cell replacement are governed by the endocrine system. The life cycle of most organisms is quite complex and beyond the scope of this book. Suffice it to say that the whole process of biomass production is a carefully choreographed dance involving thousands of ballerinas (tissues), hundreds of instruments (hormones), and a many-armed conductor (the hindbrain). The composer, evolution, worked out the score and how this dance was to be performed on a wide variety of stages and for different audiences. And in the case of the human animal, there is a critic (the forebrain), a powerful one, who through behavioral choices (like smoking or drinking or clean living) can have an impact on the performance as well.

13.3.2.4 Distribution

Both plants and animals have several vascular systems that convey materials and energy in fluids, like sap or blood. Throughout the tissues of the bodies of plants and animals there is an intercellular aqueous environment (water and dissolved organic and inorganic molecules) that supports between cell communications and transport of materials.

13.3.2.5 Consumption

As is the case with metabolism we don't have a clear case of internal consumption of products as such. The whole body, of course, consumes energy or rather degrades high-potential energy to waste heat. Much of that energy is supplied by the liver converting glycogen to glucose for distribution in animals. In both animals and plants (in respiration) energy is used, and degraded to waste heat in the work of anabolism, carried out in cellular metabolism. Tissues, in bodies, produce heat which must be conveyed away.

We don't actually find consumption the way most of us think of it, until we get to the social economy. We'll consider that below.

13.3.2.6 Exporting Products and Wastes

In multicellular organisms, various cells have become specialists in that they produce export products that are of use to other cells in other tissues, either as primary resources or as signaling inputs. These cells turn out a volume of such products, which they export to the environment, but as specialists they often require specific input resources from other cells/tissues that they cannot manufacture themselves, which leads us back to resource importing functions and how metabolism is an ongoing cyclical process. In a multicellular organism we find that the network of economic activities has now spanned multiple cell/tissue types that must cooperate in producing and exporting very specific resource products. The physiology of a multicellular organism is the next outer level of economics in which producers and consumers along with their own HCGS produce a coordinated CAS capable of sustaining itself for a time scale much longer than the time constants for activities at the subsystem level.

13.3.3 Family Economics

Throughout the world, humans have adopted a basic unit of organization that provides the basis for reproduction of the species—the family. The actual forms that families have taken can vary somewhat from society to society, but the standard model involves an adult male and an adult female who are, at least for some time, mated and responsible for the "manufacture" of more human biomass—children. Families may be extended, involving several generations of grandparents or closely related aunts, uncles, cousins, etc. There is no simple model of composition and in modern societies one even finds families based on homosexual partners who adopt children or have them via a surrogate mother. The only fundamental is that a society requires enough standard model families to assure replacement of those who die off.

Families do not exist in a vacuum of course. They aggregate in groups or societies. But within the family, there is an economic system that constitutes the subprocesses found in the socio-economic supra-system. We can, thus, analyze a family as an intermediate unit between the simpler physiological/metabolic economies and the larger group economy.

13.4 The Economics of a CAS/CAES

We now examine how the generic economic subsystem situates within the context of a CAS/CAES system.

13.4.1 Generic Economic Archetype

Before we get into the decomposition of a "typical" energy subsystem let us turn briefly to the perspective taken from a systemic biological analysis. Figure 13.6 depicts a much-generalized model of a biological entity (a single individual). As shown in the figure, we have included strategic management. This is more representative of the situation for human beings who have the ability to do some strategic thinking.

Our contention is that this model depicts a basic metabolic process. Take out the strategic management process and you have a general model of all such systems, including the economic system of the HSS. Here we are not showing replicated functions (work processes) but such functions are generally replicated in living systems. For example, one important work process in a living cell is the work of producing proteins done by the ribosomes. There are many ribosomes in a cell, each continually using messenger RNA (mRNA) archetypes to construct chains of amino acids into polypeptides and proteins that then become the work horses (like people supplying labor) for the metabolic activities in other parts of the cell.

A significant difference between organelle in a cell, and their contributions to metabolic activity, and entities in the economy is that organelle are not driven by competition. Their abundance and activities are driven strictly by demand. If the nucleus is pumping out more mRNA, then the cell will turn its attention to producing more ribosomes so as to meet demand. When demand drops, some of the ribosomes will be reabsorbed to supply the raw components (e.g., ribonucleic acid molecules) for other or later use. In the economy, as we understand it, driven by the idea of companies as producing profits for the owners, those companies will adapt

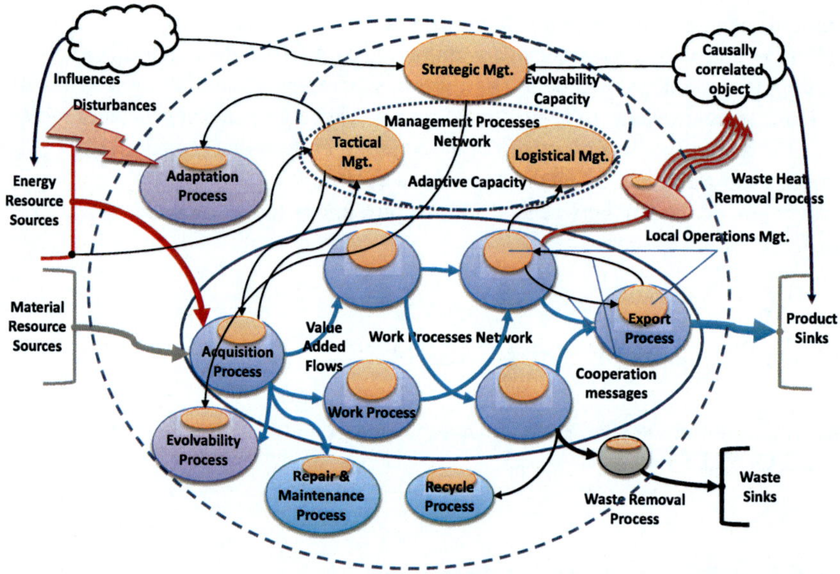

Fig. 13.6 A whole complex, adaptive and evolvable system is organized to transform resource inputs into products, expelling waste materials and heat. The system is sustainable against the vagaries of the environment by virtue of possessing both adaptive and evolvable processes (purple). Blue arrows represent value-added flows of materials transformed by work. Thin black arrows represent the flow of messages; only a small fraction of such flows is shown. The large, outer, dashed oval is the boundary of the whole system. General work processes are grouped in the center as a network of processes with value added moving from left to right. Governance processes (information and knowledge processes) are pink ovals. The whole system governance, coordination and strategic (if present), is grouped together in the smaller dashed oval. All of the other processes are auxiliary and are responsible to things like internal maintenance or recycling materials. See the text for the explanation of the "cloud" objects

or evolve to try and maintain their profit-producing function. Competition, rather than cooperation, is the rule of the day. If demand for a product decreases in the human economy, a company will attempt to adapt, perhaps by shifting its efforts to other, still-in-demand, products. Or it may try to evolve to become a different kind of company with a better position in a market. In the economy, the operating unit has become important to preserve in the sense that it is self-serving. This may be attributed to the idea that operating companies have become the main mechanism for providing support to employees through wages and profits to owners. Given the model of entities as serving the greater needs of the whole system, as organelle do, one might well ask if the economic entity model is appropriate for the working of the economy of the HSS.

Let us take a closer look at the various entity subsystems represented in the figure and begin to relate them to the economic system.

13.4.1.1 Primary Work Processes Network (Blue Ovals)

All processes do physical work on materials adding value to inputs that become resources for downstream work processes. In the figure, we identify various work processes that have special relevance to the whole system. Special processes that interface with the environment sources and sinks are the acquisition (or import) and the export processes under coordination by the tactical management agent (see below). All other work processes provide successive transformations of materials that collectively produce the product(s) of the CAES. All processes produce some waste heat and materials. The work processes usually account for the bulk of material wastes. In general net flows go from the input streams (left side of figure) to output streams (right side of figure). Internal feedback loops may also exist (e.g., some material wastes may actually be recycled).

13.4.1.1.1 Core Value-Added Work Processes

A typical work process in a living cell is the ribosome which assembles polypeptides and proteins from amino acids that are available due to the digestion of, for example, denatured proteins no longer able to do their own work. The ribosome is composed of ribonucleic acid (RNA) chains and some structural proteins. They receive instructions from the genes in the nucleus or in mitochondria (see below) in the form of a long messenger RNA chain that is used as an archetype for the construction of proteins, for example, to be used as enzymes. The amino acids are attached to transfer RNAs that carry a specific three codon marker at one end that signifies the kind of amino acid (there are 20 amino acids but triplet codes of four nucleic acids gives a total of 64 possible codes, so many of the proteins have more than one code) being conveyed into the ribosome. The triplet codes match up to their counterparts in the mRNA strand and so the ribosome can bring in the right amino acid in the right sequence. And, through the work process, new proteins are assembled for use in other parts of the cell's machinery.

There are a number of work processors in cells that conduct this kind of work. Metabolic work processes are the fundamental basis of all living systems as shown in Fig. 13.3.

Surrounding the living cells in a multicellular organism is the physiological economy (again, Fig. 13.3). Physiological work processes are conducted by various tissue clusters, usually, but not always, found within various "organs." The liver in animals, for example, is a major chemical work process involved in many transformations. For example, it converts energy storage molecules into long-term storage (i.e., glycogen) or from long-term to short-term fuel (glucose). It also works to recycle and detoxify a number of chemical substances by breaking them down into non-toxic components. The heart is a mechanical work process devoted to pumping blood (carrying nutrients, oxygen and other gasses, and chemical signals, e.g., hormones) throughout the body. Most of us are very familiar with the body's various work processes to one degree or another since they are front-and-center in our daily lives. But most of us do not necessarily realize that these comprise an economy in the sense being discussed here. Work gets done, value gets added, and products (replacement biomass) get produced by this body economy.

At this point, the reader may be sufficiently cognizant of the archetype model of economy to be able to identify work processes at the family level, in Fig. 13.3, and certainly at the social level—the subsystem we generally mean by "the economy."

13.4.1.1.2 Acquisition Processes

Raw materials and energy have to be acquired through active processes such as eating, extraction of metal ores, or cutting timber. Material resources are valuable because of how they might be worked on to produce something useful such as timber being sawed to produce lumber of specified dimensions for building construction. Timber (trees) are, in principle, a renewable resource, meaning that new trees will grow to replace those extracted. But the rate of extraction cannot exceed the rate of renewal, otherwise the resource begins to act like a finite fixed one. Metal ores and fossil fuels, on the other hand, are finite and fixed resources that can get used up after extraction so that eventually that resource will run out of stock. While metals can potentially be recycled internally, this is not the case for energy which gets degraded to waste heat in doing the work. Once you burn the gasoline in your car, it is gone forever.

The value of these raw materials in or on the ground depends on the amount of work that has to be accomplished in retrieving them. Fossil fuels provide an example of how a commodity's value may be reduced. Acquisition processes such as well drilling, pumping will invariably target the easiest to obtain oil or gas. This is known as the "Best First" principle. It is simply the result of the business of oil companies that are driven by the profit motive. It gets progressively more expensive to drill and pump reserves that are deeper in the ground, or under the ocean. At the same time, our civilization is absolutely dependent on fossil fuels to power industry and homes.

13.4.1.1.3 Export Processes

These are the processes that push waste materials and heat out into the Ecos environment. Or an export process for most CAESs that are subsystems of a larger supra-system might be to push products to "customers." At present we can find very little evidence that the HSS and its economy are producing anything of particular worth to the other subsystems in the Ecos; our economic products are aimed at serving our own consumption only. Nevertheless, the possibility for doing so exists. For example, the economy could do work to improve the biological support quality of soils, not just for farming but for natural ecosystems that our prior activities had degraded. We will discuss this more in the next part.

13.4.1.2 Management Processes Network (Orange Ovals)

The major management processes are grouped together in this network. These are the agent processes responsible for directed evolution (strategic), coordination with the entities in the environment (tactical), and coordination of the internal processes (work and auxiliary).[27] Within all of the other processes we represent the local operational management that is in communications with each other as well as with the coordination management network. The figure only shows a few message pathways to avoid clutter. Shown are cooperation messages passing between a generic work process and the export work process (right side of figure). Tactical messages are shown between the tactical manager and the acquisition process (right side of figure) as well as from the tactical manager to the adaptive process (explained below). Finally, a message pathway is shown between the strategic management and the evolvability process (an auxiliary process).

The strategic management process is observing external entities (cloud shapes) that are not directly interacting with the system but are interacting and having influence on those entities that are (e.g., energy source and product sink). This is explained in Chap. 12 regarding the hierarchical cybernetics model.

[27] Unfortunately, there is no good way to cover the subject of management without a full explication of the hierarchical cybernetic governance system model being used. That means we have to refer to Chap. 12 for a more complete handling of the model. Or we can refer to Mobus and Kalton (2015, Chap. 9). The reader is encouraged to take a pause here and refer to those chapters since we will be referencing concepts explicated there (This is the curse of systems science. Every concept depends on every other concept so we are constantly having to reference multiple chapters. Hopefully the reader will indulge us this quirk.)

13.4.1.3 Auxiliary Processes

These are support processes that provide services to the internal processes. Since other processes are subject to degradation or accidental structural damage a repair and maintenance process (which can repair itself!) operates on any process that incurs such degradation or damage. A waste removal management process ensures that material wastes do not clog any of the value-added channels. A similar process is needed to dispel waste heat through the system boundary. Since some materials that end up as waste might be recycled economically, some amount of recycling process is included.

The two processes that make the system adaptive and evolvable (aside from the management capabilities to make these decisions) are work processes in that they are responsible for the work accomplished in adapting or evolving a system. Adapting means modifying an existing subsystem process to accommodate short-term changes in its activities. For example, an adaptive process would be when the internal mechanical shop expands the capacity of an inventory cage in response to increases in material flows. Evolving means constructing a whole new work process to accomplish some entirely new activity in order to take advantage of long-term changes in the environment, such as entering a new product market.

13.4.2 Comparisons: Natural Versus Human Economies

The economy of the HSS is encapsulated within this model of a metabolic process of a living entity. But there are differences in the way we think of an economic system and the metabolic system. Therein lay the problems with our understanding of the former. What is shown in Fig. 13.6 is the layout of a *sustainable* CAES. We assert that the economic system within the HSS should support a sustainable CAES. But we note that our common understanding of the economy from classical economic theory fails to recognize this and, as a result, is producing "predictions" that are wide of the mark. When we look closely at the biophysical aspects of the real economic system, as we did in Chap. 9, we will discover very different projections for the future of the HSS.

Consider a singular example of the discrepancy between the common understanding of economics as the social science of human economy and the systems economy perspective. In the former a "goal" of the system is to grow the production of wealth, and to do so exponentially. Growth of an assumed measure of production or income, the gross domestic product (GDP), is couched in terms of a percentage increase per annum. Like compound interest this kind of growth means the stock of wealth is consistently getting larger with each passing year. Classical economists argue that this is a necessary condition for the maintenance of the HSS. Yet, from the physics and biology of reality we know that infinite exponential growth is impossible in a finite world. Rather, a CAES grows to a limit imposed by the constraints of its supra-system and then operates in a more-or-less steady-state dynamic from that time on. The question raised by a systems perspective of economics is: what is the limit of the size of the HSS, the limits of the use of finite resources, and the limits of dumping wastes into that supra-system that will allow the HSS to function as a sustainable subsystem of the Ecos? Classical economics not only doesn't ask those questions, it doesn't even recognize the need to ask them.

The assumed need for growth begs another question: Why? Why would the economy need to grow exponentially forever? Part of the answer comes from the biological mandate and the removal of some constraints on population, namely death rates keeping the stock of people within a carrying capacity range, as with other biological systems. Technology, especially the technology of agriculture (and especially that of the "Green Revolution") along with better ways to keep ourselves safe from predation and treat the causes of diseases, has led to a seeming removal of constraints such that the population stock has grown from an estimated size of around 5 million at the dawn of the First Agricultural Revolution (also called the Neolithic Revolution)[28] to around 1 billion in 1800 to 7.5 billion today. Estimates on what the carrying capacity of the planet for human beings range from several million (as in the Neolithic) to no more than 2 billion (meaning that our current population is nearly four times too large). It is not really clear that the concept of carrying capacity, developed to measure populations of minimally adaptive creatures, is all that relevant to human populations. But one thing is clear. All of those people have to eat, they have to have shelter, and they have to have meaningful employment—a way to make a living. The demand for resources applied to producing food and shelters is directly related to the size of the population.

So, one answer to the question of why the economy has to grow is that it is basically the source of jobs in the modern world and each year more people are in need of jobs. There is another, more modern impetus for growth that is much more complicated. This will be discussed below as we look at the modern energy sector dynamics and how it affects the economy.

But it isn't that simple. The biological mandate doesn't just drive reproduction. It also drives the psychological propensity to want to maximize income when times are good as a hedge against when times are hard. Animals store fat when food is plentiful so they have a reserve of energy when it isn't. That is a basic biological urge. But just as technology threw off the shackles of population growth, it also contributed to eliminating or at least reducing the bad times on net. Thus, the agglomeration of wealth during plenty was less and less offset by times when that wealth had to be consumed for survival. At some point in the history of modern cultures, the agglomeration of wealth for its own sake coupled with increasing kinds and numbers of material goods (especially stylish goods!) became recognized as the "Profit Motive." The idea that one should constantly and forever maximize any kind

[28] The Neolithic Revolution involved more than just the discovery of agriculture; it involved nearly every aspect of culture in general. There have been two additional agricultural revolutions in the west. The second or British Agricultural Revolution and the third was the Green Revolution noted in the text.

of gain from economic activities such as trade came to be the dominant theme in those activities.

Living systems as CAESs only grow while maturing. Except for cancers they do not grow beyond a limit that has been established by evolutionary pressures. This doesn't mean that the biological mandate goes away. What it means is that life has found mechanism for internal regulation of growth. Tissues need to be repaired, of course. Reproduction itself is a kind of growth—but again, in nature there are external constraints that keep most biological systems in check once a population reaches its carrying capacity. Within our bodies, cells are signaled to reproduce by elaborate systems that monitor the state of maturity of the whole organism, or the states of its tissues with respect to normal functions. When growth needs to be curbed, those signals, like growth factors, are turned off and the cells obey.

Nor do cells seek to continue ingesting food (e.g., glucose) beyond their metabolic needs. It is true that specialized cells store fat but not for themselves as much as for the whole body, again as a means to survive hard times for all. Once more, there are internal signals and controls that keep living systems in balance with its environment. But the HSS has not evolved to this stage. What drives modern commerce is profits. This will be demonstrated to be one of the prime reasons for major dysfunctions in economic systems.

13.4.2.1 Markets in Natural and Human Social Economies

In this section, we examine some comparisons between natural and human-constructed economies that highlight the significant differences and reasons why the modern human economy is not really suited to perform the functions of a true CAES economy.

13.4.2.1.1 Demand Driven Production Versus Speculative Production

The counterpart to a free-market capitalist economy is not a planned economy as tried under twentieth century communism/socialism. Rather, it is a demand-driven *non-profit* economy in which the market acts as a local optimization mechanism and work processes operate on an as-needed basis somewhat like modern just-in-time supply chains. This is the case for living systems in which any excess production is either stored as a buffer against bad times or is degraded and the components are recycled to where they might be needed for actually demanded production.

In the human economy, a different mechanism underlies how a market operates. Producers are dependent on income from the sale of their products and that income is the basis of their livelihoods. They cannot afford to be unproductive just because there is a lull in the demand (though of course, they may end up being so once their product inventory fills up). Laying off workers is a poor option but necessary under these circumstances. In the worst case, a firm goes out of business for lack of sales. Living systems work processes operate under a different paradigm. Processes are supported at a minimal subsistence when demand for their products is low. Muscles may atrophy but they never actually go away. As soon as demand picks up, these processes rebuild capacity in proportion to the demand (both in terms of magnitude and longevity).

Another complication in the human marketplace is the requirement for firms to be profitable. Not only must they use their sales amounts as a signal signifying demand, they must include some kind of price premium that goes above and beyond covering their costs of production. This is the capitalist free-market where competition to supply demand leads to more than just redundancy. It can lead to wastes. It is also the force underlying the impetus to use additional communications channels—advertising—to boost demand.

13.4.2.1.2 Communications Between Customers and Producers

In natural economies like metabolism the communications between customers, like the ribosome, and the producers, like the mitochondria, is direct and based on exact demand messages (i.e., ADP recycling). There is little opportunity for distorting noise, and certainly no deception (false information). In the modern human economy, very often, there are multiple "middle-men" between producer and customer. Each of these agents is guided by a profit motive which adds to the prices the customer pays. Therefore, the main signal, the price, is often distorted and generally not in a consistent way. Once, long ago, when human's transacted trade in local (community) markets, both buyers and sellers met face-to-face and both had some insights into the actual cost of production, thereby having some basis for negotiating a price. Those markets were more like the natural markets of metabolism or physiology, indeed being extensions of them.

The question arises, if direct and non-distorted communications are an essential feature of a sustainable economy, what does this imply for a future human social economy? We will address this question, to some degree, in Part 4 of this book. Stay tuned.

13.4.2.2 The Role of Energy

In any economy the general pattern is for the system to import relatively high entropy (low organization) materials, low entropy (high potential) energy, and along with messages from sources and sinks that inform the system of the state of the environment. The energy is consumed in the process of doing transformative work on the materials, reducing their entropy by increasing their organization. Iron ore, coke, and other raw materials are mined and eventually out comes an automobile. Amino acids or low weight polypeptides are ingested by a bacterium and assembled into functioning proteins in the cell. In the process, the energy is degraded (becomes more entropic) in the form of unusable waste heat that is carried off to the environment.

The second law of thermodynamics is not violated in any of these processes because the waste heat contributes to the rise in entropy of the larger supra-system (or the Universe as a whole).

Looked at in another way, the production of wealth (cars and proteins) requires the expenditure of energy on work processes throughout the CAS/CAES. The imported energy must be at the right potential and of the right form (high temperature heat versus electricity) for specific kinds of work processes (internal combustion engine versus electric motor).

The central problem for an economy is how to channel and direct the right amount of energy to the right points of use at the right times. For example, the human economy electrical distribution system uses wires from the power source to the homes and businesses that use electricity. Power companies have to maintain peak capacity generation in order to handle peak loads that can occur at the points of use, sometimes unpredictably.

As an economy operates over time, decisions need to be made about the distribution of the appropriate power capacities. Work processes (points of use) operate at rates that are often dictated by demand for their products or services. That, in turn, translates into demand for power to drive those processes (e.g., surges and wanes). The demand put on the work process needs to be forwarded and translated into demand on the energy flows to support the changes in work rates. That means an information system is needed to communicate changes to the power producers.

In metabolism, the main product of the citric acid cycle (also known as the Krebs cycle) is the molecule adenosine triphosphate (ATP) which can be thought of as portable batteries that distribute by diffusion along concentration gradients to the various other work processes throughout the rest of the cell such as the ribosome discussed above. The cycle takes place largely within the membrane bound mitochondria—the power plant for cells.[29] At the point of use the ATP molecule gives up a phosphate becoming adenosine di-phosphate (ADP) and supplying 2 hydrogen ions per molecule. The local concentration of ATP declines while the concentration of ADP increases. This is what drives the gradient diffusion of these molecules cycling between the point of production and the point of use. The electron transported are the energy carriers in setting up a current that is coupled to the work process. ADP is then recycled back to the mitochondria where a phosphate is added by pumping energy into the reaction—the same energy that is removed later at another point of use.

Depending on how quickly the cell types must accomplish work there could be hundreds to thousands of mitochondria throughout the cell. The mitochondria are stimulated primarily by the concentration of ADP in their vicinity and the availability of sugar molecules. So the demand-driving work process will produce more ADP the more work it does, and this signals the mitochondria in the neighborhood

[29] Green plant cells also have chloroplasts which capture sunlight converting water and carbon dioxide into sugars, which are then the main inputs to the mitochondria. Animals have to eat plants and other animals, digesting their more complex molecules down to sugars for input to their mitochondria.

to produce more ATP. The ATP/ADP currents constitute a signaling system, a communications process.

Energy flows are handled by energy-carrying channels and supply lines such as gas pipes and electricity wires. The flows are controlled by information influencing actuators, amplifiers, restrictors and other components. The information is generated by various agents that receive information, 12.g. of demand for an increase in a flow, from recipient agents. In the case of the ATP/ADP coupled production/use, the mechanism is simple concentration gradients and diffusion. However, there are more complex demand-driven examples of active transport of macromolecules within the cytoplasm instigated by a counter transport of a molecular signal that activates the one conveying the macromolecule. In the body of a multicellular organism, the examples of informational signals generated by decision agents that then cause recipient agents to apply control activation to material or energy flows are abundant. For example, the increased activities of neurons in a specific region of the brain supplies a cascade of molecular signals to the local blood supply which responds by increasing the flow of oxygen and glucose to that region to support its increased activity.

In the human economy, we also find a signaling system needed to guide the flow of energies to appropriate work processes. And like the ATP/ADP cycle (which is its own signal), it is driven by supply and demand. But an external messaging process is used. This is what we call money. Money in some nominal form is used to make purchases, which is actually a signal to the producer that their product or service is in demand. This kind of signaling and response is ubiquitous but not universal. There are several alternative models of economies being more direct and based on different premises other than, for example, profit making. Sharing or gift economies are also able to manage the distribution of goods and services (and energy) without the intermediary role of money. However, a monetary signaling economy seems to be by far the most prevalent so we will primarily focus on those (e.g., in Chap. 9).

13.4.3 Competition and Cooperation

In natural economies, metabolism and physiology, and families in Pleistocene times, there is a constant tension in economic activities involving, on the one hand, competition between processes and cooperation between neighboring processes. This is the problem that the coordination layer in the HCGS strives to solve.

Input resources may be less than the whole system needs from time to time, leading to competition between processes that use them. The coordinator needs to allocate those resources according to some optimization scheme that keeps the whole system working as best it can. At other times, resources might be abundant and the coordinator can sequester excesses in buffers.

In general, there exists a dynamic tension in the network of processes between cooperation at a local neighborhood scale, and competition between processes that are further apart. That dynamic tension can be seen in the metabolism economy as anabolism goes on at the same time as catabolism for instance. They work together to continually renew structures while breaking down complex molecules to components that can then be fed back into construction.

In body physiology, this same effective tension is seen in the ebbs and flows of hormones or the effects of the autonomic nervous system's sympathetic and parasympathetic systems.

That tension also applies to the social system economy at every scale. Within the commerce economy, it is widely held that competition is a driver of innovation. Certainly, many examples can readily be found where companies competing with one another in the same market are constantly trying to come up with novel features for their product to best their competitors and thus outsell them. However, it is questionable whether this kind of innovation is actually of great value in the long run. A certain popular smartphone company that keeps "innovating" every year to drive sales results in a lot of e-waste going into landfills or causing health concerns for workers who attempt to extract the precious metals from them. That is clearly not a sustainable model.

Competition seems to be the overwhelming force in the modern capitalist, free-enterprise economy. There have been instances of cooperation between otherwise competing firms, for example in working together to formulate standards for engineering processes. But overall, firms seek to out-compete the other firms in the same market in order to maximize profits and, in the last several decades, send their top managers home with fat pay packages.

The amount and kind of coordination in social economies is also highly questionable. Governments may attempt to regulate business practices or influence economic outcomes with various policies, especially tax policies. But the problem with this is that governments do not have a sound model of economics to work with, so formulate policies in a trial-and-error manner and are mostly reactive agents. In the United States, the Federal Reserve Bank is charged with regulating tradeoffs between full employment and low inflation through interest rates on borrowing. As of this writing it is unclear that their leverage actually is responsible for maintaining the balance.

The biggest problem with governments providing coordination services is that the model they universally subscribe to is that of growth and expansion. There are probably hundreds of reasons that people in government (and academia) believe this, and there have been an endless stream of books and journal papers devoted to explaining it. But the laws of physics as we currently understand them prohibit endless exponential growth. We'll leave it at that.

13.5 Conclusion

A CAES economy is the engine that keeps the entropy within the system as low as possible, that is, the organization is maintained, the functions are maintained, and the whole system is kept far from an equilibrium state. It achieves this by extracting high-entropy materials from its environment and capturing and using very low-entropy energy to drive the work of converting the materials into useful things like structures and machines that support the organization. A living cell's metabolism is an economic system. A multicellular body (plant or animal) has physiology as its economy. A family has its capacity to obtain food, water, shelter, and other supports to shelter the family and keep it producing more human biomass. And a human social system has its production-consumption system of conducting transactions between sellers and buyers with or without money. All of these economies are nested, metabolism the innermost and the baseline, physiology next out, and so on. Each inner one relies on the activities of the one more outward.

The human economy is similarly nested within the Ecos economy. What is different about the Ecos is that all of the material resources needed to produce low-entropy structures and functions are completely contained within the planet itself. These are available in the huge reservoirs of the major geospheres. The vast majority of high-potential (low entropy) energy comes from the sun and the waste heat that the Ecos economy produces, which is now dominated by the human subsystem's production, radiates back to space. The entropy of the Universe is increased overall, in compliance with the 2nd Law, but the entropy within local pockets on the planet is diminished. Even so, those systems may just as easily revert to entropy maximization if the Ecos, HSS, or biological economies ever fail.

Part IV
Designing CAES Artifactual Systems

In this final part of the book, we consider aspects of using what we have covered in the prior three parts to build viable CAESs that are not just the physical artifacts themselves, but artifactual systems that include humans as components. Engineers have been grappling with how to use systems science to do systems engineering. Primarily they were initially looking at the design of extremely complex systems like jumbo jetliners. But with time and experience, and some help from systems science with regard to recognizing the boundary problem, they began to realize that the physical artifacts were not by themselves the 'real' system. They began recognizing that you have to include the human beings who would interact with the physical artifacts, such as the pilots and other airline personnel. Then there where the customers to consider. Before long systems engineers realized that the problem of system design is so much more than hardware or software design. It included elements of social system design.

Today, as the world recognizes the need to think systemically about the whole world and the human social system's place in it, we have come to understand that we humans can no longer pursue willy-nilly the production of physical artifacts (or services) based on profit motives because we can now see that that approach is contributing to our global predicaments. We have to adopt a world-wide systems approach. This means we need to put effort and resources into the analysis, the deep analysis, of the Ecos, the HSS, the internal health maintenance subsystems within the HSS, and so on down to our metabolism.

With that in mind, this last part will provide a mere glimpse of what tackling the design of a viable HSS would look like. We start with a definition of what we mean by an artifactual system (Chap. 14) and, generally speaking, the design and engineering process that should be adopted to enable us to intentionally construct a viable CAES-based HSS. Then in Chap. 15 we go into greater details about that process in which we use the CAES archetype models to guide the architecture, design, and engineering of one of the most basic subsystems important to the HSS of the future, the food-producing social module.

We end the book with a look forward to what a whole HSS that is viable within the context of the Ecos might look like. This brings all aspects of the CAES archetype to bear on the question of how should society organize itself and, most importantly, what purpose with respect to the health of the Ecos should the HSS adopt? The HSS cannot be merely a consumer of finite resources, as seems to be the case now. It cannot be to replace a substantial part of the biosphere with human (and farm and pet) biomass. The proper design of any to-be-viable CAES must include determining what the outputs of the subsystem should be to benefit the larger embedding supra-system. That will be where we leave it for now.

Chapter 14
Complex Artifactual Systems

Abstract Artifactual systems are complex, adaptive, and evolvable systems that are composed of humans and artifacts in multifarious relations. Sometimes called socio-techno systems, emphasizing the human-physical artifact nature, and sometimes referred to as human-activity systems, to emphasize the human-process nature, these systems are integral parts of human social system cultures. Artifacts themselves can be CAESs. And the special relation between humans and artifacts that is their conception, design, and construction is shown to be the result of a new kind of ontogenesis, one that involves intentional organization and intentional selection replacing auto-organization and "natural" selection as operated on material systems prior to the evolution of human consciousness. We focus on CAS/CAESs and how they are conceived and designed. The design/engineering process is shown to be an evolutionary one even though intentions replace the much more stochastic processes involved in ontogenesis prior to human intentional approaches. We also argue for the inclusion of evolutionary approaches in the design/engineering process.

14.1 The Purpose of This Chapter

The concern for the next three chapters is to explore the meaning of the larger human social system (HSS, introduced in Chap. 7) which includes not just human beings but a vast and complex array of artifacts created by them forming what we generally think of as human cultures. The term "cultural artifact" is used in fields such as archeology, anthropology, and sociology to mean any object that is the result of human production in any period prehistorical to modern times. An *artifactual system* is a set of interconnected cultural artifacts and the humans that did the production and those who use the artifacts to accomplish other work. The emphasis is on the systemness of the objects, their producers, and users.[1] We will offer a definition of what is meant by the term "artifact" below.

[1] Artifacts have relations to humans and to each other in a complex entanglement. For a thorough treatment of this see Hodder (2012).

G. E. Mobus, *Systems Science: Theory, Analysis, Modeling, and Design*,
https://doi.org/10.1007/978-3-030-93482-8_14

We will also offer a new approach to understanding artifactual systems and the cultures in which they are embedded. This larger-scale systems perspective is needed in order to better understand the "context" or environment of artifactual systems. The intent is that we will develop a new approach to conceiving, designing, and engineering such systems by using the concepts and procedures developed in this book, derived from our understanding of ontogenesis in Chap. 3.

Artifactual systems represent a new "state" for matter in the Universe in the same way that life itself is now seen by some researchers as a new state (Smith and Morowitz 2016, see Sect. 1.3.3.2 for an introduction). That is artifacts are the product of a wholly new kind of dynamic in the course of ontogenesis and they are obtained from a phase transition from unconstructed form to structures that could not have obtained by some energy minimization process. Prior to human intentions and interventions in the process of combining material or organizing social structures such as communities or institutions, matter had been subject to auto-organizing (e.g., as in chemistry) and phase transitions between states such as solids, gases, liquids, and plasma. And all systems were subject to the laws of thermodynamics and mechanics. Living systems have been characterized as resulting from a phase transition by virtue of the seeming ability to go against the Second Law as dissipative systems (Smith and Morowitz 2016; Prigogine and Stengers 1984). It is the new kind of organization, behavior in obtaining free energy, and the capacity to dissipate waste heat to the environment (thus not really going against the Second Law after all) for material objects that have led to the characterization of life as a new state of matter. We assert that the same kind of logic applies to artifactual systems. Artifacts, as defined here, did not exist anywhere until humans applied intelligence and creativity as well as imagination to the combining of components to obtain a new object (system) that served a purpose. Human consciousness was itself a phase transition of living matter, networked brains, from which emerged the capacity to imagine and consider the future and mentally manipulate objects with various forms of affordances to create new objects, more complex, able to be used in new ways.

Artifact-hood cannot be taken for granted. This new state of matter exists because a new kind of organizing process has emerged in the evolution of human consciousness; we replace the notion of auto-organization with *intentional organization* and natural selection with *intentional selection*. That is, the combining and recombining of components to create new systems are done with foresight and purpose and the ongoing replay of doing so is done based on a human perception that the new system successfully fulfills that purpose.

There may be good arguments for considering some kinds of artifacts as not really being systems and not even atomic components as described in Chaps. 3 and 4. For example, a marble statue is clearly an artifact having symbolic meaning, but it does not process energy or material so far as meets the eye. It would at first glance appear not to have inputs and outputs so therefore not qualify under our definition of a system. The statue appears to be inert. However, on closer inspection, we find that the marble is itself a composite of minerals and those of atoms. So, from the hierarchical organization perspective, it embodies, albeit in an essentially static way, that nested quality of subsystems. Moreover, the statue exists in an ambient environment that can include, for example, temperature variations, meaning that heat energy is being absorbed when the temperature is higher and radiated when it is lower. In other words, energy, though of a low quality, is indeed flowing in and flowing out as conditions change. These flows, in turn, act on the marble minerals by changing the energy modes (or at least the vibrational mode) of the atoms in the minerals. One effect of these changes is that minerals tend to break down over long enough time scales and cycles of heating and cooling. The statue will last a long time relative to the human lifetime, but eventually, it too will return to dust. So, is the statue an artifactual system? We would argue, yes, but clearly a very simple system in terms of, say, behavior.

Yet a statue, like all forms of artwork, literature, poetry, even architecture, even though static from the perspective of human perception, is the product of human intention and imbued with human-relevant meaning. It has a purpose, if nothing more than the communication of the affective state of the artificer. For now, we will accept all such objects as artifactual systems and remain open to counter arguments.

In this chapter, we have several main goals. The first is to understand what we mean by artifacts and how they are the product of human intentions—this is the mechanism by which a phase transition to a new state of matter is achieved. The second is to understand what is meant by the relations between artifacts and human beings, as well as between artifacts and artifacts—the nature of artifactual systems. And the third is to consider artifactual systems in the mold of the CAES model introduced in Chap. 10 (and Part III). This latter is the setup for the next chapter in which we introduce the methods for designing and engineering CAES-based artifactual systems, especially the human social system itself. In the final chapter, we demonstrate the application of all of this thinking to the actual design of an HSS for the future of the planet.

The purpose of this chapter is several-fold. We first need to establish what we mean by artifacts in general and then how artifacts play into human culture and its evolution. We then need to address how artifacts come into existence in our current ways of designing and engineering[2] them. We then turn to specific aspects of the design and engineering of CAS/CAESs. We lay out a new way of thinking about systems engineering that will be needed to proceed forward into the realm of artifactual CAS/CAESs.

Finally, we will be proposing a radical new approach to the design and engineering of complex adaptive and sometimes even evolvable artifacts. This new approach is based on the subject of Chap. 3—Ontology, specifically the nature of ontogenesis and the coming into existence of new systems. We will explain the rational for adopting an ontogenic process for design and engineering when it comes to CAS and CAESs. We argue that Nature long ago worked out the process by which these systems come into being and achieve sustainable existence; they persist in

[2] Recall that we use the term "engineering" in a more holistic, broader sense to include any intentionally created artifact using principled approaches. In traditional engineering, for example, we use mathematical procedures based on measurement and physics, writ large. But designing and implementing a policy within an organization is also an act of engineering.

otherwise chaotic environments by being "fit" and adaptive to changes. We also argue that this process will be found to be essential to the success of design and engineering of seriously complex systems as the human social system (HSS) moves into the future. This chapter will introduce the concepts of ontogenic artifice and the next chapter will go into the details of the process of designing based on the CAES archetype model (model-based design) as used to "design" a particular social module (a food-producing domicile) which is an example of a CAES. The final chapter of the book will extend the example of the social module design to the design of a complete HSS that, because it is designed as a CAES from the start, will be sustainable.

The process we propose is a definite departure from classical engineering approaches. It does not abandon certain elements of those approaches entirely but, rather, incorporates them within a larger holistic framework based on ontogenesis. Prior to the emergence of the human mind able to invent purposeful tools, Nature operated on the basis of auto-organization (as in Chap. 3), the not entirely random, but stochastic (chance) coming together of simpler systems that could form strong associations from which emerged new systems forms and functions that none of the components could demonstrate before their combination.

Subsequent to the emergence of early human minds, the organizing process entered a new dynamic. Rather than chance encounters between components, humans for a variety of reasons (described more fully below), began to *intentionally* bring components together. Even in auto-organization, the encounters of components, bringing them close enough that free energy might generate a coupling, is seen to not be completely random. In Mobus and Kalton (2014, Chap. 10), the process that gives rise to encounters and auto-organization is described as resulting from directed energy flows (e.g., convective cycles) that can also determine sorting mechanisms (e.g., a convective cycle operating in a gravitational field) that increase the probability of certain components having encounters.[3] So, auto-organization, pre-humans, had a larger stochastic element, but it was from the start not purely random (e.g., a uniform distribution across pair-wise encounters of all types of components). There were already organizing constraints on what sorts of encounters would happen. Geometry and boundaries greatly influenced what could transpire. For example, the nature of a rotating planetary body orbiting in the "Goldilocks zone" and bathed in sunlight provided the necessary conditions for the auto-organization of the various geospheres and the prelife molecular systems. And, eventually, life itself.

Intentional organization is a far more directed process than those found in more physical versions of auto-organization. Nevertheless, mental intentions are not, themselves, wholly deterministic. Human intentions retain an element of stochasticity (for example imperfect memory recall can result in mistakes).

[3] This is opposed to a "purely" random process in which there is no directed flow of energy, just an ambient temperature. For example, an adiabatically isolated system such as an insulated flask of gas molecules will be subject only to random encounters and chance combinations.

Another big difference between intentional- and auto-organizing processes is that the former is based on "teleological" outcomes. That is, a human mind often desires a particular result from the act of organization. Whereas the latter do not involve any particular insights into outcomes, desired or otherwise. This is the distinction between living, purposeful, systems and merely mechanical/chemical systems. Life, as a state of matter and energy, has an overall objective function that leads to perpetuation of form and function. It is teleonomically purposive.[4]

Human artifice is an ontogenic process that introduces an element of cognitive purposiveness. Humans, by virtue of their ability to construct complex models of things in their mental representations (Mobus and Kalton 2014, Chap. 1, Principles 9–12), are able to "imagine" what a new or improved thing could be. They can test, to some degree, manipulations of the thing mentally before actually building that thing. The evolution of animal life led to the emergence of brains able to conceive of new arrangements of matter and energy in virtual reality (that is mental space). The conception may be motivated by some desire to achieve a goal—the new thing may be instrumental in solving a problem—and the mental manipulations are essentially simulations of how well the new thing will work toward that end. Those mental simulations may result in further modifications of the "design" of the thing. Motivation and mental simulations contribute to the meaning of the word "intentional" coupled with "organization."

In this chapter, we explore this new kind of ontogenic cycle, involving intentional organization (including the selection processes that lead to extinction or further improvements) and purposive construction of artifacts. We bring everything together to describe the general principles and approaches to systems design and engineering of complex adaptive and evolvable artifacts.

14.1.1 What Is an Artifact?

What do we mean by an artifact? Fundamentally, anything that comes into being by the devices of human artifice, anything conceived of and constructed by human effort, and anything that could not exist or come into being by mere physical or biological evolution is what we shall call an artifact.[5] This includes not only physical devices of various kinds but also the design of sequences of actions, procedures, methods, etc. that produce other artifacts. A computer program that results in a physical output (however transient it might be) and its underlying algorithm(s), is an example. We also include organizations and institutions such as commercial companies or the market economy. We include governments and city- or nation states.

[4] Teleonomy, as opposed to teleology, recognizes an Aristotelean final cause in the sense of an end purpose. In universal evolution, we see that the Universe appears to produce increasingly complex, hierarchically organized entities—systems.

[5] This includes nonuseful by-products of human artifice such as sawdust piles or rock chips resulting from flacking a stone to make an arrowhead.

In other words, everything that humans have created from their attempts to improve their existence counts as artifactual.[6] Chimpanzees have social organization as a natural consequence of their social psychological natures (De Wall 2005, 2010, 2014, 2016; Tomasello 2014, 2016, 2019). They didn't invent them. Humans also once had naturally evolved social organizations (i.e., the tribe), but the various forms of contemporary societies demonstrate the inventive urge to do things "better," which, of course, is a relative term.

The sum of artifacts constitutes the cultural milieu above natural human sociality and moral behaviors.

Various kinds of artifacts we include:

- Physical objects, tools, toys, etc.
- Methods and procedures—ways of doing something, including algorithms for performing computations, manufacturing process instructions, and interpersonal interactions, such as rules of manners.
- Institutions and organizations—social structures that facilitate cooperative behavior (e.g., moral and ethical norms), production of artifacts, governance, and economic exchanges. This includes societies with cultures and social norms, governments, and economic structures like markets.
- Policies—abstract rules for regulating the above.

In this chapter, we will be interested not only in artifice itself, but how it is accomplished and how it has evolved with the evolution of human knowledge of how the world works—or, in other words, with systems knowledge. Since the invention of writing and mathematics, humans have accumulated impressive knowledge about materials, mechanics (including electronics), and conceptual architecture/design. This knowledge is passed down to new generations along with the methods for using the knowledge in creating new things. Culture has been growing in "bulk" and complexity ever since.

14.1.2 Artifact/Cultural Evolution

Initially humans probably did more tinkering and used trial and error methods to see if their "invention" worked as intended. When they didn't work, humans wondered why, which led to investigations and more trials employing variations. In this tinkering with reflection on results, they learned what variations worked and which ones didn't (so-called best practices). Human understanding of causal relations in the world of artifacts had to go through a "bootstrap" procedure, the gradual

[6] Artifacts are still considered "natural" in the sense that what humans are and do is still the result of natural evolution, so by extension, their artifacts are natural outcomes of the ongoing process. The human capacity to invent and construct artifacts represents a major transition in the senses used by Maynard Smith and Szathmáry (1995) and Morowitz (2002).

accumulation of knowledge of how things worked and didn't (c.f. Carey 2009 for a thorough treatment of the origin of and changes to concepts.)

The accumulation of knowledge through the Paleo- and Mesolithic was slow; cultures remained remarkably similar with only minor improvements, for example in the knapping of stone tools. Then in the Neolithic and early Bronze Age, people began to take things like measuring time and distance seriously, recording the results of various "tinkering" and using those more formal characterizations to learn to reason about the rules for how the world works. The Agricultural Revolution (approximately 10,000 to 12,000 years before the present) also saw the invention of recording symbolic messages on external media (clay jars and tablets). The emergence of civilizations (particularly in the Middle East, roughly modern-day Iraq) was possible because of the inventions of measurement, recording (numbers and names of commodities and owners, etc.), and accounting (Nissen et al. 1993).

Humanity embarked on a scientific procedural formulation of how to more rigorously conduct those investigations. As scientific knowledge grew so too did the application of that knowledge to the prior design and engineering of artifacts. And those artifacts became more complex as well as more functional and esthetic over time. Humans learned to live in settled locations where they could practice agriculture (Scott 2017). They became "entangled" in ever more complex webs of relations and dependencies on the things they created (Hodder 2012).

In the modern, highly technological age, human beings have begun to design and engineer extremely complex artifacts. Specifically, in the age of so-called smart devices that are connected to one another, we have reached a position of trying to engineer CASs and CAESs. This is having powerful consequences on the process of design and engineering itself. We need to re-envision the process in light of emergent behaviors that inevitably arise from the construction and operation of adaptive networks of adaptive agents (e.g., The World Wide Web as an evolvable system).

Thus, we will introduce a new way to look at design and engineering of artifacts (including the socio-techno-economic systems we call modern societies). This new way derives directly from the prior chapters defining and demonstrating how we should use principles of systems science to deeply understand CAS/CAESs. And, in particular, it derives from our current understanding of ontogenesis from Chap. 3 wherein auto-organization is replaced with intentional organization.

Engineering has been perceived and characterized as a positivistic and deterministic process. Apply the mathematical models from nature to the design specification and you will be able to produce an artifact that does precisely what was intended. When designing and engineering various "complicated" machines in the late nineteenth and early twentieth centuries, this actually worked fairly well. On the other hand, the experiences of the last half-century, particularly in the realm of software engineering, demonstrate that this deterministic attitude is not correct. Recall the discussion of what goes wrong in Sect. 2.5. As we develop more and more complex artifacts (hardware, software, mechanical, etc.) we begin to see too many things going awry with them, too many bugs and glitches, too many failures to operate properly even when the artifact was built to specifications. Even more importantly, we witness the emergences of behaviors unintended and unpredicted from the poorly understood interactions between complex components operating in multiple time domains. And it suggests that the process of designing and engineering them is fundamentally flawed, or at very least missing something important. The reason that our approach to design and engineering is no longer serving us well is that the overall complexity of our artifacts has reached a tipping point. High complexity in systems breeds emergent properties and behaviors, as we have seen. When our machines were merely complicated (like an automobile of the last decades of the twentieth century) the deterministic approach to design and engineering worked very well. But even as this is being written, we are embarking on a new path for creating truly complex machines (let alone organizations) in which we seek to embed adaptive components (so-called smart technologies). We are now experimenting with self-driving vehicles and are being surprised by unexpected behaviors. A whole new field called "Systems of Systems" has emerged in which embedded devices interact with each other, a stochastic world, and human (uncertain) behaviors. From systems science, and especially from the nature of evolutionary development, we know that we should expect many surprises as we continue to push for increasing complexity toward adaptive and evolvable systems.

This, in turn, suggests that it is time for a new paradigm for design and engineering. We will suggest that this new paradigm involves the explicit role of evolution (and the ontogenic cycle) as being a fundamental part of artifice. Culture, including its artifacts, evolves. The forms and functions of artifacts evolve from generation to generation already.

The key difference between biological evolution and cultural evolution is that the latter includes a brand-new factor, itself having been a product of biological evolution, which changes the dynamics of increasing complexity in ontogenesis. That new factor is *intentionality* as previously mentioned. Human minds do experiments (both as mental models and as tinkering) in an attempt to solve a problem with some kind of artifact. They intend to improve their own conditions by altering the environment through the use of artifacts. Cultural evolution is still Darwinian-like evolution but with different mechanisms for replication, variation, and selection (Volk 2017). The development of all the components of cultures is still explained by the ontogenic cycle. But now we need to recognize the role of human intentionality as a factor in determining the various stages of ontogenesis. Humans direct the organization process (replacing auto-organization with intentional organization) even though it still retains an element of nondeterminism depending on the degree of complexity. They are still doing some degree of tinkering. And humans act as selection forces in several different ways. They are involved in the use of artifacts and determine their desirability or utility. Those deemed less fit will succumb to obsolescence being replaced by newer, better versions.

14.1.3 The Artifact Creation (or Improvement) Process

Human beings, and to a limited extent some other species, employ behaviors intent on changing some aspect of their environment in the service of their own advantage. Beavers construct complicated dams that alter the flows of streams to create ponds that act as moats for their protection while raising their pups. Bower birds build elaborate bowers to attract mates. Often this takes the form of using some kind of "tool" to acquire otherwise inaccessible resources. Crows, for example, are known to use twigs to dislodge food from tight places. Chimpanzees fashion fishing sticks to obtain termites.

Affordance is the capability of an animal to "imagine" the use of some existing element, like a twig, to serve their own purposes. Second-order affordance is the ability to see how to modify an element in order to be better able to serve. In humans, we witness third-order affordance in their abilities to see how to combine several existing elements, naturally occurring or previously manufactured, in new ways (invention) to accomplish those ends. And this is the beginning of design and engineering.

Consider a stone ax as developed during the late Pleistocene era by early *Homo sapiens*. Humans had already been fashioning and using stone-cutting tools with some heft to them that could be used to butcher large prey or cut down trees (slowly). They had also already been using wooden clubs to beat rivals (as depicted in the 1968 film *2001, A Space Odyssey*, where a primitive hominid primate conceives of the use of a femur bone to bash an opponent from another clan) or prey. They also knew how to use sinew from animal skins to bind things. It was, however, a major advancement in technology to recognize that a new kind of tool could be made by combining these three elements in the right fashion to produce a much more effective wood-cutting tool. This is a third-order affordance and the basis of the human ability to conceive of and construct complex artifacts.

But another cognitive factor has to enter the picture before a human being conceives of how to construct an artifact for use, namely the conceptualization of the need. Needs are the result of the biological mandate that motivates growth (and reproduction) against the problems posed by an environment with limited resources, namely energy and materials. What all biological systems seek is the accomplishment of getting food and avoiding being food. Anything that will help get these done more efficiently, which is with a lower energy expenditure per unit of growth, is to be pursued. For humans, this is experienced in the form of three related motives.

Convenience means that a task or chore can be done by an artifact, generally a machine, rather than requiring human labor. Buying groceries at a store means that people don't have to spend time and effort growing, harvesting, and preserving foodstuffs themselves. Related to convenience is the ability to do things quicker by using an artifact. Going to the grocery store in an automobile is much faster than going in a horse-drawn wagon. And even the latter is faster than walking into the field to hunt a rabbit. All of this reduces the amount of effort that an individual or society needs to expend on the tasks that can be accomplished "better" using artifacts (like stores and cars). Humans, like all other living systems, seek the minimum energy state of affairs, looking for leverage to conserve biological energy. Where humans have transcended the usual biological constraints is the use of energy-channeling artifacts and the supply of high-powered energy sources other than their own. The result is a positive feedback loop for generating ever newer, more efficient, and faster artifacts (and here we mean primarily machines and technologies) that provide more convenience and save time.

We should note that the ability to produce more goods and services faster also allows humans to spend time on creative endeavors such as painting and movie making. Here too, though, improvements in artifacts (like CGI for motion pictures) form part of a positive feedback that promotes an ever-expanding cycle of ontogenesis and increasing complexity.

14.1.4 The Artifact Cycle

The artifact cycle, we will argue, is just another version of the ontogenic cycle introduced in Chap. 3. Recapping that chapter, the cycle involves auto-organization of elements combining to form more complex entities under the influences of energy flows and geometric constraints (boundaries). Those entities display new emergent properties and behaviors. These emergent properties and behaviors are then subject to selection forces in the extant environment, including interactions with the other new entities just emerging. Said new interactions, due to emergent potentials, lead to the next round of the cycle. Since each round of combining and interaction leads to more complex entities, we can think of the overall process of ontogenesis as a spiral that grows on the complexity scale over time, what Teilhard de Chardin called "complexification" (recall Fig. 2.6).

The artifact cycle is fundamentally the same but with the twist as noted above; the introduction of intentionality in the organization phase.

The organization process is not as subject to random encounters as, say, is the case in chemical interactions. This is not to say that the combination mental modeling is not subject to some degree of randomness, or is in some ways stochastic. The human brain is not a deterministic modeling engine like a digital computer. It does not have detailed representations of the components it seeks to combine. It makes mistakes in judging interactions, for example, not completely understanding the "personalities" of the various components. Moreover, it might not have a complete grasp of the problem it is attempting to solve—that, after all, was the point of using the deep analysis methodology promoted in Chap. 6.

Even so, the artifact cycle introduces an element of teleology, an end purpose, into the ontogenic cycle. Prior to the introduction of human purpose, evolution was nonteleological. It was "blind" to purpose, purely experimental in the search space of biological possibilities. If it hit on a "configuration" or "behavior" that made the phenotype more fit, then that genetic endowment was passed on and the successors became more fit than their conspecifics leading to their out-reproducing the others and the new trait stuck in the species. It was mostly blind luck.

Artifact evolution is the result of ongoing attempts to invent new combinations of component artifacts in the cycle and to improve on existing designs of artifacts, which often involves an organization stage itself (recall Principles 11 and 12 in Chap. 2).

Cultural evolution combines artifact evolution and mental model evolution, which is to say what humans believe about themselves and their world. These two domains of evolution are locked into a kind of dance, coevolution, which results in increasing complexity in both. As per the principle of complexity described in Sect. 2.3.5 (Chap. 2), the increase in complexity of a system is achieved by modularization and increasing the depth of a hierarchical network. It turns out that how the brain represents concepts exactly replicates this phenomenon (Carey 2009). The human mind is capable of representing an incredibly complex world through the modularization of concepts and the organization of hierarchies of categories and kinds, for example, my pet is a dog, a dog is a mammal, a mammal is an animal.

Complexity itself cannot grow without boundaries. The human brain has limitations in terms of what it can ultimately represent about the world. As that world becomes increasingly complex the human mind may rebel at having to learn so much new stuff and at increasing rates. This is just as true for artifacts as for natural systems. At some point, our artifact designs are more complex than any one mind can represent in totality. After a brief survey of the formal design and engineering procedures that we currently adopt to produce artifacts, we will consider the implications of the increase in complexity rendered by designing CASs and CAESs. We are just at the beginning of building limited CASs on purpose, that is intentionally. As we proceed down that route, and the motivation for doing so will be explained below, the nature of the design and engineering procedure must itself become more complex! Complexity begets complexity for a purpose. And then the world will determine whether the increase was worth it.

With the advent of the industrial age, the artifact cycle has been formalized into what we will call the design-engineering (DE) process where intentions themselves are encoded into forms that can be manipulated directly, and modified. At first, the DE process was applied only to physical artifacts, machines, bridges, buildings, etc. Organizations, institutions, and nations (generically, societies) were generally left to form and organize at the whims of the people doing the organizing.[7]

The DE process actually includes the up-front mental modeling, these days assisted by a number of formal tools (e.g., computer-based modeling languages), as well as a more formal approach to testing, deployment, and as-used feedback (for guiding improvement in future versions). The artifact cycle is employed in the creation of new physical products but also in the creation of new policies and procedures in organizations. The basic process is outlined in the next subsections.

[7] Chapter 16 will more deeply address the design and engineering of societies.

14.1.4.1 Realization of a Need

Also called recognizing a *problem.* A problem exists when a perceived or desired need cannot be met by any existing artifacts. It could be that no relevant artifact exists or it could be that current artifacts need refinement or adjustment in order to meet the need. Our first furless ancestors experienced a need when the climate in East Africa began to cool and they found themselves without protection. They would surely have understood that they were experiencing the colder weather and needed more than a campfire to keep warm.

14.1.4.2 Conceptualization

Seeking a mental model of what kind of artifact could possibly alleviate the need is called conceptualization. This is the use of imagination and affordance in the construction of an idea of what the artifact should be and how it should fulfill the purpose ascribed to it. Returning to our ancestors, one or more, at some point, noted that the animals they hunted that had furry coats seemed to be immune to the chill. They imagined using their hides (which were basically inedible) to cover their own bodies. They conceived of clothing.

14.1.4.3 Architecture

Architecture[8] is generally considered a transition activity moving from concept to design. This phase involves larger-scale aspects of the artifact but particularly how the artifact will interact with its environment (e.g., serving the needs of users). The "*architecture*" can be considered as a kind of first pass at design, establishing the major aspects of the to-be-created new artifact. Several activities, such as modeling and applications of patterns, are similar to what happens in design (see below) but at an abstract level. In systems terminology, an architecture corresponds with the system of interest looked at as a whole or possibly the first one or two levels in the hierarchy depending on how complex the artifact is in fact. The design, then, will correspond to the rest of the analytic hierarchy deeper into the system. Sticking with the example of early human clothing, the architectural phase would identify the main components, say breeches, torso covering, etc.

[8] For an excellent reference on the concept of architecting see Sillito (2014). He also provides a broad view of the whole complex system creation process.

14.1.4.4 Design

Conceptualization and architecture provide only a fuzzy representation of the artifact. A person can create a mental picture of what the artifact should be like and what it should do, but then comes the act of formulating an actual design (also called a "detailed design"), that is the design of all of the subsystems down to the level of components. In the modern world, so much knowledge of what works and what doesn't work in specific circumstances and for specific needs has been accumulated that design (and engineering too) is approached based on existing models or patterns of structures and functions. The basic patterns of the design, for example, of a jet aircraft engine, are well known and available to base new designs on. Whatever conditions exist that are new, for example, heavier aircraft to propel, can be handled by modifications to the basic patterns, for example, by scale modifications such as has been the case for reduction in scale of electronic components on integrated circuits. Opportunities for including new features show up during the process so modifications in the basic patterns can be handled. With modifications, new patterns in the family of patterns for that artifact are created and added to the family. Thus, knowledge of design approaches increases.

Once again, to belabor the example, the design of cloths would include "specifying" the layout of the legs and considering how long they should be.

14.1.4.5 Engineering

Using the best scientific knowledge of how things work, in general, engineering is the activity that formulates a design specification for a particular artifact so that it fulfills its purpose, that is attaching numbers to the dimensions, flows, and capacities. An engineered design ensures that all the parts fit together and perform their functions according to the needs of the whole system. Whereas architecture and design specify what the parts should be and what they should do, engineering, based on detailed knowledge of the physical properties of the components and how they will interact with one another, assigns numerical values to the specifications.

The animal-hide pants need to be sized to fit the "customer" and the method of stitching with the sinew of a certain thickness determined.

Engineering also considers the mechanical aspects of how the artifact is to be constructed, tested, deployed, and used in the field.

14.1.4.6 Construction, Deployment, and Use

Artifacts are built through various construction procedures dependent on what sort of artifact it is. Obviously, computing systems differ from propulsion systems and all physical hardware systems differ from policy implementation procedures and institutions. These latter are just as "constructed" as are machines so this formal procedure applies in some form to them as well.

The aforementioned pants are constructed by cutting the skins to the right size and sewing the legs to form column-like extensions from an abdomen-skirting base. The customer can then try them on!

Deployment refers to the actual placement of the artifact within its environment of use. This may involve installment, training of personnel, and other subsidiary operations needed to join the interfaces between the entities in the environment and the artifact, for example training pilots to fly a new aircraft.

A complete, closed-loop, procedure also involves collecting data on performance and as-used in the field. Data on problems encountered during use should be analyzed and fed back to one of the previous stages depending on the nature, for example, perhaps a major subsystem turns out to not be appropriate to the overall performance or mission of the artifact and needs to be rearchitected. The ongoing performance and uses of the artifact will also be used as a feedback input to the next round of the DE procedure when a new generation of artifact is needed.

The next phase of artifact design, however, will transcend the human monitoring and improving cycles. Today we are seeking to produce artifacts that can adapt to variations in their environments to remain stable and perform properly. Additionally, we seek artifacts that can learn from experience, approaching an ability to evolve with changed conditions in their environment. Ultimately, we look at the need for artifacts to emulate biological-like evolution, allowing internal modifications to take place in the process of seeking increased viability and sustainability. Our cultural systems, such as organizations and governments, are already evolvable, but not always with insights that prevent or at least lessen unintended consequences. We need to better understand intentional organization and evolvability to do a better job of designing such systems. That will start with an approach to understanding the application of the CAS/CAES archetypes to the DE process.

14.2 CAS/CAES Artifacts

The current DE approach to engineering has worked reasonably well for many kinds of complex artifacts such as machines and business processes and with humans in the loop to detect problems and reactively mitigate them the artifacts tend to perform reasonably well over the long term. Human beings are the components of such systems that provide the "adaptivity" aspect. That is, the artifact in combination with users and maintainers constitutes a CAS capability. But when we enter the realm of complex adaptive artifacts in which adaptivity is to be inherent in the design of the artifact and not just dependent on the human factor, the situation changes quite drastically.[9]

[9] Indeed, the motivation for putting "smarts" into our artifacts is in order to get humans out of the loop! Machines that have internal subsystems that can adapt to changes in their environment should, in principle, not be subject to human foibles such as falling asleep at the wheel.

Artifactual systems need to be designed to incorporate adaptive components because they are becoming so complex that humans in the loop cannot be expected to understand all of the operating parameters sufficiently to know what to do in case of deviations from the norm in operations. Moreover, and more importantly, humans cannot react fast enough to some kinds of environmental changes that can disrupt normal operations. Think of the situation with a nuclear power station in the face of tsunami flooding as happened at the Fukushima Daiichi Nuclear Power Plant in Japan. Human reactions might not be fast enough in many time scales to keep the station from catastrophe. Or consider a pilot of a commercial jet when confronted with unusual turbulence while on auto-pilot. If their training is adequate and they are awake at the yoke things might turn out OK if they react in time. But what if they don't?

At the time of this writing, we are exploring many artifacts that will require built-in adaptivity. There are a number of autonomous mobile machines that will need to be able to react to unanticipated conditions in their environments. But real adaptive response or even anticipatory actions, as described in Chap. 10 will require much more than mere "learning" (really training) no matter how "deep" it might be. At a minimum to achieve real adaptivity, it will require an agent with the ability to modify its learned models with the accumulation of real-time experiences. Methods for real-time, online learning will be needed as well as a principled approach to concept modification.[10] One promising area for this is the use of Bayesian networks. See Sect. 14.2.1.3 below.

Truly evolvable artifacts, as of now, are found only in social organizations where human beings exercise some forms of strategic management and construct entirely new subsystems in response to perceived permanent changes in environmental conditions. See Sect. 14.2.1.4 below for a discussion on persistence to be achieved in CAESs.

In the next chapter, we will cover the methods for DE based on the CAS/CAES archetype along with the other component archetypes. In what follows we will assume the use of those models to guide the DE process.

14.2.1 Design and Engineering for Increasingly Complex Artifacts

In this general discussion of how artifacts come into existence, and evolve over time, we collapse the various stages of the DE procedure into a single overarching category, "engineering," since the kinds of activities outlined above are similar in all of them varying only in details and precision. They are stages of refinement of the detailed design of an artifact so we will collapse them into that single rubric.

[10] Here we only consider modification of existing models (concepts) as opposed to construction of new concepts, which is considered under the heading of evolvability.

Systems engineering (henceforth, SE) has come into focus from the realization that many artifact projects are becoming increasingly about complex systems. We've mentioned a few in the opening chapters of the book. But even complex systems such as a passenger airliner or a supercollider can be developed using traditional design/engineering methods augmented with systems-level purview. As mentioned above, as long as human operators are in the loop to provide the adaptivity component these complex systems generally work. The same is true for business enterprises in which talented, smart people are manning the desks and sales counters. The same may be argued for local, smaller, governments, such as cities and counties. It is becoming increasingly evident that when the scale of government exceeds some level (say a state like California) it works less well. Human decision makers cannot adequately compensate for poor mechanisms of governance. This is not just a matter of the heterogeneity of the population and range of physical characteristics, economic interests, industries, etc. It is now appearing to be a core problem with the way governments are "designed." Recall the discussion in Chap. 12 of the nature of governments of various scales.

At present, a systems engineer[11] is someone, usually coming from one of the traditional disciplines, who possesses a high-level vision of the whole project and manages a number of data tools that keep track of components and processes that involve multiple traditional disciplinary engineers.[12] In the typical process, none of those specialists need understand or even know much about the engineering of other components from other disciplines. The electrical engineer need not know what the metallurgist does about the wing material. But someone sitting high above the whole operation needs to know what needs to be known, by whom, and when to apply that knowledge.

Even in the realm of merely complex systems, like moon shots, the role of the systems engineer is formidable. To date, there seems not to be universal agreement on what it means to be a systems engineer.[13] Nor is there total agreement on what a systems engineer does. Different organizations adopt various standards and practices with, naturally, variations in effectiveness.

In this book, we have made the case repeatedly that at least part of the problem has been the seeming lack of a real science of systems (though see comments below on the reality of systems science in practice) as evidenced by the fact that there are

---

[11] A reminder that by "systems engineer" we include people who design and develop processes and policies. In the discussion here, we mostly describe engineering in the traditional sense, for example, mechanical, since this involves the most rigorous design and engineering standards.

[12] Things are slowly starting to change in the USA. There are an increasing number of mechanical and industrial engineering degree programs that include some systems engineering in their curriculum. Additionally, the Accreditation Board for Engineering and Technology, Inc. (ABET) is planning to develop a set of accreditation standards for any academic program that has "systems" in its degree title.

[13] Though note that several engineering professional organizations have attempted to codify standards for processes and data tools. For example, the International Council on Systems Engineering (INCOSE) has published an extensive document, the *Systems Engineering Body of Knowledge* which can be found at the INCOSE website.

not very many academic programs in purely system science. Mechanical engineering is informed by several areas of knowledge in physics. So too is electrical engineering. But systems engineering (and its companion design process) has very little systems science informing it. To be fair the systems engineering community is acutely aware of this problem and some are trying to rectify the situation. There are several groups and projects within the community trying to define systems science so that they can get on with the work at hand. This author has participated in several of these but remains somewhat skeptical as to the progress or the results to date. The "ordinary" sciences were not "defined" by engineers but grew in understanding as a result of scientific investigation. Being fair, we should note that engineering practices did often provide useful information to the science practices, as when experiences with designing more efficient heat engines in the eighteenth century provided useful data in formulating the theories of thermodynamics.

Consensus among systems scientists about what systems science *IS* is also not fully at hand. Indeed, there is no real consensus as to what it means to be a systems scientist. Is a systems scientist someone who uses systems science concepts and principles to guide otherwise ordinary science (for example a systems biologist)? Or are there scientists who are devoted to researching the very nature of systemness itself without regard to producing domain-specific knowledge? Within the systems science community, the majority of people today[14] are involved with using "systems thinking" to guide their application of systems concepts to specific problem domains, they are like systems engineers, users of systems concepts but are not engaged in systems science itself. This, too, is problematic since these "systems thinking" concepts have not yet been formalized (as was suggested in Chap. 4 of this volume) in a way that can be readily applied to applications. Thus, at present, the basis for systems design and engineering (as well as more general problem solving) is sketchy at best. This is indeed a problem since the continuing trend toward producing more complex systems continues unabated.

Consider, however, that if the systems design and engineering process for merely complex systems is still wide of the mark, what then will be the case for complex adaptive and evolvable systems? Throughout this book, we have argued that such systems can be understood deeply through the application of an adequate formal process (Chaps. 6 and 8) and an underlying formalism (given in Chaps. 3 and 4). We looked at examples from the biosphere and the noosphere. Socio-techno-economic systems along with their governance infrastructures are CAESs of the highest order. Moreover, we have noted that it might be useful to consider actively working to design and engineer such systems so that they actually do the job of keeping human beings alive and thriving on the planet.

When systems engineers complain that they need a systems science to point to as the basis of their practice, perhaps the problem is not that there is no systems science (Mobus and Kalton 2014 argues vehemently that there definitely IS a systems

[14] Ascertained from an informal observation of the interest groups and kinds of papers/presentations given at the annual conference of the International Society for the Systems Sciences (ISSS).

science). Perhaps it is just because the systems engineers and other systems practitioners have simply failed to grasp the nature of that science and are largely unfamiliar with the various aspects of it that could be applied to systems practice. In terms of its nature, the term systems science is misleading. It implies there is some coherent set of methodologies that produce new knowledge in the same way physics, chemistry, and biology do. Systems science is not like the other sciences that mostly still operate on a reductionist paradigm. In fact, it is *meta-science*. It is above and beyond ordinary science. It is the new "metaphysics" of our age, a metaphysics that is eminently more practical in terms of providing a reasonably sound basis for ontology and epistemology as we attempted to show in Chap. 3. Engineers may have a hard time seeing how to use a meta-science directly in their work. Let's review the approach called for in this book. Figure 14.1 summarizes the conceptualization of the "aspects" of a CAES.

By aspect we mean a particular view of a system through a specialized lens. Here we show ten different aspects that collectively and holistically define systemness. All of these aspects need to be taken into consideration simultaneously in the design and engineering of a CAES (or even a CAS). Also shown is the process of systems analysis (Chap. 6) that gets to the internals and externals (environment) of the system. Our claim is that in order to apply systems science to the engineering of CAESs the engineers need to use these aspect lenses in all considerations of their designs. How to do this is, at first glance, problematic.

The reason is that these aspects taken collectively involve a huge amount of knowledge that would be needed. We don't expect engineers, even systems engineers, to be polymathic. On the other hand, without considering each and every

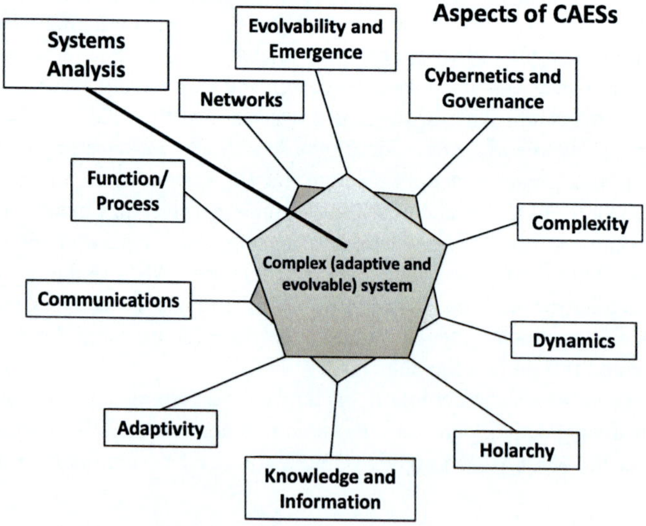

Fig. 14.1 All CAESs can be viewed from different but deeply related aspects derived from the principles given in Mobus and Kalton (2015)

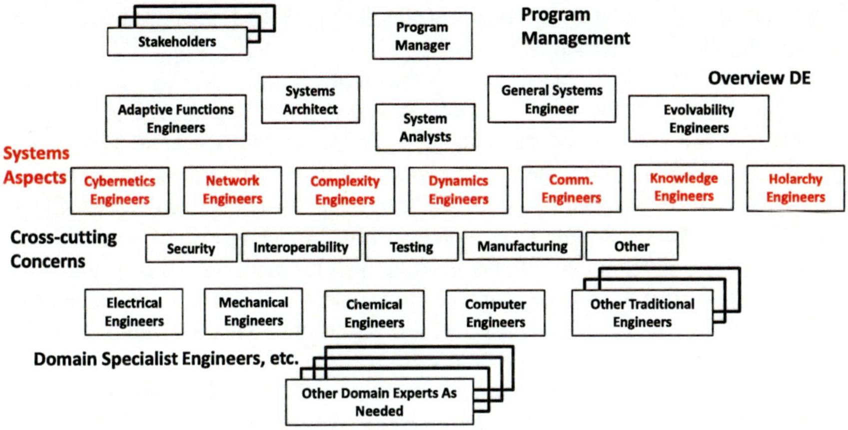

Fig. 14.2 An organizational structure for systems engineering projects. Not all of the systems aspects (third level from the top) are shown

aspect in concert there is every likelihood of failing to take some aspect into adequate account as they are all interrelated and interdependent.

Here we will suggest an organizational structure for systems engineering projects that distributes the work of designing and testing a CAES among numerous specialists, for example, traditional engineers and other domain specialists, but under the auspices of specialized systems experts. We use the breakdown of systems aspects and principles shown above and following Mobus and Kalton (2014) and Chap. 2 (see Fig. 2.3). Figure 14.2 shows a four(+)-tiered organization in which systems aspects engineers are responsible for their particular concerns for the system being designed. This structure follows from the layout of systems aspects (principles) shown above.

This organization is conceived of as following more-or-less traditional development organization structure and processes with a few notable exceptions to ensure the incorporation of systems aspect.

As with any design activity, there needs to be an overall project management that interfaces with higher-level entities (like marketing, finance, etc. not shown). The activities of the management are fairly well known but many organizations may not adhere to the HCGS architecture, which can be problematic.

Between the management and the level of systems aspects engineers are functions such as systems architecture, which implements a major part of tactical coordination and general systems engineer, who acts as the global internal logistics coordinator. These are the functions that have the whole picture in mind and coordinate the activities of all the levels below.

At a similar global level are the "adaptive functions engineers" and the "evolvability engineers." They, too, need to have a more holistic view of the project so that they can best see where adaptivity or evolvability is needed/warranted. Since these functions can affect almost any aspect of the design, these functions sit above the specific aspect of engineering functions.

What is different about this structure is the middle layer in which specialist engineers are responsible for specific aspects of systems. These are engineers highly schooled in systems science but who focus on particular aspects. We show, for example, a "cybernetics engineer" who would be in charge of all of the activities associated with control, management, and governance systems as covered in Chap. 12 within the overall design. They would review the designs of all modules in a system design and determine, for example, the various management decision types (from Chap. 12) that would be needed and how to implement the decision agents to give them all of the needed agency (requisite variety) to do the job right. In similar fashion, the "network engineer" would examine and model the various relational aspects between and within modules, all flow connections between modules within the system and flows from and to external entities as per Chaps. 4, 6, and 8. A subspecialty of the network engineering is that of interfaces engineering, wherein the engineers are concerned with the details of the interfaces, the endpoints of the flows and connections. Network engineering in this case means the connectivity of all processing modules, whether material, energy, or messages.

All of these engineers should have global knowledge and understanding of all of the other aspects so that they can cooperate on specific cases where any two or more of these aspects causally affect one another. A system is a system only if we account for all aspects simultaneously. All of these specialist engineers will need to be fluent in the special language of all the other specialist engineers. There is really no other way that we could call the activity of the whole group systems engineering unless there is seamless communication between all of them.

These aspect specialists need not have detailed knowledge of all of the traditional engineering work (see next paragraph) but they need to have a basic understanding of the main concepts in each. For example, a computation engineer (not shown in the figure) need not be a computer engineer, per se. The latter is shown as a traditional engineer. Rather they would be more like a computer scientist, but one who also understands the purposes of computing in many forms, analog, digital, and biological, for example. The computation engineer would most often work in association with the cybernetics engineer (see Mobus and Kalton 2014, Chaps. 8 and 9) perhaps, but also with all of the message processing interface engineers as well. Incoming and outgoing data through any of the communications links will need some form of information processing. They would also work with the knowledge engineer in this regard. CAESs will have learned memories that will hold a range of knowledge types, from explicit (facts) to episodic (events) to tacit (skills and generalized models of the world).

At the lowest level, the "operational" level, the majority of the detail design takes place. It is presumed that each different aspect of the systems design layer may employ any number of "traditional engineering" professionals such as electrical, mechanical, etc. The total number and kind of specialist engineers will depend on the nature of the project. "Other domain experts," for example, might include psychologists or human factors engineers. All of these "detail" specialists, while possessing all of the knowledge expected by traditional standards, will nevertheless need to understand systemness too. They need to understand how the designs of modules they work on *fit* into the *whole*, at least into the whole subsystem they work on. They must speak the general language of systems (Chap. 4) so as to communicate with their systems aspect colleagues as well as with one another. They cannot work in isolation any more than the systems aspect specialists can. Certainly, they will produce the detailed designs within their profession, but they need to know how those designs fit into the other design aspects of the module. This is not particularly new. This already is the case for the merely complex systems we have today. All of our electronic devices involve not just electronic designs but also mechanical, human factors, and esthetic designs. All of these specialists need to talk to one another. The difference in systems engineering is that all of these specialists are also systems engineers to a degree.

The major point of this structure and the way it can function is that there is always an emphasis on the application of principles and concepts from systems science to the design and engineering process. This is what is missing in systems engineering at present.

What is not shown in the figure is that within each box (function) there may be many other boxes arranged in a hierarchy that resembles that of the nested modules in the design. That is the organizational structure within each functional unit resembles the system tree diagram such as in Fig. 6.2 (the system tree) and Figs. 7.6, 7.9 (a biological model), and 7.13 (another biological model) generated from the knowledgebase produced from the deep analysis.

So, for example, for the group of traditional engineers working on a module at, say, level i, there will be a traditional engineer overseeing the whole module from the perspective of level $i - 1$ and a set of systems aspect engineers assigned to monitor the work at level i as well as a group at level $i - 1$. This may sound complicated, but then the whole point is that the design and engineering are for a highly complex system—the design and engineering process will reflect the complexity of that which is being designed and engineered. In fact, this isn't all that different from current organizational practices for merely complex systems. The management structure for the project will be similar to that of the current structures used to manage the design and engineering of such systems. Once again, the difference is the insertion of systems aspect engineers into the functional structure to oversee the implementation of systems concerns.

We now turn to several important objectives that must be part of the systems DE process.

14.2.1.1 DE for Purpose

Every system interacts with other systems in the embedding meta-system. Some of the outputs of the SOI should be of value to some other systems, that is, they should be products that serve the purposes of the whole by being of value. In general, the system to be designed must serve this larger purpose. It will do so by internalizing the larger purpose into its own governance system (see below).

Value, however, needs to be carefully examined. Humans as recipient systems have valued faster, more powerful vehicles for transportation, so the designs of, for example, cars that consumed larger amounts of fossil fuels were previously deemed designed for that purpose. Now we come to realize that that purpose is antithetical to a healthy planet. SE cannot just look at the immediate "customers" of the system's outputs. They need to examine a much larger environment that will be impacted by the operations of the system. They must consider longer time scales and issues, such as resource depletion as well as consequences of waste and by-products. They must consider the rates of flow versus the volume of stocks relative to the larger environment's capacities. These are, of course, not simple issues to consider. And they could bring into question the original purpose of the system.

14.2.1.2 DE for Resilience: Adaptability

As we seek to deploy our artifacts in environments in which numerous factors demonstrate high variability in values affecting the system, we will want to design adaptability into the system that is not necessarily dependent on human capabilities.

The various forms of adaptivity were covered in Chap. 10, under that heading. These models of adaptivity were derived from living systems examples but have been explored in artifact autonomous agents (Mobus 1999; Mobus and Kalton 2014, Chap. 9, Sect. 9.6.3.1).

There have been a number of engineering forays into adaptive systems that will continue to be important for systems, such as adaptive filters and adaptive controls. Thus far, these explorations have not extended to the wider variety of adaptivity mechanisms as seen in living systems, but we expect that with the use of the systems approaches advocated in this work, there will be an array of opportunities and methods that will allow the DE of more broadly adaptive autonomous mechanisms (modules) in future designs. It bears stressing here, however, that the current spate of artificial intelligence (AI) applications, using "Deep Learning" neural networks, does not produce autonomous adaptivity. Such "AIs" learn through supervision and only laboriously (many iterations of exposure to target patterns). The deployed AI-based agent cannot adapt to conditions for which it was not trained; such conditions should be expected in the real world. The author (Mobus 1999, 2000) has provided a model of autonomous adaptivity in an artificial agent that is more like that found in animals. More advanced AIs might be able to use that or similar methods to become truly adaptive.

Systems engineering is challenged with the need to build resilience into designs that will increasingly rely less on human reactions to changing conditions and more on the system's own ability to adapt. Reactive capability is already being designed into artifacts; think adaptable cruise control slowing down if the car in front slows down, or windshield wipers adjusting to the intensity of rain. These kinds of devices are not much more than a proportional response "thermostat." They respond to variable demand with a proportional response. Real adaptivity is another matter. That requires the system have a modifiable memory of the trend in conditions.

Adaptive response is the ability for a system to change its response "curve" as a result of ongoing experience (Chap. 10 and Mobus and Kalton 2014). Note that what we mean by adaptive response is not the same thing as merely response to changes in environmental conditions (e.g., ordinary homeostasis). Essentially the phenomenon is demonstrated in biological systems. For example, consider what happens when a person decides to work out with weight lifting to build muscle mass. Initially, lifting weights takes considerable effort and requires more energy to accomplish. But after repeatedly and often exercising, the muscles grow in bulk and the energy required to lift the same amount of weight in any one repetition is less. The body has adapted to the demand for doing more work than had been normal by building the capacity to do so. The muscles constitute a memory of past demands and will be ready for future demands. It turns out that the amount of energy needed to maintain the additional capacity is less than the energy needed to respond to the demands when those demands are made relatively frequently over time.

In the realm of artifacts, we are just entering an era of adaptive response designed into machines. Specifically, computational systems implementing forms of reinforcement learning have the ability to modify their responses to changing demands (c.f. Mobus 2000). It will be interesting to see if the adaptive response can be developed for mechanical aspects of artifacts, emulating the muscle development example. More likely it will involve a combination of mechanical and computational processes as in the Agent model described in Chap. 11.

For social systems such as organizations, the capacity for adaptive response already exists but is not generally recognized as such and thus not particularly well managed. Organizations do adapt to trending changes. For example, when the demand for a product increases, the manufacturer will increase capacity, perhaps through capital investments in more equipment or space. What is less clear is how organizations respond to fluctuations in demand, as in a nonstationary environment. What happens when the demand for a product declines and investments have already been committed? In some sectors, we see an increasing reliance of "flexible production cells," reconfigurable robotic systems that can readily be reprogrammed to produce other products as demand changes. Another example of adaptive response in the business world is the supply chain. Of course, in this instance, it is the final retailer that benefits from an ability to flexibly switch from one supplier to another as demand changes, not necessarily the suppliers.

14.2.1.3 DE for Sustainability: Governance

In Mobus (2017), the author lays out the requirements for achieving a sustainable, that is, a lasting, CAES. Every complex system requires management mechanisms in order to assure continued reliable performance of the parts, interactions between the parts, and between the whole system and its environmental conditions (as they change).

The design of a CAES is as much about the design of an appropriate hierarchical cybernetic governance system (HCGS), covered in Chap. 12, as it is about the functionality of components. Our experience with designing organizations (and maintaining their functioning) has provided a number of examples of this and should be considered as a prototype for the designs of all CAESs.

14.2.1.4 DE for Persistence: Evolvability

Artifacts are constantly subject to modification in design followed by a selection process (for example how well the artifact does in the market—how well it sells). Later in this chapter, we describe a process of intentional organization, mentioned above, that replaces the auto-organization phase in the human development of artifacts. So, in a larger sense in which human designers are a part of the ongoing evolution of artifacts, ontogenesis is at work. But what we seek is something that is, like the adaptivity described above, autonomous. That is, the artifact is, itself, capable of evolving to be sustainable and persistent in the face of a changing environment.[15]

As already pointed out, our ordinary machines are far from evolvable. The first artifact that humans have created that approaches evolvability is the Internet and the World Wide Web (WWW). But even here, the evolution depends on some human intervention. For example, the Web has been subject to revisions in its underlying protocols and the browser technology (software) has been developed to allow more interactivity, etc. The most one can claim about the autonomous evolvability of the Internet/WWW is that the network of nodes and links (e.g., web pages and hyperlinks) can be characterized as an evolving graph (Barabási 2003).

So, what would it take to design and engineer a truly autonomous artifact that was able to evolve in response to changes in the environment? This is quite obviously an open question. But it is a question that deserves attention as our artifacts are becoming increasingly complex and expected to do significant work without humans in the loop.

The epitome of CAESs is those that can alter themselves in response to significant long-term changes in the environment. As already noted, this goes beyond the notion of adaptivity, which is just a system's built-in capacity to respond to short-term changes that lie within an operational range. As demonstrated in the muscle-building example above, the body can respond to increased demand for muscle power to lift heavier weights, but within limits. What the body can't do is grow a whole new muscle group to handle demands beyond that range, or to address a completely new kind of movement. This would require evolution. Many kinds of systems are evolvable. We count the human brain, by virtue of the neocortex's

[15] However, we need to proceed with caution and wisdom in this regard. We should take the warnings in movies like *The Terminator* (Skynet) or *2001 A Space Odyssey* (the HAL 9000 computer) seriously. Already a number cyber-ethicists are exploring the implications of giving machines too much autonomy, as in combat robots (already a thing) given discretion to open fire as they deem needed.

ability to represent what might be considered an infinite variety of concepts, as an evolvable system even if the body is not. Organizations have a capacity for evolving in the sense of being able to learn from experience (Senge 2006) though most organizations seem not to be able to do this very well.[16] To make evolution and learning realizable requires a working strategic management layer in the HCGS subsystem. Some organizations have cognizant CEO's, the ones who are nominally tasked with strategic management. But, unfortunately, most heads of organizations do not really understand the concept, as presented in Chap. 12.

Biological genera are evolvable, i.e., speciation in response to changed environments. So too are human organizations. They can make varying scales of changes internally to accommodate changes in their environments. This ability requires the most advanced aspects of the HCGS and governance capacity, strategic management, and highly adaptive tactical management. The organization has to possess the ability to acquire new resources and their needs to be an internal mechanism for implementing new processes/subsystems to carry out the strategic directives.

In the design of autonomous CAESs, the designers will need to understand the need for and implementation of strategic and adaptive governance subsystems in order to achieve full CAES capacities. They will need to design and engineer for all of the above capabilities. This is "virgin" territory in so far as the DE enterprise is concerned. It extends far beyond mere complexity. It is needed for the "management" of complexity. For autonomous systems, those not relying on human intervention to cover adaptive responses, this designing of an adequate strategic layer in the HCGS will be a major challenge and a focus of needed research. Below we will address some ways to consider evolvability in CAESs through the modeling process. Essentially this involves (as noted in Chap. 6) the tagging of certain components as "evolvable" and providing parameters for the ranges of variations (of different kinds). In simulation runs, we introduce new elements or changes in conditions beyond the range of adaptivity which then trigger an evolvable response to the affected components. That response may be stochastic, replicating genetic mutation, for example, but it still involves intentional response.

The most advanced artifacts that humans have yet built (but not necessarily designed or engineered) are our societies with their various cultures and economic subsystems. These are systems in all of the senses developed in this work. They are the product of species-wide auto-organization processes with intentions generally limited to local problem-solving. Let's designate such systems as socio-techno-economic artifactual systems (STE or cultures for short). We assert without an attempt to prove it here that the natural auto-organization and subsequent emergence/selection processes of ontogenesis have created something of a conundrum. Our societies are operating in ways that are a major threat to the Earth's biosphere

[16] In fairness this subject is actually well understood by, for example, organization systems scientists like Peter Senge (1990) who has characterized the "learning organization." The subject is often encountered in business school curriculum regarding management theory. However, critiques of how well the concepts are adopted have shown that real organizations, particularly those that are profit-oriented, have great difficulty implementing the requisite mechanisms and culture.

(see Crist 2019, esp. Chap. 1). Perhaps it is time to think about systems design and engineering being applied to the design of societies that can achieve the above requirements for autonomous CAESs embedded in a super complex world. This is, in fact, what we will outline in the final chapter of this book.

14.2.2 *Design Considerations*

Here we review several DE considerations that have appeared throughout this book. These are overall architectural aspects of all CAESs. They are essential to incorporate into any CAES artifact. In what follows we assume the reader understands that the analysis of the CAES SOI has been (or is being done) in accordance with the procedures given in Chap. 6. That is, the system as a whole is the original SOI and is analyzed in terms of its environment and interactions with elements of the environment followed by recursive decomposition. This procedure ensures that functional and structural modules (e.g., work processes) will be properly identified. It also ensures that the complex networks of material, energy, and information flows are properly mapped within the SOI. The recursive decomposition will reveal the hierarchical network structure of subsystems. Finally, the capture of dynamic properties at each level in the hierarchy ensures that appropriate time scales for behaviors are recognized. Lower-level component subsystems operate on smaller time scales than higher level ones and it is necessary to get this right in order to avoid some of the more typical blunders that have been seen in too many systems engineering efforts.

14.2.2.1 Structural and Functional Modularity

Throughout the book, we have emphasized that complex systems are composed of sets of interacting modular components. Modularity is one aspect of the management of complexity. By designing modules that are functionally and structurally "limited" we keep them, relatively speaking, simple. It is in the composition of supramodules, that is, those higher function modules that contain the lower-level modules as components, that complexity rises yet is managed.

As demonstrated in Chap. 6 modularity does not necessarily mean hard boundary conditions for subsystems. We saw several examples of fuzzy boundaries characterized by the fact that different module components can participate in the activity of the module on a temporally differentiated basis. That is a component may participate for some period of time (in the appropriate time domain for the module's level) and then move to a different module to participate for a complementary amount of time. The example from Chap. 9 was the human being participating for some period in a work process (employee), in the home (parent), or, periodically, in a congregation.

A related phenomenon for some kinds of modules is that of a "subroutine." The concept name is taken from software systems in which some particular code modules are called from different places in the program, being passed parameters and returning a result value. The internal work is the same for all such calls, only the parameter values and return value are particular to the specific call. But consider a rental shop that rents lawnmowers to different people in the area. One lawnmower can be used by different customers at different times. The same is true for the services of a barber and, indeed, all service providers. The barber, giving a haircut to one customer at a time, is a social subroutine. No household needs to have a dedicated barber in residence! Finally, think of people as subroutines that participate in various social organizations and families. They, of course, play different roles in each but they can be thought of as very talented subroutines, each with its own set of sub-subroutines.

14.2.2.2 Hierarchical Network Structures

As we saw in Chap. 6, the relations between subsystems discovered within any SOI at any level above the atomic component level are based, functionally, on the flows of materials, energies, and messages (information). That is, we expect to map out a flow network within the module under examination. Those flows enter the various modular processes, get transformed, and exit as products and wastes observing the rules of conservation (matter energy/information knowledge). Then, within the module, after doing a decomposition on it, we find that the input/output flows of the module are distributed among the component modules forming another, lower level, network. Thus, we conceptualize the system as a network of subnetworks.

When we are doing recursive decomposition of a CAES system to be designed and engineered this concept should be foremost in the mind(s) of the architect/designer/engineer. It should guide their analysis and discovery process. Fortunately, network theoretic work, based on graph theory mathematics, provides a rich assortment of analytic tools for "checking" the validity of designs, such as ensuring that all inputs and outputs are accounted for (the above-mentioned conservation principles).

14.2.2.3 Time Domains and Levels of Organization

As noted, the time domains of dynamics are based on the level in a hierarchy, lower levels have time constants that are small relative to those of higher levels. By following the methods of functional/structural decomposition from Chap. 6, and identifying the transformation functions (dynamics) one has the basis for ensuring the coordination among modules at any level, thus avoiding things like parasitic lag times within a level.

14.2.3 Design for Emergence

This is a consideration, which is aligned with evolvable systems, that is just beginning to be tackled in systems engineering circles. By a common definition of emergence, that the behavior of a whole complex system cannot be predicted by the behaviors of its components, it would appear on the surface that it would not be possible to intentionally design emergent properties or behaviors into a CAES. However, our position is that that definition is a bit of a cop-out. It is true that as we have entered the realm of designing very complex systems, we have been often surprised by behaviors that were not predicted in the DE phase. Defining emergence in this way can be more of an excuse for missing something potential when combining components. This is especially the case when, for example, positive feedbacks or nonlinear behaviors produced unintended consequences (in behavior).

By the arguments in Chap. 3 regarding ontogenesis, however, it should be realized that there is a possibility to design so as to encourage the emergence of positive properties and behaviors. What the procedures of Chap. 6 show us is that by doing a complete analysis of the environment of a CAES, including, especially, entities that can potentially couple with the SOI as sources or sinks, we are in a better position to specify internal modules/processes that would be needed to realize such a coupling. Assuming that the coupling is desirable, the specification becomes part of the design for emergence, i.e., a new property or behavior. By the arguments for generating models given in Chap. 8, with respect to the sufficiency of analysis of the environment, it should be possible to simulate a wide range of environmental conditions that would elicit emergent behaviors.

This is actually not new for artifact designs. Common examples of it are just not always recognized in engineering circles. For example, consider the nature of entrepreneurship. Someone has a vision for a product or service that has a potential for serving a market need and constructs a proforma design of a company to actualize this (the artifact). What emerges is an actual organization with couplings to venture capital and, if successful in producing the product or service, results in a coupling with customers in an emerging market. Note that in advance of the actualization of the firm, no one can predict the exact results. There is still a large element of nondeterminism involved in the enterprise. Moreover, the actualization may reveal additional opportunities (or problems) that were not anticipated originally but in hindsight seem perfectly logical. Entrepreneurism is an act of intentional organization with stochastic variability in specifics. It does, of course, depend on human adaptability and foresight to succeed, but we can see the effects of emergence in the results.

We should ask, then, how do we apply this notion of emergence to designs of all CAESs? At present this is not an easy question to answer. We do see emergent properties and behaviors in some complex systems that have been designed already. One very interesting example of this may be the design of autonomous vehicles (cars and trucks). Already, even at an early stage of experimentation with self-driving cars, for example, we are witnessing the emergence of interactions between the use of these vehicles and the need for roadways and streets to include extra cue signage for both cars and pedestrians to improve the safety issues.[17]

Ultimately, design for emergence is coupled with the last two considerations of designs for CAESs, adaptability, and evolvability. A system that is composed of adaptable components may be expected to exhibit some forms of emergent behavior, that is, behavior not necessarily predicted in advance (recall the discussion of the "drunken-sailor-walk" emergent behavior in Chap. 11, Sect. 11.5). The possibility of emergence requires that the simulations of the system should be done in as wide a range of environmental variations as possible to encourage such emergence to obtain. This, of course, requires that a substantial analysis of those environmental conditions (level − 1) will have been undertaken.

14.3 Intentional Ontogenesis and Evolution

We will describe the process by which human beings, both as individuals and as collectives in a socially cooperative fashion, conceive of and construct complex artifacts. We contend that the process by which human beings bring new artifacts into existence is still part of the ongoing Universal evolution, the ontogenesis, and development of things. But with the human mind and especially the social mind, we replace the chance and necessity dynamics of auto-organization with what we have called "intentional"-organization; the bringing together of components to form a whole artifact intended to solve a problem.

Artifacts are objects or processes that come into being as a result of intentional construction (conception, design, and implementation), be this by non-human animals (chimpanzee modifying a twig for the purpose of "fishing" for termites) or human artifice. We normally think of the latter as being of "artificial" origin, but here we consider a larger scope of how things come into existence in reflection of the ontogenic process described in Chap. 3. However, in our current sense, we consider a new process involved in bringing things together in new configurations. Instead of auto-organization, or the stochastic coupling of entities with predisposed personalities and the energy sources to achieve those couplings, we introduce the notion of the organization of an artifact resulting from the construction of a mental model in which an overall function that is desired can be simulated.[18] This is the bringing together of entities for a purpose or the shaping of systems to solve a problem. This is what human beings have brought to the ontogenic table.

[17] See the article "Driverless Vehicles Set to Change the Way We Design Our Roadways?": http://www.sehinc.com/news/future-what-do-driverless-cars-mean-road-design for background. Accessed 8/13/2019.

[18] Today, these mental models are most often transferred via a formal language into a computer-based mathematical model for greater precision. Even so, the model originates in the mind of one who can envision how a construction could solve the problem at hand.

But no artifact is ever born in some kind of final state. We know that artifacts themselves evolve over time, incremental improvements in form and function continually reshape the artifact so that it serves its purpose better (e.g., more efficient, more capacity, more esthetic). If an artifact type is considered useful, then that type, like a biological genus, is kept in play (is fit). Over time, as humans discover ways to modify the artifact to be better at serving its function it morphs and may eventually give rise to a whole new kind of artifact that, perhaps, serves a different or expanded purpose. Just think of the evolution of the automobile over the century.

Occasionally wholly new artifacts do come into existence. From time to time we are faced with a new problem that cannot be addressed merely by modifying an existing artifact. Or, somewhat more rarely, someone serendipitously discovers (invents) something that allows performance of a task that has rewarding consequences. In the former case, the mind is challenged to envision some kind of artifact (and remember, artifacts are not just instruments but also procedures and other abstractions) that could possibly be used to solve the problem. Necessity is the mother of invention, after all.

14.3.1 Solving Problems, Expanding Capabilities

In what follows we use a single word, "problem," in a more generic sense than in more common usages. A problem exists whenever a cognitive entity seeks a goal state for themselves that is different from the current state or from a likely or possible future state. This definition differs from standard dictionary definitions but covers the range from mathematical problems, in which the mathematician seeks to, for example, prove a theorem, or social problems in which a group perceives that they are not able to achieve a desired condition under the current state of affairs. Individuals are almost constantly presented with problem situations from, for example, navigating through a complex terrain, getting from point A to point B, to finding food, to obtaining a sufficient education in order to get a good job. Most of the time individual problem solving involves the application of learned procedures, such as following a trail through the terrain, unconsciously. But frequently enough the problem solving involves conscious and cognitive processes, such as with the mathematician.

One particular kind of problem can be characterized as "a desire to be able to do something not previously possible to do." Another variation on this is "a desire to be able to do something faster or more efficiently (or both)." We call this "expanding capabilities" and it seems to be a universal drive in human beings. Whereas all other non-human animals seem content, generally, to live easily within their econiches and never do much in the way of the invention of artifacts that would "improve" their conditions, early human species began on an evolutionary course based on altering aspects of their environments to gain some kind of advantage, become more fit, as it were. Starting with the control of fire to reshape landscapes and later actively reshaping stones to serve purposes like butchering carcasses, they continued to seek ways in which they could be better off. And this invariably involved the creation of artifacts.

There are three main factors driving this tendency that need to be considered. These are the underlying motivations of a human or group of humans to want "improvement," the capacity to recognize and grasp a "problem," and the autonomy of thought and imagination to formulate a solution in the form of an "invention" of an artifact. We will examine each of these briefly.

14.3.1.1 Motivations

All animals possess innate drives to perform behaviors that support their living, reproduction, and, in many cases, supporting their offspring until they are able to do so on their own. These drives are motivation to action and to make decisions accordingly. Animals that have rich social lives, such as most primates, have evolved more complex behaviors to enable the maintenance of that social life. Humans have evolved extremely complex behaviors and conceptual modeling, for example, having a theory of mind (Tomasello 2019).

As we saw in Chap. 9, a main motivation for human beings is to obtain resources (as supplies to their ability to satisfy the drives/needs they experience) at the least cost in terms of energy and time. That is, they are always looking to "pick the low hanging fruit" in the fastest time possible.

One reason that this might be the case, as opposed, for example to the motivations of other great apes with more-or-less stable (and steady-state) lifestyles, is that from the initial emergence of the genus *Homo*, there appears to have been a tendency to expand territory and numbers. It is not clear why this is the case; however, it is clear that the species (indeed the genus) is incredibly adaptable. The genus is marked by a significant increase in brain size relative to body size as compared with other apes. This may have contributed to an ability to conceptualize living in new kinds of environments and experimenting with new kinds of foods. We also see, not long after the advent of the genus, the ability to use fire which, through cooking foods, provided more energy per unit weight of foodstuffs and setting up a positive feedback loop to support the further evolution of brain size. In any case, *Homo* began to explore more broadly and showed a tendency to be more acquisitive than their cousin genera.

Once this tendency started the genus became motivated to "find more." The basic animalistic balance between exploitation of existing resources and exploration (e.g., a predator trying to find a new waterhole to stake out for prey) had shifted toward higher exploration. As a consequence, early humans expanding outward from their origin region in East Africa kept encountering new challenges and thus had to develop the capacity to overcome them in order to survive and thrive. Clearly, they became very successful at doing so. So much so that the motive to expand and explore became the main theme of life.

14.3.1.2 Recognizing a Problem

A problem exists when a human mind recognizes that a drive cannot be satisfied readily, or it could be satisfied at a lower cost, *If Only…*

Another form of this problem recognition takes the form of: "I see X and recognize X as a useful resource but I am blocked from acquiring X because…" A problem, in the abstract, is a situation that blocks one from achieving a goal state, fulfilling a need, or more generally, a want.

Given the nature of the motivations described in the prior section, human beings immediately set out on a quest to discover a means for "solving the problem." Given great cleverness, resourcefulness, and desires to conquer obstacles (which seems to include other humans) members of our species seek to solve the problem.

When problems were basically simple (where can I find more food?) solutions were relatively easy to come to mind. As culture evolved and societies became more complex, as with the advent of machinery and the Internet, problems were still "felt," but not necessarily recognized for what they were. Complex problems are often not well understood; they are often seen as open-ended or wicked. As a result, the rush to find a solution to a poorly understood problem generally leads to new problems (Tainter 1988) and the cycle perpetuates.

As a rule, the failure to understand complex problems is the result of not recognizing the "problems" are really just parts of a larger embedding suprasystem. What needs to be done is to treat the "problem" as the system of interest and then use the reverse recursive analysis of Chap. 6 to better understand the environment in which the problem exists (often giving us a clue as to why it is a problem in the first place). This almost never happens. Most problems, even complex ones, are seen as local phenomena and short-term solutions are sought to speed the results. Inventors are often individuals that operate in this mode (usually with a reward in mind).

When groups are involved in recognizing (and working toward solving) a problem, the focus tends to shift from local and short-time scales due to the introduction of multiple perspectives having to be considered and integrated. This is a good thing, though it often drives diehard progressives insane since it tends to enforce a more conservative dynamic on the process. So too, marketing managers are frustrated by the perceived "slowness" of projects carried out by large teams where those projects are part of the profit-making goals of the organization. Even so, groups too often ignore important environment dynamics when seeking a solution.

14.3.1.3 Autonomy

The human mind may be constrained to thinking in certain ways by virtue of its design and its life experiences, but the imagination is incredibly free to pursue seemingly boundless possibilities. Consider, for example, the nature of language. With a limited set of phonemes (or written characters), a human can invent an uncountable number of words, especially if there are few constraints on the length requirements. Every time a human mind encounters a "new" semantic, it is possible to name it. It is true that random combinations of phonemes do not work for a variety of reasons and so the actual number of "legitimate" words is probably countable, even so, a likely infinite number of sentences can be constructed from those words.

So, it is with the combinations of things or procedures. Initially, we imagine humans only sought to combine things that were at hand, sewing skins together, attaching a stone ax to a wooden shaft. But as humans learned to shape those things at hand so that they fit together better we were off in the adventure of the exploratory invention coupled with intentional organization.

Human beings have an unprecedented level of mental autonomy in the construction of concepts (those mental models that turn into artifacts). This can be seen as a good thing when looking through the lens of innovation today, with the sentiment leaning toward novelty and serving wants. It is definitely a positive factor in the exploration of possibility space; some amount of exploration is needed to prevent getting stuck in a local and nonoptimal minimum. But, on the other hand, without constraints that reflect reality, it can lead to creations that have numerous and negative unintended consequences. It can be argued that we are seeing many more of these kinds of artifactual creations today. Autonomy in individual and group problem-solving cannot be unlimited or unconstrained by taking into account the larger suprasystem. The embedding governance system must establish some rules by which exploration is kept in keeping with the nature of reality. The overarching question is: "Just because we can do a thing, does that necessarily mean we should do that thing?"[19]

14.3.1.4 Mental Affordance

We've used the term "affordance" to mean a potential interaction between two components or systems. The human mind has proven supreme in recognizing affordances in objects and methods and is the basis for the capacity to recognize a potential solution to a problem by using an artifact. But it goes far beyond mere recognition. A chimpanzee can recognize the utility of a stick for fishing out termites. With human cognition, we find an ability to construct an artifact in our minds, a mental simulation, and check its affordance against the problem we are trying to solve.

Mental affordance is the phenomenon, coupled with the autonomy of mind covered above, which is at the base of intentional invention and design. We humans are certainly not the only problem solvers in the animal kingdom. Crows and octopuses have been able to cleverly solve problems (puzzle-boxes) involving obtaining food.

[19]This version of the question may be attributed to the Star Trek movie, The Undiscovered Country. It is one of many variants on the ethical dictum: Just because you can do something doesn't mean you *should* do that thing.

But it is not contested that the human capacity for recognizing other kinds of problems and then considering new ways to solve those problems seems limitless. However, it may be that it only *seems* so. There exist physical constraints on the feasibility of potential solutions. One might imagine a lever (having the affordance of lifting a heavy weight) that could lift the world. But where would one put the fulcrum?

14.3.1.5 Constraints

It is incumbent on us to recognize that our ability to imagine artifacts (and artifactual systems) that solve some particular problem has to be constrained by the laws of nature and situational realities.

For example, currently humanity is recognizing that the burning of fossil fuels has led to the untenable situation of global warming leading to climate chaos. We burned fossil fuels to solve one kind of problem—driving economic growth. But now, faced with the existential predicament that this has caused (not just for the human species but for the vast majority of other species) some of the more physics-naïve of us have locked onto a “solution” that involves transferring to so-called renewable energy sources and a phase-out of fossil fuels. This is based on the assumption that we can and should continue the kind of industrial civilization that we have had and that all that is necessary is to transition to “clean” energy sources. This is woefully ignorant of physical realities. The power factor of fossil fuels far exceeds that of solar energy, for example. Even though the total energy in sunlight that falls on the Earth daily exceeds that of that available in the daily burning of fossil fuels currently, the power density of sunlight is so small that we would have to convert enormous tracts of sunlit land areas to provide an aperture large enough to collect the same amount as well as a method of concentrating that relatively low-power energy into the high-power form that is needed to drive the economy. Alas, the “Green New Deal” has not really taken into account this thermodynamic fact. It is more wishful thinking than aspirational.

There are real and necessary constraints on human invention and artifact creation when it comes to solving societal problems. But at the same time, we need to recognize that the recognition of a problem is often an arbitrary situation. Humans, on the whole, want progress and see anything blocking their path forward as a problem. So, part of the “problem” is actually human perception of the situation and the needs for the future. Suppose humans recognized that continued growth (population and consumption) was not desirable. Then the loss of high-powered energy from fossil fuels would no longer be a problem, but a limitation on the numbers and permitted consumption per capita. We could seek other solutions to the problem of sustainable existence than trying to grapple with maintaining our current requirements.

14.3.2 Incremental Improvement as Evolution

Even the human endeavor to imagine a new artifact or a significant improvement on an existing design is part of an evolutionary process. As in biological evolution, the process involves incremental innovations over existing structures and functions and not, as might be sometimes imagined, the whole cloth invention of something completely new. Even the electric light bulb was proceeded by various fuel-burning methods of supplying concentrated light. So, the notion of providing a point-source of lighting was not something Edison came up with out of the blue. What he did was, knowing that electric current flowing through a resistive filament induced high heat and radiation of light as a result. Putting together the idea of, say an oil lamp wick, and an electric filament to produce a useful source of light was the act of intentional organization. After that, he had to begin a search for a filament that had the right properties and longevity to count as an economic replacement for the oil lamp, an act of intentional selection. He had not been living in a world where the darkness of night was never chased away by some kind of intense energy illumination and then suddenly thought of inventing a point-source of lighting. His ideas were based on previous inventions, a series of them, in fact.

Biological evolution demonstrates how incremental improvements on designs work in general. Variations in physiology, anatomy, or behaviors that get selected because they are fortuitous to fitness generally account for the majority of improvements. Occasionally, a major innovation in either form or function (or both) proves to be very favorable and slowly takes over in the gene pool. But more generally genetic drift guides the pathways followed in speciation.

Human innovation follows a similar pattern. Most advances in culture take the form of incremental changes in existing designs that are then subject to selection in the noosphere.

14.3.3 Admitting Evolution into the Design and Engineering Process

The engineering tradition has placed a lot of weight on the notion of formal design processes that essentially ignored the idea that DE was really an exploratory, evolutionary process. The worldview of DE has largely been derived from scientism and positivism, the traditions, themselves derived from the Enlightenment, which held that the world could be understood in strictly deterministic terms as long as we could measure the states of the world absolutely. Over the last half-century of experience, and especially over the last few decades, the realization that we really can't know anything absolutely has entered both scientific and engineering thinking. In Sanskrit, from the Vedic tradition, there is a term, "Lesha Avidya," which means the remains of ignorance that persists after learning (or even enlightenment!) takes place. In Mobus and Kalton (2014), in Fig. 7.2, we show how knowledge grows only asymptotically toward absolute ($K = 1$). The implication is that a human mind may get marginally closer to the "truth" but will always have some uncertainty about what that might be.

There will always be room for some uncertainty in any model of real systems. And, with respect to CAES designs, we have to be ready to admit that uncertainty about the future should be accounted for in our DE processes. We have to admit evolution into our DE process so that our artifacts can change themselves to meet previously unrecognized needs within the artifactual system.

Designing for evolvability begins with the identification of component subsystems that are candidates for structural/functional modifications should the environment change in unanticipated ways that exceed normal adaptive response. Of course, with intentional selection in mind, the kinds and ranges of modification would need to be anticipated and, in effect, designed into the component subsystem. The CAES in Chap. 10 included a module that is capable of taking in a problem statement and construct a new component. In the next chapter, we will show how this can be done in the design phase employing evolvable simulations.[20]

14.4 Conclusion

Artifactual systems are those that involve both artifacts such as hardware, buildings, and procedures, and humans who use, live in, and follow the artifacts. These systems can run the gamut from relatively simple (someone using a hammer to drive a nail) to complex (a commercial enterprise) to ultracomplex (human society). Some authors refer to these kinds of systems as human-activity systems with emphasis on the human components. What we have tried to do is recognize the equality of roles of both humans and the artifacts that they design and use.

The design and engineering of artifacts are based on the central idea that artifacts help humans solve various kinds of problems. But then humans come to depend on those artifacts. Let your internet service go down for an hour and you will see just how dependent we become. Our artifactual systems are an entangled mess of mutual dependencies (after all the internet could not survive without our continually managing it). Even so, the concepts of CAESs and a systems approach to deep understanding provide guidance for effectively working with and living within these systems.

We also need to admit that the DE process is just another kind of evolutionary process. It involves intentional organization, putting pieces together in ways that produce a tool for accomplishing a purpose. We humans have the mental facilities to do this in mental models, imagination, as well as produce physical models experimentally. With the advent of human mentation, ontogenesis passes from

[20] For example, we will look at the field of genetic algorithms and evolutionary programming as early examples of this kind of problem solving applied to the DE process.

dependence on the chance aspect of auto-organization to the directed aspect of intentional organization. We are not guaranteed that our designs will always work as intended, of course. There is still an element of uncertainty involved, especially in the realm of longer-term unintended consequences. But that just means there will always be new problems to solve.

Chapter 15
Designing Complex, Adaptive, and Evolvable Systems

Abstract Complex artifactual systems are realizations of the CAES archetype model. The design or improvement of these systems is based on using the CAES model for guidance in acquiring understanding of the actual system. We elaborate on the SDE process introduced in the last chapter, showing how to use the CAES model to analyze the to-be-designed system in the abstract. We then show how using the analytic and synthetic techniques covered in Chaps. 6 and 8 are used to refine and elaborate the design in terms of specific subsystems. This chapter ends with an example of a very important CAES for human society, a module that produces food in a sustainable process.

15.1 The Purpose of This Chapter

In the previous chapter, we introduced systems design and engineering (SDE) as the process that produces artifactual systems. We explained that current SDE concepts are missing the guidance of systems science resulting in an inability to tackle a real systems science-informed approach. We then suggested a new overall scheme for accomplishing SDE based on the inclusion of "systems aspect" engineers who would be concerned with the systemness of the whole design and interact with more traditional discipline engineers who would do the detail designs. In this chapter, we are going to elaborate on some additional aspects of SDE to take into account. And then we turn to the actual SDE process for CAS/CAESs using a model-based approach. That is, we show how to use the archetype models given in Part 3 to guide the analysis and design of these super complex systems—the systems of the future.

To summarize the purpose of this chapter, it is to elaborate and tie together these points:

1. A new approach to systems DE organization/process that includes system aspect engineers along with traditional discipline engineers as introduced in Chap. 14.
2. Using the deep systems analysis approach to develop the design of a new system as covered in Chaps. 6 and 8.

G. E. Mobus, *Systems Science: Theory, Analysis, Modeling, and Design*,
https://doi.org/10.1007/978-3-030-93482-8_15

3. Admitting *intentional* ontogenesis as a guiding process –evolutionary engineering as discussed in Chap. 14.
4. And using the CAS/CAES archetype and the subarchetype models as template guides for elaborating the system design.

That last point will be demonstrated at the end of the chapter with the process of designing a human activity system—an eco-agricultural system for producing adequate food for a community without generating nonrecyclable wastes. We won't design such a system; we will just show how the above four points work in an actual design and engineering project.

The word "design" used in this chapter is meant to cover the broad spectrum of thought activities that result in a functional design of an artifactual system of arbitrary complexity. The word as typically used within the systems engineering community often refers to the act of specifying components and their layouts to implement *an* artifact. But in this chapter, we are concerned with artifactual systems, that is systems comprised of both artifacts and humans/human organizations and interfacing with a larger environment/supra-system. The word "architecture" and the gerund "architecting" are used to distinguish the act of conceptualizing and specifying general properties of the artifactual system as opposed to the detailed design work. What we are interested in in this chapter is the process by which the results of a detailed deep systems analysis as described in Chap. 6, and having captured the knowledge in the knowledgebase described in Chap. 8, are the basis for the design of complex artifacts and artifactual systems as covered in the previous chapter.

Structured, deep systems analysis gives us "Design for Free," that is, the end result of following the procedures of Chaps. 6 and 8 results in a complete structural and functional breakdown of the whole system. In the case of SDE for a "new" system, the systems analysis is conceptual and virtual. That is, we are not analyzing an existing system, but one that we intend to bring into existence. The system to be designed exists in the abstract. So, also, is the case for modifying an existing system. In this case, we combine the analysis of an existing system and then extend the analysis in the virtual domain as we project the modifications to be made. In other words, the methods provided in this volume can be used globally in the evolution of artifacts and artifactual systems.

We showed in Chap. 9 the analysis of an existing artifactual system, the economic subsystem of the HSS, and quickly found problems with its design. Those problems were shown to arise as the result of an immature intertwined evolution, that is, through chance and necessity as in biological evolution, and intentional evolution. That is, humans made choices about how an economic process should work based on both what had evolved so far as a result of natural selection (e.g., trade of goods) and what they thought should be involved in the process (e.g., the use of money).

The way in which our cultural artifacts come into being is a multistep process. First, someone conceives of a way to solve some problem by having a "thing"[1] that works to do so (see section Conceptualization, below, for an aside about the nature of "problems"). Next, they envision what that "thing" will look like and how it will work in practice. This is what we mean by architecting. The next steps involve the actual working out of the details of the "thing." This is the DE activity. This was, historically, an empirical process in which one would try to construct something based on experience and "rough" ideas about what components to put together and see if it worked the way intended. Usually that process would require multiple iterations to get to an acceptable level of performance. The invention processes that Thomas Edison, for example, used before there was an accumulation of knowledge in physics and engineering. Over history, however, we have learned a lot about transforming designs into implementations. We've learned how to "engineer" a thing, that is, we can elaborate the design on paper (or in a computer tool) and create a set of specifications that when implemented provide a more sophisticated "thing." Engineering is a structured, formal, and principled method of converting a design into a working "thing."

Our focus, however, will be on the design of CAESs as covered in Chap. 10. We leave the detailed design and engineering of components, etc. to the specialists; they already know what to do once a specification is worked out. Systems engineering[2] is increasingly called upon to tackle extremely complex problems and design extremely complex systems to solve those problems. In fact, the whole conceptualization of complex problems has evolved as more experience has been gained, particularly in the realm of artifactual systems (also called techno-socio systems), those that involve human components. The fact that humans learn new things, and think new thoughts, therefore being evolvable systems in their own right, means that we need to bring evolutionary principles directly into the design arena. This makes systems engineering very different from the traditional approach to design and engineering.

The approach we will use is to base our designs on the archetype models of Part 3, Chaps. 10, 11, 12, and 13. This is not a new approach within the engineering profession. But the term, "model-based design," has two senses. One is that a design is done by building models first. The other is that designs should be based on general models of the category or kind of thing being designed. This latter sense is sometimes described as using "design patterns," reusable patterns that have been used to solve problems previously. The methods we describe in this chapter employ both senses. The archetype models presented in Part 3 are generalized patterns of subsystems found in all CAESs and we will be using these as the core design

[1] Actually, the word "thing" here doesn't just apply to physical objects. Developing procedures or a set of steps to follow to accomplish a goal, is also a design-engineer process.

[2] We've used the term design-engineering (DE) and systems engineering (SE) almost interchangeably. The reason is that at the systems level of engineering (SE) we are still involved in system design. Therefore, we will continue to use SE and DE as essentially the same activity. This is not the case at the component level where design and engineering are generally more distinguishable.

models for the design of CAES artifacts. And the designs we develop are immediately available as models for simulation and testing.

Note that this approach is actually the same one used by our brains in constructing models of systems in the world. Recall from Chap. 4 that our brains come prewired with systemese (from Chap. 4), including notions of agents and agency, and a built-in hierarchical cybernetic governance system that allows us to interact with the things in the world that we have constructed models of in our heads. We construct elaborate models by reusing elements of systemese and recursively combining them into complex concepts. Recall, also, that this process is both experimental and intentional. We build theories of what a thing is and how it behaves and then test the theory against experience. Gradually a veridical concept serves us to "predict" or "anticipate" how our interactions with the thing will go in the future. However, our concepts are subject to changes based on changes in the world. That is, our concepts can evolve as a response to how the world around us is evolving. Our brains are evolvable systems.

Thus, in our formal engineering process, we incorporate evolvability directly into our modeling process as well as into the designs of the system components themselves.

CAES model-based design involves using the items in Table 15.1 as overall guides to elaborating actual designs. Additionally, the designs of specific agents involved in the decisions in the governance subsystem are employed.

The level 1 decomposition will take each of these and consider how they are to be composed of subsystems (many being, themselves, CAESs). For example, as shown in Fig. 10.2, all of the processes are shown with their own operational management sub-processes. Generally speaking, at this level in a system of systems, this will be the case.

Table 15.1 The list of major CAES component subsystems

Work process network	
	Acquisition processes (under cooperative logistic and tactical management)
	Export processes (under cooperative logistic and tactical management)
	Work processes (under cooperative, logistic, and operational management)
	Value-added flows
Management processes network	
	Logistical management
	Tactical management
	Strategic management (where applicable)
Peripheral processes	
	Adaptation processes
	Evolvability processes (where applicable)
	Repair & Maintenance processes
	Recycle processes

15.2 Process of System Design and Engineering

In this section, we will outline the process for designing a CAES-model system, in preparation for what is to be presented in the next chapter, the systems design of society. In the next chapter, we will be introducing the modularity principle applied to social units. Many modules at the lower levels of organization will be essentially identical across regions, however some modules, as we will show here, may have some specialization due to factors like resource type distributions or having craft persons in particular trades.

Let us imagine a situation in which there are just two modules that are situated relatively close together. One module occupies a prime food-producing area, and we'll assume a reasonably stable (though likely different) climate. The other module cannot produce all of the food it needs but has access to metals and quality clays so is capable of manufacturing various artifacts, including those needed for farming. Thus, the food-producing module supplies its excess food production to the manufacturing module and the latter supplies implements and clay containers to the former. In this section, we will focus on the design process itself and then in the following section, we will apply the process to the food-producing module.

15.2.1 How Do We Design Complex Systems?

We now provide a brief overview of the design process.

15.2.1.1 Purpose of the System

Every CAES serves a purpose with respect to the larger embedding suprasystem. Some of that suprasystem constitutes the direct environment of the CAES itself. By purpose, we mean that the products of the CAES of interest are resources to other sibling systems that, in turn, are producing products for still other systems. Indeed, the CAES of interest may be a reciprocal beneficiary of one of those other systems precisely because they are all part of the economy of the suprasystem, or they may be part of the governance structure of that larger system. As long as the CAES of interest continues to produce those products as needed, it is fulfilling a "purpose" of the whole suprasystem.

In naturally evolved CAESs, we also note that the suprasystem provides the necessary resource inputs to the system in order for it to continue its functions. There exists a dynamic, often fluctuating balance between the interests of the CAES in terms of sustainability and persistence and the long-term interests of the suprasystem. This has to be the case since if the suprasystem fails it will entail the failure of the CAES of interest.

The same rules apply to human-created artifacts and artifactual systems. Artifacts have to serve a beneficial purpose with respect to the artifactual system in which they exist and the artifactual system has to serve a beneficial purpose with respect to the suprasystem in which it exists. And the largest artifactual system, the human social system (humans + cultures) has to play by these rules of serving a purpose with respect to its embedding suprasystem, the Ecos.

Thus, from artifacts themselves, CAS/CAES or merely complex, to the human systems in which those artifacts exist, all must be evolved or designed to provide benefit to the larger world. It should be clear that not all of human inventions have followed this dictum. Many artifacts have served some narrow aspects of perceived human needs but the human social system using those artifacts ended up causing destruction to the Ecos. We are all familiar with the saying that many artifacts meant to serve humanity are like two-edged swords—they cut both ways, for good (of the whole system) or destruction.

Intentionally designed CAS/CAESs need to serve a purpose that promotes the wellbeing not just of humanity (or a small segment of humanity) but of the Ecos itself. Part of the SDE process, at the very start, is to thoroughly analyze the wide-scale and long-term consequence of the system seeming to fulfill its purpose. The key questions about unintended consequences or outright Ecos destroying effects need to be asked at the very beginning. This has not been the normal procedure (especially when the ultimate purpose of an artifact is to enhance profits).

If it is determined that a CAES purpose is life-supporting in the long run, the questions about what products the CAES of interest will produce are nearly answered.

15.2.1.2 Conceptualization

In general, we think of the SDE process being employed in order that a new artifact or artifactual system will solve a problem.

15.2.1.2.1 An Aside on the Nature of Problems

A problem is said to exist when one or some of us seeks to achieve a new state of affairs but cannot see how to do so. The situation may be that something has gone wrong and is leading to dysfunction. Or it may simply be that someone sees the possibility of making "progress," that is making the world a better place, if only we could achieve this new state of affairs. A major social problem going far back in our prehistory was how to improve the facilitation of trade/transactions to eliminate the need for bulky and burdensome barter. Money was invented as a form of representing physical wealth that could be easily carried and transferred to different owners who could then use the tokens to purchase a different commodity. Problem solved. Similarly, inventions of faster transportation (e.g., wheeled carts drawn by horses or oxen) could make it possible for goods to be taken farther and result in more profit.

Note how these and many more examples of problems needing solutions are actually examples of creating an artifactual system not based on supporting the needs of the embedding Ecos, but rather created to solve a perceived local problem, that is, a problem within the human social system. It may be fair to claim that humanity has always gone about invention and design of artifacts and artifactual systems to satisfy its own perceived needs without ever considering the long-term impacts on the rest of the world. Whenever the motivation for invention is to solve a local, immediate problem, characterized in this way, there is no accounting for what the consequences will be for the Ecos. History is replete with examples of early grain states, civilizations, expanding and depleting local resources to meet internal wants and then collapsing once the costs of going further afield to obtain additional resources exceeded the value of the resource—exactly the same situation society finds itself in currently due to overexploitation of fossil fuels as we saw in Chap. 9 (Diamond 2005; Homer-Dixon 2006; Morris 2010, 2013, 2015; Tainter 1988).

The human enterprise will need to thoroughly reexamine its motivations and conceptualization of just what constitutes a "problem."

15.2.1.2.2 Forming the Initial Concept of the System

After recognizing the nature of a complex (generally wicked) problem and understanding the context or (total) environment in which the problem exists the designer sets about considering ways in which the problem can be solved, what sort of artifactual system might address or resolve the problem. Using the CAS/CAES archetype model helps with this activity as it is a starting place for conceptualizing the solution. The process is exactly the same as the top-down decomposition process except that we replace the SOI with the CAS/CAES archetype knowing that when the designers go from level 0 to level 1 they will be concerned with determining the subsystems as outlined in Part 3. At the conceptualization stage, it is not necessary to do a thorough environmental analysis nor be concerned with metrics of flows. These are left as abstract concepts which will be refined later in the design analysis.

In this phase, for example, when we start with the "product" outputs (and this means services as well). These are the outputs that interact with the entities in the environment that are parts of the problem(s), as best we understand it, so as to change the suprasystem's behavior and reduce or eliminate the problem(s).

The designer needs to take into account concepts such as time (duration), how the environment might change over time, how those changes might affect the artifactual system and its life-cycle, and so on. These need early consideration as they will affect how the CAS/CAES will be expanded during the analysis phase. They will determine how the subsystems need to be identified along with their relations with one another. Such considerations need input from historical perspectives, that is, from the histories of the environments themselves, how they have changed over time in the past. In later stages of analysis and design, we will see how explicit models of both the developing system and its environment can be refined to test the hypotheses generated during this conceptual stage.

The main objective of conceptualization is to develop a sense of how some artifactual system can improve the human condition without causing harm to the system's environment. However, it may be the case that the artifactual system might be of positive benefit to the environment. An example might be the development of methods for removing waste plastics from the environment, an effort that would include humans and machines working to improve the health of the Earth.

15.2.1.3 Architecture

Table 15.1, above, provides a rough list of the architectural features that will go into the design. Figure 10.2 (Chap. 10) provides the general layout for all of these components and their flow relations. The conceptual model of the system should have resolved the basic questions about how and how much the environment may change over the life of the system. This will determine the inclusion of the adaptivity and evolvability processes. If it is determined that the environment will remain reasonably stable over the life span of the system, then it is likely not necessary to include explicit evolvability processes. For example, in a manufacturing facility that is expected to produce the same product throughout its lifetime, provisions in the design might include adaptive capabilities to adjust for production rates if demand for the product increases or decreases. But if it is anticipated that other products might be produced in the future then the facility would include an internal design capability as well as an expanded machine shop that could design and build whatever additional equipment would be needed. In addition, the upper management of this facility would need to have all of the capabilities of a strategic management process in place to get information from the environment about possible new customers for new products and provide the tactical and logistical management layers with directions on what needs to be done to evolve the organization so as to fill that future need.

The general architecture of the system would then start with a Fig. 10.2 layout giving names to the high-level processes pertinent to the particular system.

15.2.1.4 Submodule Specification

Figure 10.2 is a very abstract representation of the CAS or CAES artifactual system to be designed. It does provide a way to focus on the different functions that need to be considered. The next step is to use the same methods described in Chaps. 5, 6, and 8 and expanded below in section CAES Archetype as Template, 15.2.2, to analyze each of the functional modules in the figure based on what they will actually do in the context of the whole system and its environment. The recursive procedure described in Chap. 6 will then allow the designers to analyze the various submodules down to lower levels in the organization hierarchy. This process of downward analysis produces the "design-for-free." At the leaf nodes of the system tree, the designers are ready to develop specifications for what and how each leaf-node module should function. For the artifact portions of the system, this would resemble classical engineering and require traditional specialist engineers to write specifications for the machinery, as it were. This was shown in Fig. 14.2 in the last chapter.

Since we are talking about an artifactual system we need to also take into consideration the way that human beings fit into the scheme, that is, specifying their role as submodules. But this is actually already done in principle because we have already identified the human role in decision-making, being agents, and having agency in governance and economic subsystems. Humans are needed where automated systems are still inadequate for filling this role. These days, with advances in cyber-physical systems that make them able to handle most low-level decision processes, humans would tend to fill the logistical and tactical as well as strategic management levels.[3] The point of analysis of all agents, cyber-physical or human, would be to understand the decision requirements and provide an adequate decision model along with the degrees of freedom or requisite variety needed to have effective agency.

If the procedures outlined in Chap. 6 are followed rigorously, and the capture of data needed to fill the knowledge base ala Chap. 8 is seen to, then at the end of the analysis the designers have enough knowledge of the system and its environment to generate functional models for the next step.

15.2.1.5 Simulation

The knowledge base contains both the structural and functional aspects of a system. From this knowledge, the designer may generate a simulation model at, in principle, any resolution and for any portion of a system. That is, the designer may call for a simulation of any subsystem at any level of the hierarchy. This is possible because the structure of the knowledgebase, derived from Eq. 4.1 and its recursive use, contains not just the knowledge of the subsystems but also the knowledge of the dynamical behavior of the sources and sinks for any subsystem. Within a system level, for any subsystem chosen, all other relevant subsystems at that level are treated as environmental entities as sources and sinks.

The simulation models are similar in nature to a systems dynamics (SD) model except that the systems are treated as bounded (modular) processes. And their subsystems are internal bounded processes so that a simulation of a system is a recursive simulation of lower modules each at their level's appropriate time scale. At least, again, in principle, a simulation of the entire system along with the behaviors of its environmental entities is accomplished by recursively calling the simulations

[3] There is, currently, much talk about how robots/machines have become smart enough to replace humans in many operational level positions. This includes skilled labor jobs in both blue- and white-collar tasks. But the new story is how AI, with deep learning, will be increasingly replacing middle management jobs, those that require higher levels of autonomy and education.

of its subsystems and of their sub-subsystems down to the last internal node of the system tree. We say "in principle" because the computational power needed for such simulations would be immense for extremely complex CAESs. However, as an aid to the design process and to test design operations, simulation of the pieces in a bottom-up manner should help resolve issues about input/output designs (interfaces) and the overall dynamical behavior of the modules. This might seem to be a retreat to reductionist approaches, but it is not. The designs fell out of a top-down holistic analysis process which retained the information regarding relations and interactions of all of the modules, so the running of simulations of those modules from the bottom up is consistent with an integrative systems approach.

Additionally, recall from Eq. 4.1 that the model of any system at any level in the hierarchy includes a summary transition/transformation function that is, in essence, the summary of the subsystem's model. It provides a more abstract representation of the dynamics of the subsystem and so, can be used in lieu of recursive calls to subsystem full simulations in a top-down simulation of a parent system. This addresses the issue of computational complexity, the time it would take to do full top-down recursive simulations. It is possible because in the process of decomposition as each new subsystem was analyzed, its transition/transformation function was determined and tested in simulation with validation and verification.

We suspect that as systems design and engineering become more endowed with experience in the design approach we advocate here, designers and engineers will begin adopting so-called Big Data and distributed computing methods to apply to this approach of simulation. Since the designs are modular as a result of following the analysis methods of Chap. 6, they are naturally represented in distributed simulations.

The hypothesized simulation engine, described above as a process-based systems dynamics modeling tool, should also include provisions for the three submodels in the CAES archetype, the agent, the economy, and governance.

15.2.2 CAES Archetype As Template

The design of an artifactual system presupposes that the system will be a CAES (or if presumed to be a CAS, sans the evolvability capacity). It will have a basic internal economy, a governance subsystem, and include provisions for adaptivity and evolvability. We will therefore proceed with using a CAES archetype model as a template for this design. Using the elements (subsystems) given in Chap. 10 we will erect a scaffold and placeholders for each of those elements. Below we will show one way to parse the whole system.

In this section, we provide an outline of the SDE process conducted by the organization, the SDE team, as depicted in Fig. 14.2 in the previous chapter. Our intent is not to provide a set of detailed standards and procedures to be adopted by any particular SDE team; those would depend on the specifics of the kind of CAES being produced. The team organization depicted in that figure is merely representative of the skill sets needed to ensure the following procedures capture the whole set of system aspects that need to be taken into account when attempting to design truly complex systems. Thus, in what follows, the reader should keep in mind that this team is constantly interacting across the "Systems Aspects Layer" shown in the figure to give due attention to each of those identified aspects that make a system a system.

The procedure for architecting, designing, and engineering an artifactual system based on the CAES archetype fundamentally follows the same approach as the system decomposition process followed in Chap. 6 and with the same knowledgebase capture covered in Chap. 8. SDE is essentially deep systems analysis but of a virtual system as opposed to an actual system. The virtual system exists in the minds of the SDE team and stakeholders as an abstraction. They know what the new system is supposed to do in general and need to drill down, as it were, on the internal mechanisms that will affect the doing.

Thus, we start with level 0, the system of interest, the CAES to be designed and engineered, and follow the procedure. Starting with an analysis of the inputs and outputs and the environment from which they come and go, level -1, we proceed just as in Chap. 6, but now with guidance from the CAES archetype model.

15.2.2.1 Start with the Environment

We follow the same algorithm for analysis as given in Chap. 6. We start by analyzing the environment of the artifactual system. This actually begins with an understanding of the nature of inputs and outputs from the system, which, in turn, means having a firm notion of where the boundary between the insides of the system and its environment lies. At the conceptual stage, this is assumed to exist. As the analysis proceeds and specific inputs and outputs are identified, and their interfaces for import/export are determined, the "location" of the boundary will become clearer.

15.2.2.1.1 Entities As Customers—Sinks for Products

Products that the system is to produce to satisfy its fitness with its environment, what do the customers need from the system? At what rate should each product be produced to satisfy the needs of the customers? This is sometimes referred to as stakeholder analysis; the identification of various other entities who benefit from the proper behavior of the system. Note that at times stakeholders can also be considered as components of the system itself, as, for example, when an employee of an enterprise (the system) is also the recipient of a salary in the role of a supplier of labor to the system. Recall that this relation requires a fuzzy systems analysis and specification.

15.2.2.1.2 Entities As Suppliers of Inputs—Resource Sources

What resources, that is, energy, material, and information, will the system need to obtain in order to produce these products? Once identified, the next question is "where do they come from?" Note that in some sense, sources, such as suppliers, may be considered a kind of stakeholder as well, since resources are to be obtained with a commensurate reward to the supplier, either directly, as when a corporation pays a supplier for their products, or indirectly when some larger feedback loop through the environment provides it. For example, when a charity provides supports for those who cannot support themselves, they receive grants and donations from others in their environment. In a well-functioning suprasystem, the environment of all of the subsystems, such feedbacks of outputs from some entities being inputs to other entities, will be in balance over the long haul.

15.2.2.1.3 Sinks for Waste Products

Whatever material inputs are not embodied into the products, and have not been recycled internally, have to be considered wastes and disposed of. Where will they go? What possible damage may accrue to the suprasystem as a result of their output and accumulation? Can we find entities in the environment that could actually use these wastes as resources?

We now know very well what the effects of not answering these kinds of questions are, and they are not good. Thus far, the HSS has not much concerned itself with making sure that waste products like CO_2 and plastics are either fully recycled or fed to entities able to do the recycling for us. We now know the harm that leaving this important aspect out of our designs is doing significant damage to the Ecos and will have to be rethought.

15.2.2.2 Consider the Economy

The economic subsystem is where the export products will be produced. As recommended in Chap. 6 and amplified below, we start our analysis with the outputs of the system. But the economy is also where the material and energy inputs are imported, so we will have nearly covered the boundary set with this approach. What will not yet be covered are the message inputs and outputs, for example, messages received by an acquisition process.

It is always a good approach to start with the internal economic subsystem since this is where the material work is accomplished and the workflow needs to be deeply understood before it is determined what sort of coordination management is going to be needed. Figure 15.1 depicts the essential parts of an economy subsystem (extracted from Fig. 10.2) starting with resource inputs and ending with product outputs. In the center, we represent a single generic work process (giving off waste

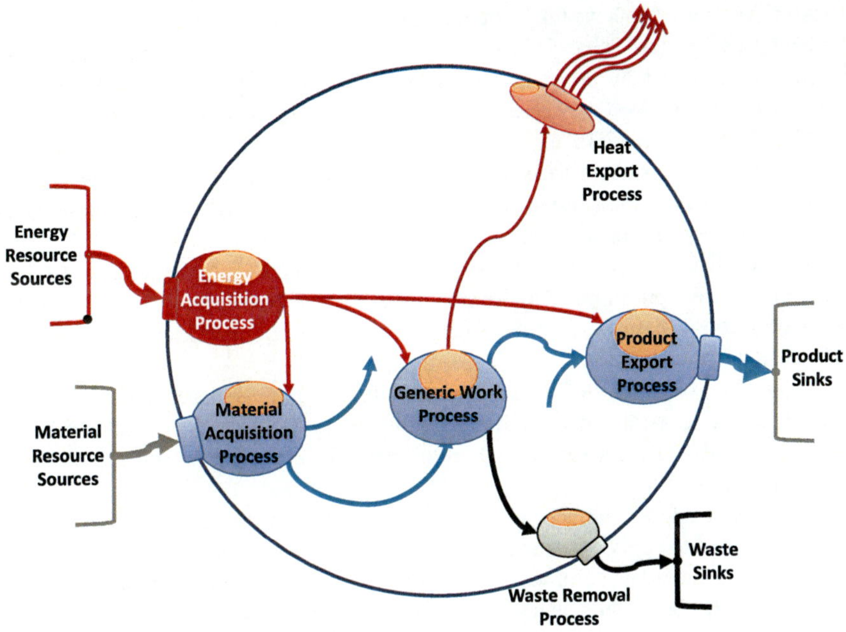

Fig. 15.1 A CAES economy archetype

heat and waste materials). Each of the ovals inside the boundary is a starting placeholder for its particular purpose in the whole system.

In all of the ovals, we see an inner orange oval that represents the local operational management agency. As per Chap. 12, these agents monitor the activity in the work process, making adjustments when needed, and report to a coordination agent. Also shown are the sources of resources and the sinks for products and wastes, again as in Chap. 10. Additionally, to emphasize their importance, we have shown the major interface subsystems interacting with the external environment.

Using this template, we use the methods of Chap. 6 to analyze and decompose each of these template objects. All of these are absolutely necessary to produce a complete economic subsystem for the CAES. The only difference between the analysis of an existing system and the analysis of a to-be-designed system is that the system is virtual in the latter case and the designer(s) determine what is to be present.

Using this template, we use the methods of Chap. 6 to analyze and decompose each of these template objects. All of these are absolutely necessary to produce a complete economic subsystem for the CAES. The only difference between the analysis of an existing system and the analysis of a to-be-designed system is that the system is virtual in the latter case and the designer(s) determine what is to be present.

All CAESs need to have high-potential energy acquisition processes, here represented by a single (red) oval. Over the course of the analysis of work that takes place in the economy, it may become evident that multiple energy sources are needed. Chapter 6 methods, you may recall, provided for adding subsystems as needed.

All CAESs also need to acquire medium-entropy material resources, so we put in a placeholder for such process (blue oval with an interface for receiving material). As with the energy inputs, if we discover later that multiple types of material inputs are needed, they can readily be added.

Finally, all CAESs need to expel waste materials and export waste heat to the environment. We put in placeholder processes representing these as well. We will not worry about these further; this exercise is just to demonstrate how the model provides guidance in design.

As per the guidelines in Chap. 6, we start with the outputs, particularly the product. In actual CAESs, there will generally be multiple products (or services) exported, so the product sink may be replicated for each one as will the product export process.

Once the products and waste exports have been determined and characterized, we can proceed to apply the deep analysis process to the generic work process, splitting it into as many level 1 processes as will be involved in producing the products. Once they have been delineated the analysis of their inputs will tell what resource inputs need to be imported, so we apply the same logic to the acquisition processes, both material and energy. Not shown in the figure are the various communications channels that will be needed for the import/export processes to coordinate with their respective sources/sinks.

15.2.2.3 Develop the Governance Subsystem

Once the details of the economic subsystem are worked out, we can begin consideration of the governance subsystem needed to manage that economy (and the agents). This, at minimum, means adding coordination-level agency for both tactical and logistical decisions. Recall that the purpose of logistical coordination is to ensure that all of the work processes work together and that there is a smooth flow of adding value to products as well as ensuring completion rates.

Figure 15.2 shows the addition of coordination level managers and some of the communications channels that allow them to regulate the work processes and cooperate with one another with respect to coordinating the flows of inputs from the acquisition process and to export processes. We include an adaptation process that is capable of operating on, in this example, a material acquisition process that is being disturbed by an environmental factor, for example, disruption to the flow of input from a source.

As with the analysis of the work processes themselves in the economic subsystem, the need for management proceeds to determine the governance architecture, that is, the number of levels of management needed to manage the complexity of the system. The guiding principle is just that. The greater the structural complexity of

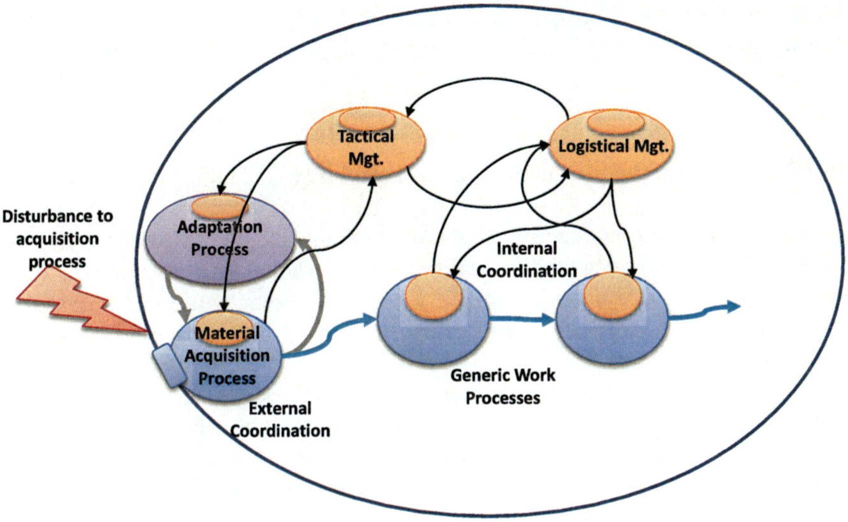

Fig. 15.2 A CAES governance archetype (without strategic management)

the economy, the more coordinators are going to be needed and that means constructing a hierarchy of management (recall Fig. 12.11).

15.2.2.4 Bring in the Agents

In both the economic and governance systems, it is time to look at all of the decisions that need to be made. In Fig. 15.2, we show the orange ovals representing agents in the governance subsystem along with communications channels between agents in the work processes of the economy and the coordination management level. At this point, the analysis turns to the agents themselves. And, as per Chap. 12, we design them according to the decision types they need to make. We identify their sources of information, for example, each operation level agent (shown in the work processes) needs real-time information on the behavior of its work process, such as the quantity and quality of the product being produced and feedforward information regarding the flow of inputs.

As the design analysis works its way down the hierarchy of subsystems the nature of the decisions becomes simpler. That is, the number of variables to be monitored and regulated decreases within the boundaries of any particular subsystem. Decision models, too, become simpler. Operational decision models, as we've seen, are generally simple error minimization calculations. Of course, the operational subsystem, if higher in the hierarchy, say, than a single variable operation, may involve multiple outputs to be monitored and the error minimization across the vector of variables may involve internal logistical (optimization) decision models. We apply the CAES archetype recursively, as we've seen, to complex operations. As the decomposition proceeds, we eventually arrive at atomic processes where the decision models are quite simple. Usually, these days, of course, these are the decision agents that are automated.

15.2.2.5 Develop the Mechanisms for Adaptability

Adaptability is a capacity needed by almost every module in a CAS/CAES. While there may be general adaptability facilities for the whole CAES as shown in Fig. 10.2, in fact, each subsystem within a very complex system, where the subsystems are, themselves, CAS/CAESs will contain its own internal mechanisms for adaptivity.

As an example, a common form of adaptivity at an operational level subsystem is the ability to increase or decrease the size of an internal buffer or reservoir based on the time-averaged flow demands; too much resource flowing in or not enough product flowing out for short time periods. Buffers with expandable or contractable boundaries, like an inventory cage that can be enlarged to hold excess incoming inventory, represent a very simple form of adaptivity.

More generally speaking, adaptivity involves the transformation function, $\boldsymbol{T}$, for any subsystem. Adaptivity types were covered in Chap. 10, Sect. 10.3.2.1, and we refer you to that section. The implementation of the three types covered there will be very different for each kind of mechanism involved in transformations (e.g., electronic, mechanical, computational).

Of course, one of the most important components that could benefit from a capacity for adaptivity would be a decision agent, especially in the coordination layer of the governance structure. For all of human history, the adaptivity of any artifactual system has lain in the mental capacities of the human agents operating machinery or managing the operations/organization. A historical problem, which persists today, is the mismatch between a human agent's ability to adapt their behavior in light of environmental changes versus their agency capacity, the power to actually change the behavior of the machinery or processes they are attempting to control. In the era of fixed machinery, they could only control the machine what it was designed a priori to do. They could not cause the machine to be adaptive on their own. In cybernetic circles, this is known as lack of requisite variety, a well-known phenomenon.

In the future, and actually starting now, every adaptive agent will also need the capacity to cause the process over which they are providing agency to also adapt its capacity in response to changes. Machines are still fairly restricted in this capacity, however, more research, especially in the area of adaptive materials, may start to change this for physical devices. In the realm of information processes, the capacity already exists and is routinely implemented in the form of software updates and machine learning in the loop.

As long as humans are incorporated as agents in the design of a CAES, the route to adaptivity is through ongoing education and life-long learning. Systems (e.g., organizations) that employ human agents should take this into account and make provisions for frequent development opportunities for their people.

15.2.2.6 Develop the Mechanisms of Evolvability

Evolvability can be achieved in two ways. In systems such as an ecosystem or even the Earth as a whole, where there is no explicit strategic management, evolution comes about essentially in the Darwinian way, through differentiation (e.g., mutations) of components followed by selection from the environment. Ecosystems are evolvable through either the evolution of member species or by invasion of new species that compete with prior members, subject to environmental changes such as climate.

In intentional systems, human-based organizations, societies, and so on, evolvability is accomplished with intentional modifications to components in response to some alteration in the environment. This is the strategic management of such a system covered in Chap. 12. Of course, accidental modifications in components of a system can still happen and cause the system to function differently. As with genetic mutations, most of these will prove detrimental or at least reduce the efficiency/effectiveness of whatever subsystem the modification takes place. For example, when an employee tasked with a routine procedure decides to take shortcuts without realizing the consequences. They can go on for some time without anyone noticing that something subtle but vital to the health of the organization is missing. This sort of degradation of processes is, unfortunately, quite common. And the larger the organization the more it seems to happen. But every once in a while, an improvement in a process is discovered (an employee has a good idea) that boosts the overall efficiency/effectiveness in the given environment.

Designing a nonintentional CAES for evolvability requires considerable anticipatory strategic thinking by the designers. The system is intended to operate autonomously and survive changes that *might* occur in its environment. Imagine a mobile, autonomous mining system operating on, say, Mars. The system would need to be at least adaptable; Mars has seasons. But if the system is to be long-lived it may need some capacity to evolve, say as when it discovers some new kind of ore that needs different extraction or processing techniques.[4] It would require a capability for modifying its current mechanisms. We will not dwell on the design of nonintentional CAESs. One way to think about how this is achieved is to consider the strategic management subsystem shown in Fig. 15.3 below as being "outside" the

[4]Least the reader thinks this example might be a stretch, please realize that the design of a true CAES has not yet been attempted. We don't really have any examples of making a nonintentional system evolvable! Putting humans in the loop (making an intentional system) provides the capacity to evolve the system in which they are embedded. The point of discussion on nonintentional but evolvable systems is to demonstrate that it is, at least in principle, possible.

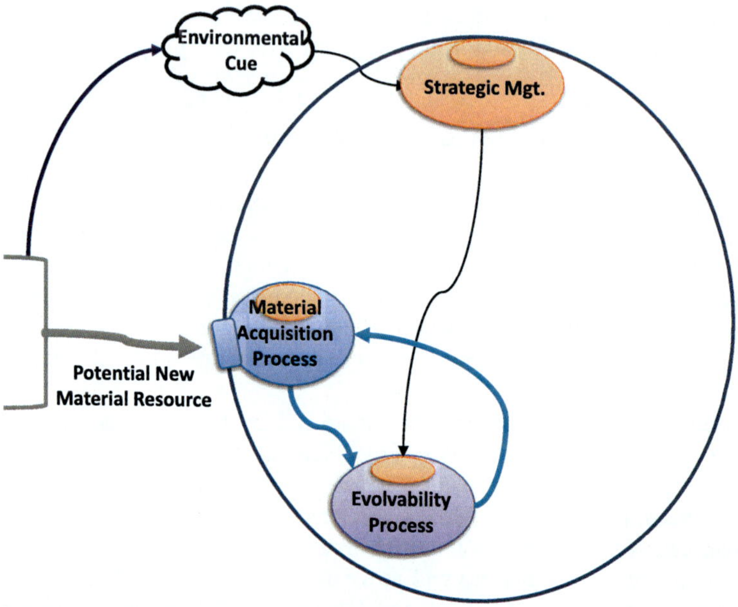

Fig. 15.3 Evolvable CAES with strategic management

system. Our main concern now is the design of intentional systems, socio-technical artifactual systems that involve humans filling the roles of agents (or at least most of them). In this figure, we show the situation when humans are in the loop and can provide strategic management.

Strategic evolution requires the inclusion in the design of processes that are able to alter other processes. As shown in the figure, a new material resource has been identified by the strategic agent from cues in the environment. But the current material acquisition process is not designed to import that material. This requires that the process be altered so that it may import the resource. The evolvability process is a peripheral subsystem that has the capacity to operate on other processes. This is represented by the blue arrows, one from the material acquisition process to the evolvability process as the latter takes the existing former in as an input, does work on it to implement the modification, and then outputs the modified process.

Evolvability processes like this may have themselves evolved from earlier repair and maintenance processes in living systems, discussed in Chap. 10. Indeed, the adaptation processes (in Fig. 15.2) may have had the same origins. Repair and maintenance processes, such as autopoiesis in living systems, require there be tools and materials in reserve that can be brought to bear when needed. Many of the same mechanisms used to repair, say a machine, could also be used to build a new kind of machine.

15.2.2.7 A Note on the Role of the History Object, H, in Eq. 4.1

Up until now little has been said about the role of the H object given in Eq. 4.1 and described briefly in Chap. 4. Simple systems may not have need of a history or memory of prior states of themselves and their environments. Atoms do not retain a memory of their past interactions as far as we know. Bridges, on the other hand, while simple in nature, are subject to wear and tear, stresses on their mechanical members that can lead to fatigue and potential failure. With modern monitoring equipment, it is possible to detect the results of their history in those fatigue signatures. Complex systems, such as airplanes, do need to have an associated record of histories, for example, to keep track of maintenance requirements. Methods similar to testing strains on structural members in bridges can detect fatigue in things like wings and tail sections, etc., but an explicit record of the airplane's history is a necessary part of the airplane as a system (along with the human interpretation and any subsequent decisions on repair if needed).

Complex adaptive systems where the adaptivity is in the form of adaptive response (described in Sect. 10.3.2.1.2 Adaptive Response, and addressed in Sect. 11.2.2 on agent autonomy) require an explicit history summarization or memory of experiences as laid out in Chap. 10. That is, some version of reinforcement learning is needed with provisions for "forgetting" or memory decay as in the Adaptrode model (Mobus 1994, 1999).

15.2.2.8 Rinse and Repeat

By rinse we mean increment the level index by 1 and perform the same analytic approach for each of the subsystems discovered in the previous level. From level 0, the SOI and its subsystems we proceed to level 1 and repeat the process for each subsystem identified at level 0. The procedure continues its recursive dive into each subsystem to discover their sub-subsystems and so on. At each recursion, the SDE team gets closer to the details of design that will eventually lead to a fully specified, elaborated CAES that meets the requirements and all of the processes (the components) will be identified and interconnected appropriately in the *N* and ***G*** networks at all levels. What remains is for the traditional disciplinary engineers to apply their skills and knowledge to make sure the atomic components meet performance requirements.

15.3 Designing the Food-Producing Module

In this section, we consider the design of human living unit that is capable of producing most of its own internal needs in order to import as few nonrenewable resources and export *no* naturally unrecyclable wastes. We assume that the human social system has arrived at an understanding that it needs to exist as a *contributing subsystem* of the Ecos and will go about reorganizing its living arrangements accordingly[5]. In the next chapter, we will describe one possible model of this living arrangement, an HSS, its interactions with the rest of the Ecos, its economy, and its governance structures needed to maintain the arrangements and interactions for the benefit of the whole Ecos.

At the most fundamental level, all living systems need a source of matter and energy—food.

We will look at the SDE process as it can be applied to the design of a basic unit of the HSS, a food-producing module or agri-village.[6] This is the basic unit that provides the food resources needed to sustain the HSS, clearly a basic requirement for living beings. In point of fact, the design of such a module has already been attempted many times, everything from a homestead to a modern industrial farm attempts to provide some food excesses (that cannot be eaten by the working participants) that can be exchanged for other food types or artifacts needed to maintain operations. As we now know all of these designs have proven problematic in multiple ways.[7]

In what follows we will not be attempting a detailed design and engineering process. Rather we will address the design considerations from the systems perspective as developed in this book. This will be a conceptual sketch of the SDE process applied to one of the most fundamental requirements for sustainable living in the human social system, the production of food.

One design approach, already in existence, has promise and we will refer to it as a basic design that provides a starting point. In the spirit of Principle #12—Systems can be improved—we start with the concept of a permaculture-based community. We will see that these modules are adaptable so that they can be situated in a wide variety of ecozones with varying resources and fluctuating climates. They are also evolvable in being able to reconfigure themselves as a result of nonstationary changes that occur in their environments. In the next chapter, we then tie these modules into a web of relations that constitutes the social systems of larger scales.

---

[5] Is this a reasonable assumption? Most evidence at present suggests probably not. As far as the vast majority of humanity is concerned, the purpose of the human social system is to produce increasing physical wealth to be consumed by humans with little concern for what the system does to benefit the larger Ecos. However, the basis for this assumption is predicated on a small but perceptible change in understanding of the need for humanity to live in balance with the productive capacities of the Ecos to support all of its various subsystems, and in particular the biosphere. As the world faces increasing threats from the global predicaments created by human commercial activity, this understanding has lately been garnering a wider audience. Realization of the impact of humanity upon the Ecos may be rapidly expanding. A tipping point in the majority of humans understanding the situation may yet be reached generating a phase change in consciousness. It is not, thus, a totally unwarranted assumption.

[6] We assume that not every village or living module will be devoted to agricultural production. Thus, those villages that are thus devoted need to produce excess foodstuffs to export to other non-food villages.

[7] For example, we now understand the pollution produced by the industrial farm as well as erosion of soils and the risks of diseases or pestilences to food crops and livestock.

However, whatever we do in terms of a social system design, it all starts with the ability to feed ourselves sustainably.

15.3.1 The Agri-Village

In the context of social modules (units) to be presented in the next chapter, we consider the unit we call a "community," which is an artifactual system involving humans and their cultural artifacts. This is a system of households or what James Scott (2017) calls the "domus,"[8] in which people reside. Domus[9] are living quarters (houses), and surrounding land and utility buildings along with gardens, orchards, water catchments, and other components for a semi-autonomous (i.e., near self-sufficient) family existence. We say semiautonomous in that the horticultural/animal husbandry capabilities as well as the capacity to make necessary artifacts of each domus are likely not able to provide total self-sufficiency. Domus (pl) may resemble "homesteads" that are largely self-sufficient but nevertheless coupled with other sources and sinks. We posit, however, that a community is a set of domus that collectively come much closer to self-sufficiency through cooperative interactions with other nearby units forming the community level of organization. Generally, it is not possible for any community to achieve complete autonomy in that there are always going to be resources that they may not have access to or products they cannot produce that they could obtain, say, from other not-too-far-away communities.[10]

In keeping with the CAES model, every community obtains resources and produces a product of some kind that can be used by other modules as well as internally consumed products used to maintain the integrity of that community. Our focus will be on a community that produces agricultural products to be used by, generally speaking, other nearby communities. Internally, each domus will produce a significant amount of its own food requirements, but each will supplement their foods with products produced by the whole community (e.g., wheat flour) in sufficient excess of the community's own needs such that the product can be exported to other communities in exchange for products that those other communities produce that are not produced locally. Thus, this type of module is both semi-self-sufficient and

[8] Scott uses this word, which comes from Latin, meaning a certain style of home in Rome, in a much more general sense of not only a residence for people but also a surrounding agricultural enterprise, much like we think of small, self-sufficient farms. He describes these as the original family sedentary homes when the domestication of plants and animals was first beginning, around 6–8 thousand years ago (first in Mesopotamia). Scott elucidates this pattern of the origins of "organized" agriculture for several additional "grain-centers" in Asia and Meso-America.

[9] The term domus is both singular and plural!

[10] In this section, we assume that humans in these communities will live with some modicum of comfortable living standard. We do not assume the culture will be a high-energy consuming throw-away materialistic one. But neither do we envision humanity returning to the Stone Age.

specializes in some particular foodstuff for trade. It is able to export some agricultural products. We'll call it an "agri-village" to evoke an image that is, in essence, a permaculture community. The term "permaculture" as originally coined, meant "permanent agriculture" but has more lately been taken to mean "permanent culture." This meaning includes all aspects of living, including all normal activities in the domus and communities. Much of the guiding ideas were influenced by systems ecology, particularly the work of Howard Odum (1983, 1994, 2007; also, Odum and Odum 2001) and developed into a holistic set of design patterns (Holmgren 2002; Mollison 1997).

Our task is to design such a community using the methods of analysis of Chaps. 6 and 8 and using the CAES archetype (and its submodels) to guide us. What follows will be a brief overview just to show the considerations that would be made in full analysis and design. We will basically discuss the level 1 decomposition and a few instances of a deeper decomposition where it will shed further light on the process.

There are many permaculture communities in existence already around the world so the concept is well established. These communities generally follow the three core ethical tenets of permaculture but can vary in terms of the degree to which they adhere. For example, the "fair share" tenet holds that individuals should take no more of the community wealth than what they need and by implication, all individuals in the community have equal participation. To some, this will seem like socialism and compare unfavorably to capitalism in terms of personal freedom to accumulate as much wealth as they can, and accept the notion of ownership and private property. Indeed, many of the communities operate as communes. On the other hand, most of these existing communities operate within the matrix of the larger capitalistic economies and have to interact with those economies; very few are truly self-sufficient. What we envision here is a world in which all communities are based on a basic permaculture approach whether they are agri-villages or manufacturing villages.

15.3.2 Purposes

All CAES designs begin with a statement of the system's purpose or what function it is supposed to perform within the context of its suprasystem. There are actually two kinds of purposes with respect to the suprasystem. The first is to produce products that are useful to the suprasystem, the obvious purpose. The second, however just as important, is the importing of resources from other entities (sources) in the suprasystem. When overall flow balance is achieved in the suprasystem, this act of importing other entities' outputs prevents those outputs from becoming unused wastes. So, the system of interest persists and thrives when it is fulfilling both of these functions at the same time that the larger suprasystem benefits.

Here we will focus on the production purpose and assume, for the moment, that the SOI is importing the right resources and in the right amounts per unit of time.

Only when the component system is producing something of value to the whole suprasystem will the whole reciprocate with support for the component system. In the case of an agri-village, as already noted, the main export will be some foodstuff that is produced in excess of the needs of the community and supplied to other communities. However, the purpose is more complex than just producing food. As will be discussed in the next chapter (which describes the larger social system that is part of the agri-village's suprasystem), every community contributes to the collective knowledge and to the governance of the whole social system. Thus, we can characterize the purpose in terms of producing a useful material/energy product as well as meaningful messages.

In addition to these productions, the community, in autopoietic fashion, must replenish its own human and physical capital. As will become clear in the next chapter this does not mean "growing" the population or increasing individual consumption. Indeed, for the human social system to be sustainable in the very long run, it will be necessary to maintain a stable population across the various module types and hierarchical levels. We will not speculate on the nature of the mechanism for establishing and managing a stable population size; many such mechanisms are well known and we will assume that societies will choose those most in consonance with its group values.

Below we will identify the product outputs that define the overall purposes of the system. We will consider the transformation function (Sect. 4.3.3.4. Transformations) for the whole SOI but in general terms. We need to determine the inputs needed in order for the system to produce these transformations. One important difference between the use of the top-down decomposition process for design as opposed to the analysis of an existing system is that the transformation function in the latter can readily be approximated from obtaining input/output data over some time frame and then using machine deep learning (or preferably causal learning, Mobus 1994) provide a starting point for modeling the SOI, noting that this approximation will be refined during the process of decomposition and the refinement of transformation functions at lower levels in the organization hierarchy. In the design of a new SOI, the transformation function is to be approximated by early specifications given the understood needs of the suprasystem (the products to be exported to sink entities). These specifications may change as the environment (i.e., level-1) becomes better understood. But these specifications provide requirements and constraints on the subsequent design of the internal modules (level 1) once the flows into and out of the SOI (level 0) are better understood.

The whole process of design is generally an iterated recursive "dive" into the SOI and its subsystems making refinements and adjustments to the assumptions made in the specifications.

15.3.3 Environment—Level-1

In keeping with the methods of Chap. 6, we begin our analysis of the system of interest with an examination of the environment or context in which the SOI operates, followed by an analysis of the inputs and outputs crossing the boundary. The former will establish how the system will serve its purposes by identifying the relevant source and sink entities and the milieu in which the system operates.

15.3.3.1 Milieu

For an agri-village module, there are several important milieu factors operating on different scales of time and space and having their own dynamics. We mention a few of the more important factors here.

15.3.3.1.1 Overall Climate

One of the more relevant factors conditioning the long-term (and thus evolutionary) aspects of an agri-village will be the climate of the region, which determines the set of feasible crops that can be grown, for example. At the time of this writing, the climate is being rapidly driven from the norms of the last ten thousand years (or so) through anthropogenic warming of the atmosphere and hydrosphere, with the acidification of the later (oceans) with increased CO_2 injections.

Climate considerations will condition all aspects of the design of an agri-village, down to the details of building construction and soil management. The reason is that the climate conditions the weather and as the former is trending toward a warmer world the weather patterns (already dynamically chaotic) are becoming increasingly unpredictable.

This speaks to the need for designing these systems and subsystems for adaptivity and evolvability. It is clearly the case that the future climate chaos will be a nonhomogeneous, nonstationary milieu. Thus, the long-term viability of a food-production system must include the ability to evolve and its short-term viability will depend on its capacity to adapt to shorter-term shifts in the climate.

15.3.3.1.2 Seasonal Variations

As the climate warms, we are now beginning to experience a number of changes in weather patterns. Among the effects we are seeing are increasing variances in seasonal conditions such as average diurnal temperatures. These changes are having an effect on the kinds of crops a particular region can grow. There are also changes in the timing of plant growth cycles. Agri-village designs will need to take this into consideration as the system may not be able to merely adapt to intermittent variations, but evolve to trending variations.

15.3.3.1.3 Storms and Other Disturbances

The climate change that is developing has also had an impact on the severity and frequency of storms such as hurricanes, floods, and tornados. Designs of buildings, as one example, should take this into consideration. It is unfortunate that the most productive soils are found in flood plains. Designs would have to either be for flood-resistant housing or possibly transporting the rich soils to villages situated on hillsides with terraced fields such as found at Machu Picchu, Peru, and elsewhere in hilly or mountainous terrains.

15.3.3.2 Entities, Sources, and Sinks

There are several principal input and output requirements for an agri-village. Perhaps the most important is solar energy, both real-time, that is daily, and accumulated embodied energy in the form, for example, of fibers for various purposes, and wood for timber and heating. As well, the rainfall in the watershed is produced by solar energy evaporation of water. The input of solar energy depends on the "aperture" or square area of land where plants are grown. Below we address the issue of this land area in terms of how much is needed to support a community.

In a permaculture-based agri-village, the forest areas are included in the landscape and are, to some degree, cultivated or at least replanted as wood is harvested. So, the forest along with the food crop fields are the main input interfaces with the sun as the source of energy. There are additional ways to capture solar energy as will be discussed below.

In addition to local solar influx, communities are expected to obtain specialized resources from other communities such as metals or manufactured goods.

Finally, agri-villages and other kinds of communities may be expected to obtain resources from "nature". The example of rainfall is a good one. The moisture in air that condenses as precipitation in the watershed will come from more than just the area of the community territory. Similarly, the food supply may be supplemented with meat from hunting. The populations of prey animals are the sources to consider.

The principal material output from the agri-village is the surplus foodstuffs it exports to customer modules, such as the one described above. In this instance, the customer is also a source for the metals and clays used in the agri-village, part of the economy of the tribal module that encompasses the communities.

Of equal importance in designing a module is the minimization of waste products that cannot be used internally and have to be exported to the environment. Every effort must go into the design of processes, the outputs of which, can be recycled internal to the module. But when this is not possible then the design needs to consider sinks in the environment that could reasonably handle the effluents. For example, composting toilets can recycle much of human waste as fertilizer thus keeping it out of the output stream (i.e., the water flowing through the community). Animal wastes can be similarly managed and recycled as fertilizer too. Solid wastes should be minimized. We imagine that the use of plastics that cannot be immediately reused or recycled will be abandoned. There is probably no way to make such plastics in any case since they require oil derivatives. In fact, about the only waste product that cannot be internally recycled (or at least not easily) is the carbon dioxide exhaled by the people and animals. If the human social system is in balance with the Ecos, this will be recycled through plant photosynthesis. The ideal design will have almost no need for waste sink services.

15.3.3.3 Boundary Conditions and Interfaces

In Sect. 4.3.3.3 Boundary, we provide guidance in how to consider the boundary of an SOI by examining the physical aspects of the boundary itself, for example, a physical container or internal constraints such as stronger bonds between components than between components and external entities. This also includes the degree to which the boundary is able to exclude penetrations by substances (porosity)[11] and control inputs and outputs through "formal" interfaces.

For purposes of description of the design process, we will make the assumption that an agri-village as envisioned will not be enclosed in a physical boundary. The nature of a permaculture design covers extended territory across many different landscapes such as fields, forests, watersheds. Rather we imagine the boundary as being the territorial extent, or occupancy footprint, of the community, that is, the amount of surface area, regardless of topographical features, possessed by the people in the community. We say possessed as opposed to "owned" since this surface area may actually need to change with changing conditions. There should probably be a wide buffer zone between neighboring communities that allow for shifts growing outward or shrinking inward. In all cases, the area possessed should never be more than a community needs to maintain self-sufficiency. Territorial recognition by neighboring modules (see the governance architecture discussion in the next chapter) and ongoing cooperation in the management of the buffer commons[12] are essential.

There are at least two ways to think about the boundary of an agri-village. One way is to consider a boundary around the "village" itself, excluding the resources that the village needs to import. Alternatively, we might consider the "territory" including the sources of resources that are obtained by the village directly, as

[11] An important aspect of porosity for human communities is the infiltration by pathogens. See the section below on the repair and maintenance functions of the SOI for a short discussion on the health care and public health functions.

[12] In truth, management isn't quite the right term here. Buffer zones would not really be "commons" in the usual sense, but rather a general kind of asset (nature) that profits the surrounding communities simply by existing and regulating its own self in the way ecosystems do.

opposed to inputs from other villages. In the former case, we need to consider those sources, the area of land and water that is "controlled" by the village. This is no easy problem. Many resources spread over a wide area may be shared by multiple villages (a commons) without conflict as long as there is monitoring of the usage rates and a coordination function making sure none of the communities participating take more than their fair share. The details will likely depend on the distribution of resources within a particular landscape. For this exercise, we will consider a boundary around a particular community and define (design) specific external sources (and sinks) and interfaces with this larger suprasystem entity (sources and sinks). In this case, what is inside the system is the village, its cultivated lands, fields used for livestock, and water reservoirs directly used by the village. Thus, all other flows are inputs to or outputs from the community system.

15.3.3.3.1 Interfaces

The boundary is defined, in part, by a list of interfaces for these inputs and outputs. Here we mention a few examples. These will segregate, of course, into input and output categories, which will be discussed below, but also into classes of each category. For example, input interfaces will include the obvious subcategories, energy, material, and messages and output interfaces will include the subcategories of product, material wastes, and waste heat. In this last case, we recognize that waste heat will radiate or be removed by convection (air currents) from every subsystem so we need not consider a special heat removal process. Each of the subcategories of inputs and outputs will further be differentiated by types of energy, types of material, types of messages, types of products, and types of wastes. Figure 15.4, below, is a starting sketch of the environmental, input/output, and interface analysis of the agri-village, completing a level-1 analysis.

For example, if a village receives a kind of food staple from a different village, one the receiving village does not itself produce, then there must be an interface for either picking up the food from the producing village or receiving a shipment from that village (who does the transporting?). Food is an interesting case where both material and energy are in a coupled state as they enter the system and both must be accounted for.[13] The energy is used to support the living entities and the waste products, CO_2, urine, and manure (and eventually the no longer living biomass of the deceased) have to be considered. All but the first should be considered as recyclable.

Other interfaces for receiving or exporting materials may include, for example, a sawmill for receiving and processing wood products obtained from a forest.

[13]This is not much different for the case of systems that, for example, burn fossil fuels to obtain energy. The outputs of oxidized carbon and hydrogen (along with things like sulfur dioxide and trace metal pollutants) and leftover ash, if any, are waste materials. The inputs are the fuels as material substances that contain the sought energies in the form of exothermic covalent bonds as well as oxygen and some initial energy to start the combustion.

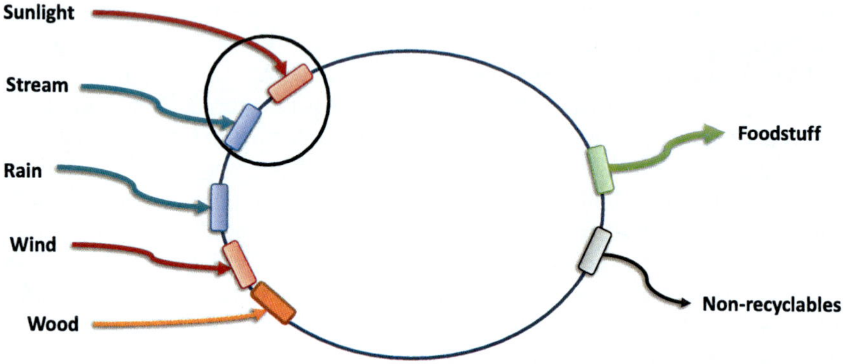

Fig. 15.4 Start with a placeholder oval representing the agri-village and begin to identify the sources, sinks, inputs, and outputs. Sketch in the interfaces that will later need to be specified

15.3.4 Input/Output—Level 0

In this section, we focus on the food-producing relevant inputs and outputs in particular. Other inputs, such as imported tools or other supplies and products from other modules will be further explored in the next chapter.

Of note here, the general requirement for inputs and outputs is that there is a material balance to be maintained. It is necessary to measure, as a time series, the mass outflows, of foodstuffs, both plant and animal products, and the inflows of organic and mineral wastes (coming from other non-food-producing modules or from surrounding land) to be used to renew what was lost from the soils in the production of the food. This means that careful measurements and control mechanisms must be in place to regulate the system's inputs/outputs over time so as to maintain a steady-state condition. We will not attempt a detailed description of these mechanisms; we only note that they are a part of the HCGS system needed to manage the module.

15.3.4.1 Export Processes, Interfaces, and Sinks

15.3.4.1.1 Main Product—Calories Out

The main product of a food-producing modular community would be nutritious calories, meaning vital nutrients and bioenergy. It is expected that to some degree or another every community in the future world will be responsible for some aspects of food production (and preserving) but some communities may be tasked with other forms of production (such as making tools) and have a lesser capacity to be completely self-sufficient. They will be customers of excess production by communities that are primarily food producing.

The design would call for careful consideration of the methods and timing of exporting food products to other communities. The interfaces could be in the form of markets as is the case today, but markets depend on a certain amount of competition which would defeat the purpose of establishing a more global system of trade. More appropriately, the design might include direct trade agreements (protocols) between communities, where the food-producing community agreed to produce certain food products at a certain rate and deliver those products at agreed upon times and places.

Since the export of foodstuffs generates a continual loss of minerals and nutrients from the soils, it will be important for the food-producing module to import organic wastes and other soil amendments from other modules (see below).

15.3.4.1.2 Waste Products

As mentioned, a well-designed permaculture community would not produce wastes that would have to be relegated to a midden or dump. Organic wastes, such as table scraps and even human excreta would be composted and recycled. Worn out or broken metal objects such as tools or glass containers could be exported to modules that were producing those objects in the first place where they would have the capacity to recycle them. The design of the system should consider every kind of material that is used in its internal processes and determine how these are to be recycled or reused.

Waste heat cannot be recycled. In general, heat will dissipate into the atmosphere which acts as a nearly infinite reservoir. The agri-village is an internally semiclosed system with respect to material cycles, and open with respect to energy flow-through. Since this system is, itself, a subsystem in a larger social system (as described in the next chapter) where flows of materials and energies from and to other subsystem modules balance in the long run, the larger suprasystem is also a semiclosed system with respect to material flows and open with respect to energy flows. For example, other non-food producing modules may send their organic wastes back to the food-producing module for composting and reuse in the soils.

15.3.4.2 Import Processes, Interfaces, and Sources

15.3.4.2.1 Energy

The main energy input of a food-producing module is sunlight. It is transformed into organic chemical energy in the form of foodstuffs by the process of photosynthesis. Every plant grown participates in the import of energy in this way. Other energy import processes involve the mechanical energy in wind and water flows through the module. The interfaces involve windmills and waterwheels respectively. Much of this energy would most likely be used in its mechanical form, though some may be transformed to electricity for various other work processes in the module, such as lighting.

Since the human and animal labor (work) done in the module is supplied by energy from food produced, largely, locally, we do not consider human effort as an input.

15.3.4.2.2 Materials

Some of the primary material imports to a food-producing module will include organic waste matter from other modules and additional soil amendments from surrounding natural areas (e.g., forest floor liter). These materials are essential to replace what is lost from the soils due to the export of foodstuffs. The actual weight of edible foodstuffs compared with total plant weight for most foods is a small percentage, but not inconsequential. Therefore, in order to maintain soil health and, thereby, plant health, these materials need to be composted and worked into the soils in an ongoing seasonal process. For animal products exported, such as eggs and meats, more organic waste material must be imported than for compensating exported plant foodstuffs. This is because more plants, such as grains and grasses, need to be grown to support the animal stock. A food-producing module needs to collect wastes from several other non-food-producing modules to achieve a balance.

Other material inputs, such as, for example, lumber or special tools needed for operating the module will be considered in the next chapter.

15.3.4.2.3 Messages

The main kinds of messages received from the world external to the food-producing module involve information on weather conditions but primarily messages from other modules to coordinate the flows of materials and foodstuffs from sources and to sinks. As shown throughout the book, communications channels are established from source export processes to the system's import processes and from the system's export processes to sink import processes for the purpose of signaling flow requirements and adjustments.

15.3.5 Internal Processes—Level 1

15.3.5.1 Work Processes and Economy

Most of the work processes at level 1 will be the domus and the agricultural process. However, the community must also be capable of producing clothing and other artifacts as needed. These will be considered under the Peripheral Processes below.

Agriculture involves the preparation of suitable land (soil) for the planting and nurturing of food crops. Humans are able to consume a huge variety of plant foods such as grains, vegetables, fruits, nuts, and herbs. The specific varieties of these various types will depend on the climate and environment of the domus. One of the major adaptations that human families will need to make is in growing and consuming plant foods that are able to grow in a particular environment. Our current horticultural practices have bred many varieties of many different plants that are adapted to a variety of climate conditions. For example, we have tomato varieties that are hardy enough to grow and produce fruits in higher latitudes. However, it is important to recognize that these are variants and their seeds may be subject to regression unless carefully curated through ongoing horticultural practices. Within permaculture, there are a number of practices that are geared toward selection of food crops that are highly resilient in various climate zones. Of course, this means that domus (pl) in different situations will need to acclimatize their consumption habits to what is possible locally.

The idea is to minimize transportation of foodstuffs between modules and definitely between regions. Communities in more northern latitudes will need to forego having tropical fruits, for example. This restriction will become clearer as we examine larger social systems in the next chapter.

The soil must be prepared, seeds planted, plants nurtured, and foodstuffs harvested. After harvest, the soil must be allowed to regenerate and helped to do so with various practices. These are the normal sequences of work process that produce foodstuffs. Some food items will be further processed for storage for consumption during the nongrowing season (e.g., winter).

The processes of growing, processing, and storing food in buffers (e.g., grain silos) have a 10,000+ year history of refinement. Our knowledge of how to do this is deep and extensive and with the formalization of systemic processes should be sustainable for an indefinite future time. What needs to be added are the mechanisms for adaptation and evolvability. Farming practices, as understood today, have evolved but under conditions of long-term climate stability. They have not needed to emphasize adaptability explicitly. Nor have they recognized a need to provide for evolvability. As an example of adaptive agriculture, consider the process of plant husbandry needed to develop new strains of crops as climate conditions change. A tremendous amount of research is currently going into this in the face of global climate change. The methods developed will need to be incorporated directly into the management of the domus and the community. Similarly, evolvability, such as in finding completely new food crops as conditions change, will need to be explicitly provided for within the community as a whole entity.

But the operation of agriculture is only one aspect, albeit a very important one, of the food-producing module. Each domus will require more than just food to support it. Within any community, for example, some domus (pl) may become specialists in producing clothing or shoeing horses and thus a basis for trade and transactions arise in the community as a local economy.

15.3.5.2 Management Processes and Governance

In the following chapter, we describe the basic management/governance architecture for the social modules. This architecture implements the governance archetype model specifically for human organizations and society in general. Here we will only mention a few level 1 requirements in terms of what needs to be managed and governed within the agri-village module.

First, of course, is the agriculture production. The distributed operational management relates to the several independent sub-processes such as soil quality monitoring and amendment application. This can be viewed as part of an import process as described above and thus will be coordinated by a tactical decision agent, a person trained in soil management. The actual measuring of soil conditions, such as pH, moisture, carbon content, and testing nutrient levels is under operational management, that is, the worker knows what data needs to be collected and is self-governing to do the job. Similarly, the adding of amendments, like composted chicken manure, is an operational process; the workers know what needs to be done and how to do it. The tactical coordinator takes care of obtaining the amendments in the right proportions and at the right time. The logistics coordinator analyzes the collected data and determines the work that needs to be done to maintain the soil at maximum health.

The same basic pattern is applied to other agricultural tasks such as planting and harvesting. The logistics managers determine the timing of the work and how many workers will participate. Take note that, as has been pointed out in discussions of human activity systems, one person can perform multiple agent tasks, being the operations worker, the tactical, and the logistics coordinator. They are making different kinds of decisions at different times.

Strategic decisions with respect to the agricultural process would involve things like deciding which crops to plant, taking into consideration climate conditions such as draughts or cooler-than-normal temperatures. Such decisions have to take into account also the demand from other non-food-producing modules. The strategic agent is responsible for evolving the agricultural process as needed to adapt to long-term changes in the environment such as setting up a breeding program to select a variety of a plant that is better suited for the changing climate.

The community of domus modules, likewise, needs a governance process involving all three levels of decision types. Each household has its own internal processes according to each family's needs/desires/options. Each is self-governing. Between domus (pl) activities would involve cooperation between near neighbors and some amount of coordination over the entire community. In the next chapter, we will discuss a model of this form of governance.

15.3.5.3 Peripheral Processes

An agri-village, a food-producing-module, should be thought of as an autopoietic system. It must contain processes that provide repair and maintenance capabilities for all of the physical facilities of the village, such as houses, barns, warehouses, tools. The people who work in these processes have to have the necessary skills, of course.

All internal work processes as well as the import/export processes should have some capacity to adapt to short-term changes in the village's environment. For example, if it is determined that other non-food-producing villages need more of a particular kind of food, the amount of growing space needs to be expanded to accommodate that.

On the other hand, if it is determined that this increase may be permanent or even increase in the future, village may need to convert previously undeveloped land to that purpose, that is, evolve.

The design of various peripheral processes should allow the community to be as self-sufficient as possible so as to minimize dependence on other communities for necessities. This ideal may not be realized, of course, but the design should seek as close to the ideal as possible.

15.3.6 Discussion

Current permaculture design principles (Holmgren 2002) guide the detailed designs at lower levels in the hierarchy of organization. The major factor in, for example, sizing anyone subprocess, say the number and range of chickens, is the trophic flows. Starting from the caloric and nutrient needs of the human occupants/workers, we work backward through the various subsystems, for example, the "chicken" system, to determine requirements. Tradeoffs will always exist and so balancing needs is part of the engineering activity in the SDE process.

There is likely no single design that will serve all food-producing modules at the detailed level. Each will require a deep analysis in order to find the balances needed. But that at least down to level 1 all agri-villages will follow a basic pattern (some of which discussed above) should be obvious.

We have provided this rough sketch of the design of a food-producing module as just a cursory glimpse of the application of systems design and engineering, based on a model of a CAES.

15.4 Conclusion

The SDE process using the CAES archetype model as a guide can be used to design a large variety of systems fit for their environments, most especially and importantly human social systems.

The key is having a comprehensive general system archetype that tells us, in advance, what kinds of internal capabilities the system to be designed should possess. The CAES provides this basic model. But, in addition, we need to follow the kind of deep systems analysis described in Chap. 6 to provide us with a way to elaborate the internal details of how each of the CAES modules will be implemented in the real systems.

The most complex systems in our experience are artifactual systems such as neighborhoods, cities, organizations/enterprises, countries, and the whole human social system. All of these can be modeled as CAESs and the CAES archetype can be used as a starting point for the architecting, design, and engineering of all such systems. With appropriate consideration for stochastic processes (inherent uncertainty), ambiguity (e.g., the membership function of a component), and other fuzzy aspects of these systems, all can be modeled with systems concepts as derived from the ontology developed in Chap. 3. And following the principle of system purpose reflecting the needs of the suprasystem to achieve fitness, there is no necessary reason that we humans cannot engage in intentional organization as a design and engineering process with fit systems being the product. Including adaptivity and evolvability into those designs ensures that the systems will endure in a changing world.

Chapter 16
Societal Systems

Abstract This chapter will provide a glance at the application of systems theory (Chaps. 3 and 4) to the analysis (Chaps. 6 and 8) of the most complex adaptive and evolvable system of greatest import to humanity, the human social system (HSS). In previous chapters, we analyzed bits and pieces of the HSS, especially Chaps. 7 (as a part of the Ecos) and 9 (the biophysical economy) as it exists. We found that the HSS of the modern world is highly dysfunctional in numerous ways. Then, in Part III, we examined the archetype models of real-world systems, the CAES as a whole and the agent, governance, and economy models that combine to make up a functional CAES. These are the models that are to provide guidance in the analysis and design of an "ideal" HSS that can operate within the context of the Ecos. We start by considering the deviations in the implementation of the current global HSS from the archetype models in Part III. Then we turn attention to what a design for a functional, that is sustainable, persistent, and stable HSS would look like if we follow the pattern/model-based design approach given in the last chapter.

16.1 Purpose of This Chapter

Arguably, a modern, technologically enabled, culturally diverse society is the most complex artifact that humans have created. It is made more so given that humans are not only the constructors but are components of that artifact. The global HSS is a complex adaptive and evolvable system to be sure, but is it a good one? The word "good" has a specific interpretation based on systems principles. The HSS is now understood as a subsystem of the Ecos. A "good" subsystem is one that provides benefits to the supra-system in which it is embedded. A subsystem that fulfills its purpose, relative to the supra-system's needs is a good one.

What purpose does the HSS have? More importantly, what about the HSS is good for the planet as things stand today. Many people both laypeople and specialists from multiple disciplines are rapidly coming to the conclusion that mankind is in the process of harming the Ecos.

In Chap. 9, we examined some subsystems in such a society, largely focused on the neoliberal capitalist-based economy as it is found today. That chapter revealed

G. E. Mobus, *Systems Science: Theory, Analysis, Modeling, and Design*,
https://doi.org/10.1007/978-3-030-93482-8_16

some significant deviations of the modern society's economic subsystem from the archetype model presented in Chap. 13, most particularly the reliance on finite stocks of hydrocarbon fuels for the majority of high-power inputs as opposed to the real-time flows of solar energy.

The current HSS is unsustainable.

Its form, a set of nation-states along with a few peripheral units such as aboriginal tribes, each pursuing its own agenda and most based on the pursuit of a growing economy even as the availability of natural resources is declining (and pollutants are accumulating) is the result of a mix between auto- and intentional-organization. The latter aspect, an example being the development and adoption of constitutions, is also a mix of short- and long-term thinking. But the latter is too often trumped by the need, seemingly, to solve short time-frame problems.

The disparate forms of current governance and economics of the HSS are the result of the same kinds of evolutionary dead ends that mark the extinction of species in the tree of life. Evolutionary adaptations that seemed appropriate in one environment turned out to be non-adaptive as the environment changed more rapidly than the species could evolve. Similarly, the various societal systems that co-exist today have evolved to fit one set of circumstances, say the availability of "cheap" fossil fuels and through unintentional niche-creation (depleting resources and polluting the atmosphere) have sown the seeds of their own maladaptation.

The irony of the human condition is that after evolving a kind of intelligence and creativity that could learn virtually any concepts and solve many problems, and applying that cleverness to finding new ways to make a living, we humans rushed to choices, in terms of ways of governing our social systems and conducting economic activities, that, in retrospect, seem incredibly foolish. As revealed in Chap. 9, for instance, adopting the assumption that markets could solve all economic "problems," and as shown in Chap. 13 not to be the case. Or, in a similar vein, adopting the notion that a capitalist, growth-oriented economy (with accompanying growing population) was the best way to provide for the common good has turned out to be at the base of the causes of our current predicaments (Klein 2015; Kolbert 2015).

There are a number of explanations for the reason that human societies have become maladapted to the modern world that they, themselves, created. One proposal by this author posits that human evolution that had produced an incredibly clever primate at the end of the Pleistocene era got sidetracked, so to speak, when that clever species invented agriculture (Mobus 2012, 2019). The theory, in brief, was that human beings were the first species to evolve a brain significantly able to provide strategic thinking capabilities. Evolution had selected for beings able to think about alternate futures and to enter into the intentional-organization ontological cycle (Chap. 10).

However, we also recognize that human beings started with no concept of how complexifying culture would eventually affect their successors. The knowledgebase of humanity was limited to natural ecosystems and how to construct primitive tools at the end of the Paleolithic age. It would take many thousands of years to accumulate more knowledge of the physical and biological worlds, let alone the psychological world. Fundamentally, the search for a workable form of society has been an evolutionary process with auto-organization replaced with intentional-organization, but intentions based on best guesses and not any definitive knowledge of humans as individuals (e.g., psychology) or in social organizations (social psychology). We were ignorant of what the effects of scale or complexity might be. It has taken thousands of years of experience to begin to accumulate an understanding sufficient to consider a systems engineering project to realize a new social contract and social structure that could achieve a sustainable society.

But, the discovery of agriculture provided another effect with respect to the evolution of the average human psyche. A careful examination of the management needs and practices of early agriculture indicates that the greater emphasis was on logistical and tactical thinking with a greatly reduced need for strategic thinking (partly a result of living in fixed settlements). Thus, selection for greater capacity for strategic thinking was diminished and for the last 10,000 years or so the vast majority of the gene pool of humanity has been subject to selection for logistical and tactical thinking, both as individuals and as groups.

Logistical thinking (making sure the crops were irrigated) is very short-term in scope. Tactical thinking (protecting the crops from marauders) is not much longer-term. In the short run, the application of cleverness to immediate problems took precedence over thinking about the consequences of over-extraction of resources (like forests, groundwater, and soils) in the future. Humans were ignorant of deep scientific knowledge. They were motivated to acquire all the "wealth" they could in as short a period of time as possible. They invented artifacts that allowed them to literally rape the earth. Historical baggage in the form of cultural practices ("we've always done it that way"), lack of systems knowledge, expansionism, acquisitionism, all of these factors have carried into the present (Crist 2016). The consequences of these practices look dire at present. They will need to be abandoned in favor of more sustainable ways.

The question we address in this chapter is: If we were to tackle the design and engineering of an HSS based on systems principles and especially the CAES archetype, what would that system look like? How would it work? In this chapter, we will tackle the systems design and engineering of the human social system (HSS) based on the CAES archetype (including the three sub-model archetypes). The reader should recall Fig. 7.2 in which the HSS is shown embedded in the whole Earth ecosystem.

This may not be a useless exercise. As noted, the current forms of civilization are not sustainable. This means, quite simply, that the global civilization could very well collapse. Given the recent revelations regarding the rate of climate change and the severity of climate chaos, the risk of collapse seems not at all untoward. We could lose everything that constitutes what we have come to accept as a normal society. In that event, wouldn't it be good to have a design for a human society that might actually be sustainable as a plan for recovering an HSS.

As with any CAES embedded in a larger meta-system, the HSS will need to meet a set of criteria for success in order to be long-term sustainable (Mobus 2017). As outlined in the reference, these conditions are:

1. Fulfil a purpose—produce valuable outputs.
2. Receive rewarding (essential) inputs.
3. Be adaptable.
4. Be evolvable.
5. Be lucky.[1]

The first condition is a statement about the fitness of a CAES, we brought this up in Chap. 14. The supra-system is comprised of many other systems entities that may cooperate with or compete with the system of interest, in this case, the HSS. If the HSS produces one or more outputs that are of value to the Ecos as a whole, then cooperation with those other entities leads to stability and rewarding feedbacks.

16.2 Systems Design for the Human Social System

Here we will tackle the most important and easily argued most difficult (complex) design task put before us, the design of a more systems-rational social system for humanity. We do not assert that what we present here is THE design. Rather what we present below is more on the order of design challenges, but with hints of how those challenges might be pursued given the systems archetypes for CAESs.

As discussed above, the design of a system starts with the knowledge captured in the knowledgebase, per Chap. 8, as a result of having done a deep systems analysis, per Chap. 6. Here we run into a peculiar situation. The HSS already exists. Moreover, the various social sciences have already uncovered significant normative knowledge about human behaviors, especially with respect to their proclivities in being agents, the economic systems that are extant in the world, and the various forms of governments also extant in the world. We could hypothesize that if all of the knowledge thus far gained could be realigned with systems' principles and theories, then the knowledgebase needed for modeling and design would be at our fingertips. The presumption might be that all we need to design are "interventions" that would mitigate, correct, or otherwise replace faulty functions in the current social system.

[1] There is good reason to believe that the Universe is fundamentally non-ergodic. That is, at each instant, the Universe enters a new state of existence and passes from a prior state that it will never enter again. Thus, all processes and phenomena are fundamentally non-stationary with respect to the relations between existing systems. This being the case, there is never any guarantee that an ultimate stable and persistent condition will ever obtain. Long-life depends on the environment of the system to remain sufficiently stable that adaptation or evolvable reconfigurations will sustain it. That is not a given, hence, the system relies on a certain amount of luck. Massive, overwhelming changes, such a being struck by a comet or meteorite, are always possible.

16.2.1 *Considering the Current Realization of CAES Subsystems*

However, intervention is not a real option. The current realization of the HSS is so far from the fundamentals of the archetype sub-models of the CAES that the likelihood of finding the right intervention or leverage points is next to nil. Recall from Chap. 9 that we used deep systems analysis to understand the economic subsystem (or at least an important piece of it) and discovered several disturbing pathologies with respect to how the nature of the neoliberal capitalist world view, coupled with beliefs in a "free market" and unlimited profit motives are acting more like cancers than healthy sustainable economic activities. As another example, from Chap. 13 we learned that the role of a "currency" in an economy is to control the flow of energy available to do useful work. But, as mentioned in Chap. 9 we saw how the monetary system in the modern market economy has almost completely lost this role, being treated as a commodity (Ferguson 2008; Polanyi 2001) rather than a standard measure of work potential.[2] The result of this loss of true information about the operating health of the underlying work processes and distribution of produced wealth has resulted in uncontrolled oscillations in the economy as a whole.

These examples of dysfunction, or rather the loss of healthy function in the economy, are just one aspect of the complete social system and if we were to do a similar analysis of the governance systems actually in use, we suspect we would discover many other dysfunctional aspects.[3] Indeed, the argument that governance and economics are strongly coupled subsystems, as was done in Chap. 10, leads us to conclude that both systems are deeply flawed with respect to achieving sustainability for the HSS in the long run.

On top of that, we have to take note of the distance from the perfection of the agents[4] that operate in both systems—human beings. Even the best-intentioned and

[2]The Gold Standard adopted by many nations seeking to participate in the international market served as a kind of standard for the creation of money. There were, however, still problems with using a non-intrinsically valued metal—except perhaps for its ornamentation value—to represent the value of work and assets produced. See the two references for analysis of these problems. The standard was abandoned for a time due to these problems and reinstated after WWII until the United States abandoned it as a basis for printing money. Now the different national currencies relative values float on a commodity market and the relation to actual production potential is lost.

[3]This is almost a laughable understatement as of this date (September 2019). National governments all over the world are easily seen to be dysfunctional. Some of this is due to the inherent inconsistencies in their designs (with respect to the governance archetype) so are endemic to the systems. More and more, however, we can see the dysfunctions with respect to those of the agents themselves—human beings—as corruptible and greedy.

[4]See the discussion of perfect decision agents in Chap. 11. Human decision makers, even when rightly motivated to make good decisions for the benefit of the system, are still plagued by a number of faulty factors such as incomplete or noisy information, lack of computing resources, and insufficient time. Then on top of that add psychological factors such as greed, desire for power, and hubris (among others) and we quickly realize that human decision makers at the heads of major corporations and governments are generally found to be far from perfect agents.

smartest, most knowledgeable humans are prone to make mistakes in judgments. Most are still subject to hidden biases and worldview beliefs that cloud their capacity to see the relevant situation and conditions that require veridical decisions. The higher one goes in the governance infrastructure of a large organization or government, the more damage they can do via errors in judgment or dishonesty. We will take a brief tour of the main deviations from the model archetypes developed in Part III.

In Chap. 7, we provided a preliminary top-down analysis of the human social system, starting with the embedding Ecos (whole Earth and particularly the biosphere). Now, using the methods of Chap. 15 we will provide a preliminary design. First, however, we need to understand why the current realization of the HSS as a CAES is far from ideal.

16.2.1.1 From the CAES Archetype: Purpose of the HSS

As noted in Chap. 7 the current realization of the HSS is fundamentally flawed as a CAES because it has no real purpose with respect to serving the good of the larger embedding supra-system, the Ecos. The human economy does not produce a product that serves the needs of other subsystems or the whole. Instead, the HSS as currently instantiated seeks to serve only its constituent members seen as "consumers" without regard to the larger Ecos. It extracts, forcibly, natural resources at a rate that exceeds their replacement, in stocks, or their feasibility in flows. And, the HSS produces by-products that are effectively toxic to other subsystems at rates that exceed the ability of the Ecos to denature them. Plastics, in particular, but also a number of chemicals such as drugs and biocides, are proving to be poisoning not only other life forms but humans as well. In its current form and function, the HSS is a blight on the planet.

The first question our analysis needs to ask is: Can there be, or should there be, a purpose served by humanity that supports the Ecos and its long-term sustainability as a living system? In the sections below, we attempt to answer this question in the affirmative. We will assert that a society of human beings, systemically conceived and instantiated, can fulfill a role for the planet very much *similar to that filled by the brain* for individual beings. But for this to become a reality we need to grasp what the various deviations from the CAES archetype are. We address this overarching flaw in the sections below.

16.2.1.2 From the Economic Archetype

In Chap. 9, we discovered a number of deviations in the form of a systems economic architecture in the way the current neoclassical, neoliberal capitalistic economy fails to account for biophysical realities such as energy and other finite resource limitations on growth and profits. From Chap. 13, we saw that a system-based economic architecture not only takes these issues into account but also explains why institutional views of things like "free markets" cannot be the full basis for a functioning economy for the HSS.

What we do now is ask what an HSS economy (and its governance) would look like if we followed the principles of systems science and utilized the lessons of natural CAS/CAESs in these archetypes. The central question might be posed as: "How can the HSS economic subsystem be architected/designed to meet the requisites of a sustainable HSS?" Recall from Chaps. 9 and 13, and as stated above, that a sustainable CAS/CAES needs to produce an output that is useful to the supra-system—in this case the Ecos—and thus be rewarded by the supra-system providing sustenance to the CAES. The HSS is completely embedded in the Ecos, and until we figure out a way to abandon this planet for a better existence, we had better determine how to best fulfill our role of benefactor of the larger Earth system. The HSS economy needs to be organized so as to use resources at a sustainable rate, recycle its material resources, and not produce toxic by-products in the process. And it must serve a purpose.

The Economic Archetype shows us that our role in the Ecos is to take in "reasonable" amounts of resources needed to sustain our own existence and still produce an output (product and/or service) that benefits the Ecos. It also shows that our waste products must be recycled and/or usable by some subsidiary processes that can ultimately benefit from those wastes (and as a result recycle positive benefits to the Ecos).

We have seen how the economic archetype is a sequence of value-adding processes that use high-potential energy resources and high (or medium) entropic raw materials to produce consumable goods and services within the economic system but also output useable products or services to the supra-system. Previously we noted in Chap. 9 that, currently, the HSS does not really produce anything of benefit to the Ecos. And, indeed, to the contrary, produces unrecyclable wastes and consumes all available resources to the detriment of the biosphere. The systems view suggests that this is unsustainable. How, then, should we design an economic system that is capable of supporting the HSS and be integrated with the Ecos? In the sections below, we will attempt to show how to begin answering this question.

16.2.1.3 From the Governance Archetype

The economic subsystem is strongly coupled with the governance subsystem. The latter is, of course, more than just the regulation of economic activities. It includes judicial oversight on misbehaviors as well as interpretation of and enforcement of agreements (contracts) between transaction partners. But the architecture of the governance subsystem is coupled with the architecture of the economic system so we must look at how this is accomplished.

The objective of this aspect of the governance subsystem is to ensure that the economic subsystem operates cleanly and efficiently. The governance of independent entities within the economy has been adequately covered in Chap. 12. In this chapter, we focus, rather, on the coordination aspects of how multiple economic entities within the HSS can be seen to cooperate or compete without damaging the overall HSS economic subsystem. That some form of coordination, and indeed, some form of strategic management, is needed to make a large-scale economic system work will be unwelcome to some worldviews of economics (and human proclivities) is foreordained. But if we stick to a systems perspective (and keep open minds regarding the significance of the findings), we may come to see that some ideologies are misplaced with respect to the optimal operation of an HSS economy (and its governance).[5]

The problem with the current political process—the process for choosing how we are going to be governed and by what principles—is that it has largely devolved into a highly divisive kind of popularity contest and has been deeply corrupted by the intrusion of market-based thinking (i.e., money as it is currently understood) rather than being a process of reasoned collective decision-making (c.f. Brennan 2016). Thus, from the outset, we point out that while the design of a functional economic system, filling the role of metabolism for a society, is completely feasible following the systems approach, it cannot yield success without simultaneously redesigning the political process to match.

And none of these can be successful without a systems approach to governance. As delineated in Chap. 12, a governance subsystem that fulfills the purpose of regulating a CAES from within is possible and, for purposes of sustainability, necessary. Translated into more common terms, this means that an HSS governance superstructure needs to be designed based on the hierarchical cybernetic governance model—in effect a new kind of constitution needs to be drawn up based on systems principles. Then the political process (subsystem) can be designed to select capable agents to fill the governance positions as described in Chap. 12. This would definitely not involve political "parties" driven by political ideologies, but rather be akin to public debates over substantive issues that would need to be addressed to keep the governance subsystem functioning in light of whatever changed environment the Ecos has become.

The potential design of a governance subsystem has been covered in Chap. 12. A political subsystem designed to select the agents to operate in that subsystem is, unfortunately, more than can be covered in this book. We are left to assert that such a co-design, with the economic subsystem, would be necessary in order that the latter would succeed. Our objective in this chapter is to show how a systems approach to the design of an important subsystem of the HSS might be approached and to offer insights into how such a design might lead to an exo-metabolic process that would keep the HSS being a productive, purposeful, subsystem of the Ecos. We will, for the time being, assume that the co-design of the political process is accomplished apace. In other words, for the moment, we will, for the moment, assume that a process that selects capable decision agents in the governance structure, as it pertains to economic matters, is a given. Later we will consider the more realistic case

[5] Specifically, we refer to liberalism and libertarian sentiments that espouse a minimal role for governance in economic activities and social lives. Put bluntly, these are worldviews that are contrary to a systems perspective.

where less than perfect decision agents are emplaced in positions of responsible decision-making.

16.2.1.4 From the Agent Archetype

The most difficult factor to approach in the design of a well-functioning HSS is the human mind as a decision agent. Classical economics required that human beings are rational decision makers, *Homo economicus*, that is they make economic decisions in their own best interest. But in the new field of behavioral economics, the evidence has been amassed that shows humans to be "victims," as it were, of biologically-evolved biases and using heuristic thinking that is "rational" in the context of making fast and mostly useful decisions in the world of the Late Pleistocene but fails to provide sufficient guidance in modern complex situations (Gilovitch et al. 2002; Kahneman 2011). This is just one of the many theoretical bases that neoclassical economics gets wrong, but it is possibly the most important error. Human cognition is evolved to deal with survival in a world of social dependence and support; as Polanyi (2001) points out, human beings were not individuals competing in a "labor market" and weren't subjected to becoming a commodity until the dawn of the Industrial Revolution. They were members of a collectivist society which worked to produce the subsistence necessaries of life. Social agency and economic governances were one and the same, in a sense. Working and relating in a tribal-like group were all part of the same process of living. And human agency prior to the Agricultural Revolution was adequate to the needs of decisions that needed to be made.

The ideal agent computes a decision based on a model and, possibly, a knowledgebase of both explicit and implicit knowledge (Mobus 2019). Unless an environment is shifting rapidly such decisions would tend to be reasonably good (veridical and sufficient). There are no guarantees of absolute correctness, of course. There are always problems with, for example, communications or sensor noise or data occlusions that reduce the validity of any given decision. But such an agent is not motivated to cheat or deceive other agents. Humans, unfortunately, seem to have this problem. It is likely that a good part of possessing non-pure motives is the fact that in the modern world (and for many centuries now) humans have been subject to inordinate psychological stress that acts to increase the noise level. That most certainly can explain part of the problem. But at the same time, humans, as mentioned at the start of this chapter, seem not to have evolved an adequate cognitive capacity for gaining and using wisdom (see more discussion in the next section). A chief component of the psychological construct of wisdom is strategic thinking, a mode we seemed to be heading toward as a result of group selection operating on the brain's capacity to obtain and use implicit knowledge of how the world worked to make judgments about how to handle complex social problems (wise grandparents and tribal councils). We have asserted that it is the lack of adequate wisdom learning that is at the root of poor decision-making in the modern society (Mobus 2019).

Regardless of the mechanisms involved, humans are notoriously non-humane in making both economic and social decisions, tending toward corruption and abuses of power positions to elevate their own wealth and status. It might go without saying that the current implementations of both economic and governance designs, even if they were closer to the archetype models, would necessarily fail just due to the "human factor" alone. Fortunately, there may be a way to circumvent this conclusion. Humans might be expected to do much better at decision-making under the right conditions. This will be a major design consideration as we proceed.

16.2.1.5 Why the Deviations?

How many thousands of history and psychology books have been devoted to wondering why our present world is the way it is? Many, many authors have wondered why. Many philosophers have questioned the nature of humans and why our evolving civilizations have followed the paths they have; civilizations have come and gone and recent finds in archeology suggest there are underlying similar patterns to the process of growth and decline/collapse (c.f. Diamond 2005; Harari 2011; Homer-Dixon, 2006; Tainter 1988).

In this section, we will provide a synthesis of some of the major ideas that have emerged in recent decades that provide a more systemic view of why our current world civilization is so dysfunctional.

16.2.1.5.1 Human Evolution: Toward Wiser Beings

There are strong reasons to believe that humans were evolving into hyper-social animals some 180–200,000 years ago (Donald 1991; Gazzaniga 2005; Nowak and Coakley 2013; Sober and Wilson 1998; Tomasello 2014, 2016; Wilson 2013) when the modern version of *Homo sapiens* emerged in East Africa. Furthermore, evidence suggests that what made humans sapient (according to Carl Linnaeus[6]) is their capacity for making wise choices. The full nature of sapience is provided in Mobus (2018) but includes conscience (or moral sentiments), tacit knowledge-based judgment, systemic, and strategic thinking. Humans are the first animal to display a capacity for genuine strategic thinking.

This is to say that the evidence for humans becoming better decision agents evolutionarily suggests that they were on an evolutionary trajectory toward becoming wiser beings, likely by virtue of being hyper-social. The subject is far beyond the scope of this section, let alone the chapter, but it is an important part of the pivotal argument. Mobus (2019), Chap. 5, provides a more comprehensive set of arguments

[6] Carl Linnaeus (1707–1778) gave us the Binomial Nomenclature system of naming species. His choice of "sapiens," meaning "wise" might be considered suspect in light of choices humans tend to make.

regarding the evolution of wisdom capacity in early humans prior to the discovery of agriculture.

The nature of human societies and concepts of household and tribal "economics" and "governance" before the advent of the agricultural revolution and the transition to early civilizations was much different than has evolved culturally since (c.f. Scott 2017). However, the argument from social psychology is that human cognition is essentially the same as it was in the Late Pleistocene so that our modern cultural innovations, begun with the revolution and elaborated into the present, especially large-scale societies, are anathema to human mental well-being. That argument is likely cogent with one exception. Mobus (2018, 2019) has argued that the agricultural revolution and the subsequent inclination to settlement life brought about an end to the selective forces favoring increasing strategic thinking. The emphasis, instead, was on logistical and tactical thinking, which had unfortunate consequences for the further coevolution of humans and their cultures.

16.2.1.5.2 Agriculture and the Re-emphasis on Logistic/Tactical Thinking

Humans evolved as groups/bands of hunter-gatherers a lifestyle in which there was an emphasis on strategic thinking to anticipate the locations and timings of game and plants, especially in the face of other groups would be doing the same. The travels of a group depended on considerations of the resources to be gained and in terms of competition for those resources. This was not dissimilar to the strategic moves of a chess game. And it had been the main selective pressure on humans up until they became much more dependent on agricultural produce and animal husbandry. This required a stronger attachment to a specific area and led to a more settled life. The domestication of plants and animals and the living style required thus lessened or even removed the need for strategic thinking.

The agricultural revolution was the first time that humans consciously sought to manage the acquisition of energy and physiological matter (food) as opposed to simply searching for it. Humans, like all animal life, had always been dependent on solar energy but had not previously sought to "regulate" its acquisition and storage (Crosby 2007; Scott 2017). With the advent of horticulture, humans were able to seriously reduce the uncertainty associated with hunting and foraging. But it required a different set of thinking skills oriented toward logistics (preparing fields, planting, cultivating, harvesting, etc. and their seasonal timing) which were managed by a small segment of the population. The majority were engaged in the labor of growing, harvesting, and preparing food. Human numbers increased in settlements and the need to organize and manage labor led to the hierarchical arrangements that we still see in today's governments and organizational management. These needs led to a new way of communicating as well—the systems of writing and numeracy that grew out of simple marks on clay jars (Nissen et al. 1993).

In the rise of agricultural-based societies, humans didn't just domesticate plants and animals. They domesticated themselves (Scott 2017).

This new living style also led to conflicts between settlements and the remnants of nomadic hunter-gatherers, raiders, who found the settled peoples easy targets. The need for tactical thinking developed, the heads of the hierarchies had to not only manage agriculture they had to take responsibility for protecting what they had.

So, starting about 10,000 years ago, humans became socially organized into pyramidal strata. Workers at the bottom of the hierarchy needed only operational thinking skills, managers in the middle needed mostly logistic thinking skills, soldiers and their leaders needed some tactical thinking skills, and chiefs on the top needed occasional strategic thinking skills. Note that the majority of the population has been at the bottom of this structure ever since, and by the logic of evolution and selection for the specific thinking skills needed at these levels, humans have, on average, been tending away from the average capacity to think strategically and toward the more immediate thinking needed for solving little and local problems of operations (Mobus 2019).

While this argument may carry a fair amount of weight in terms of explaining a lot about current human psychology (specifically, a broad lack of wisdom) 10,000 years is a short time in evolutionary terms. It is not yet the case that humans lower in the social hierarchy have lost all capacity for, say, tactical and even some strategic thinking. But it is likely that these capacities are greatly diminished from where we were as a species in the Late Pleistocene.[7]

The irony is that in a complex rapidly changing environment the need to make logistical, tactical, and strategic decisions is greater. The current global civilization has come about a few hundred years after the discovery of the use of fossil fuel (coal) and the heat engine coupled to machines—the Industrial Revolution. This is a time scale far too short to allow selection for these thinking skills to re-emerge. Hence, we find human decision-making generally inadequate for the needs of this environment. The general result is that most people when faced with a complex situation needing some kind of decision, resort to guessing (gut reactions) and faking competence.

16.2.1.5.3 Technology, Convenience, and Power

The use of technology to perform work and solve problems has always been a two-edged sword. Starting with the use of fire to spook game into snares, to cook food for extra calories, and to shape the landscape in favor of grasses, humans are unique in their ability to control external energy to do work (c.f. Harari 2011, pp. 13–14) perhaps 200,000 years ago. With the development of stone implements (even

[7] It should be pointed out that the same selection forces, or lack thereof, were not the case for peoples still engaged in hunter-gatherer lifestyles. The numbers of such people far exceeded those in agricultural-based societies for much of the history of the species. However, in terms of differential reproduction rates, domesticated humans outpaced the non-domesticated. And over the last several centuries, especially due to colonialism, domestication has spread to nearly every corner of the globe.

earlier) humans gained the advantage of leverage (e.g., skinning the hide from a prey animal and butchering it). Recall that in Chap. 9, Sect. 9.5.2.1, we explained this in the context of the physics of economy. Tools make life easier, increase the efficiency of operations, especially the gain of calories, and give human users increased power to dominate.

Tools, in all their various forms, are part of the fabric of culture, which includes collective beliefs and various institutions. All of these latter are, in some sense, also tools in that they help increase the effectiveness and efficiency of human activities. Whether those activities are ultimately beneficial to the whole of society may be questioned, of course. The old saw, "The right tool for the *right* job," might be invoked. A systems analysis of what the job is, meaning looking at the products the job produces, would include determining the worthiness of that particular activity.

It isn't too outrageous to claim that humans are spoiled and have become more so over several centuries. Over the course of the Industrial Revolution, in particular, the culture has progressively moved from mostly utility considerations, such as having the willingness to "darn socks" to keep them in service, to the desire for convenience; now we throw the socks away when they get threadbare. This is partly from the fact that consumed goods have, relatively speaking, become cheaper as the capacity to manufacture and the availability of cheap energy has increased through technological innovations. But it is also partly from the attitudes of people who we now call "consumers." People are expected to consume, use up, and replace rather than conserve and repair. The modern manufacturing process assumes continuous consumption, designs products to effectively not be repairable, and all so that sales don't sag.

The continuous throughput of material can only be sustained by the continuous, and often wasteful, use of substantial power flows. Moreover, in a growing population and increasing individual consumption habits, the amount of power consumed has to increase.

Technology is always a two-edged sword. It may provide convenience to individual consumers but at the cost of externalities that end up impacting society as a whole.

16.2.1.6 The Modern Conundrum in the Social-Political-Economy

There is a strongly held belief in Western nations that a free-market economy is self-regulating through the price adjustment process under the forces of supply and demand. Polanyi (2001) described the history of the emergence of a market-based economic model at the time and in conjunction with the emergence of a higher-powered industrial economy in England in the late eighteenth and early nineteenth centuries. Adam Smith (1723–1790) famously observed in "An Inquiry into the Nature and Causes of the Wealth of Nations" (1776) that merchants, craftsmen, and other participants in economic exchanges performed their works and traded cooperatively and that the whole affair was guided not by government intervention or control, but as if by "an invisible hand." Each participant, he concluded, in following his own self-interests (in making a living), pursued a course of action that tended to optimize the production of wealth for the whole society. This has been interpreted largely as an economic agent, following his own selfish desires, somehow makes life better for all. This is the basis for the notion of maximizing profits. It is the argument advanced to justify markets, their failures from time to time, and the accumulation of excessive wealth by capitalists (Piketty 2014).

At least a part of the reason that people are inclined to believe in the self-regulating market as the best way to conduct economic business is that history is replete with examples of human agents in governing positions, with respect to economies, making tremendously unwise decisions, either from ignorance, stupidity, or malevolence (e.g., greed and corruption). If humanity cannot trust itself to govern an economy, the network of extractors, producers, and consumers, because its agents, no matter how being in a position of regulation, are prone to corruption, then the idea that a market can regulate itself for, if not optimal, then reasonable results is quite easy to adopt. The problem is that the complexities of HSS economies, along with the dysfunctions noted above, make the price mechanism an unreliable method of comparing values.

Put simply, humans can't trust money, prices, most institutions, and especially other humans in the modern conundrum.

16.2.1.7 A Future?

If the current situation of the HSS is untenable then what is to be done? There is a very good argument for our global civilization to collapse as a result of the many dysfunctions in economies and governance (Homer-Dixon 2006; Tainter 1988).[8] What if we had a clean slate on which to design a functional HSS on the basis of the architecture of a CAES? Could we then design a workable, sustainable HSS? In principle, yes.

Here is the strategy.

All of our sciences, particularly the social sciences, medical sciences, psychology, anthropology, etc., have already acquired considerable knowledge about the human condition but it exists in disjointed structures, books, journals, and education curricula. Employing the method of deep analysis (Chap. 6) and organizing the information already made available in all of the sciences into a global knowledgebase (Chap. 8), we would be able to discover the holes in our knowledgebase. Much as nineteenth and twentieth century chemists used the periodic table to "predict" the existence of yet undiscovered elements, we could identify the areas of knowledge of humanity that needed exploring through further analysis. In many ways, this process is already underway in some key areas. For example, neuroscience and psychology are becoming more integrated by the discovery of correlates between neural activities in the brain with cognitive and affective mental states. This became

[8] This is especially relevant given the threats of climate chaos and energy depletion.

possible with the advent of a structural framework (macro- and micro-anatomy of the brain) provided by network analysis (Seung 2013; Sporns 2016) that gave researchers a way to organize knowledge about neuronal and neural network functions. This process gives us confidence that with the right structural framework all of the human knowledge, to date, can be organized and understood. This is what can be accomplished with the application of systems science and the language of systems.

16.2.2 An Architectural Design for the HSS

If we were starting from scratch, what kind of social system would we design? Imagine that humans, toward the end of the Pleistocene, had further evolved their capacity to acquire and use veridical wisdom, that when they discovered the methods of agriculture, they did not unquestioningly simply expand their numbers and start massive acquisition of natural resources. Imagine further that humans reflected on what they were doing and noticing how their activities impacted their environments. Of course, this isn't what happened, obviously. But if it had, what kind of society would befit human beings and benefit the Ecos as a subsystem?

In Chap. 7, we spent no small effort and space on the description of the analysis of the brain as a representative CAES. That effort was in anticipation of this chapter. Consider some interesting isomorphic "resemblances" between neurons and human beings, and between networks of neurons and networks of human relations, and between the information processing capabilities of the brain and that produced by the larger network of human interactions. The brain might be described as a network of functional modules, each a network of networks (e.g., the cortical columns), responsible for computing relevant inputs in coordination such that the organism is aware of its environment and produces responses (and plans) that support its existence tactically. Neurons are the main computational and memory trace encoding components that participate in a more global computational process resulting from the ways in which the modules and sub-networks are interconnected. Human individuals in society also participate in global computations, though, obviously in far more complex ways.

Here we propose an approach to a model-based design of a human social system architecture that could, in principle, be fit in the context of the whole Ecos and provide a valuable service to it as one of its subsystems. The archetype model is, of course, the CAES described in Chap. 10. More specifically, however, we propose that the architecture of the brain provides the scaffolding for elaborating a design for humanity. We will describe that scaffold below as a basic model framework and then apply some of our system design approaches outlined in the last chapter to develop more specific suggestions as to how to proceed.

16.2.2.1 Society

It is probably a good idea to start with what we know about the nature of human social units in the late Pleistocene since this was considered a natural state that had been a part of human evolution. Specifically, we know that humans formed small groups, extended family, and non-related members, numbering up to around 150 to 200 individuals of all ages. In fact, much of human evolution of the language and planning facilities are owed to a process of group selection—cooperation within a group and (sometimes) competition between near neighbor groups (Bourke 2011; Buller 2005; Calcott and Sterelny 2011; Coen 2012; Deacon 1997; Donald 1991; Geary 2005; Harari 2011; Mithen 1996; Sober and Wilson 1998; Tomasello 2014; Westneat and Fox 2010; Wilson and Sober 1994; Wilson and Wilson 2008; Wilson 2013). Group selection operating on a population in the context of the Late Pleistocene was putting favor on groups that had members better capable of strategic and systemic thinking (Mobus 2019).

It was only after the advent of larger-scale agriculture, particularly when the need for irrigation arose, that groups began to coalesce and expand in numbers (Scott 2017).[9] Settlements turned into large villages and some of these turned into towns. Civilizations arose as populations increased and agriculture demanded societies to become organized around the labor and management of it and land. Within the span of only a few thousand years, the human social system went from small group dynamics to large metropolitan interactions. Humans are adaptable beings, but adaptation does not mean peaceful accommodation. Population densities convey stresses that may be adjusted for, but not relieved. The story of a domesticated and crowded man is not one of simple adaptation. It is one of accumulating mental stress and the emergence of bad behavior.

Suppose that we reverted to a group size and organization that resembled the Pleistocene social module. Suppose that we organized societies to minimize population density stress. But also suppose that we carried forward our hard-won knowledge of how the world works, how we work. We might be organized in tribal bands, but not ignorant of why this form of organization is beneficial to our psyches. We will pursue this idea below when we get to the actual design of a social system. For the moment we note that that design is based on a hierarchy of social modules, with the basic module, a domicile, and an aggregate of a small number of such domiciles creating a community module. Higher-level modules involve the hierarchical governance architecture.

[9] The pattern of forming settlements organized around the growing of grains or similar staples may have started in the Middle East (the Fertile Crescent) around 10,000 years ago but was repeated in other regions of the world independently. We infer that the form of the new kind of social organization that developed and was fundamentally the same in these regions was a natural consequence of the needs to organize society around the agricultural mode of life.

16.2.2.2 Economy

The basis for an economy of society is the acquisition of necessary resources, food, shelter, clothing, etc. The economy is the means of producing two basic forms of wealth. The first form is the wealth that directly supports human life, of course. The second form is the wealth that is exported to the Ecos; the economy must produce a product that makes it worthy of being a subsystem of the Ecos. We need to use the economic archetype model to design an economy for human society that fulfills this requirement.

Considerations for such an economy include the necessity of not overexploiting resources and not overproducing wastes that cannot be processed by the Ecos so as to not become toxic. Ideally, all outputs from economic activity would be beneficial to some part of the Ecos, just as biological wastes, feces, and urine, for example, are resources to decomposers.

Each social module will have an economy designed on the basis of its level in the hierarchy. The domicile module will be the basis of work processes with an emphasis on operational management. The community module will act as the primary level at which a CAES economy will operate. Higher modules will be supported, economically, by the community modules and their purpose is primarily coordination management.

16.2.2.3 Governance

Below we will describe some details regarding the implementation of the HCGS architecture explained in Chap. 12. The governance of a human social module requires some specific elements beyond the plain vanilla version of the governance archetype described in that chapter.

Here we note only that the governance of this new concept of a social module is deeply intertwined with the management of the economic activity and includes a reflexive governance of the decision agents—people—who make decisions outside of the economic subsystem sphere that, nevertheless, have impacts on every aspect of the society.

16.2.2.4 Technology

What makes human cultures significantly different from other animals is the presence of tools and other objects that are the result of technological development and their use is the economy of resource acquisition, production, and consumption. In the kind of HSS to be described below, we want to emphasize that the kinds of technologies that will be applicable will be limited in terms of their energy consumption (i.e., electric motors) and employed only in higher priority work processes. As will be explained below, technologies such as computing and communications may serve an important role in managing the whole HSS and it fulfilling its CAES purpose vis-à-vis the Ecos.

Otherwise, we envision that most technology will be focused on the improvement of manually operated tools that will be used to do most of the work in the economy.

16.3 Designing a Human Social System as a Complex Adaptive and Evolvable System

In Chap. 7, we began the top-level analysis of the HSS (recall Fig. 7.17). Then in Chap. 9, we further analyzed the existing economic subsystem, at least insofar as the energy acquisition and conversion to free energy was concerned. That analysis revealed some serious deficiencies in the current fossil fuel-based industrial economy. In both of these chapters, we mentioned problems with current forms of governance, especially of the economy and especially with respect to humans as agents.

Then in Chaps. 10, 11, 12 and 13, we introduced the archetype CAES model and its primary sub-models, agents, economies, and governance. These, we claimed, are found in naturally evolved complex systems such as living systems from cells to ecosystems. We also made the claim that society is a CAES information through a combination of auto- and intentional-organization and that the problems that the HSS has experienced were the results of insufficient knowledge had among the human designers of historical experiments in social systems. Put simply, the search for an ideal society has been based on best (not always good) guesses about how a society might achieve a stable, resilient, and persistent existence. That search started in ignorance of systems principles and led down many a blind (and destructive) alley.

Now, knowing considerably more about all of the various components of human nature and the sciences that gave us our modern technologically based culture, might it not be possible to suggest the way forward in designing a true CAES-based human social system? In this section, we seek to start that process. It should be obvious that one author, in one book, could not hope to get very far into the analysis/design of a total system. All we can offer here is a glimpse of what that process might entail. Even so, we think the glimpse will be instructive. At some time in the future, when there is a body of systems designers who have taken the procedures and visions of this volume to heart, perhaps a full-blown deep systems analysis of the needs of a properly functioning HSS can be undertaken. In the meantime, here are several of the aspects of such a system we have considered.

16.3.1 Designing for Faulty Agents

Humans make fundamentally three kinds of decisions. One kind of decision involves interpersonal relationships and is largely driven by emotional/affective mental processes. These include the observation of social norms and commonly held beliefs, such as religious convictions. They are the evolved responses to living in social groups. We will not dwell on how these decisions are made or their consequences except to note that these kinds of decisions are driven by subconscious, affective factors that are subject to all kinds of mistakes (see Mobus 2012, 2019, for a deeper analysis of affective decision-making).

Governance decisions are related to regulating social and economic processes. The former concerns laws, mores, etc., and their enforcement, acting as the explicit, conscious mechanisms overriding the affective-driven decisions just described above. Social governance attempts to reduce affective mistakes as well as extend the beneficial forms of interpersonal and social behaviors. The latter concerns operations of and coordination between work processes.

Economic decisions are related to obtaining resources and using them wisely, or should be. That use includes determining what products need to be produced to support human life and wellbeing. And decisions need to be made regarding how the produced wealth is to be distributed equitably in the population. As mentioned in Chap. 9, currently there is a widespread belief that something called the "market economy" somehow magically solves all of these problems effectively and efficiently. This belief persists in spite of both theoretical (Polanyi 2001) and empirical evidence (c.f. Keen 2011, for explication) that this is far from true. Yet economic actors make decisions and behave as if it is true, and that is a fundamental problem.

As has been mentioned, humans as agents have some serious faults. When the nature of the overall system is relatively simple, small in scale, not involving too much technology, and does not involve the abstraction of "money" as wealth, then humans can be reasonably effective decision makers, as when they lived tribal lives in the Late Pleistocene. But, when the situation is as it is now, size scale beyond anyone's comprehension, technological and relationship complexities also not grasped by the average human, the capacity for veridical decision-making has become seriously compromised.

Thus, the design of any form of HSS must include serious simplification of the social milieu in order to greatly reduce the psychological stresses that lead to selfish, self-centered, defensive behaviors. This implies returning to a social structure based on small tribe-like units. We will further discuss this idea below in Sect. 16.3.3, Social Organization. For now, we will simply note that human beings living and working in smaller groups do perform much better. Operating in much less complex social situations, they are far more prone to make decisions for the benefit of the group as opposed to for their own gain. Complex technological environments also press people into more anti-social behaviors. Observe the rash of online bullying among students on so-called social media.

While humans may behave better, and make more rational decisions in a simpler and smaller group environment, they are still capable of deviations from time to time. They are still able to make mistakes for a variety of reasons not just caused by stressful conditions. For a very long time now we have understood a basic concept of third-party auditing as a means to catch mistakes (or fraud) and report these to "higher management" or external enforcement to activate correction procedures. The audit function can be found in all-natural CAESs. For example, in the maintenance of the genetic code in chromosomes, there are proteins tasked with monitoring the "health" of the strands of DNA. If an anomaly is detected, such as a point mutation in a somatic cell's gene, an error correction mechanism is activated that repairs the site.

The design of an HSS governance system should include provisions for an audit function for all activities. For example, in our current system, we need a policing function and a justice system to enforce laws relating to interpersonal interactions. Humans had evolved cognitive means for detecting dishonesty (cheating) in members of a tribe and retain these abilities today. The policing function we find necessary today was fulfilled by that system of auditing behaviors and public rebuking or expulsion from the group for guilty parties. The design of the human social unit should allow the re-emergence of this kind of audit function in interpersonal relations.

The problem of faulty agents in government has always been a problem in civilizations. Societies, especially in response to growing larger and more complex, evolved various forms of distributed governance. Even autocratic government found it necessary to offload various decisions to "underlings," particularly those involving more operational and logistical problems. Eventually, the problem with faulty agents in positions of authority provided a selective pressure favoring what might be seen as the ultimate form of distributed governance that depended not on one agent in charge, but a collective of agents, a forum, or a legislative body. Democracy is presumed to be a method for countering the effects of agents in power positions from abuses of that power. But it is found that democracies have their own problems when the right to vote is restricted to a non-representative set of citizens, or when the citizenry is largely ignorant of the governance issues and policies (Brennan 2016). The more complex a society becomes the more the notion of egalitarian (liberal) participation in governance falls down. Yet the notion that governance of a social unit, a module of "reasonable" size and complexity, should involve a plenum of members seems worth pursuing. In the design of life-critical control systems engineers often use redundancy to guard against failure of the system. For example, a troika of automatic (embedded) controllers use the same or similar sensor data to make a control decision. The rule used is that if two out of the three come to the same decision that is the one taken even if the third controller comes to a different decision. Critical decision points are handled by a council rather than a single all-powerful controller. It is believed this was the principle at work in the evolution of tribal councils. Below we consider this principle in the design of a social governance system.

Ultimately, given the psychological nature of modern human beings, there is probably little that can be done to prevent corruptibility. However, we suspect that actual corruption is propelled by cultural drives rather than a natural tendency for humans to be selfish and self-centered. It is our modern social milieu that promotes corruption rather than an overarching propensity to cheat and steal and lie (everyone else is doing it so why shouldn't I). Our thought is that given the design of a social system as described herein, where the systems at the modular level are kept simple and scaled to human capacities, the more dominant propensity of cooperation and eusociality will prevail. And while cheaters may always emerge, that they can be dealt with in the traditional fashions of shaming and ostracizing the culprit.

16.3.2 Designing a Sustainable Economy

The job of an economy is the production or acquisition of energy and material resources, the production of useful wealth, and the equitable distribution of that wealth. Just as importantly, however, the economy must export products and byproducts to the environment. Internally, recycling as much as possible should reduce the output of waste products. All other work processes are ancillary. For example, the construction of farming equipment, requiring the extraction of metals, working them into shape, etc., is driven by the needs of food production.

However, as noted earlier, a CAES embedded within the matrix of a larger supra-system must produce a product that will benefit that larger system. That is what makes the system fit and ensures that the larger system will provide sustaining support (resources) to the system. Every CAES has a purpose. As previously noted also, the HSS currently does not produce any useful product for the planet. But that need not be the case.

We will, briefly, analyze the HSS as a sustainable, resilient, and persistent CAES.

16.3.2.1 Boundary Analysis

As recommended in Chap. 6, we will start with the examination of the boundary and the flows of energy, materials, and messages into and out of the system, and the environment, the sources, sinks, and milieu. Initially, this involves the delineation of the boundary conditions, describing the human occupancy of the land and the interfaces between the HSS and the rest of the Ecos. These will include input/output interfaces. In the society of today, these would include interfaces such as farms to capture solar energy and convert it into calories for human consumption and fossil fuel-burning processes (such as internal combustion engines) for output of carbon dioxide into the atmosphere. But what we seek in our utopian society are input/output flows that are in balance with the natural capacities of the Ecos to supply inputs and process outputs. As we will describe below, the HSS needs sufficient energy to support human life and activities. But the amount of power consumed cannot exceed the thermal balance with the Ecos. Essentially this means real-time solar input flows.

The boundary of the HSS is fuzzy and highly porous. For example, many of the potentially toxic chemicals we put into the environment are now being found in food and water supplies around the world; they are ending up in our bodies without any control. A seemingly clear boundary condition is that humans are a single species, so have essentially the same biological needs. But humans reside in domiciles (families) that include non-human components vis-à-vis vegetation, domesticated animals, and myriad artifacts. Each domicile will occupy some amount of space and with persistence for an extended time. The collection of domiciles we will call a "community" (to be detailed below) plus fields and connecting pathways can be similarly considered as the space that the HSS occupies and thus in the HSS. However, the physical boundary will vary in composition and extent depending on local conditions. In our consideration of what is in the HSS we shall take the strength of connections between human beings and the "things" with which they interact on a more or less daily basis. We consider what is outside as the rest of the Ecos and natural resources that may be extracted periodically or on an as-needed basis.

As demonstrated in Chap. 6, our boundary analysis should look first at the outputs of the HSS and then the inputs. This will put us in the position of tracing the internal flows and processes that need to be in place to produce the outputs from the inputs.

16.3.2.1.1 Outputs of the HSS

Let us start with the output of a "product," that is, we ask what is the purpose of the HSS with respect to the Ecos and what product output would serve that purpose? What should that be and what other subsystems of the Ecos benefit? The brain-architecture mentioned above gives us something to consider. Pushing the analogy further, if the HSS is effectively operating as a brain for the planet, then the main output might well be expected to be the management of the planet (c.f. Grinspoon 2016 for a perspective on this idea) in the same way the brain manages the body. There will, no doubt, be an immediate negative reaction to this notion. The argument will be that humans have failed to be decent stewards of the planet thus far, why should they now try to do so? The answer is that if the HSS is designed to be a sustainable CAES, employing a planet-wide HCGS subsystem (which is the brain-like aspect), then humans will have the social structure needed to be better decision agents (as per the section above regarding designing for faulty agents). Furthermore, implicit in this line of argument, we seek to design the HSS so that it benefits the planet.

Consider an example. The current HSS has the science and sensing capacity to monitor large parts of the Solar System, in particular the asteroids and comets that inhabit it. We have the capacity, presently, to compute the trajectories of many bodies that might be a threat to the Earth. And, though probably nascent in effectiveness we have some means for intercepting and redirecting any bodies that might threaten to collide with the planet and reek the same kind of havoc as with the End Cretaceous Event that did in the dinosaurs. This would constitute the tactical management of the planet's interactions with other bodies in the Solar System.

As an example of logistical management, we have gained extensive knowledge of ecosystems dynamics and mechanics. We actually know what would need to be done, for example, to restore populations of plants and animals that have been decimated by a human overshoot. However, in the long run, the planet had done a reasonable job of managing the logistics. The Gaia hypothesis holds that the biosphere has performed as both operational and logistics level regulators for maintaining the conditions of the other "spheres" to support life (Lovelock and Margulis 1974). It may be the case that once humans have helped restore balanced ecosystems that there would not be much needed in the way of management of the non-human systems.

16.3.2.1.2 Extraction of Resources: Inputs to the HSS

The HSS, as any real system, requires inputs from the surrounding environment such as energy, minerals, and forest products, among others. Farming provides a major source of energy flow in the form of food. Mining operations extract mineral resources.

We should assume that the extraction of fossil fuels will have come to an end, due as much to the low energy return on energy invested as to the need to not add any more carbon to the atmosphere. It is uncertain, at present, that photovoltaic or wind energy can capture any significant power, in all likelihood not enough to run our modern technological society. Dams and nuclear power plants actually require significant power to maintain. So, we are faced with a social system that must learn to live off of the real-time flow of solar energy primarily through a greatly reduced energy-consuming agriculture.

The majority of the work to be done in these various extraction activities will be through manual labor; indeed, the vast majority of work processes will involve manual labor (or animal labor).

This fact need not mean that there could not be any technology. It isn't that a future society will need to revert completely to, say, a Middle Age feudal form. There are several different kinds of modern technology that would be useful to maintain and thus provide for some extra energy inputs (i.e., electricity). The communication capabilities of an Internet-like network would be near essential to maintain coordination between far-flung modules. In the arena of agriculture, it would be useful to maintain some kinds of metal working (from recycled metals) for producing efficient hand tools and plows, etc. The vision of the early nineteenth century blacksmith comes to mind.

It isn't yet clear that solar photovoltaic energy extraction and conversion to electricity can effectively be self-reproducing (recall Sect. 9.6.2.1). The same is even more true for wind turbines. However, in a society that does not demand as much electric power as our current western ones do (indeed, hardly any at all), and if solar panels in the future can be simplified and built primarily from recycled materials, it is conceivable that some form of direct solar energy might be viable. This is made more so if the production processes can be scheduled in coordination with the availability of sunlight, reducing the need for extensive storage.[10] As long as the electric power is reserved for powering the important technologies (e.g., communications) and self-regeneration, then it is conceivable that this form of real-time energy capture could be practical.

16.3.2.2 Production of Necessaries and Niceties

The model of the blacksmith may be thought of as a kind of archetype for the work processes that produce goods in a low-energy society; manual labor, skilled in various trades. The emphasis has to be on the notion of "necessaries." One can imagine a rank ordering in priorities from the preservation of foodstuffs without refrigeration or chemicals, to clothing, to housing, with supporting tool manufacture accompanying. What needs will be fulfilled will depend entirely on exosomatic energy production and the availability of free energy of sufficient power. As indicated above, for example, electricity production should be reserved for powering important (to society) technologies. Otherwise, the production of, say, clothing can be done completely manually using early twentieth century treadle sewing machines similar to the one the author learned to sew on. Similarly, the production of fibers can revert to spindles, weaving on looms, knitting, etc.

A key to understanding the ability to revert production activities to manual labor with late nineteenth century tools (with any improvements like saw blades sharpened to late twentieth century standards) is that with a much smaller population not engaged in population growth, a much simpler lifestyle, and an economic system running on an as-needed basis, there is no need for continuous production at volume and buffering excesses (inventories). One model for this kind of economy is that of the Amish (and related sects) colonies in the mid-eastern USA. These German settler-derived societies have lived successful agrarian lives, eschewing modern technology. So, we know that a much simpler lifestyle is not only possible, it is also likely to contribute to greater mental health.

The kind of economy being suggested does not mean "subsistence," or constant toil just to stay alive. Using our best knowledge of organic agriculture and production practices, and assuming social modules are situated in areas not hyper-adversely impacted by climate change-related calamities (e.g., the coasts), a reasonable proportion of living may be devoted to aesthetic pursuits—to niceties that enhance

[10]The efficiency of both solar panels and batteries is dependent on some rare or expensive materials and more complex designs. Real costs for these materials and manufacture are currently hidden in externalized costs rather than reflected in the price of solar energy, which gives the illusion that solar is more economic than it is in reality. Simpler, less efficient designs might mitigate these real costs but would obviously cause the cost per kilowatt hour to increase. There is a balance point between efficiency and sufficiency that needs to be investigated.

mental health and experiences of joyfulness. The objective of the kind of social system being described is to provide a modular environment in which individuals may achieve the higher levels of Abraham Maslow's hierarchy of needs, esteem, and self-actualization.[11] Early humans decorated cave walls with paintings, tools with adornments, and even created sculptures in clay. There is no reason to believe that this creative urging cannot be continued if the kind of social and economic milieu is much greatly simplified in the fashion being suggested here.

16.3.2.3 Recycling of Waste Products

Every CAES will produce outputs and the HSS is no exception. Ideally, those outputs are taken up as inputs to other entities in the supra-system or degrade to harmless materials. The rate of output flows must not exceed the rate of uptake or degradation; otherwise, the intermediate products will accumulate in the environment and potentially poison some other processes.

To minimize the potential of accumulation, it is imperative that the HSS make every effort to recycle unusable but low entropy materials, such as scrap metals, glass, and building materials as well as high entropy organic wastes. In principle, the recycling of materials reduces the need for importing/extracting raw materials or the effort required in exporting wastes so the energy that might have been expended doing the latter input/output processes is better applied to doing the former.

16.3.2.4 Proper Role of a Market

We have established in Chaps. 9 and 13 that markets as generally understood in modern economics are only able to operate over short distances, when full costs of production are visible to all parties, and where speculative behaviors are disallowed. A CAES market is one where energy, materials, and messages can flow between near neighbors in a stable "price" setting. Below we discuss the proper role of money as a signaling device, but here we point out that the use of money in a properly organized market is simply a convenient means of communication between buyers and sellers.

In fact, nothing like our current concept of markets (e.g., financial, labor, real estate) can exist in a CAES-based HSS. The only markets that make sense are the kind that operates today in so many parts of the world; farmers' markets where food, clothing, furnishings, physical assets, and commodities can be exchanged. Money can provide a convenient mechanism as it does now, but the exchange value of transactions would be equal for both parties. A community market, for example, could

[11] Maslow's famous model of positive psychology has been criticized on several counts, primarily as being overly simplistic. However, it still represents an understanding that mental health includes things like self-esteem and the esteem of others in the group as well as a personal sense of being embedded in the world.

provide a mechanism for exchanges between domiciles as well as a social "event," say once a week. It would be an occasion not just for the exchange of goods, but of information and strengthening of friendships.

In a low-energy economy distance trading of various products would likely be limited to essentials that cannot be obtained locally within a module. This would operate at most scales, though with lesser volume as we go up the hierarchy due to the greater distances that need to be traveled.

The scale of any market is kept small by the modular and hierarchical design of the HSS. Transactions would mostly be point-to-point, that is buyer and seller, directly with an emphasis on eliminating so-called middle-persons (no wholesale or retail as we now know it). These mechanisms are generally seen as supporting a mass marketing, growing economy, which the CAES-based HSS would definitely not be.

16.3.2.5 Proper Role of Money

We have already made comments on the role of money in an economy, both in Chap. 9 (with a nod to the history of money) and in Chap. 13 regarding the economy archetype model. In the latter, we argued that economic agents make decisions about acquisitions and production and signal those decisions, to suppliers or consumers, respectively.

At the level of metabolism there are various signaling mechanisms depending on the kind of work that needs to be done (say protein production at the ribosome), but a basic currency is the flow of ATP to the work process to supply energy and ADP back to the energy supply facility, the mitochondria. This constitutes both an energy flow and the signaling process itself in terms of requirements for additional energy flow. What causes the ribosome, as a work process and consumer of energy, to do its work in the first place is the receiving of messenger RNA molecules (mRNA) that activate the production of specific proteins. The DNA gene that codes for that particular protein, was, itself, activated to produce copies of mRNA by other signals from the cytoplasm generated because of a sensed shortfall in the concentration of that particular protein. In other words, there is a cascade of messages from the initial recognition of a deficit (or need) to the production center (ribosome) and then on to the energy supply center (mitochondria) that in the end controls the flow of energy and materials into work processes.

In bodily physiology, there is a separation between signaling and energy flow (as well as material flow control). Various low-weight molecules, like hormones, are released and responded to in order to control the flow of energy molecules (e.g., glucose). The nervous system helps to coordinate the endocrine system signaling.

In the human economy of the modern world, the role of a signaling mechanism is taken on by money. At one time money was based on commodity value (e.g., the gold standard). But today that approach has been abandoned in favor of treating money as just another commodity (see Polanyi 2001 for the historical record of this abandonment). A commodity-backed monetary system had the virtue of basing the amount of money, either fiat or bank-based, on the volume of value in the commodity. This helped to stabilize economic activity to some degree and facilitate market transactions. But the problem was the choices of commodities. When that commodity backing was based on the amount of grain (e.g., in barter systems some amount of grain could be worth the value of a cow) then there was a direct tie to the amount of energy (and matter) needed to support various work processes (basically, food for human workers and draft animals). Later, after the introduction of precious metal coinage, the notion that the metal somehow represented, similarly, an amount of energy became very fuzzy. The metal started to take on a value of its own.

Today, with the advent of money-as-a-commodity with a floating rate of exchange, and with the advent of financialization of the economy (the equivalent of a casino) money has pretty much lost its original purpose as a mechanism for signaling to control the flows of energy and matter, and the work that should be accomplished. We will examine in greater detail what role money can serve in a human economy based on the archetype model.

16.3.2.5.1 Energy Standard Basis for Monetary Instruments

The production of true wealth, food, tools, and other productive assets, takes free energy. That is, the energy that is of the right potential and kind (e.g., electricity) and flows at a sufficient rate (power) to drive the work process that transforms the inputs into outputs. The extraction and movement of raw materials, and the movement of products, that is, all material flows, require the expenditure of free energy. Therefore, there is an energy equivalency of material available to an economy. In other words, all aspects of economic activity require flows of free energy.

We use this principle to suggest that the fundamental basis of economic activity is free energy itself. This corresponds with the original role of money in human economic activities. When a buyer pays for a product, they are signaling that energy should be applied to the production and transportation of that product. In this sense money is a form of information about the amount of energy that should be used for different economic activities. In effect, this is another form of commodity money (e.g., the free energy content of a barrel of oil). Rather than base monetary denomination value on an abstract notion of the worth of an ounce of gold, the monetary unit (e.g., a dollar) would be based on an amount of free energy measured in a common unit of energy such as the joule. This is justified on the basis that there is a direct link between the production of wealth and the availability of energy to do useful work.

The translation in monetary policy is straightforward. The "amount" of money in units is exactly corresponding to the amount of free energy available to an economy. There can only be as many units of money (dollars say) as there are units of free energy backing that money. We can determine the amount of free energy available (from the raw energy resources in hand) from thermodynamics and engineering knowledge of the power requirements of work processes. The accounting systems we employ now would become meaningful in measuring true costs and revenues.

This has significant consequences for economies. In aggregate, the total production of wealth is tied to the availability of free energy. If, for some reason, the input flow of free energy declines, then the economy is, by definition, in a contraction; much less wealth can be produced. Contrariwise, if the flow of free energy increases, say due to increased efficiency in the extraction of energy from a flow-limited source, or just increased efficiency in the work processes themselves, then the production of wealth may expand accordingly.

Money, in this sense, is used to regulate the use of energy and the production of wealth for the HSS. The role of the governing process is to monitor the availability of free energy and regulate the money supply accordingly. Elsewhere, we consider the sorts of transactions that use token money (dollars) to affect those transactions. But the overall pattern is not dissimilar to what happens now (review Fig. 9.14, the messages used to regulate flows). Two parties enter into an agreement to trade value (a product for a sum of money), but the price is determined by the energy control transfer. The buyer agrees to pay an amount of money equivalent to the emergy value[12] of the product, plus, possibly, a small increment representing use or esthetic value premium (that which is known presently as "profit"). Below we will discuss these values and their relations to energy flows.

16.3.2.5.2 Measure of Values

A key part of economic theory (including neoclassical economics) is the notion of the value of an asset. In Chap. 13, we noted the idea that a "tool" has an inherent economic value because it allows the user to increase the availability of free energy to a work process by virtue of reducing the total amount of raw energy needed to accomplish a given task. The tool, therefore, has a value that exceeds its emergy value (the amount of energy that was required to form the tool). Here we will explore several aspects of the basis of value to "users" that contribute to variations on what "price" a buyer might be willing to pay for an asset. This is important since a price system needs to incorporate the total free energy flow in the whole economy. The first and second Laws of Thermodynamics demand it.

16.3.2.5.2.1 Base Cost to Produce

As mentioned above, emergy is a measure of the amount of energy that was consumed in the production of an asset. If the monetary system of an economy is based on energy, then this represents a base cost of production. It is a starting place for assessing the value of an asset product. The producer must add up all relevant energy costs including labor and externally supplied energies. Presumably, the labor costs include the producer's own energy (that is their energy used not only to do the work

[12] Emergy, a term introduced by Howard Odum, is the "embedded energy" in a low-entropy material object, that is, the amount of free energy that was used to produce the object. Odum developed a complete energetics model of the economy. See Odum (2007).

but also to support their own family). All free energy inputs to the production process should be considered. These form the basis of computing a true cost in monetary units and set the basis for a price to be used in establishing an archetype market price for the asset. Essentially this is similar to Karl Marx's and David Ricardo's "labor theory of value," except that the actual work processes in production today rely on machinery using high-power energy and are merely controlled by or augmented with human labor.

The base cost is just the first step in establishing a fair price in an archetype economy market. Premiums for the extended value of a product might be established beyond the base cost, as discussed above. These are the use or utility, esthetic, and need value evaluations. In a human economy, there are factors above and beyond the mere cost value that we see expressed in, for example, the metabolism exchanges discussed in Chap. 13. The human economy allows for functions beyond mere energy flow that support aspects such as development (improvement/evolvability) and psychological support (esthetics and the hierarchy of needs).

16.3.2.5.2.2 Use Value and Utility

Let's say that every plow that was ever produced in pre-industrial civilization took the exact same amount of energy to make (emergy value). The plow produced by blacksmith Bob was fundamentally the same as that produced by blacksmith Ray. Then in a balanced economy the total production of Bob and Ray output would equal the total needs of the society they served. But suppose Bob invented a new shape for a plow that allowed a single ox, as opposed to two oxen, to do almost the same amount of work in the same amount of time, that is, be more efficient for the farmer. Bob is probably justified in asking a higher price for his plows than is Ray. And farmers that can afford to pay that higher price (because they saved more of their earnings in the past) can pay the premium. It may be the case that Bob's emergy investment is no different from Ray's, but the end result, the production costs of a crop, result in a net saving to society. In other words, Bob's efforts (and inventiveness) increased the total free energy to society.

The use value, and the use value premium, is a result of how much total free energy is added to the society (system) as a result of any advantage that asset (tool) provides to the production processes that use it. The price charged for an asset that increases the overall free energy in the system might be viewed as a kind of profit, but it is not the same as in a current market—what the buyer is willing to pay. The price in this case is established by the benefit to society, not a whim of buyers and possible deceit by sellers. There must be a quantifiable benefit to society in the form of increased usefulness in the form of increased free energy available.

The benefit to society can be realized either in the form of reduction of work effort for the same production, or increased development of culture (i.e., more art). Society does not "need" to grow as a result. It can enjoy more leisure or more esthetics as a result of increasing the amount of free energy available to a particular level of production.

16.3.2.5.2.3 Esthetic Value

Human beings enjoy life. We cannot begin to address the effects of affective influences on human productivity. But there is now a body of evidence that humans who engage in esthetic activities (especially group activities) are ultimately more productive in work activities. People who have recreational time simply do work more efficiently. So even though esthetics seem to be counter-economic, in fact, they are contributory to enhanced economic outcomes.

This extends to assets that have not just functional value but esthetic value as well. We can see this in early humans' design of ornamental accouterments to tools like knife handles carved from ivory or bone. Some of the earliest artifacts of humanity include seemingly purely esthetic relics such as the cave paintings in Paleolithic Europe or the fertility figures of that time. Clearly esthetic qualities have been an important value to humans for a very long time.

It is not surprising then that those esthetic qualities factor into the valuation of assets. Today humans seek all kinds of artifacts that are not only functional but "beautiful" as well. We suggest that this tendency supports psychological well-being and as such has economic value. Happy humans are more social (and social organization is at the base of the success of the species) and more productive. They, therefore, tend to optimize the use of free energy for the good of the whole society.

16.3.2.5.2.4 Need and Exchange Value

There is another aspect of value above and beyond the utility or esthetics used in the calculus of price. People trade assets because they have a need for what the seller is selling and less need for the asset they seek to trade. In a happy situation, the various values of the to-be-traded assets mentioned above will be essentially the same and the trade will make both parties prosper. The problem comes in when, say, the use value of one asset exceeds that of the proffered asset. The need doesn't go away so the buyer and seller may have to negotiate. In today's market economy, this can lead to sellers trying to hide real costs of production or promoting a higher utility than will actually be realized so as to boost the price. In a more ideal situation, the buyer should understand what the value of the asset to-be-bought is and realize they may need to sweeten the deal with some additional asset to exchange so as to even the transaction.

16.3.2.5.3 On the Consideration of Value in Economics

On the basis of the above analysis, we have a basis for assigning value to economic products in the human economy. Value is based on what maintains or improves human life, not in terms of (as is the case now in Western societies) growth of possessions and in the overall economy but in terms of what contributes to human psychological well-being. Wealth, then, is that which supports human happiness and contentment, not that which may seem to promote status on the basis of ownership of frivolous assets. Status enhancement has always been a human desire, of course, but one would hope that humans build their status on the basis of their deeds rather than their possessions.

Mitochondria do not compete with one another for status. Nor do ribosomes. Nor do lungs or any other living organs or organisms. Competition is based on energy flow and selection.

16.3.2.6 Proper Role of Savings and Surpluses

Wealth accumulation for its own sake has become, mixing source languages, both the sine qua non and *raison d'être* in neo-liberal capitalism. However, it should be noted that the definition (or conception) of wealth includes the monetary value recorded on some account (today digitally) and not necessarily true physical wealth. Even real-estate values are merely estimates of asset holdings in monetary units and are based on some market-based perceptions. A house is worth absolutely nothing in monetary terms unless there is someone to buy it. The same goes for the value of corporate stocks. This is "paper" wealth that is meaningless without both a market and gullible agents to buy and sell. All such market-based mechanisms for valuation are subject to bubble phenomena and crashes—then where does the wealth go?

Real wealth means the holding of physical assets the valuation of which is based on the criteria given above (e.g., use value). In a sustainable CAES economy, the values are established by the real criteria. Prices of units of the assets do not go up because there is a higher demand (though the amount of energy expended to produce the assets will go up). There are no winners or losers in competition to acquire necessaries as in current market-based exchanges.

Some commodities, such as foodstuffs, may from time to time be produced in surpluses above immediate demand. Care must be taken (logistical management) to optimize production. For example, in the case of some foodstuffs (e.g., grains), excesses in 1 year can be saved as insurance against a bad year following.

Money, while not actually a commodity, might also form a surplus if a seller, say of some commodity or esthetic objects, enjoys a particularly efficient production and sells more than in a normal year. A bank can provide a service to hold such surpluses as savings. Again, the purpose of such a buffer is to provide insurance against future poor years or catastrophic asset losses (like a house burning down).

Both grain (as an example) and money savings might accumulate to a point that either the granary or the bank can "loan" some of the accumulation to help a member of the community improve a process or even start a new enterprise within the constraints of the module. This is a basic form of fractional-reserve banking in which a small portion of all account holders' savings are pooled for the loan. Unlike modern fractional-reserve banking, wherein banks, for example, are allowed to loan out larger than safe proportions of savings and where the effect is often said to increase the money "supply," this use of savings should be completely transparent with the account holders being made aware of and approving the loans to be made. Also, the loans would be considered only for improvements, as in the potential increase in free energy as described above in the section on Use Value, 16.3.2.5.2.2. The manager of the granary or the banker may charge a small fee to account holders and borrowers for their services just as any other provider of a service but no "interest" would be paid to account holders nor to the bankers and granaries.

The purpose of surpluses (when they occur) and savings should strictly be viewed as a buffer against unforeseen circumstances that invoke, for example, repair activities. Another use of excesses would be to support non-producing members of the community, such as the elders participating in councils (see below).

16.3.3 Social Organization

Related to the design for faulty agents' concepts given above, the actual physical layout of a social unit[13] will do much to lessen the kinds of stresses on individuals that allow faulty decision-making to dominate humans as agents. A first principle here is that a social unit should not have more people in it than a normal person should interact with on a daily basis. For example, Robin Dunbar (1992) suggested that there is something like an optimal number of individuals in a group. He used the size of the neocortex of various primates and correlated this measure with the mean number of social contacts in the "troop." According to his index, human social units should have a mean size of around 150 individuals. Other anthropological studies of prehistoric and contemporary tribes do show that the number of people grouped together does tend to be in a range of 100 to 200, so Dunbar's number, as it is called, seems about right.[14] The psychological theory behind Dunbar's number and the size of the neocortex is that that is the number of close personal contacts that the human brain evolved to process. One individual may be able to handle, on average, that number of mental models of other people.

16.3.3.1 Design Considerations: Modularity, Network, and Hierarchy

Briefly, we consider several factors that will be important in the design of a social system. These will be expanded later below. What we will describe is an architecture that actually reflects the natural one that humans originally had at some stage of their cultural evolution, what is called, by Daniel Quinn (2000), a "new tribalism." Quinn sees this as the next stage in cultural evolution, restoring human relations in social units of appropriate size and complexity. This basic organization of indigenous cultures arose naturally as human populations increased and spread around the globe. It was a successful strategy for governing larger numbers of

[13] A "unit" of society takes on somewhat different meanings based on the level in the hierarchy as explained below. The term "module" is basically synonymous with the unit.

[14] Other researchers, using different approaches, have come up with numbers much larger but not even an order of magnitude so.

ethnically or genetically related people. And it carried through the early stages of the Agricultural Revolution into the era of what we now call civilizations. It evolved in a bottom-up fashion as would be expected for an ontogenic process.

One of the most successful ways to manage complexity and reduce the impacts of occasional dysfunction in some components in a system is to design component subsystems as "modules," that is, bounded subsystems.[15] As we saw in Chap. 3, this is how nature evolved the complex systems we see today. Semi-independent modules may be linked by the kinds of connections (flows) demonstrated in Chap. 4 to form supra-modules. We will explore the various scale modules, but we can readily picture the way this might work for human social systems. The lowest level module can be considered a "family," however that is constituted, or, more generically, a "domicile." Groups of domiciles may constitute the next higher-level module, for example, a village or community using Dunbar's number as a guide to a reasonable size for a community module.

Likewise, a multiplicity of communities may be grouped into a larger-scale module (system) that might be called a "tribe.[16]" These tribal units may be further grouped as will be discussed below. What is being described is a hierarchical network, modules composed of submodules down to a basic fundamental element; a system of systems. All modules starting at the very bottom of the hierarchy are connected by flows of communications and distribution of assets within the module. And modules are governed by a common hierarchical cybernetic structure as well as sharing principal aspects of culture. The flows are designed to allow specialization of modules at any level, and the distribution of goods and services in the same sense as we currently think markets work. The governance architecture can be the same for all modules within a given level and up through all the levels.

Figure 16.1 shows a schematic representation of the modular design approach. It shows the basic structure of a social module composed of submodules, communications, and asset flows. And it hints at the governance architecture we will be advocating (and mentioned above regarding a brain-like one). The figure includes the idea of a module wherein submodules may act as interfaces for the import of asset resources and export of product/byproduct assets as previously described. These flows would represent internal flows at the next higher level in the hierarchy.

Also shown in the figure is a hint of the governance architecture in the form of a module council. The council, to be explained below, is the coordinator for the submodules and has the responsibility for strategic management of the module.

We can now outline several modules at increasing levels in the hierarchy. Later we will describe their nested relations in the hierarchy of modules.

[15] In designs of mechano-electrical equipment, modular design makes it possible to easily replace a faulty module. This is not a particularly useful benefit for social modules, but in the event of a disease or other calamity striking a module, one can imagine seeding a new module as needed.

[16] The term here is used to designate a grouping that is very similar in nature to social systems found in many indigenous peoples, including the American natives.

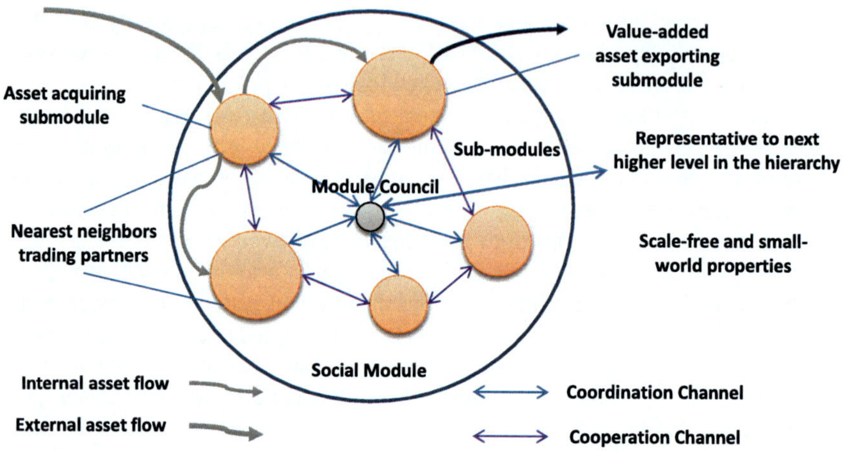

Fig. 16.1 A social module (at any level in the hierarchy) is composed of submodules that can share/trade assets and communicate with one another

16.3.3.1.1 Domus

It might be useful to think of individuals as atomic processes. The lowest-level module we will consider we're calling a "domus" or household (as defined in the previous chapter). As stated above this is likely synonymous with "family" so long as we don't think of strict blood relations defining the bounds. A domus[17] is also a basic unit of work and reproduction.

While it is not likely that domus could be self-sustaining units, in a low-power world, most food production will likely be done by them, as farming and food preserving processes. They would also be the locus of population renewal. As elders die off, babies would be conceived for replacement in a steady-state population.

Domus (pl) are the locus of operational-level governance, with logistics relating to farming and other domestic tasks. Tactical governance involves interactions with other domus (pl) in the community.

16.3.3.1.2 Communities

A collection of interacting domus (pl) constitutes a community. The physical arrangements of domus (pl) within a community could vary widely, from a spread-out farming community to a clustered set of domus (pl) like a village. The

[17] As explained in the last chapter, owing to a quirk of Latin, domus is both singular and plural, leading us to adopt a clumsy notation. When we say "domus" we mean a singular household plus its surroundings that make it a system. When we say domus (pl) we mean the plural as in what you would think would be domii! Being no linguist and not knowing Latin outside of its use in biological nomenclature, the author apologizes in advance.

determinant would be based on the geography and main sorts of activities of the community. As a subsystem of the next level, the tribe, communities may very well engage in activities other than just farming, such as crafts and products meant for trade with other communities. Since communities within a tribe may be relatively near in space, the opportunities for communities to take advantage of local specific resources to create (or grow) commodities that other communities would value. Thus, the activities of communities are similar to the metabolism of cells (see Chap. 13) that produce export commodities/products.

Communities need not be completely self-sufficient either. However, the collective production of food and artifacts of several communities should provide a level of self-sufficiency to the next level up.

Communities are also the locus of coordination governance for the domiciles. The community council (see below) would be expected to help the community to achieve a relatively high level of self-sufficiency. The reason is that in a changing world (e.g., climate change), there are always threats to domiciles and whole communities. In the event of a calamity to one community, there should be sufficient resilience in the others to not be impacted by the loss.

Additionally, communities are the providers of infrastructure services, such as inter-domicile communications and education, at least through what we would call "high school."

16.3.3.1.3 Tribes

The collection of some number of communities (see below for an analysis of numbers at each level) can share a larger geographical region and many aspects of culture. We will use the term tribe to describe this modular unit.

There is any number of structural organizations that a tribe might embody. For instance, one community in the tribe may act as a kind of capital or central governance specialist where the tribal council can meet. Presumably, such a community would be like a village in which the domus (pl) contribute to the support requirements for representatives from the various other communities living for extended periods. We might also expect the other clerical trappings of a government center for record-keeping and analysis of data gathered from the other communities for purposes of coordinating their activities.

The tribal level might also be expected to be responsible for organizing the higher education needs of young adults (colleges). This could be accommodated in the same community as the governance but could also have its own specialized community.

16.3.3.1.4 Nations

The logic of building up larger collections of modular units to handle the coordination among the lower-level modules continues with the collection of some number of tribes as a nation. Whereas tribes probably share a similar environment over some geographic range, or bioregion, it is likely that different tribes will experience different ecological conditions due to a much wider separation in geographic space. At the same time, a mechanism for providing coordination among tribes would be useful for communications and cultural exchanges. Tribes may undergo independent cultural evolution (but within the global guidelines for sustainability discussed below) and be able to share ideas and, especially, new knowledge. The amount of commodity trade between tribes would probably be more limited since the distances between them would likely only allow small material flows owing to the limits of energy available to support long-distance transportation.

As with the previous units, the governance of a nation is through a council of representatives from the various tribes. The same basic structure (which now could be described as "fractal") of specialist tribal units for centers of governance and higher education (e.g., universities) would allow for the centralized processes to be tightly coupled while maintaining an overall distributed architecture.

16.3.3.1.5 Federations

In all likelihood, the population distributed among nations might need an additional level of coordination that could be described as a federation (collection) of nations. Again, the considerations are based on geographic distributions, for example, continent-scale. As with the prior modules the general pattern of the modular design, including its governance, will be based on a council of national representatives and any additional coordination facilities that would be needed. As with the above descriptions, the main flows between subsystem modules would be messages for coordination and sharing new knowledge.

16.3.3.1.6 Global HSS

We imagine that the whole hierarchical structure terminates in a global HSS having the self-same architecture of those below.

16.3.4 Governance

The governance archetype model of Chap. 12 will provide us guidance for developing a workable governance system for the HSS. Most of the current governance systems are too monolithic and overly complex no matter the underlying political philosophies. The constitution that worked well for the early United States as a representative democracy seems to be failing the current actual polity (Brennan 2016). The design of three branches of government does follow from the three primary duties of a government, but the implementation of the legislative, executive, and judicial branches has long been problematic and becomes more so as the overall size of the country, in population, and the complexity due to cultural and technological innovations increases with time. This model does not scale well. In the USA, the central, federal government is not an adequate coordinator for the various state governments (that take on the same form as the federal government). Superficially there appears to be a hierarchy in the sense of the HCGS, but in fact, this is not the case. Even less so for the various local governments, counties, and cities.

We will not dwell on the inefficiencies and ineffectiveness of the current broken system. What we will do is explore a completely different vision of what governance of the HSS should be. Let us call this vision a distributed HCGS wherein true political power comes from the lower modules in the hierarchy. This will be a true democracy.

16.3.4.1 Democratic Sapiocracy

Democracy as currently understood is the idea that all citizens of a polity should be able to participate in decisions about who should attain positions in government (and of "power") and thus affect the kinds of policies that are adopted by the society.

The problem with this conception is that it depends critically on both the definition of "citizen" and on the notion that whoever is a citizen is sufficiently educated (as well as thoughtful) to understand the workings of government (as in "civics") and the norms and mores of the society. As has come to light in recent years, in some of the purported "greatest democracies," such as the United States of America where suffrage is supposed to be universal, two critical issues are faced. First, there are known ways to suppress voting by certain groups, thus diminishing the idea of universal suffrage. Second, a significant proportion of the voting public is extremely ignorant in civil matters (Brennan 2016).

Political philosopher Jason Brennan (2016) has proposed that a much more logical and efficient approach to the political process is to adopt an "epistocracy," or only allow those citizens who demonstrate a basic knowledge of civics and awareness of potential policy options and candidate qualification to vote and hold office. He provides a number of arguments, especially the lack of knowledge by average voters, as to why democracy does not and cannot work in a society that is so huge and complex. But he thinks the solution is to require this basic level of knowledge (episteme) of citizens who want to participate. Given the recent events in the American political process, there may be much to be said for this. Thomas Jefferson said, "A Nation's best defense is an educated citizenry." And, the story goes, that as Benjamin Franklin, after participating in the writing of the US constitution, was asked by a woman, "Well, Doctor, what have we got, a republic or a monarchy?" He replied, "A republic, if you can keep it." He had been very worried that the electorate might not be able to sufficiently understand the political requirements needed of an informed citizenry. The electoral college was created as a safeguard against a majority of ignorant citizens electing “the wrong kind of president.”

However, knowledge by itself is a very tricky thing and can be interpreted in several ways, such as “facts” which are subject to interpretation from different perspectives, for example, evolution. Facts alone are not sufficient. This whole book has been about a very different kind of knowledge—understanding. This requires not just knowing facts but also knowing how things work, how those facts interrelate. This is the domain of wisdom or sapience (Mobus 2012, 2019). The proposal advanced here is that, just as groups and tribes of old were governed by a council of wise elders, the human social units should also be guided by councils composed of the wisest (which will generally mean the more senior members of the group). Wisdom is the accumulation of tacit knowledge gained over the lifetime of an individual providing they have the neurological equipment and mental capacity to acquire it. It is the basis of veridical judgments and intuitions guiding complex social decisions/problem-solving.

Let us call this concept sapiocracy (from the Latin “sapio,” meaning discerning, wise, and judicious, and the name given to our species, *sapiens*). That is the governance, especially at the strategic level, is provided by a small group of wise elders. Somewhat similar to Brennan’s concept of epistocracy, those who are chosen to lead are selected not just on the basis of their explicit knowledge but on their holistic tacit knowledge about how the world works, including how people work.

There is an age-old method for recognizing the wisdom in people. You ask others in the social group who have had an ongoing relationship and experience with the candidates (Sternberg 1990a, b). People, even young people, are often very good at recognizing wiseness (or sagacity) in other people. Within a domicile this could be as simple as the eldest member, say a grandmother, would represent the domicile in the community council. Then, in a community, the council members could, through deliberation, decide whom they should send to the tribal council. This form of government and political process for selecting leaders is more democratic than the current examples of, say, representative democracies, especially those in which the wealthy have a disproportionate influence on the campaigning and voting mechanisms.

It is important to note that this is not meritocracy in the traditional sense. Leaders are not chosen on the basis, for example, of how successful they have been in life pursuits, nor the presumption of how “smart” they are. What is needed in leadership is wisdom above cleverness.

16.3.4.2 Hierarchical Network of Modules and Councils

The hierarchy of social modules is depicted in Fig. 16.2 below. The number of levels depicted is not suggested to be some absolute number. The point of this hierarchy is to keep each module at all levels manageable by limiting the number of participating submodules. So, if a community contains roughly 15 domiciles (with

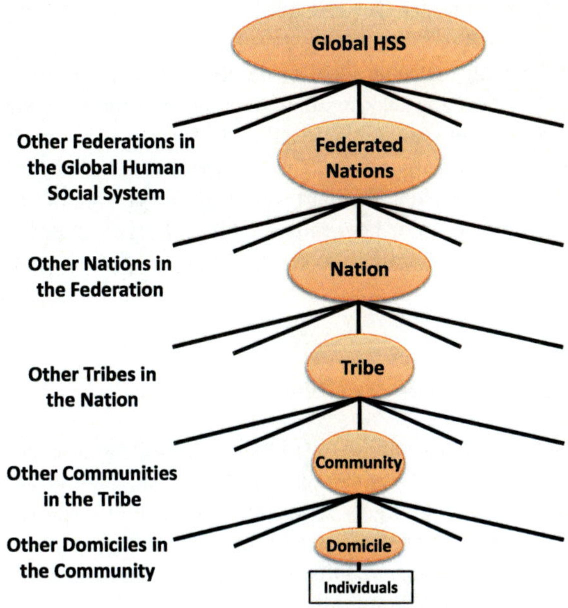

Fig. 16.2 The global HSS is based on a hierarchy of modular social units, as depicted in Fig. 16.1 above, starting with domiciles (families and a few other individuals), which are collected into higher-level modules, communities, and those in turn are collected into tribes and so on

10 individuals participating in an average domicile), which keeps the community population at Dunbar's number, and then we can use similar logic in determining the number of communities that come under a single tribe. Using the Dunbar number as a guide, suppose the average number of communities in a tribal module is kept somewhere below 150. This is to keep the tribal council size at or below Dunbar's number. The total population of the tribe is 150 × 150 or approximately 22,500 individuals. Of course, this should probably be viewed as an upper limit for the purposes of keeping the size of a tribal module manageable. The number of communities in a tribe would more likely reflect the overall conditions of the geographical area, the local ecosystems, resource availability, etc. The total territory of a tribe, likewise, would be determined by the environmental conditions supporting the population.

This design principle is extended up to the collection of the whole HSS population.

With appropriate population sizes in these modules distributed over sufficient land areas, the various communities can be spaced such that interactions between individuals on a daily basis would be kept to the level that does not put an overburden on a person's mental capacities. Similarly, the distribution of tribes, say by regional considerations would further buffer interactions between members inter-nationally.

The biggest advantage of this hierarchical, distributed modular, network design is that local processes are the dominant mode of operations and management. As opposed to a top-down government such as is seen in most countries today, the main form of governance starts at the community level.

16.3.4.3 Local Governance

In keeping with the ideal of democracy or universal participation in political and policy decision-making, the governance of modules is basically local. Thus, effective decision-making is vested in the local council, meaning the council of the module type in the hierarchy. The vast amount of the power to affect life is vested at the community level. The purpose of higher councils is not to tell communities what they should do or how to live their collective lives. It is purely for the purpose of providing a coordination service, that is logistical support and appropriate tactical "advice."

16.3.4.4 Global Governance

From the standpoint of the global HSS, the role of governance is to observe both the Ecos and the HSS, the long-term behaviors, actions taken and consequences resulting, and attempting to understand the dynamics of the whole system. The ultimate role of the global council is to consider strategic moves that are meant to benefit the planet as well as humanity. This may manifest in the form of suggestions for tactical moves for the federations.

Likewise, the federation councils are monitoring some of the same considerations but for the scope of the federation of nations under their purview. The difference between the global and federation governance is mostly scaled in space and time; the time scales of lower modules are shorter than for higher modules as a rule.

This pattern is recursively applied down to the community level. Upper councils are concerned with the sustainable operation of the whole module unit and provide logistic support, for example, maintaining any needed infrastructure to allow sub-modules to cooperate and interact, and tactical suggestions (not laws and coercion) to the sub-modules in the interest of facilitating the whole module's interactions with other peer modules.

16.3.4.5 Module Governance Architecture

16.3.4.5.1 Integrated Governance: Social Behavior and Economy

Once upon a time, the study of political and governance processes was integrated with the study of economic processes—political-economy. Today, mainstream political science and economics are often studied almost independently and certainly with differing world views. This is remarkable since governments routinely intervene in economic processes in a number of ways. As is often observed, the so-called free market has never actually been free of government interventions (Polanyi

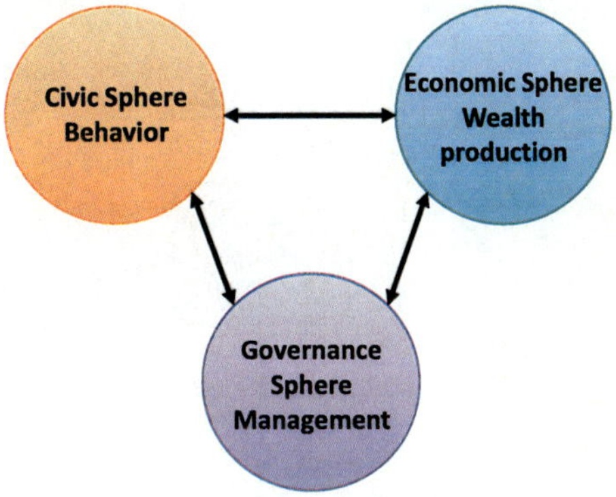

Fig. 16.3 Civic activities (interpersonal behaviors), economic activities, and governance are all interrelated and need to be considered as a whole

2001).[18] Similarly, the economic processes are the sources of "revenue" for the support of governments. So, it is odd to think of the two as separate domains. We did provide two archetype models, one of the governance mechanisms (Chap. 12) and economy mechanisms (Chap. 13), but this was to show the particularities of two different kinds of processes. We also showed how these two models are related, or interrelated, and how governance applies to both civic and economic decisions and management.

Figure 16.3 depicts this co-equal relation between the three spheres of social activities.

In what follows we consider that the governance of a social module is embodied in a government structure that serves several necessary functions.

16.3.4.5.2 Functions of Government

We turn attention to the architecture of a government that fulfills the various functions of a government over both civic and economic activities. As with all systems we have examined in this work, we consider the subsystems and the flows of influence (messages) between them. Figure 16.4 provides a general systems overview of a government system. There are three basic work processes and two repositories/decision models.

16.3.4.5.2.1 Establishment Rules: Governance System Inputs

Unlike many system inputs from an environment the "establishment rules" or "constitution" are a result of a, generally, one-time flow. A society establishes its set of values and mores that *should* govern the body politic and establish a set of rules for the organization and general operation of the government structure itself. Such rules

[18] And as noted in several places, particularly Chap. 13, we have established the thesis that markets are not able to self-regulate as supposed by neo-classical and capitalistic views. So, the notion that a free market could, in principle, govern itself without government intervention is a pure myth.

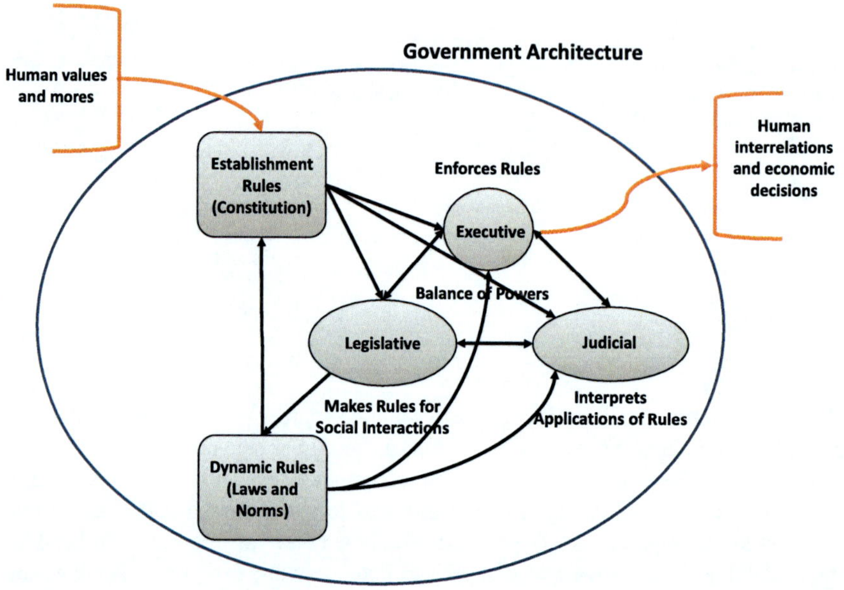

Fig. 16.4 The general architecture of government used by current political units

are considered stable for very long time frames, say the lifetime of the social module itself. The formal constitutions of many nation-states today reflect these establishment rules.

While these rules are generally viewed as stable (and stationary), practical experience in an evolving society shows that the rule base needs to also be an evolving, or learning, system. Changes to a constitution are rare in practice, but sometimes necessary to accommodate new conditions not anticipated by the originating values/mores that gave rise to them initially. There are several mechanisms used by nation-states to achieve these changes. In the figure, we show that such changes are guided by the body of "dynamic rules" (below).

16.3.4.5.2.2 Dynamic Rules: Governance Output to the Human Subsystem

Societies have been, historically, and are likely to continue to be evolving systems. Moreover, the establishment rules are often too general to handle day-to-day or even year-to-year changes in the system. As such they need to have a mechanism for producing and codifying rules that can, themselves, be changed dynamically. The repository shown in Fig. 16.4 as "Dynamic Rules" constitutes the laws and evolving mores of the society over time. These constitute the basic behavioral decision models used by individuals and institutions to conduct normal operations. These rules are subject to nearly constant revision, addendum, and deletions as the conditions within the module change over time. However, it should be noted that in a stable society, the rate of changes to rules, would be anticipated to be much lower than is observed today in most of the world's nation-states.

Laws are generated by legislative action (below). They are the result of intentional design (through deliberation generally). Mores and norms are adopted more or less unconsciously by the spread of social memes. Once again, for a stable society in which novelty is encountered but slowly, these various rules of conduct would not be expected to change rapidly.

16.3.4.5.2.3 Executive

In most governments, the executive branch (whether a dictator or a president) is required to "enforce" the rules, both establishment and dynamic. What enforcement entails today is dependent on the form of a political model. In a dictatorship, this might mean coercion under the threat of personal harm. In a democracy, it could mean incarceration (for breaking the rules) or at least a stiff monetary penalty (for cheating in the economic system).

In the envisioned social system, under the assumption that with a much simpler and stable society people will be less prone to try to cheat, executive actions might look similar when cheating does occur, but more broadly it might be viewed as providing helpful corrective advice when citizens or institutions make mistakes. This follows from the philosophy that authority does not mean punitive power but higher wisdom and understanding.

16.3.4.5.2.4 Legislative

The role of the legislative branch of a government is to assess the societal situation and conditions and address any disparities that might be developing through the construction (systems engineering) of appropriate rules. This is fundamentally the modification of the decision models used by various decision agents in the civic or economic spheres. This is the adaptive process of a society, applied only when needed to adapt to changes in the environment.

In our current society, we see this adaptive process working not just to adapt to external changes but to internal evolutionary changes within the society itself. For example, congresses need to adopt new laws to regulate the uses of new technologies or materials that have come into existence due to innovations within the society itself. For example, the establishment of the Environmental Protection Agency (EPA) in the USA and the writing of environmental protection legislation have been the result of the development of numerous "unnatural" chemicals, like plastics and biocides that have been shown to threaten the natural environment. We might call this second-order adaptation, but it is really just a form of evolvability.

16.3.4.5.2.5 Judicial

Judgment of the "correctness" of behaviors is a constant requirement of designing for faulty agents (as described above). All governments include a judiciary function for this purpose. There will always be differences of opinion and perspective among human agents and so there needs to be a mechanism for adjudicating (that is resolving) those differences when they arise. Of course, for the current species of humans that includes a variety of anti-social behaviors that may arise as a result of the stress levels generated by the scale and complexity of industrial or even agricultural societies. We will not attempt to psychoanalyze modern society, but simply assert that there is ample evidence that crime and anti-social behaviors are high and that the complexity and failures of modern societies might be to blame. The competition that is fostered by a capitalistic society is a major source of stress and disagreements. In the envisioned social system, there is no need for that kind of competition.

A judicial function will probably always be needed in any governmental system. However, in a scaled modular social unit, it is anticipated that the need for adjudication of either civic or economic misbehavior will be much less than we currently experience.

16.3.4.5.3 Governance in the Social Module

What, then, would the governance of a systemic social module look like? In Fig. 16.5 we offer one possible model. The social module already consists of a set of operating sub-modules (the various trades in a community or specializations in a tribe, etc.) and these are governed by a modular government. Here we show how the functions discussed above are handled and add a crucial new function.

In this figure, we show the governing council as discussed previously, but having a specific structure that allows for the implementation of some of the functions described above. We also show the judicial function as separate from the council and a new function that is analogous to the "fourth estate" in democratic governments, what we have labeled as the "audit" function.

The council has been organized as "inner" and "outer" councils with different roles as explained below. As the reader may recall, councils at all levels will consist of from 50 to 150 members (depending on the level in the hierarchy and the number of actual sub-modules). The design suggested in Fig. 16.5 assumes that some smaller number of, presumably, the wisest members of the council are elected to the inner council and that that unit is responsible for the strategic decision-making for the whole module and for the executive functions of government. The outer council would fulfill the role of a legislative body and be concerned with tactical governance of the module. The judicial function could be handled by a rotating sub-council of the regular council members. They would recognize their role for the period during which they served in that role.

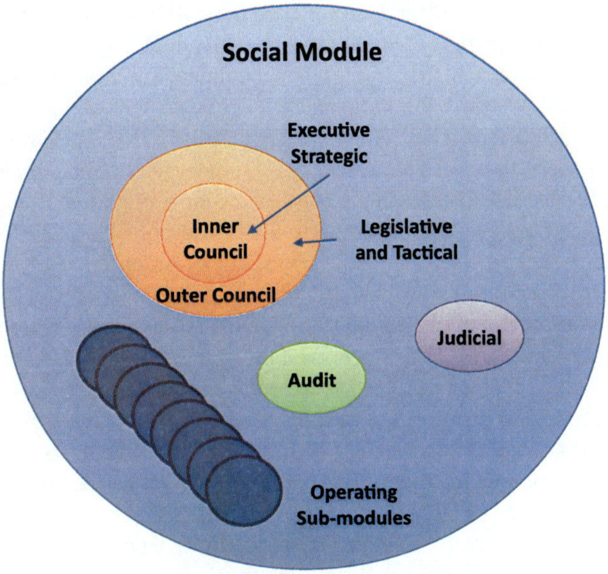

Fig. 16.5 A general architecture for a social module would provide for the four main functions of government and cover the operating (work processes) submodules

What is labeled as the "audit" function in this model is essentially the role supposedly played by the fourth estate or the "press" in modern societies. In this case, the role is closer to that of internal auditors in large corporations and other institutions. The audit role is to monitor the actions taken by the other three branches of government (assuming complete transparency) and report these to the rest of the module citizens. This is the basis for citizens to consider who among their members are truly wise. That is, citizens would have complete access to information regarding how their governors were performing and be able to thus make judgments on how well their current set of leaders are serving the whole module. This is another aspect of designing around potentially faulty agents.

16.4 Conclusion

The human social system that evolved at the end of the Pleistocene epoch provided a natural structure and function for the general well-being of individuals. That doesn't mean life was not 'hard' in the sense that survival was not assured and much work needed to be done to stay alive. But humans had also evolved some unique talents that made them significantly different from even their nearest cousins, the other great apes. They had language, collective intentionality, and a capacity to imagine and construct more complex tools out of at-hand materials. Starting from having nothing but innate concepts of how the world worked (i.e., physics) they gradually developed the ability to invent ever more clever tools and methods for living. During the whole time leading up to, roughly, the Upper Paleolithic humans had been evolving sapience, the capacity to accumulate a wealth of implicit knowledge, strategic and systemic thinking, and a heightened sense of moral sentiment. Then, around 10,000 years ago something happened. Humans discovered agriculture and everything changed. The further development for strategic thinking would seem to have been sidetracked in favor of short-term operational and logistical thinking. And the bulk of the populations that arose out of nascent civilizations tended to be selected precisely for their lack of higher-order thinking.

The modern, technologically enhanced, the world is the outcome of ten millennia of cultural evolution, intentional-organization, and various selection mechanism for determining the 'successes' of artifacts. The most potent selection force of the last two to three hundred years has been the mental state induced by capitalism, the desire to accumulate wealth for its own sake. As a result, the vast majority of people, today, believe that endless economic growth is a "good" when in fact it is the thing that will likely destroy civilization and is the cause of what has been called the Fifth Mass Extinction.

In this chapter, we have considered how systems science, its major principles, and systems design and engineering might be applied to the human social system. Using the archetype model of a complex adaptive and evolvable system, a system that is designed to be sustainable, persistent, and stable in the face of changes in climate and biota, we found that the structure of a global HSS to be essentially like the one that existed at the end of the Upper Paleolithic. That structure involves modular groups in a hierarchy of modular units, similar to tribal organizations of so-called primitive aboriginal peoples.

The purpose of this exercise was dual. First, we wished to show the contrast between what a CAES-based HSS and our modern nation-state-based organizations would be in the hope that more rational decision makers, if they exist, could consider what would it take to transition from that in which we live now to that which would fulfill the CAES ideal. Second, in the event that the current system implodes and civilization as we know it crashes, the ideas promulgated here might serve as a blueprint and guide for what to do to recover humanity.

If these pages survive.

Bibliography

References and Further Readings

Abdou K, Shehata M, Choko K, Nishisono H, Matsuo M, Muramatsu S, Inokuchi K (2018) Synapse-specific representation of the identity of overlapping memory engrams. Science 360(6394):1227–1231

Ackoff RL (1971) Towards a system of systems concepts. Manag Sci 17(11):661–671

Alkon DL (1987) Memory traces in the brain. Cambridge University Press, Cambridge, UK

Alperovitz G (2005) America beyond capitalism: reclaiming our wealth, our liberty, and our democracy. Wiley, Hoboken

Alur R (2015) Principles of cyber-physical systems. The MIT Press, Cambridge, MA

Anderson WT (2004) All connected now: life in the first global civilization. Westview Press, Boulder

Arsenault JS, Buchsbaum BR (2015) Distributed neural representations of phonological features during speech perception. J Neurosci 35(2):634–642

Ashby WR (1958) Requisite variety and its implications for the control of complex systems. Cybernetica 1(2):83–99. Available on-line: http://pespmc1.vub.ac.be/Books/AshbyReqVar.pdf. Accessed 28 Nov 2016

Baars BJ, Gage NM (eds) (2007) Cognition, brain, and consciousness: introduction to cognitive neuroscience. Elsevier, Boston

Barabási A-L (2003) Linked: how everything is connected to everything else and what it means for business, science, and everyday life. Penguin Group, New York

Bateson G (2000) [1972] Steps to an ecology of mind: collected essays in anthropology, psychiatry, evolution, and epistemology, University of Chicago Press, Chicago

Beer S (1959) Cybernetics and management. English University Press, London

Beer S (1966) Decision and control: the meaning of operational research and management cybernetics. Wiley, New York

Beer S (1972) Brain of the firm: a development in management cybernetics. Herder and Herder, New York

Boulding KJ (1956) General systems theory – the skeleton of science. Manage Sci 2:197. Reprinted in Klir, 2001

Bourke AFG (2011) Principles of social evolution. Oxford University Press, Oxford

Box GEP (1976) Science and statistics. J Am Stat Assoc 71(356):791–799

Brennan J (2016) Against democracy. Princeton University Press, Princeton

Brodt S, Gais S, Beck J, Erb M, Scheffler K, Schonauer M (2018) Fast track to the neocortex: a memory engram in the posterior parietal cortex. Science 362:6418. 30 November, 2018, pp 1045–1048

G. E. Mobus, *Systems Science: Theory, Analysis, Modeling, and Design*,
https://doi.org/10.1007/978-3-030-93482-8

Buller DJ (2005) Adapting minds: evolutionary psychology and the persistent quest for human nature. The MIT Press, Cambridge, MA

Burton RA (2008) On being certain: believing you are right even when you're not. St. Martin's Griffin, New York

Calcott B, Sterelny K (eds) (2011) The major transitions in evolution revisited. The MIT Press, Cambridge, MA

Camazine S, Deneubourg J-L, Franks NR, Sneyd J, Theraulaz G, Bonabeau E (2001) Self-organization in biological systems. Princeton University Press, Princeton

Capra F (1975) The Tao of physics: an exploration of the Parallles between modern physics and eastern mysticism. Shanbala, Boston

Capra F (1982) The turning point: science, society, and the rising culture. Bantom Books, New York

Capra F (1996) The web of life: a new understanding of living systems. Random House, New York

Capra F (2002) The hidden connections: integrating the biological, cognitive, and social dimensions of life into a science of sustainability. Doubleday, New York

Capra F, Luisi PL (2014) The systems view of life: a unifying vision. Cambridge University Press, Cambridge, UK

Carey S (2009) The origin of concepts. Oxford University Press, Oxford

Carroll SB (2006a) Endless forms most beautiful: the new science of Evo Devo and the making of the animal kingdom. W. W. Norton & Company, New York

Carroll SB (2006b) The making of the fittest: DNA and the ultimate forensic record of evolution. W.W. Norton & Company, New York

Carroll S (2016) The big picture: on the origins of life, meaning, and the universe itself. Dutton, New York

Carter R (1999) Mapping the mind. University of California Press, Berkeley

Casti JL (1994) Complexification: explaining a paradoxical world through the science of surprise. HarperCollins, New York

Charette RN (2005) Why Software Fails: We waste billions of dollars each year on entirely preventable mistakes. IEEE Spectrum.

Checkland P (1999) Systems thinking, systems practice, includes a 30-year retrospective. Wiley, Chichester

Churchman CW (1960) Prediction and optimal decision. Prentice Hall, Englewood Cliffs

Churchman CW (1968a) Challenge to reason. McGraw-Hill, New York

Churchman CW (1968b) The systems approach. Delacorte Press, New York

Churchman CW, Ackoff RL, Arnoff EL (1957) Introduction to operations research. Wiley, New York

Clegg B (2019) Dark matter & dark energy: the hidden 95% of the universe. Icon Books Ltd., London

Coen E (2012) Cells to civilizations: the principles of change that shapes life. Princeton University Press, Princeton

Costanza R, Cumberland J, Daly H, Goodland R, Norgard R (1997) An introduction to ecological economics. St. Lucie Press, Boca Rotan

Crist E (2019) Abundant earth: toward an ecological civilization. University of Chicago Press, Chicago

Crosby AW (2007) Children of the sun: a history of humanity's unappeasable appetite for energy. W. W. Norton & Company, New York

Cuban L (1999) How scholarship trumped teachers: change without reform in university curriculum, teaching, and research, 1890–1990. Teachers College Press, New York

Daly HE, Farely J (2010) Ecological economics, principles and applications, 2nd edn. Island Press, Washington, DC

Daly HE, Townsend KN (1993) Valuing the earth: economics, ecology, ethics. The MIT Press, Cambridge, MA

Damasio AR (1994) Descartes' error: emotion, reason, and the human brain. G.P. Putnum's Sons, New York

Damasio AR (2000) The feeling of what happens: body and emotion in the making of consciousness. Mariner Books, New York

De Waal F (2005) Our inner ape. The Berkley Publishing Group, New York

De Waal F (2010) The age of empathy: nature's lessons for a kinder society. Broadway Books, New York

De Waal F (2014) The bonobo and the atheist: in search of humanism among the primates. W.W. Norton & Company, New York

De Waal F (2016) Are we smart enough to know how smart animals are? W.W. Norton & Company, New York

Deacon TW (1997) The symbolic species: the co-evolution of language and the brain. W. W. Norton & Co., New York

Dehaene S (2014) Consciousness and the brain: deciphering how the brain codes our thoughts. The Penguin Group, New York

Diamond J (2005) Collapse: how societies choose to fail or succeed. Penguin Group, New York

Donald M (1991) Origins of the modern mind: three stages in the evolution of culture and cognition. Harvard University Press, Cambridge, MA

Donald M (2001) A mind so rare: the evolution of human consciousness. W.W. Norton & Co., New York

Dretske FI (1988) Explaining behavior: reasons in a world of causes. MIT Press, Cambridge, MA

Dretske FI (1999) Knowledge and the flow of information. CLSI Publications, Palo Alto

Dunbar RIM (1992) Neocortex size as a constraint on group size in primates. J Hum Evol 22(6):469–493. https://doi.org/10.1016/0047-2484(92)90081-J

Ehrlich PR, Ehrlich AH (2002) The dominant animal: human evolution and the environment. Island Press, Washington, DC

Ferguson N (2008) The ascent of money: a financial history of the world. The Penguin Press, New York

Findeisen W, Bailey FN, Brdys M, Malinowski K, Tatjewski P, Wozniak A (1980) Control and coordination in hierarchical systems. Wiley, New York

Flyvbjerg B, Bruzelius N, Rothengatter W (2003) Megaprojects and risks: anatomy of ambition. Cambridge University Press, Cambridge, UK

Fodor J (1975) The language of thought. Harvard University Press, Cambridge, MA

Ford A (2010) Modeling the environment. Island Press, Washington, DC

Forrester J (1961) Industrial dynamics. Pegasus Communications, Waltham

Fox RF (1988) Energy and the evolution of life. W.H. Freeman & Company, New York

Friendshuh L, Troncale L (2012) SoSPT I.: Identifying fundamental systems processes for a general theory of systems (GTS). In: Proceedings of the 56th annual conference, International Society for the Systems Sciences (ISSS), July 15–20, San Jose State University. [Electronic Version: http://journals.isss.org/index.php/proceedings56th]

Fuller B (ed) (1968) Operating manual for spaceship earth. University of Southern Illinois Press, Carbondale

Fuller B (1970) Utopia or oblivion: the prospects for humanity. Allen Lane, London

Fuller B (1982) Critical path, 2nd edn. St. Martin's Griffin/McMillan Press, New York

Galbraith JK (2014) The end of normal: the great crisis and the future of growth. Simon & Schuster, New York

Gazzaniga MS (2005) The ethical brain. Dana Press, New York

Geary DC (2005) The origin of mind: evolution of brain, cognition, and general intelligence. American Psychological Association, Washington, DC

Geiger RL (2004) Knowledge & money: research universities and the paradox of the marketplace. Stanford University Press, Stanford

Gilbert D (2006) Stumbling on happiness. Alfred A. Knopf, New York

Gilovitch T, Griffin DW, Kahneman D (2002) Heuristics and biases: the psychology of intuitive judgment. Cambridge University Press, Cambridge, UK

Gleick J (1987) Chaos: making a new science. Penguin Books, New York

Goldstein J (2020) Money: the true story of a made-up thing. Hatchet Books, New York

Gordon RJ (2016) The rise and fall of American growth: the U.S. standard of living since the civil war, The Princeton economic history of the western world. Princeton University Press, Princeton

Gould SJ, Eldredge N (1977) Punctuated equilibria: the tempo and mode of evolution reconsidered. Paleobiology 3:115–151. Paleontological Society, Bethesda

Gribbin J (2004) Deep simplicity. Random House, New York

Grinspoon D (2016) Earth in human hands. Grand Central Publishing, New York

Grisogono A-M (2017) (How) Did information emerge. In: Walker S, Davies P, Ellis GFR (eds) From matter to life: Information and causality. Cambridge University Press, Cambridge, UK, pp 61–96

Hall CAS, Klitgaard KA (2012) Energy and the wealth of nations: understanding the biophysical economy. Springer, New York

Hall CAS, Cleveland CJ, Kaufmann R (1986) Energy and resource quality: the ecology of economic process. Wiley, New York

Hanson JS, Olson CR (eds) (1990) Connectionist modeling and brain function: the developing interface. The MIT Press, Cambridge, MA

Hansson SO (1994) Decision theory: a brief introduction. Department of Philosophy and History of Technology, Royal Institute of Technology (KTH), Stockholm. Available: https://web.archive.org/web/20060705052730/http://www.infra.kth.se/~soh/decisiontheory.pdf. Accessed: 27 Nov 2016

Harari YN (2011) Sapiens: a brief history of humankind. Vintage Books, London

Harari YN (2017) Homo Deus: a brief history of tomorrow. HarperCollins Publishers, New York

Harnad S (1990) The symbol grounding problem. Physica D 42:335–346. Available online in Cogprints, http://cogprints.org/3106/. Accessed 1 June 2017

Harold FM (2001) The way of the cell. Oxford University Press, Oxford

Hartley RVL (1928) Transmission of information. Bell Syst Tech J 7(3):535–563, (July 1928)

Hawkins J (2004) On Intelligence: How a New Understanding of the Brain Will Lead to the Creation of Truly Intelligent Machines. Times Books, New York

Hawkins J, Blakeslee S (2004) On intelligence, Times Books/Henry Holt and Company, New York

Heylighen F, Rosseel E, Demeyere F (eds) (1990) Self-steering and cognition in complex systems: toward a new cybernetics. Gordon and Breach Science Publishers, New York

Hidalgo C (2015) Why information grows: the evolution of order, from atoms to economies. Basic Books, New York

Hinton GE, McClelland JL, Rumelhart DE (1987) Chapter 3: Distributed representations. In: Rumelhart DE, McClelland JL (eds) Parallel distributed processing, vol. 1: foundations. The MIT Press, Cambridge, MA

Hodder I (2012) Entangled: an archaeology of the relationships between humans and things. Wiley, New York

Hogarth R (1988) Judgement and choice, 2nd edn. Wiley, New York

Holland JH (1998) Emergence: from chaos to order. Addison-Wesely Publishing Company, Inc., Reading

Holland JH (2006) Studying complex adaptive systems. J Syst Sci Complex 19(1):1–8

Holmgren D (2002) Permaculture: principles and pathways beyond sustainability. Holmgren Design Services, Hepburn

Homer-Dixon T (2006) The upside of down: catastrophe, creativity, and the renewal of civilization. Random House Canada, Ontario

Hughes TP (2004) Human-built world: how to think about technology and culture. The University of Chicago Press, Chicago

Huth A, Nishimoto S, Vu AT, Gallant J (2012) A continuous semantic space describes the representation of thousands of object and action categories across the human brain. Neuron 76:1210–1224. Elsevier, Inc. New York

Ivenenko VI (2009) Decision systems and nonstochastic randomness. Springer, New York

Jackson T (2009) Prosperity without growth: economics for a finite planet. Earthscan, London

Judd SJ (1991) Neural network design and the complexity of learning. Bradford Books, New York

Kahneman D (2011) Thinking fast and slow. Farrar, Straus, and Giroux, New York

Kauffman S (1993) The origins of order: self-organization and selection in evolution. Oxford University Press, Oxford

Kauffman S (1995) At home in the universe: the search for the laws of self-organization and complexity. Oxford University Press, Oxford

Kauffman S (2000) Investigations. Oxford University Press, Oxford

Keen S (2011) Debunking economics – revised and expanded edition: the naked emperor dethroned? Zed Publishing, New York

Klein N (2015) This changes everything: capitalism vs. climate. Simon & Schuster, New York

Klir GJ (2001) Facets of systems science, 2nd edn. Kluwer Academic/Plenum Publishers, New York

Koestler A (1967) The ghost in the machine (1990 reprint ed.). Penguin Group, New York

Kolbert E (2015) The Sixth Extinction: An Unnatural History. Picador, London

Koshland DE (1980) Bacterial chemotaxis as a model behavioral system. Raven Press, New York

Koshland DE (1981) Biochemistry of sensing and adaptation in a simple bacterial system. Annu Rev Biochem 50:765–782

Kosko B (1990) Fuzziness vs. probability. Int J Gen Syst 17:211–240

Kraus N, Cheour M (2000) Speech sound representation in the brain. Audiol Neurootol 5:140–150

Krugman P, Wells R (2012) Economics, 3rd edn. Worth Publishers, Duffield

Kuppers B-O (1990) Information and the origin of life. The MIT Press, Cambridge, MA

Laszlo E (2002) The systems view of the world: a holistic vision for our time. Hampton Press, Inc., Cresskill

LeDoux J (2002) Synaptic self: how our brains become who we are. Viking, New York

Lehrer J (2009) How we decide. Houghten Mifflin Harcourt, New York

Leontief WW (1986) Input–output economics, 2nd edn. Oxford University Press, New York

Leshem D (2016) The origins of neoliberalism: modeling the economy from Jesus to Foucault. Columbia University Press, New York

Libet B, Gleason CA, Wright EW, Pearl DK (1983) Time of conscious intention to act in relation to onset of cerebral activity (readiness-potential). Brain 106(3):623–642

Lombardo T (2006) The evolution of future consciousness. AuthorHouse, Bloomington

Losos JB (2017) Improbable destinies: fate, chance, and the future of evolution. Riverhead Books, New York

Lovelock JE, Margulis L (1974) Atmospheric homeostasis by and for the biosphere: the Gaia hypothesis. Tellus 26(1–2):2–10. Series A. Stockholm: International Meteorological Institute

Mackintosh NJ (1983) Conditioning and associative learning. Oxford University Press, Oxford

Marcus G (2008) Kluge: the haphazard construction of the human mind. Houghton Mifflin Co., New York

Margulis L, Sagan D (2000) What is life? University of California Press, Los Angeles

Marieke LS, Howarth C, Attwell D (2008) The cortical energy needed for conscious perception. NeuroImage 40(4):1460–1468

Maslow AH (1943) A theory of human motivation. Psychol Rev 50(4):370–396

Maynard Smith J, Szathmáry E (1995) The major transitions in evolution. Oxford University Press, Oxford

McGilchrist I (2009) The master and his emissary: the divided brain and the making of the Western world. Yale University Press, New Haven

Meadows DH (2008) Thinking in Systems: A Primer, Chelsea Green Publishing. White River Junction, VT

Meadows DH, Meadows DL, Randers J, Behrens WW III (1972) The limits to growth. Potomac Associates, Washington, DC. Available online at: http://www.donellameadows.org/wp-content/userfiles/Limits-to-Growth-digital-scan-version.pdf. Retrieved 20 Oct 2016

Meadows DH, Randers J, Meadows DL (2004) Limits to growth: 30-year update. Chelsea Green Publishing, White River Junction

Meuleman L (2008) Public management and the meta governance of hierarchies, networks and markets: the feasibility of designing and managing governance style combinations. Springer, New York

Micklethwait J, Wooldridge A (2014) The fourth revolution: the global race to reinvent the state. The Penguin Press, New York

Miller JG (1978) Living systems (1995 edition). University of Colorado Press, Niwot

Miller GA (1995) WordNet: a lexical database for English. Commun ACM 38(11):39–41

Miller JH, Page SE (2007) Complex adaptive systems: an introduction to computational models of social life. Princeton University Press, Princeton

Minsky M (1986) The society of mind. Simon & Schuster, New York

Mitchell M (2009) Complexity: a guided tour. Oxford University Press, Oxford

Mithen S (1996) The prehistory of the mind: the cognitive origins of art, religion, and science. Thames and Hudson, London

Mobus GE (1994) Toward a theory of learning and representing causal inferences in neural networks. In: Levine DS, Aparicio M (eds) Neural networks for knowledge representation and inference. Lawrence Erlbaum Associates (now Taylor & Francis), New York

Mobus GE (1999) Foraging search: prototypical intelligence. In: The third international conference on computing anticipatory systems, Liege, Belgium, August, 1999. Available on-line: http://faculty.washington.edu/gmobus/ForagingSearch/Foraging.html

Mobus GE (2000) Adapting robot behavior to a nonstationary environment: a deeper biologically inspired model of neural processing. In: Sensor fusion and decentralized control in robotic systems III. SPIE – International Society for Optical Engineering, Boston. Available on-line: http://faculty.washington.edu/gmobus/Mavric/Nonstationary/spie-paper.html

Mobus GE (2012) The evolution of wisdom. In: Science, wisdom, and the future: humanity's quest for a flourishing earth. Collins Foundation Press, Santa Margarita, pp 83–89

Mobus GE (2015) A systems science framework for understanding the nature of governance. In: Conference: governing the anthropocene, International Society for the Systems Sciences, Berlin, Germany, August 2–7

Mobus GE (2017) A framework for understanding and achieving sustainability of complex systems. Syst Res Behav Sci 34:5. https://doi.org/10.1002/sres.2482

Mobus GE (2018) A systems-based architecture/design for the human social system. In: Systems analysis in economy. Moscow Finance and Economics University, Moscow. ORCID: https://orcid.org/0000-0001-8145-0908

Mobus GE (2019) A theory of sapience: using systems science to understand the nature of wisdom and the human mind, Currently available online, hosted by The Millennium Alliance for Humanity and the Biosphere at Stanford University, https://mahb.stanford.edu/library-item/theory-sapience-using-systems-science-understand-nature-wisdom-human-mind/. Accessed 7 Oct 2019. [In preparation for print edition]

Mobus G, Anderson K (2016) System language: understanding systems. In Conference: realizing sustainable futures in socio-ecological systems, International Society for the Systems Sciences, Boulder, July 24–30, pp 76–77

Mobus GE, Fisher P (1994) "MAVRIC's brain", technical paper. Paper presented at Seventh international conference on Industrial & Engineering Applications of Artificial Intelligence & Expert Systems IEA/AIE'94, University of North Texas, San Diego. Available on-line: http://faculty.washington.edu/gmobus/Mavric/MAVRICS_Brain_rev.html

Mobus GE, Fisher P (1999) Foraging search at the edge of chaos. In: Levine D, Brown VR, Shirey VT (eds) Oscillations in neural networks. Lawrence Erlbaum Associates, Hillsdale. Available on-line: http://faculty.washington.edu/gmobus/Mavric/chaos.html

Mobus GE and Kalton MC (2014) *Principles of Systems Science*, Springer, New York

Mobus GE, Kalton MC (2015) Principles of systems science. Springer, New York

Mollison B (1997) Permaculture: a designers' manual. Ten Speed Press, Berekley

Monod J (1971) Chance and necessity: an essay on the natural philosophy of modern biology. Alfred A. Knopf, New York

Montague R (2006) Why choose this book? Penguin Group, New York
Montgomery DR (2007) Dirt: the erosion of civilizations. University of California Press, Berkeley
Morowitz HJ (1968) Energy flow in biology. Academic Press, Cambridge, MA
Morowitz HJ (1992) Beginnings of cellular life: metabolism recapitulates biogenesis. Yale University Press, New Haven
Morowitz HJ (2002) The emergence of everything: how the world became complex. Oxford University Press, Oxford
Morris I (2010) Why the west rules – for now: the patterns of history, and what they reveal about the future. Farrar, Straus, and Giroux, New York
Morris I (2013) The measure of civilization: how social development decides the fate of nations. Princeton University Press, Princeton
Morris I (2015) Foragers, farmers, and fossil fuels: how human values evolve. Princeton University Press, Princeton
Newman F, Couturier L, Scurry J (2004) The future of higher education: rhetoric, reality, and the risks of the market. Jossey-Bass, San Francisco
Nicolis G, Prigogine I (1989) Exploring complexity: an introduction. W.H. Freeman & Company, New York
Nissen HJ, Damerow P, Englund RK (1993) Archaic bookkeeping: early writing and techniques of economic administration in the ancient Near East. The University of Chicago Press, Chicago
Nowak MA (2006) Evolutionary dynamics: exploring the equations of life. The Belknap Press, Cambridge, MA
Nowak MA, Coakley S (eds) (2013) Evolution, games, and god: the principle of cooperation. Harvard University Press, Cambridge, MA
Odum HT (1983) Systems ecology: an introduction. Wiley, New York
Odum HT (1994) Ecological and general systems: an introduction to systems ecology (Revised edition). University Press of Colorado, Boulder
Odum HT (2007) Environment, power, and society for the twenty-first century: the hierarchy of energy. Columbia University Press, New York
Odum HT, Odum EC (2001) A prosperous way down: principles and policies. University Press of Colorado, Boulder
Orr DW (2004) Earth in mind: on education, environment, and the human Prospect. Island Press, Washington, DC
Patt, YN and Patel, SJ (2004) Introduction to Computing Systems: From Bits & Gates to C & Beyond. McGraw Hill, Boston
Pazzani MJ (1990) Creating a memory of causal relationships. Lawrence Erlbaum Associates, Hillsdale
Pfeifer R, Scheier C (1999) Understanding intelligence. The MIT Press, Cambridge, MA
Piketty T (2014) Capital in the twenty-first century, (translated by Goldhammer, A). The Bellknap Press of Harvard University Press, Cambridge, MA
Pinker S (1997) How the mind works. W. W. Norton & Company, New York
Pinker S (2002) The blank slate: the modern denial of human nature. Penguin Books (Paperback Ed.), New York
Pinker S (2007a) The language instinct: how the mind creates language. Harper Perennial, New York
Pinker S (2007b) The stuff of thought: language as a window into human nature. Viking, New York
Polanyi K (2001) The great transformation: the political and economic origins of our time. Beacon Press, Boston
Potter, VR (1971) Bioethics: bridge to the future. Prentice-Hall, New York
Prandoni P, Vetterli M (2008) Signal processing for communications. EPFL Press, Lausanne. Available in .pdf at no cost: http://www.sp4comm.org/. Accessed: 13 Aug 2017
Prigogine I, Stengers I (1984) Order out of chaos: man's new dialogue with nature. Bantam New Age Books, New York

Primack JR, Abrams NE (2006) The view from the center of the Universe. Riverhead Books, New York

Qualman D (2019) Civilization critical: energy, food, nature, and the future. Fernwood Publishing, Halifax

Quinn D (2000) Beyond civilization: humanity's next great adventure. Broadway Books, New York

Radzicki MJ, Taylor RA (2008) Origin of system dynamics: Jay W. Forrester and the history of system dynamics. In: U.S. Department of Energy's introduction to system dynamics. http://www.systemdynamics.org/DL-IntroSysDyn/start.htm. Accessed 12 Jan 2017

Reich D (2018) Who we are and how we got here: ancient DNA and the new science of the human past. Pantheon Books, New York

Rhodes FHT (2001) The creation of the future: the role of the American University. Cornell University Press, Ithaca

Rosen R (1985) Anticipatory systems: philosophical, mathematical and methodological foundations. Pergamon Press, Oxford

Rosen R (1991) Life itself. Columbia University Press, New York

Rothschild M (1990) Bionomics: economy as ecosystem. Henry Holt and Company, New York

Rousseau D, Billingham J, Wilby J, Blachfellner S (2016) In search of general systems theory. Systema 4(1):76–99. Special issue – General systems transdisciplinarity

Rousseau D, Wilby J, Billingham J, Blachfellner S (2018) General systemology. Springer, New York

Ruhla C (1989) The physics of chance. Oxford University Press, Oxford

Rumelhart DE, McClelland JL (1987) Parallel distributed processing, vol. 1: foundations, PDP research group, Bradford books. The MIT Press, Cambridge, MA

Sapolsky RM (2017) Behave: the biology of humans at our best and worst. Penguin Books, New York

Sawyer RK (2005) Social emergence: societies as complex systems. Cambridge University Press, Cambridge, UK

Sayama H (2015) Introduction to the modeling and analysis of complex systems. Open SUNY Textbooks, Geneseo

Scalaidhe SPÓ, Wilson FAW, Goldman-Rakic PS (1999) Face-selective neurons during passive viewing and working memory performance of rhesus monkeys: evidence for intrinsic specialization of neuronal coding. Cereb Cortex 9(5):459–475

Schneider ED, Sagan D (2005) Into the cool: energy flow, thermodynamics, and life. University of Chicago Press, Chicago

Schwartz B (2004) The paradox of choice: why more is less. Harper Perenial, New York

Schwarz E (1992) A generic model for the emergence and evolution of natural systems toward complexity and autonomy. In: Proceedings of the 36th annual meeting of the international society of systems science, Denver, Vol. II

Schwarz E (1997) Toward a holistic cybernetics: from science through epistemology to being. Cybern Hum Know 4:17–50

Scott JC (2017) Against the grain: a deep history of the earliest states. Yale University Press, New Haven

Senge P (2006) The fifth discipline: the art and practice of the learning organization, 2nd edn. Doubleday, New York

Seung S (2013) Connectome: how the brain's wiring makes us who we are. Mariner Books, New York

Shannon CE, Weaver W (1949) The mathematical theory of communication. University of Illinois Press, Champaign

Sillito H (2014) Architecting systems: concepts, principles, and practice. College Publications, London

Simon HA (1957) A behavioral model of rational choice. In: Models of man, social and rational: mathematical essays on rational human behavior in a social setting. Wiley, New York

Simon HA (1991) Bounded rationality and organizational learning. Organ Sci 2(1):125–134

Simon HA (1998) The sciences of the artificial, 3rd edn. The MIT Press, Cambridge, MA
Smith SW (1997) The scientist and engineer's guide to digital signal processing, California Technical Pub. California. Free, on-line version: http://www.dspguide.com/. Accessed 13 Aug 2017
Smith E, Morowitz HJ (2016) The origin and nature of life on earth: the emergence of the fourth geosphere. Cambridge University Press, Cambridge, UK
Smolin L (1997) The life of the cosmos. Oxford University Press, New York
Sober E, Wilson DS (1998) Unto others: the evolution and psychology of unselfish behavior. Harvard University Press, Cambridge, MA
Sporns O (2016) Networks in the brain. The MIT Press, Cambridge, MA
Squire LR, Kandel ER (2009) Memory: from mind to molecules, 2nd edn. Roberts & Company, Greenwood Village
Staddon JER (2001) Adaptive dynamics: the theoretical analysis of behavior. The MIT Press, Cambridge, MA
Sternberg RJ (ed) (1990a) Wisdom: its nature, origins, and development. Cambridge University Press, New York
Sternberg RJ (1990b) Wisdom and its relations to intelligence and creativity. In: Sternberg RJ (ed) Wisdom: its nature, origins, and development. Cambridge University Press, New York, pp 142–159
Stewart J (2000) Evolution's arrow. The Chapman Press, Canberra
Streeck W (2016) How will capitalism end? Essays on a failing system. Vesro, New York
Striedter GF (2005) Principles of brain evolution. Sinauer Associates, Inc., Sunderland
Suddendorf T (2013) The gap: the science of what separates us from other animals. Basic Books, New York
Surowiecki J (2004) The wisdom of crowds. Doubleday, New York
Sutton RS, Barto AG (2017) Reinforcement learning: an introduction (2nd edn. In draft). The MIT Press, Cambridge, MA
Tainter JA (1988) The collapse of complex societies. Cambridge University Press, Cambridge, UK
Taso DY (2019) Face values. Sci Am 320(2):22–29
Tomasello M (2014) A natural history of human thinking. Harvard University Press, Cambridge, MA
Tomasello M (2016) A natural history of human morality. Harvard University Press, Cambridge, MA
Tomasello M (2019) Becoming human: a theory of ontology. Harvard University Press, Cambridge, MA
Troncale LR (1978a) Origins of hierarchical levels: an emergent evolutionary process based on systems concepts. In: Ericson R (ed) Proceedings of the 22nd annual meeting of the Society for General Systems Research (SGSR), pp 84–94. [Note: This Society has been renamed: International Society for Systems Science, ISSS]
Troncale LR (1978b) Linkage propositions between fifty principal systems concepts. In: Klir GJ (ed) Applied general systems research: recent developments and trends: N.A.T.O. conference series II. Systems science. Plenum Press, New York, pp 29–52
Troncale LR (1993) Qualitative and quantitative research on hierarchies as a case study of general systems research. In: Ethical management of science as a system, 37th annual proceedings of the international society for systems science, International Society for the Systems Sciences
Troncale LR (2006) Toward a science of systems. Syst Res Behav Sci 23:301–321. Wiley Interscience http://onlinelibrary.wiley.com/doi/10.1002/sres.729/abstract;jsessionid=A67D28A263EB8BE7D9E3A8F62ADBF27E.f03t01. Accessed 13 July 2017
Türke R-E (2008) Governance – systemic foundation and framework. Springer, New York
Unger RM, Smolin L (2015) The singular universe and the reality of time. Cambridge University Press, Cambridge, UK
Varela FG (1991) The embodied mind. MIT Press, Cambridge, MA
Varela FG, Maturana HR, Uribe R (1974) Autopoiesis: the organization of living systems, its characterization and a model. Biosystems 5(4):187–196
Vignieri S, FahrenKamp-Uppenbrink J (2017) Ecosystem earth. Science 356(6335):258

Volk T (1995) Metapatterns across space, time, and mind. Columbia University Press, New York

Volk T (1998) Gaia's body: toward a physiology of earth. Springer, New York

Volk T (2017) Quarks to culture: how we came to be. Columbia University Press, New York

von Bertalanffy L (1968) General system theory: foundations, development, applications. George Braziller, New York

von Foerster H (1974) Cybernetics of cybernetics. University of Illinois, Urbana. OCLC 245683481

von Foerster H (1981) Observing systems. Intersystems Publications, Seaside

von Foerster H (2003) Understanding understanding: essays on cybernetics and cognition. Springer, New York

von Neumann J, Morgenstern O (1944) Theory of games and economic behavior. Princeton University Press, Princeton

Wahl DC (2016) Designing regenerative cultures. Triarchy Press, Axminster

Walker SI, Davies PCW, Ellis GFR (2017) From matter to life: information and causality. Cambridge University Press, Cambridge, UK

Wallerstein I (2004) World-systems analysis: an introduction. Duke University Press, London

Warfield JN (2006) An introduction to systems science. World Scientific Publishing Company, Singapore

Washburn J (2005) University Inc.: the corporate corruption of higher education. Basic Books, New York

Westneat DE, Fox CW (2010) Evolutionary behavioral ecology. Oxford University Press, Oxford

Wiener N (1950) The human use of human beings. The Riverside Press (Houghton Mifflin Co.), New York

Wiener N (1961) Cybernetics: or the control and communication in the animal and the machine, 2nd edn. Martino Fine Books, Eastford

Wierzbicka A, Goddard C (2014) Words and meanings: lexical semantics across domains, languages and cultures. Oxford University Press, Oxford

Wilson EO (2013) The social conquest of earth. Liveright, New York

Wilson DS, Sober E (1994) Reintroducing group selection to the human behavioral sciences. Behav Brain Sci 17(4):585–654

Wilson DS, Wilson EO (2008) Evolution 'for the good of the group'. Am Sci 96(5):380–389

Wolf M (2012) Computers as components: principles of embedded computing system design. Elsevier, New York

Wolpert D, Libby E, Growchow JA, DeDeo S (2017) The many faces of state space compression. In: Walker S, Davies P, Ellis GFR (eds) From matter to life: Information and causality. Cambridge University Press, Cambridge, UK, pp 199–243

Wymore AW (1967) A mathematical theory of systems engineering – the elements. Wiley, New York

Yogi MM (1974) Maharishi International University Catalogue. MIU Press, Los Angeles